D1560032

engineering thermodynamics

RICHARD E. BALZHISER
Electric Power Research Institute

MICHAEL R. SAMUELS
E. I. DuPont and Company

Prentice-Hall, Inc., Englewood Cliffs, New Jersey 07632

Library of Congress Cataloging in Publication Data

BALZHISER, RICHARD E
Engineering thermodynamics.

Includes indexes.
1. Thermodynamics. I. Samuels, Michael R., joint author.
III. Title.
TJ265.B23 621.4'021 76-25440
ISBN 0-13-279570-1

10 9 8 7 6 5 4 3 2 1

Printed in the United States of America

PRENTICE-HALL INTERNATIONAL, INC., *London*
PRENTICE-HALL OF AUSTRALIA PTY. LIMITED, *Sidney*
PRENTICE-HALL OF CANADA, LTD., *Toronto*
PRENTICE-HALL OF INDIA PRIVATE LIMITED, *New Delhi*
PRENTICE-HALL OF JAPAN, INC., *Tokyo*
PRENTICE-HALL OF SOUTHEAST ASIA PTE. LTD., *Singapore*
WHITEHALL BOOKS LIMITED, *Wellington, New Zealand*

contents

SIX THE PROPERTY RELATION AND THE MATHEMATICS OF PROPERTIES 181

SEVEN THERMODYNAMICS OF FLUID FLOW, COMPRESSION, AND EXPANSION 251

EIGHT THERMODYNAMICS OF ENERGY CONVERSION 310

NINE REFIGERATION, HEAT PUMPS, AND GAS LIQUEFACTION 376

TEN MULTICOMPONENT SYSTEMS 405

preface

While we may appear a bit presumptions in generating still another introductory thermodynamics text, we assure the reader that it is with both humility and purpose that we have undertaken the project. It is our firm belief that future introductory courses in thermodynamics must include an adequate treatment of chemical thermodynamics as well as the classical material if tomorrow's engineers are to have the understanding, as well as the tools, to deal with such increasingly important and complex matters as energy and the environment. As members of the University of Michigan and University of Delaware faculties, we have taught such courses to students from all engineering disciplines and concluded that the chemical thermodynamics derived from the general chemistry courses typically taken by most engineering students fail to provide the applications orientation required by today's engineers in dealing with the study of energy. Most general thermodynamics texts have failed to include a sufficiently well-developed treatment to provide the engineer with the ability to apply it to practical problems. Our subsequent experience in industry has reinforced these convictions.

We firmly believe that energy availability will be the primary forcing function in changing societal development patterns for many decades to come. Our ingenuity as scientists and engineers in finding new sources and more efficient methods for utilizing all forms of energy will have as profound an impact on the future as did the industrial revolution that created our energy intensive society. Engineers, whether trained for the energy producing industries, the energy consuming industries or the manufacturing or construction industries, require a basic understanding of thermodynamics principles, including those of chemically reacting systems.

Combustion products, our most serious environmental pollutants today, are the result of complex chemical reactions occurring in boilers, automobiles, aircraft, furnaces, smelters and other industrial processes. Likewise, in assessing the many approaches to using coal most efficiently as an energy or raw material resource, one must consider not only direct combustion, but its possible coversion into a clean liquid or gaseous form, useable as a fuel or as a source of synthetic hydrocarbons. These latter processes consume a portion of the coal's energy in the conversion process, and an overall thermodynamic assessment must include these losses as well as those inherent in converting thermal energy to a mechanical form. Fuel cells enable us to convert chemical energy directly to a mechanical form and may some day become as important in power production as they have been to the space program. All of this simply demonstrates that we believe today's engineer should be as well-grounded in the chemical side of thermodynamics as in the thermal side. Specialization should follow, for we agree that engineering schools must produce energy doers rather than conversationalists.

In adapting our earlier material* for a general engineering course, we have assumed that the student's chemistry background includes no more chemistry than typically required for enrollment in an engineering college. At the same time our approach to the subject is sufficiently different so that it is not likely to duplicate material covered in any chemistry courses which the student may have taken. For this reason we believe the text will serve the needs of chemical engineers as well as others. We have supplemented our earlier energy conversion with additional material on thermodynamic cycles in Chapter 8. This chapter also includes a considerable amount of descriptive material on energy which we believe is valuable in giving the student a contemporary perspective on energy, its future importance in society, and the importance of thermodynamics to his education.

The first six chapters develop the basic thermodynamic and property relationships. Emphasis is placed on the fundamental concepts: energy, entropy, and equilibrium, their interrelations, and the engineering relationships to which they give rise. Considerably more attention is given to entropy than is characteristic of most undergraduate texts; this topic is introduced early in the book as an indicator of the effectiveness with which man utilizes his energy reserves. The concept of lost work is used in developing the entropy generation term to further emphasize the significance of dissipative processes that degrade energy potentials.

Chapters 7, 8, and 9 are applications oriented and treat compression and expansion processes, fluid flow, energy conversion cycles, and refrigeration. Chapters 10 and 11 are devoted to phase equilibrium and chemically reacting systems. The text can be used to accomplish various objectives, depending on the curriculum and the instructor's desires. Sufficient material exists for a two-term sequence designed to provide the student with a broad background in applied thermodynamics at the undergraduate level. Such a student should

* R. E. Balzhiser, M. R. Samuels, and J. D. Eliassen, *Chemical Engineering Thermodynamics*, Prentice-Hall, Inc., Englewood Cliffs, N. J., 1972.

be well-equipped for either a career in industry or enrollment in any graduate thermodynamics course. A one-semester (15-week) course will require selective pruning by the instructor to meet his objectives. We believe Chapters 1 through 6 are essential for students with no background in the subject. Chapters 7 through 9 are organized so that continuity is not sacrificed if many sections are skipped. Sections 1 through 6 of Chapter 10, and 1 through 8 of Chapter 11 provide the student with the fundamentals of phase behavior and chemical thermodynamics.

The student has been foremost in the minds of the authors throughout the preparation of this text. Sufficient detail and sample problems are used in the development of important principles that we believe it is possible to rely heavily on the text, rather than lectures, for development of the principles, allowing the instructor to devote class time to problem analysis and discussion of student questions. Sample problems have been selected for their instructional value and should be studied carefully by the student to derive maximum benefit.

A variety of engineering oriented problems that can be assigned for homework are found at the end of each chapter. Their complexity varies from simple discussion questions to several worthy of a computer solution. Many of the problems are suitable alone as a homework assignment. We have resisted the temptation to incorporate comprehensive property tabulations and plots which would compete with handbooks. We have included sufficient data and information for most problems in the text, but have not felt it inappropriate to expect the student to use other reference material in the library on occasion. We believe the data included will permit the student to derive the full educational value from the text without forcing him to lug a data bank around campus.

The preparation of this manuscript was aided by our experience with *Chemical Engineering Thermodynamics*, and the constructive comments of both students and faculty. Jack Eliassen, our co-author on the first book, made numerous contributions to our development of the subject-matter which have been useful in preparing this text. We are also grateful to Professors William Bathe and Ephraim Sparrow for their reviews of the manuscript and their help in adapting it to a larger cross-section of undergraduate engineering needs. We quickly learned to appreciate the opportunity that academia provides for scholarly pursuits as contrasted with the demands of government and industry. At the same time, without the enlightenment that we have gained by our associations with E. I. DuPont de Nemours, the Electric Power Research Institute and the White House Office of Science and Technology, we probably would not have felt as compelled to pursue this project.

Last but most important in this endeavor has been the willingness of our families to spare us the evenings and weekends that have been required over the last several years to prepare the manuscript. In appreciation of their encouragement and understanding, we join in dedicating this book to our wives, Chris Balzhiser and Sue Samuels, and our children; Gary, Bob, Patti and Cheryl Balzhiser and Amy and Lisa Samuels.

RICHARD E. BALZHISER
MICHAEL R. SAMUELS

ONE

engineering, energy, entropy, and equilibrium

Introduction

For many years the developed countries of the world, and particularly the United States, whose energy consumption rate approximates 35% of the world's total usage, have taken their energy supply for granted. However, a series of events, commencing with the major Northeast blackout in 1965 and followed by regional brownouts, natural gas curtailments, and finally the long gasoline lines as a result of the Arab oil embargo in early 1974, began to sensitize citizens and their governments to the finite nature of our energy resources and the essential role energy plays in modern society. Since then, conservation and improved efficiency in energy conversion and utilization have received increasing attention among legislators, regulators, manufacturers, consumers and energy suppliers.

The importance of energy conversion processes is clearly shown by examining actual energy flows for the United States in 1970 (Figure 1-1) and projected flows for 1980 (Figure 1-2). Less than half the energy consumed in 1970 was productively used and the picture as projected will worsen by 1980, largely due to the lower thermal efficiencies of nuclear power plants, which will become an increasingly important source of electrical power during the decades ahead.

These figures also show the relative importance of the fossil fuels (coal, oil and gas) in our present and future energy systems, even with the rapid growth of nuclear and geothermal power. In 1970 24% of the total energy consumed was used to generate electrical energy and 26% to meet the nation's transportation needs. By 1980 these two categories are projected to utilize 31% and 28% respectively. It should be noted that these are the two forms of energy utilization which contribute most significantly to our energy losses. Only about 36%[1] of the energy consumed in the electrical generation process is

[1]This figure represents a 1970 average for the entire generating mix, including fossil steam, hydroelectric, nuclear, geothermal and gas turbine generating capacity.

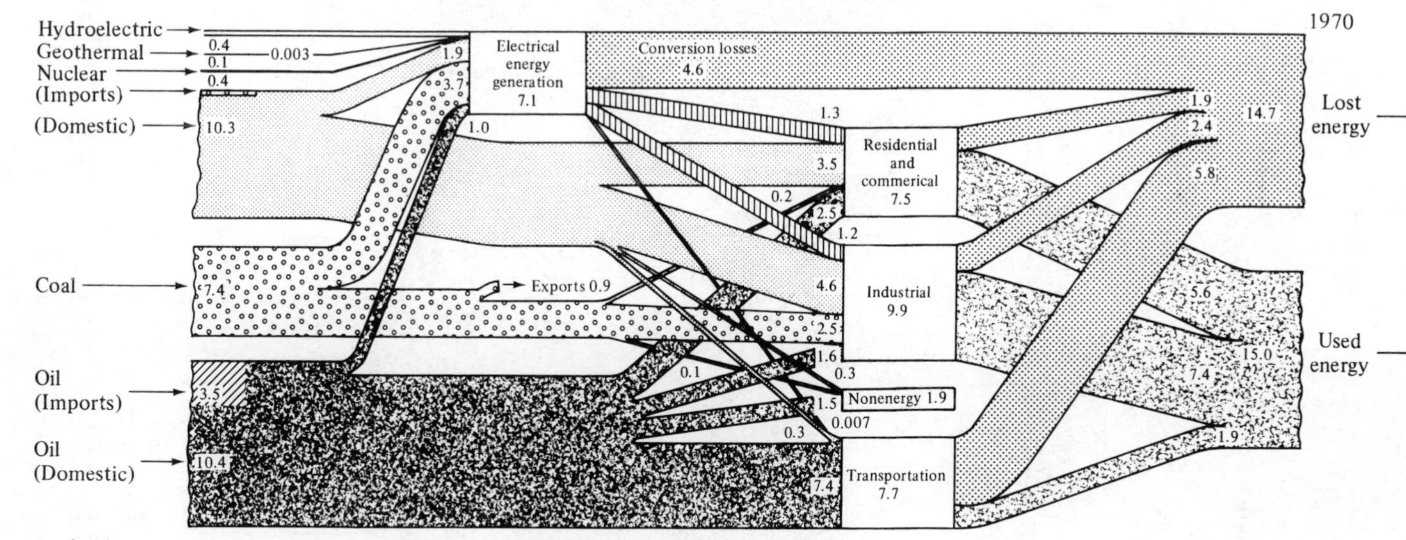

Source: "Understanding the 'National Energy Dilemma'." JCAE, 1973

Figure 1-1. The actual energy flow pattern for the United States-1970.

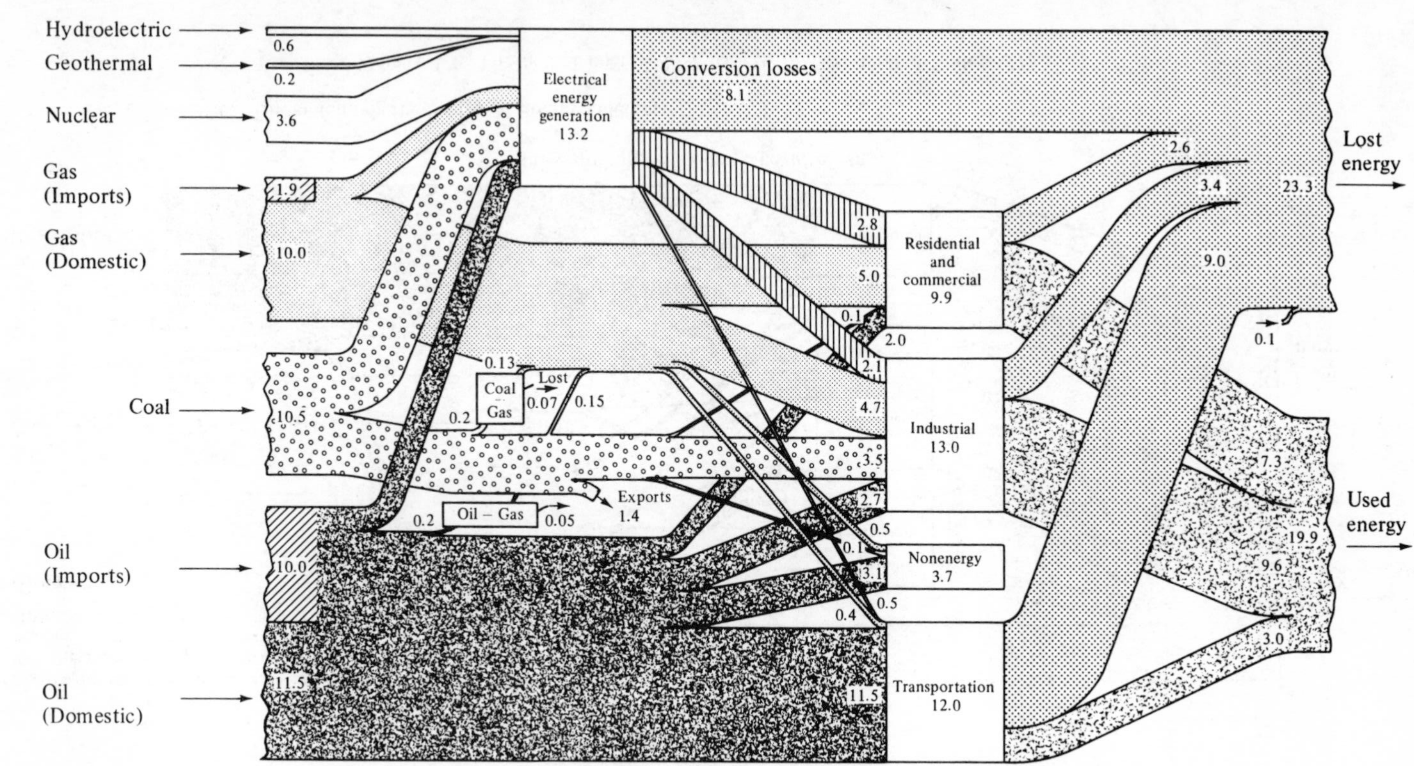

Source: "Understanding the 'National Energy Dilemma'." JCAE, 1973

Figure 1-2. The projected energy flow pattern for the United States-1980.

converted to electricity and the transportation sector productively uses only about 25%[2] of the energy which it consumes. In both cases the waste heat resulting from engine inefficiencies, friction, etc., are discharged to the lakes and streams or the atmosphere, where they sometimes create a local thermal pollution problem. It should also be noted that electrical energy generation and transportation are the two consuming sectors projected to grow most rapidly over the decade of the seventies.

From a resource point of view it is becoming increasingly evident that the world's petroleum and natural gas resources are too valuable to burn under boilers for generating electrical energy. Fortunately, the United States has enormous coal and nuclear resources which can meet these needs in the future if the safety and environmental issues which have caused so much concern in the past can be overcome. The abundance of these resources and the strong probability of resolving the environmental uncertainties makes it likely that electrical energy will become increasingly attractive and significant in future years.

Electrical energy is the cleanest and most versatile form of energy available today. It can be used for heating, lighting, or operating motors, radios, TV's, computers, appliances and vehicles. Demand for electricity is doubling virtually every decade. For this reason it is important that the conversion processes by which it is produced be operated as efficiently as possible if the nation is to make maximum use of its finite (by present technology) fuel resources. It thus becomes an immense challenge to future engineers to improve upon the performance of today's electrical generation and transmission technology so that the world's finite resources can be used most efficiently under conditions where the public's health and safety are not compromised.

Transportation systems consume approximately 25% of the energy used in the United States, making the U.S. population by far the most mobile society in the world. Virtually all of this energy is currently provided by petroleum, a resource which the U.S. can no longer expect to obtain from domestic production alone. As our dependence on foreign sources grows, an increasing incentive to improve the efficiency of propulsion systems will result along with greater interest in unconventional systems which use alternative fuels and energy sources. These include considerations of hydrogen, methanol and LNG (liquified natural gas) as fuels and the development of battery powered vehicles for which the primary energy source could be coal or uranium with the eventuality of using solar or fusion energy a real possibility.

The imposition of stringent environmental standards on emissions from propulsion systems has created a major perturbation in engine and vehicle

[2]A 1970 average for all modes of transportation including rail, automobile, truck, ship, and airplane.

design considerations which could lead to major system redesign in the years ahead. Internal combustion engines have served as the backbone of the U.S. transportation system. However, these engines are a principal source of the air pollutants, carbon monoxide, unburned hydrocarbons, and nitrogen oxide in our major urban areas today. Early attempts to reduce the amounts of these pollutants in vehicle emissions have markedly reduced vehicle performance, particularly with respect to fuel consumption.

The pyramiding of these environmental concerns, as the nation becomes more urbanized and its transportation system grows, in combination with possible energy shortages, is leading to a major reexamination of propulsion systems. The internal combustion engine, much like the conventional steam power plant, has revolutionized the industrialized societies of the world. They both represent highly developed and reliable technologies that will remain tough competitors for newly emerging challengers in their respective areas of application. However, emerging energy and environmental constraints portend sufficiently significant perturbations to our industrialized society that new propulsion technology may well emerge in mass produced vehicles during the balance of the twentieth century.

The responsibility for developing the technology to use existing energy resources safely and efficiently and to harness fusion and solar energy economically will fall largely on the emerging generation of engineers. Thermodynamics, the science of energy and its interconversion, will represent one of their most valuable analytical tools. The material in this text is developed with its engineering application foremost in mind. For this reason we have included in the next section of this introductory chapter a discussion of engineering as a prelude to the three basic E's of thermodynamics—energy, entropy and equilibrium.

1.1 Engineering

It has been said that politics is the bridge across which ideas march to meet their destiny. It seems equally appropriate to characterize engineering as the bridge which links the laboratories of the scientist with the products of industry. Indeed, as we examine more carefully the role of the engineer in society, we become increasingly aware of the sizable gap that must be spanned by the engineering profession. In a technical sense the engineer represents the interface between the scientific community and the business world. When it is economically feasible, the engineer translates the scientists' successes in the laboratories to the production operations of industry. This "when" is an important one, which requires the engineer to have both technical and economic training to successfully perform his role.

The engineer's responsibilities range from devising a process complete with the specification of operating conditions and the type and size of all equipment needed, to figuring the depreciation, cash flow, and profitability associated with a manufacturing process. He draws heavily on the principles of mathematics, science, economics, and common sense in performing such a role. However, these alone are insufficient to handle all aspects of his professional responsibilities. Equations and formulas can seldom completely replace experience and intuition in the engineering scale-up of a process. The latter commodities would be difficult to produce on a scale that would match the needs of a highly industrialized society if the engineering profession itself had not advanced from an "art" to a science in its own right.

By sharing and correlating the "experiences" of investigations through the years, an improved understanding of many physical phenomena has been realized. Nevertheless, many seemingly simple processes such as the flow of a fluid through a pipe or the flow of heat across an interface defy complete understanding. To circumvent his inability to describe completely certain processes, the engineer will commonly choose to approximate the process by a simplified model. By identifying the key parameters which affect the process and then varying them in an ordered way, the engineer is frequently able to develop an expression that relates all of the important parameters.

The flow of a fluid in a pipe serves as an excellent illustration of the power of this approach. By varying pipe diameter, pressure drop, and fluid properties while determining flow rates, it is possible to devise an empirical correlation that relates all these parameters. Thus, although we may not completely understand turbulence or the resistance to flow occurring at the wall, we are able to use such a correlation to size the pipe and pumps for a given application. Pooling of all such information soon permits refinements in the correlating expressions which can then be used with reasonable confidence for fluids and conditions beyond those specifically studied.

Such procedures have been used over and over by the engineer to organize and coordinate the experiences of many for the collective good of the profession. Postulate, experiment, correlate, refine—that is the sequence by which the engineering profession has developed the tools of the trade. The procedure is not unlike that used by Mendeleev and other chemists in devising the periodic chart of the elements. In spite of all the observations made to date with respect to elements and their chemical behavior and in spite of the fact that the system has permitted man to predict the existence of elements before they were experimentally detected, the concept of the atom, on which the periodic chart is based, could be completely erroneous.

Such is often the case when extending the frontiers of technology. The process is one of trial and error. Each observation consistent with the model tends to increase the likelihood of its aptness and hence increases its utility and reliability. Thus, although many "laws" are considered as absolute truths, they could in fact become fiction as investigators continue their explorations.

Many engineering relationships have been devised by these procedures. They have offered great utility in making reliable engineering estimates of certain process specifications. Frequently it is deemed desirable to incorporate a safety factor in finalizing a design. This is merely the engineer's way of compensating for the reliability of the procedures available to him. Similarly, efficiencies are often used to relate actual performances to a highly idealized model of a process that is not capable of exact description. These techniques are all part of an increasingly sophisticated array of procedures available to the engineer.

Included in this collection of engineering "tools" is thermodynamics, which is often referred to as an engineering science. This designation arises in part from its origin in the sciences and in part from the increasing sophistication and utility of the relationships developed. Frequent references are made to the laws of thermodynamics. In reality they are laws only in the sense that no one has yet disproved them. Their formulation followed the pattern (described earlier) that has been used by scientists and engineers to expand progressively their understanding of the physical world. Energy and entropy are in effect "models" of a conceptual sort; both have been refined and expanded since their initial introduction many years ago.

Additional changes may well occur in the future as man continues to open new frontiers of understanding. In this development an attempt will be made, while building on the concepts as presently understood, to illustrate to the student where progress in future years is likely to lead. The areas of irreversible and statistical thermodynamics are two such possibilities. Both are intimately tied to the concepts of entropy and equilibrium. Although somewhat tangential to a basic treatment of the subject for the undergraduate, both will likely play an increasingly prominent role in graduate thermodynamics courses.

1.2 Energy

The relationships between heat and temperature were poorly understood until the early 1800s. Indeed, the thermodynamic concept of heat itself had not developed until that time. Evidence of this confusion between the concepts of heat and temperature is still encountered occasionally, in expressions of the following type: "The body was brought to a high *heat* (instead of *temperature*) through contact with the flame." The proper relationship between heat and temperature was not fully recognized until the concept of energy was developed.

The observation that a body's temperature could be changed by contact with a hotter (or colder) body is almost as old as mankind itself. Since the body's temperature has changed, scientists have concluded that something must have been gained or lost by the body. Around the middle of the 1700s this something was termed "caloric" and was considered to be a massless, volumeless substance that could flow between bodies by virtue of a temperature difference. According to the caloric theory, the higher the caloric content of a body, the higher its temperature.

In the late 1700s, however, Count Rumford (a native of Woburn, Mass., who was working for the Bavarian government at the time) observed that the temperature of cannon barrels became quite high when the barrels were subjected to the mechanical action of boring tools. Rumford questioned how this heating effect was produced, and he finally concluded that the "heating" was brought about by friction from the boring tool and that it would continue as long as the boring bits were kept in motion.

The consequences of Rumford's observations had a profound influence on the scientific thinking of his time. In a relatively small number of years James P. Joule had conclusively shown that many forms of mechanical work (that is, the effect of a force moving through a distance) could be used to produce a heating effect. Joule also observed that a given unit of mechanical work, regardless of its form, always produced the same temperature rise when applied to the same body.

In an effort to unite these apparently unrelated phenomena, the concept of energy evolved. The basis of this concept assumes that a body may contain several forms of energy. From physics we are familiar with the observation that a body may possess potential energy by virtue of its position within a force field (such as height within a gravitational field), and kinetic energy by virtue of its average bulk motion. In each of these instances the energy is determined by the position or velocity of the center of gravity of the mass, not by the microscopic particle motions or interactions that occur on an atomic scale. Although molecular motions and interactions are excluded from our conventional concepts of potential and kinetic energy, they must clearly be accounted for in any overall discussion of energy. *Internal energy* is defined in Chapter 2 to include all forms of energy possessed by matter as a consequence of random molecular motion or intermolecular forces.

We shall divide energy in transit between two bodies into two general categories: *Heat* is the flow of thermal energy by virtue of a temperature difference; *work* is the flow of mechanical energy due to driving forces other than temperature, and can be completely converted (at least in theory) by an appropriate device to the equivalent of a force moving through a distance. Included in the broad class of mechanical energies are kinetic, potential, electrical, and chemical energies. Thermal energy, on the other hand, is unique in that there is no known device whose sole effect is to completely convert thermal energy into a mechanical form. Therefore, we are led to the conclusion that heat, a flow of thermal energy, cannot be completely converted into work, a flow of mechanical energy. This statement expresses the essence of what is frequently called the "second law of thermodynamics" and will be considered in great detail later.

As originally envisioned, energy was a conservative property. That is, it was assumed that the energy content of the universe remained constant—energy could change its form, but the total amount always remained fixed. Findings by nuclear physicists in relatively recent times have shown this assumption to be incorrect. Their findings suggest that it is mass and energy together which are conserved. They even succeeded in relating mass to energy through the Einstein relation $E = mc^2$. For our purposes as thermodynamicists the conversion of mass to energy by nuclear reactions is of little current interest, and for applications outside the realm of nuclear processes, it is possible to think of energy itself as being conserved.

As an example of the restrictions just discussed, let us consider a roller coaster as it accelerates down a hill. Since its height is decreasing, the roller coaster is losing some of its potential energy. However, since its velocity is increasing, the kinetic energy is also increasing. If the roller coaster is frictionless, the decrease in potential energy is exactly counterbalanced by the increase in kinetic energy. On the way up the hill, this kinetic energy is reconverted into potential energy as the roller coaster returns to its original height. Thus we find that kinetic and potential energy are directly and completely interconvertible—a characteristic common to all forms of mechanical energy. We call such a process *reversible* since the whole universe may be returned to its starting condition.

If, on the other hand, the roller coaster is not frictionless, some of its potential energy will be converted to thermal energy by the rubbing action of bearing surfaces. Consequently, less potential energy gets converted to kinetic energy and the kinetic energy at the bottom of the hill is less than the original potential energy (by an amount equal to the amount of thermal energy produced). Thus the roller coaster is unable to return to its original height (or potential energy), since the thermal energy it now possesses leads only to a higher temperature and is of no value in returning the roller coaster to its original position. Only by performing work (pushing on the cart) can you return it to its original elevation.

At this point it may be asked: Couldn't this work be supplied by converting the thermal energy generated by friction back into work? The answer to this is: Not completely. We have already indicated that it is not possible to *completely* convert thermal energy into mechanical energy. Thus, although we could theoretically recover some of the frictional losses, we could never recover all of them. Frictional effects reduce forever the amount of energy in the universe which is capable of transformations that produce work. Thus, the occurrence of friction, or the degradation of mechanical energy to a thermal form, produces a process from which the whole universe cannot be returned to its starting condition. We call this kind of process *irreversible*.

As we will see, the distinction between reversible and irreversible processes is extremely important. Indeed, it is just this distinction and our need to *quantitatively* describe the level of irreversibility encountered during a given process which lead us into our discussion of a second fundamental property of matter—*entropy*.

1.3 Entropy

The importance of the entropy concept can best be illustrated by an example. Consider a gas as it flows through a wind tunnel. The gas molecules possess kinetic energy, a portion of which is random and a portion of which is ordered and contributes to the bulk velocity of the gas as it moves through the duct. The ordered portion is similar to the kinetic energy of any macroscopic object and is mechanical in form. As such it is capable of being converted to work by an appropriate device such as a turbine or windmill. Extraction of this ordered kinetic energy as work by a perfectly designed turbine would reduce the overall velocity of the gas and hence its kinetic energy, but would not affect the random behavior of the collection of molecules as the gas passed through the blades.

The random contribution to the total energy of the gas is effectively superimposed on the oriented flow. It contributes nothing to the energy flow to the turbine blade, as its random character produces as many collisions which tend to prevent the turbine from rotating as those which would assist it. Thus the random, or thermal, component does not decrease through its interaction with the turbine. Theoretically one could extract all the oriented kinetic energy possessed by the gas and leave only the thermal component. Such a total conversion would require many stages and an unreasonably large device in proportion to what the latter stages of the conversion would yield. Nevertheless, this process represents (in theory) the most efficient use of the energy available and provides the thermodynamicist with a standard with which he is able to compare less ideal conversions. In addition, this change is reversible since the gas can be reaccelerated to its initial kinetic energy, by using the work which was obtained from the turbine during the deceleration.

Contrast this process in which the ordered kinetic energy of the gas is completely converted to work with the condition that would exist if the inlet and outlet ducts of the turbine were suddenly closed. Clearly the total energy of the gas trapped inside would remain unchanged, as it would become essentially isolated. As the flowing molecules strike the closed outlet they rebound and interact with one another such that after a short period of time all kinetic energy will be random in nature. The extent of randomness within the collection will have increased significantly, as the original thermal component will now have added to it a thermal component equal to the kinetic component of the oriented flow.

Although the total energy remains unchanged, any attempt to convert any portion of this energy to work with *the turbine* is now impossible. We will show in Chapter 5 how a fraction of this thermal energy may be converted to work. However, this fraction is always less than unity and, under the circumstances pictured, would actually be quite close to zero. Thus, in fact, little work can be recovered from the mechanical energy once it has been converted to the thermal form. The change that took place in this element of gas by simply trapping it inside the turbine is quite revealing. The net effect was to leave its total energy unchanged but to convert a portion of the mechanical form (kinetic) to the thermal form with an accompanying loss in the ability of the gas to convert its energy to work. (The gas will also experience an increase in pressure as a result of the flow stoppage; since this is mechanical in form, it could be used to convert a portion of the internal energy of the gases to a mechanical form at a subsequent time. However, the amount of mechanical energy gained during the pressure rise will be quite small in comparison with the kinetic energy lost, so we still experience a net loss in the usefulness of our original energy.) Since mechanical energy has been degraded to the thermal form during this process, the process is irreversible. We will define *entropy* as a measure of the level of irreversibility associated with any process. If the process occurs reversibly, then the total entropy of the interacting bodies remains constant. If the process is irreversible, then the total entropy of the interacting bodies increases, the severity of the irreversibilities being directly proportional to the entropy increase.

Thus we may picture the entropy as providing a measure of the effectiveness with which energy transfers occur. Although in theory all mechanical forms of energy are completely interconvertible or transferable as work, in practice these ideal conversions cannot be realized. Frictional and other dissipatory effects inevitably lead to a downgrading of the available energy resources to a thermal form from which it can *never*, even by an ideal process, be completely reconstituted to a mechanical form. When the mechanical forms of energy possessed by a body are permitted to degenerate by any process whatsoever to the thermal (or random) form, the entropy of the body is increased. Thus entropy is also a measure of the extent of randomness within a system and thus provides an accurate indication of the effectiveness of energy utilization.

If we were to watch the gas molecules of the previous example, we would find that the likelihood that the random molecular motions will ever reorient themselves without some external input is extremely small. That is, the random thermal energy will not freely revert to a mechanical form. Since the entropy of a substance is related to its randomness, the entropy will not decrease without some external interaction. However, the only manner known to man to reduce molecular randomness is to transfer the randomness to another body, thereby increasing the randomness and entropy of the second body. Thus, as thermal energy is transferred from one body to another, entropy is effectively transferred. The body receiving the thermal energy experiences an increase in entropy while the body releasing the thermal energy experiences a reduction in entropy. The transfer of thermal energy (or randomness) as heat in this fashion is the *only* manner by which it is possible to reduce a body's entropy. Since, as we shall see later, this process at best results in a constant amount of randomness in the universe (and generally produces an increase), the entropy of the universe is forever increasing. Unlike energy, it is nonconservative, and therein lies much of the mystery that continues to perplex students. Hopefully the development of the concept in Chapters 2 and 5 and Appendix H will help clarify students' understanding of this most important and useful concept.

1.4 Equilibrium

A body at equilibrium is defined to be one in which all opposing forces, or actions, are exactly counterbalanced (subject to the restraints placed upon the system), so that the macroscopic properties of the body are not changing with time. Experience tells us that all bodies tend to approach an equilibrium condition when they are isolated from their surroundings for a sufficient period of time. For example, if a ball is placed on a surface as shown in Fig. 1-3, it tends to settle in the lowest portion of the surface, where the force of gravity is exactly counterbalanced by the supporting force of the surface. Thus the ball has settled in its equilibrium position—subject to the constraint that it remain in contact with the surface.

The equilibrium condition described in Fig. 1-3 is termed a *stable equilibrium* because the ball will always return to this condition after it has been

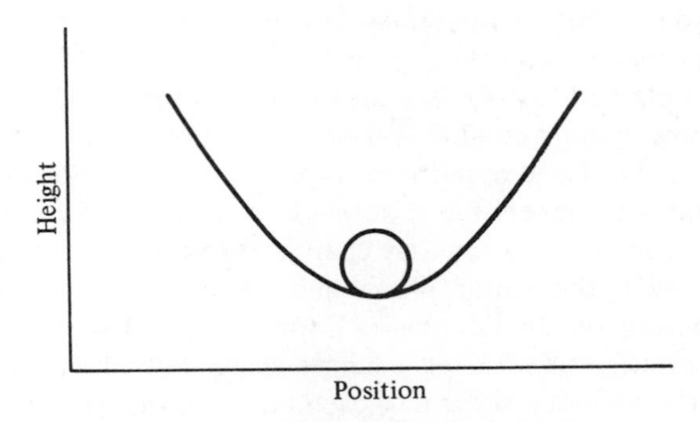

Figure 1-3. The equilibrium condition.

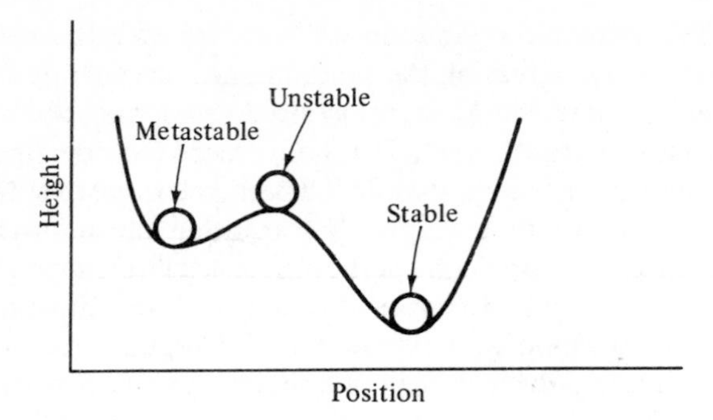

Figure 1-4. Types of equilibrium.

moved away (or disturbed) from the equilibrium position. In addition to stable equilibrium conditions, we may have metastable and unstable equilibrium conditions, as shown in Fig. 1-4.

A metastable equilibrium is one in which the system will return to its original state if subjected to a small disturbance, but which will settle at a different equilibrium condition if subjected to a disturbance of sufficient magnitude. For example, a mixture of hydrogen and oxygen can remain unchanged for great periods of time if not greatly disturbed. However, if the mixture is disturbed sufficiently, say by an electrical spark or a mechanical shock, then a violent reaction between the hydrogen and oxygen can be expected to occur.

An unstable equilibrium condition is one in which the system will not return to its original condition whenever it is subjected to a finite disturbance. For example, if we carefully balance a dime on its edge, we have a body that is essentially in unstable equilibrium, because a small disturbance will cause the dime to topple.

Although the three kinds of equilibrium have been illustrated in the previous paragraphs, we shall now restrict the remainder of our consideration to the discussion of *stable equilibrium*, as this is the condition to which most systems will ultimately move.

For the simple ball-and-surface illustration, we may picture equilibrium to be a static unchanging condition. However, for many of the problems we shall encounter, this static description is far too simple. For example, if we examine the equilibrium between vapor and liquid water on a molecular scale, we would find constant motion and change: Molecules from the liquid are constantly entering the vapor phase and molecules from the vapor are constantly entering the liquid phase. Equilibrium between the liquid and vapor phases occurs not when all changes cease, but when these molecular (or microscopic) changes just balance each other, so that the macroscopic (or gross) properties remain unchanged. As seen in this broader context, equilibrium is actually a dynamic process on a microscopic scale, even though we treat it as a static condition in macroscopic terms.

All spontaneous (naturally occurring) events tend toward more probable and more random molecular configurations. (If the new configuration were not more probable, the system would not tend toward it spontaneously.) Thus, if a body is isolated from its surroundings and allowed to interact with itself, all changes in the body's properties must lead to more probable and random configurations. When the body finally attains its most probable configuration, it can undergo no additional change. Since the body may undergo no further change in its properties, we observe that it must then be in equilibrium with itself, or simply in *equilibrium* (subject, of course, to the constraint of isolation from its surroundings). Conversely, if an isolated body is in its equilibrium condition—so that its properties do not change with time—the body must be in its most probable and random configuration. (If it were not in the most probable configuration, it would tend to move toward that configuration.) As we have shown, the more random or probable a configuration, the higher is the entropy of the material in this arrangement. Since the equilibrium condition of an isolated body corresponds to the most probable conditions, the equilibrium conditions must correspond to the conditions of maximum entropy—subject to the constraints placed upon the body.

In macroscopic terms the equilibrium condition requires that all energy potentials, such as temperature and pressure (which measure the availability of energy), be uniform throughout the body. If this were not true, energy flows would occur and could theoretically be used to produce work if the flow were channeled through an ideal engine. Thus a body in equilibrium (with itself) can be described as one from which no work can be derived if any part of the body is allowed to communicate with any other part through an ideal engine. Later we shall deal with many types of problems in which the criteria for equilibrium will be extremely difficult to specify. In these cases we shall find it useful to rely on this last observation to develop other useful criteria for equilibrium.

1.5 System—Definition and Characterization

A thermodynamic *system* may consist of any element of space or matter specifically set aside for study, while the *surroundings* are thought of as representing the remaining portion of the universe. The system *boundary*, which may be real or imaginary, separates the system from the surroundings. A system that is not permitted to exchange mass with the surroundings is termed a *closed system*, whereas a system that exchanges mass with the surroundings is called an *open system*. A closed system that exchanges no energy with its surroundings, either in the form of heat or work, is called an *isolated system*. Examples of a closed system might include a block of steel or a fixed amount of gas confined within a cylinder; a pipe or a turbine through which mass is flowing are examples of open systems. However, if one were to identify a given mass of fluid and follow its passage through a pipe, this fixed amount of mass, surrounded by an imaginary boundary, would constitute a closed system. Two blocks of steel at different temperatures but perfectly insulated from their surroundings constitute an isolated system.

The thermodynamic system will provide a basis for analysis in subsequent chapters of this text. In the solution of any problem the first step will always involve a clear definition of the system under consideration. Such a choice is often arbitrary, because several possibilities generally exist. Experience demonstrates that in some cases an open system will have inherent advantages, whereas in other cases a closed system might be preferable.

The condition in which a system exists at any particular time is termed its *state*. The state of a system is defined by specifying the numerical value of its properties such as pressure, temperature, and density. As we will soon see, thermodynamic analyses also use other properties which have yet to be developed. Indeed, much of the discussion in later chapters will revolve about the definition and use of the so-called thermodynamic properties internal energy, enthalpy, and free energy.

A change in the state of a system brought about by some interaction with its surroundings always results in the change of at least one of the properties which describe the state. However, if by a series of interactions with the surroundings the system is restored to its original state, then all the properties by which that state was originally characterized must return to their original values. A true *state property*, is one whose value corresponds to a *particular state* and is completely independent of the sequence of steps by which that state was achieved. Many of the properties with which we are familiar, such as pressure, temperature, and density, are state properties.

A system may consist of one or more *phases*. A phase is defined as *a completely homogeneous and uniform state of matter*. (This definition is strictly valid only for an equilibrium phase. However, since all phases encountered in this text are at equilibrium, no ambiguity will result if the term "phases" is used alone.) Although both ice and water have a uniform composition, they do not have a uniform consistency or density and would therefore be considered as two different phases. On the other hand, two immiscible liquids would possess different compositions and thus regardless of densities would be considered as separate phases.

In general, matter may exist in gaseous, liquid, or solid phases. For single-component systems no more than three phases can exist at one time, and that is possible only under a unique set of conditions. For multicomponent systems it is possible to have additional phases present. For systems containing more than one phase, a complete description of the system must also include the relative amounts of the phases present, as well as the specification of a sufficient number of properties to fix the state of the individual phases present.

1.6 Units

Any physical measurement must be expressed in units. For example, if the length of a football field was expressed as 100 long, it would convey no meaning. However, if we add a unit and say the field is 100 yards long, the statement has meaning. The term "yard" is a unit of length whose value is defined in terms of a reference length which is kept at the National Bureau of Standards. There are many units which the engineer is likely to encounter. Some of these units consist of groupings of other units. In certain cases different units may be used to express the same physical property (for example, "foot" and "yard" are both units of length) and therefore are interconvertible.

The subject of units has been greatly confused by the use of different sets of units in the scientific and engineering communities in the United States and several other countries. The scientific community generally uses the metric system; the engineering community uses the engineering, or British, system. Although the engineering and metric systems appear at first glance to be basically different, we shall see that they are actually quite similar. In any case, use of the engineering units is fading. The United States is the only major country still using the British system. Most of the rest of the world uses the *Metric System Internationale d'Unites*, or S.I. units, and it appears likely that the U.S. will begin switching in the near future. In order to illustrate both the similarities and differences between the engineering and S.I. units, we will use both liberally throughout the text.

A vast number of physical units exist. However, the great majority of these units can be expressed in terms of the four fundamental dimensions of mass, length, time, and temperature. The dimensions for both force and energy can be expressed from these four fundamental dimensions. Most of the difficulty encountered in the use of dimensions and units arises because the relation between the dimensions (and units) of mass and force is not properly understood. This problem is compounded greatly by the different mass–force relations used in the engineering and metric systems.

Let us now examine the similarities and differences between the engineering and S.I. systems of units. In the S.I. system the unit of length is the meter; in the English system the unit of length is the foot. In both systems the basic unit of time is the second. The S.I. unit of mass is the kilogram; the English unit of mass is the pound (abbreviated lb_m, for pound mass). Up to this point the S.I. and engineering systems are very similar, except for the numerical factor for converting feet to meters or lb_m to kg. However, as we stated before, the unit of force is a derived quantity and the S.I. unit of force is defined in a slightly different manner from the English units of force. A second metric system, the CGS system, is widely encountered in the U.S. physical sciences. In the CGS system the basic units are the centimeter (length), gram (mass), and second (time). The CGS system is very similar in all respects to the S.I. system and we will treat both of them as more or less comparable metric systems. It appears, however, that use of the CGS system is likely to fade as the S.I. system becomes more entrenched in the U.S.

The unit of force may be derived from Newton's law of acceleration, which says that the force, F, necessary to uniformly accelerate a body is directly proportional to the product of the mass of the body and the acceleration it undergoes. That is,

$$F \propto M \cdot a \tag{1-1}$$

or

$$F = \frac{Ma}{g_c} \tag{1-2}$$

where g_c is the universal conversion factor, whose magnitude and units depend on the units chosen for F, M, and a.

For example, in the metric system the unit of force is either the newton (S.I.) or the dyne (CGS). The newton is defined as the force needed to accelerate a 1-kg mass at 1 m/sec^2; the dyne is the force necessary to accelerate a 1-g mass at 1 cm/sec^2. Substitution of these values into equation (1-2) gives

$$1 \text{ newton} = \frac{1 \text{ kg} \cdot 1 \text{ m/sec}^2}{g_c} \tag{1-3}$$

and

$$1 \text{ dyne} = \frac{1 \text{ g} \cdot 1 \text{ cm/sec}^2}{g_c} \tag{1-4}$$

We may now solve for g_c as

$$g_c = 1 \frac{\text{kg m}}{\text{newton sec}^2} = 1 \frac{\text{g cm}}{\text{dyne sec}^2} \tag{1-5}$$

Thus, in the metric system g_c has the value unity, and the dimensions (where a dimension is a whole class of units)

$$g_c [=] \frac{\text{mass} \cdot \text{length}}{\text{force} \cdot \text{time}^2} \tag{1-5a}$$

where [=] represents "has the dimensions of."

It would seem natural to define the units of force in the engineering system in a manner similar to that of the metric system, that is, as the force necessary to accelerate a 1-lb_m mass at 1 ft/sec^2. The *poundal* is, in fact, defined in just this manner. However, the poundal has never received great acceptance as a unit of force and is hardly ever seen. The pound force, lb_f, is the most frequently used unit of force in the engineering system; it is defined as the force necessary to accelerate 1 lb_m at 32.174 ft/sec². (It should be noted that 32.174 ft/sec² is the acceleration of the earth's gravity field at the equator.) Therefore, in the engineering system g_c is defined as

$$g_c = \frac{1\ lb_m \cdot 32.174\ ft/sec^2}{1\ lb_f} \tag{1-6}$$

or

$$g_c = 32.174 \frac{lb_m\ ft}{lb_f\ sec^2} \tag{1-6a}$$

The weight of an object can be calculated from equation (1-2) by remembering that the weight of a body is identical to the force necessary to accelerate the body at the same rate it would accelerate during free fall in a vacuum. Therefore, the weight of a body is expressed as

$$W = \frac{Mg}{g_c} \tag{1-7}$$

where g is the acceleration of gravity and W the weight.

In the engineering system of units we may express the weight of a 1 lb_m as

$$W = \frac{1\ lb_m \cdot 32.174\ ft/sec^2}{32.174\ lb_m/lb_f \cdot ft/sec^2}$$
$$= 1\ lb_f \tag{1-8}$$

That is, in the engineering system of units the magnitudes of the weight and mass of a body are identical at sea level, where $g = 32.174$ ft/sec², when the weight is expressed as lb_f and the mass as lb_m, and the acceleration of gravity is 32.174 ft/sec². Thus the lb_f might alternatively have been defined as the force necessary to support a 1-lb_m mass against the forces of gravity at the equator.

The weight in newtons of a 1-kg mass is given by

$$W = \frac{1\ kg \cdot 9.8\ m/sec^2}{1\ kg \cdot m/sec^2} \cdot \text{newtons} \tag{1-9}$$

where $g = 9.8$ m/sec², or

$$W = 9.8 \text{ newtons} \tag{1-9a}$$

Thus in the metric system the weight of a mass does not have the same numerical value as its mass, whereas in the engineering system it does. On the other hand, in the metric system the magnitude of $g_c = 1$; in the engineering system the magnitude of g_c = the magnitude of g at the equator.

Since most scientific fields prefer to use metric, rather than engineering, units, the omission of the term g_c causes no serious problem, and the practice of not including g_c in the equations involving the conversion of mass units to force units is almost universal. *However, in the engineering system neglect of g_c can be catastrophic, because the magnitude of g_c is not equal to unity.* Thus it is of primary importance that the proper use of g_c be fully understood.

Table 1-1 lists the more commonly used systems of units and their respective mass, length, time, and force conversion.

Table 1-1 Common Systems of Units

System of Units	Unit of Length	Unit of Time	Unit of Mass	Unit of Force	g_c	Definition of the Force Unit
Engineering	ft	sec	lb_m	lb_f	$32.174 \frac{lb_m\text{-ft}}{lb_f \sec^2}$	Force needed to accelerate a 1-lb_m mass at 32.174 ft/sec^2
〃	〃	〃	lb_m	poundal	$1 \frac{lb_m}{\text{poundal}} \frac{\text{ft}}{\sec^2}$	Force needed to accelerate a 1-lb_m mass at 1.0 ft/sec^2
Metric (CGS)	cm	sec	gm	dyne	$1 \frac{\text{g}}{\text{dyne}} \frac{\text{cm}}{\sec^2}$	Force needed to accelerate a 1-g mass at 1.0 cm/sec^2
Metric (S.I.)	m	sec	kg	newton	$1 \frac{\text{kg}}{\text{newton}} \frac{\text{m}}{\sec^2}$	Force needed to accelerate a 1-kg mass at 1.0 m/sec^2
Combined	cm	sec	gm	g_f	$\frac{980 \text{ g cm}}{g_f \sec^2}$	Force needed to accelerate a 1-g mass 980 cm/sec^2
Combined	m	sec	kg	kg_f	$\frac{9.8 \text{ kg m}}{kg_f \sec^2}$	Force needed to accelerate a 1-kg mass 9.8 m/sec^2

Let us now examine the dimensions and units of some of the quantities most frequently encountered in engineering studies. Since all units that have identical dimensions must be interconvertible, conversion factors between these units must exist. Lists giving the commonly needed conversion factors and some other useful constants are presented as Tables 1-4 and 1-5.

1. *Work* has the dimensions of force · length. In the engineering system, this is ft-lb_f. In the S.I. system, the unit of work is the newton-meter, or joule. We shall see in later chapters that because both work and heat are energy terms, they must have the same dimensions. In the engineering system heat is expressed in British thermal units (Btu's). The conversion factor between

ft-lb_f and Btu is approximately

$$1 \text{ Btu} = 778 \text{ ft-lb}_f \tag{1-10}$$

2. *Kinetic energy*, being a form of energy, must also have dimensions of force · length. Kinetic energy is evaluated from the formula

$$\text{KE} = \frac{1}{2}\frac{Mu^2}{g_c} \tag{1-11}$$

In the English system

$$M = \text{mass, lb}_m$$
$$u = \text{velocity, ft/sec}$$
$$g_c = 32.17 \text{ lb}_m \text{ ft/lb}_f \text{ sec}^2$$
$$\text{KE } [=] \text{ ft-lb}_f$$

In the S.I. system

$$M = \text{mass, kg}$$
$$u = \text{velocity, meter/sec}$$
$$g_c = 1 \text{ kg-m/newton sec}^2$$
$$\text{KE } [=] \text{ N m, or joule}$$

SAMPLE PROBLEM 1-1. Calculate the kinetic energy, in Btu, of a 4000-lb_m car traveling at 60 mph. Repeat using S.I. units.

Solution:

$$60 \text{ mph} = 88 \text{ ft/sec}$$
$$\text{KE} = \frac{1}{2}\frac{4000 \text{ lb}_m (88 \text{ ft/sec})^2}{32.17 \text{ lb}_m \text{ ft/lb}_f \text{ sec}^2}$$
$$= 4.8 \times 10^5 \text{ ft lb}_f$$
$$= 618 \text{ Btu}$$

In S.I. units we get

$$u = 60 \text{ mph} = 26.82 \text{ m/sec}$$
$$M = 4000 \text{ lb}_m = 1816 \text{ kg}$$
$$\text{KE} = \frac{1}{2}\frac{1816 \text{ kg } (28.62 \text{ m/sec})^2}{1 \text{ (kg m/newton sec}^2)}$$
$$= 7.44 \times 10^5 \text{ Nm} = 7.44 \times 10^5 \text{ J}$$

---

3. *Potential energy* also has the dimensions of force · length and for gravitational potential energy is evaluated from the formula

$$\text{PE} = \frac{MgZ}{g_c} = WZ \tag{1-12}$$

where M = mass
Z = height
g = local acceleration of gravity
g_c = universal conversion constant
W = weight

The units of the various terms in the English and S.I. units are summarized below:

Quantity	*English*	*S.I.*
Mass	lb_m	kg
Height	ft	m
g	$32.17\ ft/sec^2$	$9.8\ m/sec^2$
g_c	$32.17 \dfrac{ft\text{-}lb_m}{lb_f\text{-}sec^2}$	$1.0 \dfrac{kg\ m}{N\ sec^2}$
W	lb_f	N
PE	$ft\text{-}lb_f$	Nm or J

4. *Pressure* has the dimensions of force/area. Often hydrostatic pressure is measured by the formula

$$P = \frac{\rho g h}{g_c} \tag{1-13}$$

where ρ = density
g = acceleration of gravity
h = height of liquid
g_c = universal conversion constant

The discussion of the fourth fundamental dimension, temperature, is taken up in Chapter 2. Its role in thermodynamics is of paramount importance, and a more detailed analysis of its true meaning is imperative.

SAMPLE PROBLEM 1-2. A normal atmosphere will support a column of mercury 0.760 meter high. Calculate the pressure exerted by this atmosphere in both S.I. and English units. The density of mercury is 13.60 g/cm³.

Solution: We use equation (1-13) to calculate the pressure:

$$P = \frac{\rho g h}{g_c}$$

We begin in metric units:

$$\rho = 13.60\ g/cm^3 = 1.36 \times 10^4\ kg/m^3$$

$$g = 9.80\ m/sec^2$$

$$h = 0.76\ m$$

$$g_c = 1.0 \frac{kg\ m}{N\ sec^2}$$

Therefore

$$P = \frac{(1.36 \times 10^4)(9.8)(0.76)}{1}\left[\frac{\text{kg}}{\text{m}^3}\frac{\text{m}}{\text{sec}^2}\text{m}\bigg/\left(\frac{\text{kg m}}{\text{N sec}^2}\right)\right]$$

$$= 1.013 \times 10^5 \text{ N/m}^2 = 1.013 \times 10^5 \text{ pascal}$$

In English units:

$$\rho = 13.60 \text{ g/cm}^3 = 847.3 \text{ lb}_\text{m}/\text{ft}^3$$

$$g = 32.17 \text{ ft/sec}^2$$

$$h = 0.76 \text{ meter} = 2.49 \text{ ft}$$

$$g_c = 32.17 \text{ lb}_\text{m} \text{ ft/lb}_\text{f}\text{-sec}^2$$

Therefore

$$P = \frac{(847.3)(32.17)(2.49)}{(32.17)}\left[\frac{\text{lb}_\text{m}}{\text{ft}^3}\frac{\text{ft}}{\text{sec}^2}\text{ft}\bigg/\left(\frac{\text{lb}_\text{m}\text{ ft}}{\text{lb}_\text{f}\text{-sec}^2}\right)\right]$$

$$= 2109.8\frac{\text{lb}_\text{f}}{\text{ft}^2} \equiv 14.7\frac{\text{lb}_\text{f}}{\text{in.}^2} \quad \text{or} \quad 14.7 \text{ psi}$$

The previous examples described some of the physical quantities whose units can be expressed in terms of the fundamental units of length, mass, and time. Many other such quantities exist. A summary of the more commonly encountered physical quantities and their derived units in both the English and S.I. systems are shown in Table 1-2. Also listed in Table 1-2 are the special names and symbols which have been given to many of these quantities in the S.I. system. Table 1-3 summarizes the names and units of the basic electrical and magnetic quantities in the S.I. system.

Table 1-2 Summary of Derived Quantities

	English Units		S.I. Units			
Quantity	Expression in Terms of Other Units	Expression in Terms of Base Units	Name	Symbol	Other Units	Base Units
Area		ft^2				m^2
Volume		ft^3				m^3
Velocity		ft/sec				m/s
Acceleration		ft/sec^2				m/s^2
density		$\text{lb}_\text{m}/\text{ft}^3$				kg/m^3
frequency		sec^{-1}	hertz	Hz		s^{-1}
force	lb_f	$\text{ft lb}_\text{m}/\text{sec}^2$	newton	N		m kg/s^2
pressure	$\text{lb}_\text{f}/\text{ft}^2$	$\text{lb}_\text{m}/\text{sec}^2\text{ ft}$	pascal	Pa	N/m^2	$\text{kg/s}^2\text{m}$
energy or work	ft-lb_f	$\text{lb}_\text{m}\text{ ft}^2/\text{sec}^2$	joule	J	Nm	$\text{kg m}^2/\text{s}^2$
power	$\text{ft-lb}_\text{f}/\text{sec}$	$\text{lb}_\text{m}\text{ ft}^2/\text{sec}^3$	watt	W	J/s	$\text{kg m}^2/\text{s}^3$

Table 1-3 Summary of Electrical and Magnetic Quantities in S.I. Units

			S.I. Unit	
Quantity	Name	Symbol	Expression in Terms of Other Units	Expression in Terms of S.I. Base Units
Electric Current	ampere	A	Base unit in S.I. system	
Quantity of electricity, electric charge	coulomb	C	A·s	s·A
Electric potential, potential difference, electromotive force	volt	V	W/A	$m^2 \cdot kg \cdot s^{-3} \cdot A^{-1}$
Capacitance	farad	F	C/V	$m^{-2} \cdot kg^{-1} \cdot s^4 \cdot A^2$
Electric resistance	ohm	Ω	V/A	$m^2 \cdot kg \cdot s^{-3} \cdot A^{-2}$
Conductance	siemens	S	A/V	$m^{-2} \cdot kg^{-1} \cdot s^3 \cdot A^2$
Magnetic flux	weber	Wb	V·s	$m^2 \cdot kg \cdot s^{-2} \cdot A^{-1}$
Magnetic flux density	tesla	T	Wb/m^2	$kg \cdot s^{-2} \cdot A^{-1}$
Inductance	henry	H	Wb/A	$m^2 \cdot kg \cdot s^{-2} \cdot A^{-2}$

For convenience we have listed several of the more commonly used conversion factors in Table 1-4. Table 1-5 gives conversion factors between the most commonly encountered work or energy units.

Table 1-4 Miscellaneous Factors

1. Miscellaneous Units

Multiply	*by*	*to get*
ft^3	7.48	U.S. gal
ft^3	28.3	liters
ft^3	0.0283	m^3
U.S. gal	3.79	liters
lb_m	0.45359	Kg
in.	2.54	cm
ft	0.3048	m
m	3.2808	ft
kg/m^3	0.0624	lb_m/ft^3
atm	1.013×10^5	pascal
lb_f	4.448	newton

2. Pressure

$$1 \text{ atm} = 760 \text{ mm Hg} = 14.7 \text{ psia} = 0.00 \text{ psig}$$
$$= 33.9 \text{ ft } H_2O = 29.9 \text{ in. Hg} = 1.013 \times 10^5 \text{ Pa}$$

3. Temperature

$$0°C = 32.0°F = 273.15°K† = 491.67°R†$$

†These can be rounded to 273°K and 492°R.

Table 1-5 Energy-Conversion Factors†

	joules = 10^7 ergs	$kg_f m$	$ft\text{-}lb_f$	kw hr	hp hr	liter atm	kcal	Btu	g cal
joules	1	0.102	0.738	2.77×10^{-7}	3.73×10^{-7}	9.69×10^{-3}	2.39×10^{-4}	9.48×10^{-4}	0.239
kg_f m	9.80	1	7.23	2.72×10^{-6}	3.65×10^{-6}	9.68×10^{-2}	2.34×10^{-3}	9.29×10^{-3}	2.344
$ft\text{-}lb_f$	1.36	0.14	1	3.77×10^{-7}	5.05×10^{-7}	1.34×10^{-2}	3.24×10^{-4}	1.29×10^{-3}	0.324
kw hr	3.6×10^{6}	3.67×10^{5}	2.66×10^{6}	1	1.34	3.55×10^{4}	8.61×10^{2}	3.41×10^{3}	8.61×10^{5}
hp hr	2.68×10^{6}	2.74×10^{5}	1.98×10^{6}	0.746	1	2.65×10^{4}	6.42×10^{2}	2.55×10^{3}	6.42×10^{5}
liter atm	1.01×10^{2}	10.3	74.7	2.82×10^{-5}	3.77×10^{-5}	1	2.42×10^{-2}	9.60×10^{-2}	24.2
kcal	4.18×10^{3}	4.27×10^{2}	3.09×10^{3}	1.17×10^{-3}	1.56×10^{-3}	41.3	1	3.97	1×10^{3}
Btu	1.06×10^{3}	1.07×10^{2}	7.78×10^{2}	2.93×10^{-4}	3.93×10^{-4}	10.4	0.252	1	2.52×10^{2}
g cal	4.18	0.427	3.09	1.16×10^{-6}	1.56×10^{-6}	4.13×10^{-2}	1×10^{-3}	3.97×10^{-3}	1

†To find a value in units of those along the top row, take the value you have in units of the left column and *multiply* by the number in the corresponding box of that row and column. Example: Btu × 778 = $ft\text{-}lb_f$.

Table 1-6 summarizes the accepted prefixes for scale changes in the S.I. system.

Table 1-6 S.I. Prefixes for Scale Factors

Factor	Prefix	Symbol	Factor	Prefix	Symbol
10^{12}	tera	T	10^{-1}	deci	d
10^{9}	giga	G	10^{-2}	centi	c
10^{6}	mega	M	10^{-3}	milli	m
10^{3}	kilo	k	10^{-6}	micro	μ
10^{2}	hecto	h	10^{-9}	nano	n
10^{1}	deka	da	10^{-12}	pico	p
			10^{-15}	femto	f
			10^{-18}	atto	a

Problems

1-1. A 1-kg mass is accelerated with a force of 10 lb_f. Calculate the acceleration in ft/sec^2 and cm/sec^2.

1-2. What is the weight of a 50-g mass (in CGS units)?

1-3. CO_2 is contained in a vertical cylinder at a pressure of 30 atm by a piston with a mass of 57 lb_m. If g is 32.4 ft/sec^2 and the barometric pressure is 29.7 in. Hg, what is the area of the piston?

1-4. A spring scale is calibrated to read pounds mass at a location where the local acceleration of gravity is 32.40 ft/sec^2. (That is, 1-lb mass placed on the scale will give a scale reading of 1, a 2-lb mass will give a reading of 2, etc.) If the scale is moved to a location where g is 32.00 ft/sec^2, what will be the mass, in pounds mass, of an object which, when placed on the scale, gives a reading of 20.00? Justify your method of calculation.

1-5. A $\frac{1}{2}$-lb_m ball traveling at 60 mph is struck by a bat. If the bat imparts an impulse of 4 lb_f sec on the ball, what is the velocity in ft/sec at which the ball leaves the bat? (*Hint:* You should remember from physics that the impulse is equal to the change in momentum—both vector quantities.)

1-6. Determine the amount of energy (in joules, calories, watt-seconds, and Btu) that the bat transmits to the ball in Problem 1-5.

1-7. A 500-lb_m pile driver hammer freely falls a distance of 10 ft before impacting a wooden pile. Determine the velocity of the driving hammer at the instant it hits the pile. What is the kinetic energy of the driver at this instant. Express your answer in Btu and joules.

1-8. A space station in the form of a large donut is 1000 ft in diameter. The space station is rotated at a rate of 1 revolution per minute to generate an artificial gravity. What is the artificial acceleration of gravity so generated? Determine the "weight" of an astronaut whose mass is 150 lb_m.

TWO

thermodynamic properties

2.1 Measurable and Conceptual Properties

The description of the state of matter in terms specific enough to permit a unique characterization is an important first step in the development of thermodynamics. Certain parameters, because of their fundamental significance and our ability to measure them, have long been used to describe a system. Mass, composition, temperature, pressure, and volume are such properties. The mass of a substance can be determined by measuring its interaction with the earth's gravitational field, either with a simple spring scale or by comparing its mass with a known mass by means of a balance. Composition and volume can also be determined by fairly standard procedures. In the case of temperature and pressure, we are familiar with instruments or devices that can provide us with an accurate measure of these properties over a large range of each parameter.

Experience has shown that these properties are not sufficient to describe completely all the transformations that matter might undergo. It was in response to such a need that the concepts of energy and entropy were introduced. Neither energy nor entropy can be measured directly on an energy or entropy meter. Values are usually expressed in relation to an arbitrary reference state, as is the case with the steam tables, where all values are calculated relative to liquid water at 32°F. Thermodynamics has been called by many "the science of energy and entropy." Such a statement seems appropriate in that much of the effort in this book will be devoted to giving the student a better understanding of each of these conceptual properties and the laws and principles that have been developed to relate them to physical changes in the world.

Our inability to measure energy and entropy directly frequently leads to apprehension on the part of the student about accepting them as he does temperature, pressure, or volume. He generally has fewer reservations about energy than entropy, because he has discussed the former in relation to heat, work, motors, and engines in the home or in earlier courses. Thus he has at least a superficial understanding of energy in a practical, if not a thermodynamic, sense. The biggest challenge for any instructor or textbook is to develop a similar appreciation for the role of entropy.

2.2 Intensive and Extensive Properties

The properties we have mentioned above can be subdivided into two categories: Those which relate to the amount of matter present and those which do not. If matter existed in a uniform state and one wished to characterize it with the properties just discussed, the composition, pressure, and temperature would be the same regardless of whether one were to describe all the matter or any part of it. Properties such as these, which are independent of the amount of matter, are termed *intensive* properties.

On the other hand, the mass or volume of the matter would vary directly as the amount is changed, even though the state remains unchanged. Properties that are dependent on the extent or size of the system are termed *extensive* properties. In addition to mass and volume, energy and entropy also fall into the extensive category, as our later discussions will demonstrate.

Extensive properties can be converted to a unit-mass basis by simply dividing the extensive property by the mass of the system. Extensive properties presented on a per-unit basis are termed *specific* properties and are represented by a lower case italic letter. Thus for volume, $v = V/M$. Since the specific properties are independent of the amount of matter considered, they are also intensive in nature. Thus it is possible to convert an extensive parameter to an intensive form.

2.3 Mass and Volume

The *mass* of a system is a measure of the amount of matter present and is directly proportional to the number of molecules in the system (the proportionality constant being the molecular weight divided by Avogadro's number). The standard unit of mass is the kilogram in the metric (S.I.) system and the pound mass, lb_m, in the engineering system. The mass of an unknown body may be determined by comparison with standard masses on a balance, or by use of a scale. Normally the symbol M is used to express mass.

Volume is a measure of the physical size of system and is defined as that portion of space occupied by the system. The volume of a body may be determined directly by measuring its physical dimensions or indirectly by measuring the amount of fluid (say water) that the body displaces. The units of volume are cubic meters in the metric (S.I.) system and cubic feet in the engineering system. The symbol V is used to indicate volume.

2.4 Pressure and Temperature

The pressure exerted on or by a system is defined quite directly as the normal force exerted per unit area of surface. If the normal force is uniformly distributed across the entire surface, then the pressure is simply given as the ratio of the force to the total surface area:

$$P = \frac{F_n}{A} \tag{2-1}$$

If the normal force is nonuniformly distributed across the system's surface, then the pressure is not constant. In this case, the local pressure is simply the ratio of the local normal force to the increment of area of over which that force is exerted.

Most pressure gauges read the difference in pressure between the body under consideration and the surrounding atmosphere. Such readings are called *gauge pressures* and are positive for pressures above atmospheric and negative for pressures below atmospheric. (Such readings are often referred to as *vacuum pressures*). In thermodynamics, however, we are always interested in the *absolute pressure*. This is the pressure that would be read if the atmospheric pressure were identically zero. In the general case, the absolute and gauge pressures are related by

$$P_{\text{abs}} = P_{\text{gauge}} + P_{\text{atm}} \tag{2-2}$$

A simple quantitative definition of the property *temperature* is unfortunately a little more difficult. The concept of temperature evolved historically from man's wish to quantify the physical sensation of hotness or coldness. It had been empirically observed that various substances appeared to change their volumes more or less in relation to the hotness of a body with which they were in contact. Thus the earliest *thermometers* (or temperature-measuring devices) were constructed by placing a column of mercury (or other fluid) into two baths whose temperatures were arbitrarily fixed. A freezing mixture of ice and water and a boiling mixture of steam and water were the most common such baths. The heights of the mercury column in these two reference baths were marked and recorded, and the remainder of the temperature scale was formed by uniformly dividing the distance between the hot and cold reference works into as many subdivisions as desired. The familiar Fahrenheit and centigrade scales are established when the ice and boiling points are chosen as indicated below:

	Fahrenheit	*Centigrade*
Ice temp.	32	0
Boiling temp.	212	100

Unfortunately, the temperature scale so defined is totally arbitrary and has no fundamental significance. To make matters worse, if we build a second thermometer with a different working fluid (alcohol, for example), calibrate the ice and boiling points of water, and then divide the scale into the same number of increments as the mercury scale, we find that the mercury and alcohol thermometers yield slightly different temperatures when both are used to measure the temperature of a given body. Which, if either, thermometer would be "correct"? In point of fact, neither thermometer would be *exactly* correct, although both will be close enough for the large majority of applications.

To develop a thermometer scale which is both of fundamental significance and independent of the choice of working fluid, we must first examine the P–v–T behavior of gases at low pressures. In 1660 Robert Boyle, while investigating vacuums and the behavior of gases under vacuum, discovered that the pressure and specific volume of a gas were inversely related under isothermal conditions. That is:

$$Pv = f(T) \tag{2-3}$$

It was not until about 100 years after Boyle's experiments that Jacques Charles observed that, when a quantity of gas was heated at constant pressure, the increase in volume was always directly proportional to the temperature change as shown in Fig. 2-1.

By extrapolating to the point where $v = 0$, it is possible to define a new temperature scale T' such that

$$T' = T + \Delta \tag{2-4}$$

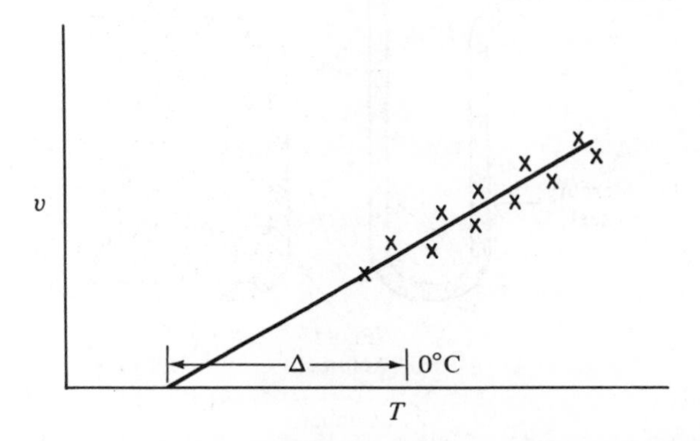

Figure 2-1. Volume–temperature relationship observed by Charles.

In the new temperature scale $T' = 0$ when $v = 0$, and the volume and temperature are directly proportional:

$$v \propto T' \tag{2-5}$$

Combining (2-5) and (2-3) we then obtain the so-called ideal-gas equation of state:

$$Pv = RT' \tag{2-6}$$

where R is the proportionality constant between Pv and T'. (For future reference we drop the prime and simply refer to the ideal gas temperature as T.)

Thus we may construct an ideal-gas temperature scale, and an ideal-gas thermometer with which to read it, as suggested by equation (2-6). The thermometer could either be a constant-pressure or constant-volume arrangement, although experimental factors usually dictate constant volume. The constant-volume thermometer consists of a gas bulb, seal arrangement, and pressure sensor. The seal and pressure sensor are commonly a U-tube manometer. A sample thermometer might be arranged as shown in Fig. 2-2.

Once the bulb is filled with the desired quantity of gas, mercury from the reservoir is forced into or out of the U tube until the mercury column in the right leg is exactly at the reference mark. The pressure in the gas bulb is then read directly from the manometer. A single fixed (or reference) temperature is used to establish the proportionality factor between pressure and temperature, and the thermometer is ready for use.

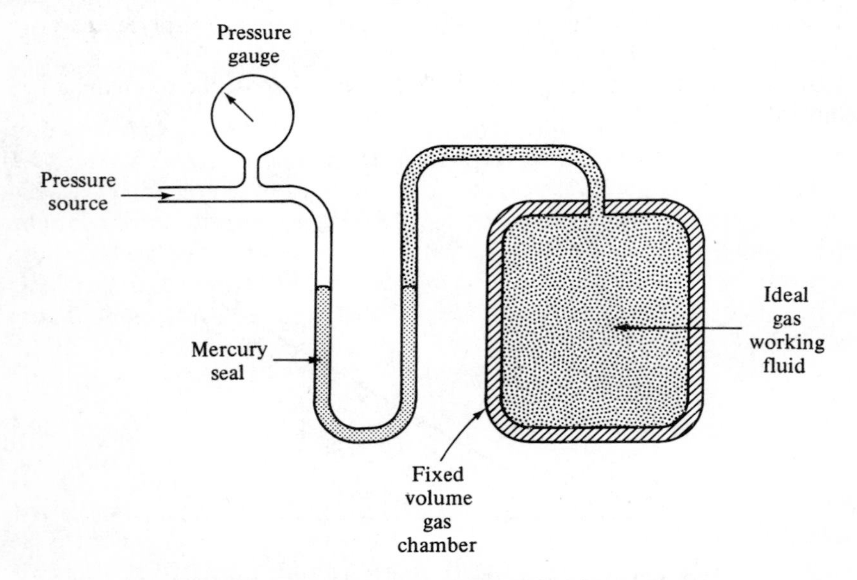

Figure 2-2. The constant volume idea-gas thermometer.

The ideal-gas thermometer and its ideal-gas temperature scale have several distinct advantages over the older mercury thermometers with Fahrenheit or centigrade temperature scales:

1. The zero temperature has a special significance—it is the temperature at which an ideal gas would have a vanishing specific volume. (Clearly, no real gas satisfies this condition since the atoms making up the gas have some nonzero volume. But this "molecular volume" is small enough to be negligible at all but the lowest temperatures when the pressure is low enough.)
2. The temperatures obtained from an ideal-gas thermometer are independent of the working fluid, provided the working fluid is at a low enough pressure.
3. Lastly, a minor point, the thermometer requires only a single arbitrary reference point. (By convention the triple point of water is assigned a temperature of 273.16° Kelvin or 491.69° Rankine on the ideal-gas temperature scale.)

The ideal-gas temperature scale we have just described is termed an *absolute temperature scale* because the point of zero temperature has a rather fundamental significance independent of the fluid used in the thermometer. We will show later that this temperature scale is also equivalent to the *thermodynamic temperature scale* which will be described in our discussion of the entropy function.

Although at first glance the ideal-gas thermometer appears to be simple and straightforward to use, in practice its use is extremely difficult and time-consuming. In addition, it is not useful at very low temperatures, where no gas behaves ideally, or at very high temperatures, where the confining vessel would melt. To overcome these difficulties, the International Committee of Weights and Measures adopted in 1968 a new temperature scale known as the International Practical Temperature Scale of 1968 (IPTS 68) and a specific set of thermometers for measuring temperatures according to this scale. The new scale is based on a series of accurately reproducible fixed points whose temperatures have been accurately established by means of an ideal-gas thermometer. The practical scale then consists of a series of interpolating procedures for measuring temperatures between the fixed points.

2.5 Internal Energy and Enthalpy

The energy directly associated with the existence of matter is termed its *internal energy* and is unrelated to its bulk position, or velocity (and hence the potential or kinetic energy of the total mass). Although the internal energy is not associated with the potential or kinetic energy of the total mass, the same is not true for the atomic and subatomic units from which the mass is comprised. Indeed, a significant portion of a mass's internal energy is associated with the kinetic and potential energy of its atomic and subatomic constituents.

In an effort to better understand the basis of internal energy, let us consider matter from a microscopic point of view. Inasmuch as internal energy is related to the structure of matter, it seems reasonable to assume that an absence of mass in a given system would result in an absence of what we define as internal energy. Although such an assumption fixes a zero value for internal energy, the problem of determining the internal energy when several molecules are then added to a system is extremely perplexing. Let us now discuss the various types of energies we may associate with atoms and their interactions.

Translational Energy

In the gaseous phase, molecules are known to move about randomly. Each molecule possesses kinetic energy by virtue of its finite mass and velocity. In the liquid state, we find the molecules somewhat more restricted in their movement. Thus the contribution of translational energy is reduced. However, since clusters of molecules can still undergo short-range random motion, translational effects are not completely absent. In the solid state, the lattice structure essentially eliminates any possibility of significant translational contributions to internal energy.

Rotational Energy

Consider the CO_2 molecule, which has a dumbbell configuration as shown in Fig. 2-3. In addition to its translational energy, the molecule may also rotate about some axis through its center of gravity as indicated in Fig. 2-3. Just as a rotating flywheel possesses rotational energy, so does a rotating molecule. Rotational energy can be exchanged with neighboring molecules or converted to translational energy by means of molecular collisions. For liquids and solids, rotational contributions are less important because molecular motion is restricted by interactions with other nearby molecules.

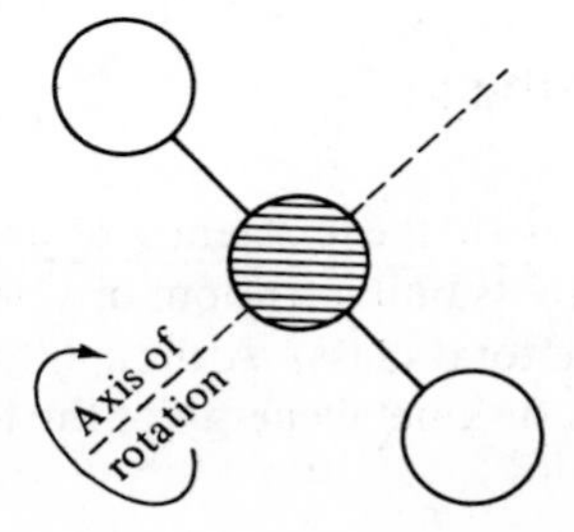

Figure 2-3. CO_2 molecule.

Vibrational Energy

Further consideration of the CO_2 molecule suggests the possibility of a vibrational motion within the molecule. Strong bonding forces, which act like springs, hold the carbon and oxygen atoms together. When excited, the oxygen atoms can vibrate with respect to the central carbon atom as shown in Fig. 2-4. Since these oscillations are capable of activating other molecules through molecular collisions, they, too, must be considered in computing the total internal energy of a substance. In the gaseous phase, both vibrational and rotational contributions are associated only with polyatomic molecules. In the solid state the lattice structure virtually eliminates the possibility of translational or rotational energy, but the vibrational contribution of atoms about their lattice points is one of the principal contributions to the internal energy of matter.

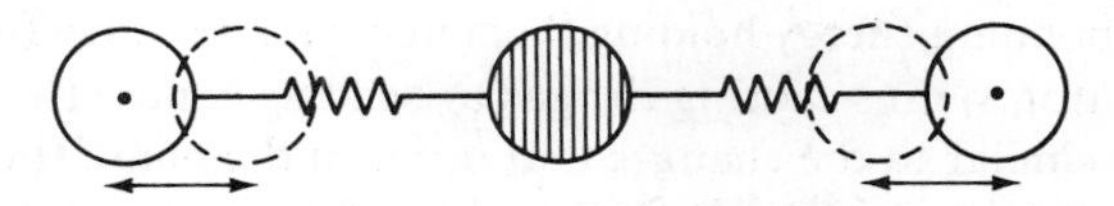

Figure 2-4. Vibrational energies.

Atomic and Molecular Interactions

When molecules, or atoms, approach each other, strong intermolecular, or interatomic, forces come into play. Although these forces are repulsive at very short intermolecular spacing, they rapidly become attractive as the spacing increases. Since the energy levels associated with intermolecular forces are quite significant, these forces exert a controlling influence on the state of matter. As long as the molecules possess sufficient translational energy to overcome these attractive forces, matter remains in the gaseous state. As the energy content is reduced, the translational energy may no longer be sufficient to overcome the intermolecular forces; then a change from the gaseous to the liquid state occurs.

In the liquid state the molecules continue to translate, although not nearly as freely as in the gas phase. As the energy content is further reduced, intermolecular forces become more important, and all translation ceases as the liquid state transforms to the solid state.

Nuclear and Electron Contributions

In attempting to account for the internal energy possessed by matter, we have thus far examined only the modes of energy that atoms or molecules might possess by virtue of their relative motion or position. An attempt to further subdivide matter into the more fundamental particles poses great difficulties. Many of our analogies with macroscopic phenomena vanish and quantum mechanical principles must be introduced if these subatomic contributions are to be considered. Until recently thermodynamics did not concern itself with processes that disturbed the atomic makeup of matter. The atomic structure was thus assumed to remain constant, and internal energies have always been referred to a reference state that ignored the enormous energy content of the atom itself. Processes were always assumed to involve only changes in the molecular-energy modes previously cited.

In recent years the plasma (highly ionized gas) has become of engineering significance; therefore, the energies associated with the plasma state warrant some discussion at this point. Modern atomic theory pictures the atom as highly concentrated nuclear matter surrounded by orbiting electrons. Because of the positively charged nucleus, the electrons experience an attractive force that tends to hold them in orbit. If the internal energy of a gas is raised to a sufficiently high level, the outer electrons are energized to a point where they overcome the bonding energy holding them in orbit and move freely throughout the sea of atoms, thus causing the gas to become conductive. This ionization process is similar to the changes that occur in the phase transitions from solid to liquid to gas and thus is frequently said to lead to a fourth state of matter. The movement of electrons from higher or excited levels down to the lower or more stable levels occurs with a release of energy. Thus it can be seen that the energy possessed by electrons within an atom can also contribute to the internal energy of matter.

Finally, the advent of fission has demonstrated that the atom itself possesses a quantity of energy which dwarfs all contributions cited so far. Inasmuch as our concerns in thermodynamics will not involve nuclear reactions, no attempt is made beyond the Einstein relationship to account for this form of internal energy. Thermodynamic analyses generally are concerned with processes producing changes in the state of matter which do not affect its atomic makeup. Thus only those energy changes arising from the translational, rotational, vibrational, and intermolecular modes will be of concern to us when treating systems of fixed composition. If processes are to be studied in which matter experiences a change in its chemical composition, the energy changes associated with chemical bonding must also be considered.

Calculation of Internal Energy

The internal energy, then, describes the total quantity of energy matter possesses by virtue of its existence. This energy is observed to be a function only of the state that the matter exists in, and not a function of how the matter got to that state. Thus the internal energy is a state function, or *property*, whose value is determined once the state of the system is specified. Since the energy associated with matter is clearly a function of the mass under consideration, the internal energy is an extensive property, to which we give the symbol U. The specific internal energy, u, is the intensive property obtained by dividing the internal energy by the total mass:

$$u = \frac{U}{M} \tag{2-7}$$

Therefore, the internal energy (or more exactly, the specific internal energy u) takes its place alongside T, P, v, etc., as one of the properties that is fixed by, or may be used to fix, the state of the system.

In thermodynamics we are seldom interested in absolute values of internal energy. The working relationships that will be developed in later chapters call for dU or ΔU, the differential or finite internal energy change between two states. Since it is the difference that is of interest to us, we can use any arbitrary reference state as a basis for calculating internal energy in all other states. Since the reference state cancels out when we difference the internal energy between two states, the value we assign to U in the reference state is unimportant. Typically, the reference state is assigned a value of zero at a convenient temperature and pressure. The reference conditions are commonly, but by no means universally, chosen to be 32°F and 1 atm.

If, for example, the reference state of water was specified to be liquid water at 32°F, its internal energy could be evaluated in higher and lower energy states by evaluating the changes in the various energy modes which contribute to the internal energy. These changes can be determined experimentally through calorimetric measurements (as discussed in Chapter 4) or, for certain special molecules, theoretically using statistical mechanics and the kinetic theory. The calorimetric measurements being *macroscopic*, or bulk, measurements do not allow us to determine the changes in the various energy modes, but give us only the total energy change.

Enthalpy

In much of our discussion in Chapter 4 we will find it convenient to define a new property, which we will call the *enthalpy*, H, by combining the internal energy and the product of the pressure times the volume:

$$H = U + PV \tag{2-8}$$

The enthalpy is also an extensive property whose value is completely determined once the state of the system is fixed. The specific enthalpy, h, is obtained by dividing the enthalpy by the total mass:

$$h = \frac{H}{M} \tag{2-9}$$

Both internal energy and enthalpy are energy quantities and hence have the dimensions of energy—force $\times$ length. The units are typically Btu in the engineering system and joules in the metric or S.I. system.

2.6 Entropy

In Chapter 1 we observed that all forms of mechanical, chemical, electrical, magnetic, and any other nonthermal energy can be completely converted from one form to the other. We also observed that any of these forms can be converted to a thermal form. However, no one has yet devised a means for completely converting thermal energy to any of the interconvertible (mechanical) forms. Thus, when mechanical energy is degraded to the thermal form, the energy experiences an irreversible change after which it is less useful to us. Although energy is always conserved, its usefulness may not be.

Entropy, first suggested by the German physicist Clausius, is a quantitative measure of the degradation which energy experiences as a result of changes in the universe. Like energy, entropy is a conceptual property which cannot be measured directly. If the entropy is to be of any quantitative value to us, it must be a function only of the state of the system, rather than its history. Thus, as we develop the concept of entropy, we note that entropy will be a state property. Also, since entropy is to measure the degree of irreversibility encountered during the change in a system and/or its surroundings, the total entropy of the universe remains constant during a reversible change and increases during an irreversible one.

Although the preceding discussion has shown how the total entropy of the universe changes, it has not dealt with the entropy of a single body. We can, however, use the facts already developed to relate the entropy of a single system to its state. Consider an isolated system containing, for example, two separate sections containing gas at different pressures P_{I} and P_{II}. If the

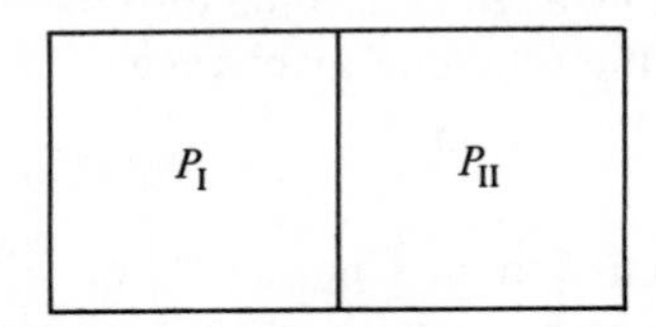

partition between the sections is removed, a highly irreversible change in the system occurs as the two pressures equilibrate. Since the change is irreversible, the total entropy of the universe increases. However, the surroundings have undergone no change during the pressure equalization, and thus their state and entropy remain constant. Thus we conclude the entropy of an isolated system increases when the system undergoes an irreversible change.

Let us consider now a slightly different change in the isolated system considered above: Assume that a small gas turbine connects the two chambers. As gas flows from the high-pressure chamber to the low-pressure one, a portion of the gas's thermal energy is converted to mechanical energy by the gas turbine. The mechanical energy is stored by having the turbine lift a weight. Under the appropriate frictionless conditions such a change can be completely reversible since it is possible (in theory at least) to run the turbine backwards as a pump with the falling weight suppling the mechanical energy needed to run the pump. Since this change is reversible, the total entropy of the universe remains constant. As before, the surroundings undergo no change so their entropy remains fixed. Thus, as an isolated system undergoes a reversible change its entropy remains constant.

Now suppose that the mechanical energy stored during the reversible pressure equalization is degraded to thermal energy which is transferred to the gas within the isolated system. Since this is clearly irreversible, the total entropy of the universe increases—the magnitude of this increase being directly proportional to the amount of mechanical energy degraded to the thermal form. As before, the surroundings remain unchanged, so their entropy is constant. Thus the entropy of the system increases in direct proportion to the amount of mechanical energy degraded into the thermal form.

Let us return to the above example at the point where the pressures are equal, but mechanical energy is still stored in the lifted weight. This time, let the surroundings transfer into the system an amount of thermal energy equivalent to that stored in a mechanical form by the raised weight. At the end of both sequences the thermal energy content of the systems are equivalent. However, so are their entropies. The equivalence of their entropies can be demonstrated by observing that the only difference between the two final states is the presence of the stored mechanical energy in the previous example. However, we have shown that mechanical energy can be (reversibly) transferred to the surroundings without changing the entropy of the system.

Thus we learn that the addition of thermal energy to a system—either by internal dissipation of mechanical energy or by transfer of thermal energy from the surroundings—always leads to an increase in the entropy of the system. In addition, the systems' entropy increase is directly proportional to the amount of thermal energy added to the system. On the other hand, we have shown earlier that the (reversible), flow of mechanical energy into or out of a system has no such entropy change directly associated with it. Thus we conclude that flows of thermal and mechanical energy have strikingly different effects on the entropy of a system and its surroundings.

We may now summarize the characteristics we associate with the property we have called entropy:

1. The entropy is a quantitative measure of the degradation which energy experiences as a result of changes in the universe.

2. For a reversible change, the total entropy of the universe remains fixed. For an irreversible change, the total entropy of the universe increases in direct proportion to the amount of mechanical energy degraded to a thermal form.

3. Reversible transfer of mechanical energy to or from a system may be

accomplished without changing the entropy of either the system or surroundings.

4. The transfer of thermal energy into (out) of a system leads to an entropy increase (decrease) in the system which is directly proportional to the amount of thermal energy transferred.

The concept of entropy originally evolved in a purely macroscopic sense and, as we have shown, was quantitatively related to mechanical and thermal energy transfers. However, recent advances in our understanding of the molecular (microscopic) nature of matter and the application of statistical and quantum mechanics to the structure of matter have led to a microscopic description of the property entropy which is in many ways considerably more revealing than that given by macroscopic considerations alone. A simplified semiquantitative microscopic development of entropy is included in Appendix G for those students who desire another view of the entropy concept and its relationship to some of the other basic properties of matter—volume, mass, and energy.

We will return in Chapter 5 to our development of the concept of entropy. At that time we will show that the proportionality constant between entropy changes and thermal energy flows is the reciprocal of the absolute temperature ($1/T$). Although we have not yet verified this important result, it does indicate that entropy changes associated with thermal energy flows are dependent on the temperature at which the energy flow occurs—the higher the temperature, the lower the entropy change, and vice versa.

Although the third law of thermodynamics establishes an absolute reference state for the entropy—the entropy of a pure crystalline solid at a temperature of absolute zero is zero—most property tables and charts are based on some arbitrary reference state. In the thermodynamic calculations of interest to us in this text, absolute values of the entropy are not essential—only entropy changes (from which the reference state entropy drops out) are needed.

2.7 Interrelationships Between Properties

The properties discussed in the previous sections are interrelated with all other physical properties for every state in which matter exists. The study and development of these relationships has been, and continues to be, the subject of much research. We stated previously that each state of matter is characterized by a unique set of properties. We observe, for example, that when water is taken at atmospheric pressure and 70°F, it always possesses the same density or specific volume. Since the state of a substance is not related to the mass of the substance considered, we correctly note that v, the specific volume, or $1/v$, the density, both intensive properties, are parameters that are functions of the temperature and pressure of the substance. Exactly the same observation could be made for steam at a given pressure and temperature, or for any single-phase substance: If v is measured at a given value of pressure and temperature, it will always have the same value.

Suppose we ask ourselves if the same value of v would result if only one property, such as temperature, were specified while pressure was permitted to change. If the pressure on the water was increased to 10 atm and the density very carefully determined, a slight change could be observed. The amount of such a change for liquids and solids, both relatively incompressible phases of matter, would be small but finite. For the gaseous phase, steam, an increase in the pressure on a fixed mass produces an obvious reduction in the volume and hence v. Thus it is clear that specification of temperature alone does not result in a uniquely defined state for any single-component, single-phase system. Rather, experience indicates that two properties are required to specify a third property for a single-component, single-phase system. However, the choice of properties is somewhat arbitrary. For example, if one chose to fix T and v for these systems, the same value of P would always result.[1]

In each of these instances we have observed that specifications of two intensive properties was sufficient to define uniquely a third intensive property. Although pressure, temperature, and volume have been used in this illustration, any other intensive property would behave in an analogous fashion. The specific internal energy for either the water or steam would always be the same, regardless of how many times one were to measure it for given values of P and T. The same would be true for entropy, enthalpy, and the free energies (these properties will be introduced and discussed later) if measured on a unit-mass basis. We thus conclude *on the basis of many observations* that *the specification of any two independent intensive properties is sufficient to determine all other intensive properties and hence the state of a single-phase, single-component substance.*

If we perform a series of experiments in which two phases of a single component are held in equilibrium with each other, we observe that specification of only a single property of one phase completely specifies the states of the individual phases. For example, specification of the temperature of a steam–water mixture fixes the pressure exerted by the mixture, as well as the density, entropy, and other properties of the *individual phases*. If we examine an equilibrium mixture containing three phases of a single component, we would find that no properties can be arbitrarily specified without losing one

[1] This uniqueness may occasionally break down with two or three possible values of temperature yielding the same density at a given pressure—for example, the density of liquid water is a maximum at +4°C at 1 atm. Thus for each temperature below 4°C there is another temperature above 4°C where the two densities are identical at 1 atmosphere pressure.

phase. The requirement of three coexisting equilibrium phases for a single component automatically fixes all the properties of the individual phases.

Thus we see for a single-component system, experience indicates that the number of degrees of freedom, f (that is, the number of properties of the individual phases that can be arbitrarily specified before fixing the states of the individual phases), is given by the expression

$$f = 3 - P \tag{2-10}$$

where P is the number of coexisting phases.

Equation (2-10) is a special case of the *Gibbs phase rule*, which expresses the degrees of freedom for a multicomponent-multiphase mixture:

$$f = C + 2 - P \tag{2-11}$$

where C is the number of components.

Equation (2-11) tells us that a two-phase mixture of a single component can exist only at a single pressure (temperature) once the temperature (pressure) of the mixture has been specified. However, equation (2-11) offers no clue about the form of this relationship. The equilibrium-phase diagram illustrated in Fig. 2-5 allows us to convey concisely the equilibrium pressures and temperatures for which various pairs of phases may be in equilibrium with each other.

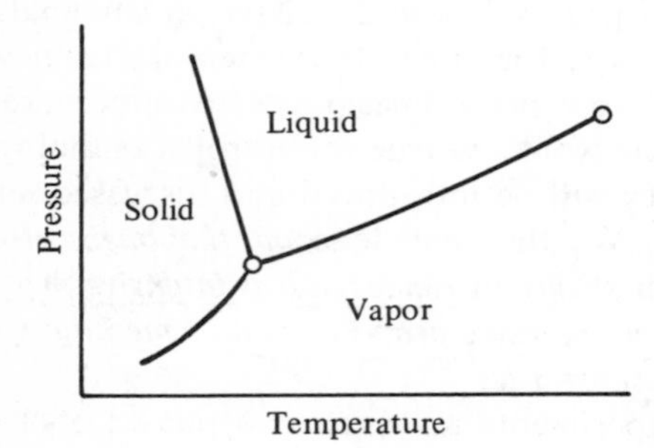

Figure 2-5. Equilibrium-phase (or P–T) diagram for water.

We may determine the equilibrium phase (phases) of a system by plotting its pressure and temperature on the phase diagram. If the point falls within a single-phase region, the system will exist in that phase. If the point falls on the dividing line between two phases, these phases *may* coexist at equilibrium (a phase that can exist at equilibrium with one or more other phases is called *saturated*). Thus the solid lines represent the pressures and temperatures at which two *saturated phases* may coexist. (There is, however, *no guarantee* that both phases *do exist*.) The intersection of the three lines gives the *triple point*, the point at which three saturated phases *may* coexist.

The line separating the solid and liquid regions on the P–T diagram is the *freezing-point line* and shows the effect of pressure on freezing temperature. For almost all substances this line is nearly vertical and indicates that the freezing temperature is relatively independent of pressure. Since the freezing temperature normally increases with pressure, the slope of the freezing-point line is usually positive. Water, however, behaves in the opposite direction and has a negatively sloped freezing-point line, because the freezing temperature of water decreases with increasing pressure. Indeed, it is just this phenomenon that is involved when an ice skate melts a thin film of water under the skate blade. This film of water acts as a lubricant and allows the skate to glide smoothly over the ice. In very cold areas the temperature may drop so low that the skate blade cannot melt the ice. When this occurs, conventional ice skating is not possible.

The line that separates the liquid and vapor phases (and represents the pressures and temperatures at which *saturated* liquid and vapor may coexist) is known as the *vapor–pressure curve*. As opposed to the freezing-point line, which may have a positive or negative slope, all vapor-pressure curves have a positive slope (the same is true for the line that separates the solid and vapor regions).

Since the vapor-pressure line has a positive slope, we know that an increased pressure is needed to liquefy a gas when its temperature is increased. At higher pressures the gaseous phase becomes more dense, and eventually it is impossible to distinguish between the gaseous and liquid phases. The point on the vapor-pressure curve where the two phases become indistinguishable is known as the *critical point*. The pressure and temperature at the critical point are known as the *critical pressure* and *critical temperature*, respectively. The critical temperature is the highest temperature at which a liquid phase may coexist in equilibrium with a separate vapor phase. The critical point is frequently determined by plotting some property (such as density) of the vapor and liquid phases against the equilibrium pressure or temperature and extrapolating to the point where the properties of the phases are equal. In this manner it is possible to determine the critical conditions without having to observe the disappearance of the liquid phase.

The complete volumetric behavior of a substance may be represented by its P–v–T surface. An example of such a surface for a substance with a normal freezing curve is shown in Fig. 2-6.

Plotting any two of the three properties (P, v, T) determines a point on the surface and thus fixes the state of the material. (If the point so located falls on the phase boundaries indicated by the P–T plot, then the properties of the mixture of phases that may result can be obtained from the properties of the individual phases as indicated in the next paragraph.) The P–v–T surface may be divided into several readily classifiable regions: Central to the diagram is the *saturation dome* which encloses the *solid–vapor* and *liquid–vapor* two-phase regions. When the substance falls within this region it is in reality composed of a mixture of saturated vapor and saturated liquid (or solid). In a similar manner, the *solid–liquid* region to the left of the diagram indicates the region where saturated solid and liquid can coexist. To the left of the

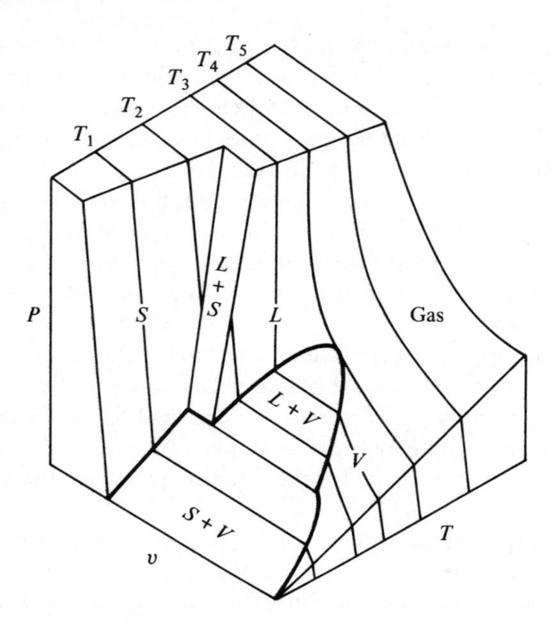

Figure 2-6. The *P-v-T* surface.

solid–liquid region we find *pure subcooled solid.* The *line of constant temperature,* or *isotherm,* which passes through the critical point is known as the *critical isotherm.* The region between the critical isotherm and the solid-liquid region is a *subcooled liquid* (except, of course, right at the phase boundary lines where a saturated liquid exists). At temperatures above the critical temperature we have what is called a gas. Although this gas may be under extremely high pressure and possess a density higher than some liquid densities (especially those near the critical point), the gaseous material will always completely fill its confining vessel and shows no meniscus. Thus the material is not a liquid, and we call it a gas, or *permanent gas.* The region to the right of the saturation dome, but below the critical isotherm, also represents a *gaseous* material. However, this material is termed a *vapor* to indicate that it may be condensed to a liquid via an isothermal pressure increase, as opposed to the gaseous material which cannot be liquified via any isothermal change.

The *P–T* diagram of Fig. 2-5 may be obtained from the *P–v–T* surface by simply examining the diagram end-on from the temperature surface. The sharp breaks in the *P–v–T* surface give the phase boundaries directly. In a similar manner, if we look at the diagram from the volume surface we obtain the *P–v* diagram of the substance. To make the *P–v* diagram more useful in a quantitative sense, the volumetric behavior along various selected isotherms is frequently added. The resulting *P–v* diagram is then as illustrated in Fig. 2-7.

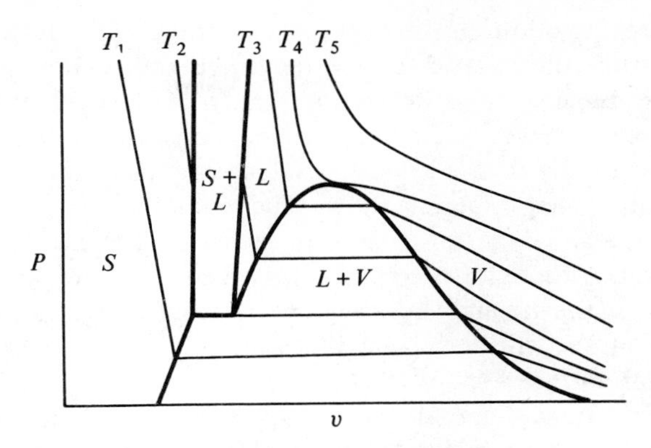

Figure 2-7. The *P*-*v* diagram.

When a system contains a mixture of two saturated phases, it is necessary to know the relative amount of each phase before the total properties of the mixture can be calculated. For example, the specific volume, v, of a mixture is given by the total volume of the mixture divided by the total mass of the mixture. The volume of the mixture is the sum of the volumes of the individual phases, as, of course, is the mass:

$$V_{m \cdot x} = \sum_{i=1}^{P} M^i v^i \tag{2-12}$$

where i is the phase counter. The specific volume of the mixture is then

$$v_{\text{mix}} = \frac{V_{\text{mix}}}{M_{\text{mix}}} = \sum_{i=1}^{P} \frac{M^i}{M_{\text{mix}}} v^i = \sum_{i=1}^{P} x^i v^i \tag{2-12a}$$

where x^i is the mass fraction of mixture in phase i. Thus the specific volume of mixture is simply the weighted average of the specific volumes of the individual phases.

Although equation (2-12a) has been derived for the specific volume, it is equally applicable to any specific property of the mixture, e_{mix}:

$$e_{\text{mix}} = \sum_{i=1}^{P} x^i e^i \tag{2-13}$$

2.8 Equations of State

The equation of state expresses a relationship between two or more thermodynamic properties. For single-component, single-phase systems the equation of state will always involve three properties, any two of which may be assigned values independently. Although, in principle, functional relationships involving any three thermodynamic properties (such as T, P, v, internal energy, and entropy) might be constructed, analytical expressions of the interrelationships among properties have been almost completely limited to

P, T, and v. Because of our incomplete understanding of the interactions that occur among molecules, particularly in the liquid and solid states, empirical methods have been used in developing many of the widely used P–v–T equations of state. Since pressure, volume, and temperature can all be measured directly, the data needed to evaluate the constants in such equations can be obtained experimentally. The equation of state for P–v–T behavior is thus an analytic expression for the P–v–T surface discussed in the previous section. Although the commonly used equations of state are widely applicable only in the gas and vapor regions, a few are useful a short way into the liquid region. No currently available equation of state is applicable for the entire P–v–T surface.

Properties such as internal energy, u, or entropy, s, are not directly measurable and must be calculated from other properties and appropriate thermodynamic relations. Equations of state in which u or s appear as variables have not been independently developed, although we will see in Chapter 6 how these may be derived from the P–v–T equation of state. (We shall henceforth restrict our use of the term "equation of state" to mean only a relation among P, v, and T.)

An equation of state may be long and complicated, sometimes involving up to 15 terms—as in the Martin-Hou equation[2]—or short and simple, with as few terms as the one term of the ideal-gas equation. The choice of which equation to use in a given application depends greatly on the desired accuracy and the endurance of the user. Since the coefficients of almost all equations of state must be evaluated by fitting the equations to various experimental P–v–T data, the equations can never be more accurate than the data they represent. However, in many instances the equation of state cannot adequately represent the available P–v–T data, and thus limits our accuracy. This is particularly true when the simpler equations are applied in the region of the critical point. We shall now examine a few of the simpler equations of state and their limitations. For a more detailed discussion the interested reader is referred to *Chemical Engineering Thermodynamics*.[3]

As we discussed previously in our development of the ideal-gas temperature scale, the early experiments by Boyle and Charles indicated that the pressure, volume and temperature of gases (or vapors) at low pressures could be represented by the *ideal-gas equation of state*

$$Pv = RT \tag{2-14}$$

or

$$PV = nRT \tag{2-14a}$$

where R is a constant and T is the ideal-gas temperature. The constant R is

[2] J. J. Martin, and Y. C. Hou, *Amer. Inst. Chem. Engrs. J.*, **1**, 142 (1955).

[3] R. E. Balzhiser, M. R. Samuels, and J. D. Eliassen, *Chemical Engineering Thermodynamics*, Prentice-Hall, Inc., Englewood Cliffs, N. J., 1972.

the ideal-gas constant and has different numerical values, depending on the units of P, V, n, and T. Frequently used units of R, and its corresponding numerical values, are presented in Table 2-1.

Table 2-1 Values of the Ideal-Gas Constant

Value of R	Units
8.317×10^7	ergs/(g-mol °K)
1.9872	cal/(g-mol °K)
8.3144	J/(g-mol °K)
0.082057	(liter atm)/(g-mol °K)
82.057	(cm^3 atm)/(g-mol °K)
62.361	(liter mm Hg)/(g-mol °K)
0.0848	(kg_f/cm^2 liter)/(g-mol °K)
998.9	(mm Hg ft^3)/(lb-mol °K)
1.314	(atm ft^3)/(lb-mol °K)
1.9869	Btu/(lb-mol °R)
7.805×10^{-4}	(hp hr)/(lb-mol °R)
5.819×10^{-4}	(kw hr)/(lb-mol °R)
0.7302	(atm ft^3)/(lb-mol °R)
555	(mm Hg ft^3)/(lb-mol °R)
10.731	(psi ft^3)/(lb-mol °R)
1545	(lb_f-ft)/(lb-mol °R)
1.851×10^4	(lb_f-in.)/(lb-mol °R)

Equation (2-14) is a formal statement of the ideal-gas equation of state. Since it was derived from experimental observations of real gases at low pressures, the ideal-gas equation of state is widely used for calculations involving gases and vapors under these conditions. Unfortunately, as the pressure is raised above 5 to 10 atm ($5 - 10 \times 10^5$ N/m^2), most gases no longer behave according to the ideal-gas equation of state, and more complicated equations are needed to obtain reasonable engineering accuracy.

It is interesting to note that the ideal-gas equation of state may be derived from statistical mechanics if the gas molecules are assumed to be infinitesimal spheres which occupy no volume and which undergo completely elastic collisions with each other and with the walls of the container. It must also be assumed that no intermolecular attraction or repulsion occurs between the gas molecules.

The van der Waals Equation

In an effort to correct the ideal gas equation of state for its two worst assumptions—infinitesimal molecular size and no intermolecular forces—van der Waals proposed the following relation:

$$\left(P + \frac{a}{v^2}\right)(v - b) = RT \tag{2-15}$$

The term b is meant to account for the finite size of the gas molecules and is sometimes referred to as the "molecular volume." Its value depends on the size and nature of gas molecules. The term a/v^2 is a correction which was meant to account for the attractive forces that exist between molecules. This attractive force tends to increase the effective pressure on the gas, and therefore is added to the external pressure to get the total effective pressure.

The two constants in the van der Waals equation may be chosen to fit experimental P–v–T data in any small region. However, since there are only two constants in the equation, we would not expect it to accurately describe P–v–T data over a great range of P, v, or T. However, the van der Waals equation does allow for greater accuracy than the ideal gas equation, and therefore may be used when a simple equation of state with somewhat greater accuracy than the ideal gas equation is needed.

It is possible to determine the a and b of the van der Waals equation without recourse to specific P–v–T data from the following general observation: Experimental P–v data on all real substances *at their critical temperature* indicates that the P–v isotherm goes through a horizontal inflection point at the critical pressure of the substance, as shown in Fig. 2-7. That is, both the first and second derivatives of P with respect to v vanish at the critical conditions:

$$\left(\frac{\partial P}{\partial v}\right)_T = \left(\frac{\partial^2 P}{\partial v^2}\right)_T = 0 \qquad (2\text{-}16)$$

when $P = P_c$ and $T = T_c$.

Substituting these relations into the van der Waals equation we find that a and b are related to the critical P and T of the material by the following relations:

$$a = \frac{27}{64} R^2 \frac{T_c^2}{P_c}$$
$$b = \frac{RT_c}{8P_c} = \frac{v_c}{3} \qquad (2\text{-}17)$$

and

$$\frac{P_c v_c}{RT_c} = \frac{3}{8} \qquad (2\text{-}18)$$

The critical constants (and van der Waals constants calculated from them) for several common gases are presented in Table 2-2.

However, it must be stressed that these values will not produce results nearly as accurate as those found when a and b are determined from fitting experimental P–v–T data over a small region. In fact, the critical volume is, in general, not predicted accurately by these constants.

As we have indicated, real-gas equations of state frequently get considerably more complex than the van der Waals equation presented above. However, in the interests of brevity we will cease our discussion at this point.

Table 2-2 Critical† and van der Waals Constants for Several Gases

Gas	P_c, atm	T_c, °R	a, atm ft⁶/(lb-mol)²	b, ft³/lb-mol
O_2	50.1	278.6	349	0.510
N_2	33.5	227.1	346	0.618
H_2O	218.3	1165.3	1400	0.486
CO	34.5	240	374	0.630
CO_2	72.9	547.5	924	0.685
CH_4	45.8	343.9	579	0.684
C_2H_6	48.2	549.8	1410	1.04
C_3H_8	42.0	665.9	2370	1.45
C_4H_{10}	37.5	765.2	3670	1.94
NH_3	111.3	729.8	1080	0.598
H_2	12.8	59.9	63	0.427

†Critical constants taken from K. A. Kobe and R. R. Lynn, Jr., *Chem. Rev.*, **52**, 47–236 (1953).

Those interested in further discussion of this still evolving area, are referred to *Chemical Engineering Thermodynamics*.[3]

2.9 The Law of Corresponding States

If the constants expressed by equation (2-17) are substituted in the van der Waals equation, an equation of state in terms of *reduced variables* is obtained:

$$\left(P_r + \frac{3}{V_r^2}\right)(3V_r - 1) = 8T_r \tag{2-19}$$

where $P_r = P/P_c$ = reduced pressure
$T_r = T/T_c$ = reduced temperature
$V_r = v/v_c$ = reduced volume

Although we recognize that equation (2-19) is not an accurate representation of the volumetric behavior of real gases, it does suggest that it may be possible to develop a "generalized" equation of state in terms of reduced properties that would be applicable to all real gases. That is, if we specify any two reduced properties (P_r, T_r, or V_r), the third one is fixed. Since pressure and temperature are the easiest properties to arbitrarily fix, these are the properties usually specified. Gases at the same reduced pressure and temperature are said to be in *corresponding states*. The *law of corresponding states* postulates that *all fluids in corresponding states should have the same reduced volume.*

We now introduce the compressibility factor,

$$Z = \frac{Pv}{RT} \tag{2-20}$$

as a measure of deviation of real gas behavior from ideal-gas behavior. The compressibility may be expressed in terms of reduced variables as

$$Z = \frac{P_r V_r}{RT_r} \frac{P_c v_c}{T_c} = \frac{P_r V_r}{T_r} \frac{P_c v_c}{RT_c} \tag{2-20a}$$

The term $P_c v_c/RT_c$ is simply the compressibility at the critical point, Z_c, so equation (2-20a) becomes

$$Z = Z_c \frac{P_r V_r}{T_r} \tag{2-21}$$

Since the law of corresponding states indicates that V_r is a universal function of P_r and T_r, then equation (2-21) suggests that the compressibility factor should also be a universal function of P_r and T_r for all gases that have the same critical compressibilities. Examination of the critical properties (P_c, v_c, and T_c) of most real gases indicates that their critical compressibilities fall within a fairly restricted range $0.20 \leq Z_c \leq 0.31$. Thus the law of corresponding states indicates that the compressibility factors for all gases should be essentially a universal function of P_r and T_r. Shown in Fig. 2-8 are some experimental curves of compressibility factors as functions of P_r for various values of T_r.

Examination of the data in Fig. 2-8 shows that it is possible to adequately represent the various Z versus P_r isothermal data by single curves. For many engineering applications, the law of corresponding states expressed in terms of compressibilities, with constant Z_c, is a useful method for correlating the P–v–T data of many gases onto a single curve. A generalized compressibility chart for $Z_c = 0.27$ is presented in Appendix C. For the quantum gases hydrogen and helium, it is empirically observed that the generalized charts correspond more accurately to the real-gas behavior if the critical constants used to evaluate T_r and P_r are modified as shown below:[4]

$$T_r = \frac{T}{T_c + 8^\circ\text{K}}$$
$$P_r = \frac{P}{P_c + 8\text{ atm}} \tag{2-22}$$

Examination of the generalized compressibility chart in Appendix C shows several important features of the volumetric behavior of real gases: At reduced temperatures above 2 and reduced pressures below about 6, the volumetric behavior of the real gases are described within about 5 percent by the ideal-gas equation of state. For reduced temperatures below 2, the ideal-

[4] R. H. Newton, *Ind. Eng. Chem.* **27**, 302 (1935).

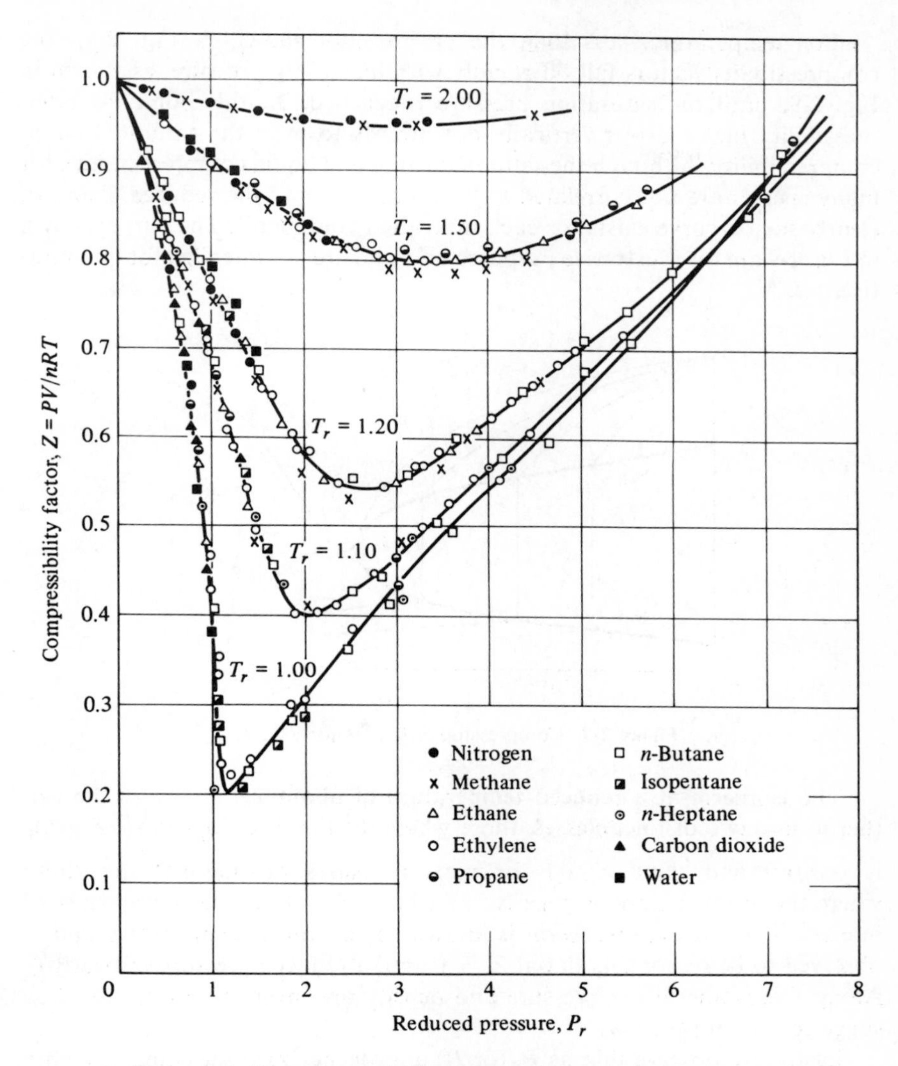

Figure 2-8. Gas compressibilities as functions of reduced pressure and temperature.

gas equation of state will seriously overestimate the specific volume of a gas at known P and T. This overestimation gets continuously more pronounced as the reduced temperature is decreased at constant pressure, but it shows a decided maximum (minimum in the curve) as the pressure is increased at constant temperature. At reduced pressures greater than about 8, the ideal-gas equation of state will underestimate the specific volume for all temperatures greater than the critical.

For temperatures less than the critical (not shown in Fig. 2-8), the compressibility factors fall off rapidly with increasing pressure, as shown in Fig. 2-9, until the saturation pressure is reached. At this point the compressibility factor drops vertically downwards to meet the saturated liquid compressibility. Although the saturated vapor and liquid compressibilities for many species are not correlated well in terms of only the reduced T and P, clearly such a curve exists for each individual compound. The curves shown in Fig. 2-9 are for illustrative purposes only and are not intended for quantitative use.

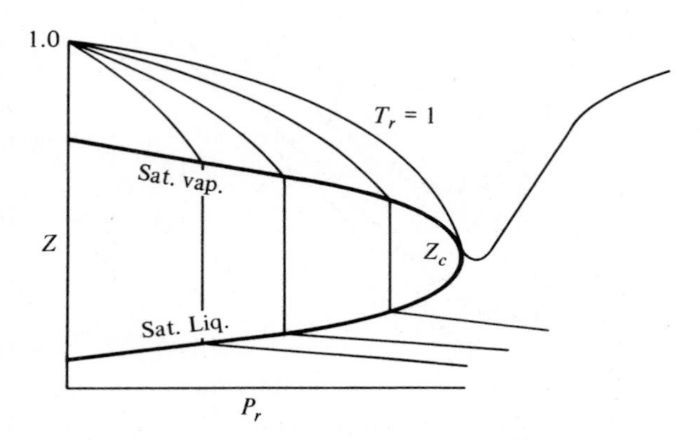

Figure 2-9. Compressibility factors for $T < T_c$.

The isotherm at a reduced temperature of about 2.4 separates the isotherms into two distinct classes, those where the limiting slope $\lim_{P_r \to 0} (\partial Z/\partial P_r)_T$ is positive and show $Z > 1$ for most pressures of interest, and those where the limiting slope is negative and have $Z < 1$ for most pressures of interest. This dividing isotherm is known as the *Boyle* temperature and is observed to be reasonably flat at $Z = 1$ until a reduced pressure of nearly 6. Along this isotherm the pressure and density are inversely proportional as suggested by *Boyle's law.*

Lastly, we observe that as P_r (or P) approaches zero the compressibility factor monotonically and uniformly approaches unity irrespective of the reduced temperature. Thus the basis of our original observation that the volumetric behavior of real gases is correctly described by the ideal-gas equation of state provided that the pressure is low enough.

SAMPLE PROBLEM 2-1. Natural gas is to be transported by pipeline from gas fields in Texas to major markets in the Midwest. The gas, which is essentially pure methane, enters the pipeline at a rate of 70 lb_m/sec at a pressure of 3000 $lb_f/in.^2$ and a temperature of 65°F. The pipeline has a 12-in. inside diameter.

Calculate the inlet density expressed as lb_m/ft^3 and the initial velocity expressed in ft/sec, assuming that methane obeys

(a) The ideal-gas equation of state.
(b) The generalized compressibility factors.
(c) The van der Waals equation of state.

Solution: (a) We begin by discussing the direct solution based on the ideal-gas equation of state:

$$Pv = RT$$

Since $\rho = 1/v$, the equation of state becomes

$$\rho = \frac{P}{RT}$$

Before attempting to substitute numbers into this equation, we must examine the units of the various terms to ensure that ρ will have the proper units. Although the unit problem may seem trivial (and actually is quite simple) for the ideal-gas equation, we shall find that the unit problem is more significant when we attempt to deal with the van der Waals equation.

It has been the experience of the authors that the simplest way of overcoming unit problems is simply to convert the units of all physical quantities into a consistent set of units before any calculations are attempted. In this way the need for conversion factors in the governing equation is eliminated and the chance of error significantly reduced. Since the engineering system of units is still used in this country, we shall use this system in about half of the illustrative problems. The remaining problems will be solved using S.I. units to familiarize the students with this unitary system. For this problem the engineering system will be used. Therefore, all dimensions will be expressed in terms of the units feet (ft), pounds force (lb_f), pounds mass (lb_m), seconds (sec), and degrees Fahrenheit (°F) or degrees Rankine (°R).

In the equation

$$\rho = \frac{P}{RT}$$

we express

$$P = 3000 \text{ psi} = 4.32 \times 10^5 \text{ lb}_f/\text{ft}^2$$

$$T = 65°\text{F} = 525°\text{R}$$

$$R = 1545 \frac{\text{ft-lb}_f}{\text{lb-mol °R}}$$

but the molecular weight of methane is 16 lb_m/lb-mol, so the ideal-gas constant (in mass units) is expressed as

$$R = 1545 \frac{\text{ft-lb}_f}{\text{lb-mol°R}} \frac{\text{lb-mol}}{16 \text{ lb}_m} = 96.7 \frac{\text{ft-lb}_f}{\text{lb}_m \text{°R}}$$

and the density is given by

$$\rho = \frac{4.32 \times 10^5 \text{ lb}_f/\text{ft}^2}{[96.7 \text{ ft-lb}_f/(\text{lb}_m{}^\circ\text{R})] \cdot 525^\circ\text{R}} = 8.5 \frac{\text{lb}_m}{\text{ft}^3}$$

Note that the units on density automatically give lb_m/ft^3 when all other quantities are expressed as shown.

The entering velocity may be calculated from the expression

$$\dot{M} = \rho V A$$

$$V = \frac{\dot{M}}{\rho A}$$

$$\dot{M} = 70 \text{ lb}_m/\text{sec}$$

$$\rho = 8.5 \text{ lb}_m/\text{ft}^3$$

$$A = \frac{\pi D^2}{4} = \frac{\pi (1 \text{ ft})^2}{4} = 0.785 \text{ ft}^2$$

Therefore,

$$V = \frac{70 \text{ lb}_m/\text{sec}}{8.5 \text{ lb}_m/\text{ft}^3 \cdot 0.785 \text{ ft}^2} = 10.5 \text{ ft/sec}$$

(b) To use the generalized charts we must first evaluate the reduced temperature and pressure of the incoming methane:

$$T_r = \frac{T}{T_c}, \qquad P_r = \frac{P}{P_c}$$

but

$$T_c = 343.9^\circ\text{R}$$

$$P_c = 45.0 \text{ atm} = 9.70 \times 10^4 \text{ lb}_f/\text{ft}^2$$

Thus,

$$T_r = \frac{525^\circ\text{R}}{343.9^\circ\text{R}} = 1.53$$

$$P_r = \frac{4.32 \times 10^5 \text{ lb}_f/\text{ft}^2}{9.70 \times 10^4 \text{ lb}_f/\text{ft}^2} = 4.46$$

According to Fig. 2-8, the compressibility factor for these conditions will be $Z = 0.82$. The density is then given by:

$$\rho = \frac{P}{ZRT} = \frac{1}{0.82}\left(\frac{P}{RT}\right) = \frac{1}{0.82}(8.5 \text{ lb}_m/\text{ft}^3) = 10.3 \text{ lb}_m/\text{ft}^3$$

The velocity is then

$$V = \frac{70 \text{ lb}_m/\text{sec}}{(10.3 \text{ lb}_m/\text{ft}^3)(0.785 \text{ ft}^2)} = 8.6 \text{ ft/sec}$$

Under these conditions the ideal-gas equation of state underestimates the density by about 18 percent and overestimates the velocity by a comparable amount.

(c) van der Waals equation:

$$(P + a\rho^2)\left(\frac{1}{\rho} - b\right) = RT$$

We obtain values of a and b for methane from Table 2-2:

$$a = 579 \text{ atm ft}^6/(\text{lb-mol})^2$$
$$b = 0.684 \text{ ft}^3/\text{lb-mol}$$

P and T are given by the problem statement:

$$P = 3000 \text{ lb}_f/\text{in}^2 = 204 \text{ atm}$$
$$T = 65°\text{F} = 525°\text{R}$$

Since the units of the van der Waals constants are given in atm and ft³, we should use the same units for the ideal-gas constant. In this way no internal unit conversions will be necessary. The appropriate value of R can be found in Table 2-1:

$$R = 0.730 \text{ atm ft}^3/\text{lb-mol°R}$$

Thus the density of methane at the pipeline inlet can be obtained by solving:

$$\left(204 \text{ atm} + \frac{579 \text{ atm ft}^6}{(\text{lb-mol})^2}\rho^2\right)\left(\frac{1}{\rho} - 0.684\frac{\text{ft}^3}{\text{lb-mol}}\right) = (0.730)(525)\frac{\text{atm ft}^3}{\text{lb-mol}}$$

Multiplication of both sides by ρ then gives:

$$\left(204 + 579\frac{\text{ft}^6}{(\text{lb-mol})^2}\rho^2\right)\left(1 - 0.684\rho\frac{\text{ft}^3}{\text{lb-mol}}\right) = 384\rho\frac{\text{ft}^3}{\text{lb-mol}}$$

Multiplication and collection of terms then gives:

$$396\rho^3\frac{\text{ft}^9}{(\text{lb-mol})^3} - 579\rho^2\frac{\text{ft}^6}{(\text{lb-mol})^2} + 523\rho\frac{\text{ft}^3}{\text{lb-mol}} = 204$$

or

$$\rho^3 - 1.46\rho^2 + 1.32\rho = 0.492$$

where ρ is expressed in lb-mol/ft³.

We now solve for the density via a direct trial-and-error solution—we use the ideal-gas equation of state density as a first guess:

i	ρ_i	ρ_i^3	$-1.46\rho_i^2$	1.32ρ	Σ	*Comment*
1	0.53	0.15	−0.41	0.7	0.44	Try larger ρ
2	0.56	0.18	−0.46	0.74	0.46	Try larger ρ
	0.60	0.22	−0.53	0.79	0.48	Try slightly larger ρ
	0.62	0.24	−0.56	0.82	0.50	Slightly too large

Although the last guess is too large, it is only slightly so, and the final result is very close to:

$$\rho = 0.615 \text{ lb-mol/ft}^3 = 9.85 \text{ lb}_m/\text{ft}^3$$

which corresponds to a compressibility factor of

$$Z = \frac{P}{\rho RT} = 0.87$$

The inlet gas velocity is then given by:

$$V = \frac{\dot{M}}{\rho A} = \frac{70 \text{ lb}_m/\text{sec}}{9.85 \text{ lb}_m/\text{ft}^3 \cdot 0.785 \text{ ft}^2} = 9.05 \text{ ft/sec}$$

Experimental data on the compressibility of methane reported by Su[5] indicate a compressibility factor of 0.80 for methane under these conditions. This value is about halfway between that predicted by the ideal-gas and van der Waals equations of state. For pressures this high it is somewhat fortunate that either the van der Waals or the ideal-gas equations of state give reasonable values—this result is not to be expected generally.

2.10 Other Property Representations

In theory equations of state can be developed which relate any property of a system to any two other properties. However, the extremely complicated nature of many of these equations make their use extremely cumbersome even where computers are available to handle the calculations. To present property information in an accurate but convenient form for hand or graphical calculations, various property charts and tabulations were developed. Frequently these charts and tables are prepared from calculations based on one of the complex equations of state. However, once tabulated or plotted, these data are readily available.

Probably the simplest form of presenting property data is through the property table, or tabulation. In these tables we usually find h, v, and s (where we use the symbol S for *entropy*) presented as functions of P and T. Since the property tables must be of finite size, h, v, and s can only be presented at a certain number of values of pressure and temperature. If the properties at an intermediate P and/or T are needed, then some form of interpolation must be used. Property tabulations usually allow greater precision than charts, but the charts have the distinct advantage of presenting the data in a somewhat more convenient and readily usable form.

Property tabulations that cover large regions of P and T are available for only a very limited number of gases and liquids. Some of the more readily obtainable property tabulations are those for steam, Freon 12 and Freon 22,

[5] G. J. Su, *Ind. Engin. Chem.*, **38**, 803 (1946).

NH_3, SO_2, H_2, O_2, and air. See, for example, *Mechanical Engineers' Handbook*, R.H. Perry et al., *Chemical Engineers' Handbook* (McGraw-Hill, New York, 4th ed., 1963), and E.I. Du Pont de Nemours, *Freon Property Tabulations*. Probably the most widely used of these tables are those which apply to steam. Since we will have frequent need for these tables, they have been reprinted in part and included with this book as Appendix B. The appendix is divided into three smaller tables. In Table 1 the properties of saturated liquid and vapor (as well as the change on vaporization which is simply the difference between the vapor and liquid values) are presented at uniform increments of temperature. The properties listed are (from left to right): saturation pressure, specific volume, enthalpy, and entropy. In Table 2 essentially the same information is presented, except now for uniform increments in saturation pressure. In this case the first entry to the right is simply the saturation temperature. For two-phase mixtures (or single *saturated* phases) the phase rule tells us that a single property of one of the phases present is sufficient to fix the state of all the phases which may coexist under these conditions. Thus, for example, if we have saturated liquid water at 14.696 psia pressure, Table 2 tells us that its temperature must be 212°F, the specific volume of the liquid will be 0.01672 ft^3/lb_m, and its specific enthalpy 0.3120 $Btu/lb_m{}^\circ R$. Likewise, saturated vapor at 14.696 psia will also be at 212°F, its specific volume will be 26.83 ft^3/lb_m, specific enthalpy 1150.4 Btu/lb_m, and specific entropy 1.7566 $Btu/lb_m{}^\circ R$. For mixtures of saturated liquid and vapor, the mixing rules discussed in Section 2-7, can be used to determine average mixture properties. For temperatures or pressures that do not lie exactly on one of the entries in the table, linear interpolation would normally be used. A comparable set of tables in S.I. units has also been included.

SAMPLE PROBLEM 2-2. A mixture of liquid and vapor water at 325°F is found to contain 32 percent vapor. Determine its average specific volume, density, specific enthalpy, and specific entropy.

Solution: If the mixture contains 32 percent vapor, it must also contain 68 percent liquid. The mixture specific properties can then be obtained from the mixing rule

$$e_{\text{mix}} = x^L e^L + x^V e^V$$

but

$$x^L = 0.68, \qquad x^V = 0.32$$

Thus determination of the saturated liquid and vapor specific properties is all that we need to determine the specific mixture properties. This procedure will not work for the density (why not?). For the mixture density we must use the reciprocal of the mixture specific volume.

The saturated liquid and vapor properties are then read from Table 1 in Appendix B:

	$P_{sat.}$, psia	$v_{sat.}$, $\frac{ft^3}{lb_m}$	$h_{sat.}$, $\frac{Btu}{lb_m}$	$s_{sat.}$, $\frac{Btu}{lb_m°R}$
Sat. Liq.	110.31	0.01782	305.91	0.4835
Sat. Vap.	110.31	11.039	1189.1	1.5949

Thus the mixture specific properties are given by:

$$v_{mix} = [(0.68)(0.01782) + (0.32)(4.039)]\ ft^3/lb_m = 1.305\ ft^3/lb_m$$

$$\rho_{mix} = 1/v_{mix} = 0.7665\ lb_m/ft^3$$

$$h_{mix} = [(0.68)(305.91) + (0.32)(1189.1)]\ Btu/lb_m = 588.5\ Btu/lb_m$$

$$s_{mix} = [(0.68)(0.4835) + (0.32)(1.5949)]\ Btu/lb_m°R = 0.8391\ Btu/lb_m°R$$

SAMPLE PROBLEM 2-3. At a pressure of 50 psia a mixture of liquid and vapor water is found to have specific enthalpy of 925 Btu/lb_m. Determine the moisture content of this mixture.

Solution: According to Table 2, Appendix B, the saturated liquid and vapor enthalpies at 50 psia are 250.09 Btu/lb_m and 1174.0 Btu/lb_m, respectively. The mixture specific enthalpy can be expressed in terms of the saturation enthalpies as

$$h_{mix} = x^L h^L + x^V h^V$$

but

$$x^V = 1 - x^L$$

so that

$$h_{mix} = x^L h^L + (1 - x^L)h^V = h^V + x^L(h^L - h^V)$$

Rearranging and solving for x in terms of the known enthalpies yields:

$$x^L = \frac{h^V - h_{mix}}{h^V - h^L} = \frac{(1174.0 - 925)\ Btu/lb_m}{(1174.0 - 250.1)\ Btu/lb_m} = 0.270$$

or 27 percent moisture.

Table 3 of Appendix B describes the properties of superheated steam. In the superheated region only the gaseous phase can exist and thus two properties are needed to fix the state of the steam. Since the most commonly (but not exclusively) known properties are pressure and temperature; the other properties (v, h, and s) are tabulated as functions of P and T. Note that for each value of P and T, four variables are listed. The first variable is labeled

"Sh" and is the "superheat" of the steam. The quantity is simply defined as the actual temperature minus the saturation temperature at the pressure of interest. The remaining three variables have the meanings: v = specific volume, ft^3/lb_m, h = specific enthalpy, Btu/lb_m, and s = specific entropy, $Btu/lb_m{}^\circ R$. The first entries to the right of the pressure readings are the properties of the saturated liquid and vapor and duplicate exactly the information presented in Table 2. To determine the properties of steam at a given pressure and temperature we first locate the pressure of interest along the left-hand side of the page, and then proceed along this row until coming to the column which contains the desired temperature. Thus, for example, at 45 psia and 500°F the steam would have 225.55°F of superheat, a specific volume of 12.57 ft^3/lb_m, a specific enthalpy of 1283.0 Btu/lb_m, and an entropy of 1.7990 $Btu/lb_m{}^\circ R$. As in the previous discussion, properties at pressures and/or temperatures not specifically listed in the tables would usually be found by linear interpolation (higher order interpolation may occasionally be used when extreme precision is required).

SAMPLE PROBLEM 2-4. Determine the specific enthalpy of steam at 36 psia and 430°F.

Solution: Since neither 36 psia or 430°F correspond to entries in the table, we will need to perform a double interpolation. We will begin by evaluating the enthalpy of steam at 36 psia and 400°F by linear interpolation between 35 psia and 40 psia at 400°F.

$$(h)_{36\,\text{psia},\,400^\circ\text{F}} = \frac{(4)(h)_{35\,\text{psia},\,400^\circ\text{F}} + (1)(h)_{40\,\text{psia},\,400^\circ\text{F}}}{5.0}$$

$$= \frac{[(4)(1236.9) + (1)(1236.4)]\ \text{Btu/lb}_\text{m}}{5.0} = 1236.8\ \text{Btu/lb}_\text{m}$$

Likewise,

$$h_{36\,\text{psia},\,450^\circ\text{F}} = \frac{(4)(h)_{35\,\text{psia},\,450^\circ\text{F}} + (1)(h)_{40\,\text{psia},\,450\,\text{psia}}}{5.0}$$

$$= \frac{[(4)(1260.1) + (1)(1259.6)]\ \text{Btu/lb}_\text{m}}{5.0} = 1260.0\ \text{Btu/lb}_\text{m}$$

The desired enthalpy at 36 psia and 430°F is now found by the second interpolation:

$$h_{36\,\text{psia},\,430^\circ\text{F}} = \frac{(2)(h_{36\,\text{psia},\,400^\circ\text{F}} + (3)(h_{36\,\text{psia},\,450^\circ\text{F}})}{5.0}$$

$$= \frac{[(2)(1236.8) + 3(1260.0)]\ \text{Btu/lb}_\text{m}}{5.0} = 1250.7\ \text{Btu/lb}_\text{m}$$

As an exercise the student should attempt the interpolation in the reverse order, that is, interpolate temperatures first at P = 35 psia and P = 40 psia, and then interpolate on the pressures. Do the two answers agree? Are you surprised? Is this a general result?

SAMPLE PROBLEM 2-5. Steam initially available at 200 psia and 400°F is expanded to atmospheric pressure in such a fashion that (a) its enthalpy remains constant (an *isenthalpic* expansion) or (b) its entropy remains fixed (an *isentropic* expansion). Determine the properties of the resulting material.

Solution: Since we are told that the final pressure is 1 atm (14.696 or 14.7 psia) the determination of any other property is sufficient to fix the state, and hence all other properties, of the outlet stream.

(a) In this part we are told that the expansion takes place at constant enthalpy. The inlet enthalpy is obtained from the inlet pressure and temperature:

$$(h)_{200\text{ psia},\,400^\circ\text{F}} = 1210.8\text{ Btu/lb}_\text{m}$$

Thus the outlet enthalpy is also 1210.8 Btu/lb$_\text{m}$. Now we turn to the $P = 14.696$ row of the steam tables. For this pressure, at 300°F $h = 1192.0$ Btu/lb$_\text{m}$, while at $T = 350$°F $h = 1215.4$ Btu/lb$_\text{m}$. Thus the final temperature is between 300 and 350°F. The actual temperature is found by interpolation.

$$T = 300^\circ\text{F} + 50^\circ\text{F}\left(\frac{1210.8 - 1192.0}{1215.4 - 1192.0}\right) = 341^\circ\text{F}$$

The specific volume and entropy are now found by interpolation using the 341°F final temperature:

$$v = \left[30.52 + \frac{41}{50}(32.61 - 30.52)\right]\frac{\text{ft}^3}{\text{lb}_\text{m}} = 32.20\frac{\text{ft}^3}{\text{lb}_\text{m}}$$

$$s = \left[1.8148 + \frac{41}{50}(1.8446 - 1.8148)\right]\frac{\text{Btu}}{\text{lb}_\text{m}{}^\circ\text{R}} = 1.8685\frac{\text{Btu}}{\text{lb}_\text{m}{}^\circ\text{R}}$$

(b) In this part, the entropy remains constant. The outlet entropy is then equal to the inlet entropy:

$$(s)_{200\text{ psia},\,400^\circ\text{F}} = 1.5599\text{ Btu/lb}_\text{m}{}^\circ\text{R}$$

Thus, the outlet entropy is also 1.5599 Btu/lb$_\text{m}$°R. Again we turn to the $P = 14.696$ row of the steam tables. At 250°F the lowest temperature listed the entropy is still above 1.5599 Btu/lb$_\text{m}$°R. Even the entropy of the saturated vapor is above this figure. However, the entropy of the saturated liquid is not. Thus we conclude that the outlet material must be a mixture of saturated liquid and vapor. The moisture content may be obtained as illustrated in sample Problem 2-3:

$$x^\text{L} = \frac{s^\text{V} - s_\text{mix}}{s^\text{V} - s^\text{L}} = \frac{1.7566 - 1.5591}{1.7566 - 0.3120} = 0.136$$

The specific volume and enthalpy are then obtained from the mixing rules:

$$v_{\text{mix}} = x^L v^L + x^V v^V = v^L + x^V(v^V - v^L)$$
$$= [0.0167 + 0.864(26.828)]\ \text{ft}^3/\text{lb}_\text{m} = 23.2\ \text{ft}^3/\text{lb}_\text{m}$$
$$h_{\text{mix}} = h^L + x^V(h^V - h^L)$$
$$= [180.1 + 0.864(1150.4 - 180.1)]\ \text{Btu/lb}_\text{m} = 1018\ \text{Btu/lb}_\text{m}.$$

Because they are so convenient and simple to use, property charts are perhaps the most widely used method of presenting property data. Two properties are plotted along the horizontal and vertical axes with constant values of additional properties superimposed as lines to complete the chart.

We shall now examine some of the more commonly encountered property charts.

P–v Diagram

The pressure–volume diagram is a graphical representation of the P–v–T behavior of a given substance as illustrated in Fig. 2-7. In the figure are shown lines of *constant temperature*, or *isotherms*. Movement along an isotherm corresponds to an *isothermal* (that is, constant temperature) compression or expansion and gives the density or specific volume as a function of pressure for any given temperature.

Several regions of interest occur in the P–v diagram: The region above the line $T = T_c$ corresponds to temperatures greater than the critical temperature, and therefore represents a fluid state. The region below the line $T = T_c$ corresponds to temperatures below the critical temperature, where it is possible to have more than one phase in equilibrium. The saturated liquid and saturated vapor lines represent the specific volume-pressure relations for the saturated liquid and vapor, respectively. The point where the saturated liquid and saturated vapor lines meet is, of course, the critical point. Note that the critical isotherm is tangent to the saturated-vapor–saturated-liquid line at the critical point.

The region below the saturated-vapor—saturated-liquid line is the two-phase region. In this region specification of P and T will not fix v of the system. v may, however, be determined, as previously demonstrated, from v^{liq} and v^{vap} and the mass fraction of liquid.

The region to the left of the saturated liquid line and below the $T = T_c$ isotherm corresponds to a liquid phase; the region to the right of the saturated vapor line and below the $T = T_c$ isotherm corresponds to a gaseous or vapor phase.

The P–v diagram like all other property diagrams is used in a manner similar to that of the property tables. The state of the material in question is located by plotting the values of the two known properties. The values of any other properties may then be obtained by reading the value of the property line in question which passes through the state point. If the state point does not fall exactly upon a listed property line for the property in question, graphical interpolation is used. Since the property lines are usually closely spaced, this graphical interpolation is often accomplished by "eyeballing it." It is this ability to do interpolation more or less instantly that makes use of the property charts so much quicker than a comparable property tabulation. The price that is obviously paid for this simplicity is precision—while a property tabulation may be precise to four or five significant figures, it is rare to find a property chart that is good to more than two significant figures.

T–s Diagram

The temperature–entropy diagram is one of the more useful of the property diagrams. It finds its greatest application in the analysis of heat and power cycles, and therefore is most readily available for those materials which are used in these cycles. An example of a typical *T–s* diagram is shown in Fig. 2-10. The dome enclosed by the saturated-liquid–saturated-vapor line represents the two-phase region. The lines $P =$ const represent lines of *constant pressure*, or *isobars*. The line $P = P_c$ is the *critical isobar* and is tangent to the saturation dome at the critical point. Lines of *constant enthalpy*, or *isenthalps*, and/or lines of *constant volume*, or *isochores*, are also frequently included on *T–s* diagrams.

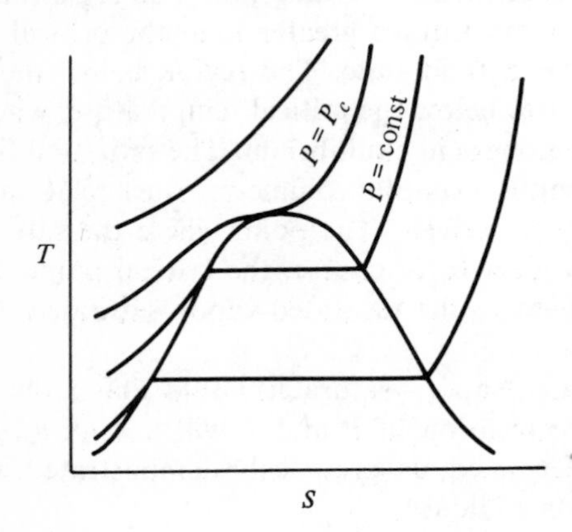

Figure 2-10. Temperature-entropy diagram.

Mollier, or h–s, Diagram

The Mollier diagram is an *enthalpy–entropy* diagram with lines of constant temperature and pressure (and others on occasion) added for extra usefulness. It is commonly used for calculations involving heating, cooling, expansion or compression, and so on. Complete Mollier diagrams are

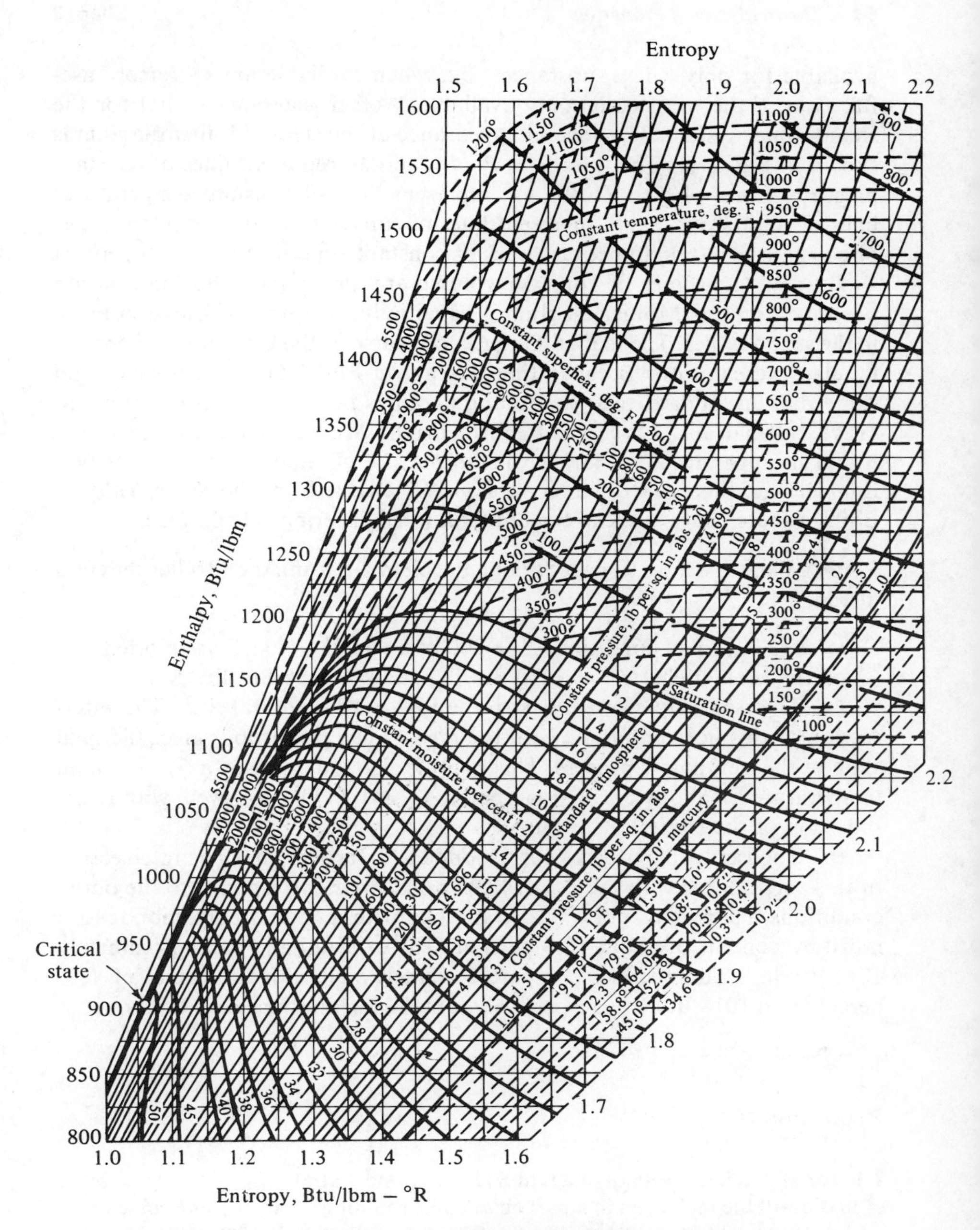

Figure 2-11. Mollier diagram for steam (H_2O). (From Joseph H. Keenan and Joseph Keyes, *Thermodynamic Properties of Steam*, John Wiley & Sons, Inc., New York, 1936, by permission.

available for only a few substances, but when available are extremely useful. One of the most commonly available Mollier diagrams is that for the steam–water system. The general appearance of the steam Mollier diagram is shown in Fig. 2-11. The lines $P = P_1$, P_2, and P_3 represent lines of constant pressure. Lines of $T = T_1, T_2, \ldots$ represent lines of constant temperature; lines of $X = X_1, X_2, \ldots$, represent lines of constant moisture content (lines of constant *quality*). Sometimes lines of constant superheat (where superheat $= T - T_{sat}$) and lines of constant volume are also given. The information portrayed on this Mollier diagram is essentially the same information given in the steam tables. The main difference, of course, is the graphical rather than tabular presentation. Because the diagram in Fig. 2-11 is fairly small and somewhat out of date, this particular diagram is not really useful for quantitative calculations. Larger and more current Mollier charts for steam are found in appendix A. A larger version may be obtained from Combustion Engineering, Inc.[6] A still larger version is available from the *Steam Tables*.[7] Both of these charts are extremely useful in engineering calculations.

SAMPLE PROBLEM 2-6. Repeat Sample Problem 2-5 using the Mollier diagram in Fig. 2-11.

Solution: The inlet conditions $P = 200$ psia and $T = 400°F$ are plotted as a ⊙ on the Mollier chart. $h = 1210$ Btu/lb$_m$ and $s = 1.56$ Btu/lb$_m$°R.

(a) The isenthalpic process corresponds to a horizontal line. The intersection of this horizontal line with the $P = 14.696$ psia isobar gives the final state of the isenthalpic process. This point is also indicated by a ⊙. The final temperature is about 335°F (compared to 341°F before—well within the precision of the charts).

(b) The isentropic process corresponds to a vertical line. The intersection of this vertical line with the $P = 14.696$ psia isobar corresponds to the outlet conditions again. This point is also marked with a ⊙ and corresponds to a moisture content of about 13.8 percent. The exit enthalpy is approximately 1015 Btu/lb$_m$. These values compare extremely well with the values of 13.6 percent and 1018 Btu/lb$_m$, respectively, obtained from the steam tables.

f= C+2-P

Problems

2-1. (a) If you had 1 lb-mol of gas at STP (32°F and 1 atm), what is the probability of finding all the molecules in a particular cubic foot of gas? In any one cubic foot?

(b) If it could be assumed that the life of a microstate is 10^{-6} sec, how many years would likely pass (on the average) between such occurrences? Would it make

[6]Combustion Engineering, Inc., Windsor, Conn. 06095

[7]J. H. Keenan, F. G. Keyes, P. G. Hill., and J. G. Moore, *Steam Tables*, John Wiley & Sons Inc., New York, N.Y., (1969)

an appreciable difference if we made the life a microstate 10^{-100} sec? Could you detect such an occurrence if it lasted 10^{-6} sec? 10^{-100} sec?

2-2. Pure nitrogen and pure oxygen in two compartments of a cylinder are separated by a membrane as shown in Fig. P2-2. The membrane is broken and the two gases allowed to mix. Has the entropy of the universe increased, decreased, or remained the same?

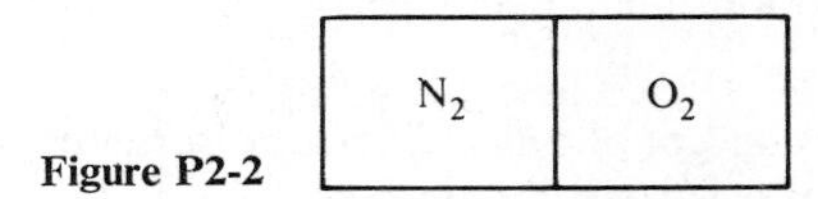

Figure P2-2

2-3. Two blocks of copper are initially at different temperatures. The two blocks are held together until the temperatures in both blocks are the same.

(a) What has happened to the entropy of the block that initially was hotter?

(b) Is this a reversible process?

(c) How do we reconcile our answer to (a) with our statement that during an irreversible process the entropy always increases?

(d) What has happened to the entropy of the block that was initially cool?

(e) How does the magnitude of the change of entropy for the hot block compare with the magnitude of the change in entropy for the cold block?

2-4. Let us consider two similar heat exchangers, both working with the same fluids and the same rates. In one exchanger there is an average temperature difference of 10°F between hot and cold streams. In the second exchanger the temperature difference is 20°F. Since the same amounts of heat are to be transferred by each exchanger, the unit with the 20°F ΔT is only half as large as the unit with 10°F ΔT and therefore will cost considerably less to install. However, there are other factors involved in choosing the actual ΔT that will be used.

(a) Will either exchanger operate reversibly?

(b) Which exchanger will have the greater entropy production?

(c) Which exchanger will be more efficient in a thermodynamic sense? (That is, which exchanger will recover more of the available energy from the high-temperature stream?)

This problem should serve to illustrate that with every irreversible process there is a loss in potential and a corresponding increase in the entropy of the universe. The more irreversible a process is (that is, the higher the rate of the process), the greater this loss in potential and the higher the increase in entropy.

2-5. Suppose we have two connected cylinders as illustrated in Fig. P2-5. Initially the cylinder on the left is filled with a gas at a pressure, P. The cylinder on the right is completely evacuated. Suddenly the valve is opened. Explain, using three different concepts, whether there is a change in the entropy of the system and, if so, whether it is an increase or decrease.

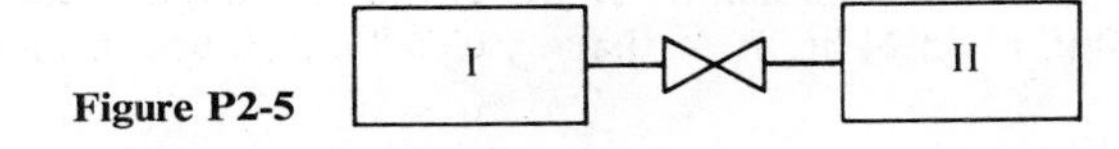

Figure P2-5

2-6. A blob of putty is sitting on the top of a wall. If somebody pushes the putty off the wall so that it falls to the ground, what has happened to the entropy of the universe?

2-7. If we had 6×10^{20} particles in two boxes,

(a) What do you think the probability of having exactly 3.0×10^{20} particles in any one box is?

(b) What do you imagine to be the probability of finding a "more or less" even distribution where no more than 50.001 percent of the particles were in any one box?

2-8. Let us extend our concept of ordered versus disordered motion from speaking of atoms to speaking of electrons. Suppose we have a stream of moving electrons (an electric current) in a conductor.

(a) Can you describe some methods in which usable energy could be obtained from these electrons?

(b) If the electron stream passes through a resistor so that its potential (voltage) is decreased, what has happened to the energy of the electron stream?

(c) What has happened to the entropy of the universe?

2-9. We have stated that the entropy of an isolated system at equilibrium must be maximized (for the total volume and energy of the system). Show that this statement is in agreement in at least two different ways with our description of entropy.

2-10. Sugar placed in a cup of hot coffee slowly dissolves. What has happened to the entropy of the universe? Justify your answer in two ways. What happens to the entropy of the coffee–sugar mixture as the mixture cools to room temperature? Justify your answer.

2-11. The Clausius statement of the second law of thermodynamics may be phrased as follows: It is impossible to construct a device that operates in a cycle and produces no effect other than the transfer of heat from a cooler body to a hotter body. Bear in mind that a device operating a cycle periodically returns to its original state and thus exhibits no net change in its properties with time. Show that the Clausius statement is consistent with our statements regarding entropy.

2-12. During the passage of a strong pressure (or shock) wave through a polyatomic gas, it is found that the energy which the shock wave transfers to the gas appears mainly in the translational and rotational forms. Very little increase in the vibrational energy is noted immediately after passage of the shock wave. However, as time passes, random molecular collisions transfer energy from the translational and rotational modes to the vibrational modes, and equilibrium is finally achieved.

(a) Describe the entropy change in the universe which occurs during the period after the passage of the shock wave, but before the vibrational energy approaches its equilibrium value.

(b) Justify your answer to (a) with three different arguments.

2-13. It has been suggested that methane in pressurized cylinders be used as an emergency fuel for a plant heating system which normally uses natural gas (largely methane). A sufficient stockpile of gas cylinders must be kept on hand to provide 100,000 Btu/hr for 24 hr. If methane yields 175,000 Btu/lb-mol upon burning and is available in 2.0-ft^3 cylinders at 3000 psia and 70°F, how many cylinders must be kept in the stockpile? Obtain predictions based on each of the following equations of state:

(a) The ideal gas equation.
(b) The van der Waals equation.
(c) The law of corresponding states.

2-14. Throttling devices are usually assumed to be adiabatic, so the flow through them is treated as being isenthalpic. However, these devices do not operate in an isentropic manner. If the state function enthalpy (h) is constant across the throttle and the throttle operates adiabatically, why isn't the entropy, also a state function, constant?

2-15. A steam/water mixture originally contains 15% water at a pressure of 20 Bars. To what pressure must the mixture be expanded *isenthalpically* to produce a single phase material? What is that phase? To what pressure must the mixture be expanded *isentropically* to increase its moisture content to 20%? What is the new temperature of the mixture?

2-16. Steam originally available at 400°F and 350 psi is heated at constant pressure until the enthalpy increases 50 Btu/lb_m. What is the new temperature? How much has the entropy of the steam changed? Has it increased or decreased?

2-17. 2 lb_m of steam originally available at 300°F and 2 atmosphere pressure is compressed until its volume is $\frac{1}{4}$ its original value. Determine the final T, P, h, and s if the compression occurs at constant:

(a) Temperature
(b) Pressure
(c) Enthalpy
(d) Entropy.

2-18. A 3 m^3 tank originally contains 5 kg of water at 5 Bar and 200°C. How much, if any, of the water is in the liquid phase? If half of the mixture leaks out of the tank determine the final T, P, h, and s if the leakage occurs at constant:

(a) T
(b) P

THREE

processes—the interaction of a system and its surroundings

3.1 Definition of a Process

Changes in the physical world are brought about by *processes*. In the thermodynamic sense a process represents a change in some part of the universe. This change may affect only a single body, as in the approach to equilibrium of an isolated system that was not initially in its equilibrium state, or a process may involve changes in both system and surroundings. The expansion of a gas as it flows through a turbine to produce work is an example of a process in which work is developed by utilizing a pressure difference to extract energy in the form of work from the gas. A thermodynamic analysis might consider either a fixed mass of gas (*closed system*) or the turbine (*open system*) as the system. In either of these cases the process involves an interaction between the system and its surroundings.

The fixed mass of gas expands against its surroundings, a part of which is represented by the turbine blades. As it impinges, the gas exerts a force that causes the shaft to rotate. The rotating shaft, if attached to an appropriate device, can do useful work in the surroundings. In this instance both the system (the gas) and the surroundings experience a change in their states as the process proceeds.

In the choice of the turbine as the system, the nature of the interaction between system and surroundings is different, although the net effect on the universe remains unchanged. The interaction of the turbine and the surroundings is represented by passage of mass into the system at a given energy level, followed by an internal conversion of the kinetic energy of the gas to shaft work, and the subsequent discharge of a lower-energy gas at the turbine's exhaust. In this case the surroundings are being changed, whereas the system itself may operate indefinitely with no change in its state.

This example illustrates the relationship of a system to a process. The process is defined in terms of a specific change to be accomplished (such as the production of shaft work by expanding a gas from a high pressure to a low pressure); however, several different systems might be specified in a thermodynamic analysis of the problem. The system which is chosen must involve the process of interest if the analysis is to yield meaningful results. In some instances the change may occur totally within a system such that a redefinition of the system is essential to permit a more meaningful analysis.

As an illustration of this point, consider a process in which two metal blocks are originally at different temperatures. If the two blocks are brought together, we observe that the temperatures of the two blocks approach the same value. The process involves energy transfer from the hotter block to the colder one. If one defines the system to be both blocks, the process occurs within the boundaries of the system and no net change in the energy content takes place in either the system or the surroundings. If one desires to know the amount of energy transferred, such an isolated system is a poor choice. On the other hand, if one chooses either block as the system, the process would involve a flow of energy between the system and part of its surroundings. Physically the choice of system makes no difference in the final state of the universe, that is, both blocks end up at the same temperature regardless of which system is specified.

Because a process must involve some change in the universe, the description of a process usually reduces to describing the changes that occur as part of the process under consideration. The science of thermodynamics deals primarily with the effects of energy transfers upon the properties of the system and its surroundings. Thus, as our discussion unfolds, we will consider in great depth those processes in which energy transfers are of central importance. Of particular interest to us will be those energy transfers which occur *across the system boundaries*, that is, those energy transfers which occur between the system and its surroundings. We shall find it useful to separate energy transfers into two distinct categories: (1) thermal energy transfers and (2) mechanical energy transfers. Thermal energy transfers across the system boundaries will be called *heat*, or *heat flows*, and mechanical energy transfers across the system boundaries will be termed *work*. As we will soon learn, determination of the quantity of heat and work transferred is almost always a major portion of the description of a specific process.

Frequently we find it useful to consider processes in which one of the system properties remains constant throughout the whole process. Because of the frequency that these processes are encountered, we have developed a short-hand notation for describing these processes:

1. A process that occurs at constant temperature is called *isothermal.*
2. A process that occurs at constant pressure is called *isobaric.*
3. A process that occurs at constant volume is called *isochoric.*
4. A process that occurs at constant enthalpy is called *isenthalpic.*
5. A process that occurs at constant entropy is called *isentropic.*

3.2 Simple Processes—The Evaluation of Work and Heat

Work

The dictionary defines work in several ways. It may be referred to as either an exertion directed to produce or accomplish work, toil, or employment, or it may be referred to as the result of an exertion, labor, or activity. Unfortunately neither of these descriptions provides the quantitative description that we as engineers demand. To provide this quantitative description we turn to the physicists' definition of work: Work is the product of an applied force, F, and the distance, ΔX, through which the force moves, multiplied by the cosine of the angle, θ, between the force and the displacement it causes. In thermodynamics we shall restrict our definition of work to include only energy in transition across the system boundaries by virtue of an energy potential other than temperature (for example, pressure or voltage). *Energy transfers that occur wholly within a system are not considered as work*, much as they will not be considered heat when thermal energy is transferred internally.

Let us consider a man who supports a 100-lb_f weight without raising it. He has transferred no work to the weight, because the force he applies was not accompanied by motion. On the other hand, a man who lifts a 1-lb_m book (taken as the system) and places it on a table has transferred work to the system, since this force is accompanied by motion.

When either the applied force or the angle between the force and the displacement changes during the process, we may evaluate the work performed during the whole process as the sum of a number of smaller processes during which both F and θ are constant. In the limit this sum becomes an integral, and we obtain the relation

$$W = \int_{X_1}^{X_2} |F| \cos\theta \, |dX| \tag{3-1}$$

The following sign convention will be adopted in regard to work: Work will be considered positive when done by a system on its surroundings, negative when done by the surroundings on the system. Thus it is seen that a choice of system and surroundings must be made before a sign can be given to a unit of work.

Examination of equation (3-1) indicates that the units of work (or for that matter any energy term) are force $\times$ length. In the engineering system this would normally be expressed as ft-lb_f. In the S.I. system the unit of energy is the newton-meter, or joule, where

$$1 \text{ joule} = 1 \text{ newton-meter} \tag{3-2}$$

As we will shortly see, another energy unit is commonly encountered in the engineering unitary system. This is the British thermal unit, or Btu. The

Btu was originally defined as the quantity of energy needed to raise the temperature of 1 lb_m of water from 59.5 to 60.5°F. However, as more refined experiments became available, it was necessary to continuously revise the conversion factor between the Btu and the ft-lb_f. Since this is a fundamental conversion factor it is rather inconvenient to have its value subject to constant revision. Therefore, the Btu has been redefined directly in terms of the ft-lb_f:

$$1 \text{ Btu} \equiv 778.16 \text{ ft-lb}_f$$

Although 1 Btu of energy will not raise the temperature of 1 lb_m of water from 59.5°F to exactly 60.5°F, it will very nearly do so, and for all but the most precise applications the two definitions of the Btu are analogous.

A similar unit of energy exists in the metric system. It is called the calorie, or cal, and was originally defined as the amount of energy needed to raise the temperature of 1 g of water from 14.5°C to 15.5°C. The calorie, however, has also been redefined directly in terms of the joule:

$$1 \text{ cal} = 4.1861 \text{ joule} \tag{3-3}$$

Although 1 cal of energy will not raise the temperature of 1 g of water exactly from 14.5 to 15.5°C, it nearly will and for all but the most precise applications the difference in definitions is again negligible.

Let us now examine the quantitative calculation of the work performed in several commonly encountered processes.

Friction Processes

Suppose a block rests on a flat surface as illustrated in Fig. 3-1. When the force F exerted on the block (the system) is large enough to overcome the frictional resistance, the force causes a horizontal motion through a differential distance dX. The differential work this force has performed on the system is then

$$\delta W = F \cos \theta \, dX \tag{3-4}$$

If the force continues to move the block until it has undergone movement through a total distance $X_2 - X_1$, we may evaluate the total work performed on the system by integrating equation (3-4) to yield

$$W = \int_{X_1}^{X_2} \delta W = \int_{X_1}^{X_2} F \cos \theta \, dX \tag{3-5}$$

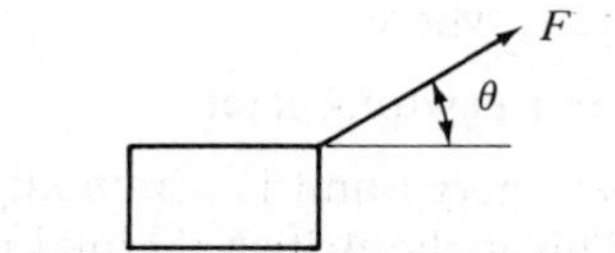

Figure 3-1. Work against friction.

If F and θ remain constant, equation (3-5) yields

$$W = F\cos\theta(X_2 - X_1)$$

or (3-6)

$$W = F\cos\theta(\Delta X)$$

where the Greek letter Δ (capital delta) signifies the difference between a property at the end and the beginning of a process. This convention for the symbol Δ is followed throughout this text.

Lifting Process

If a constant force F is applied to lift a weight (the system) through a distance $\Delta Z = Z_2 - Z_1$, the work required to lift the system is

$$W = \int_{Z_1}^{Z_2} F\cos\theta\,dZ = F\int_{Z_1}^{Z_2}\cos\theta\,dZ \qquad (3\text{-}7)$$

However, $\theta = 0°$, because F is directed parallel to the direction of the motion it produces. Therefore,

$$W = F\,\Delta Z \qquad (3\text{-}8)$$

and the work done on the system has resulted in an increase in its potential energy.

Accelerating Processes

If a system with mass M is accelerated from a velocity V_1 to a velocity V_2, we may evaluate the work necessary to perform this acceleration as follows:

$$W = \int_{X_1}^{X_2} F\cos\theta\,dX \qquad (3\text{-}5)$$

But $\cos\theta = 1$, and from Newton's law, $F = Ma/g_c$, where a is the acceleration of the body.

However, by definition,

$$a = \frac{dV}{dt} \qquad (3\text{-}9)$$

and

$$V = \frac{dX}{dt} \qquad (3\text{-}10)$$

Therefore,

$$F = \frac{M}{g_c}\frac{dV}{dt} = \frac{M}{g_c}\frac{dV}{dX}\frac{dX}{dt} = \frac{M}{g_c}\frac{dV}{dX}V \qquad (3\text{-}11)$$

so that

$$W = \frac{M}{g_c}\int_{X_1}^{X_2} V\frac{dV}{dX}\,dX = \frac{M}{g_c}\int_{V_1}^{V_2} V\,dV \qquad (3\text{-}12)$$

since M is constant. Upon integrating,

$$W = \frac{M}{2g_c}(V_2^2 - V_1^2) = \frac{M}{2g_c}\Delta V^2 \tag{3-13}$$

where the work has resulted in an increase in the system's kinetic energy.

Spring Processes

Suppose we examine a weightless, frictionless, ideal spring (the system) as illustrated in Fig. 3-2. The spring is attached to a solid wall at one end and is allowed to float freely with no constraint on the other end. The no-load position of the free end is chosen as a coordinate reference point to which the value $X = 0$ is attached. If a force F is applied to the spring, it will be compressed and shortened according to Hooke's law,

$$F = KX \tag{3-14}$$

where K is the spring constant of the spring, F the force applied to the spring, and X the displacement of the spring caused by the force F (Fig. 3-3).

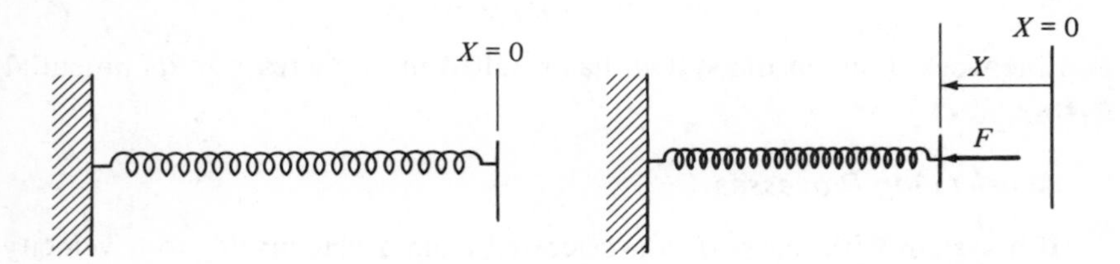

Figure 3-2. Uncompressed spring. **Figure 3-3.** Compressed spring.

The work necessary to compress the spring from the point $X = 0$ to $X = X$ may be evaluated from

$$W = \int_{X=0}^{X=X} F\,dX = \int_{X=0}^{X=X} KX\,dX \tag{3-15}$$

or

$$W = K\int_{X=0}^{X} X\,dX \tag{3-16}$$

which gives

$$W = \left(\frac{KX^2}{2}\right)_{X=0}^{X=X} = \frac{KX^2}{2} \tag{3-17}$$

The work done on the system in this case results in an increase in the system's internal energy.

Sample Problem 3-1. A spring with a spring constant of 500 newtons/meter is originally compressed 10 cm shorter than its uncompressed length. Determine the work required to compress the spring to 25 cm less than its uncompressed length. Express your answer in joules, calories, newton-meters, ft-lb_f, and Btu.

Solution: We showed that the work needed to compress a spring is given by:

$$W = K\int_{X_1}^{X_2} X\,dX \quad \text{or} \quad W = \frac{K(X_2^2 - X_1^2)}{2}$$

The problem statement tells us that $K = 500$ newtons/meter, $X_1 = 10$ cm and $X_2 = 25$ cm. For convenience we convert all numbers to a single unitary system. In this case we'll use newtons and meters:

$$K = 500 \text{ newtons/meter}$$

$$X_1 = 0.1 \text{ meter}$$

$$X_2 = 0.25 \text{ meter}$$

Thus

$$W = \frac{500(0.25^2 - 0.1^2)}{2}\frac{\text{newton}}{\text{meter}} \cdot \text{meter}^2$$

$$= 13.1 \text{ newton-meters}$$

But a newton-meter equals a joule. Thus

$$W = 13.1 \text{ joules}$$

Conversion to other units is now simply a matter of applying the appropriate conversion factors from Table 1-2:

$$W = 13.1 \text{ joules} = 9.67 \text{ ft-lb}_f = 3.13 \text{ cal} = 1.24 \times 10^{-2} \text{ Btu}$$

Compression or Expansion Processes—Closed Systems

Assume that a gas (the system) of volume V and pressure P is contained within a cylinder and piston as illustrated in Fig. 3-4. The piston is assumed to move frictionlessly in the piston and has a force F applied to it in order to constrain the gas within the cylinder. If the force F is increased so that the gas is compressed, we may evaluate the work of compression performed by the surroundings on the system as follows:

$$W = \int_{X_1}^{X_2} F_{res}\,dX \tag{3-18}$$

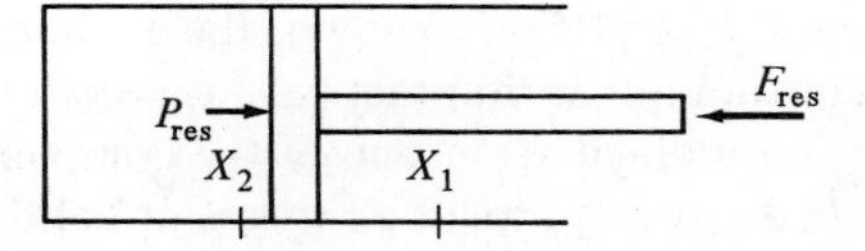

Figure 3-4. Work of compression.

where F_{res} represents the *resisting force exerted on the system boundary.* But the force exerted on the system boundary can be converted to an effective pressure by defining P_{res} such that

$$F_{res} = P_{res}A \tag{3-19}$$

where A is the area of the piston and P_{res} is the resisting pressure on the system boundary. Substitution of equation (3-19) into (3-18) then gives

$$W = \int_{X_1}^{X_2} P_{res}A\, dX = \int_{X_1}^{X_2} P_{res}A\, dX \tag{3-20}$$

Since $A\, dX = dV_{sys}$, the volume swept out by the piston as it advances (or equivalently, the change in volume of the system), equation (3-20), can be rewritten as

$$W = \int_{V_1}^{V_2} P_{res}\, dV_{sys} \tag{3-21}$$

During the compression the volume of the gas is decreasing, so dV_{sys} is a negative quantity. Therefore, the amount of work transferred by the system to the surroundings is negative. That is, the surroundings are transferring work to the system. If the gas were expanding, the volume change of the system would be positive, so W is positive and the system is transferring work to the surroundings.

If the piston moves frictionlessly within the cylinder and the net forces on the piston are balanced so it does not undergo a finite acceleration, then the resisting pressure on the system must be identical to the pressure within the system, and we may replace P_{res} with P_{sys} in equation (3-21)

$$W = \int_{V_1}^{V_2} P_{sys}\, dV_{sys} \tag{3-21a}$$

This replacement results in a major simplification in the evaluation of W, since W is now related only to system properties. On the other hand, the requirements of frictionless and nonaccelerating motion are extremely restrictive and effectively remove all real processes from rigorous application of equation (3-21a). Although most real processes do not meet one or both of the restrictions implicit in the use of equation (3-21a), we frequently use this relation because it is the only choice we have. However when we use such an approximation it is important to realize that we are *using an approximation* and the results must be viewed with some degree of honest skepticism.

As we shall see later, the evaluation of the integral in equation (3-21a) is often an extremely difficult task, because the relation of P_{sys} to V_{sys} may not be explicitly known. Even in some cases where an analytical expression between P_{sys} and V_{sys} is known, the evaluation of equation (3-21a) may be time consuming or impractical—especially if the expression relating P_{sys} to V_{sys} is obtained from one of the complex equations of state. However, if a plot of P_{sys} vs V_{sys} for the process at hand can be constructed as shown in

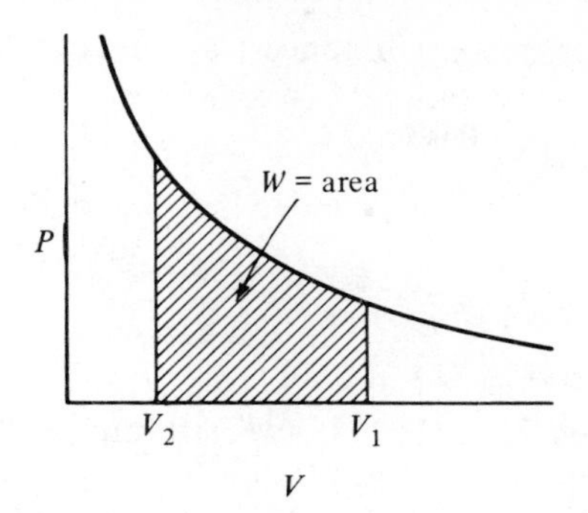

Figure 3-5. Work of compression—P–V diagram.

Fig. 3-5, the evaluation of $\int_{V_1}^{V_2} P_{sys}\, dV_{sys}$ is reduced to the evaluation of the area under the curve between the limits of V_1 and V_2. For example, in Fig. 3-5 the work done on the gas in compressing it from V_1 to V_2 is equal to $P_{sys}\, dV_{sys}$, which is identically equal to the shaded area under the curve. It should be noted that the area is negative if the integration is from state 1 to state 2. Since in many cases the $P_{sys} - V_{sys}$ curve for a gas may be available or readily plotted, evaluation of the area under the P–V curve (called *graphical integration*) is often a far simpler and quicker procedure than analytical evaluation of the integral.

For many processes involving ideal gases, the pressure and volume of the gas are related according to an expression of the form:

$$PV^k = \text{constant} \tag{3-22}$$

Such processes are termed *polytropic* and will be encountered frequently in our future discussions. If $k = 1$, the polytropic process reduces directly to an isothermal process

$$PV = \text{constant} \tag{3-22a}$$

where the constant is given by nRT. For nonisothermal processes, k may have a value anywhere between 1.0 and 2.0, depending on the gas and the nature of the process under consideration.

SAMPLE PROBLEM 3-2. One lb-mol of an ideal gas at 273°K and 1 atmosphere pressure is compressed to 10 atmospheres pressure in a frictionless piston-cylinder arrangement as shown in Figure 3-4. Determine the amount of work needed if the compression is a polytropic process with $k = 1.4$.

Solution: We begin by determining the initial volume of the gas. This we obtain from the ideal gas equation of state:

$$V = \frac{nRT}{P}$$

For simplicity we will convert all units to lb-mol, lb_f, ft, and °R. In these units.

$$n = 1 \text{ lb-mol}$$
$$R = 1545 \text{ ft-lb}_f/\text{lb-mol°R}$$
$$T = 273\text{°K} = 492\text{°R}$$
$$P = 1 \text{ atm} = 2.12 \times 10^3 \text{ lb}_f/\text{ft}^2$$

Thus,

$$V = \frac{(1)(1545)(492)}{2.12 \times 10^3} \frac{(\text{lb-mol})[\text{ft-lb}_f/(\text{lb-mol°R})](\text{°R})}{(\text{lb}_f/\text{ft}^2)}$$

or

$$V = 359 \text{ ft}^3$$

We now evaluate the constant for the polytropic process from:

$$\text{K} = PV^k$$

where k is now the exponent of the polytropic process. Since PV^k is constant throughout the polytropic process we may determine the value of K from the initial conditions:

$$\text{K} = P_1V_1^k = (2.12 \times 10^3)(359)^{1.4} \frac{\text{lb}_f}{\text{ft}^2} (\text{ft}^3)^{1.4}$$
$$= 7.85 \times 10^6 \text{ lb}_f\text{-ft}^{2.4}$$

We are told that the piston-cylinder arrangement is frictionless, and will assume that the compression occurs without acceleration of the piston. Thus equation (2-57a) will be used to evaluate the work of compression:

$$W = \int_{V_1}^{V_2} P_{\text{sys}}\, dV_{\text{sys}}.$$

However, the system pressure and volume are related by the polytropic conditions:

$$P_{\text{sys}} = \frac{\text{K}}{V_{\text{sys}}^k}$$

Thus:

$$W = \int_{V_1}^{V_2} \frac{\text{K}}{V_{\text{sys}}^k}\, dV_{\text{sys}} = \text{K} \int_{V_1}^{V_2} \frac{dV_{\text{sys}}}{V_{\text{sys}}^k}$$

for $k \neq 1.0$ the integral becomes:

$$W = \frac{\text{K}}{1-k}\left(\frac{1}{V^{1-k}}\right)\Bigg]_{V_1}^{V_2} = \frac{\text{K}}{1-k}[V_2^{k-1} - V_1^{k-1}]$$

This expression may be simplified as follows:

$$W = \frac{1}{1-k}[\text{K}\,V_2^{k-1} - \text{K}\,V_1^{k-1}] = \frac{1}{1-k}[P_2V_2^kV_2^{k-1} - P_1V_1^kV_1^{k-1}]$$

or

$$W = \left[\frac{P_2V_2 - P_1V_1}{1-k}\right]$$

All that is needed to allow evaluation of W are the final conditions. The final pressure is specified to be 10 atm or 2.12×10^4 lb_f/ft^2. The final volume is obtained from the polytropic relation:

$$V_2 = \left(\frac{K}{P_2}\right)^{1/k} = \left(\frac{7.85 \times 10^6}{2.12 \times 10^4}\right)^{1/1.4} \text{ft}^3 = 68.5 \text{ ft}^3$$

Thus:

$$W = \left[\frac{(2.12 \times 10^4)(68.5) - (2.12 \times 10^3)(359)}{(1-1.4)}\right] \frac{\text{lb}_f}{\text{ft}^2} \text{ft}^3$$

or

$$W = -1.73 \times 10^6 \text{ ft-lb}_f$$

where the negative sign indicates that work is required to perform the compression.

SAMPLE PROBLEM 3-3. Steam initially available at 800°F and 900 psia is expanded isothermally in a frictionless piston cylinder to a pressure of 100 psia. Determine the quantity of work liberated per pound of steam expanded.

Solution: We are told that the piston movement is frictionless, and again we will assume that the motion is slow enough to neglect acceleration effects. Under these circumstances, the work is evaluated from equation (3-21a)

$$W = \int_{P_1}^{P_2} P_{sys}\, dV_{sys}$$

The work liberated per lb_m of steam expanded is obtained by dividing the work expression by the total mass of steam (which is constant) and recognizing that $W/M = w$ and $V_{sys}/M = v_{sys}$. Thus, the specific work is obtained from

$$w = \int_{P_1}^{P_2} P_{sys}\, dv_{sys}$$

If we had an analytical expression for the equation of state of steam we could obtain the P–V relation for the steam at 800°R and use this to evaluate the required integral. Although we do not have the equation of state of steam, we do have a great quantity of volumetric data in the steam tables. From the data in the steam tables the 800°R isotherm for steam may be plotted on a P–v diagram as shown in Fig. SP 3-3. The integral is now obtained by

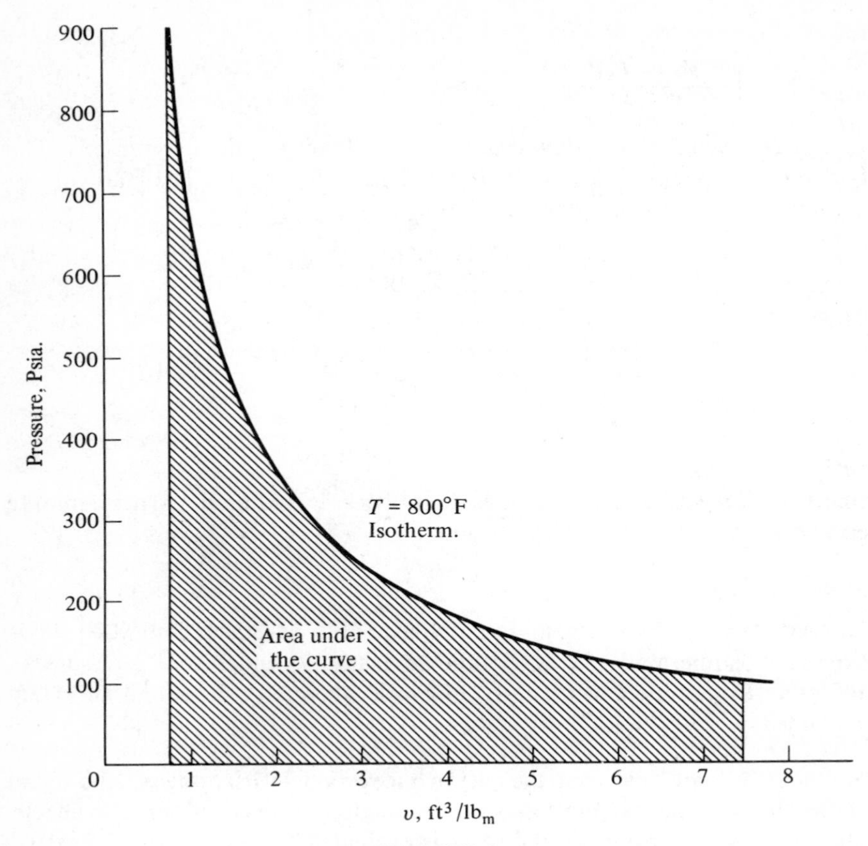

Figure SP3-3

graphical integration of the P–v curve between the limits of $100 \text{ psia} < P < 900 \text{ psia}$.

Integration yields the value:

$$w = \int_{P=900\text{ psia}}^{100\text{ psia}} P_{sys}\, dv_{sys} = +1640\,(\text{lb}_f/\text{in}^2)(\text{ft}^3/\text{lb}_m) = 2.36 \times 10^5 \,\frac{\text{ft} - \text{lb}_f}{\text{lb}_m}$$

$$= 3.03 \times 10^2 \text{ Btu/lb}_m.$$

Heat

By the late eighteenth and early nineteenth centuries our present view of *heat as energy in transit by virtue of a temperature difference* had pretty well evolved. Although it is clearly possible to have energy transfers wholly within a system by virtue of internal temperature differences, we shall not refer to these energy flows as heat. Rather we shall restrict our definition of heat (in a thermodynamic context) to include only *those energy transfers which occur between the system and its surroundings by virtue of a temperature difference*. We use the symbol Q to denote a heat flow. Thus to determine the existence or nonexistence of heat flow, we must examine the boundary between the system and its surroundings. It is also meaningless to attempt to describe heat by reference to an isolated system: Heat can be transferred only between a system and its surroundings. Once heat has crossed the boundary between the system and surroundings, it is no longer considered heat but becomes part of the thermal component of the receiving body's energy. A process during which no heat is transferred between the system and surroundings is termed *adiabatic*.

Heat is measured in terms of the energy changes it causes in the system and surroundings. In the early development of the heat-flow concept these energy changes were most readily described in terms of the temperature changes that the heat flows caused. As we have indicated, the two commonly encountered thermal energy units were originally defined as follows:

1 Btu = amount of energy needed to raise the temperature of 1 lb_m of water from 59.5 to 60.5°F at a constant pressure of 1 atm.

1 cal = amount of energy needed to raise the temperature of 1 g of water from 14.5 to 15.5°C at a constant pressure of 1 atm.

However, these definitions have proven to be somewhat inconvenient since it was necessary to continuously revise the conversion factor between "thermal" and "mechanical" energies as more accurate experiments were performed. Therefore, the calorie and Btu have been defined directly in terms of the mechanical energy units:

$$\begin{aligned} 1\ \text{Cal} &= 4.1861\ \text{joules} \\ 1\ \text{Btu} &= 778.16\ \text{ft-lb}_f \end{aligned} \tag{3-23}$$

Although these definitions do not agree exactly with the original definitions, the differences are extremely small and may be neglected for all but the most precise applications.

3.3 Reversible and Irreversible Processes

One objective of thermodynamics is to describe the interactions between system and surroundings (such as heat and work) which take place as a system moves from one (equilibrium) state to another. For instance, suppose that the gas in a well-insulated cylinder (the system) expands against a piston and transfers mechanical energy to the surroundings. If the piston moves frictionlessly within the cylinder and slowly enough so that viscous losses can be neglected, the work done by the gas will be just equal to the mechanical energy received by the surroundings. Moreover, the mechanical energy received by the surroundings can be stored (by the raising of a weight, for example) and used to return both the system and the surroundings to precisely their original states.

On the other hand, if there is friction between the piston and the cylinder, a part of the work done by the gas in its expansion will be converted to thermal energy and cannot be stored in the surroundings in a mechanical form. Similarly, on the return stroke of the piston, not all the work done by the surroundings will be transferred to the gas, but some will be converted to thermal energy by the friction. Thus, if the piston–cylinder arrangement is to be returned to its initial state, the surroundings will have to supply more work to the gas during the compression than was received during the expansion. Similarly, the gas would have to transfer an equivalent amount of heat to the surroundings (in order to return to the original energy level). Thus the overall effect of the frictional expansion–compression cycle is a net transfer of mechanical energy into the gas and an equivalent net transfer of thermal energy back to the surroundings. However, as we have previously indicated, it is impossible to conceive of a device whose only effect is to completely convert thermal energy back into mechanical energy. Thus we find that there is a net change in the universe which can never be completely reversed. A similar result would have been obtained for a frictionless piston if the expansion were allowed to take place so rapidly that nonuniformities in pressure could occur within the gas in the cylinder. The dissipation of a shock wave is an extreme example of this type of nonuniformity.

In order to provide a criterion by which to distinguish between the two types of processes discussed above, we define a *reversible process* as one which occurs in such a manner that both the system and its surroundings can be returned to their original states. *This condition cannot be realized unless all changes associated with the process (both in the system and surroundings) occur without frictional effects.* Any process that does not meet these stringent conditions is called an *irreversible process.*

An irreversible process always involves degradation of an energy potential without producing the maximum amount of work or a corresponding increase in another energy potential other than temperature. The degradation results from an imbalance of mechanical-energy potentials, as in the case of the free expansion. We can generalize this observation to state that a process will be irreversible if that process (by virtue of a finite driving force) occurs at a rate that is large compared to the rate at which molecular adjustment in the system can occur. Since molecular processes occur at a finite (even if rapid) rate, a truly reversible process will always involve an infinitesimal driving force to assure that the energy transfer occurs without degradation of the driving potential. Hence they will take place at an infinitesimally low rate. Under such conditions it is always possible for a system to readjust on the molecular level to the process change such that the system moves successively from one equilibrium state to another. Such processes are called *quasi-static* or *quasi-equilibrium* and are frequently amenable to analysis by much the same technique as reversible processes. Thus a reversible process would always qualify as a quasi-static process, but the reverse need not be true.

For a reversible process it is a straight-forward matter to completely describe the interaction of system and surroundings. This behavior is a major virtue of reversible processes. For an irreversible process, such as the frictional expansion, it is not such a straightforward matter to describe the interactions between system and surroundings. For example, knowledge of either the amount of mechanical energy transferred from the surroundings to the piston or that transferred from the piston to the gas does not give us the other unless we also know the amount of mechanical energy dissipated in overcoming the friction between piston and cylinder. Thus for the frictional process, not only do we need to describe the changes that take place in the system, but we need to know how these changes are transferred to the surroundings before our description of the process is complete. This additional complexity of irreversible processes carries over into other types of systems as well and will be the subject of much discussion in the next several chapters.

Before leaving the topic, let us examine some of the general conclusions that can be reached concerning reversible and irreversible processes. Since it is much easier to enumerate those things which make a process irreversible than vice versa, the following is a discussion of the sources of irreversibility and their consequences.

We have already noted that mechanical friction leads to irreversibility because the mechanical energy dissipated in overcoming friction is transformed into thermal energy. Similar irreversibilities are exhibited by a large class of processes, of which the following are examples:

1. The flow of a viscous fluid through a pipe. System: the fluid.
2. The transfer of electricity through a resistor. System: the resistor.
3. The inelastic deformation of a solid material. System: the solid.

Each process has in common the fact that a part of the work provided by the surroundings to carry out the process can be stored and is recovered when the direction of the process is reversed, but part of the work is lost to friction (molecular and electronic in these examples) and is converted to thermal energy. Thus the system and surroundings cannot both be returned to their original states and these processes are irreversible.

Not all irreversibilities involve the degradation of work to thermal energy immediately, although it can be shown that restoration of the system to its original state will eventually lead to such a conversion. Consider the following processes in isolated systems:

1. A sealed and insulated cylinder containing one high-pressure chamber and one vacuum chamber with the connecting valve then opened (free expansion). System: the gas.
2. A hot and a cold block of metal insulated from their surroundings but brought into thermal contact with each other. System: both blocks.

To restore the gas to its original pressure the gas must be compressed using work supplied by the surroundings. During the compression an equivalent amount of heat (thermal energy) must be transferred back to the surroundings to restore the gas to its original energy level. Thus the surroundings undergo a net change that cannot be reversed, and the process is irreversible.

To restore the second system to its original state, the temperature of one block must be increased by transferring thermal energy to it from a high-temperature reservoir in the surroundings. The other block must be cooled by transferring thermal energy to a low-temperature reservoir. As we shall show later, it is not possible to transfer thermal energy from the low-temperature reservoir to the high-temperature reservoir without supplying work. Thus, although the system can be returned to its original state, the surroundings cannot; the process is again irreversible.

The common factor in these two processes which can be generalized is that energy has been transferred from one part of the system to another and an energy potential (pressure difference or temperature difference) has been reduced without the production of work. If, in the first case, the energy potential (pressure difference) had been reduced by expanding the gas against a frictionless piston until its volume equaled that of the combined chambers, the work produced in the surroundings would have been just that necessary to recompress the gas. Although the final state of the work-producing system would not be the same as for the free expansion, this process, in contrast, is reversible.

In the second case a heat engine (for example, a steam engine) might have been employed to produce work during the transfer of thermal energy from the high-temperature block to the low-temperature block. This work could then be used to drive a heat pump (refrigerator) which could drive thermal energy from the low-temperature block back to the high-temperature block. Under ideal conditions it is possible to return both blocks to their initial states, thereby reversing the original process.

Although it is generally much easier to describe a system's behavior under reversible conditions rather than irreversible ones, most processes of interest are irreversible. However to facilitate thermodynamic analysis we often invent reversible processes that closely approximate actual process behavior and yet provide some basis with which to compare irreversible processes between the same end states.

In practice, we cannot afford to operate processes at an infinitely slow rate, because our output and hence profit would also be infinitesimal. Thus we intentionally sacrifice our energy potentials to accomplish immediate change. However, many process changes, while occurring at a finite rate, do not occur so rapidly that the system is unable to adjust on a molecular level (owing to the very rapid rate at which molecular processes occur). Thus, the assumption of reversibility, although not perfect, is often a useful engineering approach.

3.4 State and Path Functions

We have defined work as equivalent to a force moving through a distance. Any time we have motion of the boundary of a system against a restraining force, work is performed. Earlier it was shown that the amount of work performed by a closed system is given by

$$W = \int_{V_1}^{V_2} P_{\text{res}}\, dV_{\text{sys}} \tag{3-21}$$

where P_{res} is the effective pressure exerted on the system boundary. For simplicity, let us again consider our system to be the gas enclosed within the cylinder of Fig. 3-6. The piston is assumed to be weightless and considered as part of the system. However, the piston is *not* assumed to be frictionless. The process is assumed to be slow, so the pressure within the gas is at all points equal. When the piston moves, the force of friction (when it is present) will prevent the pressure of the gas within the system from equaling that of the surroundings. Thus, when we evaluate the work performed by the system on the surroundings, we must be very careful to specify only that (mechanical) energy which is actually transferred to the surroundings. That portion which might have been transferred to the surroundings but is dissipated by friction within the system is termed *lost work* and will be discussed in some detail in Chapter 5.

The work that is actually transferred to the surroundings is W and is evaluated from

$$W = \int_{V_1}^{V_2} P_{\text{res}}\, dV_{\text{sys}} \tag{3-21}$$

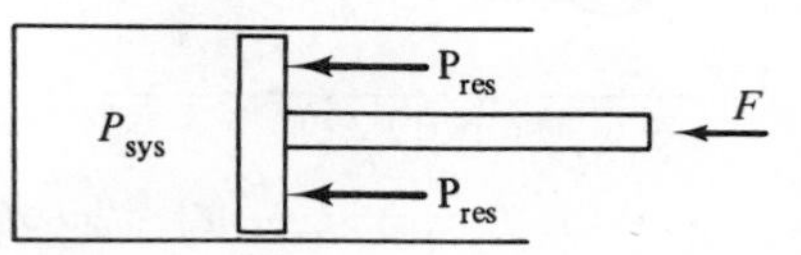

Figure 3-6. Piston–cylinder confinement of a gas.

Thus we see that the work done on, or by, a system may very well be a function of things not associated with the terminal states of the system, as, for example, the effects of friction. We find, then, that in going between any two states the work a system can produce may have many different values depending on the path the system or the surroundings follows between the two states. Therefore, we say that the work a system can produce in going between two states is a *function of the path* or *a path function*. Similarly, it may be demonstrated that the heat absorbed by a system in going between two states is also a path function.

In addition to path functions, which depend on the path followed between states, there are many system properties that are independent of path. These properties are called *state properties* or *functions of state*. All physical properties—density, pressure, temperature, viscosity, etc.—are examples of state functions. In addition, any property that can be expressed uniquely as a function of only these properties must itself be a function of state. Thus we find that all the thermodynamic properties $u, h, s, \ldots$ which may be expressed as functions of temperature and pressure alone are functions of state!

The distinction between functions of path and functions of state may be quantitatively described by looking at the integral of the function in question around a closed path such as that shown in Fig. 3-7. If the function P in question is a state function, the closed-path integral $\oint dP$ will equal zero independent of the path traversed between a and b or between b and a. If a single closed path can be found for which the closed-path integral is not identically zero, the function is a path function.

Let us examine several ways in which we may frictionlessly expand an ideal gas contained within a piston–cylinder (system: the gas) between states 1 and 2 on the P_{sys}–V_{sys} diagram shown in Fig. 3-8. For illustration purposes, let $T_1 = T_2$. Since we have assumed that the process is performed without frictional losses, $P_{res} = P_{sys}$. Thus, the work performed along each leg of path may be given by

$$\delta W = P_{sys}\, dV_{sys} \tag{3-24}$$

For the isothermal portion of the process, the ideal-gas equation of state

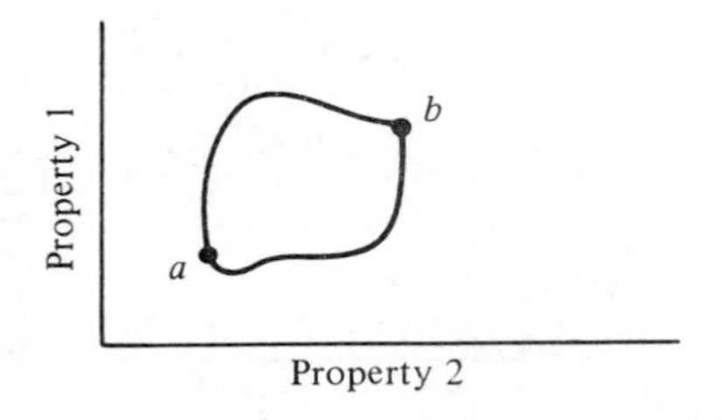

Figure 3-7. The path integral.

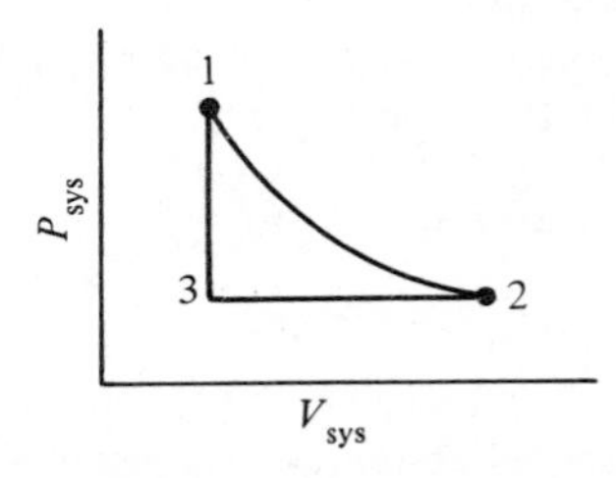

Figure 3-8. Path integral on a P–V diagram.

may be used to eliminate P_{sys}:

$$P_{sys} = \frac{nRT}{V_{sys}} \tag{3-25}$$

so that

$$\delta W_1 = nRT_1 \int_{V_1}^{V_2} \frac{dV_{sys}}{V_{sys}} = nRT_1 \ln \frac{V_2}{V_1} \tag{3-26}$$

Along the indirect path 1-3-2, the gas is assumed to undergo a frictionless two-step process. During the first step (1-3) the gas is cooled at constant volume until the pressure drops to the outlet pressure, P_2. In the second step (3-2) the gas is reheated, but at constant pressure, back to the inlet temperature, T_1. Since we are assuming that both steps are frictionless, $P_{res} = P_{sys}$, and the work transferred to the surroundings is given by

$$W_{1\text{-}3\text{-}2} = W_{1\text{-}3} + W_{3\text{-}2} \tag{3-27}$$

$$= \int_{V_1}^{V_3} P_{sys}\, dV_{sys} + \int_{V_3}^{V_2} P_{sys}\, dV_{sys} \tag{3-27a}$$

Since no volume change occurs in the first step, $W_{1\text{-}3}$ is zero, and

$$W_{1\text{-}3\text{-}2} = \int_{V_3}^{V_2} P_{sys}\, dV_{sys} \tag{3-28}$$

Since P_{sys} is constant at P_2 during the second step, the integral reduces to

$$W_{1\text{-}3\text{-}2} = P_2(V_2 - V_3) = P_2(V_2 - V_1) \tag{3-28a}$$

Examination of the two work expressions, $W_{1\text{-}2}$ and $W_{1\text{-}3\text{-}2}$, clearly indicates that they are different—even though both processes are frictionless. *Thus we must conclude that work is a path function whose value depends on the path chosen between the initial and final states.* Similarly, if we examine the closed-path integral we find that the integral does not vanish since $W_{1\text{-}2}$ does not equal $W_{1\text{-}3\text{-}2}$. Once again we are led to the conclusion that W is a path function.

The mathematical differences between path and state functions will become more significant when we deal with the mathematics of properties. We know that changes in state functions may be uniquely and conveniently expressed in terms of exact differentials. Such expressions are not possible for path functions which are inexact differentials.

3.5 System Analysis—The Concept of Accountability

In thermodynamics the behavior of a system is studied by monitoring the changes in state that it experiences during a process, and the flows of mass or energy that *cross the boundary*. An effective accounting scheme is necessary so that changes may be carefully monitored and analyzed. For example, let

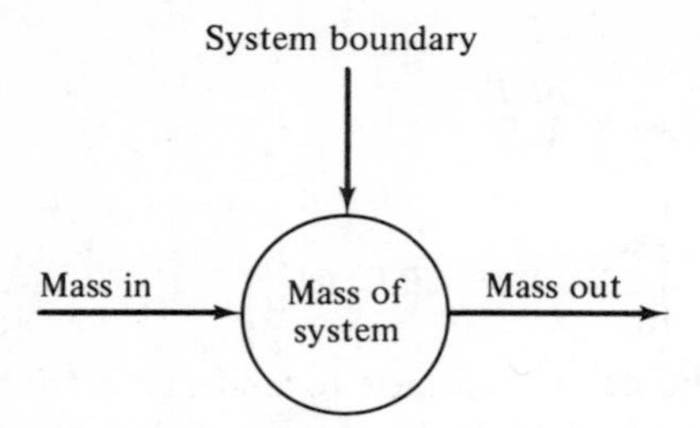

Figure 3-9. Open system mass flow.

us consider a simple open system into and out of which mass is flowing as shown in Fig. 3-9. We may monitor the accumulation of mass in the system by observing the amounts of mass that cross the systems boundary by virtue of flow into and out of the system.

$$\text{Mass in} - \text{mass out} = \text{mass accumulated} \tag{3-29}$$

Since total mass is conserved in all nonnuclear processes, such an accounting scheme is sufficient for the total mass. However, such an accounting scheme is not sufficient to monitor the accumulation of individual chemical species in a chemical reactor. For example, suppose hydrogen and oxygen are mixed and burned in a torch to produce water. Clearly, our accounting scheme must be extended to allow for the generation (or disappearance) of individual species within the system. Thus if we are attempting to monitor the accumulation of hydrogen within the torch we would write:

$$\begin{gathered}\text{Hydrogen in} - \text{hydrogen out} + \text{hydrogen generated by the} \\ \text{chemical reaction} = \text{hydrogen accumulated in the torch}\end{gathered} \tag{3-30}$$

or in short hand:

$$(H_2)_{\text{in}} - (H_2)_{\text{out}} + (H_2)_{\text{generated}} = (H_2)_{\text{accumulated}}. \tag{3-30a}$$

In the particular problem described above the amount of hydrogen generated would be a *negative number* since hydrogen is actually consumed in the reaction. On the other hand, if we monitor water accumulation by means of this accounting scheme, the amount of water generated would be a *positive number* since water is actually formed by the chemical reaction.

Although equation (3-30a) has been developed to describe the accumulation of hydrogen in a chemically reacting system, a similar relation can be used to describe the accumulation of any quantity which flows into and out of a system and/or is internally generated (or consumed) within the system:

$$\text{In} - \text{out} + \text{generation} = \text{accumulation} \tag{3-31}$$

In our later discussion we will frequently need to describe the infinitesimal changes that occur within a system due to infinitesimal flows into and out of the system. For such instances we use the differential form of equation (3-31).

Thus if we are describing infinitesimal flows of a quantity B into and out of a system, we would write a differential balance expression such as:

$$(\delta B)_{in} - (\delta B)_{out} + (\delta B)_{gen} = d(B)_{sys} \tag{3-32}$$

where the symbol δ is used to indicate either a differential flow, or a differential generation, and the symbol d is used to indicate the differential change in the amount of B within the system. The mathematical difference between the δ and d differential operators becomes important when we consider their integrals: The integral of δ is simply the total quantity that either flows across the system boundary or is generated within the system. The integral of $d(B)_{sys}$, on the other hand, is the finite change in the quantity of B within the system:

$$\int d(B)_{sys} = \Delta(B)_{sys} = (B_{sys})_{end} - (B_{sys})_{beginning} \tag{3-33}$$

where the symbol Δ is used as a short-hand notation for the total change.

This same accounting scheme will be used later to analyze changes in mass, energy, and entropy experienced by a system during various processes. Although the time rate of change of these variables is not of particular concern in a thermodynamic analysis, one could incorporate the notion of rates by simply dividing each term by a time factor. The same accounting scheme will also be used to formulate basic rate expressions for different processes in later courses.

SAMPLE PROBLEM 3-4. In the production of alcoholic beverages one starts with a mixture of starches, sugars, and yeast. (The starches and sugars are usually obtained from one or more different kinds of grain or fruit—the kind of grain or fruit used will, in large proportion, determine the kind of beverage produced.) The yeast biologically degrade, or ferment, the starches and sugars to produce carbon dioxide and ethyl alcohol. For noncarbonated beverages (that is, most wines and all liquors) the CO_2 is simply allowed to escape. Unfortunately, when the alcohol content reaches about 10 percent by weight, the yeast cease their activity and the fermentation ceases. To produce a product of higher alcohol content than this, the alcohol–water mixture must be separated into an alcohol-rich stream and a water-rich stream. This separation is normally accomplished by a technique known as fractional distillation. 1000 lb_m/hr of this 10 percent by weight alcohol and water solution is to be separated into a mixture that contains 40 percent by weight alcohol and one that contains 2 percent by weight alcohol. Determine the product rates for each of these streams assuming that no alcohol or water is accumulating in the distillation equipment.

Solution: We may picture the process described as shown in Fig. SP 3-4. The streams have been labeled 1, 2, and 3.

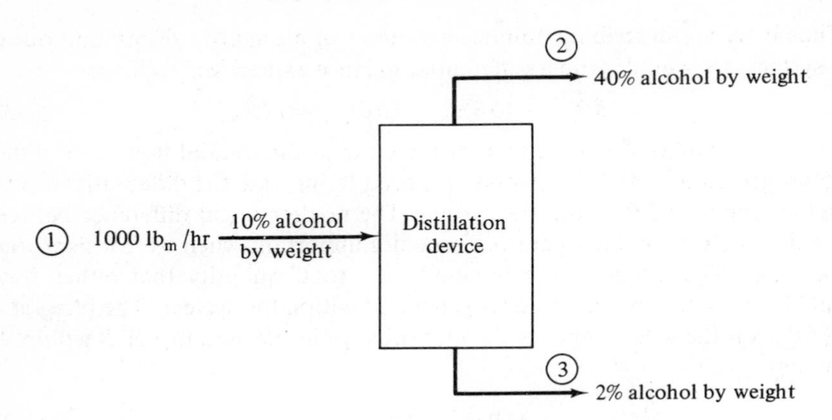

Figure SP3-4

Let us write an overall mass balance around the distillation device:

$$\delta M_1 - \delta M_2 - \delta M_3 = (dM)_{\text{dist.device}}$$

Likewise we may also write a component mass balance for alcohol around the device:

$$(\delta M_a)_1 - (\delta M_a)_2 - (\delta M_a)_3 + (\delta M_a)_{\text{gen}} = (dM_a)_{\text{dist.device}}$$

However, we are told the device is not accumulating either alcohol or water so that $(dM)_{\text{device}} = (dM_a)_{\text{device}} = 0$. Also there is no chemical reaction occurring within the device so that $(\delta M_a)_{\text{gen}} = 0$. Thus the overall and component mass balances reduce to:

$$\delta M_1 - \delta M_2 - \delta M_3 = 0$$

$$(\delta M_a)_1 - (\delta M_a)_2 - (\delta M_a)_3 = 0$$

However, the quantity of alcohol in any given stream is simply given by the total mass flow in that stream times the mass fraction of the stream that is alcohol. That is, $\delta M_a = x_a \delta M$. Thus the alcohol component balance reduces to

$$(x_a)_1(\delta M_a)_1 - (x_a)_2(\delta M_a)_2 - (x_a)_3(\delta M_a)_3 = 0$$

We may put both the overall and component mass balances on a per-unit-time basis by dividing by dT and observing that the ratio $\delta M/dT$ is simply the time *rate* of flow of mass which we denote by a dot over the M:

$$\dot{M} = \frac{\delta M}{dT}$$

The time-rate equations for the overall and component mass balances then reduce to:

$$\dot{M}_1 - \dot{M}_2 - \dot{M}_3 = 0$$

$$(x_a)_1 \dot{M}_1 - (x_a)_2 \dot{M}_2 - (x_a)_3 \dot{M}_3 = 0$$

The problem statement tells us that

$$\dot{M}_1 = 1000 \text{ lb}_m/\text{hr}$$

$$(x_a)_1 = 0.10, \qquad (x_a)_2 = 0.40$$

and $(x_a)_3 = 0.02$. Thus we now have two equations in the two unknowns $\dot{M}_2$ and $\dot{M}_3$. We proceed to solve for $\dot{M}_2$ and $\dot{M}_3$ as follows:

$$\dot{M}_2 = \dot{M}_1 - \dot{M}_3$$

which may be used to eliminate $\dot{M}_2$ from the component mass balance. This elimination yields:

$$\dot{M}_1[(x_a)_1 - (x_a)_2] - \dot{M}_3[(x_a)_3 - (x_a)_2] = 0$$

or

$$\dot{M}_3 = \frac{\dot{M}_1[(x_a)_1 - (x_a)_2]}{[(x_a)_3 - (x_a)_2]} = 1000 \text{ lb}_m/\text{hr} \frac{(0.10 - 0.40)}{(0.02 - 0.40)}$$

$$= 790 \text{ lb}_m/\text{hr}$$

$\dot{M}_2$ is then obtained from the overall mass balance:

$$\dot{M}_2 = \dot{M}_1 - \dot{M}_3 = (1000 - 790) \text{ lb}_m/\text{hr} = 210 \text{ lb}_m/\text{hr}.$$

Problems

3-1. A baseball weighing $\frac{1}{2}$ lb$_m$ is travelling at 50 ft/sec before it is struck by a baseball bat. If the collision between the ball and bat is perfectly elastic, how much energy must the bat transmit to the ball if the ball leaves with a velocity of 150 ft/sec?

3-2. If the ball of problem 3-1 is popped straight-up how high will it rise (neglect viscous drag from air) before returning to earth?

3-3. A space craft weighs 15,000 kg. To what velocity must the craft be accelerated (at ground level) if it is to escape the earth's gravitational field? Assume the following:

(1) The earth has a diameter of 13,000 km

(2) The gravitational attraction of the earth falls off as $1/r^2$, where r is measured from the center of the earth.

3-4. A block weighing 10 N compresses a horizontal spring with a constant of 25 N/m. If the spring is compressed $\frac{1}{4}$ m and the block released, what is the velocity of the block when it leaves the spring? Neglect frictional effects.

3-5. A car radiator contains 3 kg of an antifreeze mixture which contains 30% by weight ethylene glycol. It is desired to raise the concentration of glycol to 50%. How much pure ethylene glycol must be added

(a) if no liquid is removed from the radiator first?

(b) if sufficient liquid is first removed from the radiator so that the final mixture still weighs 3 kg? How much liquid was withdrawn?

3-6. A wine originally contains 10% by volume of ethyl alcohol. What volume of a 50% (vol) alcohol in water mixture must be added to fortify the wine to 20% alcohol (vol)? Assume that all volumes are additive.

3-7. A swimming pool originally contains 100 m^3 of sea water which contains 3.5% (by weight) NaCl. In an effort to reduce the salt concentration, fresh water is added to the pool and an equivalent amount of salty water removed. If the water in the pool is well mixed, and fresh water added at a rate of 1 kg/sec how long will it take to reduce the salt concentration in the pool to 2%?

FOUR

the energy equation

Introduction

In this chapter we shall develop and apply the energy equation—a mathematical relationship expressing the conservation of energy, a special case of which is sometimes called the first law of thermodynamics. The application of the energy equation to typical thermodynamic processes is illustrated in the remainder of the chapter. We will discover how this equation can be used to relate the changes of state of a system to the energy exchanges between the system and its surroundings. We will also learn that, in making this connection for a given process, some shrewdness and experience in selecting the elements to be identified as the system can greatly facilitate the computations.

Examination of the general balance equation shown in Chapter 3 reveals that the energy equation will incorporate groups of terms representing (1) the addition of energy to the system, (2) the removal of energy from the system, and (3) the accumulation of energy in the system. We examined in Chapter 3 the concepts of heat and work and showed that these are two of the ways in which energy can be exchanged between the system and its surroundings. We will examine later in this chapter the ways in which a system may accumulate energy and the ways in which energy may flow into a system by virtue of a mass flow. These terms will all be combined to yield the desired energy equation. We begin our discussion by extending our previous discussion of energy transfers as work to include frictional processes as well as those involving finite accelerations.

4.1 Evaluation of Work for Complex Processes

In Chapter 3 we introduced the quantitative definition of work and showed how the work term could be evaluated for several simple processes. Before proceeding with the development of the energy balance, it is appropriate to expand this discussion to consider the evaluation of the work term for several more complex processes. For purposes of simplicity we will subdivide this discussion into two parts.

Processes with Infinitesimal Driving Forces

Suppose a piston and cylinder are arranged as shown in Fig. 4-1. The piston is assumed to move frictionlessly within the cylinder. A force F is

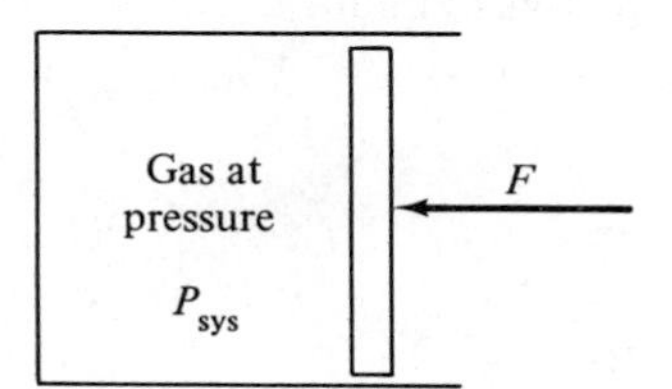

Figure 4-1. The piston-cylinder confinement of a gas.

applied to the piston and is assumed to exactly counterbalance the force the gas exerts on the piston head. Thus the pressure within the cylinder is always equal to the resisting pressure on the system boundary. If the force exerted on the piston decreases by a differential amount, the differential pressure difference that now exists will cause the piston to move to the right, so the volume of the gas within the cylinder increases by a differential amount dV. This is termed a *quasi-static process*, since all motion takes place *infinitesimally slowly*. The work performed by the gas within the cylinder on the surroundings may then be evaluated from

$$\delta W = P_{res}\, dV_{sys} \tag{4-1}$$

where the subscript "sys" refers to the system. However, for the frictionless nonaccelerating process,

$$P_{res} = P_{sys} \tag{4-2}$$

Therefore,

$$\delta W = P_{sys}\, dV_{sys} \tag{4-3}$$

As the force is continually lowered, the gas continues to expand. The total work performed by the gas on its surroundings may then be determined by integrating equation (4-3):

$$W = \int_{V_1}^{V_2} P_{sys}\, dV_{sys} \tag{4-3a}$$

Thus for a frictionless process we may evaluate the total work performed as a function only of system properties—a relation we shall find extremely useful.

The integration of equation (4-3a) requires two other pieces of information—an equation of state for the substance under consideration, and a relation between the temperature of the system and either its pressure or volume. With these additional items we may then integrate equation (4-3a). As may be expected, these two pieces of information will, in general, *not* be available —particularly the pressure (or volume)–temperature relation of the process.

SAMPLE PROBLEM 4-1. An ideal gas is to undergo isothermal reversible compression in a frictionless piston-cylinder from 1 Bar to 10 Bar. Calculate the initial and final molar volumes of the gas and the work necessary to perform the compression if the gas is initially at 25°C.

Solution: The ideal-gas equation of state is given by

$$Pv = RT$$

or

$$v = \frac{RT}{P}$$

where $T = 298°K$
$R = 8.314$ kJ/kg-mol°K

That is,

$$v = \frac{8.314 \times 10^3(298)}{P}\left(\frac{Nm}{\text{kg-mol}}\right)$$

$$= \frac{2.48 \times 10^6}{P}\left(\frac{Nm}{\text{kg-mol}}\right)$$

For $P = 1$ Bar $= 10^5 Nm^{-2}$:

$$v = 24.8 \frac{m^3}{\text{kg-mol}}$$

At $P = 10$ Bar $= 10^6 Nm^{-2}$:

$$v = 2.48 \frac{m^3}{\text{kg-mol}}$$

Since the compression is reversible, the work of compression may be obtained from the integral

$$w = \int_{v_1}^{v_2} P_{\text{sys}}\, dv_{\text{sys}}$$

But for an ideal gas, $P_{\text{sys}} = RT/v_{\text{sys}}$. Therefore,

$$w = \int_{v_1}^{v_2} \frac{RT}{v}\, dv = \int_{v_1}^{v_2} RT \frac{dv}{v}$$

But we have been told that the process is isothermal; therefore, T is a constant and RT may be removed from under the integral sign, to give

$$w = RT \int_{v_1}^{v_2} \frac{dv}{v} = RT \ln \frac{v_2}{v_1}$$

or

$$w = (8.314)(298) \ln\left(\frac{24.8}{2.48}\right) \frac{\text{kJ °K}}{\text{kg-mol °}K}$$

$$= 5710 \frac{\text{kJ}}{\text{kg-mol}} = 5.71 \times 10^6 \frac{Nm}{\text{kg-mol}}$$

SAMPLE PROBLEM 4-2. If the piston–cylinder of Sample Problem 4-1 is made adiabatic, that is, perfectly insulated against heat transfer, it can be shown that the following relation exists between the pressure and temperature of the gas within the cylinder (see Section 6.7 for a derivation of this result):

$$\frac{T}{P^{1-(1/k)}} = \text{constant}$$

Using this relation, obtain an expression for the amount of work needed to adiabatically and reversibly compress an ideal gas from 1 atm and 100°F to 10 atm. The coefficient k is a constant whose value depends on the ideal gas in question.

Solution: From the ideal-gas equation of state we may eliminate temperature from the above equation to give

$$\frac{Pv}{RP^{1-(1/k)}} = \text{constant}$$

which simplifies to

$$P^{1/k}v = C'$$

or

$$Pv^k = C$$

The value of C is obtained from the inlet conditions

$$C = P_1 v_1^k$$

The pressure–volume relation developed above is of the general class we have defined as polytropic. We have shown in Sample Problem 3-2 that the work requirements for a quasi-static, polytropic compression are given by:

$$w = \frac{P_2 v_2 - P_1 v_1}{1 - k}$$

which is the desired relation. The final specific volume, v_2, is determined from the relation

$$v_2 = v_1 \left(\frac{P_1}{P_2}\right)^{1/k}$$

Processes with Finite Driving Forces

Even if the piston of Fig. 4-1 is not frictionless, or if more than a differential pressure difference exists between the gas confined by the piston and the effective pressure of the surroundings, we may still be able to substitute P_{sys} for P_{res} in the determination of W provided we choose the system carefully enough. For example, let us examine Fig. 4-2. Initially the effective pressure

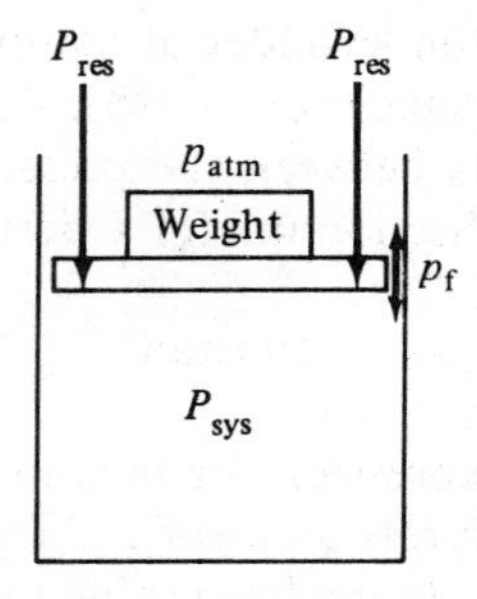

Figure 4-2. Expansion with finite pressure difference.

of the piston and weights (P_{surr}) is identical to the pressure of the system, where the system is chosen to be the cylinder and gas *but not including the piston.* Let us for the moment assume that the piston is frictionless. If we suddenly remove a portion of the weights on the piston, the effective pressure of the surroundings instantaneously drops before the volume of the system has had a chance to adjust itself (see Fig. 4-3). However, if we examine a force balance around the system boundary, we see that P_{res} must still equal P_{sys} if no unbalanced force is to exist across the system boundary. (Since the boundary has no mass, it cannot support a finite pressure imbalance.) On the other hand, a finite pressure difference now exists across the piston

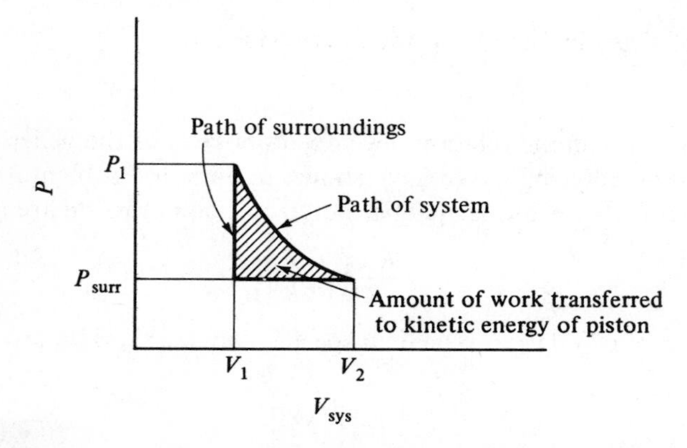

Figure 4-3. The variation of P_{sys} with P_{res}.

resulting in an unbalanced force on the piston, and Newton's law tells us that this unbalanced force will cause the piston to accelerate. If this acceleration is not too rapid, we may still assume that the gas within the cylinder is undergoing a quasi-static process. Therefore, the work that the system (the gas) does is identical to the amount of work the gas did in the previous case (after all, the gas cannot tell what it is pushing against as long as it is undergoing a quasi-static process). That is, the work done by the gas can be expressed as

$$W = \int P_{sys}\, dV \tag{4-4}$$

However, of this only

$$W_{surr} = \int P_{surr}\, dV \tag{4-5}$$

goes to pushing back the surroundings. What has happened to the rest? Very simply, it has gone into increasing the kinetic energy (acceleration) of the piston! It is now a simple matter to show that as long as the piston is frictionless, once set in motion like this it will continue to move for time immemorial (provided it does not shoot out of the cylinder), in a periodic oscillation whose frequency and amplitude will depend on the mass of the piston and its weights and the initial difference between P_{sys} and P_{surr}.

If the piston is now assumed to be frictional, we have a situation very similar to that in the previous case. Suppose a frictional force F_f must be overcome to move the piston up or down in the cylinder. Then we can say that a frictional pressure drop $P_f = F_f/A$ is required before motion can occur. That is, unless the absolute magnitude of $|P_{sys} - P_{surr}| > P_f$, no motion can occur. When this inequality is satisfied, the piston will move (in the direction toward the lower pressure). Suppose $|P_{sys} - P_{surr}| \gg P_f$. Then, as before, the piston will experience an imbalanced force and will accelerate. If the acceleration is slow enough to assume that the gas within the cylinder is undergoing a quasi-static process, then once again the *work performed by the gas* is given by

$$W = \int P_{sys}\, dV \tag{4-6}$$

The work needed to push back the surrounding atmosphere is given by

$$W_{atm} = \int P_{atm}\, dV \tag{4-7}$$

The work lost through friction is

$$W_f = \int P_f\, dV \tag{4-8}$$

and the remainder $(W - W_{atm} - W_f)$ goes into accelerating the piston head. Because of the effects of friction, the piston in this case will not continue to move indefinitely but will, after a time, settle at some position where $|(p_{sys} - P_{surr})| \leq P_f$.

SAMPLE PROBLEM 4-3. A 1440-lb_m piston is initially held in place by a removable latch above a cylinder as shown in Fig. SP4-3. The cylinder has an area of 1 ft^2; the volume of the gas within the cylinder initially is 2 ft^3 and at a pressure of 10 atm. The working fluid may be assumed to obey the ideal-gas equation of state. The cylinder has a total volume of 5 ft^3 and the top end is open to the surrounding atmosphere, whose pressure is 1 atm.

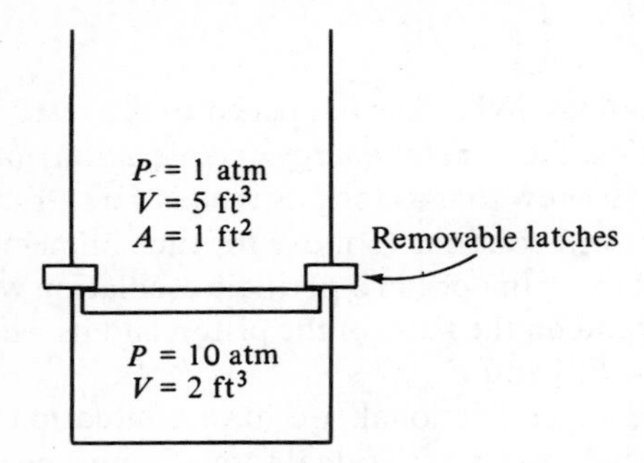

Figure SP4-3

(a) If the piston rises frictionlessly in the cylinder when the latches are removed and the gas within the cylinder is always kept at the same temperature, what will be the velocity of the piston as it leaves the cylinder?

(b) What will be the maximum height to which the piston will rise?

Solution: (a) First we must find the effective pressure of the surroundings:

$$P_{surr} = P_{atm} + \frac{\text{weight of piston}}{\text{area of piston}}$$

$$= 14.7 \text{ psia} + \frac{1440 \text{ lb}_f}{144 \text{ in.}^2}$$

$$= 24.7 \text{ psia}$$

Next, we determine the total work performed by the gas on both the piston and the atmosphere. We have shown for isothermal expansions of ideal gases that the total work performed by the gas is,

$$W = nRT \ln (v_2/v_1)$$

But

$$nRT = P_1 V_1$$

Thus

$$W = (10 \text{ atm} \cdot 2 \text{ ft}^3) \ln (5.0/2.0)$$

$$= 18.4 \text{ atm ft}^3 = 3.88 \times 10^4 \text{ ft-lb}_f$$

Of this, however,

$$W_{surr} = \int P_{surr}\, dV_{sys}$$

$$= 24.7 \frac{lb_f}{in.^2} \cdot 3\ ft^3$$

$$= 1.07 \times 10^4\ ft\text{-}lb_f$$

must be supplied to move the "effective" surrounding pressure. Therefore,

$$(3.88 - 1.07) \times 10^4 = 2.81 \times 10^4\ ft\text{-}lb_f$$

of energy is supplied to accelerate the piston. That is, the change in the kinetic energy of the piston is equal to 2.81×10^4 ft-lb_f, which tells us that

$$\frac{1}{2} \frac{MV^2}{g_c} = 2.81 \times 10^4\ ft\text{-}lb_f$$

or

$$\frac{1}{2} \frac{1440\ lb_m\ sec^2}{32.2\ ft} V^2 = 2.81 \times 10^4\ ft\text{-}lb_f$$

or

$$V^2 = 1255\ ft^2/sec^2$$

or

$$V = 35.4\ ft/sec \text{ at its exit from the cylinder}$$

(b) At the maximum rise of the piston its potential energy (PE) would be equal to the kinetic energy imparted to it. Therefore, at its maximum height,

$$PE = \frac{Mg\,Z}{g_c} = WZ = 2.81 \times 10^4\ ft\text{-}lb_f$$

But its weight is 1440 lb_f; therefore,

$$Z_{max} = \frac{2.81 \times 10^4}{1440} = 19.5\ ft \text{ above cylinder top}$$

(Checking of the units is left to the reader as an exercise).

4.2 Conservation of Energy

The fact that many forms of mechanical energy are completely interconvertible (theoretically) has been well known for many years. However, the recognition that thermal and mechanical energy are interconvertible is not so apparent. Indeed, as we have seen, it was not until the late 1700s that the correct relation between heat and work began to evolve.

As further evidence of the interconvertibility of heat and energy accumulated, several crude experiments were undertaken to determine the "mechanical equivalent of heat." In 1843 Joule performed the first accurate and reproducible determination of this quantity using the simple equipment illustrated in Fig. 4-4. The weight W is allowed to fall through a distance ΔZ, performing an amount of work $W \Delta Z$ on the paddle wheel. The paddle wheel in turn performs this same amount of work on the water, where viscous forces convert the mechanical energy into thermal energy. By measuring the temperature rise of the water, Joule was able to determine the amount of thermal energy absorbed, and, from the amount of work the weight performed, was able to determine the mechanical equivalent of heat. The best value obtained by Joule was[1]

$$1 \text{ Btu} = 772.5 \text{ ft-lb}_f$$

or

$$1 \text{ g cal} = 4.155 \text{ J}$$

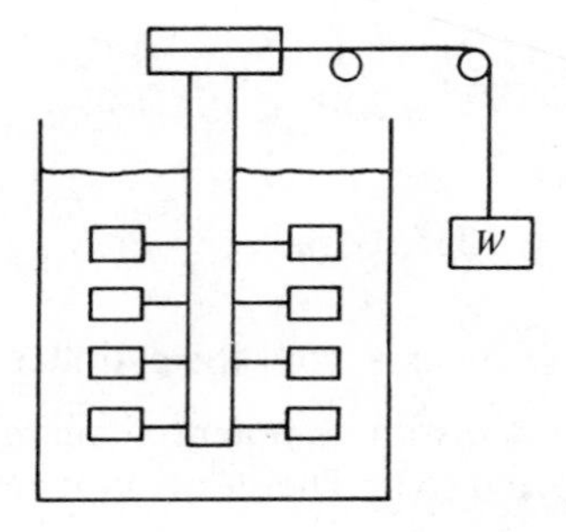

Figure 4-4. Joule's experiment.

As we have indicated, these conversion factors have now been (somewhat) arbitrarily fixed at:

$$\begin{aligned} 1 \text{ Btu} &= 778.16 \text{ ft-lb}_f \\ 1 \text{ cal} &= 4.1861 \text{ joule} \end{aligned} \tag{4-9}$$

Once it was established that mechanical energy could be reproducibly converted into thermal energy, the concept of equivalency and interconvertibility of all forms of energy quickly evolved. This concept is expressed in the *law of conservation of energy,* which states: Energy can neither be created nor destroyed—only changed in form. It would seem at first glance that atomic and nuclear processes in which mass is converted to energy according to Einstein's law, $E = mc^2$, disobey the law of conservation of energy. However, nuclear processes only indicate that the traditional concept of the separation of mass and energy is coming under attack. It now appears that mass and energy themselves are two different forms of the same fundamental phenomenon, and therefore are in many ways no more different than thermal

[1] M. W. Zemansky and H. C. Van Ness, *Basic Engineering Thermodynamics,* McGraw-Hill Book Company, Inc., New York, 1966, p. 74.

and mechanical energy. However, in this book we shall not deal with atomic or nuclear reactions and therefore will not be concerned with the interconversion of energy and mass. For our purposes we shall consider mass and energy to be separate properties, each of which obeys its respective law of conservation.

4.3 The Energy Equation

In the following sections we shall discuss the derivation and use of the energy equation. The equation is derived under very general conditions. We shall then consider some of the simplications that are commonly justified.

In formulating the energy equation for a system, the same procedure used in Chapter 3 for mass will be used. For an open system, the overall energy equation may be written as

$$\text{energy}_{\text{in}} - \text{energy}_{\text{out}} + \text{energy}_{\text{gen}} = \text{energy}_{\text{acc}} \tag{4-10}$$

In the previous section we indicated that energy can neither be created nor destroyed (at least in nonnuclear processes). Therefore, $\text{energy}_{\text{gen}} = 0$ and the energy equation becomes

$$\text{energy}_{\text{in}} - \text{energy}_{\text{out}} = \text{energy}_{\text{acc}} \tag{4-10a}$$

Let us assume that a differential amount of mass, δM_{in}, flows into the system and a differential amount of mass, δM_{out}, flows out of the system, as illustrated in Fig. 4-5. The mass entering the system has an energy per unit mass of e_{in}; the material leaving contains e_{out} energy per unit mass. Let us also assume that a differential amount of heat, δQ, flows into the system and that the system performs a differential amount of work, $\delta W'$, on the surroundings. The total energy accumulated by the system will then be $(eM)_{\text{sys}}$ at the end, $-(eM)_{\text{sys}}$ at the beginning, or $d(eM)_{\text{sys}}$ for the differential process just described. The energy equation may then be written as

$$\underbrace{(e\delta M)_{\text{in}} + \delta Q}_{\text{in}} - \underbrace{(e\delta M)_{\text{out}} - \delta W'}_{\text{out}} = \underbrace{d(eM)_{\text{sys}}}_{\text{accum}} \tag{4-11}$$

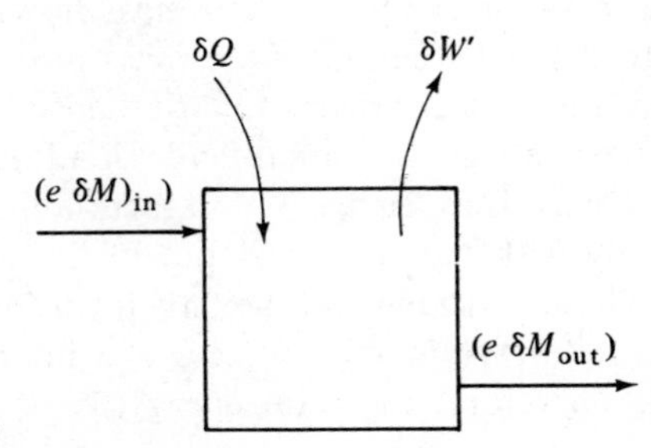

Figure 4-5. The open system.

For a process with no chemical reaction the mass balance is written as

$$\delta M_{\text{in}} - \delta M_{\text{out}} = dM_{\text{sys}} \tag{4-12}$$

Particular attention should again be given to the different meanings of the two infinitesimal operators δ and d: δ is used to indicate the transfer of a differential amount of something such as mass, heat, or work (in addition, we use δ to describe a differential generation term); d is used to signify the *change* in a (state) property [that is, the value of a (state) property now, minus its value at some previous time]. The difference between δ and d becomes especially important when they are integrated. The integral of δ is simply the total amount of whatever is entering (or leaving) the system, or the total internal production. However, the integral of d is the total change of the system property and is represented by the difference operator Δ. That is,

$$\int_{A_1}^{A_2} dA = \Delta A = A_2 - A_1$$

while (4-13)

$$\int \delta W = W$$

Sign Convention for δQ and δW

In the previous derivation of the energy equation we have assumed that heat, δQ, is being added to the system while the system is performing work, $\delta W'$, on the surroundings. That is, δQ is positive when heat is transferred into the system. If the system is losing heat to the surroundings, then δQ will be a negative number—but the sign before the δQ term in the energy balance should remain $+$. Similarly, $\delta W'$ has a positive value when the system is performing work on the surroundings. However, when the surroundings perform work on the system, $\delta W'$ has a negative value, which when combined with the minus sign in the energy balance yields a positive contribution to the energy accumulated within the system.

Total Energy, e

Let us now consider the ways by which energy may flow into the system by virtue of the mass flow: when mass flows into the system it clearly may bring with it kinetic energy if the mass possesses a finite velocity. The mass will also bring with it gravitational potential energy by virtue of its position within the earth's gravitational field. In addition to the potential and kinetic energies, a mass flow brings with it a certain amount of internal energy associated solely with the very existence of the mass itself.

We shall now *assume* that the only forms of energy that the matter entering or leaving the system will possess are internal, kinetic, and gravitational potential. Therefore, the total energy, e, is simply the sum of the internal energy u, the kinetic energy $V^2/2g_c$, and the (gravitational) potential energy gZ/g_c. Thus, e becomes:

$$e = u + \frac{V^2}{2g_c} + \frac{gZ}{g_c} \tag{4-14}$$

where all terms represent energy per unit mass (mole). In some fields of study other forms of energy such as electrical, magnetic, surface, and rotational–kinetic are also important. In these cases the definition of e can be directly expanded to include these terms as well. However, in this text we shall restrict ourselves, for convenience, to the three most frequently encountered forms of energy suggested by equation (4-14).

Now let us examine $\delta W'$, the total work transferred from the system to the surroundings. We have shown that whenever a force acts over a distance, work is performed. That is, whenever a system's boundaries move, work is done.

When we examine an open system in close detail, the following process is seen to occur at the inlet and outlet (as illustrated in Fig. 4-6). As a differential amount of mass, δM_{in}, is forced into the system, the surroundings perform an amount of work equal to $(P\,dV)_{in} = (Pv\delta M)_{in}$ on the increment of mass entering the system. Similarly, as δM_{out} is forced out of the system, the system performs an amount of work equal to $(Pv\delta M)_{out}$ on the surroundings. If an amount of work δW is performed by the system on its surroundings in addition to the work required to force mass in and out of the system (such as the work needed to lift a weight or turn a shaft), then the total amount of work performed by the system on its surroundings is

$$\delta W' = \delta W + (Pv\delta M)_{out} - (Pv\delta M)_{in} \tag{4-15}$$

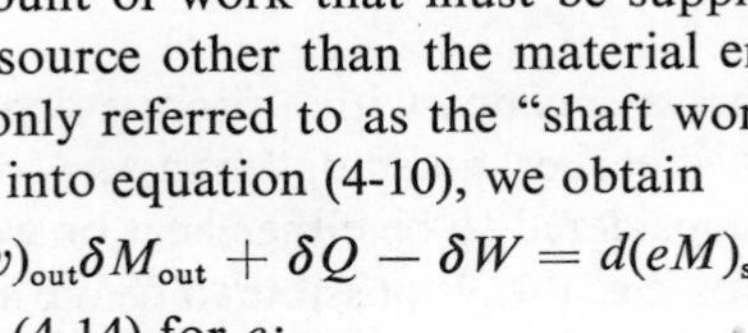

Figure 4-6. The effect of mass flows.

The term δW represents the amount of work that must be supplied to (or removed from) the system by a source other than the material entering or leaving the system and is commonly referred to as the "shaft work." If the expression for $\delta W'$ is substituted into equation (4-10), we obtain

$$(e + Pv)_{in}\delta M_{in} - (e + Pv)_{out}\delta M_{out} + \delta Q - \delta W = d(eM)_{sys} \tag{4-16}$$

or upon substitution of equation (4-14) for e;

$$\left(u + Pv + \frac{V^2}{2g_c} + \frac{gZ}{g_c}\right)_{in} \delta M_{in} - \left(u + Pv + \frac{V^2}{2g_c} + \frac{gZ}{g_c}\right)_{out} \delta M_{out} + \delta Q - \delta W = d\left[M\left(u + \frac{V^2}{2g_c} + \frac{gZ}{g_c}\right)\right]_{sys} \tag{4-17}$$

Equation (4-17) is the *open-system energy* equation. It is extremely general (although only for processes in which only internal, potential and kinetic energies changes are important). As we shall see, the energy equation can be greatly simplified for many special cases.

In equation (4-17) the term $u + Pv$ appears twice. As discussed in Chapter 3, this term is encountered frequently and has attained the status of a property. The sum $u + Pv$ is called the *specific enthalpy* and given the symbol h.

$$h = u + Pv \tag{4-18}$$

or, multiplying by the total mass, M, the *total enthalpy* is

$$H = U + PV \tag{4-18a}$$

Substitution of the enthalpy into the energy balance gives

$$\begin{aligned}\left(h + \frac{V^2}{2g_c} + \frac{gZ}{g_c}\right)_{\text{in}} \delta M_{\text{in}} - \left(h + \frac{V^2}{2g_c} + \frac{gZ}{g_c}\right)_{\text{out}} \delta M_{\text{out}} + \delta Q - \delta W \\ = d\left[M\left(u + \frac{V^2}{2g_c} + \frac{gZ}{g_c}\right)\right]_{\text{sys}}\end{aligned} \tag{4-19}$$

Since u, P, and v are functions only of the state of the material involved, h is also a state function, whose value therefore does not depend on its past history. Values of enthalpy are commonly found in tabulations of thermodynamic properties.

4.4 Flow or Shaft Work and Its Evaluation

We have said that work is equivalent to a force moving through a distance. A system is capable of performing (or consuming) work in three primary fashions: (1) the boundaries of the system may move against a restraining force; (2) a shaft may add (or remove) work through the system's boundary, or (3) energy may be transferred across the system boundary by virtue of a thermodynamic potential other than temperature (for example, an electrical potential).

The work associated with a process depends on the choice of the system. As was noted earlier, an energy transfer—regardless of the potential causing it—is not considered to be either heat or work if it occurs wholly within the system. Thus it is always possible to define a system such that all energy flows are internal and thus neither heat nor work occur in a thermodynamic sense. These observations will be examined in the discussion that follows.

Let us consider first a simple closed system as illustrated in Fig. 4-7 (choose the gas as the system). As the gas expands, it does work by pushing back the surroundings. If a shaft within the system supplies an amount of

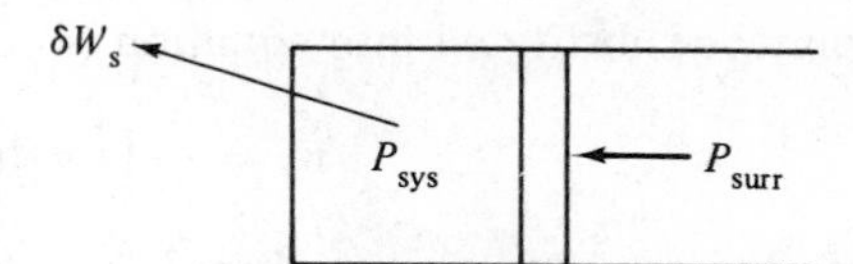

Figure 4-7. Work terms for a closed system.

work δW_s to the surroundings, then the total work may be evaluated as

$$\delta W_{\text{total}} = \delta W_s + P_{\text{res}}\, dV_{\text{sys}} \tag{4-20}$$

where $P_{\text{res}}\, dV$ represents the work needed to push back the surroundings.

Let us now consider a gas flowing through a turbine or a compressor (Fig. 4-8). Suppose we choose as our system 1 lb_m of gas as it flows through the turbine (the turbine blades and shaft are chosen to be outside the system). The gas represents a closed system. As the gas flows through the turbine the total amount of work it does is

$$w' = \int P_{\text{sys}}\, dv_{\text{sys}} \tag{4-21}$$

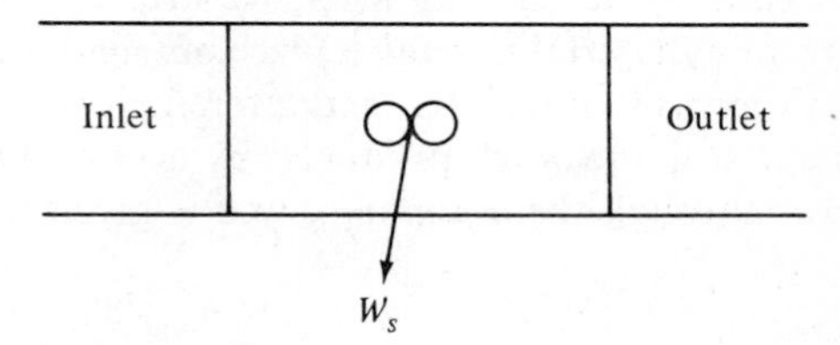

Figure 4-8. Work terms for an open system.

where the prime refers to the total amount of work delivered by the mass of gas to its surroundings. This work is divided into three parts: (1) that work which the gas does on the gas in front of it as it moves down the pipeline, given by $P_{\text{out}}v_{\text{out}}$; (2) that work which the gas receives from the gas behind as it is forced down the pipeline, given by $P_{\text{in}}v_{\text{in}}$; and (3) that work which is transmitted to the shaft, w_s. Thus

$$w' = \int P_{\text{sys}}\, dv_{\text{sys}} = w_s + P_{\text{out}}v_{\text{out}} - P_{\text{in}}v_{\text{in}} \tag{4-22}$$

Solution for the work transmitted to the shaft, w_s, then gives

$$\begin{aligned} w_s &= \int P_{\text{sys}}\, dv_{\text{sys}} - (P_{\text{out}}v_{\text{out}} - P_{\text{in}}v_{\text{in}}) \\ &= \int P_{\text{sys}}\, dv_{\text{sys}} - \Delta(Pv) \end{aligned} \tag{4-23}$$

Integration by parts of the first integral in equation (4-23) gives

$$\int_1^2 P_{\text{sys}}\, dv_{\text{sys}} = -\int_1^2 v\, dP + \int_1^2 v(Pv) \tag{4-24}$$

or

$$\int_1^2 P_{\text{sys}}\, dv_{\text{sys}} = -\int_1^2 v\, dP + \Delta Pv \tag{4-25}$$

which may be substituted into equation (4-23) to give

$$w_s = -\int_1^2 v\,dP \tag{4-26}$$

w_s is then the actual work *transferred from the gas to the shaft*. (If the shaft passes frictionlessly through the walls of the turbine, then w_s is the work transmitted to the *surroundings of the turbine*. If the shaft encounters frictional losses as it passes through the turbine housing, then the amount of work transmitted to the turbine's surroundings will be decreased by a comparable amount.)

If we now consider the turbine and its contents as an open system, then the work term in the energy balance is only that work which the system transmits to its surroundings via the connecting shaft. (Had the volume of the turbine been changing, a $P_{\text{surr}}\,dV_{\text{sys}}$ term would also be needed.) For the frictionless process this would be w_s of equation (4-26). For a frictional process it would be w_s minus the frictional losses.

Thus we learn that when we are attempting to analyze the frictionless flow of a material *through* a piece of (rigid) equipment, there are two different work terms that may be encountered, depending on how the system is chosen. The first is the work produced by a unit mass of material (the system) as it flows through the equipment and is given by

$$w' = \int P_{\text{sys}}\,dv_{\text{sys}} \tag{4-21}$$

The second is work transmitted by the gas via a mechanical linkage (shaft) to the surroundings of the equipment and is given by

$$w_s = -\int v_{\text{sys}}\,dP_{\text{sys}} \tag{4-27}$$

where the system is taken as the turbine and its contents. The two work terms differ by $\Delta(Pv)$, the net work the unit of material expends pushing fluid out of its way as it passes through the piece of equipment. Because this amount of work, $\Delta(Pv)$, is the work associated only with the flow of the working fluid, it is frequently termed *flow energy* or *flow work*. For processes in which no mass flow occurs, there is no work needed to move the fluid around, and thus such a term does not arise.

The shaft work, w_s, is the maximum amount of recoverable work that may be obtained from the *flow* of a fluid through a piece of process equipment (for irreversible flows, the actual w is less). For example, if we are analyzing the main turbine in an electrical generating station, the maximum amount of work that may be recovered from the working fluid and transmitted to the electric generators would be w_s, not w'. Similarly, if we were analyzing the work requirements of a compressor, w_s, not w', would be the *minimum* amount of work that would have to be supplied to the compressor motor in the form of electrical energy. Thus we see that, for flow processes, it is the shaft work, w_s, not w', that is the term of primary importance to us. This is the reason we have chosen to write the general energy balance in terms of w_s rather then w'. Of course, if the (flow) system boundaries are expanding or contracting against a surrounding pressure, then a work term for this must be added to w_s. Similarly, for frictional processes, the actual work transfer, w, is the term of primary importance.

4.5 Special Cases of the Energy Equation

Now let us examine some of the simplifications of the energy equation for certain special but still common and important cases.

Closed System

The gas enclosed by a cylinder and piston is a typical example of a closed system (see Fig. 4-1). Since no mass enters or leaves the closed system, $\delta M_{in} = \delta M_{out} = dM_{sys} = 0$. Therefore, the energy equation becomes

$$\delta Q - \delta W = Md\left(u + \frac{V^2}{2g_c} + \frac{gZ}{g_c}\right)_{sys} \tag{4-28}$$

If the system does not accelerate greatly its velocity, V, does not vary greatly and we may neglect changes in $V^2/2g_c$. If its height does not vary appreciably, we may also neglect changes in gZ/g_c in comparison to u, and the closed-system energy balance further reduces to

$$\delta Q - \delta W = M\, du_{sys} = du_{sys} \tag{4-29}$$

Equation (4-29) is frequently referred to as the *first law of thermodynamics* and is presented as a general form of the energy equation. However, as we have seen, equation (4-29) is merely a special case of the more general, open-system energy equation. *It may only be used for closed systems that neither accelerate nor change their height appreciably.*

Open Systems

As you will learn in later courses, most equipment and process designers strive to design equipment and processes that operate at steady state for extended periods of time. Unsteady processes are avoided because they are usually less economical than steady-state processes. In addition, flow processes are usually much simpler to operate and maintain than unsteady processes.

Since all flow processes may be considered as open systems (with the process equipment and its contents as the system), most engineering design problems may be treated as open systems—usually operating at steady state.

Of course, any process can be treated using a closed-system analysis if we choose as the system a unit of mass as it flows through the equipment. However, when this is done we must be able to evaluate the work performed on, or by, the system as it flows through the process. Unfortunately, this evaluation normally requires knowledge of all the system properties at every point during the process—knowledge that is frequently not available. The open-system approach, on the other hand, requires knowledge of only the "shaft work" supplied to, or by, the processing machinery. Usually this quantity can be easily calculated or directly measured. For this reason the flow-system approach to thermodynamic calculations is usually preferred to the closed-system approach. When sufficient data for both a closed- and an open-system analysis are available, the results from both calculations *must be identical* (except for determination of the system work and heat transfer).

A *steady-state process* is defined as one in which the system properties (either point or average properties) do not change with time. That is, if we examine either a single point in the system, or the system as a whole, its properties will not vary with time. Note, this in no way implies that the properties at all points must be identical, only that the properties of each point are invariant with time. Since the mass of a system is a property of the system, steady state implies that $\delta M_{in} = \delta M_{out}$ and $dM_{sys} = 0$. Since total energy is also a system property, steady state also implies that $d[u + (V^2/2g_c) + (gZ/g_c)]_{sys} = 0$. Thus the energy equation for a steady-state process reduces to

$$\left(h + \frac{V^2}{2g_c} + \frac{gZ}{g_c}\right)_{in} \delta M - \left(h + \frac{V^2}{2g_c} + \frac{gZ}{g_c}\right)_{out} \delta M + \delta Q - \delta W = 0 \quad (4\text{-}30)$$

We may divide equation (4-30) by δM to get

$$\left(h + \frac{V^2}{2g_c} + \frac{gZ}{g_c}\right)_{in} - \left(h + \frac{V^2}{2g_c} + \frac{gZ}{g_c}\right)_{out} + \frac{\delta Q}{\delta M} - \frac{\delta W}{\delta M} = 0 \quad (4\text{-}30a)$$

where

$$\frac{\delta Q}{\delta M} = q = \frac{\text{heat absorbed}}{\text{unit-mass flow through equipment}}$$
$$\frac{\delta W}{\delta M} = w = \frac{\text{work liberated}}{\text{unit-mass flow through equipment}} \quad (4\text{-}31)$$

Thus equation (4-30a) reduces to

$$-\Delta\left(h + \frac{V^2}{2g_c} + \frac{gZ}{g_c}\right) + q - w = 0 \quad (4\text{-}32)$$

Compressors and Expanders—Steady State

Compressors and expanders are machines that are capable of changing the pressure on a substance. Compressors (pumps) are used to increase the pressure on a substance, and consume work in so doing. There are two main classes of compressors: positive-displacement compressors, which include the piston, diaphragm, and gear compressors; and centrifugal compressors. Positive-displacement compressors are usually capable of high compression ratios per stage, but being noncontinuous they usually cannot handle high flow rates. Centrifugal compressors, on the other hand, are not capable of producing high pressure ratios per stage but can handle extremely large flow volumes, because they operate continuously. Most of the pumps and compressors used in normal process operations are of the centrifugal type—fitting into our previous observation that most processes tend to be of the continuous type.

Expanders operate in the reverse fashion from compressors—decreasing the pressure on the working fluid while producing work. All compressors can in theory be run backward to operate as expanders. However, in practice turbine expanders are by far the most commonly encountered type of expansion machines.

For most compressors and expanders it is found that the $\Delta(V^2/2g_c)$ and $\Delta(gZ/g_c)$ terms are usually quite small in comparison to the Δh term. Therefore, these terms are often assumed to be negligible and are neglected. Under these assumptions the steady-state energy equation reduces to

$$\Delta h = q - w \tag{4-33}$$

Equation (4-33) can be considered the open-system analogy to equation (4-29).

If the compressor is run adiabatically, that is, with no heat exchange with the surroundings, then $q = 0$, and equation (4-33) reduces to

$$\Delta h = -w \tag{4-34}$$

Note that this result is valid whether or not the compression is quasi-static!

Throttling Devices—Steady State

A throttling device is a device used to reduce the pressure of a flowing fluid without obtaining any shaft work. The valve that reduces the water pressure in your kitchen faucet from the pressure of the water main to atmospheric pressure is a good example of a throttling device. (See Fig. 4-9.) As with the majority of compressors and expanders, most throttling devices produce a negligible change in $V^2/2g_c$ and gZ/g_c; therefore, the energy equation reduces to

$$-\Delta h + q - w = 0 \tag{4-33}$$

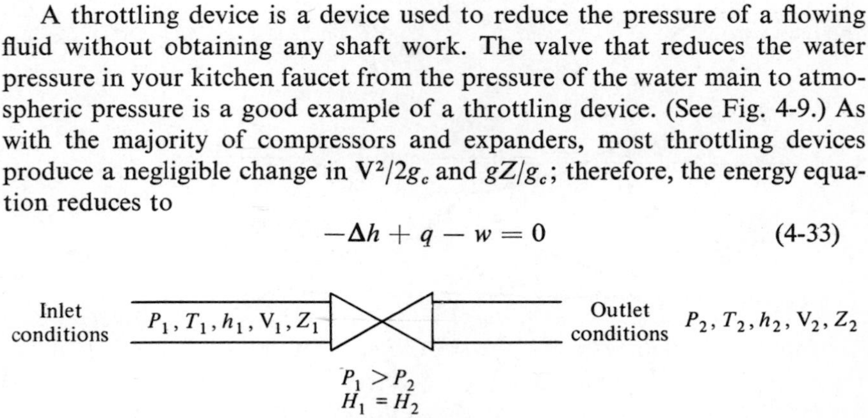

Inlet conditions P_1, T_1, h_1, V_1, Z_1 | Outlet conditions P_2, T_2, h_2, V_2, Z_2

$P_1 > P_2$
$H_1 = H_2$

Figure 4-9. Throttling devices.

However, since throttling devices provide no work to the surroundings, $w = 0$, and equation (4-33) reduces to

$$\Delta h = q \tag{4-35}$$

Usually the fluid passing through a throttling device is moving so rapidly that it does not remain within the device long enough to absorb, or give off, much heat. Therefore, for many cases it is reasonable to assume that $q = 0$ (unless special effort is taken to provide heating or cooling within the device), and the energy equation becomes

$$\Delta h = 0 \tag{4-36}$$

or

$$h_{\text{in}} = h_{\text{out}} \tag{4-36a}$$

Nozzles and Diffusers—Steady State

A nozzle is a device specifically designed to increase the kinetic energy of a high-pressure fluid at the expense of its pressure and temperature. A diffuser, on the other hand, increases the pressure and temperature within a fluid at the expense of its kinetic energy. Since a nozzle (diffuser) is designed to increase (decrease) the velocity of the working fluid (and often to a very high degree), we cannot neglect the $V^2/2g_c$ term. However, except for extremely long vertical nozzles, the gZ/g_c term is still small, and the q is usually neglected for the same reasons as with the throttling devices. A simple nozzle or diffuser is illustrated in Figure 4-10. Normally no shaft work is produced, so $w = 0$ and the energy balance reduces to

$$\Delta h + \Delta\frac{V^2}{2g_c} = 0 \tag{4-37}$$

or

$$\Delta h = -\Delta\frac{V^2}{2g_c} \tag{4-38}$$

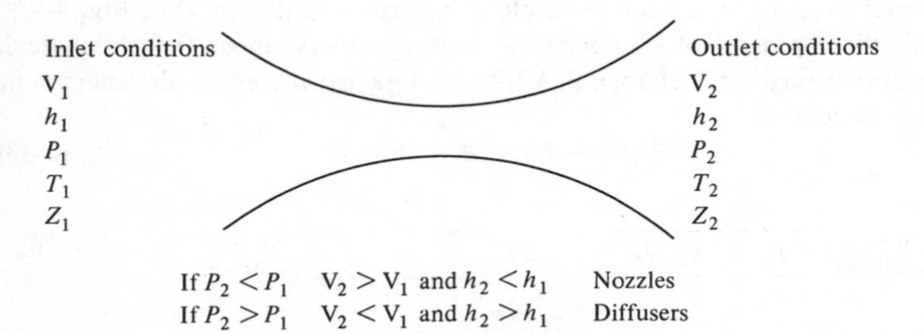

Figure 4-10. Nozzles and diffusers.

4.6 Heat Capacities

We now define the constant-volume and constant-pressure heat capacities:

$$C_V = \left(\frac{\partial u}{\partial T}\right)_V, \quad \text{the constant-volume heat capacity} \tag{4-39}$$

$$C_P = \left(\frac{\partial h}{\partial T}\right)_P, \quad \text{the constant-pressure heat capacity} \tag{4-40}$$

Since the heat capacities are defined solely in terms of state properties, they are state properties themselves. The heat capacities are extremely valuable properties because they can be measured directly and, once available, allow calculation of many other useful thermodynamic properties.

The relationship between the heat capacities and other directly measurable quantities is developed as follows.

Consider as a system a unit of mass as it proceeds through a given process. Since the system is closed, its mass is constant and the energy balance reduces to:

$$\delta Q - \delta W = Md\left(u + \frac{V^2}{2g_c} + \frac{gZ}{g_c}\right) \tag{4-41}$$

If we assume that potential and kinetic energies of the system are constant, then equation (4-41) simplifies to:

$$\delta Q - \delta W = M\,du \tag{4-42}$$

Now, if the process is performed at *constant volume* (or *specific volume* since M is constant), the work term ($\delta W = P_{\text{res}}\,dV_{\text{sys}}$) vanishes and we get:

$$(\delta Q)_v = M(du)_v \tag{4-43}$$

however, from the definition of C_V:

$$(du)_v = C_V\,dT \tag{4-44}$$

Thus:

$$\delta Q = MC_V\,dT \quad \text{at constant volume} \tag{4-45}$$

If, on the other hand, the process is performed at constant pressure—as in a flow calorimeter—the system's pressure is the same as the gas that surrounds it. Thus, $P_{\text{sys}} = P_{\text{surr}}$ and

$$\delta W = P_{\text{res}}\,dV_{\text{sys}} = P_{\text{sys}}\,dV_{\text{sys}} \tag{4-46}$$

Since P_{sys} is a constant, it may be moved within the differential sign to give:

$$\delta W = d(PV)_{\text{sys}} = M\,d(Pv)_{\text{sys}} \tag{4-46a}$$

Substitution of the work expression into the simplified energy balance yields

$$\delta Q - M\,d(Pv)_{\text{sys}} = M\,du \quad \text{at constant } P \tag{4-47}$$

Rearranging and collecting terms then gives:

$$\delta Q = M[du + d(Pv)] = M\,d[u + Pv] = M\,dh \tag{4-48}$$

at constant pressure. However, from the definition of C_P

$$(dH)_P = MC_P\,dT \tag{4-49}$$

Thus

$$\delta Q = MC_P\,dT \quad \text{at constant pressure} \tag{4-50}$$

Heat capacities (either C_P or C_V) are, in general, functions of both T and P, or V. However, over small temperature ranges C_P and C_V are usually relatively constant, so that finite heat transfers can be expressed as

$$Q = MC_P\,\Delta T \quad \text{for constant pressure}$$

or

$$Q = MC_V\,\Delta T \quad \text{for constant volume} \tag{4-51}$$

provided that the temperature differences are "small enough" that the heat capacities are relatively constant. For larger temperature ranges we must integrate $C_P\,dT$ (or $C_V\,dT$) or use an average heat capacity

$$(C_P)_{\text{av}} = \frac{\int_{T_1}^{T_2} C_P\,dT}{T_2 - T_1}$$

$$(C_V)_{\text{av}} = \frac{\int_{T_1}^{T_2} C_V\,dT}{T_2 - T_1} \tag{4-52}$$

Heat capacities are usually measured directly in an instrument known as a calorimeter. Two types of calorimeters are commonly used: the constant-volume (or bomb) type for measuring C_V; and the constant pressure (or flow) type for measuring C_P. The calorimeter experiment is extremely simple (in theory): A known amount of thermal energy is supplied to a known mass of fluid at constant pressure (or volume) and the temperature change is measured. The average heat capacity over the temperature range is then given by solving equations (4-51) and (4-52) for the average heat capacities:

$$(C_P)_{\text{av}} = \frac{Q}{M\,\Delta T} \quad \text{for constant pressure}$$

$$(C_V)_{\text{av}} = \frac{Q}{M\,\Delta T} \quad \text{for constant volume} \tag{4-53}$$

By taking the limit as ΔT approaches zero the temperature-dependent values of C_P and C_V may be determined. (Since it is not an easy matter to deliver a precisely known amount of thermal energy to a mass of fluid, these experiments are extremely tedious and difficult in practice.) Thus we find that the constant volume, or pressure, heat capacities can be determined directly in terms of measurable quantities.

We will show later that the heat capacities of gases which obey the ideal-gas equation of state have two rather useful properties:

1. C_P and C_V are independent of P and v at constant temperature, although they may still vary with temperature.
2. C_P and C_V are related through:

$$C_P = C_V + R \tag{4-54}$$

In addition, it is frequently assumed that C_P and C_V are independent of T for a gas which obeys the ideal-gas equation of state. However, it must be remembered that this is a *separate assumption* which is not required by the ideal-gas equation of state. Lastly, we will show in Chapter 6 that the constant pressure and volume restrictions in equations (4-39) and (4-40) may be dropped for fluids which obey the ideal-gas equation of state.

SAMPLE PROBLEM 4-4. A turbine is driven by 5000 kg/hr of steam, which enters the turbine at 45 Bar and 450°C with a velocity of 65 m/sec. The steam leaves the turbine exhaust at a point 5 m below the turbine inlet with a velocity of 400 m/sec. The shaft work produced by the turbine is measured as 600 kw and the heat loss from the turbine has been calculated to be 10^5 kJ/hr.

A small portion of the exhaust steam from the turbine is passed through a throttling valve and discharges at atmospheric pressure. Velocity changes across the valve may be neglected.

(a) What is the temperature of the steam leaving the valve?

(b) What is the quality or degrees of superheat (whichever is applicable) of the steam leaving the throttling valve?

Solution: The process is as illustrated in Fig. SP4-4. Since we know the pressure of the gas leaving the throttling valve (point 3), determination of one other property of the throttled stream will serve to fix its state and thus its temperature and quality (or superheat). Since we are supplied with considerable information about the energy requirements of the turbine, it is logical that the property of the gas at point 3 we should attempt to determine is the enthalpy.

We begin our determination of the enthalpy at point 3 by considering an energy balance around the throttling valve:

$$\left(h_2 + \frac{V_2^2}{2g_c} + \frac{gZ_2}{g_c}\right)\delta M_{\text{in}} - \left(h_3 + \frac{V_3^2}{2g_c} + \frac{gZ_3}{g_c}\right)\delta M_{\text{out}} + \delta Q - \delta W$$
$$= d[M(e)_{\text{valve}}]$$

The valve is operating at steady state, so $d(Me) = 0$ and $\delta M_{\text{in}} = \delta M_{\text{out}}$. We are told to neglect $\Delta(V^2/2g_c)$ and recognize that ΔZ will be small enough to neglect also. Therefore, the energy balance becomes

$$(h_2 - h_3)\delta M + \delta Q - \delta W = 0$$

However, we showed earlier that $\delta W = 0$ for a throttle valve, and for most cases δQ is also equal to zero. The energy balance then gives

$$h_3 = h_2$$

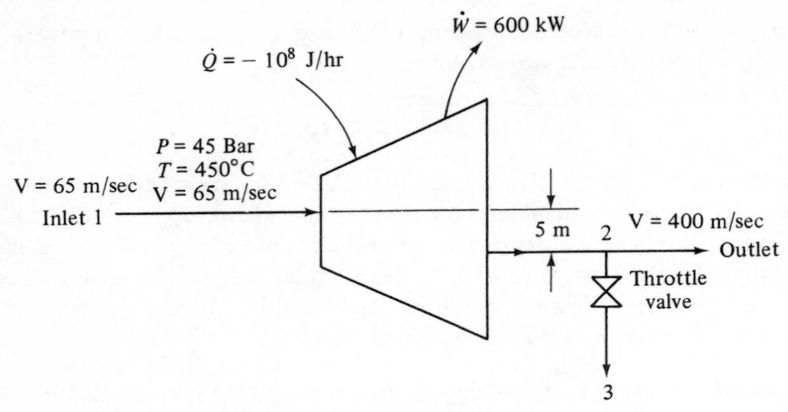

(The dot above a symbol represents "per unit time.")

Figure SP4-4

Thus, if we can determine h_2, we will know a second property, h_3 at point 3. h_2 is determined from an energy balance around the turbine:

$$\left(h_1 + \frac{V_1^2}{2g_c} + \frac{gZ_1}{g_c}\right)\delta M_1 - \left(h_2 + \frac{V_2^2}{2g_c} + \frac{gZ_2}{g_c}\right)\delta M_2 + \delta Q - \delta W$$
$$= d(Me)_{\text{turb}}$$

Base all future calculations on $\delta M_1 = 1$ kg.

Since the turbine is operating at steady state, $d(Me)_{\text{sys}} = 0$ and $\delta M_2 = \delta M_1$. Also, since the turbine exit is only 5 m below the inlet, we know that the $g\Delta Z/g_c$ term will be small and may be neglected. The energy balance then becomes

$$\left(h_1 + \frac{V_1^2}{2g_c}\right) - \left(h_2 + \frac{V_2^2}{2g_c}\right) + q - w = 0$$

or

$$-\Delta h - \Delta V^2/2g_c + q - w = 0$$

But from the data given in the problem statement:

$$q = \frac{\dot{Q}}{\dot{M}} = \frac{-10^5 \text{ kJ/hr}}{5 \times 10^3 \text{kg/hr}} = -20 \text{ kJ/kg}$$

$$w = \frac{\dot{W}}{\dot{M}} = \frac{600 \text{ kJ/sec}}{5 \times 10^3 \text{ kg/hr}} = \frac{(600 \times 3600)}{5000}\frac{\text{kJ}}{\text{kg}} = 432\frac{\text{kJ}}{\text{kg}}$$

$$h_1 = 3230 \text{ kJ/kg}$$

$$\frac{\Delta V^2}{2g_c} = \frac{[(400)^2 - (65)^2]\text{m}^2/\text{sec}^2}{2(1 \text{ kg m/N sec}^2)} = 77900 \text{ Nm/kg} = 77.9 \text{ kJ/kg}$$

(A quick examination of the $\Delta V^2/2g_c$ term indicates that the 65 m/sec term accounts for only about 3 percent of the total change in kinetic energy, or about 2 kJ/kg. That is, for all practical purposes the kinetic energy associated with velocities below 65 m/sec is negligible in comparison with the enthalpy or internal energy.) Thus the energy balance gives

$$\Delta h = -77.9 \text{ kJ/kg} - 20 \text{ kJ/kg} - 432 \text{ kJ/kg} = -530 \text{ kJ/kg}$$

but

$$\Delta h = h_2 - h_1 = -530 \text{ kJ/kg}$$

Therefore,

$$h_2 = h_1 - 530 \text{ kJ/kg} = (3230 - 530) \text{ kJ/kg}$$
$$= 2700 \text{ kJ/kg}$$

but

$$h_3 = h_2 = 2700 \text{ kJ/kg}$$

We may now examine the steam tables to locate the point where $P = 1$ atm (≈ 1 Bar) and $h = 2700$ kJ/kg. This point is slightly above the saturation dome, and therefore we know that the gas must be superheated—about 12°C in this case, since $T_3 \approx 112$°C.

SAMPLE PROBLEM 4-5. Steam flowing through an insulated 0.02-m diameter pipe has a pressure of 60 Bar and a quality of 95 percent. If the steam is adiabatically throttled to 1 atm through a horizontal valve in the line, what should the diameter of the downstream line be if there is to be no change in the velocity of the steam?

Solution: The process under consideration is illustrated in Fig. SP4-5. We may write a mass balance around the throttling valve as follows:

$$\dot{M}_{\text{in}} = \dot{M}_{\text{out}} = \dot{M}$$
$$\dot{M} = \rho_1 \text{V}_1 A_1 = \rho_2 \text{V}_2 A_2$$

where ρ_1 = density at point 1
V_1 = velocity at point 1
A_1 = area of pipe at point 1

However, we are asked to find the downstream diameter such that $\text{V}_1 = \text{V}_2$. Therefore,

$$\rho_1 A_1 = \rho_2 A_2$$

or

$$\rho_1 \pi D_1^2 = \rho_2 \pi D_2^2$$

or

$$D_2 = D_1 \sqrt{\frac{\rho_1}{\rho_2}} = D_1 \sqrt{\frac{v_2}{v_1}}$$

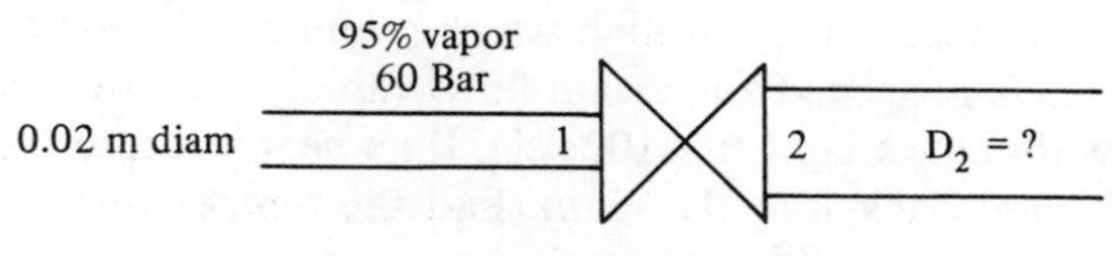

Figure SP4-5

since $\rho = 1/v$. Therefore, our problem reduces to determining the downstream specific volume, since we can determine the upstream volume from the saturation pressure and moisture content.

If we can determine one more property of the downstream fluid (other than its pressure), we can fix the state of the fluid, from which we can determine its specific volume. An energy balance around the throttling valve will supply us with the needed property value.

Assuming horizontal, nonaccelerating, adiabatic, steady-state flow through the throttling valve, the energy balance reduces to

$$h_2 = h_1$$

but

$$h_1 = 2705 \text{ kJ/kg} \qquad \text{(from Mollier diagram)}$$

Therefore,

$$h_2 = 2705 \text{ kJ/kg}$$

Knowing P_2 and h_2, we can find T_2 and v_2 from the steam tables.

$$T_2 = 115°\text{C}$$

$$v_2 = 1.77 \text{ m}^3\text{/kg (from } P_2 \text{ and } h_2)$$

Now let us determine the inlet volume, from which we can evaluate D_2:

$$v_1 = X_1 v^V + (1 - X_1)v^L$$

where X_1 = vapor fraction at 1. But at 60 Bar the saturation properties of vapor and liquid are

$$v^V = 32.4 \times 10^{-3} \text{ m}^3\text{/kg}$$

$$v^L = 1.32 \times 10^{-3} \text{ m}^3\text{/kg}$$

$$X_1 = 0.95 \text{ mass fraction vapor}$$

Therefore,

$$v_1 = [(0.95)(0.0324) + (0.05)(0.00132)] \text{ m}^3\text{/kg}$$

$$= 0.0308 \text{ m}^3\text{/kg}$$

$$D_1 = 0.02 \text{ m}$$

but

$$D_2 = D_1\sqrt{\frac{v_2}{v_1}}$$

$$= 0.02 \text{ m}\sqrt{\frac{1.77}{0.0308}} = 0.149 \text{ m}$$

Sample Problem 4-6. An evacuated chamber with perfectly insulated walls is connected to a steam main through which steam at 100 psia and 350°F is flowing. The valve is opened and steam flows rapidly into the chamber until the pressure within the chamber is 100 psia. If no heat is lost to the surroundings or transferred back into the main, find the temperature of the steam in the chamber when the flow stops.

Solution: When the valve is opened, steam will flow into the chamber until the pressure within the chamber is equal to that in the line (Fig. SP4-6a).

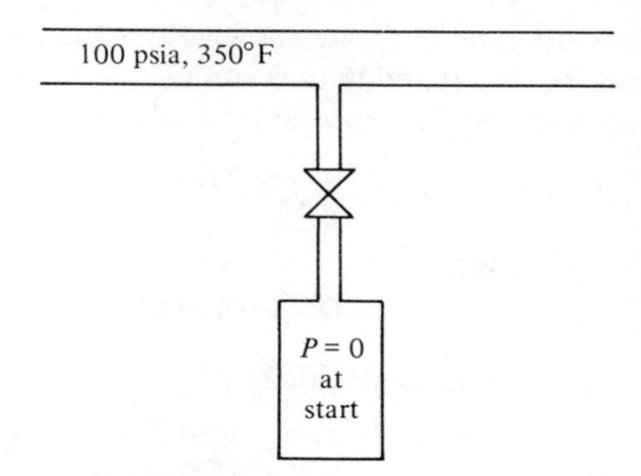

Figure SP4-6a

However, since no heat is transferred from the chamber, the temperature within the chamber may not (indeed will not) be equal to that in the main. Therefore, it is necessary to determine another property of the steam in the chamber before the state can be fixed and its temperature determined. Thus we should like to know the enthalpy (or, as we shall see, more conveniently the internal energy) of the material in the tank at the end of the process. At this point we expect the energy equation will probably be needed. But before we apply the energy equation we must choose the system. At least two systems are possible:

1. We could choose the closed system: all the gas that ends up in the chamber at the end of the process, or
2. We could pick the open system: the chamber, the valve, and their contents at any moment.

For this illustration we shall choose the open-system approach. (The student should solve this problem using the closed-system approach to prove that both methods of attack yield the same solution.) With the open-system approach we write the energy equation for the chamber and its contents as

$$\left(h + \frac{V^2}{2g_c} + \frac{gZ}{g_c}\right)_{\text{in}} \delta M_{\text{in}} - \left(h + \frac{V^2}{2g_c} + \frac{gZ}{g_c}\right)_{\text{out}} \delta M_{\text{out}} + \delta Q - \delta W$$
$$= d\left[M\left(u + \frac{V^2}{2g_c} + \frac{gZ}{g_c}\right)\right]_{\text{sys}}$$

However, no mass flows out of the system; that is, $\delta M_{\text{out}} = 0$. Also, we may neglect $V^2/2g_c + gZ/g_c$ in relation to h and u. Therefore, the energy equation simplifies to

$$(h\,\delta M)_{\text{in}} + \delta Q - \delta W = d(Mu)_{\text{sys}}$$

But $\delta Q = 0$ because of the insulation, and $\delta W = 0$ because the system boundaries do not move and no shaft brings work through the system boundaries. The energy equation now becomes

$$(h\,\delta M)_{in} = d(Mu)_{sys}$$

We must now integrate the energy equation over all the mass that enters. Since H_{in} is a function only of the T and P in the main, it does not vary with mass and may be removed from under the integral sign, to give

$$h_{in} \int \delta M_{in} = \int d(Mu)_{sys}$$

or

$$(hM)_{in} = \int d(Mu) = (Mu)_2 - (Mu)_1$$

But M_{in} = total mass added to the system = $M_2 - M_1$. Therefore,

$$h(M_2 - M_1) = (Mu)_2 - (Mu)_1$$

But $M_1 = 0$ since chamber was initially evacuated; therefore, the integrated energy equation becomes

$$h_{in}M_2 = M_2 u_2$$

or

$$h_{in} = u_2$$

That is, the internal energy of the material in the chamber at the end of the process is equal to the enthalpy of the incoming material.

Now we must find that temperature at which the internal energy of steam at 100 psia is equal to the enthalpy of the incoming steam, which is h_{in} = 1200 Btu/lb$_m$. Unfortunately, the Mollier diagram and steam tables do not have entries with the internal energy. Therefore, we must determine u from the relation

$$u = h - Pv$$

and the given values of h and v at various temperatures (Table SP4-6; $P = 100$ psia). Now we must interpolate between the values of u to find T_2. The interpolation is performed in Fig. SP4-6b, from which we see that the final temperature is 570°F. Of course, if our Mollier diagram or steam tables had tabulated values of the internal energy (and some do), then this interpolation would be unnecessary.

Table SP 4-6

T, °F	h, Btu/lb$_m$	v, ft^3/lb$_m$	Pv, Btu/lb$_m$	u
450	1253.7	5.266	97.2	1156.5
500	1278.6	5.589	103.5	1175.1
600	1327.9	6.217	115.0	1212.9
700	1377.5	6.836	126.5	1251.0

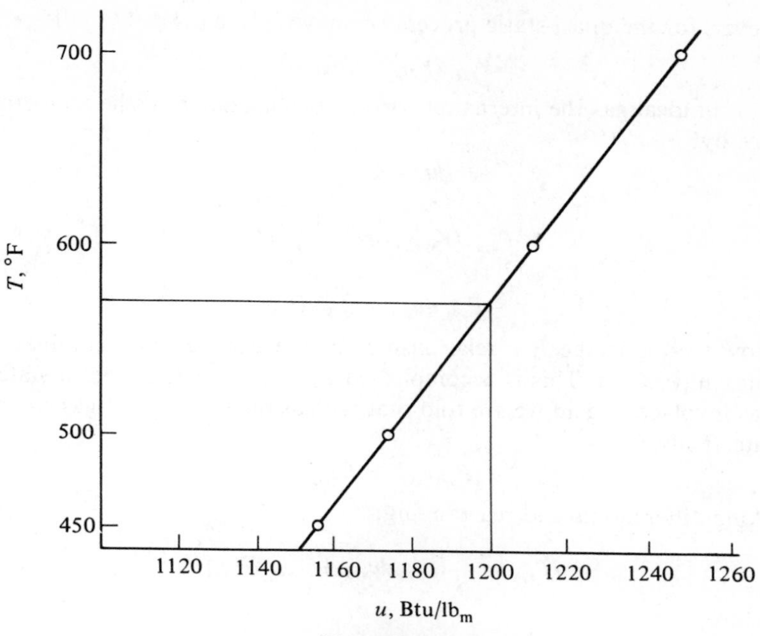

Figure SP4-6b

SAMPLE PROBLEM 4-7. Develop the P–v relation for a gas which obeys the ideal gas equation of state as it undergoes an adiabatic quasi-static compression. Assume that the heat capacities are independent of temperature.

Solution: We begin by choosing a system for thermodynamic analysis. Since we are interested in describing the Pv relationship of the gas under compression, we shall choose a unit mass of gas as it proceeds through the process. We may write the energy equation for any portion of the compression as given below:

$$\left(h + \frac{V^2}{2g_c} + \frac{gZ}{g_c}\right)_{\text{in}} \delta M_{\text{in}} - \left(h + \frac{V^2}{2g_c} + \frac{gZ}{g_c}\right)_{\text{out}} \delta M_{\text{out}} + \delta Q - \delta W$$
$$= d\left[M\left(u + \frac{V^2}{2g_c} + \frac{gZ}{g_c}\right)\right]_{\text{sys}}$$

The unit mass of gas comprises a closed system in which $\delta M_{\text{in}} = \delta M_{\text{out}} = dM_{\text{sys}} = 0$. Since we are told that the process is adiabatic, $\delta Q = 0$. Lastly, we will assume that potential and kinetic changes in the gas are negligible so that the energy equation reduces to:

$$-\delta W = M_{\text{sys}}\, du_{\text{sys}}$$

However, for the quasi-static process we may replace δW by $P_{sys}\, dV_{sys}$:

$$-P_{sys}\, dV_{sys} = M_{sys}\, du_{sys}$$

For an ideal gas the internal energy change is related to the temperature change by:

$$du = C_V\, dT$$

Thus

$$-P_{sys}\, dV_{sys} = M_{sys}\, C_V\, dT_{sys}$$

or

$$-P_{sys}\, dv_{sys} = C_V\, dT_{sys}.$$

Now to obtain the P–v relationship for the gas we must eliminate the system temperature. This is accomplished by using the equation of state for the gas involved. Again we are told that the gas obeys the ideal-gas equation of state. Thus:

$$P_{sys} v_{sys} = RT_{sys}$$

or taking differentials and rearranging:

$$dT_{sys} = \frac{1}{R}[P_{sys}\, dv_{sys} + v_{sys}\, dP_{sys}]$$

Therefore,

$$-P_{sys}\, dv_{sys} = \frac{C_V}{R}[P_{sys}\, dv_{sys} + v_{sys}\, dP_{sys}]$$

or

$$-RP_{sys}\, dv_{sys} = C_V[P_{sys}\, dv_{sys} + v_{sys}\, dP_{sys}]$$

This is rearranged to give

$$-P_{sys}\, dv_{sys}[R + C_V] = C_V v_{sys}\, dP_{sys}$$

but for an ideal gas $R + C_V \equiv C_P$, so that:

$$-C_P P_{sys}\, dv_{sys} = C_V v_{sys}\, dP_{sys}$$

Division of this result by $P_{sys} v_{sys}$ puts the differential equation into a separated form which may be easily integrated:

$$-C_P \frac{dv_{sys}}{v_{sys}} = C_V \frac{dP_{sys}}{P_{sys}}$$

or

$$-\frac{C_P}{C_V} d \ln v_{sys} = d \ln P_{sys}$$

Since the heat capacities are independent of P and v this result may be integrated from the initial pressure, P_1, and volume, v_1, to the pressure and volume at any general point within the compression:

$$-\frac{C_P}{C_V} \ln \frac{v}{v_1} = \ln \frac{P}{P_1}$$

or upon exponentiating:

$$\left(\frac{v}{v_1}\right)^{-C_P/C_V} = \frac{P}{P_1}$$

which upon rearrangement yields the desired form:

$$Pv^{C_P/C_V} = P_1 v_1^{C_P/C_V} = \text{constant}$$

Note that this is the form we used in Sample Problem 4-2. The adiabatic, quasi-static compression (or expansion) of a gas which obeys the ideal-gas equation of state is thus seen to be a polytropic process. We will show in Chapter 6 how the P–v relationship derived above can be obtained in a considerably more direct fashion.

SAMPLE PROBLEM 4-8. Self-Contained Underwater Breathing Apparatus (or SCUBA) tanks must be refilled with compressed air before each use. After each previous use the tanks would normally contain air at approximately 50 psia and a temperature of 75°F. At a particular filling station compressed air is available from an air pipeline at 1000 psia and 120°F. The SCUBA tanks are refilled by connecting them to the pipeline and allowing compressed air to flow into each cylinder until the pressure in the tank is 1000 psia. Each tank normally has a volume of approximately 1 ft^3.

(a) If the tanks are filled slowly so that the air is at room temperature (75°F) at the end of the filling, how much air has been added to each tank? How much air does each tank contain at the end of the filling operation?

(b) If the tank is filled rapidly so that the process is essentially adiabatic, how much air is added to each tank? How much air is contained in the tank at the end of the filling operation?

Although the pressures are quite high, *assume* that air obeys the ideal-gas equation of state under these conditions. (The answers will be off by 5–10 percent because of this assumption.) The constant-volume heat capacity of air may be taken as

$$C_V = 4.4 + 1.5 \times 10^{-3}\,T \text{ cal/g-mol °K}$$

where $T =$ °K for the temperature range of interest.

Solution: (a) We begin by determining the mass of air remaining in the tank at the beginning of the filling:

$$M = \frac{V}{v}$$

where v is obtained from the ideal-gas equation of state and the known initial pressure and temperature:

$$v = \frac{RT}{P}$$

where $R = 1545$ ft-lb$_f$/lb-mol °R
$P = 50$ psia $= 7.20 \times 10^3$ lb$_f$/ft^2
$T = 75°F = 535°R$

Therefore,

$$v = \frac{(1545)(535)}{7.20 \times 10^3} \frac{(\text{ft-lb}_f)\ °\text{R}/(\text{lb-mol}\ °\text{R})}{(\text{lb}_f/\text{ft}^2)}$$
$$= 115\ \text{ft}^3/\text{lb-mol}$$

To express the results in lb$_m$ rather than lb-mol we must divide by the molecular weight of air: 29.0 lb$_m$/lb-mol:

$$v = 115\ \text{ft}^3/\text{lb-mol} = 3.96\ \text{ft}^3/\text{lb}_m$$

Thus the amount of air left in each tank is:

$$M = \frac{1\ \text{ft}^3}{3.96\ \text{ft}^3/\text{lb}_m} = 0.254\ \text{lb}_m$$

At the end of the filling operation

$$P_{end} = 1000\ \text{psia} = 1.44 \times 10^5\ \text{lb}_f/\text{ft}^2$$
$$T_{end} = 75°\text{F} = 535\ °\text{R}$$

Therefore

$$v_{end} = \frac{RT_{end}}{P_{end}} = \frac{(1545)(535)}{1.44 \times 10^5}\ \text{ft}^3/\text{lb-mol}$$
$$= 5.72\ \text{ft}^3/\text{lb-mol} = 0.197\ \text{ft}^3/\text{lb}_m$$

and

$$M_{end} = \frac{1\ \text{ft}^3}{0.197\ \text{ft}^3/\text{lb}_m} = 5.07\ \text{lb}_m$$

The amount of air needed to fill each cylinder is then the difference between M_{end} and $M_{initial}$:

$$\Delta M = 5.07\ \text{lb}_m - 0.25\ \text{lb}_m = 4.82\ \text{lb}_m$$

(b) For the adiabatic filling we do not know the final temperature; thus we must attempt to determine T_{end}. Since we know P_{end} is still 1000 psia, determination of any other end property will give us the desired temperature. The additional property we will attempt to determine is the internal energy (or, equivalently, the enthalpy). We thus begin by picking a system for analysis and applying the energy equation to the system and process (as shown in Fig. SP4-8.)

As in Sample Problem 4-6, several choices of system are available for this problem. Choice of the tank, the valve, and their contents as the system will lead to the most direct solution of the problem. We now assume that potential and kinetic energy changes are negligible during the filling process. In addition, no mass leaves the system during the filling operation, so that the

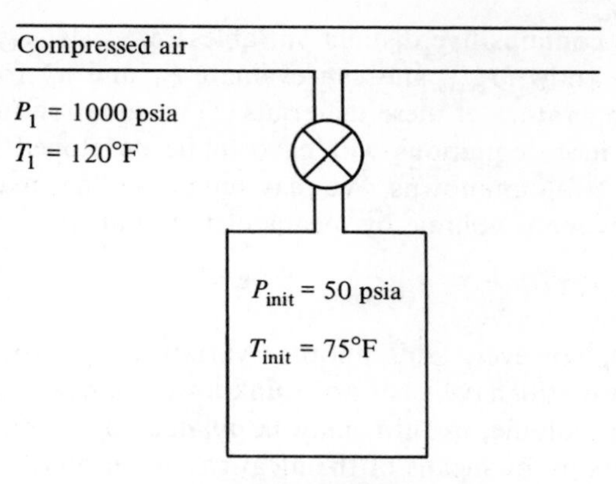

Figure SP4-8

energy equation around the system can be written as:

$$(h\,\delta M)_{in} + \delta Q - \delta W = d(Mu)_{sys}$$

However, we are told the filling operation is adiabatic, so that $\delta Q = 0$. In addition, since no shaft passes through the system boundaries, and the system boundaries are rigid and immobile, $\delta W = 0$ and the energy equation reduces to:

$$h\,\delta M_{in} = d(Mu)_{sys}$$

Now this result must be integrated over the entire filling operation:

$$\int h\,\delta M_{in} = \int d(Mu)_{sys}$$

h_{in} is the enthalpy of the air entering the system, in this case, the enthalpy of the air in the supply pipeline. Since the supply pipeline conditions do not change during the filling operation, h_{in} is constant and may be removed from under the integral of the left-hand side. The integral of δM_{in} is the total amount of mass which enters the tank. The integral of the right-hand side is the total change in (Mu_{sys}) during the process:

$$h_{in}(M_{in}) = [(Mu)_{end} - (Mu)_{begin}]_{sys}$$

or

$$h_{in}M_{in} = [M_{end}u_{end} - M_{begin}u_{begin}]_{sys}$$

M_{in} may be eliminated by means of the mass balance:

$$M_{in} = M_{end} - M_{begin}$$

Substitution into the energy balance then gives:

$$h_{in}[M_{end} - M_{begin}]_{sys} = [M_{end}u_{end} - M_{begin}u_{begin}]_{sys}$$

or

$$M_{end}[h_{in} - u_{end}] = M_{begin}[h_{in} - u_{begin}] \qquad \text{(a)}$$

Equation (a) contains five distinct variables: M_{end}, M_{begin}, h_{in}, u_{begin}, and u_{end}. We already know M_{begin} and can evaluate h_{in} and u_{in} from the known pressure and temperature of these materials. Thus equation (a) contains two unknowns, and more equations will have to be developed before we can determine the missing unknowns. We may relate the final mass of air in the cylinder to the cylinder volume by means of the final specific volume:

$$M_{end} = \frac{v_{tank}}{v_{end}} \tag{b}$$

Equation (b), however, adds another variable, v_{end}, to the system of equations, and we still have one more unknown than we have equations. The final specific volume, in turn, may be related to the final temperature and (known) pressure by means of the ideal-gas equation of state:

$$v_{end} = \frac{RT_{end}}{P_{end}\,(\text{mol wt})} \tag{c}$$

Although equation (c) adds a fourth unknown, T_{end}, to our system of equations, the closing equation is now obtained if we remember that the specific internal energy of a gas which obeys the ideal-gas equation of state is a function only of the gas's temperature:

$$u_{end} = f(T_{end}) \tag{d}$$

Since equation (d) adds no new unknowns, our system of equations is now complete: We have four equations in four unknowns. However, before solving these equations, we must develop the functional form indicated in equation (d):

As indicated in equation (4-39), the constant-volume heat capacity C_V is defined by:

$$C_V = \left(\frac{\partial u}{\partial T}\right)_v$$

As we have indicated, however, the constant-volume restiction is unnecessary for a gas which obeys the ideal-gas equation of state. Thus for this problem:

$$\frac{du}{dT} = C_V$$

or

$$du = C_V\,dT \tag{e}$$

Now let us arbitrarily assign a reference state internal energy: $u = 0$ at $T = 75°F$. The internal energy at any other temperature is then obtained from the integral of equation (e)

$$u(T) = \int_{T=75°F}^{T} C_V\,dT$$

Substituting the expression for C_V,

$$C_V = [4.4 + 8.35 \times 10^{-4}T(°R)^{-1}]\,\frac{\text{Btu}}{\text{lb}_m\,°\text{R}}$$

then gives

$$u(T) = \int_{535}^{T} [4.4 + 8.35 \times 10^{-4}T(°R)^{-1}]\, dT \frac{\text{Btu}}{\text{lb}_m°\text{R}}$$

$$= \{4.4[T - 535°\text{R}] + 4.18 \times 10^{-4}[T^2 - (535°\text{R})^2]°\text{R}^{-1}\} \frac{\text{Btu}}{\text{lb}_m°\text{R}}$$

or

$$u(T) = [4.4T(°\text{R})^{-1} + 4.18 \times 10^{-4}T^2(°\text{R})^{-2} - 2470] \frac{\text{Btu}}{\text{lb}_m} \qquad \text{(d}'\text{)}$$

The enthalpy is obtained from:

$$h(T) = u(T) + Pv$$

$$= u(T) + RT$$

or

$$h(T) = [6.4T(°\text{R})^{-1} + 4.18 \times 10^{-4}T^2(°\text{R})^{-2} - 2470] \frac{\text{Btu}}{\text{lb}_m} \qquad \text{(d}''\text{)}$$

Therefore:

$$u_{\text{begin}} = 0 \text{ Btu/lb}_m$$

$$h_{\text{in}} = h(585°\text{R}) = 1213 \text{ Btu/lb}_m$$

and we have shown

$$M_{\text{begin}} = 0.254 \text{ lb}_m$$

Thus the system of equations to be solved can now be expressed as:

$$M_{\text{end}}\left[1213 \frac{\text{Btu}}{\text{lb}_m} - u_{\text{end}}\right] = 0.254 \text{ lb}_m[1213 - 0] \frac{\text{Btu}}{\text{lb}_m}$$

or

$$M_{\text{end}}\left[1213 \frac{\text{Btu}}{\text{lb}_m} - u_{\text{end}}\right] = 296 \text{ Btu} \qquad \text{(a}'\text{)}$$

$$M_{\text{end}} = \frac{1 \text{ ft}^3}{v_{\text{end}}} \qquad \text{(b}'\text{)}$$

$$v_{\text{end}} = \frac{1545}{29} \times \frac{T}{1.44 \times 10^5} \frac{\text{ft-lb}_f}{\text{lb}_m(\text{lb}_f/\text{ft}^2)}$$

or

$$v_{\text{end}} = 3.70 \times 10^{-4}T\left(\frac{\text{ft}^3}{\text{lb}_m°\text{R}}\right) \qquad \text{(c}'\text{)}$$

and

$$u_{\text{end}} = [4.4\, T_{\text{end}}(°\text{R})^{-1} + 4.18 \times 10^{-4}T_{\text{end}}^2(°\text{R})^{-2} - 2470] \frac{\text{Btu}}{\text{lb}_m} \qquad \text{(d}'''\text{)}$$

Since equation (d‴) is nonlinear in temperature, we cannot use a simple direct solution of equations (a′) through (d‴). Rather, we will reduce the system of equations to a single nonlinear equation in T_{end}. This equation will then be solved by appropriate means.

Eliminate M_{end} between equations (a′) and (b′):

$$\frac{(1213 \text{ Btu/lb}_m - u_{end})\text{ft}^3}{v_{end}} = 296 \text{ Btu} \tag{f}$$

v_{end} is now eliminated between (f) and (c′):

$$\frac{(1213 \text{ Btu/lb}_m - u_{end}) \text{ ft}^3}{3.70 \times 10^{-4} T_{end} \text{ ft}^3/(\text{lb}_m\,^\circ\text{R})} = 296 \text{ Btu}$$

or

$$(1213 \text{ Btu/lb}_m - u_{end}) = 1.09 \times 10^{-1} T_{end} \frac{\text{Btu}}{\text{lb}_m\,^\circ\text{R}} \tag{g}$$

u_{end} is then eliminated between (g) and (d‴) to give a single equation in T_{end}:

$$[3684 - 4.4T_{end}(^\circ\text{R})^{-1} - 4.18 \times 10^{-4} T_{end}^2(^\circ\text{R}^{-2})]\frac{\text{Btu}}{\text{lb}_m} = 1.09 \times 10^{-1} T_{end} \frac{\text{Btu}}{\text{lb}_m\,^\circ\text{R}} \tag{h}$$

Although this equation is nonlinear in T_{end}, it is only quadratic, and thus the use of the quadratic formula appears the most direct means of solution. Equation (h) may be rearranged to:

$$4.18 \times 10^{-4} T_{end}^2(^\circ\text{R}^{-2}) + 4.5\, T_{end}(^\circ\text{R})^{-1} - 3684 = 0$$

Application of the quadratic formula yields:

$$T_{end} = \frac{-4.5 + \sqrt{(4.5)^2 + (4)(3684)(4.18 \times 10^{-4})}}{2 \times 4.18 \times 10^{-4}}\,^\circ\text{R}$$

or

$$T_{end} = 765^\circ\text{R}$$

Thus:

$$v_{end} = 0.283 \text{ ft}^3/\text{lb}_m$$

and

$$M_{end} = 3.53 \text{ lb}_m$$

The mass added during filling is

$$M_{end} - M_{begin} = (3.53 - 0.254) \text{ lb}_m = 3.28 \text{ lb}_m$$

Problems

4-1. (a) Write the *complete* first-law energy balance for the system shown in Fig. P4-1.

(b) If the black box is a compressor operating at steady state and streams 2 and 5 are cooling water while streams 1, 3, and 4 are the working fluid (streams 3 and 4 are at different pressure), reduce the energy balance to its simplest form.

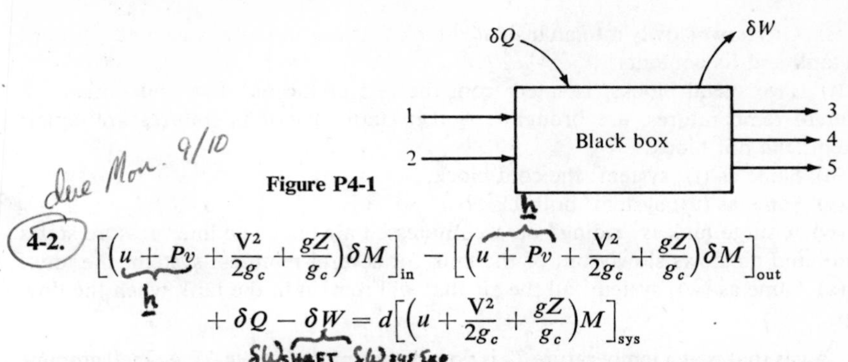

Figure P4-1

4-2.

$$\left[\left(u + Pv + \frac{V^2}{2g_c} + \frac{gZ}{g_c}\right)\delta M\right]_{\text{in}} - \left[\left(u + Pv + \frac{V^2}{2g_c} + \frac{gZ}{g_c}\right)\delta M\right]_{\text{out}} + \delta Q - \delta W = d\left[\left(u + \frac{V^2}{2g_c} + \frac{gZ}{g_c}\right)M\right]_{\text{sys}}$$

Reduce the above mathematical statement of the first law of thermodynamics to its simplest possible form for application to each of the following processes and systems:

(a) A piece of hot steel suddenly immersed in cold water; system: the piece of steel.

(b) Cold water being heated in the tubes of a heat exchanger, horizontal flow at a constant rate; system: the tubes and the water in them.

(c) A freely falling body passing through a differential increment of height; system: the body.

(d) Steam flowing steadily through a horizontal, insulated nozzle; system: the nozzle and its contents.

(e) Same as (d); system: 1 lb_m of steam flowing.

(f) A rubber balloon being inflated; system: the rubber.

(g) A storage battery discharging across a resistance; system: the resistance.

(h) An automobile accelerating on smooth, level pavement (assume frictionless); system: the automobile.

(i) A frictionless windmill driving an electric generator; system: the windmill.

(j) A tennis ball dropped from shoulder height bounces on a sidewalk until it finally comes to rest; system: the tennis ball.

(k) A bullet embeds itself in a bowling ball that is rolling frictionlessly along a horizontal surface; system: the bowling ball.

(l) Hot oil vapors are cooled at a steady rate by cooling water in a horizontal double-pipe heat exchanger that is well insulated; system: the exchanger and its contents.

(m) A heavy steel block slides slowly down an inclined plane until it comes to rest; system: the block.

(n) Water drips slowly out of a hole in the bottom of an enclosed tank; system: the tank and its contents.

(o) A gas is confined in a vertical cylinder fitted with a frictionless piston and an evacuated space above the piston. The piston rises as the cylinder is heated; system: the gas.

(p) Same as (o); system: the gas and the piston.

(q) Same as (p) except that the space above the piston is not evacuated.

(r) Water flows steadily through a long, horizontal pipe; system: the pipe and its contents.

(s) Gas flows slowly into an insulated tank that was initially evacuated; system: the tank and its contents.

(t) Two metal blocks, isolated from the rest of the universe and initially at different temperatures, are brought together until their temperatures are equal; system: the hot block.

(u) Same as (t); system: the cold block.

(v) Same as (u); system: both blocks.

(w) A surge tank is "riding" on a compressed air line. The line pressure starts to fall and air flows slowly out of the tank for several minutes; system: the tank.

(x) Same as (w); system: all the air that will remain in the tank when the flow stops.

4-3. A gas that is at a temperature T_0 is flowing in a pipe (Fig. P4-3). A small amount

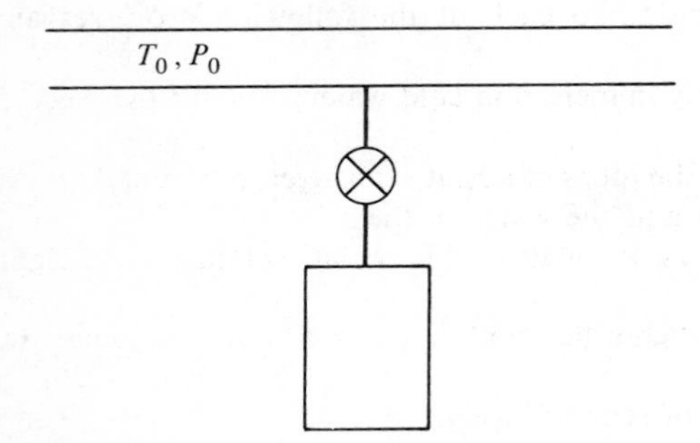

Figure P4-3

is bled off into an evacuated cylinder. The bleeding continues until the pressure in the tank is equal to P_0, the pressure in the pipe. If the cylinder is perfectly insulated and the gas is ideal (with C_P and C_V independent of temperature and pressure), find the temperature, T, which exists in the cylinder at the end of the process by two separate approaches:

(a) Assume that the cylinder and its contents constitute an open system.

(b) Assume that the mass of gas M, which will end up in the cylinder, constitutes a closed system.

You should express your answers to (a) and (b) in terms of T_0, C_P, C_V, and P_0 alone. Do the answers to (a) and (b) differ? Does this surprise you? Which approach is the simpler one to apply *in this case*, (a) or (b)?

9-17

4-4. High-pressure steam at a rate of 1000 lb_m/hr initially at 500 psia and 700°F is expanded in a turbine to produce work. Two exit streams are removed from the turbine as illustrated in Fig. P4-4. Stream 2 is at 200 psia and 400°F and has a flow

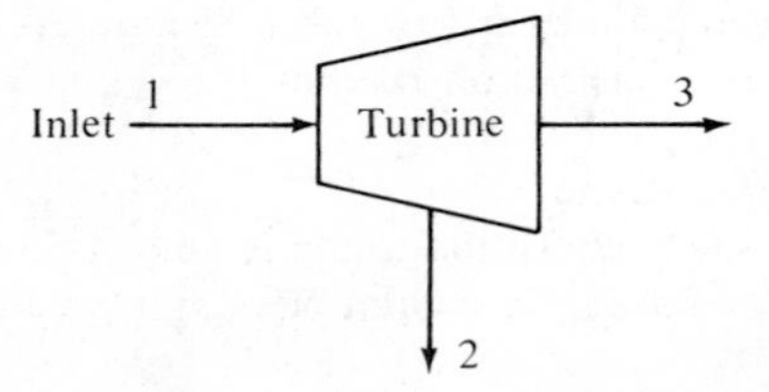

Figure P4-4

rate equal to one third of the inlet flow. Stream 3 is at 100 psia and is known to be a mixture of saturated vapor and liquid. A small representative fraction of stream 3 is passed through a throttling valve and expanded to 1 atm. The temperature after expansion is found to be 240°F. If the measured work obtained from the compressor is 54 hp, estimate the heat loss from the turbine expressed as Btu/hr.

4-5. An adiabatic steam turbine is used to drive an electrical generator, which, in turn, supplies power to four water pumps. The steam enters the turbine at 500 kg/hr with an enthalpy of 2.4 MJ/kg and leaves with an enthalpy of 1.2 MJ/kg. Each pump pumps 50 gal/min of water from an open tank 100 m above sea level to another 800 m above sea level. Kinetic energy changes and friction in the water lines may be neglected. What is the overall efficiency of the operation, that is, the fraction of the energy (enthalpy) recovered from the steam that is converted to the potential energy of the water?

4-6. A turbine is used in a nitrogen refrigeration process to derive some work from the expansion of nitrogen; 1000 lb_m of nitrogen per hour enters at 1500 psia and −40°F at 200 ft/sec. The nitrogen leaves the turbine exhaust at a point 10 ft below the inlet with a velocity of 1200 ft/sec. The measured shaft work of the turbine is 12.6 hp, and heat is transferred from the surroundings to the turbine at the rate of 10,000 Btu/hr. A small portion of the exhaust nitrogen from the turbine is passed through a throttling valve and discharged at atmospheric pressure. Velocity change in passing through the valve may be neglected. What is the temperature of the stream leaving the valve?

4-7. An insulated tank initially contains 500 kg of steam and water at 35 bar. Fifty percent of the tank volume is occupied by liquid and 50 percent by vapor. Twenty kg of moisture-free vapor is slowly withdrawn from the tank so that the pressure and temperature are always uniform throughout the tank. Analyze the situation carefully and calculate the pressure in the tank after the 20 kg of steam is withdrawn.

4-8. A well-insulated cylinder, fitted with a frictionless piston, initially contained 20 lb_m of liquid water and 1 lb_m of water vapor at a pressure of 200 psia (Fig. P4-8). The valve to the steam line was opened and 5 lb_m of superheated steam at 250 psia was admitted to the cylinder. The valve was closed and the contents of the cylinder allowed to come to equilibrium. If the final volume of the contents of the cylinder was six times the initial volume, determine the temperature of the superheated steam that was admitted to the cylinder.

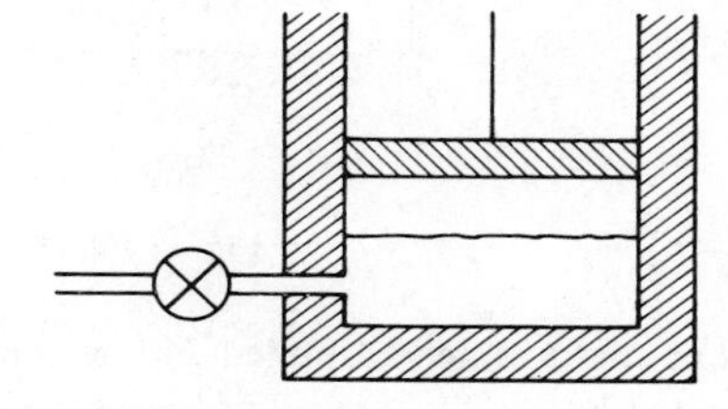

Figure P4-8

4-9. A fireless steam engine is used to haul boxcars around an explosives plant. The engine has a well-insulated 100-ft^3 tank. This tank is periodically charged with high-pressure steam which is used to power the engine until the tank is depleted. At the end of a given run the tank contains saturated steam at atmospheric pressure. It is then connected to a supply line carrying steam at 800 psia and 620°F. A valve in the supply line is opened until no more steam flows into the tank. The valve is then closed. The filling operation takes place very rapidly. Using the steam tables (and stating any assumptions), determine the amount of steam in the tank just after filling.

4-10. A well-insulated vertical cylinder is fitted with a frictionless piston and contains 1 kg of steam at 270°C and 8 bar. A line leads from the bottom of the cylinder to an insulated tank having a volume of 0.2 m^3 and initially containing 0.5 kg of steam at 3 bar. A valve is opened and steam flows slowly into the tank until the tank and cylinder pressures are equal, at which time the valve is closed. It may be assumed that the volume of the line is negligible and that there is no transfer of *heat* through the valve.

(a) How many pounds of steam flow into the tank?

(b) What is the final temperature in the tank?

(c) Calculate Q, W, and ΔU (all in J), taking as the system (1) the atmosphere, (2) the piston, (3) the piston and atmosphere, and (4) all the steam. Determine ΔH in two different ways.

(d) What is the internal energy of (1) the steam in the tank under initial conditions? (2) the steam initially in the cylinder which is to flow into the tank? (3) the steam in the tank under final conditions?

9-19 **4-11.** An insulated cylinder with a volume of 2 ft^3 contains a piston (volume negligible) attached to a shaft as shown in Fig. P4-11. The piston is initially at the extreme left of the cylinder, which contains air at 1 atm. The valve is opened and steam enters the cylinder, forcing the piston to the right. When the piston first touches the right end of the cylinder, the valve is closed. At this point 43.6 Btu of shaft work has been performed and the cylinder contains 0.5 lb_m of steam. What is the pressure of steam in the cylinder? (*Note:* The piston cannot be considered frictionless.)

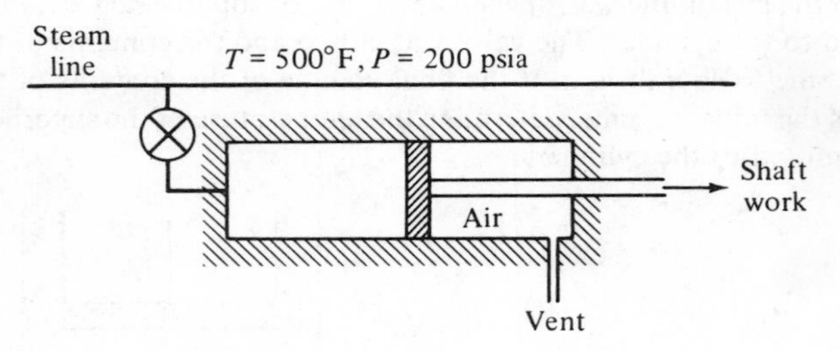

Figure P4-11

9-21

4-12. An adiabatic turbine is supplied with steam at 16 bar and 280°C and exhausts to a pressure of 1 bar. If the quality of the exhaust steam is 96 per cent and there is negligible change in velocity, compute the quantity of work done by 1 kg of steam flowing through the turbine.

4-13. The exhaust steam from a turbine is to be used in a chemical process. The turbine is supplied with steam at 15 bar and 260°C, 60 MJ of heat is lost per kg of steam supplied, and the work is 280 MJ/kg of steam. Find the enthalpy and quality of the exit steam if the exit pressure is 2.5 bar.

4-14. A 10-hp turbine is supplied with steam at 20 bar and 315°C and exhausts at 1.2 bar and 98 percent quality. Find the heat loss in Btu/hr, if 100 kg of steam are supplied per hour.

4-15. The steam–water power cycle shown in Fig. P4-15 has been recommended for the production of useful work needed elsewhere in the plant. Heat to the cycle is supplied in the boiler and superheater and is removed in the condenser. The turbine produces an amount of work W_1. However, $-(W_2 + W_3)$ of this is needed to power the pumps. On the basis of the information supplied in Fig. P4-15, determine the thermal efficiency of this cycle,

$$\eta = \frac{\text{net work produced}}{Q_1 + Q_2}$$

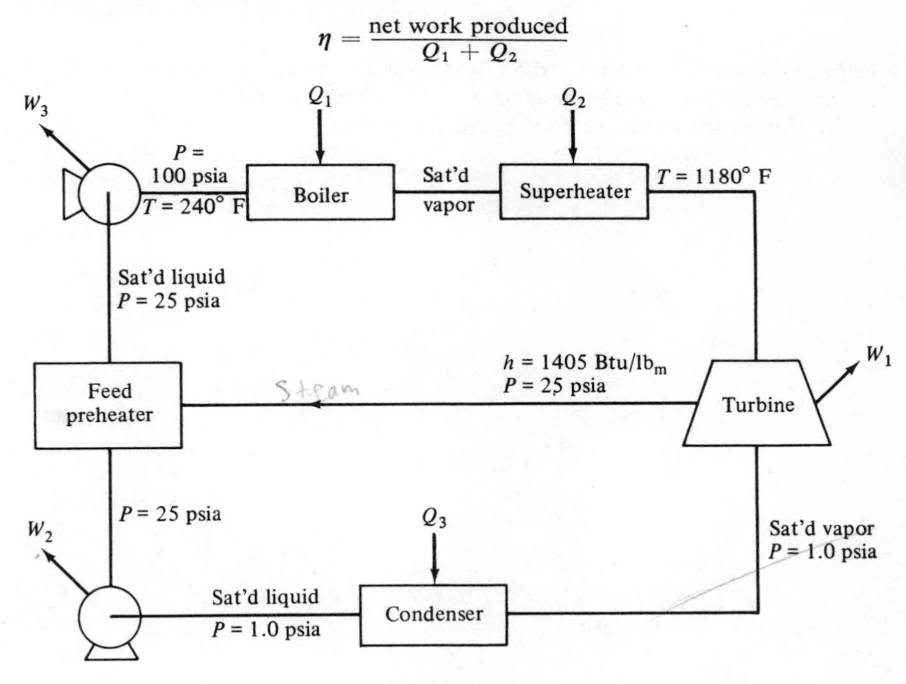

Figure P4-15

Supply all pressures, temperatures, and enthalpies which do not appear on the diagram. You may assume that there are no pressure changes across any equipment or piping except the turbine and pumps. Also you may assume v of the liquid passing through the pump to be independent of pressure.

4-16. In a certain paper mill two steam boilers are to be operated in parallel. Each has a volumetric capacity of 1000 ft³ and each contains 18,000 lb_m of steam and water. The first boiler registers a pressure of 200 lb/in.² absolute, but, owing to an error, it is connected to the second boiler when the pressure in the latter is but 75 lb/in.^{2} absolute. What will be the pressure in the system after equilibrium has been attained, on the assumption that no steam is withdrawn, that no heat is added to the system during the change, and that there is no interchange of heat between the boiler shells and their contents?

4-17. (a) One mole of gas is confined on one side of a piston at 5 atm and 200°F as shown in Fig. P4-17. If the piston has a mass of 1 lb_m and the gas expands adiaba-

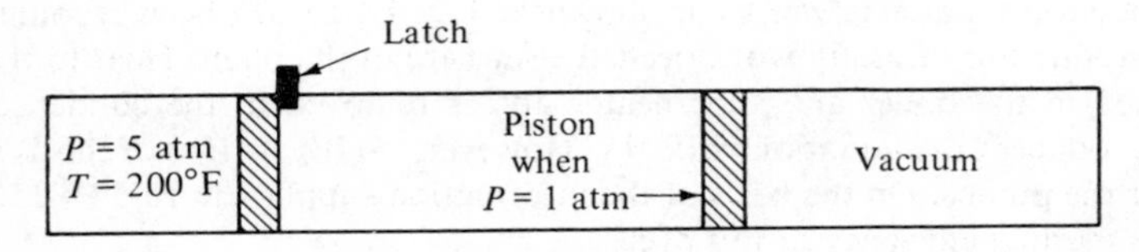

Figure P4-17

tically and reversibly, calculate the piston's velocity when the pressure has fallen to 1 atm. Assume the gas to be ideal with $C_P = 6$ Btu/lb-mol °F.

(b) How much work was done by the gas in (a)?

FIVE

the entropy equation

Introduction

In Chapter 4 the first of three basic thermodynamic relationships, the energy equation, was developed. Its utility in the analysis of engineering problems will become increasingly evident as we progress. However, the energy equation alone is insufficient for analyzing many thermodynamic systems and processes because nature has imposed certain restrictions on energy transfer and conversion over and above those embodied in the conservation-of-energy principle. The quantitative treatment of these restrictions has been facilitated by the introduction of the entropy concept.

In our earlier discussion of entropy, we emphasized its relationship to the degradation of mechanical forms of energy to the thermal form. From a microscopic treatment of entropy one can deduce certain important characteristics of the entropy property. Appendix *H* contains a simplified development of entropy from the microscopic point of view. Several significant results from that treatment were developed in the discussion of entropy in Chapter 2.

However, the utility of the entropy concept depends on relating it to the changes with which we as engineers are involved. Since our concerns are generally macroscopic in nature, it is necessary to associate entropy with macroscopic phenomena, such as heat and work, so that it can be used in the analysis of processes involving these energy flows and interconversions. Development of the entropy equation provides a quantitative relationship of considerable engineering utility, as will be demonstrated in the remainder of this chapter.

The idea of accounting for entropy changes in a system by constructing an entropy equation is one that arises rather naturally from our earlier treatments of mass and energy. Entropy, like mass and energy, is an extensive property that may be used to characterize the state of thermodynamic systems at equilibrium. A system's entropy can undergo change as a result of a process just as its energy, mass, or volume may change. The general procedure cited in Section 3-5 for monitoring such changes can be applied to entropy in the same manner as it was applied to mass or energy. Thus, if we consider an open system as shown in Fig. 5-1, we may write the entropy equation as

$$\text{entropy}_{\text{in}} - \text{entropy}_{\text{out}} + \text{entropy}_{\text{gen}} = \text{entropy}_{\text{acc}} \tag{5-1}$$

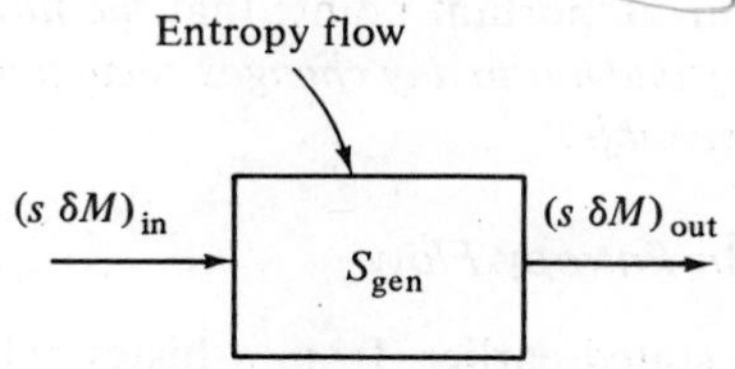

Figure 5-1. An open system.

As in the case of the energy equation, we break the inlet and outlet terms into an entropy flow that is associated with the mass flow, $s\,\delta M$, and an entropy flow that is not associated with mass flow. Since entropy is a nonconservative property, unlike mass and energy, it will be necessary to include a generation term in the entropy equation to account for entropy production arising from system irreversibilities. The accumulation term is simply the change in entropy of the system and is expressed as $d(Ms)_{\text{sys}}$. Thus the entropy equation becomes

$$(s\,\delta M)_{\text{in}} - (s\,\delta M)_{\text{out}} + \delta(\text{entropy flow}) + \delta(\text{entropy})_{\text{gen}} = d(Ms)_{\text{sys}} \tag{5-2}$$

Before we proceed, it is necessary to investigate the character of the entropy flow and generation terms.

5.1 Entropy Flow

In developing the energy equation a seemingly arbitrary distinction was made between heat and work as energy flows. Justification for this differentiation becomes apparent if we reconsider certain points made in Section 2-6 regarding the effect that these flows have on a system's entropy.

It must be remembered that both heat and work refer to flows across system boundaries. Just as we choose to designate water as rain when moving from water droplets in clouds to the earth (once it is in our rivers and lakes, it is no longer rain), we designate energy as heat or work only as it crosses the boundary of our system. Energy crosses the boundary of a system in one of these two categories, but once in or out of the system it is no longer regarded as heat or work, but simply becomes part of the system's (or the surroundings') total energy.

If we could station ourselves at the system's boundary and watch only the energy flowing across it, we would discover that we were unable to distinguish energy flows in reversible processes from those in irreversible ones. Since the irreversibilities that affect a system's entropy are those occurring within the system, we would not even see them. Furthermore, we would have no basis for relating them to our observations of energy flow at the boundary. In formulating the flow terms for the entropy equation, it is imperative that we recognize this important point: that *the flow terms relate just to what crosses the boundary and not to any changes occurring within the system, be they reversible or irreversible.*

Heat: An Entropy Flow

As was stated earlier, from a historical point of view, entropy was first thought of in a macroscopic sense as relating to heat and its conversion to work. The consistency of the more recent microscopic developments with the classical evolution of the concept was suggested when we examined the relationships among entropy, heat, and work. We observed that the addition of a small amount of heat, δQ, produces a small entropy increase, dS:

$$dS = K\,\delta Q \tag{5-3}$$

where K represents some positive proportionality factor that relates the entropy change to the flow of thermal energy.

It should be remembered that entropy as it relates to equilibrium states is a state property, that is, its value is independent of the processes by which a given state was achieved. Therefore, the difference in entropy between two states, $\Delta S = S_2 - S_1$, must also be independent of the path taken in moving from state 1 to state 2. If the equation $dS = f(\delta Q)$ is to satisfy such a condition, the right-hand side of the equation must also behave as a state variable such that $\int_1^2 f(\delta Q)$, which equals ΔS, is independent of the path chosen. Heat flow, or δQ, is not a state variable, inasmuch as different amounts of energy exchanged as heat are possible in moving between any two given states. Thus, from a mathematical point of view, the constant K in equation (5-3) must also serve as an integrating factor which converts the non-state function, Q, into the state variable, S. K can be shown to equal $1/T$, where T is the absolute temperature, by the following reasoning.

Consider the pressure–volume behavior of a substance subjected to a cyclical reversible process consisting of two isothermal steps, *A-B* and *C-D*, and two adiabatic steps, *B-C* and *D-A*, as shown in Fig. 5-2. The system begins in state A and after the four steps is returned to the same state. All state properties, including entropy, should thus return to their original values at the completion of the process.

The curves labeled t_1, t_2, and t_3 represent isotherms of progressively lower

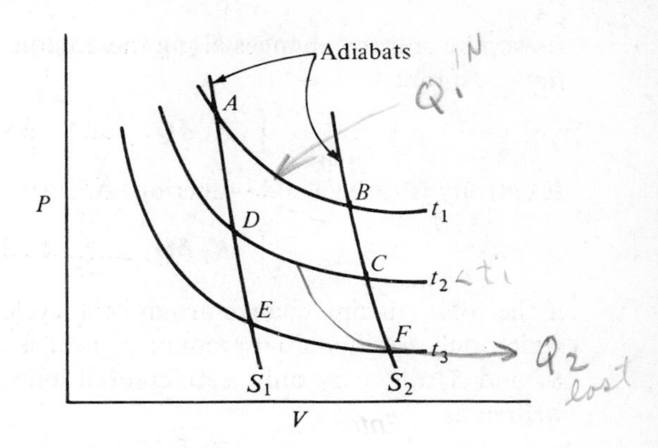

Figure 5-2

temperature (that is, $t_1 > t_2 > t_3$); the curves labeled S_1 and S_2 represent adiabatic lines which for reversible changes are also lines of constant entropy. (The function t is assumed to be related to the thermodynamic temperature but not necessarily identical to it.) As discussed earlier, in the absence of irreversibility the only means of changing a closed system's entropy is by adding or removing heat. Thus step *A-B* involves addition of heat to the system at a constant temperature, t_1; step *B-C* involves an adiabatic expansion; step *C-D* involves heat removal at a lower temperature, t_2; and step *D-A* involves an adiabatic compression that returns the system to its original state. Such a cycle represents a Carnot engine (a device for converting thermal energy to mechanical energy), which will be discussed further later in this chapter.

In Chapter 3 it was shown that the area beneath a *P–V* curve equals the work associated with the process. In a cyclical process the area enclosed by the path *ABCD* equals the net work, W, for the cycle. Application of the energy balance to the system yields

$$W = Q_1 + Q_2 \tag{5-4}$$

where Q_1 and Q_2 represent the total heat effects at t_1 and t_2, respectively, Q_1 being a positive quantity (heat absorbed) and Q_2 a negative quantity (heat rejected). Clearly, $|Q_1| > |Q_2|$ if work is produced by the system (since W is positive by convention if work is done by the system).

Consider what happens if instead of discharging heat at t_2 it is discharged at t_3. The enclosed area becomes larger; therefore, W increases and (for a given Q_1) $|Q_3|$ must be less than $|Q_2|$. In general, for a reversible, cyclic process consisting of two isothermal and two adiabatic legs, Q_i is a function of t_i. Moreover, it can be shown that $|Q_i|$ is a monotonic function of t_i ($|Q_1| > |Q_2| > |Q_3|$ in this case).

In either of the cycles *ABCD* or *ABFE*, entropy changes occur only along the isothermal legs of the path. Since the only entropy change encountered by the gas along the isothermal paths are those associated with the entropy flows, the entropy changes along these paths are simply equal to the entropy flows, so that:

$$\Delta S_{t_1} = \int_A^B K_1\,\delta Q_1 \quad \text{and} \quad \Delta S_{t_2} = \int_C^D K_2\,\delta Q_2 \tag{5-5}$$

If entropy is to be a state function, $\Delta S_{t_1} \equiv -\Delta S_{t_2}$ and:

$$\int_A^B K_1\,\delta Q_1 + \int_C^D K_2\,\delta Q_2 = 0 \tag{5-6}$$

if the total entropy change around the cycle is to vanish. For a process in which only a differential amount of heat is involved at t_1 and t_2 (and thus S_1 and S_2 differ by only a differential amount, dS), equation (5-6) can be written as

$$K_1\,\delta Q_1 + K_2\,\delta Q_2 = 0 \tag{5-7}$$

or

$$\frac{\delta Q_1}{\delta Q_2} = -\frac{K_2}{K_1} \tag{5-8}$$

or

$$\frac{|\delta Q_1|}{|\delta Q_2|} = \frac{K_2}{K_1}$$

It can now be argued that because $|\delta Q_i|$ is a function of t_i, the ratio $|\delta Q_1|/|\delta Q_2|$ is a ratio of functions of the temperatures.

$$\frac{|\delta Q_1|}{|\delta Q_2|} = \frac{K_2}{K_1} = \frac{f(t_1)}{f(t_2)} \tag{5-9}$$

Rewriting as

$$\frac{|\delta Q_1|}{f(t_1)} = \frac{|\delta Q_2|}{f(t_2)} \tag{5-10}$$

where $K_1 = 1/f(t_1)$ and $K_2 = 1/f(t_2)$, it becomes apparent that K_1 and K_2 must be a function of the temperatures t_1 and t_2. *The thermodynamic temperature scale is now defined such that*

$$T = f(t) \tag{5-11}$$

Although many alternative definitions are possible—and in fact several have been suggested—no other definition has received wide acceptance, and thus we shall continue to use the definition suggested in equation (5-11). We will show in Chapter 6 that this thermodynamic temperature is identical to the ideal gas temperature scale discussed in Chapter 2.

Substitution of equation (5-11) into the definition of K then yields

$$K = \frac{1}{T} \tag{5-12}$$

so that the entropy change associated with a heat flow is

$$\delta S = \frac{\delta Q}{T} \tag{5-13}$$

which is the desired result. Because this entropy change is directly associated with heat flows, we will term this an *entropy flow*, and give it the symbol δS. The temperature scale, which is used in calculations based on equation (5-13), must be either the Rankine or Kelvin scale, each of which is an absolute, or thermodynamic, temperature scale.

Substitution of the thermodynamic temperature scale for $f(t)$ in equation (5-10) then yields

$$\frac{\delta Q_1}{T_1} + \frac{\delta Q_2}{T_2} = 0 \tag{5-14}$$

Equation (5-14) may be generalized for a reversible process to give

$$\oint \frac{\delta Q}{T} = \oint dS = 0 \tag{5-15}$$

which demonstrates that $1/T$ is the appropriate integrating factor for relating changes in the state property, entropy, to the path function, heat.

Thus we learn that a heat flow produces an entropy flow which is given by

$$\delta(\text{entropy flow}) = \frac{\delta Q}{T} \tag{5-16}$$

where T is the absolute temperature of the system at the point where heat flow occurs. When the temperature of the system at the boundary is not constant over the areas where heat transfer is occurring, equation (5-16) must be rewritten:

$$\delta(\text{entropy flow}) = \int_{\text{surface area}} \frac{\delta q \, dA}{T} \tag{5-17}$$

where q = finite heat transferred per unit area of boundary
δq = differential heat transferred per unit area of boundary

The total entropy flow for a process attributable to heat is given by the double integral

$$\text{entropy flow} = \int_{\text{process}} \int_{\text{surface area}} \frac{\delta q \, dA}{T} \tag{5-18}$$

Work: No Entropy Flow

In the preceding discussion, entropy flows were related to heat flows. Certainly a system's energy can be altered by energy flows other than heat, and we can justifiably ask: What about entropy changes resulting from mechanical energy (work) flows? However, we showed in Chapter 3 that mechanical energy transfers do not have an entropy change directly associated with the mechanical-energy flow. Thus no *entropy flow* accompanies a mechanical energy flow. *This statement is not meant to imply that entropy changes cannot occur within a system transferring mechanical energy to or from its surroundings, only that such changes are not a direct consequence of the mechanical energy flow*. Irreversibilities occurring within a system as it transfers mechanical energy can lead to an entropy change in the system as can any thermal energy transfers which occur during the mechanical energy flow.

5.2 Irreversibilities Within a System

Suppose that a system insulated from its surroundings expands without transferring energy to its surroundings, that is, a free expansion. An example of such a process might be the gas confined by a piston assumed to have no mass as shown in Fig. 5-3. Let us assume that no resisting force restricts the motion of the piston. The first molecule that strikes the massless piston after the latch is removed causes the piston to move immediately to the far right part of the container. If the piston does not rebound, then the volume undergoes a sudden increase. When the piston is struck by subsequent molecules it is no longer moving, and thus no energy interchange occurs between the gas and the piston. The total energy of the gas remains constant while the volume has increased.

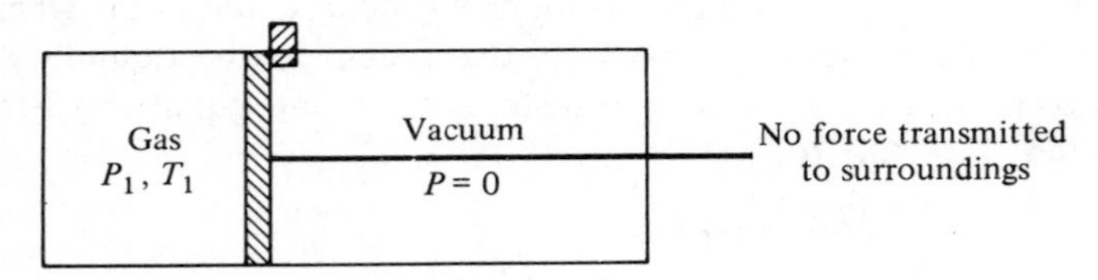

Figure 5-3. Free expansion of gas.

The process described above is clearly irreversible, because the system can never be returned to its original state without work supplied by the surroundings. Furthermore, it represents a reduction of the pressure potential ($P_2 < P_1$) of the gas without having utilized it to extract energy (as work) from the system. In effect the system, by undergoing this irreversible process, has forever lost the ability to perform a certain amount of work in spite of the fact it still has the same energy. Such a sacrifice is termed *lost work*. The symbol LW will be used to characterize this quantity, which will be considered later in this chapter. It will be observed that lost work arises any time an irreversible change within the system causes the degradation of an energy potential. Furthermore, it will always lead to an increase in entropy, just as in this illustration.

The concept of lost work proves useful in later chapters, where the efficiency of energy-conversion processes is of importance. Physically it may be thought of as work which could have been produced had the process been conducted reversibly, but which can no longer be completely realized by any process known to man. It might more appropriately be thought of as "lost potential," inasmuch as energy continues to be conserved throughout any irreversible change. It simply becomes less usable to man as he seeks to produce change in the universe.

As has been shown, all nonthermal forms of energy transfer between system and surroundings (excluding the transfer of mass) have no entropy flow associated with them. However, dissipatory processes (such as friction, turbulence, etc.), which occur simultaneously *within the system* while it is receiving or delivering work or heat, can result in entropy changes. Indeed, they can occur in a system for which neither a heat nor a work effect exist. Entropy generation by such means should not be confused with the entropy flow. However, since work and heat are defined as energy in transition across a system boundary, energy flows wholly within a system do not qualify as either heat or work, so the preceding discussion does not apply to such flows if they occur within the system as it is defined. Thus it is essential that the entropy equation contain a generation term to account for system irreversibilities.

SAMPLE PROBLEM 5-1. Two blocks that are initially at different temperatures and completely isolated from the remainder of the universe are brought together so that thermal energy flows from the hot block to the cold block. Analyze the entropy change of the universe associated with this process by choosing as your system (a) the two blocks individually, and (b) the two blocks together. Assume that the heat transfer within each individual block is reversible so that any irreversibilities which occur, occur only at the interface between the two blocks.

Solution: (a) Let us examine the individual blocks first. For each block

$$dS = \frac{\delta Q}{T}$$

Applied individually to the low- and high-temperature blocks, the above equation gives

$$dS_L = \frac{\delta Q_L}{T_L}, \qquad dS_H = \frac{\delta Q_H}{T_H}$$

The change in entropy of the universe as a result of this process then is given by

$$dS_{\text{univ}} = dS_H + dS_L = \frac{\delta Q_H}{T_H} + \frac{\delta Q_L}{T_L}$$

The total change in entropy of the universe is obtained by integrating the above expression to give

$$\Delta S_{\text{univ}} = \int \frac{\delta Q_H}{T_H} + \frac{\delta Q_L}{T_L}$$

or, since $\delta Q_L = -\delta Q_H > 0$,

$$\Delta S_{\text{univ}} = \int \left(\frac{1}{T_H} - \frac{1}{T_L}\right) \delta Q_H$$

When performing the integration, it is important to keep in mind that T_H and T_L refer to the temperatures at which δQ_H and δQ_L are transferred. Since $T_H > T_L$, our integral for ΔS_{univ} indicates that $\Delta S_{\text{univ}} > 0$.

(b) If we take both blocks together as the system, $\delta Q = 0$ because the system is isolated from the surroundings. The energy transfer from the high-temperature block to the low-temperature block is an internal transfer and not a heat flow, with the system as defined. Since the system is the only portion of the universe undergoing changes, $\Delta S_{\text{univ}} = \Delta S_{\text{sys}}$. The change in ΔS_{univ} must be the same regardless of what system we choose for the analysis. In this case the irreversible heat flow between the two blocks was totally within the system boundaries and thus the total entropy change of the universe is now associated with the internal irreversible flows.

5.3 Entropy Generation and Lost Work

The need for an entropy-generation term to account for the entropy changes brought about by irreversible energy flows internal to the system is clearly demonstrated by Sample Problem 5-1 and the preceding section. The entropy change of the universe was shown to be a positive, nonzero number when the two blocks were considered separately as systems. Since the entropy change of the universe must be independent of the way by which it is calculated, it must be the same for parts (a) and (b). When the two blocks of Sample Problem 5-1 were taken as a combined system, it became an isolated system, because no exchange of energy or mass occurred between system and surroundings. Therefore, the only way in which the entropy change of the system (and hence the universe) could be accounted for was by entropy generation within the system.

In this example we could clearly identify a flow of thermal energy from the high-temperature part of the system to the lower-temperature portion. Dissipation of the temperature potential results in a permanent decrease in the amount of work that can ultimately be produced in our universe. As shown later in the chapter, a portion of the resulting thermal energy flow could have been converted to work by a heat engine. Failure to have done so results in an increase in the system's entropy. The entropy equation will account for all transfers across the system boundary. However, our definition of the system in the second case incorporated the flow resulting from an internal, finite driving force within the system, and therefore it escaped detection. Thus we must devise a term to account for such irreversible or dissipative phenomena when they occur within the system.

Although the observation that the entropy-generation term is positive is based on a single calculation in Sample Problem 5-1, we shall soon show that this is universally true. Only in the case of a reversible process where any internal gradients are infinitesimal and friction absent can the entropy-generation term become zero; *it can never be negative.*

In the case of the gas undergoing an adiabatic free expansion, the net effect of the process was to end up with the same amount of gas at the same total energy but at a lower pressure. Since pressure is an energy potential which can cause an energy flow that results in work, it is obvious that at a lower pressure the gas is less capable of converting its energy into work than it was at its original pressure.

The term *lost work* is used to indicate this loss in the ability of the gas to produce work. It should be emphasized that the conservation-of-energy principle is not violated in such a process because the energy has not really been lost—only the ability to convert a portion of it to work. Lost work is thus defined as work that could have been performed but was not because of dissipative effects or irreversibilities. Whenever an irreversible change within a system leads to the lowering of an energy potential (such as pressure, temperature, electrical potential, etc.), without transferring as much energy to the surroundings, in the form of work, as possible, lost work results and entropy production occurs.

Friction is another example of an irreversible, or dissipatory, process which reduces a system's ability to deliver work as shown in Fig. 5-4. The system is taken to be the gas + cylinder+ piston. As the gas expands, it exerts a force $F = P_{gas} \times A_{piston}$ on the piston. In the absence of friction this force can be transferred directly to the surroundings through the shaft and will produce useful work. The presence of friction between the piston and the wall diminishes the net force that is transferrable to the surroundings and hence the maximum amount of work that may be produced. In a sense it is equivalent to a process that occurs with a finite difference in pressure if we compare the pressure of the gas, P_{gas}, with the pressure, P_{res}, which is resisting the motion of the system boundary. Thus, although the friction slows the expansion down so the process may be considered quasi-static, it requires the existence of a finite pressure difference between system and surroundings to produce an energy transfer.

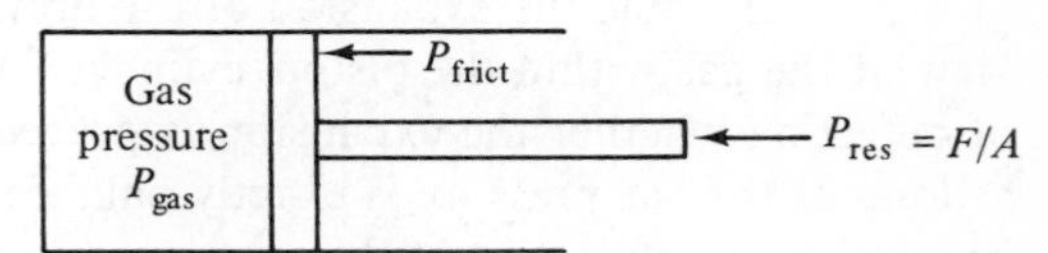

Figure 5-4. The effect of friction.

The frictional force may be thought of as a friction pressure, P_{frict}, acting to oppose movement of the piston. In this case, it would act to resist motion to the right and is thus shown as acting toward the left. The sum of the pressures opposing the motion of the piston must be less than the gas pressure, P_{gas}, if an expansion is to occur. Assuming that the piston has negligible mass, it can support only an infinitesimal pressure (or force) imbalance without experiencing an acceleration. Thus the pressures acting on the piston must be in equilibrium, so that

$$P_{\text{gas}} = P_{\text{res}} + P_{\text{frict}} \tag{5-19}$$

If the gas pressure is held constant, an increase in the frictional losses, P_{frict}, causes a corresponding decrease in the usable force, P_{res}, transmitted to the surroundings. Therefore, less work can be transferred to the surroundings.

In the expansion process pictured in Fig. 5-4 the maximum work will be transferred to the surroundings when the frictional losses vanish so that $P_{\text{res}} = P_{\text{gas}}$. Since the work delivered to the surroundings is given by

$$\int_{V_1}^{V_2} P_{\text{res}}\, dV$$

the maximum work that can be transferred to the surroundings is given by

$$\int_{V_1}^{V_2} P_{\text{gas}}\, dV$$

where V_1 and V_2 equal the initial and final volumes of the confined gas. In those instances where friction is present, $P_{\text{res}} = P_{\text{gas}} - P_{\text{frict}}$, which is less than P_{gas}. The decrease in work transferred to the surroundings is given by

$$\int_{V_1}^{V_2} P_{\text{frict}}\, dV$$

This lost work is degraded into thermal energy by friction at the interface between the piston and the cylinder. If the piston and cylinder are insulated against transfer of heat to the surroundings, this thermal energy will be transferred back into the gas, causing its entropy to increase (assuming that it is not absorbed by the piston and cylinder—that is, assuming that the piston and cylinder have either negligible masses or negligible heat capacities). This transfer qualifies as heat if we consider the gas alone as the system. However, if the system is taken as gas + piston + cylinder, the transfer, although still occurring, is internal and not considered as heat. However, in terms of this system it has clearly reduced the amount of work extractable from the expansion and therefore qualifies as lost work.

Let us examine the expansion of the previous example from the point of view of the gas within the piston–cylinder. We find that the gas has no way of detecting whether the expansion is proceeding reversibly, or irreversibly, as long as the gas pressure is exactly balanced by the sum $P_{\text{res}} + P_{\text{frict}}$. The irreversibilities that exist in the total process occur outside the gas and therefore do not affect the gas except insofar as these irreversibilities may lead to a heat flow back to the gas.

In analyzing this expansion process, we might choose to define a system such that the irreversibilities occur outside the system (that is, choose the gas alone as system) or in such a way that the irreversibilities occur within the system (for example, the gas + piston + cylinder as a system). If we assume that the piston + cylinder are perfectly insulated against heat loss to the surroundings, then the gas in either case will undergo an increase in entropy. In the former case the entropy increase results from the transfer of thermal energy (caused by friction) from the surroundings. In the latter case no heat flow is present, because the thermal-energy transfer (from piston to gas) is internal to the system and not between the system and the surroundings. In the second instance the entropy change is attributable wholly to entropy generation, which results from the internal irreversibilities.

The entropy increase in the first case can be calculated as

$$\Delta S_{gas} = \int \frac{\delta Q}{T} \tag{5-20}$$

where δQ is the heat flow to the gas as a result of the interfacial friction. As shown before, δQ is exactly equal to the mechanical energy dissipated at the wall:

$$\delta Q = P_{frict}\, dV$$

The temperature in the denominator of equation (5-20) is the system temperature at the point where the heat transfer takes place. Thus the entropy change is given by

$$(\Delta S)_{gas} = \int \frac{\delta Q}{T} = \int_{V_1}^{V_2} \frac{P_{frict}\, dV}{T_{sys}} \tag{5-21}$$

In the second case, however, there is no δQ term, but the same entropy increase must be experienced by the gas. In this case the system produced less work than it could have in the absence of friction. The lost work, LW, represents this difference:

$$LW = \int_{V_1}^{V_2} (P_{gas} - P_{res})\, dV = \int_{V_1}^{V_2} P_{frict}\, dV \tag{5-22}$$

or, upon differentiating,

$$\delta LW = P_{frict}\, dV \tag{5-22a}$$

where δ again represents the differential amount of the quantity LW.

From our earlier calculation we know that the entropy change experienced by the gas is given by

$$\Delta S = \int_{V_1}^{V_2} \frac{P_{frict}\, dV}{T_{sys}} \tag{5-23}$$

but since $\delta LW = P_{\text{frict}}\, dV$, equation (5-23) reduces to

$$\Delta S = \int_{V_1}^{V_2} \frac{\delta LW}{T_{\text{sys}}} \tag{5-24}$$

Thus the entropy change associated with an internal irreversibility or entropy generation is given for a differential process by

$$\delta(\text{entropy})_{\text{gen}} = \frac{\delta LW}{T_{\text{sys}}} \tag{5-25}$$

If irreversibilities are distributed throughout the volume of the system and the temperature varies within the system, the entropy generation must be evaluated by integrating $(\delta \boldsymbol{lw}/T)dV$ over the entire volume of the system along the entire path between initial and final states:

$$(\text{entropy})_{\text{gen}} = \int_{\text{process}} \int_{\text{volume}} \frac{\delta \boldsymbol{lw}}{T}\, dV \tag{5-26}$$

where $\boldsymbol{lw}$ = (finite) lost work per unit volume of system
$\delta \boldsymbol{lw}$ = differential lost work per unit volume of system

(*Note:* $\boldsymbol{lw}$ is used for lost work on a unit volume basis as contrasted to LW, used earlier, which is the total lost work.) The limits on the integral imply that the quantity $\delta \boldsymbol{lw}/T$ must be evaluated at all points in the system where irreversibilities occur.

Although we were successful in evaluating the lost work and entropy production for the relatively simple piston–cylinder problem, in general it cannot be precisely calculated because of the extremely complex nature of the dissipative phenomena. Although we can measure changes in other properties by which we can calculate entropy changes arising from irreversibilities, an a priori prediction of LW is extremely difficult and generally impractical.

This difficulty in treating irreversible processes in an analytic manner necessitates the use of empirical methods for predicting such effects. For example, consider the flow that occurs within a pipeline due to a pressure gradient. Since lost work (and entropy production) must be evaluated with respect to a specified system, let the pipe and its contents be our system. The drag and turbulence that occur in the pipeline lead to lost work much as friction does in the piston–cylinder example. Unfortunately, a quantitative evaluation of this dissipation is a difficult task. Fluid mechanics offers a mathematical expression for the dissipation, but its evaluation frequently requires a greater knowledge of the fluid velocity distribution than is available. Consequently, empirical procedures are generally used to estimate these effects. See, for example, Balzhiser et al.[1] for a detailed discussion of

[1] R. E. Balzhiser, M. R. Samuels, and J. D. Eliassen, *Chemical Engineering Thermodynamics*, Prentice-Hall, Inc., Englewood Cliffs, N.J., 1972.

the means by which these empirical correlations for lost work may be used in the analysis of fluid flow problems.

Direct evaluation of lost work effects in process equipment, such as turbines and compressors, is far beyond our present capabilities. Therefore, in analyzing processes involving operations of this nature, it is often necessary to approximate the real situation with a reversible one in which $LW = 0$. Past experience with many devices, such as compressors and turbines, often permits the engineer to relate performance under reversible conditions to actual operation under real conditions. The relation is usually expressed by means of an efficiency factor. For devices such as pumps and compressors which utilize work from the surroundings, efficiency is defined as

$$\text{efficiency} = \frac{W_{\text{rev}}}{W_{\text{act}}} \times 100\% \tag{5-27}$$

For turbines and other expansion devices that supply work to the surroundings the definition is inverted to give

$$\text{efficiency} = \frac{W_{\text{act}}}{W_{\text{rev}}} \times 100\% \tag{5-28}$$

In either case, W_{rev} is the work produced or consumed if the process is operated reversibly, and W_{act} is the actual work that is involved. A knowledge of the efficiency of a particular type of device, coupled with an analysis based on a reversible process, thus enables one to estimate with reasonable engineering accuracy the actual energy requirements for a specific piece of equipment.

SAMPLE PROBLEM 5-2. Consider an insulated oven with an electrical heating element. The oven is turned on, and the temperature begins to rise. Indicate whether LW or Q are present for the following systems:

(a) The heating element alone.
(b) The oven and its contents (including the element).
(c) The gas within the oven.

Solution: (a) The heating element alone: In the heating element, electrical energy (a form of mechanical energy) is transformed into thermal energy and then transferred to the surroundings as heat. Thus the heating element has a Q term. In addition to the Q term, a LW term is present because of the internal degradation of electrical energy to thermal energy resulting from the electrical resistance of the element. Thus both Q and LW are present when the heating element is chosen as a system.

(b) If we choose the oven and its contents as a system, LW is clearly present, since electrical energy is converted to thermal energy within the system. If we assume that the oven is insulated so that no heat is lost to the surroundings, then no heat transfer occurs, because the thermal energy transfers that exist are wholly within the system.

(c) If we now consider the gas within the oven as a system, Q is present in the form of the energy transferred from the heating element to the gas. If we assume that there is a negligible temperature gradient within the gas, then the gas is undergoing reversible heat transfer and no LW occurs.

Energy flowing as heat was shown to produce entropy changes within a system which were *directly attributable* only to the flow of heat, whether the overall process is reversible or not. Energy transfer as work was discussed only for a simple compression or expansion process and was *shown not to produce an entropy change* as a direct result of the energy transfer. In a more general sense, one might extend that observation to all energy-transfer processes between system and surroundings in which the energy transfer is categorized as work. In each case the energy content of the system is changed, but the change is compensated for by a change in some other extensive property, such as volume, so that no net change in entropy occurs. This statement is not intended to imply that entropy changes cannot result within a system which is either doing or receiving work, *only that such entropy change is not a direct consequence of the* (*mechanical*) *energy flow. Thus, in terms of energy flows between system and surroundings, only heat has an entropy flow associated directly with it.*

5.4 The Entropy Equation

We have shown that the entropy flow associated with heat transfer is given by $\int_{\text{surface}} \delta q/T\, dA$, and the entropy generation due to irreversibilities is given by $\int_{\text{volume}} \delta lw/T\, dV$. Incorporation of these terms in equation (5-2) yields the complete entropy equation for a general process:

$$(s\,\delta M)_{\text{in}} - (s\,\delta M)_{\text{out}} + \int_{\text{surface}} \frac{\delta q}{T}\, dA + \int_{\text{volume}} \frac{\delta lw}{T}\, dV = d(Ms)_{\text{sys}} \qquad (5\text{-}29)$$

If the system temperature is everywhere uniform, equation (5-29) reduces to

$$(s\,\delta M)_{\text{in}} - (s\,\delta M)_{\text{out}} + \frac{\delta Q}{T} + \frac{\delta LW}{T} = d(Ms)_{\text{sys}} \qquad (5\text{-}30)$$

This equation can be applied to any thermodynamic system just as the energy balance is applied. The balance has been written in differential notation with δ used to denote differential flow and production terms. Sign conventions are the same as in the energy equation; that is, $d(Ms)_{\text{sys}}$ represents the entropy change of the system over a differential time period. δQ represents heat flowing into the system (if positive) and to the surroundings (if negative) for a differential time period. δM represents a mass flow over the differential time period and is always taken as positive (since its sign is already accounted for in the "in" and "out" terms). δLW represents the energy dissipation over the differential period and will always be positive.

The temperature in the flow term $\delta q/T$ is the temperature of the system at the point along the boundary where the transfer occurs. By the same token, the temperature in the $\delta lw/T$ term is the temperature at that point within the system where the irreversibility is occurring. For isothermal (uniform temperature) systems, application of the entropy balance is greatly simplified. For nonisothermal systems it is necessary to sum (or integrate) the $\delta q/T$ and $\delta lw/T$ terms over all portions of the system where either δq or δlw is present. It is also important to stress that the terms s_{in} and s_{out} relate to the stream entropies at the point where the streams cross the system's boundary.

As with the energy equation, the entropy equation we have derived is an extremely general relation designed to fit a wide variety of processes. As such, it contains terms that may frequently be ignored in any given problem. Let us now examine some of the more commonly encountered simplifications of the entropy equation (For purposes of simplicity, we use the isothermal form of the entropy equation in the ensuing discussion. However, it must be remembered that the integral form must be used when heat transfer or lost work exists in a nonisothermal system.)

Closed Systems

For a closed system the δM_{in} and δM_{out} terms vanish, as does dM_{sys}, and the entropy equation reduces to

$$M\,ds_{sys} = \frac{\delta Q}{T} + \frac{\delta LW}{T} \tag{5-31}$$

For a reversible process $\delta LW = 0$, and we obtain a further simplification:

$$dS_{sys} = M\,ds_{sys} = \frac{\delta Q}{T} \tag{5-32}$$

Open Systems

When an open system operates at steady state $d(Ms)_{sys} = 0$ and $\delta M_{in} = \delta M_{out}$, so the entropy equation reduces to

$$(s_{in} - s_{out})\,\delta M + \frac{\delta Q}{T} + \frac{\delta LW}{T} = 0 \tag{5-33}$$

For an adiabatic process $\delta Q = 0$, whereas for a reversible process $\delta LW = 0$. These facts may be used to modify equations (5-31) and (5-33) when applicable.

SAMPLE PROBLEM 5-3. Two insulated horizontal cylinders are separated by removable insulation as shown in Fig. SP5-3. Each cylinder is divided into two parts by a freely floating, insulated piston having negligible mass. Initially, chamber A has a volume of 5 ft³ and contains steam at 490 psia and 520°F; chamber B, also 5 ft³, contains steam at 740°F; chamber C contains 10 lb_m of saturated water vapor at 320°F; chamber D contains air. The removable insulation is raised, and heat flows slowly from chamber B to C, until the steam in chamber A is a saturated vapor. At the same time, air is bled slowly out of D to maintain a constant temperature in C. What is the final pressure in C?

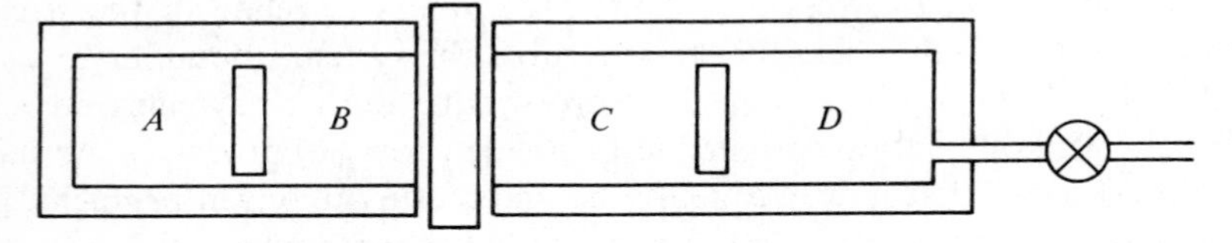

Figure SP5-3

Solution: We may summarize the changes that occur in each region as follows: Chamber A undergoes an adiabatic reversible expansion (no irreversibilities occur in chamber A) in which it performs work on chamber B. Chamber B undergoes a nonadiabatic compression during which it receives work from chamber A and loses heat to chamber C. Since the piston separating chambers A and B is frictionless, $P_A = P_B$ at all times. Chamber C undergoes an isothermal expansion at 320°F, during which it receives heat from chamber B and transfers work to chamber D. Chamber D serves to receive work from chamber C, allowing the temperature in C to remain constant. The initial conditions in the various chambers are summarized in Table SP5-3.

Table SP 5-3

Chamber	P, psia	T, °F	h, Btu/lb_m	s, Btu/lb_m °R	v, ft³/lb_m	V, ft³	M, lb_m	$u = h - Pv$, Btu/lb_m
A	490	520	1248.4	1.5116	1.0535	5	?	1152
B	490	740	1378	1.630	1.390	5	?	1252
C	90	320	1185	1.6116	4.898	?	10	Unneeded
D (air)	90							

We may calculate the mass in chambers A and B from the definition of V:

$$v = \frac{V}{M}$$

or

$$M = \frac{V}{v}$$

Thus

$$M_A = 5\ \text{ft}^3/1.0535\ \text{ft}^3/\text{lb}_m = 4.75\ \text{lb}_m$$

$$M_B = 5\ \text{ft}^3/1.390\ \text{ft}^3/\text{lb}_m = 3.60\ \text{lb}_m$$

The volume in C is calculated to be

$$V_C = v_C M_C = 48.9\ \text{ft}^3$$

We may now begin determining the conditions when the steam in chamber A reaches saturation conditions. Since the steam in chamber A is to be a saturated vapor at the end of its expansion, determination of any property will serve to fix its final state. Choosing the steam in chamber A as a system, the energy and entropy equations are written as follows (assuming negligible changes in potential and kinetic energy):

Energy:

$$(h\,\delta M)_{\text{in}} - (h\,\delta M)_{\text{out}} + \delta Q - \delta W = d(Mu)_{\text{sys}}$$

Entropy:

$$(s\,\delta M)_{\text{in}} - (s\,\delta M)_{\text{out}} + \int_{\text{area}} \frac{\delta q}{T}\,dA + \int_{\text{vol}} \frac{\delta lw}{T}\,dV = d(Ms)_{\text{sys}}$$

Since the system is closed, $\delta M_{\text{in}} = \delta M_{\text{out}} = dM_{\text{sys}} = 0$, and the energy and entropy equations reduce to

$$\delta Q - \delta W = M\,du$$

$$\int_{\text{area}} \frac{\delta q}{T}\,dA + \int_{\text{vol}} \frac{\delta lw}{T}\,dV = M\,ds$$

Since the expansion in chamber A is adiabatic and reversible, $\delta q = \delta lw = 0$, and the equations further reduce to

$$-\delta W = M\,du$$

$$0 = ds$$

Integration of the entropy balance over the entire expansion then yields

$$(s_A)_{\text{final}} - (s_A)_{\text{init}} = 0$$

$$(s_A)_{\text{final}} = (s_A)_{\text{init}} = 1.5116$$

Thus the entropy equation allows us to determine the needed property which fixes the state of the steam in A when it reaches saturation.

If we integrate the energy equation over the entire expansion we can determine the work done by the gas in chamber A as

$$-W = -\int \delta W = M[(u_A)_{\text{final}} - (u_A)_{\text{init}}]$$

$(u_A)_{\text{init}}$ is obtained from $u = h - Pv$, and $(u_A)_{\text{final}}$ can be determined from the final state of the steam in A as follows. Since the steam in A undergoes an isentropic expansion, $(s_A)_{\text{final}} = (s_A)_{\text{init}} = 1.5116$. At saturation conditions this entropy corresponds to a final state summarized as follows:

$$T = 420°\text{F}$$

$$P = 300 \text{ psia}$$

$$h = 1203 \text{ Btu/lb}_m$$

$$v = 1.54 \text{ ft}^3/\text{lb}_m$$

$$u = h - Pv = 1117 \text{ Btu/lb}_m$$

The final volume of chamber A is given by

$$(V_A) = Mv = (4.75)(1.54)(\text{ft}^3/\text{lb}_m)\text{ lb}_m$$

$$(V_A)_{\text{final}} = 7.32 \text{ ft}^3$$

The work performed by the steam in chamber A is given by

$$-W = M\,\Delta u = (4.75)(1117 - 1152) \text{ Btu}$$

or

$$W = 166 \text{ Btu}$$

We may now begin to analyze the final state of the material in chamber B. Since $P_A = P_B$ at all times, $(P_B)_{\text{final}} = (P_A)_{\text{final}} = 300$ psia. Thus determination of one other property of the system will completely specify its state. We may easily determine the specific volume of the steam in chamber B by noting that $(V_A)_{\text{final}} + (V_B)_{\text{final}} = (V_A)_{\text{init}} + (V_B)_{\text{init}} = 10 \text{ ft}^3$ and remembering that $(v_B)_{\text{final}} = (V_B)_{\text{final}}/M_B = (V_B)_{\text{final}}/3.6 \text{ lb}_m$. Since $(V_A)_{\text{final}} = 7.32 \text{ ft}^3$, $(V_B)_{\text{final}} = 2.68 \text{ ft}^3$ and

$$(v_B)_{\text{final}} = \frac{(V_B)_{\text{final}}}{3.6 \text{ lb}_m} = \frac{2.68}{3.6} \text{ ft}^3/\text{lb}_m$$

or

$$(v_B)_{\text{final}} = 0.745 \text{ ft}^3/\text{lb}_m$$

If we attempt to locate $v = 0.745 \text{ ft}^3/\text{lb}_m$ and $P = 300$ psia in the steam tables, we find that this corresponds to a mixture of saturated liquid and saturated vapor. The proportion of vapor and liquid is determined from the mixture rule:

$$v_{\text{mix}} = v^{\text{V}}X^{\text{V}} + v^{\text{L}}X^{\text{L}}$$

But

$$X^{\text{V}} = 1 - X^{\text{L}}$$

$$v_{\text{mix}} = 0.745 \text{ ft}^3/\text{lb}_m$$

$$v^{\text{V}} = 1.54 \text{ ft}^3/\text{lb}_m$$

$$v^{\text{L}} = 0.091 \text{ ft}^3/\text{lb}_m$$

or

$$0.745 = 1.54(1 - X^{\mathrm{L}}) + 0.091X^{\mathrm{L}}$$

Solving for X^{L} yields

$$X^{\mathrm{L}} = 0.52$$

$$X^{\mathrm{V}} = 0.48$$

$(u_B)_{\text{final}}$ is then determined from the mixture rules:

$$(u_B)_{\text{mix}} = X^{\mathrm{L}}(u)_{\text{sat liq}} + (1 - X^{\mathrm{L}})(u)_{\text{sat vap}}$$

$$u_{\text{sat vap}} = 1117 \text{ Btu/lb}_{\text{m}} \text{ (from chamber } A)$$

$$u_{\text{sat liq}} = (h - Pv)_{\text{sat liq}} = 393 \text{ Btu/lb}_{\text{m}}$$

so

$$(u_B)_{\text{mix}} = (0.52)(393) + (0.48)(1117) = 740 \text{ Btu/lb}_{\text{m}}$$

We are now in a position to determine the heat transfer between chambers *B* and *C*. The energy balance is written about chamber *B*. Since the system is closed, $\delta M_{\text{in}} = \delta M_{\text{out}} = dM_{\text{sys}} = 0$. Neglecting potential and kinetic energy changes, the energy equation reduces to

$$\delta Q - \delta W = M\, du_{\text{sys}}$$

This may be integrated to yield

$$Q - W = M\,\Delta u = M[(u_B)_{\text{final}} - (u_B)_{\text{init}}]$$

We know $(u_B)_{\text{final}}$, $(u_B)_{\text{init}}$, and M. In addition, the only work transferred between chamber *B* and its surroundings is the work transferred from chamber *A* to chamber *B*. Since work is considered positive when transferred from the system to the surroundings

$$W_B = -W_A = -166 \text{ Btu}$$

We may now calculate the heat transferred to chamber *B*:

$$\begin{aligned} Q_B &= W_B + M_B[(u_B)_{\text{final}} - (u_B)_{\text{init}}] \\ &= -166 + (3.6)[(740) - 1252] \text{ Btu} \\ &= -2010 \text{ Btu} \end{aligned}$$

Having calculated the heat loss from chamber *B* we can now begin to work on chamber *C*. Since chamber *C* undergoes an isothermal expansion, $T =$ const $= 320°\text{F}$. Thus once again, determination of another system property will fix the remaining properties and thereby allow us to determine the final pressure in chamber *C* as desired.

Assuming that chamber *C* is a closed system the entropy and energy equations around system *C* reduce to

$$\text{Energy:} \quad \delta Q - \delta W = M\, du$$

$$\text{Entropy:} \quad \frac{\delta Q}{T} + \frac{\delta LW}{T} = M\, ds$$

These may be integrated to yield

$$Q - W = M\,\Delta u$$

$$\frac{Q}{T} + \frac{LW}{T} = M\,\Delta s$$

Since the steam in chamber C undergoes a reversible process, $LW = 0$, so that

$$\frac{Q_C}{T_C} = M_C\,\Delta s_C = M_C[(s_C)_{\text{final}} - (s_C)_{\text{init}}]$$

But

$$Q_C = -Q_B = 2010 \text{ Btu}$$

$$T = 320°\text{F} = 780°\text{R}$$

$$(s_C)_{\text{init}} = 1.6116 \text{ Btu/lb}_\text{m}°\text{R}$$

$$M_C = 10 \text{ lb}_\text{m}$$

Thus we may solve for the final entropy in chamber C to get

$$(s_C)_{\text{final}} = (s_C)_{\text{init}} + \frac{Q_C}{T_C M_C}$$

$$= \left[1.6116 + \frac{2010}{(780)(10)}\right] \text{Btu/lb}_\text{m}°\text{R}$$

or

$$(s_C)_{\text{final}} = 1.8715 \text{ Btu/lb}_\text{m}°\text{R}$$

From the final entropy and temperature, the final pressure in chamber C is found to be $(P_C)_{\text{final}} = 10$ psia.

Sample Problem 5-4. A steam line supplies steam at 620 psia and 700°F to an adiabatic reversible turbine (expander) which discharges to an insulated

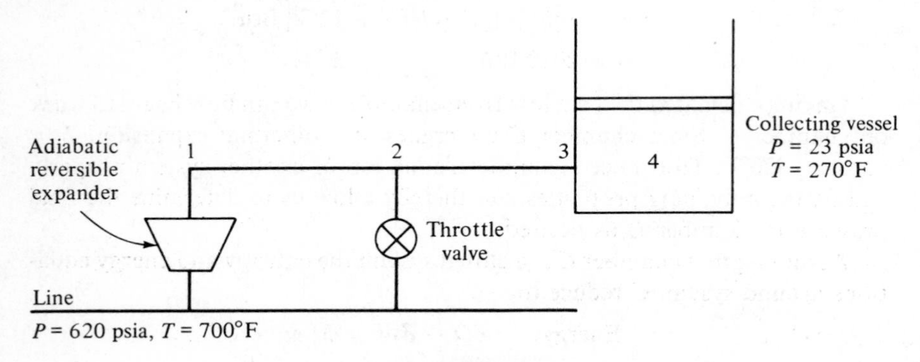

Figure SP5-4

collector fitted with a frictionless piston that maintains a constant pressure of 23 psia. Additional steam is fed to the collector via a throttling valve so as to hold the temperature in the collector constant at 270°F, as shown in Fig. SP5-4. If the collecting vessel has a cross-sectional area of 37.18 ft², how many pounds of steam flow through the turbine per foot of piston rise? Neglect potential and kinetic energy changes in the steam and heat and frictional losses in the connecting lines.

Solution: We begin our analysis by attempting to determine the state of the steam entering the collecting vessel. Since the pressure in the collector is 23 psia and there are no frictional losses in the connecting lines, the pressure of the material in the line entering the collecting vessel will also be 23 psia. Thus determination of one more property will fix the state of the entering steam. The entropy equation may be written around the collecting vessel and gives

$$(s\,\delta M)_{\text{in}} - (s\,\delta M)_{\text{out}} + \int_{\text{area}} \frac{\delta q}{T}\,dA + \int_{\text{vol}} \frac{\delta lw}{T}\,dV = d(Ms)_{\text{sys}}$$

Since no mass flows out of the system, $\delta M_{\text{out}} = 0$. The collecting vessel is well-insulated, so $\delta \boldsymbol{q} = 0$. Since no irreversibilities occur within the system (the piston and connecting lines are frictionless), $\delta \boldsymbol{lw} = 0$ and the entropy equation reduces to

$$(s\,\delta M)_{\text{in}} = d(Ms)_{\text{sys}}$$

We may integrate this expression for 1 lb_m of steam entering the collector. Since the state of both the steam entering the collector and that within it are not functions of the mass in the cylinder, the specific entropies of the entering steam and those within the collector are also independent of the mass within the collector. Thus they may be removed from under the integral sign, and the integrated entropy equation reduces to

$$(sM)_{\text{in}} = (\Delta M)s_{\text{sys}}$$

However, the conservation of mass indicates that $M_{\text{in}} = \Delta M$, so the entropy equation becomes

$$s_{\text{in}} = s_{\text{sys}}$$

That is, the entropy of steam does not change as it enters the collector. However, we have already indicated that $P_{\text{sys}} = P_{\text{in}}$, so two of the properties of the steam do not change as the steam enters the collector. Since two properties fix the state of a system, two properties remaining unchanged means that the state of the steam is unchanged and all other properties are unchanged.

We now determine the properties of the steam within the collecting vessel. We know that $P = 23$ psia and $T = 270°\text{F}$; therefore,

$$h = 1175\ \text{Btu/lb}_\text{m} \quad \text{and} \quad v = 18.6\ \text{ft}^3/\text{lb}_\text{m}$$

The cross-sectional area of the collector is 37.18 ft², so the volume increase corresponding to 1 ft of piston rise is 37.18 ft³. Since $v = 18.6\ \text{ft}^3/\text{lb}_m$, 1 ft of piston rise corresponds to $37.18/18.6 = 2.0\ \text{lb}_m$ of steam entering the piston. The enthalpy of the steam in the line entering the collector is then $h_3 = 1175\ \text{Btu/lb}_m$.

The energy balance around the mixing "T" (point 2 in the diagram) is now written as

$$\sum\left[\left(h + \frac{V^2}{2g_C} + \frac{gZ}{g_C}\right)\delta M\right]_{in} - \left[\left(h + \frac{V^2}{2g_C} + \frac{gZ}{g_C}\right)\delta M\right]_{out} + \delta Q - \delta W$$
$$= d\left[M\left(u + \frac{V^2}{2g_C} + \frac{gZ}{g_C}\right)\right]_{sys}$$

However, we may neglect $V^2/2g_C$, gZ/g_C, and δQ. $\delta W = 0$ since no motion of, or through, the system boundaries occurs. In addition, if the mixing process in the "T" is at steady state, the entire right-hand portion of the equation vanishes, and the energy equation reduces to

$$\sum (h\,\delta M)_{in} - (h\,\delta M)_{out} = 0$$

Since the enthalpies of the various streams do not change with M, the energy equation can be integrated to

$$\sum (hM)_{in} - (hM)_{out} = 0$$

or

$$\sum (hM)_{in} = (hM)_{out}$$

The incoming streams to the mixing "T" come from the turbine and throttling valve. Thus the summation reduces to

$$(hM)_{turb} + (hM)_{valve} = (hM)_{out}$$

A mass balance around the "T" gives

$$M_{turb} + M_{valve} = M_{out}$$

Since we are attempting to solve for M_{turb}, eliminate M_{valve} between the energy equation and mass balance to give

$$(hM)_{turb} + h_{valve}(M_{out} - M_{turb}) = (hM)_{out}$$

or, solving for M_{turb}

$$M_{turb} = \frac{(h_{out} - h_{valve})M_{out}}{h_{turb} - h_{value}}$$

Since the material leaving the mixing "T" is the same as that entering the collecting chamber, $M_{out} = 2\ \text{lb}_m$ per foot of piston rise, and $h_{out} = 1175\ \text{Btu/lb}_m$. Thus if we can determine h_{valve} and h_{turb}, we can solve directly for the mass passing through the turbine per foot of piston rise.

We may determine h_{valve} from an energy equation around the valve. Neglecting potential and kinetic energies across the valve and assuming steady-state operation, the energy equation around the valve reduces to

$$(h_{in})\delta M - (h_{out})\,\delta M + \delta Q - \delta W = 0$$

However, for the valve $\delta W = 0$, and if we assume that the valve is adiabatic, $\delta Q = 0$. Thus the energy equation becomes

$$(h_{in})_{valve} = (h_{out})_{valve}$$

But $(h_{in})_{valve} = 1350$ Btu/lb$_m$, so

$$(h_{out})_{valve} = 1350 \text{ Btu/lb}_m$$

However,

$$h_{valve} = (h_{out})_{valve} = 1350 \text{ Btu/lb}_m$$

We now need to determine the exit enthalpy from the turbine. Since we known that the outlet pressure from the turbine is 23 psia, determination of any other system property will fix the state of the outlet stream and give us its enthalpy. The energy and entropy balances may be written around the turbine as a system. If we neglect potential and kinetic changes and assume steady-state operation, these equations reduce to

Energy:

$$(h_{in} - h_{out})_{turb}\,\delta M + \delta Q - \delta W = 0$$

Entropy:

$$(s_{in} - s_{out})_{turb}\,\delta M + \int_{area} \frac{\delta q}{T}\,dA + \int_{vol} \frac{\delta lw}{T}\,dV = 0$$

Since the turbine is adiabatic and reversible $\delta q = \delta lw = 0$ and the entropy equation gives

$$(s_{out})_{turb} = (s_{in})_{turb}$$

Since we know P_{in} and T_{in} we can obtain s_{in} and then s_{out}. From s_{out} and $P_{out} = 23$ psia we may now look up the final enthalpy and find

$$(h_{out})_{turb} = h_{turb} = 1065 \text{ Btu/lb}_m$$

(Had we continued to work with the energy equation we finally would have obtained $-w = \Delta h$, so we could now determine the work produced by the turbine if this were desired.) Substitution of h_{turb} and h_{valve} into the expression for M_{turb} then gives

$$M_{turb} = \frac{(1175 - 1350)\text{ Btu/lb}_m}{(1065 - 1350)\text{ Btu/lb}_m}\,\frac{2.0\text{ lb}_m}{\text{ft of piston travel}}$$

Therefore,

$$M_{turb} = 1.24\,\frac{\text{lb}_m}{\text{ft of piston rise}}$$

and 1.24 lb$_m$ flow through the turbine per foot of piston rise in the collector.

SAMPLE PROBLEM 5-5. A vapor-compression desalination process will be used to produce drinking water aboard an ocean-going ship. The process is as illustrated in Fig. SP5-5a. Seawater (3.5 percent salt) enters the unit and is preheated by countercurrent contact with the drinking water and waste brine. The preheated seawater then enters the evaporator, where a portion is boiled off by condensing steam. The condensing steam is obtained by compressing the vapor formed in the evaporator. The cooled condensate is the drinking water.

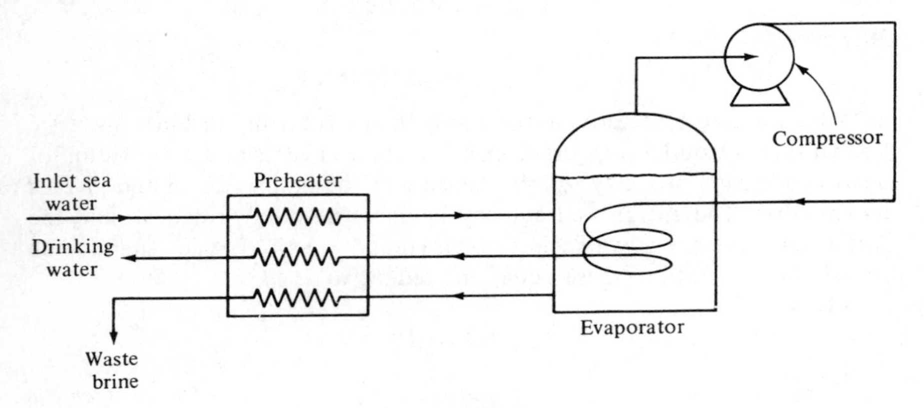

Figure SP5-5a

The evaporator normally operates at 1 atm pressure. The temperature is 105°C because of the boiling-point elevation. The vapor entering the compressor is compressed adiabatically to 1.5 Bar. If the compressor operates at an efficiency of 60 percent, determine:

(a) The outlet temperature from the compressor.
(b) The work of compression per kilogram of drinking water formed.

Solution: We may picture the compressor as a separate unit that receives super-heated steam at 105°C and 1 atm and discharges it at 1.5 Bar, as shown in Fig. SP5-5b. Choosing the turbine and its contents as a system the energy balance may be written. However, if potential and kinetic energy changes are neglected and the process is assumed to be at steady state, the energy balance reduces to

$$q - w = \Delta h = h_{\text{out}} - h_{\text{in}}$$

W

$P = 1.5$ Bar

$T = 105°\text{C}$

$P = 1$ atm

Figure SP5-5b

Since we are told that the compression is adiabatic, $q = 0$ and the energy balance becomes

$$-w = \Delta h$$

Thus evaluation of either w or Δh fixes the other. Our problem is to evaluate w and the outlet temperature T_2. However, since we already know that the outlet pressure $P_2 = 1.5$ Bar, determination of any other system property will fix the state, and hence temperature, of the outlet stream. Determination of Δh will give us the outlet enthalpy, since the inlet enthalpy can be obtained from the inlet tempeature and pressure. Therefore, our problem reduces to one of finding either w or Δh.

We are told that the compressor may be assumed to be 60 percent efficient. Therefore, the ratio of the actual work of compression to that needed if the compression were reversible is

$$\frac{w_{\text{rev}}}{w_{\text{act}}} = 0.60$$

or

$$w_{\text{act}} = \frac{w_{\text{rev.}}}{0.60}$$

If the compression were reversible, then $\delta lw = 0$ and the entropy balance (assuming steady state) could be written as

$$(s_{\text{in}} - s_{\text{out}})\,\delta M + \int_{\text{area}} \frac{\delta q}{T}\,dA = 0$$

Since the process is adiabatic (whether it's reversible or not, it can always be adiabatic), $\delta q = 0$ and the entropy balance for the reversible process reduces to

$$s_{\text{in}} = s_{\text{out}}$$

Thus the process is isentropic. From the inlet conditions we determine the inlet entropy. From the entropy and the outlet pressure we obtain the outlet conditions (assuming, of course, that the process were reversible). That is,

$$(h_2)_{\text{rev}} = 2760 \text{ kJ/kg}$$

The inlet enthalpy is

$$h_1 = 2685 \text{ kJ/kg}$$

Therefore if the process were reversible, the enthalpy change would have been

$$\Delta h_{\text{rev}} = (2760 - 2685) \text{ kJ/kg} = 75 \text{ kJ/kg}$$

Since the energy balance is independent of the reversibility of the process,

$$-w_{\text{rev}} = (\Delta h)_{\text{rev}} = 75 \text{ kJ/kg}$$

But from the efficiency of the process we know

$$w_{act} = \frac{w_{rev}}{0.60} = \frac{-75 \text{ kJ/kg}}{0.60} = -125 \text{ kJ/kg}$$

so that the actual work requirements of the compressor are -125 kJ/kg of drinking water.

Again, since the energy balance holds independent of any reversibility arguments

$$(\Delta h)_{act} = -w_{act} = 125 \text{ kJ/kg}$$

The inlet enthalpy is still $h_1 = 2685$ kJ/kg, so that the actual outlet enthalpy is

$$(h_2)_{act} = (2685 + 125) \text{ kJ/kg} = 2810 \text{ kJ/kg}$$

The actual outlet enthalpy and the outlet pressure are enough to fix the outlet state and allow determination of the outlet temperature, which is

$$T_2 = 168°\text{C}$$

So the temperature of the steam leaving the compressor is 168°C.

5.5 Work from Heat: The Carnot Cycle

Since the thermal potential, temperature, cannot be used directly to produce work, it is necessary to utilize the thermal potential to produce an increase in some other energy potential, such as pressure, before work can be extracted. For example, the heating of a closed vessel of water (producing steam) would produce an increased pressure as its energy content was increased. The higher pressure possessed by the system could then be used to push a piston or turn a turbine as illustrated in Sample Problem 5-6.

SAMPLE PROBLEM 5-6. An emergency pump in an oil refinery is to be powered by a steam turbine as shown in Fig. SP5-6a. Feedwater will enter the pump at 1 Bar and 30°C. The water is pumped to a pressure of 8 Bar and passed into the boiler, where it is converted to saturated steam. From the boiler the steam enters the turbine, where it is expanded to 1 Bar. A portion of the work produced by the turbine is used to power the feedwater pump; the remainder is the useful work of the engine. If the turbine is assumed to operate in an adiabatic and reversible fashion, determine the net work produced per unit of thermal energy supplied to the boiler.

Solution: The feedwater enters the engine (the combination of pump, boiler, and turbine is termed "the engine") at 30°C and 1 Bar. All calculations are based on 1 kg of water passing through the engine. The work required by the

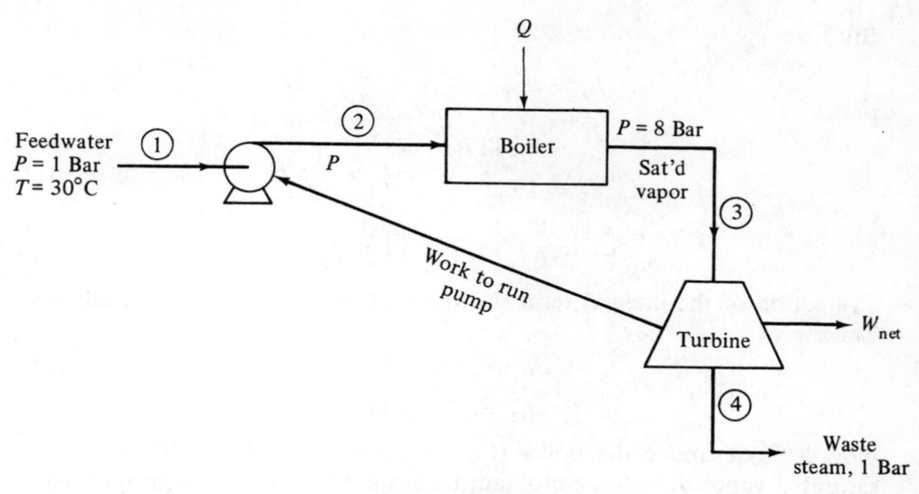

Figure SP5-6a

pump to increase the pressure of the water from 1 Bar to 8 Bar is given by

$$w_{\text{pump}} = -\int_{P_1}^{P_2} v\,dP$$

but for liquid water v is constant, and the integral reduces to

$$w_{\text{pump}} = -v\int_{P_1}^{P_2} dP = -v(P_2 - P_1)$$

$v = 0.001\ \text{m}^3/\text{kg}$ for liquid water, and w becomes

$$\begin{aligned} w_{\text{pump}} &= -0.001\ \text{m}^3/\text{kg}\,(8.0 - 1.0) \times 10^5\ \text{N/m}^2 \\ &= -700\ \text{Nm/kg} = -700\ \text{J/kg} \end{aligned}$$

An energy balance taken around the pump, assuming steady-state and negligible changes in potential and kinetic energies, gives

$$q - w = \Delta h$$

Assuming the flow of water through the pump to be essentially adiabatic, $q = 0$, and the energy balance reduces to

$$\Delta h = -w = +700\ \text{J/kg}$$

or, if we label the various positions in the engine as indicated in the diagram,

$$h_2 - h_1 = 700\ \text{J/kg}$$

Thus if we can find the enthalpy, h, of the liquid entering the engine, we can calculate the enthalpy of the liquid leaving the pump and entering the boiler. We begin by noting that we can find the enthalpy of a saturated liquid at 30°C in the steam tables. That is,

$$h_{\text{sat liq}} = 126\ \text{kJ/kg}$$

But

$$h_1 = h_{\text{sat liq}} + \int_{P_{\text{sat}}}^{P_1} \left(\frac{\partial h}{\partial P}\right)_T dP$$

$$P_{\text{sat}} = 0.042 \text{ Bar}$$

$$P_1 = 1 \text{ Bar}$$

$$\left(\frac{\partial h}{\partial P}\right)_T = v - T\left(\frac{\partial v}{\partial T}\right)_P$$

Evaluation of the integral term shows its value to be essentially negligible. Therefore,

$$h_1 = 126 \text{ kJ/kg}$$

$$h_2 = 127 \text{ kJ/kg}$$

Now let us examine the boiler (Fig. SP5-6b). Since the outlet steam is a saturated vapor at 8 Bar, we obtain from the steam tables (Appendix B)

$$h_3 = 2769 \text{ kJ/kg}$$

$$s_3 = 6.663 \text{ kJ/kg °K}$$

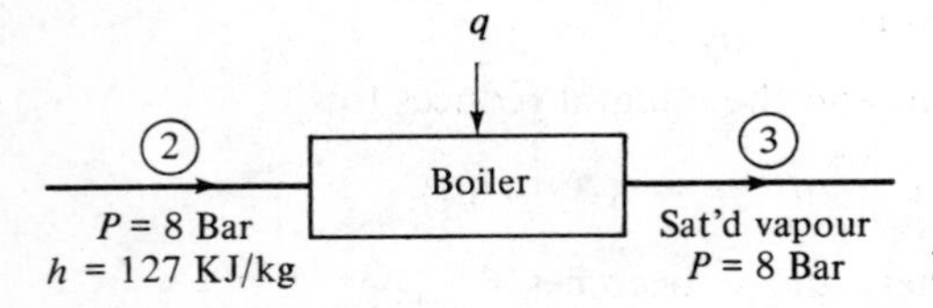

Figure SP5-6b

An energy balance around the boiler, assuming steady-state and negligible changes in potential and kinetic energies, reduces to

$$q - w = \Delta h$$

Since there is no work added or removed from the boiler, $w = 0$, and the energy balance gives

$$q = \Delta h = h_3 - h_2 = (2769 - 127) \text{ kJ/kg}$$

or

$$q = 2642 \text{ kJ/kg}$$

and represents the total thermal energy input per pound of water passing through the process.

Since the turbine is assumed to be adiabatic and reversible, the flow through it is isentropic. That is,

$$\Delta s = s_4 - s_3 = 0$$

But

$$s_3 = 6.663 \text{ kJ/kg °K}$$

Therefore,

$$s_4 = 6.663 \text{ kJ/kg °K}$$

However, we also know that $P_4 = 1$ Bar. Since we know two properties of the steam leaving the turbine, we can determine all the remaining properties. In particular we find (on the Mollier diagram, Appendix A)

$$h_4 = 2404 \text{ kJ/kg}$$

Again the energy balance around the turbine gives

$$q - w = \Delta h$$

Since the turbine is adiabatic, $q = 0$, and the energy balance reduces to

$$-w = \Delta h = h_4 - h_3 = (2404 - 2769) \text{ kJ/kg}$$

or

$$w = 365 \text{ kJ/kg}$$

However, of this, 0.7 kJ/kg is used to run the pump. Thus the net work is

$$w_{\text{net}} = (365 - 0.7) \text{ kJ/kg} = 364 \text{ kJ/kg}$$

and the ratio of net work to thermal energy input (frequently called the thermal efficiency) is given by

$$\eta = \frac{w_{\text{net}}}{q} = \frac{364 \text{ kJ/kg}}{2642 \text{ kJ/kg}} = 0.138$$

so only 13.8 percent of the thermal energy supplied to the boiler is converted to useful work. The remaining 86.2 percent is lost with the exhaust steam.

As is shown in the problem solution, only a fraction of the thermal energy is converted to work; most of it is wasted in the exhaust steam. One might question if it is necessary to discharge the steam at this relatively high energy level. Would it not be possible to recycle the exhaust steam through the engine in an effort to avoid losing this energy?

The recycling of the discharge steam requires two additional considerations. First, since the boiler operates at a higher pressure than the exhaust of the turbine (or whatever expander is used), the pressure of the working fluid must be increased before it returns to the boiler. Thus an expenditure of work is required for this step. This compression work must be subtracted from the work produced by the turbine to calculate the net work obtained from the engine (this loss is partially recovered, because no feedwater pump is required).

Second, from our earlier discussions of reversible processes it is apparent that the net work obtainable from the process will be maximized if both the expander and the compressor (pump) operate reversibly. In addition, let us assume that the pump and expander operate adiabatically; thus the flows through the compressor and expander occur at constant entropy.

Let us now apply energy and entropy balances to each of the three components our system is known to require: (1) the boiler, (2) the expander, and (3) the compressor. Assume steady state, negligible changes in potential and kinetic energies across each unit, and one unit mass of material flowing.

1. The boiler (Fig. 5-5) vaporizes the fluid reversibly at constant temperature, T_H, and pressure.

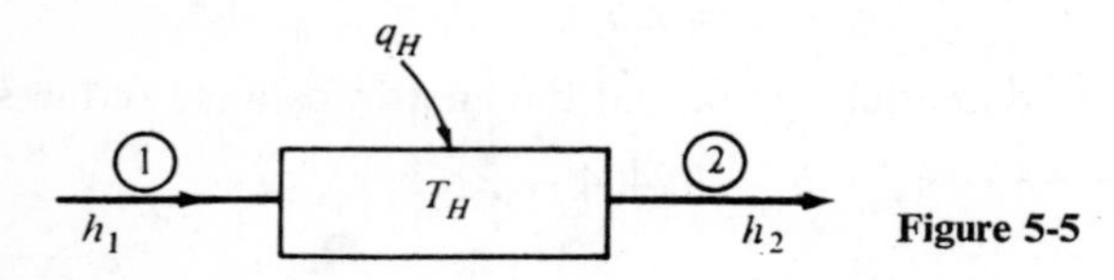

Figure 5-5

Energy equation:

$$q_H = \Delta h = h_2 - h_1$$

Entropy equation:

$$q_H = T_H(\Delta s_H) = T_H(s_2 - s_1)$$

2. The expander (turbine) operates reversibly and adiabatically (Fig. 5-6). Energy equation:

$$-w_T = \Delta h = h_3 - h_2$$

Entropy equation:

$$\Delta s = 0 = s_3 - s_2$$

3. The compressor operates reversibly and adiabatically to return fluid to the boiler (Fig. 5-7). The asterisk indicates the state in which the fluid must enter the pump if it is to be returned to state 1 by the pump.
Energy equation:

$$-w_P = \Delta h = h_1 - h_*$$

Entropy equation:

$$\Delta s = 0 = s_1 - s_*$$

Examination of the three operations shows that the pump can, in theory,

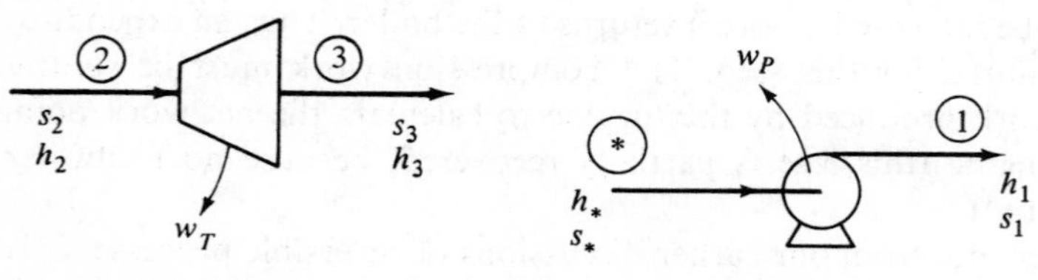

Figure 5-6 **Figure 5-7**

return the working fluid to its original energy level. However, if we examine the three entropy equations, we see that two of the processes do not affect the entropy, while the third process (the boiler) increases the entropy. This increase must be offset by a corresponding decrease elsewhere in the system if the working fluid is to complete the cycle in its original state (a necessary condition for cyclical operation—otherwise the process cannot operate at steady state).

The only method by which the fluid's entropy can be decreased in a closed cycle is by removing energy as heat. Thus it is essential to include a heat-removal step. This heat removal must not be immediately before or after that of the boiler. (If it were, its effect would have to be just opposite that of the boiler, and the fluid leaving the two units would be in the same state as that entering, and the cycle would be useless.) Therefore, the heat removal must occur between the expander and the pump. Since the temperature of the working fluid leaving the expander is less than its entering temperature, heat removal takes place at a lower temperature than heat addition in the boiler.

The low-temperature heat exchanger in this type of cycle is called the *condenser* and will be assumed to operate reversibly, isothermally, and at steady state (Fig. 5-8). It will remove sufficient heat so that the entropy of the exiting fluid is equal to the entropy of fluid entering the boiler.

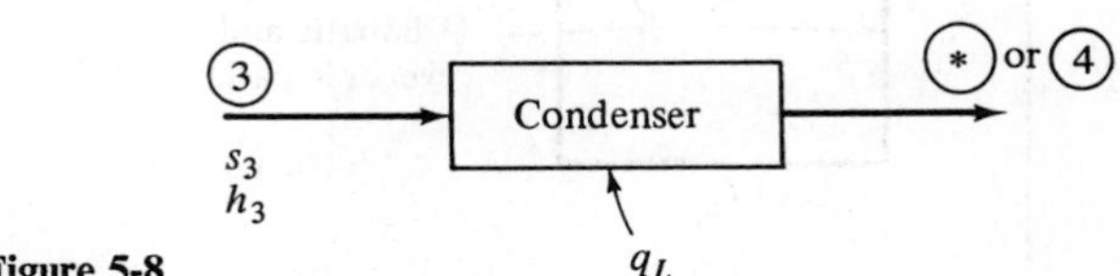

Figure 5-8

Energy equation:

$$q_L = h_4 - h_3$$

Entropy equation:

$$s_4 - s_3 = \frac{q_L}{T_L}$$

T_L refers to the temperature of the fluid at which heat is removed in the condenser. We may picture the completed cycle as shown in Fig. 5-9.

The idealized cycle we have developed, in which all processes occur reversibly and which contains isothermal heat absorption and rejection steps and isentropic expansion and compression steps, is termed a *Carnot cycle*. On a temperature–entropy diagram the fluid follows a rectangular path as shown in Fig. 5-10. As we shall demonstrate shortly, the Carnot cycle is the most efficient (efficiency $= W_{net}/Q_H$) cycle that can operate between any two given temperatures T_H and T_L.

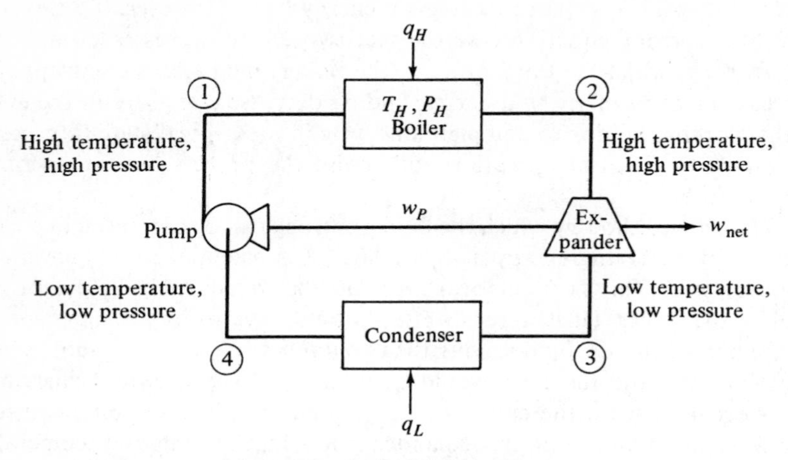

Figure 5-9. Cyclical heat engine.

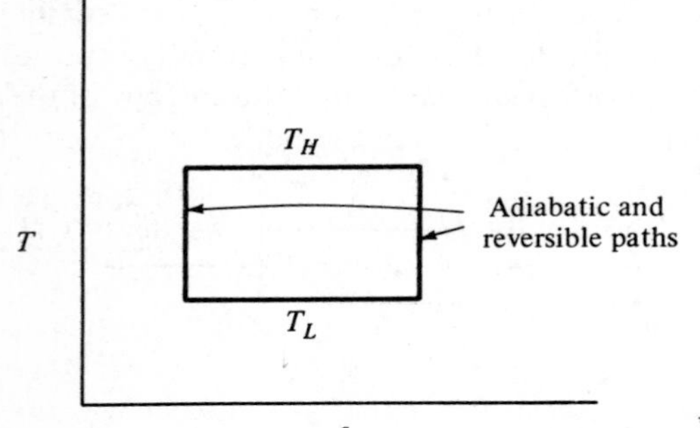

Figure 5-10. Carnot Cycle.

A practical method of achieving such a cycle is to operate within the two-phase (vapor–liquid) region as shown in Fig. 5-11. Since the operation between states 3 and 4 corresponds to a partial condensation of the working fluid, we can see why this heat exchanger is commonly termed the "condenser." This cyclic process has the ability to absorb heat from a high-temperature source and convert a portion of this energy to work, provided there is a low-temperature heat sink to which the process can discard heat.

Up to this point no mention has been made of the fluid that is to circulate in our Carnot cycle. Obviously in any real device such a consideration will play an important role. However, for the present let us assume that from the many possible working fluids, we can find a suitable choice for whatever temperature range is of interest.

The energy equation applied to the Carnot cycle as a whole reduces to the relationship

$$w_{net} = q_H + q_L \tag{5-34}$$

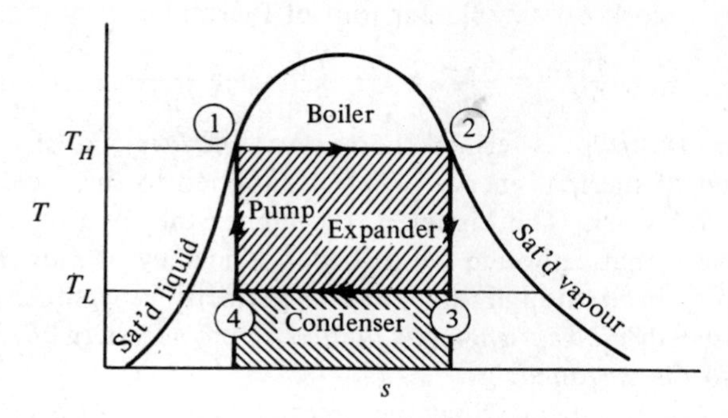

Figure 5-11

where w_{net} is the net work $(w_{turb} + w_{pump})$ supplied by the cycle per unit mass of fluid circulating. q_H and q_L are the thermal-energy transfers at the high and low temperatures, respectively, per unit mass of fluid circulating. The cycle (a closed system by definition) is assumed to be operating reversibly and in the steady state.

Since the entropy equation reduces to $\delta q = T\,ds$ for a heat exchanger operating under these conditions, application of the entropy equation to the boiler and the condenser produces

$$q_H = T_H \, \Delta s_H \tag{5-35}$$

$$q_L = T_L \, \Delta s_L \tag{5-36}$$

Thus it is seen that the areas under curves 1-2 and 3-4 in Fig. 5-11 represent q_H and q_L, respectively. The area under curve 1-2 is positive; the area under curve 3-4 is negative, because $\Delta s_{3\text{-}4}$ is negative. Addition of the two quantities, as suggested by equation (5-34), yields w_{net}. Graphically this results in a cancellation of the area beneath curve 3-4 and shows that w_{net} for a Carnot engine is represented by the enclosed area of the cycle.

Application of the entropy equation to the cycle then yields

$$\frac{q_H}{T_H} + \frac{q_L}{T_L} = 0 \tag{5-37}$$

or

$$q_L = -\frac{T_L}{T_H} q_H \tag{5-38}$$

which may be su[illegible]tuted into the energy equation, equation (5-34), to give

$$w_{net} = q_H - \frac{T_L}{T_H} q_H = q_H\left(1 - \frac{T_L}{T_H}\right) \tag{5-39}$$

where the subscript "net" indicates that this is the actual work that can be provided by the cycle to the surroundings. Equation (5-39) can be rearranged to give the net work obtainable per unit of thermal-energy input:

$$\frac{W_{\text{net}}}{Q_H} = \frac{w_{\text{net}}}{q_H} = 1 - \frac{T_L}{T_H} = \frac{T_H - T_L}{T_H} \tag{5-40}$$

The ratio (W_{net}/Q_H) is termed the *thermal efficiency*, η, of the cycle, and is the fraction of thermal energy which is supplied to the cycle that is converted to useful work. The higher η, the higher the fraction of Q_H that is converted. Note that although we call η an efficiency, it *must* have a value less than unity even though the cycle is perfectly reversible. *The thermal efficiency of any cyclic heat engine is limited by the necessity of rejecting thermal energy to the surroundings.*

Equation (5-40) indicates that the thermal efficiency of a Carnot cycle is a function solely of the absolute temperature levels at which the cycle absorbs and rejects heat. Complete conversion of heat to work is attainable only if the cycle absorbs heat at an infinite temperature or discards it at an absolute temperature of zero. Clearly neither of these conditions is attainable in any practical cycle, so thermal efficiencies of unity are not possible. Ordinarily, the low temperature is limited to ambient temperatures or to temperature levels at which large amounts of cooling capacity are available. Thus we find that power plants are usually built near rivers, lakes, or the ocean where water acts as a convenient low-temperature sink.

The maximum temperature of a cycle is generally determined by our ability to handle the fluid at the temperature involved. When water is used as the working fluid, the upper temperature is partly limited by the vapor pressure of the water. With other fluids, such as liquid metals, vapor pressure is not a serious problem, but the limitation frequently is one of finding materials that can withstand the corrosive effects of the liquid metals at high temperatures. Thus we find that it is generally impossible to fully utilize the high temperatures at which thermal energy is often available. Flame temperatures frequently exceed 2000°F (2460°R), but steam temperatures in boilers seldom exceed 1000°F. Nuclear-reactor operating temperatures are limited only by our ability to contain the fission process. (However, this limitation is normally more restrictive than the maximum boiler temperatures cited above, and thus nuclear reactors usually operate at maximum temperatures lower than conventional fossil-fuel units.)

SAMPLE PROBLEM 5-7. Preliminary cost estimates for a 1 million-kW nuclear-powered thermal power station require an estimate of the cooling-water demands of the facility. For the purposes of this estimate, the heat engine will be approximated by a Carnot cycle operating between 650°F and 150°F. If the maximum temperature rise in the cooling water is restricted to 50°F, at what rate (gal/min) must the cooling water be supplied?

Solution: The net power produced by the plant is to be 1 million kW = 10^9 W = 5.69×10^7 Btu/min. Therefore, $\dot{W}_{\text{net}} = 5.69 \times 10^7$ Btu/min, where the dot over the symbol indicates *per unit time*. The thermal efficiency of this cycle is given by

$$\eta = \frac{T_H - T_L}{T_H} = \frac{1110°\text{R} - 610°\text{R}}{1110°\text{R}} = \frac{500°\text{R}}{1110°\text{R}} = 0.45$$

but the thermal efficiency is also

$$\eta = \frac{\dot{W}_{\text{net}}}{\dot{Q}_H}$$

Therefore,

$$\dot{Q}_H = \frac{\dot{W}_{\text{net}}}{\eta} = \frac{5.69 \times 10^7}{0.45}\ \text{Btu/min} = 1.26 \times 10^8\ \text{Btu/min}$$

The heat rejected to the cooling water is obtained from the overall energy balance:

$$\dot{W}_{\text{net}} = \dot{Q}_H + \dot{Q}_L$$

Therefore,

$$\dot{Q}_L = -(\dot{Q}_H - \dot{W}_{\text{net}}) = -6.9 \times 10^7\ \text{Btu/min}$$

Since all this heat must be transferred to the cooling water,

$$\dot{Q}_{\text{cw}} = -\dot{Q}_L = 6.9 \times 10^7\ \text{Btu/min}$$

An energy balance around the cooling water gives

$$\dot{Q}_{\text{cw}} = \dot{M}_{\text{cw}}\,\Delta h_{\text{cw}} = \dot{M}_{\text{cw}}(C_P\,\Delta T)_{\text{cw}}$$

but C_P for $H_2O = 1.0$ Btu/lb$_m$ °F, $\Delta T_{\text{cw}} = 50$°F. Therefore, the flow rate of cooling water is given by

$$\dot{M}_{\text{cw}} = \frac{6.9 \times 10^7\ \text{Btu/min}}{50.0\ \text{Btu/lb}_m} = 1.38 \times 10^6\ \text{lb}_m\text{/min} = 1.66 \times 10^5\ \text{gal/min}$$

That is, 166,000 gallons of cooling water are needed every minute. On a daily basis, this becomes 240 million gallons per day. As a basis of comparison, a small city of about 10,000 normally consumes approximately 1 million gallons per day of drinking water! Consider the effect of raising the temperature of this flow rate of water 50°F on the receiving body of water. This effect is termed "thermal pollution" and can pose a serious problem for smaller bodies of water. Hence the use of cooling towers with many large power plants.

5.6 The Second Law of Thermodynamics

In our earlier discussions of entropy, we observed that spontaneously occurring processes always proceeded in a certain predictable direction. A cup of hot coffee placed in a cool room always tended to cool to the temperature of the room. Once it was at room temperature, we observed that the cup of coffee would never return spontaneously to its original hot state. These observations concerning the one-wayedness of naturally occurring processes were embodied in our statement that all spontaneous processes proceed in such a way that the entropy of the universe never decreases. The consequences of this statement are usually termed the *second law of thermodynamics*. Since our statement that the entropy of the universe never decreases is rather broad, its consequences are also far-reaching.

Although many statements of the second law have been proposed, most of these can be broadly classed as variations of two basic premises—that of Kelvin and Planck, and that of Clausius. Let us examine these statements:

1. The Kelvin–Planck statement: It is impossible to construct a cyclic process whose only effect is to absorb heat at a single temperature and convert it to an equivalent amount of work.
2. The Clausius statement: It is impossible to construct a cyclic process whose only effect is to transfer heat from a lower temperature to a higher one.

We have seen in our development of the Carnot cycle that the Kelvin–Planck statement of the second law is accurate. Although the Carnot cycle is by no means the only conceivable cycle, no other cycle has ever been devised which violates the Kelvin–Planck statement or which outperforms the Carnot cycle. Therefore, we have been unable to disprove it, and it appears that we must live within its consequences. Once the Kelvin–Planck statement is accepted, we may "prove" many other statements and corollaries of the second law.

The Clausius statement of the second law may be proved by examining the cycle shown in Fig. 5-12. Cycle 1 is a proposed cycle which violates the Clausius statement by being able to transfer a unit of heat from the low-temperature sink to the high-temperature sink without receiving work. Let the Carnot engine absorb a unit of heat from the high-temperature reservoir, convert a fraction of it to work, and reject the remainder to the low-temperature sink. Cycle 1 is now allowed to take the amount of heat rejected and transfer it back to the high-temperature reservoir, so the net effect of the two cycles is to absorb some heat from the hot reservoir and completely convert

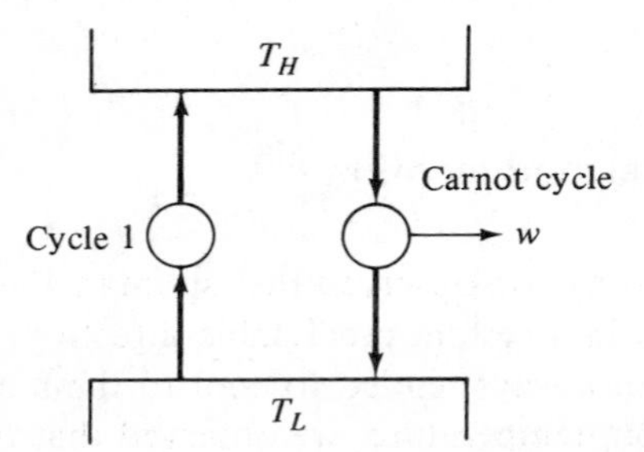

Figure 5-12. Clausius statement.

it to work—a feat that violates the Kelvin–Planck statement. Thus the scheme cannot operate as proposed. Since we know that the Carnot engine is theoretically possible, it must not be possible to construct a cycle that operates in the fashion proposed for cycle 1. Thus the Clausius statement must be true.

Using reasoning similar to that we used in the proof of the Clausius statement, we now demonstrate that the Carnot cycle is the most efficient cycle that can operate between any two thermal sinks. Before proceeding let us observe that the Carnot engine is capable of operating in reverse, such that it behaves as a heat pump rather than a heat engine. When operated in reverse, the Carnot heat pump consumes exactly the same amount of work to return a unit of heat to the hot reservior as a Carnot heat engine produces when it absorbs this amount of heat. Thus if we place a Carnot heat engine and a heat pump back to back, as shown in Fig. 5-13, no net effect is produced on the two sinks.

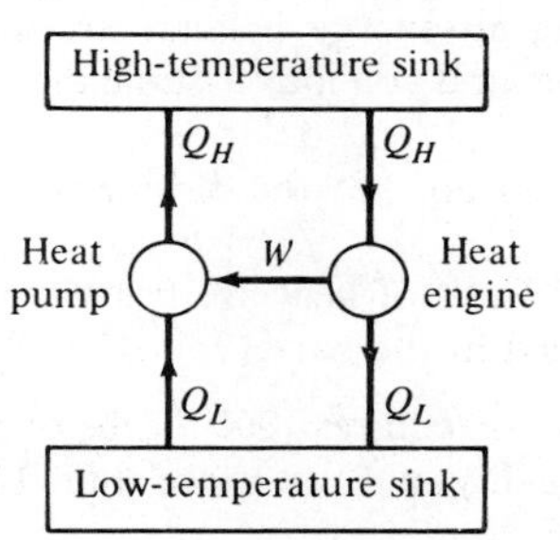

Figure 5-13. Reversible Carnot engines.

Now let us assume that some heat engine which is more efficient than the Carnot engine is substituted for the Carnot engine. Since this engine is assumed to be more efficient than the Carnot engine, it produces more work than the Carnot engine. Therefore, it provides more work than the Carnot heat pump needs to return Q_L to the high-temperature sink. Thus the net effect of the two units is to absorb heat from the high-temperature sink and completely convert it to work, a feat that again violates the Kelvin–Planck statement of the second law. Thus it must not be possible to construct a heat engine which is more efficient than a Carnot engine, and the proposition is demonstrated.

In an analogous fashion we may show that (1) a reversible cycle is always more efficient than an irreversible cycle operating between the same temperatures, and (2) a Carnot heat pump requires no more work than any other cycle to transfer a given amount of heat from a low to a high temperature.

Problems

5-1. A friend claims to have invented a flow device to increase the superheat of steam, and he solicits your financial backing. He is secretive about details but boasts that he can feed steam at 1.4 Bar and 120°C and obtain steam at 322°C and 1 bar. The device also yields liquid water at 100°C and 1 Bar. It receives no additional heat or work from the surroundings, but heat losses may be anticipated. The ratio of product steam to product water is 10:1.

(a) Will you invest your money in the project? Justify the soundness of your conclusion, on thermodynamic grounds.

(b) Describe how the device *may* operate, that is, what kind of equipment would be needed inside the box?

(c) If the ratio of product steam to product water were increased, how would this affect the operation? What, if any, are the thermodynamic limitations to such an increase?

5-2. Steam is supplied at 490 psia and 780°F to a well-insulated reversible turbine. The turbine exhausts at 40 psia, the exhaust steam going directly to the heating coils of an evaporator where the steam is condensed. The liquid condensate from the heating coils is trapped (that is, put through a trap which allows only liquid to pass) at 40 psia and flows from the trap into an open barrel, where it is weighed for the purpose of making an energy balance around the evaporator. The atmospheric pressure is 14.7 psia, and you may assume that the condensate leaves the evaporator at 212°F.

(a) Find the work done by the turbine per pound of liquid water weighed in the barrel.

(b) How many Btu's of heat are transferred in the heating coils per pound of liquid water weighed in the barrel?

5-3. A certain process requires 1000 lb_m/hr of process steam at 20 psia with not less than 96 percent quality and not more than 12°F superheat. Steam is available at 260 psia and 500°F.

(a) It has been suggested that the exhaust steam from a turbine operating from the available steam supply be utilized for the purpose. What maximum horsepower would be available from such a turbine if the heat losses from the turbine were 5000 Btu/hr?

(b) The nature of the process makes it necessary for the process steam to be available when the turbine is down for service. It has been suggested that an alternative source of process steam could be obtained by throttling the available steam supply (260 psia and 500°F) through an adiabatic throttling valve to the required pressure and then cooling to the required condition of quality or superheat. What minimum amount of heat would have to be removed from the throttled steam (Btu/hr) to attain the required conditions for the process steam?

(c) If the expansion in (a) was conducted adiabatically and reversibly, how much superheat must the steam have originally to ensure no liquid in the turbine exhaust at 20 psia if the original pressure is 260 psia?

5-4. A turbine is supplied with 630 g/sec of steam at 61 Bar and 438°C. The exhaust pressure is 5 Bar. A sample of the exhaust steam is passed through an adiabatic throttling calorimeter, where it expands to atmospheric pressure and a temperature of 115°C. Heat losses from the turbine are estimated to be 41.2×10^3 J/sec. How much work is being done by the turbine? What is the quality of the steam exhausting from the turbine? Show whether there is any lost work in the turbine.

5-5. The two cylinders in Fig. P5-5 are in a well-insulated box having a removable insulating partition C. Both cylinders are fitted with heavy, frictionless pistons of 1 ft^2 cross-sectional area each. Space A contains 1 lb_m of saturated water vapor at 500 psia and 467°F. Space B contains 50 lb_m of saturated liquid water at 50 psia and 281°F. The partition C is withdrawn and later replaced when the volume in space A is one half its original value. Assuming negligible heat capacity of the materials of construction, find Q, W, ΔU, and ΔS for:

(a) The vapor in space A as the system.

(b) The liquid in space B as the system.

(c) Both pistons and the fluids in both cylinders as the system.

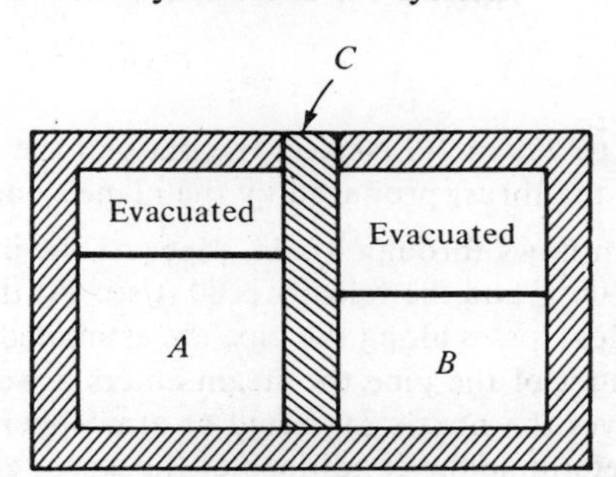

Figure P5-5

5-6. Maneuvering of an orbiting space capsule is accomplished by use of small thruster rockets mounted on the side of the capsule. In a typical design liquid hydrogen and oxygen are fed to a combustion chamber, where they combine to form hot steam. The steam is then expanded in a nozzle and produces the required thrust. A steam rocket of this type, in which the steam enters the nozzle at 700 psia and 800°F, is being designed. During ground tests the nozzle will discharge its exhaust steam at 15 psia.

(a) What is the exit velocity of the steam?

(b) If the nozzle has an exit cross-sectional area of 1 in.^{2}, what is the volumetric flow of steam into the atmosphere (in ft^2/sec)?

(c) What is the mass flow rate of steam through the nozzle (in lb_m/sec)?

You may make the following assumptions:

(1) The flow of steam through the nozzle is adiabatic and reversible (this is confirmed by experiment).

(2) The velocity of the steam entering the nozzle is negligible in comparison to the velocity of the exiting steam.

(3) The flow within the nozzle is one-dimensional, so you need use only *one* velocity component in the energy balance.

5-7. Airplanes are launched from aircraft carriers by means of a steam catapult as shown in Fig. P5-7. The catapult is a well-insulated cylinder that contains steam and is fitted with a frictionless piston. The piston is connected to the airplane by a cable. As the steam expands, the movement of the piston causes movement of the plane. A catapult design calls for 600 lb_m of steam at 2000 psia and 800°F to be expanded to 50 psia. Will this catapult be adequate to accelerate a 30-ton fighter aircraft

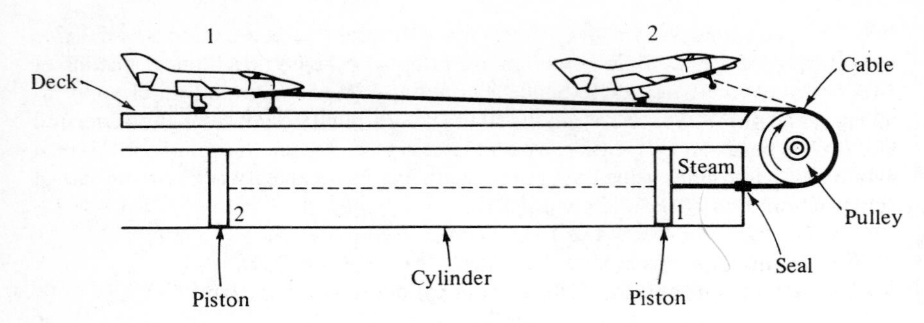

Figure P5-7

from rest to 200 mph? Neglect the mass of the piston head and connecting cables as well as the thrust produced by the plane's engines.

5-8. Steam flows through a 1-in. pipe. At the inlet of the pipe the steam is at 240 psia and 600°F and the velocity is 80 ft/sec. At the outlet of the pipe, the pressure is 60 psia. Heat losses along the pipe are estimated to be 50 Btu/lb_m of steam flowing. At the outlet of the pipe the steam enters a well-insulated, reversible nozzle. The steam leaves the nozzle saturated at atmospheric pressure.

(a) Determine the conditions of the steam entering the nozzle.

(b) What is the velocity of the steam leaving the nozzle?

Use the same assumptions as in Problem 5-6 regarding flow through the nozzle.

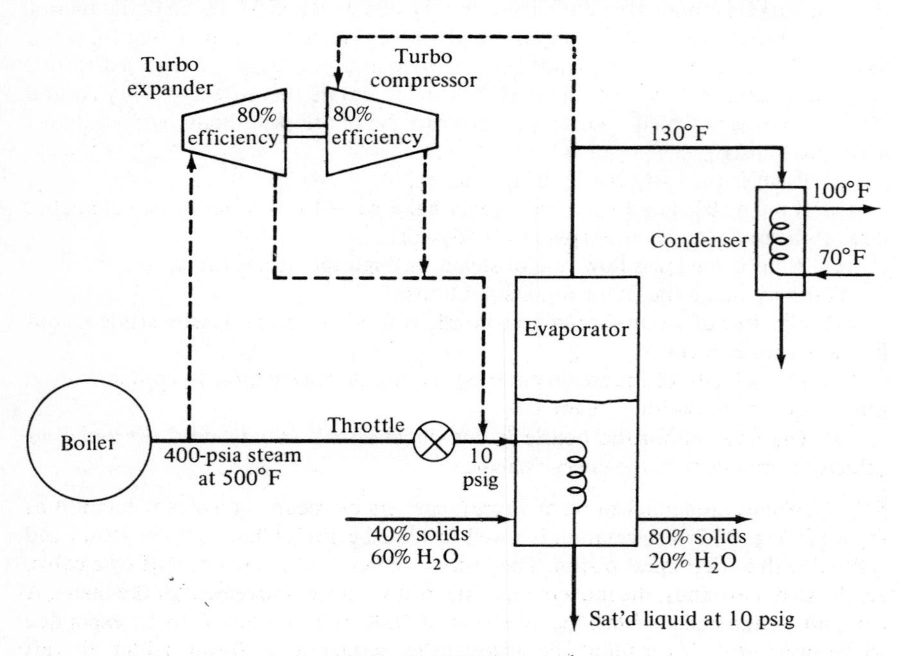

Figure P5-9

5-9. In a certain chemical plant a 40 percent solution of organic salts is being evaporated to an 80 percent solution (all percentages on a mass basis). The organic salts cause only a very small boiling-point rise, which can be considered negligible. The condenser operates in such a manner that the pressure in the vapor line to the condenser is equivalent to a temperature of 130°F. At the present time the only steam available for this operation is that which comes directly from the boiler. Since the tubes in the evaporator cannot withstand high pressure, the boiler steam is throttled to 10 psig before entering the evaporator. A manufacturer of turbo expander–compressors suggests improving the steam economy by installing one of his machines, shown by the dashed lines in Fig. P5-9. Per pound of feed solution handled, calculate the saving in boiler steam which might be made if the turbo expander–compressor operates with an efficiency of 80 percent (based on adiabatic reversible operation) at both ends.

5-10. Low-pressure steam at 100 psia and 400°F is to be used to fill an insulated tank with a movable piston whose resisting force is equivalent to 500 psia. An adiabatic compressor is used to boost the steam pressure from 100 to 500 psia, as shown in Fig. P5-10. If the tank is initially empty and the compressor operates at 70 percent efficiency, what will be the final pressure in the tank when its volume is 10 ft^3, and how much work will have been supplied to the compressor?

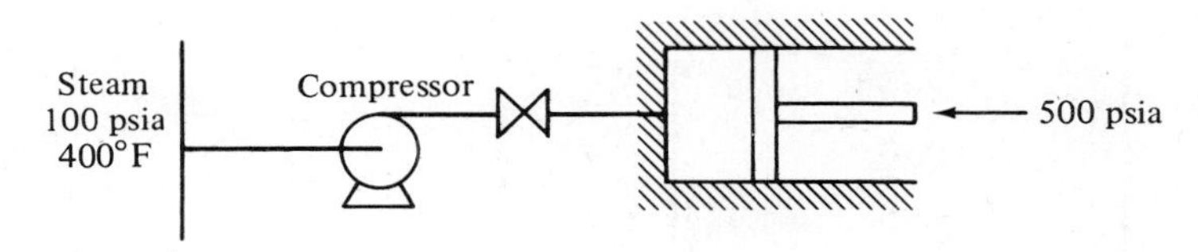

Figure P5-10

5-11. It is desired to determine the unknown volume of a tank (tank B) in a complex chemical process (Fig. P5-11). It is not possible either to measure directly the volume of the tank or to weigh its contents. However, tank B may be evacuated and then connected to a second tank (tank A) of known volume which contains steam at known P and T. (The lines and valve connecting the two tanks are assumed to have negligible volume and to be perfectly insulated.) The valve in the line connecting the two tanks is opened, and the pressures in the tanks are allowed to equilibrate rapidly. The valve is then closed (before any heat transfer can occur) and the pressure in one of the tanks is measured. Explain how you would determine the volume of tank B.

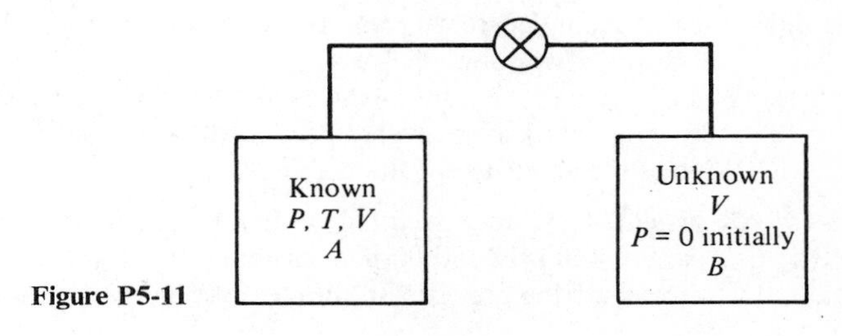

Figure P5-11

5-12. Water is circulated through a nuclear reactor (Fig. P5-12) and leaves at 1500 psia and 500°F (as *liquid*). The water is passed into a flash evaporator that operates at 300 psia. The vapor from the evaporator is used to drive a turbine (adiabatic and reversible). The condenser is operated at 1 psia. Water from the condenser and evaporator are mixed and recirculated. (Assume that all equipment is well insulated and that pressure drops in lines are negligible.) For liquid water at 1500 psia and 500°F, $h = 487.53$ Btu/lb$_m$ and $s = 0.68515$ Btu/lb$_m$ °F. Reference state: $h = s = 0$ for saturated liquid water at 32°F.

(a) How much vapor (*in pounds*) enters the turbine per pound of water leaving the reactor?

(b) How much work is obtained from the turbine per pound of water leaving the reactor?

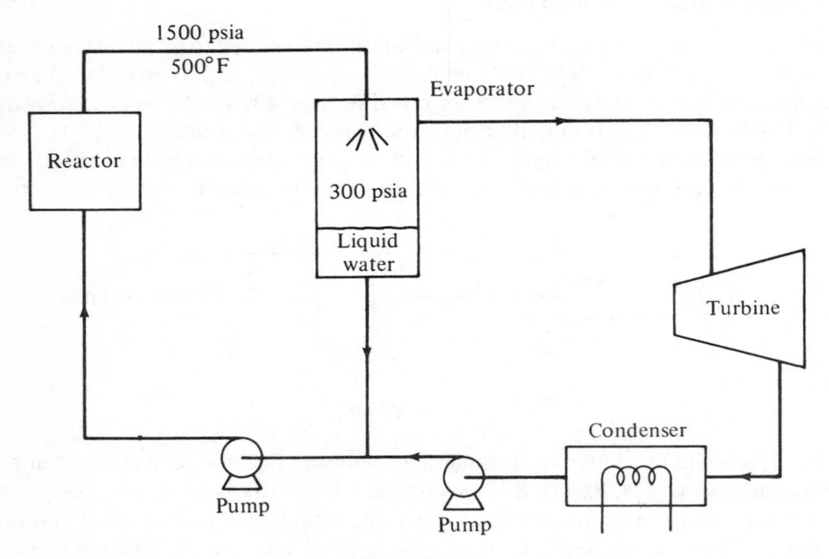

Figure P5-12

5-13. A steam turbine is being tested after installation in a refinery steam plant. Steam at 1000 psia is generated in a boiler and fed to the turbine. Just before the steam enters the turbine, a small portion is bled through a well-insulated throttling valve and expanded to 20.0 psia. The temperature of the throttled steam is found to be 740°F. The unthrottled portion of the steam is passed into the turbine, where it is adiabatically expanded to 20 psia. Tests on the turbine exhaust vapor indicate that it is a saturated vapor.

(a) What is the temperature of the steam entering the turbine?

(b) How much work is recovered per lb$_m$ of steam expanded in the turbine?

(c) What is the efficiency of the turbine?

5-14. An insulated cylinder is fitted with a freely floating piston and contains 1 lb$_m$ of steam at 120 psia and 90-percent quality; the space above the piston contains air to maintain the pressure on the steam (Fig. P5-14). Additional air is forced

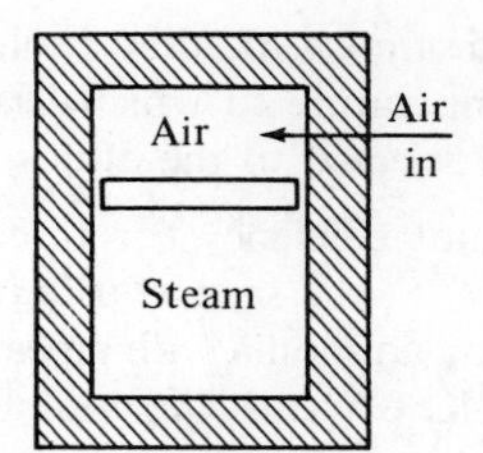

Figure P5-14

into the upper chamber, forcing the piston down and increasing the steam pressure until the steam has 100-percent quality.

(a) Determine the steam pressure at 100-percent quality.

(b) How much work must be done on the steam during the compression?

5-15. The cycle shown in Fig. P5-15 is used to convert thermal energy (in the form of heat transferred to the boiler–superheater) into work. It is assumed that the turbine operates adiabatically and reversibly and that the work consumed by the pump is small, so that $h_e \approx h_d$.

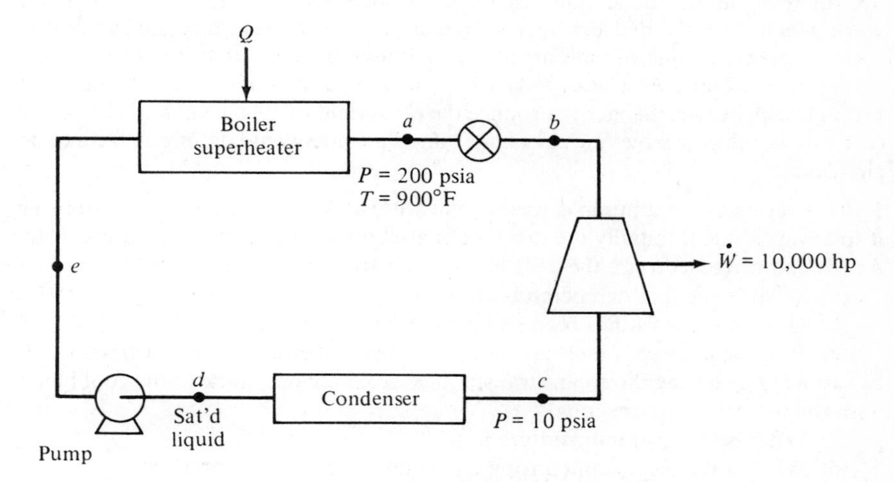

Figure P5-15

(a) The cycle is meant to operate with the valve *open* so that all the properties at point *a* equal those at *b*. (1) Under these conditions, what is the circulation rate (lb_m/sec) of water in the cycle to provide the required power output in the turbine (10,000 hp)? (2) At what rate is heat to be provided in the boiler–superheater?

(b) The valve between *a* and *b* is inadvertently left partially closed and the steam adiabatically throttled to a pressure of 120 psia at *b*. (The properties at *a*, *d*, and *e* are assumed to be unchanged.) (1) What is the water circulation rate needed to provide the required power output in the turbine (10,000 hp)? (2) At what rate is heat to be provided in the boiler–superheater?

5-16. Calculate the maximum work that could be extracted from a geothermal fluid at 200°C if the temperature to which heat can be rejected to ground water is 10°C. If heat must be rejected to the atmosphere at 25°C.

5-17. Calculate the Carnot efficiency for three cycles, one using sodium with a peak temperature of 1800°F, the second using water with a peak temperature of 1000°F and the third using ammonia with a peak temperature of 200°F. Assume all cycles reject heat at 290°K. Contrast the cooling water requirements per megawatt of power produced for the sodium and ammonia cycles if once-through cooling with a 10°F water temperature rise is used.

5-18. Prove that no refrigeration cycle can have a higher coefficient of performance than the reversed Carnot cycle.

5-19. Ocean thermal gradient power plants have been proposed as a means of using the naturally occurring temperature differences in the oceans. At some locations surface temperatures average 20°C whereas the temperature at depths of several thousand feet remain at about −6°C. Calculate the maximum work which a Carnot cycle could deliver per joule of heat addition if operated between these two temperatures. If the surface water experienced a 5°C decrease and the subsurface water a 5°C increase in the boiler and condenser respectively, calculate the maximum work which could be delivered per kilogram of water passing through the boiler. Assume the cyclic fluid in the Carnot engine is ammonia and that it boils and condenses at constant temperatures. At what rate would seawater have to be pumped through each heat exchanger to produce the equivalent of 1000 megawatts of power (the equivalent of a large nuclear plant)? At what rate would ammonia have to be circulated?

5-20. A reversible heat pump is used to extract heat from a cool block and transfer it to a warm block. Initially the two blocks are both at the same temperature, 0°C. At the end of the process, the cool block has a temperature of −200°C.

(a) What is the final temperature of the hot block?

(b) How much work has been supplied to heat pump?

(c) If the heat pump is replaced by a reversible heat engine, how much work can be extracted by letting the engine transfer heat from the hot block to the cool block until the temperatures are equal?

(d) What is the final temperature in (c)?

(e) What is the change in entropy of the universe for each process?

Assume that MC_P for each block = 1.0; all engines, pumps, and heat transfer are reversible.

SIX

the property relations and the mathematics of properties

6.1 The Property Relations

In previous sections we have developed the single-component energy and entropy balances as follows:

Energy:

$$\left(h + \frac{V^2}{2g_c} + \frac{gZ}{g_c}\right)_{\text{in}} \delta M_{\text{in}} - \left(h + \frac{V^2}{2g_c} + \frac{gZ}{g_c}\right)_{\text{out}} \delta M_{\text{out}} + \delta Q - \delta W$$
$$= d\left[M\left(u + \frac{V^2}{2g_c} + \frac{gZ}{g_c}\right)\right]_{\text{sys}} \tag{6-1}$$

Entropy:

$$(s\delta M)_{\text{in}} - (s\delta M)_{\text{out}} + \int_{\text{surface area}} \frac{\delta q \, dA}{T} + \int_{\text{volume}} \frac{\delta lw \, dV}{T} = d(Ms)_{\text{sys}} \tag{6-2}$$

These expressions may be applied to both open and closed systems through proper use of the δM_{in} and δM_{out} terms. However, these equations are frequently inconvenient to use to evaluate changes in the state variables h, u, or s. By combining these equations in such a way as to eliminate the terms that are path functions (δQ, δW, δq, and δlw) we shall describe the changes in state properties of a differential mass as it undergoes a given process in terms of other state functions, such as P, v, T, C_P, and C_V. The differential mass is a closed system so that the energy and entropy equations may be written:

Energy:

$$\delta Q - \delta W = d\left[M\left(u + \frac{V^2}{2g_c} + \frac{gZ}{g_c}\right)\right]_{\text{sys}} \tag{6-3}$$

Entropy:

$$\int_{\text{surface area}} \frac{\delta q \, dA}{T} + \int_{\text{volume}} \frac{\delta lw \, dV}{T} = dS_{\text{sys}} \tag{6-4}$$

For simplicity let us assume that $V^2/2g_c$ and gZ/g_c are negligible in relation to u and therefore may be neglected. (These assumptions are not critical, and these terms can easily be included in everything that follows.) Thus the energy balance reduces to the familiar closed-system form

$$\delta Q - \delta W = dU \tag{6-5}$$

As the state of the system changes, it may transfer work to its surroundings by expansion. Let us choose our system boundaries such that all work which is transferred to the surroundings is transferred as expansion, or $P\,dV$, work; that is, assume that no shaft crosses the boundaries of the system as it flows through the process. (This in no way says that the gas cannot expand against a shaft and do work, but only that the system boundaries be chosen to exclude the shaft from entering the system.) Thus we say that the work transferred to the surroundings is given by

$$\delta W = P_{res}\,dV_{sys} \tag{6-6}$$

Now suppose that due to the effect of friction, the pressure within the system is different from that outside the system. As an example, let us take the piston and the gas confined within a cylinder as our system. Assume that the gas is expanding against the forces of friction and the surroundings as illustrated in Fig. 6-1. P_{sys} is the pressure within the system, P_{res} the effective resisting pressure of the surroundings on the piston, and P_{loss} the effective pressure loss due to friction which equals (frictional force)/(piston area). P_{loss} is assumed to have a plus sign when in the direction of P_{res}.

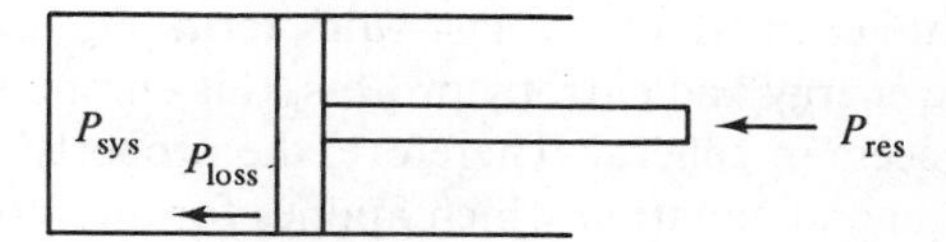

Figure 6-1. Effect of friction.

If the piston is massless, or not accelerating, a simple force balance gives

$$P_{sys} = P_{res} + P_{loss} \tag{6-7}$$

which may be substituted into equation (6-6) for δW to give

$$\delta W = (P_{sys} - P_{loss})\,dV_{sys} \tag{6-8}$$

Since we are considering a differential mass of fluid as our system, we may assume that the temperature is uniform across the system so that equation (6-4) may be rearranged to give

$$\int_{area} \delta \boldsymbol{q}\,dA + \int_{volume} \delta \boldsymbol{lw}\,dV = T_{sys}\,dS_{sys} \tag{6-9}$$

or

$$\delta Q + \delta LW = T_{sys}\,dS_{sys} \tag{6-10}$$

But as we have shown in Chapter 5,

$$\delta LW = P_{loss}\,dV_{sys} \tag{6-11}$$

which may be substituted into equation (6-10) to yield

$$\delta Q = T\,dS - P_{\text{loss}}\,dV_{\text{sys}} \tag{6-12}$$

Substitution of equations (6-8) and (6-12) into (6-5) for the work and heat terms, respectively, gives

$$[T_{\text{sys}}\,dS_{\text{sys}} - P_{\text{loss}}\,dV_{\text{sys}}] - [P_{\text{sys}} - P_{\text{loss}}]\,dV_{\text{sys}} = dU_{\text{sys}} \tag{6-13}$$

or dividing by M, the system mass, and canceling the two $P_{\text{loss}}\,dV_{\text{sys}}$ terms, we obtain

$$du_{\text{sys}} = T_{\text{sys}}\,ds_{\text{sys}} - P_{\text{sys}}\,dv_{\text{sys}} \tag{6-14}$$

Equation (6-14) is the *fundamental property relationship* and represents the third of the three basic equations of thermodynamics. (The energy and entropy balances are the other two.) Its great value lies in the fact that it relates only state functions to each other. For this reason, it is an exact differential equation and has properties that we shall find extremely useful in our work. It is left as an exercise to attempt this analysis for the case where the piston is not part of the system. (What happens to Q in this case?)

If instead of assuming that the system does work on its surroundings we assumed that the surroundings did work on the system, our development would have been identical. Since P_{loss} is positive when in the direction of P_{res}, we would have simply found that P_{loss} was a negative number. Equations (6-8) through (6-14) would not, however, have been affected in the least!

Although we have used the simple piston process to help us visualize the manner in which the lost-work term, $P_{\text{loss}}\,dV_{\text{sys}}$, is eliminated by combining the energy and entropy process, this elimination can be demonstrated for any process in general. Therefore, the property relation, equation (6-14), is itself a general equation which applies for any process in which the only forms of energy change are heat transfer and the transfer of *expansion* work to or from the surroundings. In our later work on the thermodynamics of chemical equilibria, we will see how equation (6-14) may be expanded to take into account the energy changes associated with changes in chemical composition. However, at this point we shall restrict ourselves to single-component, or constant-composition, processes. Similarly, electrical and other forms of work could have been considered and would have led to other work terms similar in form to the $P\,dV$ term in this abbreviated form of the fundamental property relationship.

6.2 The Convenience Functions and Their Property Relations

Although it is entirely possible to express all thermodynamic relations in terms of the fundamental properties u, v, P, s, and T, we find that certain combinations of these fundamental properties occur so frequently that it is to our advantage to define these groupings as new thermodynamic functions.

Since these functions are combinations of state properties, they must also be functions of state themselves.

Since these properties are not fundamental to the study of thermodynamics but only introduced for our convenience, they have been given the name *convenience functions*. We have already introduced the first of the convenience functions, the enthalpy, H. It has been defined as

$$H = U + PV \tag{6-15}$$

We now add two more convenience functions: the *Helmholtz free energy*, A, and the *Gibbs free energy*, or *Gibbs free enthalpy*, G. These are defined as

$$A = U - TS \tag{6-16}$$

$$G = H - TS \tag{6-17}$$

Equations (6-15) through (6-17) may be divided by total mass to yield their specific counterparts:

$$h = u + Pv \tag{6-15a}$$

$$a = u - Ts \tag{6-16a}$$

$$g = h - Ts \tag{6-17a}$$

We may derive property relations for H, A, and G similar to equation (6-14) as follows. Differentiate equation (6-15) to give

$$dH = dU + P\,dV + V\,dP \tag{6-18}$$

Substitute equation (6-14) for dU to give

$$dH = T\,dS + V\,dP \tag{6-19}$$

Similarly, differentiation of equations (6-16) and (6-17) followed by elimination of the dU and dH terms yields

$$dA = -S\,dT - P\,dV \tag{6-20}$$

$$dG = -S\,dT + V\,dP \tag{6-21}$$

which along with the relations for dH and dU,

$$dH = T\,dS + V\,dP \tag{6-19}$$

$$dU = T\,dS - P\,dV \tag{6-14}$$

are the desired property relations for U, H, A, and G.

Mass independent forms of the property relations may be obtained by dividing equations (6-14), (6-19), (6-20), and (6-21) by the total mass to obtain

$$du = T\,ds - P\,dv \tag{6-14a}$$

$$dh = T\,ds + v\,dP \tag{6-19a}$$

$$da = -s\,dT - P\,dv \tag{6-20a}$$

$$dg = -s\,dT + v\,dP \tag{6-21a}$$

These property relations will be of fundamental importance to us in the development of the mathematics of property changes. Since U, H, A, and G are functions of state, equations (6-14) and (6-19) to (6-21) are exact differentials and therefore are susceptible to mathematical manipulations that yield still other useful relations among properties.

6.3 The Maxwell Relations

We begin our discussion and derivation of the Maxwell equations by first discussing some of the useful mathematical properties of the state variables. We have indicated in earlier chapters that any of the intensive state variables, T, u, P, s, h, v, g, and a, can be expressed in terms of any two other state variables. In mathematical terminology we express this fact through an equation of the form

$$F = F(A, B) \tag{6-22}$$

where F, A, and B can be any three intensive state properties and $F(A, B)$ is the functional relation among the three properties. If F, A, and B were, for example, P, v, and T, then $F(A, B)$ would represent an equation of state.

Since F can be expressed solely as a function of A and B, the laws of calculus tell us that

$$dF = \left[\frac{\partial F(A, B)}{\partial A}\right]_B dA + \left[\frac{\partial F(A, B)}{\partial B}\right]_A dB \tag{6-23}$$

or dropping the (A, B) for convenience,

$$dF = \left(\frac{\partial F}{\partial A}\right)_B dA + \left(\frac{\partial F}{\partial B}\right)_A dB \tag{6-23a}$$

Since F is an exact function of A and B, equations (6-23) and (6-23a) are exact differentials. For purposes of simplicity, we now express the partial derivatives in equation (6-23a) as X and Y according to the following relationships:

$$X = \left(\frac{\partial F}{\partial A}\right)_B \qquad Y = \left(\frac{\partial F}{\partial B}\right)_A \tag{6-24}$$

Then

$$dF = X\,dA + Y\,dB \tag{6-25}$$

Conversely, if an equation of the form of equation (6-25) is encountered, X and Y must be the partial derivatives indicated in equation (6-24). For example, consider the du property relation:

$$du = T\,ds - P\,dv$$

We see that T and P are analogous to X and Y in equation (6-25) and thus must be expressible as

$$T = \left(\frac{\partial u}{\partial s}\right)_v$$
$$P = -\left(\frac{\partial u}{\partial v}\right)_s \tag{6-26}$$

Similar expressions for T may be derived from the equation for dH, and for P from dA. Thus we see that T and P themselves are not fundamental thermodynamic properties, but can be considered as convenient representations for the derivatives of the state variables u, s, and v.

However, exceedingly more useful relations than equation (6-26) can be derived from (6-25). If we cross-differentiate the variables in equation (6-24), we obtain

$$\left(\frac{\partial X}{\partial B}\right)_A = \left[\frac{\partial}{\partial B}\left(\frac{\partial F}{\partial A}\right)_B\right]_A$$
$$\left(\frac{\partial Y}{\partial A}\right)_B = \left[\frac{\partial}{\partial A}\left(\frac{\partial F}{\partial B}\right)_A\right]_B \tag{6-27}$$

However, the rules of calculus tell us that the order of differentiation of a state function cannot change the value of the derivatives. That is,

$$\left[\frac{\partial}{\partial B}\left(\frac{\partial F}{\partial A}\right)_B\right]_A \equiv \left[\frac{\partial}{\partial A}\left(\frac{\partial F}{\partial B}\right)_A\right]_B$$

Therefore, we reach the conclusion that

$$\left(\frac{\partial X}{\partial B}\right)_A = \left(\frac{\partial Y}{\partial A}\right)_B \tag{6-28}$$

Thus in the du property relation if we let $F = u$, $X = T$, $Y = -P$, and A and B equal s and v, respectively, equation (6-28) yields

$$-\left(\frac{\partial T}{\partial v}\right)_s = \left(\frac{\partial P}{\partial s}\right)_v \qquad \text{(from } du\text{)} \tag{6-29}$$

Similar expressions derived from the dh, da, and dg property relations yield (it is left to the student as an exercise to derive the relations)

$$-\left(\frac{\partial s}{\partial P}\right)_T = \left(\frac{\partial v}{\partial T}\right)_P \qquad \text{(from } dg\text{)} \tag{6-30}$$

$$\left(\frac{\partial T}{\partial P}\right)_s = \left(\frac{\partial v}{\partial s}\right)_P \qquad \text{(from } dh\text{)} \tag{6-31}$$

$$\left(\frac{\partial s}{\partial v}\right)_T = \left(\frac{\partial P}{\partial T}\right)_v \qquad \text{(from } da\text{)} \tag{6-32}$$

Equations (6-29) to (6-32) and their reciprocal relations,

$$\left(\frac{\partial v}{\partial T}\right)_s = -\left(\frac{\partial s}{\partial P}\right)_v \tag{6-33}$$

$$-\left(\frac{\partial P}{\partial s}\right)_T = \left(\frac{\partial T}{\partial v}\right)_P \tag{6-34}$$

$$\left(\frac{\partial P}{\partial T}\right)_s = \left(\frac{\partial s}{\partial v}\right)_P \tag{6-35}$$

$$\left(\frac{\partial v}{\partial s}\right)_T = \left(\frac{\partial T}{\partial P}\right)_v \tag{6-36}$$

are termed the *Maxwell equations*. The utility of these equations will become evident when we attempt to derive expressions for the evaluation of property changes in terms of the easily measured variables P, T, v, C_P, and C_V, rather than the unwieldy and nonmeasurable quantity entropy. (Mass-dependent forms of the Maxwell equations can be found by multiplying by the total mass, and introducing the total properties.)

6.4 Mathematics of Property Changes

With the derivation of the Maxwell relations we are now ready to consider other property changes. In the following sections we shall establish formulas that allow us to calculate changes in u, h, g, and a solely in terms of P, v, T, C_V, and C_P—variables that can be controlled or measured.

We begin our discussion by developing an expression for du in terms of P, v, T, C_V, and C_P. The property relation tells us that

$$du = T\,ds - P\,dv \tag{6-14a}$$

However, the occurrence of entropy in the property relation makes actual evaluation of du from this equation difficult, because we have no direct means of measuring entropies.

As we showed previously, the property relation is a specific example of the more general form

$$du = \left(\frac{\partial u}{\partial s}\right)_v ds + \left(\frac{\partial u}{\partial v}\right)_s dv \tag{6-37}$$

which implies that u is an exact function of s and v, or

$$u = u(s, v) \tag{6-38}$$

However, we also know that u can be expressed as an exact function of any two intensive state properties. Therefore, consider

$$\begin{aligned} u &= u(P, v) \\ u &= u(T, v) \\ u &= u(T, P) \\ u &= u(s, a) \end{aligned} \tag{6-39}$$

or any other combination of properties. Since we really wish to express u as a function of P, v, and T, it appears that one of the three first alternatives is most likely to yield the desired result. However, experience has shown that one of these alternatives will yield a formulation for du which is easier to evaluate than the other two; this does not mean that correct formulas cannot be developed by starting with the other two alternatives, only that these formulas will not be as useful to us as those derived from the best alternative.

As will be seen shortly, the best form of equation (6-39) is

$$u = u(T, v) \tag{6-39}$$

Thus we may write

$$du = \left(\frac{\partial u}{\partial T}\right)_v dT + \left(\frac{\partial u}{\partial v}\right)_T dv \tag{6-40}$$

However, the heat capacity at constant volume, C_V, has been defined as

$$C_V = \left(\frac{\partial u}{\partial T}\right)_v$$

which may be substituted into equation (6-40) to give

$$du = C_V\, dT + \left(\frac{\partial u}{\partial v}\right)_T dv \tag{6-40a}$$

The $(\partial u/\partial v)_T$ term in equation (6-40a) must be derived from the *du* property relation:

$$du = T\, ds - P\, dv \tag{6-14a}$$

Division of equation (6-14a) by dv at constant T then gives us

$$\left(\frac{\partial u}{\partial v}\right)_T = T\left(\frac{\partial s}{\partial v}\right)_T - P \tag{6-41}$$

Since we wish to replace all terms involving entropy with terms involving only P, v, or T, we use a Maxwell relationship to replace $(\partial s/\partial v)_T$. From equation (6-32) we observe that

$$\left(\frac{\partial s}{\partial v}\right)_T = \left(\frac{\partial P}{\partial T}\right)_v \tag{6-42}$$

Therefore equation (6-41) becomes

$$\left(\frac{\partial u}{\partial v}\right)_T = T\left(\frac{\partial P}{\partial T}\right)_v - P \tag{6-43}$$

Equation (6-43) is substituted into (6-40) to give

$$du = C_V\, dT + \left[T\left(\frac{\partial P}{\partial T}\right)_v - P\right] dv \tag{6-44}$$

Equation (6-44) is the desired relation for *du* in terms of the directly measurable properties, C_V, P, T, and v.

Had we attempted to determine *du* in terms of *dP* and *dT*, or *dP* and *dv*, rather than *dv* and *dT*, we would have found that the expression for *du* was more complicated than equation (6-44) and therefore more difficult to use. Since in the long run we shall want to integrate equation (6-44) to obtain Δu, any of the three expressions for *du* could be used (knowledge of any two of the three properties P, v, or T fixes the third through the equation of state), but clearly the simplest expression is the most desirable one.

SAMPLE PROBLEM 6-1. Express *dh* as a function of C_P, C_V, P, v, and T.

Solution: As with du, we can express dh in terms of dT and dP or dv, or dv and dP, but experience tells us that dT and dP will yield the simplest expression. The property relation for dh is

$$dh = T\,ds + v\,dP$$

Assume that

$$h = h(T, P)$$

The total differential for enthalpy may be expressed as

$$dh = \left(\frac{\partial h}{\partial T}\right)_P dT + \left(\frac{\partial h}{\partial P}\right)_T dP$$

But

$$\left(\frac{\partial h}{\partial T}\right)_P = C_P$$

Therefore

$$dh = C_P\,dT + \left(\frac{\partial h}{\partial P}\right)_T dP$$

We can determine $(\partial h/\partial P)_T$ from division of the dh property relation by dP at constant T to yield

$$\left(\frac{\partial h}{\partial P}\right)_T = \left(\frac{\partial s}{\partial P}\right)_T + v$$

From the Maxwell relations we obtain

$$\left(\frac{\partial s}{\partial P}\right)_T = -\left(\frac{\partial v}{\partial T}\right)_P$$

which is substituted into the last equation to yield

$$\left(\frac{\partial h}{\partial P}\right)_T = -T\left(\frac{\partial v}{\partial T}\right)_P + v$$

The desired expression for dH then becomes

$$dh = C_P\,dT + \left[v - T\left(\frac{\partial v}{\partial T}\right)_P\right] dP$$

Evaluation of Entropy Changes

We now turn our attention to developing expressions for changes in entropy in terms of C_P, C_V, P, v, and T. Once these expressions are found, determinations of changes in g and a can be calculated from the corresponding changes in h, u, and s.

Let us find an expression for ds in terms of dT and dv. Let

$$s = s(T, v) \tag{6-45}$$

Therefore,

$$ds = \left(\frac{\partial s}{\partial T}\right)_v dT + \left(\frac{\partial s}{\partial V}\right)_T dv \tag{6-46}$$

Since we have derivatives of entropy we first check the Maxwell relations to see if either of the terms can be eliminated. We find that $(\partial s/\partial v)_T = (\partial P/\partial T)_v$ but can find no expression for $(\partial s/\partial T)_v$. Therefore we must still find a way of evaluating this term. Let us examine the internal-energy property relation:

$$du = T\,ds - P\,dv \tag{6-14a}$$

Division of this expression by dT at constant v yields

$$\left(\frac{\partial u}{\partial T}\right)_v = T\left(\frac{\partial s}{\partial T}\right)_v \tag{6-47}$$

But $(\partial u/\partial T)_v = C_V$. Therefore, we find that

$$\left(\frac{\partial s}{\partial T}\right)_v = \frac{C_V}{T} \tag{6-48}$$

which is the necessary expression for $(\partial s/\partial T)_v$. Substitution of equation (6-48) and the Maxwell relation for $(\partial s/\partial v)_T$ into equation (6-46) yields the desired relation for ds in terms of dv and dT:

$$ds = \frac{C_V}{T}\,dT - \left(\frac{\partial P}{\partial T}\right)_v dv \tag{6-49}$$

SAMPLE PROBLEM 6-2. Find ds in terms of dT and dP.

Solution: Let $s = s(T, P)$. Therefore,

$$ds = \left(\frac{\partial s}{\partial T}\right)_P dT + \left(\frac{\partial s}{\partial P}\right)_T dP$$

but from the Maxwell relation,

$$\left(\frac{\partial s}{\partial P}\right)_T = -\left(\frac{\partial v}{\partial T}\right)_P$$

so

$$ds = \left(\frac{\partial s}{\partial T}\right)_P dT - \left(\frac{\partial v}{\partial T}\right)_P dP$$

We evaluate $(\partial s/\partial T)_P$ from the enthalpy property relation:

$$dh = T\,ds + v\,dP$$

by dividing by dT at constant P. That is,

$$\left(\frac{\partial h}{\partial T}\right)_P = T\left(\frac{\partial s}{\partial T}\right)_P$$

But

$$\left(\frac{\partial h}{\partial T}\right)_P = C_P$$

Therefore,

$$\left(\frac{\partial s}{\partial T}\right)_P = \frac{1}{T}C_P$$

so

$$ds = \frac{C_P}{T}\,dT - \left(\frac{\partial v}{\partial T}\right)_P dP$$

which is the desired relation for ds in terms of dP and dT.

It is interesting to note that the equations previously derived for du and dh in terms of dT and dv, and dT and dP, respectively, may be easily derived from their respective property and entropy relations as follows:

$$du = T\,ds - P\,dv \tag{6-14a}$$

But

$$ds = \frac{C_V}{T}\,dT + \left(\frac{\partial P}{\partial T}\right)_v dv \tag{6-49}$$

Therefore,

$$du = C_V\,dT + \left[T\left(\frac{\partial P}{\partial T}\right)_v - P\right] dv \tag{6-50}$$

Similarly for dh,

$$dh = T\,ds + v\,dP \tag{6-51}$$

But

$$ds = \frac{C_P}{T}\,dT - \left(\frac{\partial v}{\partial T}\right)_P dP \tag{6-52}$$

Therefore,

$$dh = C_P\,dT + \left[v - T\left(\frac{\partial v}{\partial T}\right)_P\right] dP \tag{6-53}$$

which are the relations we previously derived for du and dh.

6.5 Other Useful Expressions

In addition to the expressions for du, dh, and ds previously developed, we shall find use for certain other relations which may be derived using techniques similar to those used for du, dh, and ds.

Joule–Thomson Coefficient

We have previously noted that a fluid which undergoes an adiabatic throttling process at steady state, while experiencing negligible changes in potential energy and kinetic energy, has undergone an isenthalpic (constant-enthalpy) process. Very often this isenthalpic throttling process is used to reduce the temperature of the fluid as part of a refrigeration process. Therefore, it is frequently desirable to know the degree of temperature reduction that can be obtained during the expansion. A quantitative measure of this temperature change is expressed by the *Joule–Thomson coefficient*, μ, which is defined by the relation

$$\mu = \left(\frac{\partial T}{\partial P}\right)_h \tag{6-54}$$

μ may be expressed in terms of C_P, P, v, and T as follows:

$$dh = C_P\, dT + \left[v - T\left(\frac{\partial v}{\partial T}\right)_P\right] dP \tag{6-53}$$

But for an isenthalpic process $dh = 0$; therefore,

$$0 = (C_P\, dT)_h + \left(\left[v - T\left(\frac{\partial v}{\partial T}\right)_P\right] dP\right)_h \tag{6-55}$$

or

$$(C_P\, dT)_h = \left(\left[T\left(\frac{\partial v}{\partial T}\right)_P - v\right] dP\right)_h \tag{6-56}$$

Therefore,

$$\mu = \left(\frac{\partial T}{\partial P}\right)_h = \frac{T(\partial v/\partial T)_P - v}{C_P} \tag{6-57}$$

A similar expression, the *Euken coefficient*, ζ, can be derived for a closed-system process at constant internal energy:

$$\zeta = \left(\frac{\partial T}{\partial v}\right)_u = -\frac{T(\partial P/\partial T)_v - P}{C_V} \tag{6-58}$$

Variations in C_P and C_V With Respect to P and v, Respectively

We may evaluate the derivative $(\partial C_P/\partial P)_T$ as follows. The enthalpy property relation is written

$$dh = C_P\, dT + \left[v - T\left(\frac{\partial v}{\partial T}\right)_P\right] dP \tag{6-53}$$

Since equation (6-53) is an exact differential, we may apply a Maxwell-relation type of cross differentiation to obtain

$$\left(\frac{\partial C_P}{\partial P}\right)_T = \left(\frac{\partial[v - T(\partial v/\partial T)_P]}{\partial T}\right)_P \tag{6-59}$$

or

$$\left(\frac{\partial C_P}{\partial P}\right)_T = \left(\frac{\partial v}{\partial T}\right)_P - \left(\frac{\partial v}{\partial T}\right)_P - T\left(\frac{\partial^2 v}{\partial T^2}\right)_P \tag{6-60}$$

Therefore, the change of C_P with pressure is given by

$$\left(\frac{\partial C_P}{\partial P}\right)_T = -T\left(\frac{\partial^2 v}{\partial T^2}\right)_P \tag{6-61}$$

Similarly, an expression for $(\partial C_V/\partial v)_T$ can be derived from the expression for du as

$$\left(\frac{\partial C_V}{\partial v}\right)_T = T\left(\frac{\partial^2 P}{\partial T^2}\right)_v \tag{6-62}$$

These expressions may be derived alternatively as follows:

$$C_P = T\left(\frac{\partial s}{\partial T}\right)_P$$

Therefore,

$$\left(\frac{\partial C_P}{\partial P}\right)_T = T\frac{\partial}{\partial P}\left[\left(\frac{\partial s}{\partial T}\right)_P\right]_T = T\frac{\partial}{\partial T}\left[\left(\frac{\partial s}{\partial P}\right)_T\right]_P \tag{6-63}$$

but from the Maxwell relations

$$\left(\frac{\partial s}{\partial P}\right)_T = -\left(\frac{\partial v}{\partial T}\right)_P$$

Therefore,

$$\left(\frac{\partial C_P}{\partial P}\right)_T = -\left(\frac{\partial^2 v}{\partial T^2}\right)_P \tag{6-64}$$

which is identical to equation (6-61). The reader should rederive equation (6-62) in the same manner.

General Derivative Formulas

Examination of the eight thermodynamic variables P, v, T, s, u, h, a, and g indicates that there are 168 different partial derivatives (ignoring reciprocals) which involve any three of these variables. For example, $(\partial P/\partial v)_T$, $(\partial h/\partial s)_g$, $(\partial v/\partial a)_s$, . . . can be listed as just a few. As may be imagined, the number of equations relating these partial derivatives is extremely great. However, if we add the additional restriction that all derivatives be expressed in terms of P, v, T, C_P, and C_V, the number is greatly reduced.

We shall now develop techniques for expressing each of these 168 derivatives in terms of P, v, T, C_P, and C_V. We begin by stating without proof several extremely important properties of functions whose differentials are exact. (For proofs of the various properties the reader is referred to any standard textbook on advanced calculus.) Let us assume that F is an exact function of A and B:

$$F = F(A, B) \tag{6-65}$$

Then the following properties of the derivatives of F, A, and B are true:

$$dF = \left(\frac{\partial F}{\partial A}\right)_B dA + \left(\frac{\partial F}{\partial B}\right)_A dB \tag{6-66}$$

$$\left(\frac{\partial A}{\partial B}\right)_F = \frac{1}{(\partial B/\partial A)_F}$$

$$\left(\frac{\partial B}{\partial A}\right)_F = \frac{(\partial B/\partial C)_F}{(\partial A/\partial C)_F} \qquad \text{where } C \text{ is a third property other than } A \text{ or } B$$

$$\left(\frac{\partial A}{\partial B}\right)_F = -\frac{(\partial A/\partial F)_B}{(\partial B/\partial F)_A} = -\frac{(\partial F/\partial B)_A}{(\partial F/\partial A)_B}$$

The last property is frequently called the *triple-product relation* and is extremely useful because it allows us to move an unfavorable constraint (such as s, g, a, h, or u) into the derivatives.

The general technique for reducing a complicated derivative is then as follows:

1. If the derivative contains u, h, a, or g, bring these functions to the numerator of their respective derivatives. Then eliminate the u, h, a, or g via the appropriate property relation. If a u, h, a, or g still appears as a constraint on the derivative, use the triple product to bring it inside the derivatives. The offending quantity is then eliminated via its property relation as illustrated below.

$$\left(\frac{\partial T}{\partial v}\right)_u = -\frac{(\partial T/\partial u)_v}{(\partial v/\partial u)_T} = -\frac{(\partial u/\partial v)_T}{(\partial u/\partial T)_v}$$

Now substitute $du = T\,ds - P\,dv$ to obtain

$$\left(\frac{\partial T}{\partial v}\right)_u = \frac{T(\partial s/\partial v)_T - P(\partial v/\partial v)_T}{T(\partial s/\partial T)_v} = \frac{T(\partial s/\partial v)_T - P}{T(\partial s/\partial T)_v}$$

2. At this point we have eliminated all references to u, h, a, or g from the derivatives. We now eliminate the entropy. First bring the entropy to the numerator of all derivatives (the triple product may be used to eliminate entropy as a constraint). Next eliminate any entropy derivatives that appear in the Maxwell equations. Then eliminate derivatives of entropy with respect to temperature via the definitions of C_P and C_V:

$$C_P = T\left(\frac{\partial s}{\partial T}\right)_P \quad \text{and} \quad C_V = T\left(\frac{\partial s}{\partial T}\right)_v$$

Any entropy derivatives that still exist may be eliminated by introducing $1/dT$ into the numerator and denominator of the derivative (see the third property of general derivatives), and then using the definitions of C_P and/or C_V to eliminate the entropy, as indicated in the following example.

$$\left(\frac{\partial T}{\partial v}\right)_u = \frac{T(\partial s/\partial v)_T - P}{T(\partial s/\partial T)_v}$$

but $(\partial s/\partial v)_T = (\partial P/\partial T)_v$ via the Maxwell relation, and $T(\partial s/\partial T)_v = C_V$. Thus

$$\left(\frac{\partial T}{\partial v}\right)_u = \frac{T(\partial P/\partial T)_v - P}{C_V}$$

SAMPLE PROBLEM 6-3. Evaluate $(\partial h/\partial P)_g$ in terms of P, v, T, C_P, C_V, and absolute entropy, s (no entropy derivatives, however).

Solution: We begin by substituting the dh property relation for the dh term in the numerator:

$$\left(\frac{\partial h}{\partial P}\right)_g = T\left(\frac{\partial s}{\partial P}\right)_g + V\left(\frac{\partial P}{\partial P}\right)_g = T\left(\frac{\partial s}{\partial P}\right)_g + v$$

Now bring g into the derivatives by application of the triple-product expansion:

$$\left(\frac{\partial h}{\partial P}\right)_g = -T\frac{\left(\frac{\partial g}{\partial P}\right)_s}{\left(\frac{\partial g}{\partial s}\right)_P} + v$$

Now eliminate g with the dg property relation:

$$\left(\frac{\partial h}{\partial P}\right)_g = -T\left[\frac{-s\left(\frac{\partial T}{\partial P}\right)_s + v}{\left(-s\frac{\partial T}{\partial s}\right)_P + v\left(\frac{\partial P}{\partial s}\right)_P}\right] + v$$

But $(\partial P/\partial s)_P = 0$, so we get

$$\left(\frac{\partial h}{\partial P}\right)_g = -T\left[\frac{-s\left(\frac{\partial T}{\partial P}\right)_s + v}{-s\left(\frac{\partial T}{\partial s}\right)_P}\right] + v$$

$$= -T\left[\left(\frac{\partial T}{\partial P}\right)_s\left(\frac{\partial s}{\partial T}\right)_P - \frac{v}{s}\left(\frac{\partial s}{\partial T}\right)_P\right] + v$$

Now use the triple product on the first term in parentheses:

$$= T\left[\left(\frac{\partial s}{\partial P}\right)_T + \frac{v}{s}\left(\frac{\partial s}{\partial T}\right)_P\right] + v$$

But $(\partial s/\partial P)_T = -(\partial v/\partial T)_P$ from the Maxwell equations, and

$$\left(\frac{\partial s}{\partial T}\right)_P = \frac{C_P}{T}$$

Thus the derivative is finally expressed as

$$\left(\frac{\partial h}{\partial P}\right)_g = -T\left(\frac{\partial v}{\partial T}\right)_P + v\left(\frac{C_P}{s} + 1\right)$$

The absolute entropy cannot be reduced, so this is the simplest form possible.

Let us now derive one more useful expression—that for the difference between C_P and C_V. We have shown that C_V may be expressed as

$$C_V = T\left(\frac{\partial s}{\partial T}\right)_v$$

However, we also know that

$$ds = \frac{C_P}{T}\,dT - \left(\frac{\partial v}{\partial T}\right)_P dP$$

Thus the expression for C_V becomes

$$C_V = T\left[\frac{C_P}{T}\left(\frac{\partial T}{\partial T}\right)_v - \left(\frac{\partial v}{\partial T}\right)_P\left(\frac{\partial P}{\partial T}\right)_v\right] \qquad (6\text{-}67)$$

which reduces to

$$C_V = C_P - T\left(\frac{\partial v}{\partial T}\right)_P \left(\frac{\partial P}{\partial T}\right)_v \tag{6-68}$$

or

$$C_P - C_V = T\left(\frac{\partial v}{\partial T}\right)_P \left(\frac{\partial P}{\partial T}\right)_v \tag{6-68a}$$

which is the desired relation. Several other identical forms of equation (6-68a) are possible if we employ the triple product to expand one or both of the remaining derivatives. However, these forms obviously convey no new information.

For a gas that obeys the ideal-gas equation of state,

$$\left(\frac{\partial v}{\partial T}\right)_P = \frac{R}{P}; \qquad \left(\frac{\partial P}{\partial T}\right)_v = \frac{R}{v}$$

Thus:

$$C_P - C_V = \frac{TR^2}{Pv} = R \tag{6-69}$$

6.6 Thermodynamic Properties of an Ideal Gas

We have defined an ideal gas as one that obeys the ideal equation of state:

$$Pv = RT$$

Using this equation of state, we may calculate thermodynamic property changes which occur with an ideal gas as follows:

Internal Energy

$$du = C_V\,dT + \left[T\left(\frac{\partial P}{\partial T}\right)_v - P\right] dv \tag{6-50}$$

but from the ideal-gas equation of state,

$$T\left(\frac{\partial P}{\partial T}\right)_v = T\frac{R}{v} = P$$

Therefore, equation (6-50) becomes

$$du = C_V\,dT \tag{6-70a}$$

That is, the internal energy change of an ideal gas is independent of the volume change and may be expressed as a function of temperature alone.

Enthalpy

$$dh = C_P\,dT + \left[v - T\left(\frac{\partial v}{\partial T}\right)_P\right] dP \tag{6-53}$$

But for an ideal gas $T(\partial v/\partial T)_P = v$, so that

$$dh = C_P\, dT \tag{6-70b}$$

Therefore, the enthalpy, as well as the internal energy, of an ideal gas is a function of its temperature only, and not a function of its pressure.

Entropy: T and V Known

$$ds = \frac{C_V}{T}\, dT + \left(\frac{\partial P}{\partial T}\right)_v dv \tag{6-49}$$

But for an ideal gas,

$$\left(\frac{\partial P}{\partial T}\right)_v = \frac{R}{v}$$

Therefore,

$$ds = C_V \frac{dT}{T} + R\frac{dv}{v}$$

or

$$ds = C_V\, d\ln T + R\, d\ln v \tag{6-70c}$$

Entropy: T and P Known

$$ds = \frac{C_P}{T}\, dT - \left(\frac{\partial v}{\partial T}\right)_P dP \tag{6-52}$$

But for an ideal gas,

$$\left(\frac{\partial v}{\partial T}\right)_P = \frac{R}{P}$$

Therefore,

$$ds = C_P \frac{dT}{T} - R\frac{dP}{P}$$

or

$$ds = C_P\, d\ln T - R\, d\ln P \tag{6-70d}$$

Thus we see that changes in u, h, and s can be calculated easily and directly for an ideal gas provided that we know C_P or C_V (remember that $C_P - C_V = R$ for an ideal gas) and the initial and final temperature and pressure. For u and h this simplifies even further to requiring that we know only the initial and final temperatures and C_P or C_V.

Joule–Thomson Coefficient, μ

We have shown that the Joule–Thomson coefficient may be evaluated from

$$\mu = \left(\frac{\partial T}{\partial P}\right)_h = \frac{T(\partial v/\partial T)_P - v}{C_P} \tag{6-57}$$

But we have shown that for an ideal gas

$$T\left(\frac{\partial v}{\partial T}\right)_P - v = 0$$

Therefore, the Joule–Thomson coefficient for an ideal gas is identically equal to zero and indicates that an ideal gas will neither increase nor decrease its temperature during an isenthalpic pressure change.

The Euken Coefficient, ζ

The Euken coefficient is given by

$$\zeta = \left(\frac{\partial T}{\partial v}\right)_u = \frac{T(\partial P/\partial T)_v - P}{C_V} \tag{6-58}$$

However, for an ideal gas

$$T\left(\frac{\partial P}{\partial T}\right)_v - P = 0$$

Therefore, the Euken coefficient is zero, and an ideal gas can neither increase nor decrease in temperature during a constant internal-energy process.

Variation of C_P and C_V with P and v, Respectively

$$\left(\frac{\partial C_P}{\partial P}\right)_T = -T\left(\frac{\partial^2 v}{\partial T^2}\right)_P \tag{6-61}$$

$$\left(\frac{\partial C_V}{\partial v}\right)_T = T\left(\frac{\partial^2 P}{\partial T^2}\right)_v \tag{6-62}$$

But for an ideal gas

$$\left(\frac{\partial^2 v}{\partial T^2}\right)_P = \left(\frac{\partial^2 P}{\partial T^2}\right)_v = 0$$

Therefore, for an ideal gas

$$\left(\frac{\partial C_P}{\partial P}\right)_T = \left(\frac{\partial C_V}{\partial v}\right)_T = 0 \tag{6-70e}$$

Thus, we find for an ideal gas that C_P is independent of P, and C_V is independent of v. Therefore, we can say that C_P and C_V are, at most, functions only of the temperature, and not independent functions of P or v for ideal gases.

If the ideal gas is composed of rigid, noninteracting, monatomic particles, it is possible, by use of statistical mechanical arguments, to show that the heat capacities of the ideal gas are independent of temperature, as well as pressure and volume, and therefore are true constants. (For very high temperatures we must also exclude ionization effects.) However, if the ideal gas is composed of real molecules (for example, oxygen at room temperature and 1 atm) which are polyatomic, then the heat capacities will not be independent of temperature and are not true constants.

SAMPLE PROBLEM 6.4. Demonstrate that the temperature scale defined by the ideal–gas thermometer, and that defined by

$$T = \left(\frac{\partial U}{\partial V}\right)_S$$

(the so-called "thermodynamic temperature scale") are equivalent.

Solution:

The fact that the ideal-gas temperature scale is identical to the thermodynamic temperature scale may be shown as follows. For the purpose of distinguishing between the ideal and thermodynamic temperatures in the subsequent discussion, the ideal gas temperature will be represented by t and T will continue to be used for the thermodynamic temperature. Let us begin with the property relationship based on the thermodynamic temperature:

$$dU = T\,dS - P\,dV$$

which is divided by dV at constant T to yield

$$\left(\frac{\partial U}{\partial V}\right)_T = T\left(\frac{\partial S}{\partial V}\right)_T - P$$

The entropy derivative is then eliminated by using a Maxwell relationship:

$$\left(\frac{\partial S}{\partial V}\right)_T = \left(\frac{\partial P}{\partial T}\right)_V$$

to give

$$\left(\frac{\partial U}{\partial V}\right)_T = T\left(\frac{\partial P}{\partial T}\right)_V - P$$

If it is assumed that t varies monotonically with T such that for a given value of T, t remains constant, $(\partial U/\partial V)_T = (\partial U/\partial V)_t$, we can rewrite the above equation as

$$\left(\frac{\partial U}{\partial V}\right)_t = T\left(\frac{\partial P}{\partial T}\right)_V - P$$

For a gas which obeys the ideal gas equation of state, the kinetic theory indicates $(\partial U/\partial V)_t = 0$. Thus we obtain

$$0 = T\left(\frac{\partial P}{\partial T}\right)_V - P$$

The derivative term may then be written in terms of t as

$$T\left(\frac{\partial P}{\partial T}\right)_V = \frac{T}{t}\left(\frac{\partial t}{\partial T}\right)_V t\left(\frac{\partial P}{\partial t}\right)_V$$

But from the equation of state,

$$t\left(\frac{\partial P}{\partial t}\right)_V = \frac{nRt}{V} = P$$

Thus, the expression for $(\partial U/\partial V)_t$ reduces to

$$P\left[\frac{T}{t}\left(\frac{\partial t}{\partial T}\right)_V - 1\right] = 0$$

which may be rearranged to

$$\frac{\partial t}{t} = \frac{\partial T}{T}$$

or

$$\partial \ln t = \partial \ln T$$

This equation is then integrated to give

$$\ln T = \ln t + \ln k$$

or

$$T = kt$$

where k is a constant of integration. Thus the ideal gas and thermodynamic temperature scales vary by at most a multiplicative constant, and hence are equivalent.

6.7 Evaluation of Changes in *u, h,* and *s* for Ideal Gases and Various Processes

We shall now examine the techniques for integratingthe equations for du, dh, and ds. These integrations will allow us to evaluate the changes in u, h, s, g, and a which occur when a substance undergoes a given process. Since the equations for du, dh, and ds take their simplest form when applied to an ideal gas, we shall study this case first. Then we shall examine the methods used to integrate more complex equations generated when the working fluid is not an ideal gas.

In Section 6.6 we showed that for an ideal gas du, dh, and ds could be expressed as

$$du = C_V\, dT \qquad \text{(6-70a)}$$

$$dh = C_P\, dT \qquad \text{(6-70b)}$$

$$ds = C_V\, d\ln T + R\, d\ln v \qquad \text{(6-70c)}$$

$$ds = C_P\, d\ln T - R\, d\ln P \qquad \text{(6-70d)}$$

For an ideal gas whose C_P and C_V are also independent of T; C_P and C_V are true constants, and these four relations may be integrated directly to yield

$$\int_{u_1}^{u_2} du = \Delta u = u_2 - u_1 = \int_{T_1}^{T_2} C_V\, dT = C_V(T_2 - T_1) \qquad \text{(6-71a)}$$

$$\int_{h_1}^{h_2} dh = \Delta h = h_2 - h_1 = \int_{T_1}^{T_2} C_P\, dT = C_P(T_2 - T_1) \qquad \text{(6-71b)}$$

$$\int_{s_1}^{s_2} ds = \Delta s = s_2 - s_1 = \int_{T_1}^{T_2} C_V \, d\ln T + \int_{v_1}^{v_2} R \, d\ln v$$

$$= C_V \ln \frac{T_2}{T_1} + R \ln \frac{v_2}{v_1} \tag{6-71c}$$

$$\int_{s_1}^{s_2} ds = \Delta s = s_2 - s_1 = \int_{T_1}^{T_2} C_P \frac{dT}{T} - \int_{P_1}^{P_2} R \, d\ln P$$

$$= C_P \ln \frac{T_2}{T_1} - R \ln \frac{P_2}{P_1} \tag{6-71d}$$

Thus we can evaluate the change in any property directly in terms of P, v, T, C_P and C_V.

As we mentioned earlier, almost all real gases at room (or higher) temperatures, and relatively low pressures—less than 5 atm—obey the ideal gas equation of state. However, for many of these gases the heat capacities may vary appreciably with temperature (although not at all with P or v), and therefore may not be considered constants. Heat capacities at low pressures are usually distinguished from all others by the superscript*, that is, C_P^* and C_V^*. Usually $C_P^*(T)$ and $C_V^*(T)$ are presented in the form of a polynomial expansion in (T):

$$\begin{aligned} C_P^*(T) &= A + BT + CT^2 + \cdots \\ C_V^*(T) &= A' + B'T + C'T^2 + \cdots \end{aligned} \tag{6-72}$$

At all temperatures, however,

$$C_P^*(T) - C_V^*(T) = R$$

which implies that $B' = B$, $C' = C$, and $A - A' = R$.

SAMPLE PROBLEM 6-5. Derive an expression for Δu, Δh, and Δs for a gas at low pressure whose heat capacities are given by

$$C_P^* = A + BT + CT^2$$

$$C_V^* = (A - R) + BT + CT^2$$

Solution: Assume that the gas obeys the ideal-gas equation of state, so

$$\Delta u = \int_{T_1}^{T_2} C_V^* \, dT = \int_{T_1}^{T_2} (A - R + BT + CT^2) \, dT$$

Therefore,

$$\Delta u = (A - R)(T_2 - T_1) + \frac{B}{2}(T_2^2 - T_1^2) + \frac{C}{3}(T_2^3 - T_1^3)$$

$$\Delta h = \int_{T_1}^{T_2} C_P^* \, dT = \int_{T_1}^{T_2} (A + BT + CT^2) \, dT$$

or

$$\Delta h = A(T_2 - T_1) + \frac{B}{2}(T_2^2 - T_1^2) + \frac{C}{3}(T_2^3 - T_1^3)$$

$$\Delta s = \int_{T_1}^{T_2} C_P^* \frac{dT}{T} - \int_{P_1}^{P_2} R\, d\ln P$$

Therefore,

$$\Delta s = \int_{T_1}^{T_2} \left(\frac{A}{T} + B + CT\right) dT - R \ln \frac{P_2}{P_1}$$

or

$$\Delta s = A \ln \frac{T_2}{T_1} + B(T_2 - T_1) + \frac{C}{2}(T_2^2 - T_1^2) - R \ln \frac{P_2}{P_1}$$

Equations (6-71c) and (6-71d) give us some very useful information about the P–v–T relations of an ideal gas undergoing an isentropic process. Since many compressors and expanders operate in nearly an adiabatic and reversible fashion, processes that may be treated as isentropic will be encountered fairly often in our work. For an isentropic process $\Delta s = 0$. Therefore, equations (6-71c) and (6-71d) yield

$$\Delta s = 0 = C_V \ln \frac{T_2}{T_1} + R \ln \frac{v_2}{v_1} \qquad (6\text{-}73)$$

or

$$\Delta s = 0 = C_P \ln \frac{T_2}{T_1} - R \ln \frac{P_2}{P_1} \qquad (6\text{-}74)$$

Therefore,

$$\ln \frac{T_2}{T_1} = -\frac{R}{C_V} \ln \frac{v_2}{v_1} \qquad (6\text{-}75)$$

$$\ln \frac{T_2}{T_1} = \frac{R}{C_P} \ln \frac{P_2}{P_1} \qquad (6\text{-}76)$$

Equations (6-75) and (6-76) allow us to calculate the outlet temperature from an isentropic process if we know the inlet conditions (T and P or v) and the outlet P or v. Once we have determined the outlet temperature T_2, we may use equation (6-70a) or (6-70b) to determine Δu or Δh. Thus it is seen that determination of the outlet temperature is the first step in calculating the work requirement of an isentropic process. As we shall see later, this becomes an extremely complicated and tedious task when we have to deal with nonideal gases.

By equating the right-hand sides of equations (6-75) and (6-76), we can relate the pressure and volume during an isentropic process as follows:

$$-\frac{C_P}{C_V} \ln \frac{v_2}{v_1} = \ln \frac{P_2}{P_1}$$

Let $C_P/C_V = k =$ the ratio of specific heats. Therefore,

$$0 = \ln \frac{P_2}{P_1} + \ln \left(\frac{v_2}{v_1}\right)^k = \ln \frac{P_2 v_2^k}{P_1 v_1^k} \tag{6-77}$$

or

$$P_2 v_2^k = P_1 v_1^k \tag{6-78}$$

which relates the outlet pressure and volume to the inlet conditions.

The Gas Tables and Their Use with Ideal Gases

When the range of temperatures encountered in the process under consideration is large, it may not be reasonable to assume that the heat capacities are independent of temperature. For these cases the heat capacities must remain under the integrals when equations (6-70) are integrated as illustrated in Sample Problem 6-5. Although the resulting expressions for Δh, Δu, and Δs are of a closed form, they are somewhat cumbersome to use—especially if similar calculations on the same specie are to be performed repetitively. Because the thermodynamic properties of air are so commonly used, they have been tabulated for our use. These tables are termed the "gas tables" and are presented in Appendix F. The values listed in these tables have been obtained by appropriate integrations of precision heat capacity data. An explanation of the entries in these tables follows: All values are reported at a pressure of 1 atm and at the temperature indicated. It is assumed that the gas behaves according to the ideal-gas equation of state for all temperatures at this pressure. The values of u, h, and s are arbitrarily set equal to zero at $T = 0°R$ and $P = 1$ atm. The enthalpy, h, and internal energy, u, are calculated directly from the the appropriate integrals of C_P or C_V:

$$u = u_0 = \int_{T_0}^{T} C_V \, dT$$

setting $u_0 = 0$ at $T_0 = 0°R$ gives

$$u = \int_0^T C_V \, dT$$

Similarly,

$$h = \int_0^T C_P \, dT$$

The entropy at $P = 1$ atm is given by:

$$(s - s_0)_{1 \text{ atm}} = \int_{T^\circ}^{T} \frac{C_P}{T} \, dT$$

or again setting $s_0 = 0$ at $T^\circ = 0°R$ and 1 atm

$$(s)_{1 \text{ atm}} = \int_0^T \frac{C_P}{T} \, dT \equiv \phi(T)$$

Both u and h are independent of P for a gas which obeys the ideal-gas equation of state. Thus the entries for u and h in Appendix F may be used at all pressures. The entropy of a gas which obeys the ideal gas equation of state, however, varies with pressure according to equation (6-70d)

$$s(P, T) - s(P_0, T) = -R \ln \frac{P}{P_0} \tag{6-79}$$

Thus, the entropy at any general pressure and temperature may be expressed as:

$$s(P, T) = \int_0^T \frac{C_P}{T} dT - R \ln \frac{P}{P_0} \equiv \phi(T) - R \ln \frac{P}{P_0} \tag{6-80}$$

where $P_0 = 1$ atm for the gas tables of Appendix F.

SAMPLE PROBLEM 6-6. Air originally available at 1000°R and 25 atm is expanded adiabatically to 1 atm and 500°R. Determine Δs, Δh, and Δu using the gas tables.

Solution: We picture the process as shown in Fig. SP6-6. Assuming that air obeys the ideal-gas equation of state under these conditions, Δh and Δu are obtained directly from the gas tables:

$$\begin{aligned}\Delta h &= h_2 - h_1 = h(500°\text{R}) - h(1000°\text{R}) \\ &= (119.48 - 240.98)\ \text{Btu/lb}_\text{m} = -121.5\ \text{Btu/lb}_\text{m} \\ \Delta u &= u_2 - u_1 = u(500°\text{R}) - u(1000°\text{R}) \\ &= (85.20 - 172.43)\ \text{Btu/lb}_\text{m} = -87.2\ \text{Btu/lb}_\text{m}\end{aligned}$$

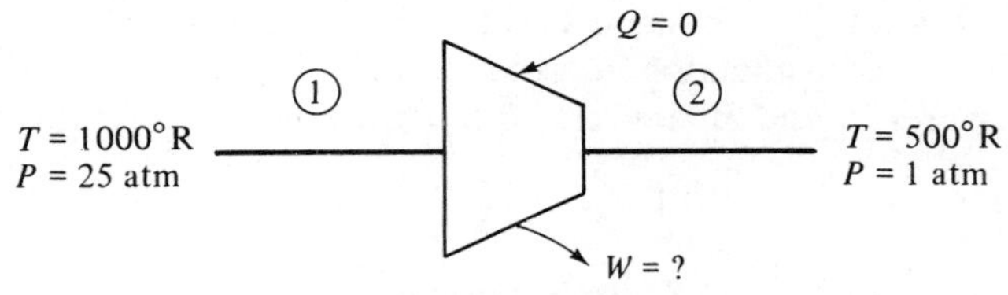

Figure SP6-6

The entropies are evaluated from equation (6-80):

$$\begin{aligned}s_2 &= \phi(T_2) - R \ln P_2/P_0 = \phi(T_2) = 0.59173\ \text{Btu/lb}_\text{m}°\text{R} \\ s_1 &= \phi(T_1) - R \ln P_1/P_0 = \phi(T_1) - R \ln 25 \\ &= [(0.75042) - 0.212]\ \text{Btu/lb}_\text{m}°\text{R}\end{aligned}$$

where $R = 0.0685$ Btu/lb$_\text{m}$°R. Thus

$$s_1 = 0.538\ \text{Btu/lb}_\text{m}°\text{R}$$

The overall entropy change is then

$$\Delta s = s_2 - s_1 = [0.592 - 0.538] \text{ Btu/lb}_m{}^\circ\text{R}$$
$$= +0.054 \text{ Btu/lb}_m{}^\circ\text{R}$$

The gas tables can also be used to determine the outlet conditions from an isentropic process if the *relative pressure* and *volume*, p_r and v_r respectively, are introduced. For an isentropic process of a gas which obeys the ideal-gas equation of state we may set $ds = 0$ in equation (6-70d). Thus:

$$\frac{dP}{P} = \frac{C_P}{R}\frac{dT}{T} \tag{6-81}$$

This equation, which relates P and T along an isentrope, can be integrated from some reference conditions P_0, T_0 on the isentrope to any other conditions, P, T which are also on the isentrope:

$$\ln\frac{P}{P_0} = \frac{1}{R}\int_{T_0}^{T}\frac{C_P}{T}\,dT \tag{6-82}$$

If we let $T_0 = 0°\text{R}$, then the integral on the right-hand side of equation (6-82) is simply $\phi(T)$. Thus:

$$\ln\frac{P}{P_0} = \frac{\phi(T)}{R} \tag{6-82a}$$

The right-hand side of equation (6-82a) is a function of T only. We now define the *relative pressure* by the expression

$$\ln p_r = \ln\frac{P}{P_0} = \frac{\phi(T)}{R} \tag{6-83}$$

Values of p_r can also be tabulated as functions of T. If we consider any two states 1 and 2 along a common isentrope, P_0 will be identical for both states and the pressures P_1 and P_2 are related to their relative pressures by:

$$\frac{P_2}{P_1} = \frac{(p_r)_2}{(p_r)_1} \tag{6-84}$$

Thus if we know P_1, P_2, and T_1 we can obtain $(p_r)_1$ from the gas tables. $(p_r)_2$ can then be obtained from equation (6-84), and T_2 is then found in the gas tables from $(p_r)_2$.

The development of a *reduced specific volume* (v_r) is similar. The ratio of relative specific volumes in an isentropic process is also equal to the ratio of the specific volumes:

$$\frac{v_1}{v_2} = \frac{(v_r)_1}{(v_r)_2} \tag{6-84a}$$

SAMPLE PROBLEM 6-7. Air originally available at 1 atm and 140°F is to be isentropically compressed to 20 atm in an open-system compressor. Determine the outlet temperature. If the compression is adiabatic, determine the work required per lb_m of air.

Solution: We may picture the compressor as shown in Fig. SP6-7. $T_1 = 140°F = 600°R$. From the gas tables $(p_1)_r = 2.005$. We get $(p_r)_2$ via equation (6-84) and the compressor's pressure ratio:

$$(p_r)_2 = (p_r)_1\left(\frac{P_2}{P_1}\right) = (2.005)(20) = 40.01$$

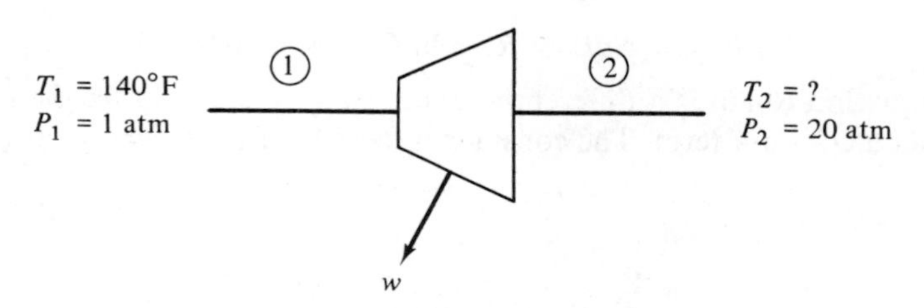

Figure SP6-7

The outlet temperature is then obtained from the gas tables:

$$T_2 \approx 1380°R = 920°F$$

For the adiabatic compression the work required is obtained from the open-system energy balance. Assuming steady-state operation and negligible kinetic or potential energy changes, the energy balance reduces to:

$$-w = \Delta h = h_2 - h_1$$

The enthalpies in turn are read from the gas tables:

$$h_2 = h(T_2) = 337.68 \text{ Btu/lb}_m$$
$$h_1 = h(T_1) = 143.47 \text{ Btu/lb}_m$$

therefore

$$-w = (337.68 - 143.47) \text{ Btu/lb}_m = 194.21 \text{ Btu/lb}_m.$$

or

$$w = -194.21 \text{ Btu/lb}_m$$

with the minus sign indicating that work must be supplied to the compressor.

6.8 Fugacities and the Fugacity Coefficient

From the Gibbs free-energy-property relation we learn that

$$\left(\frac{\partial g}{\partial P}\right)_T = v \tag{6-85}$$

or

$$dg = v\,dP \qquad \text{at constant } T \tag{6-85a}$$

For a fluid that obeys the ideal-gas equation of state, $v = RT/P$ and equation (6-85) becomes

$$(dg)_T = RT\frac{dP}{P} = RT\,d\ln P \tag{6-86}$$

For fluids that do not obey the ideal-gas equation of state, equation (6-86) is not correct. However, its form is quite useful and thus we may define a new quantity—*the fugacity*—such that the form of (6-86) is retained even for gases that do not obey the ideal-gas equation of state. The fugacity, f, is thus defined so that

$$(dg)_T = v\,dP = RT\,d\ln f \qquad \text{at constant } T \tag{6-87}$$

Since equation (6-86) is a differential expression, it can only fix the fugacity to within a constant term. The constant is fixed by also requiring that

$$f \longrightarrow P \qquad \text{as } P \longrightarrow 0 \tag{6-88}$$

or

$$\frac{f}{P} \longrightarrow 1 \qquad \text{as } P \longrightarrow 0 \tag{6-88a}$$

Equations (6-87) and (6-88) or (6-88a) then completely define the fugacity function.

The *fugacity coefficient*, ν, is defined as the ratio of the fugacity to the pressure:

$$\nu = \frac{f}{P} \tag{6-89}$$

The fugacity coefficient for a fluid may be expressed in terms of the P–v–T behavior of the fluid as follows. We begin with equation (6-87)

$$RT\left(\frac{\partial \ln f}{\partial P}\right)_T = v \tag{6-87}$$

Subtract $RT(\partial \ln P/\partial P)_T = RT/P$ from both sides of equation (6-87), and rearrange to give

$$RT\left(\frac{\partial \ln f/P}{\partial P}\right)_T = v - \frac{RT}{P} \tag{6-90}$$

But $f/P = \nu$, so (6-90) reduces to

$$RT\left(\frac{\partial \ln \nu}{\partial P}\right)_T = v - \frac{RT}{P} \tag{6-91}$$

Now integrate from $P = 0$ to the pressure of interest at constant temperature:

$$RT\int_{P=0}^{P} d\ln\nu = \int_{P=0}^{P}\left(v - \frac{RT}{P}\right)dP \tag{6-92}$$

or

$$RT\,[(\ln\nu)_P - (\ln\nu)_{P=0}] = \int_{P=0}^{P}\left(v - \frac{RT}{P}\right)dP \tag{6-93}$$

But from the definition of the fugacity, at $P = 0$, $\nu = 1$, so equation (6-93) reduces to

$$RT \ln \nu = \int_{P=0}^{P} \left(v - \frac{RT}{P}\right) dP \tag{6-94}$$

or

$$\ln \nu = \int_{P=0}^{P} \left(\frac{v}{RT} - \frac{1}{P}\right) dP \tag{6-94a}$$

For a gas that obeys the ideal-gas equation of state, $1/P = v/RT$, so the right-hand side of equation (6-94a) vanishes and we find that $\nu = 1$ independent of P or T. Thus for an ideal gas the fugacity is simply equal to the pressure. For a gas that does not obey the ideal-gas equation of state, the fugacity coefficient may be found by performing the integration in equation (6-94a).

6.9 Calculation of Fugacities from an Equation of State

Most of the useful equations of state represent pressure, P, as the dependent variable as a function of the independent variables temperature, T, and specific volume, v. Since fugacity is a state function, it should be possible to use such equations to compute fugacity as a function of v and T. Our defining equation for fugacity,

$$\left(\frac{\partial \ln f}{\partial P}\right)_T = \frac{v}{RT} \tag{6-87}$$

is not well suited to this task, however, because it treats P and T as the independent variables. We can, however, rearrange equation (6-87) to obtain the alternative expression,

$$\left(\frac{\partial \ln f}{\partial v}\right)_T = \left(\frac{\partial \ln f}{\partial P}\right)_T \left(\frac{\partial P}{\partial v}\right)_T = \frac{v}{RT}\left(\frac{\partial P}{\partial v}\right)_T \tag{6-95}$$

Thus, in principle, we can insert an equation-of-state expression for P here and, by integration, evaluate f as a function of v and T. Since $f \longrightarrow P$ as $P \longrightarrow 0$ $(v \longrightarrow \infty)$, the range of the integration should be $v = \infty$ to v. We note that, because of their logarithmic nature, some terms in this expression may not be bounded in the limit $v \longrightarrow \infty$, and thus will seek an alternative form in which these problems are eliminated. Since the fugacity coefficient, $\nu = f/P$, approaches unity (rather than zero) as $P \longrightarrow 0$, we shall attempt to derive an expression for ν rather than f directly.

First we rearrange equation (6-95) to the form

$$\left(\frac{\partial \ln f}{\partial v}\right)_T = \left(\frac{\partial}{\partial v}\right)_T \frac{Pv}{RT} - \frac{P}{RT} \tag{6-96}$$

and note that

$$\left(\frac{\partial}{\partial v}\right)_T \ln \frac{Pv}{RT} = \left(\frac{\partial \ln P}{\partial v}\right)_T + \frac{1}{v} \tag{6-97}$$

which can be rearranged to give

$$\left(\frac{\partial \ln P}{\partial v}\right)_T = -\frac{1}{v} + \left(\frac{\partial}{\partial v}\right)_T \ln \frac{Pv}{RT} \tag{6-98}$$

Substraction of equation (6-98) from (6-96) will yield the desired expression:

$$\left(\frac{\partial}{\partial v}\right)_T \ln \frac{f}{P} = \frac{1}{v} - \frac{P}{RT} + \left(\frac{\partial}{\partial v}\right)_T \left(\frac{Pv}{RT} - \ln \frac{Pv}{RT}\right) \tag{6-99}$$

Integration over the range $v = \infty$ to v, noting that in the limit $v \longrightarrow \infty$, $f/P \longrightarrow 1$, and $Pv/RT \longrightarrow 1$, produces

$$\ln \frac{f}{P} = \frac{1}{RT} \int_{\infty}^{v} \left(\frac{RT}{v} - P\right) dv + \frac{Pv}{RT} - 1 - \ln \frac{Pv}{RT} \tag{6-100}$$

Although the first term in the integral might be integrated directly, it is convenient to leave it within the integral so as to assure convergence at $v \longrightarrow \infty$.

Equation (6-100) is useful with any equation of state in which pressure is given as a function of v and T. The only limit on its application is the range of validity of the equation of state, for the derivation above has made no approximations that are not generally valid. For equations such as the Beattie–Bridgeman or Redlich–Kwong it can be used to compute vapor-phase fugacities, but for equations such as the Benedict–Webb–Rubin, which represent the liquid region as well as the vapor region, the equation can be used for vapor or liquid fugacities.

6.10 Equilibrium Between Phases

Up to this point we have talked of computing thermodynamic properties and property changes for systems that undergo no phase change. How are we to compute changes when a gas condenses? What, in fact, are the conditions at which a vapor will liquify reversibly? To answer these questions, we must establish the conditions necessary for equilibrium between phases.

We first recall that one speaks of equilibrium of a body or system with respect to the constraints that are imposed on the body. Thus a hot body may be in equilibrium if it is perfectly insulated from its surroundings. (The constraint is its isolation from any body with which it can exchange thermal energy.) If, however, the same hot body is placed in contact with a cooler body, it will come to equilibrium at a lower temperature than before. (The constraint is now the contact with another body, with which the hot body can exchange thermal energy.) Thus, as we reopen the question of equilibrium, we must be careful to specify the constraints that we wish to impose on the body.

Having specified the constraints to be imposed on a system at equilibrium, we have observed in Chapter 5 that (1) the entropy of the system and its surroundings is maximized, (2) there are no potential differences between the system and its surroundings, and (3) it is impossible to obtain work from an engine operating reversibly between the system and its surroundings. (The surroundings in each of the above cases is construed as that portion of the universe with which the system is free to interact.) These are not all independent criteria, but each provides some insight not directly available from the others.

These ideas, although useful, are not sufficiently quantitative for our purposes in describing phase and chemical equilibrium. We shall therefore make use of them to obtain quantitative criteria for equilibrium—that is, the relationship between the thermodynamic state variables of bodies in equilibrium with each other.

The ensuing development has as its immediate purpose the development of criteria for phase equilibrium of pure materials. At the outset, however, let it be noted that the treatment does not specifically rely on this circumstance. The conclusions obtained will be applicable to the seemingly more complex situations of multicomponent phase equilibrium (Chapter 10) and chemical reaction equilibrium (Chapter 11) as well.

The constraints that will be common to all considerations of phase and chemical equilibrium in this book are:

1. The absence of barriers to thermal-energy transfer between the system and its surroundings.

2. The absence of barriers to mechanical-energy transfer between the system and its surroundings through a volume change.

Except when we consider chemical reaction equilibrium specifically, we shall impose an additional constraint:

3. No chemical reaction can take place between the chemical species present.

If any of these constraints is altered, or additional constraints imposed, we shall so specify at that time.

By any of the criteria of equilibrium mentioned earlier it is evident that constraints 1 and 2 require that the temperature and pressure of the system be the same as that of the surroundings and that there be no temperature gradients within the system. Were this not so we could, for instance, obtain work from a heat engine operating reversibly between the high- and low-temperature regions or from an expansion of the high-pressure material to lower pressure. Because of these conclusions, it is common to merely specify that temperature and pressure are fixed by the surroundings—which we see is an equivalent assumption.

A further important observation about equilibrium in such systems can be made by asking what happens to the total Gibbs free energy of a material in a constant pressure, isothermal, steady-flow process which is proceeding reversibly—that is, one in which the maximum amount of work is being exchanged between the flow system and the surroundings. The flowing material might, for example, consist of liquid and gaseous A, or of liquid and gaseous mixtures of A and B, or of a gaseous mixture of chemically reacting A, B, and C. The conclusions reached will be generally valid; the above examples are offered only as an aid in visualizing the process.

For the purpose of illustration, let the flow system be divided into a sequence of smaller stages as shown in Fig. 6-2. To determine what changes occur, let us write the energy balance and the entropy balance around the first stage, remembering that reversibility is assured if $LW = 0$. (Kinetic and potential energy changes are assumed to be negligible, although a similar development can be followed for cases where they are not.)

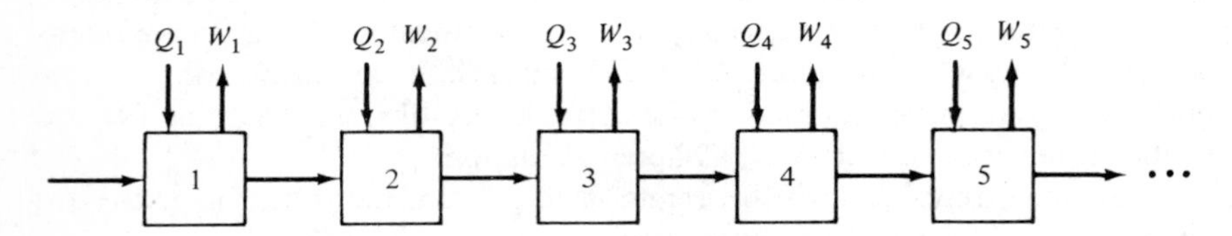

Figure 6-2. Staged, isothermal, reversible flow process.

Using the notation that H and S are the total enthalpy and total entropy of the flowing stream and that Q and W represent the heat and work transfers for the same basis, the energy and entropy balances are

Energy:

$$H_{\text{in}} - H_{\text{out}} + Q_1 - W_1 = 0$$

Entropy:

$$S_{\text{in}} - S_{\text{out}} + \frac{Q_1}{T} = 0 \tag{6-101}$$

If the two equations are combined, eliminating Q_1, we have

$$W_1 = -[H_{\text{out}} - TS_{\text{out}} - (H_{\text{in}} - TS_{\text{in}})] \tag{6-102}$$

or, since $G = H - TS$,

$$W_1 = -(G_{\text{out}} - G_{\text{in}}) = -\Delta G_1 \tag{6-103}$$

where $\Delta G_1 = G_{\text{out}} - G_{\text{in}}$ denotes the total Gibbs-free-energy change of the flowing material across the first stage. By repetition of the arguments, we see that the work produced in any portion of the steady, isothermal, reversible flow process is equal to the negative of the change in total Gibbs free energy across that portion of the process. As long as the total Gibbs free energy decreases, work can be recovered from the process.

If we replace the finite number of stages by an infinite series of infinitesimal stages, then the requirement for positive work production in each stage is simply that $dG < 0$. If $dG > 0$, then work must be supplied by the surroundings to continue the process; $dG = 0$ implies that work is neither produced nor consumed. Thus, if we plot G as a function of X, a process parameter (as shown in Fig. 6-3), we can distinguish the region in which work is produced as the region of negative slope. A region of positive slope indicates that work must be supplied to continue the process. The minimum, corresponding to $dG = 0$, is the point at which work is neither produced nor consumed. The process parameter, X, might be any useful measure of the change occurring within the system, such as overall specific volume in the case of phase equilibrium or fractional conversion in the case of chemical reaction.

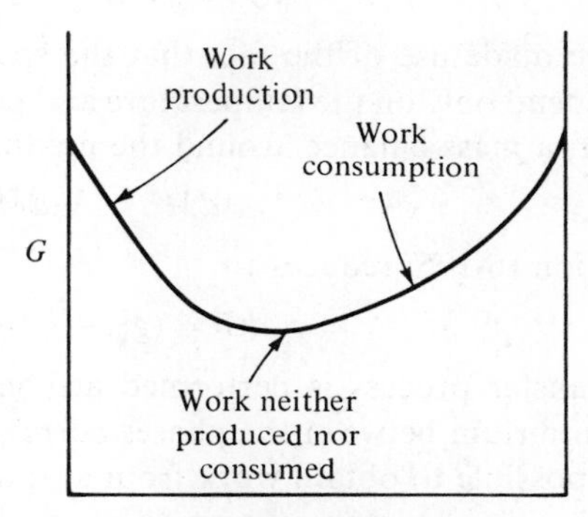

Figure 6-3. Gibbs free energy as a function of the process parameter X.

Let us now consider what happens if we simply allow the process to occur spontaneously. As long as the system is to the left of the minimum in the G–X curve, the process will proceed with a decrease in the total Gibbs free energy of the system because the reversible work for that process would be positive. When the minimum in the curve is reached the process can proceed no further, because further change can occur only if the surroundings supply work to the system. Our assumption of spontaneity implies that the surroundings cannot supply the needed work; hence the change must cease. Similarly, the process cannot proceed in the reverse direction, because this, too, would correspond to increasing G. Similarly, if the system started in a state to the right of the minimum in Fig. 6-3, it could move only toward the minimum point (or to the left) if the change is to occur spontaneously. Therefore, we find that the process will proceed only up to that point at which the minimum in the Gibbs-free-energy curve is reached. Once this point is attained, the change can proceed in neither direction, and the system will remain in this state indefinitely (unless work is supplied from the surroundings). The system at this point is in equilibrium under the constraints we have imposed, because there is no way one can obtain further work from this system. Thus we conclude that, *at fixed temperature and pressure, the Gibbs free energy of the system is a minimum at equilibrium.* For specific cases of phase or chemical equilibrium, we shall determine the conditions that correspond to the minimum in the G–X curve by setting $(dG/dX)_{T,P} = 0$.

Let us now apply this to the specific instance of phase equilibrium for pure materials. If we denote the two phases present as I and II, then

$$G = M^{\mathrm{I}}g^{\mathrm{I}} + M^{\mathrm{II}}g^{\mathrm{II}} \tag{6-104}$$

Now if we transfer a small amount of material from one phase to the other without changing the temperature or pressure, the total change in Gibbs free energy of the mixture is given by

$$dG = g^{\mathrm{I}}dM^{\mathrm{I}} + g^{\mathrm{II}}dM^{\mathrm{II}} \tag{6-105}$$

(We have made use of the fact that the specific Gibbs free energies of each phase depend only on the temperature and pressure, and hence do not change.) However, a mass balance around the mixture gives

$$dM^{\mathrm{I}} = -dM^{\mathrm{II}} \tag{6-106}$$

so equation (6-105) reduces to

$$dG = (g^{\mathrm{I}} - g^{\mathrm{II}})\, dM^{\mathrm{I}} \tag{6-107}$$

If the transfer process is performed at constant temperature and pressure, then equilibrium between the phases corresponds to that condition in which it is not possible to obtain work from a transfer of mass between the phases. Thus dG must vanish at the equilibrium conditions, and equation (6-107) becomes

$$g^{\mathrm{I}} - g^{\mathrm{II}} = 0 \tag{6-108}$$

or

$$g^{\mathrm{I}} = g^{\mathrm{II}} \tag{6-108a}$$

at equilibrium.

We can say in summary that equilibrium between different phases (of pure materials) is characterized by the equalities

$$P^{\mathrm{I}} = P^{\mathrm{II}}, \qquad T^{\mathrm{I}} = T^{\mathrm{II}}, \qquad g^{\mathrm{I}} = g^{\mathrm{II}} \tag{6-109}$$

Because of the desire to cover more general cases to come in Chapters 10 and 11 we have spoken of fixing both P and T independently. Since specification of P and T fully determines the state of a pure material, we cannot assure the existence of two phases for arbitrary choices of the two. The relationship of the points on the P–T equilibrium curve is discussed in a following section. We can say, nonetheless, that whenever two phases of a pure material are in equilibrium, all of equations (6-109) must be satisfied.

Let us now return to the original question of this section: How can we compute property changes for systems undergoing a phase change? We have learned that the phase change can be performed at constant temperature and pressure with no change in the Gibbs free energy. If we measure the heat associated with a phase change (in an isothermal flow calorimeter, for example), we can compute Δh and, since $\Delta g = 0$, $\Delta s = \Delta h/T$. If, in addition, we measure the volume change associated with the phase change, we find that $\Delta u = \Delta h - \Delta(Pv) = \Delta h - P\,\Delta v$ and, since $\Delta g = 0$, $\Delta a = \Delta g - \Delta(Pv) = -P\,\Delta v$. In short, for a pure material, knowledge of the volume and enthalpy change associated with a phase change at known temperature and pressure is sufficient to permit calculation of the change in all other thermodynamic properties of the material.

6.11 Evaluation of Liquid- and Solid-Phase Fugacities

As we will see when we deal with mixtures in Chapters 10 and 11, much useful information can be obtained from the fugacity. Accordingly, we now open the question of calculating liquid- and solid-phase fugacities.

Equation (6-87) upon integration at constant temperature becomes

$$RT \ln \frac{f^{\text{II}}}{f^{\text{I}}} = g^{II} - g^{I} \tag{6-110}$$

We have noted that when the superscripts I and II denote the properties of two phases in equilibrium, the right-hand side of this equation is identically zero. Consequently, it must also be true that the fugacities are equal at equilibrium,

$$f^{\text{I}} = f^{\text{II}} \tag{6-111}$$

This relationship enables us to compute the fugacity of pure liquids and solids assuming that we may evaluate the fugacity of a vapor which is in equilibrium with the liquid or solid under consideration. We have seen in Section 6-8 how the fugacity of a gas at elevated temperature and pressure may be obtained by integrating equations (6-95) to (6-100) at constant temperature from a low pressure reference state to the pressure in question. If the elevated pressure is the vapor pressure, P', we have shown that the gas and liquid fugacities are equal at that pressure. Since the definition of fugacity in equation (6-87) is not limited to the gaseous state, we may use that equation for the liquid state as well,

$$\left(\frac{\partial \ln f^{\text{L}}}{\partial P}\right)_T = \frac{v^{\text{L}}}{RT} \tag{6-112}$$

and continue integration into the liquid range. The computation of liquid fugacity then follows the procedure outlined by the following integrated form of equation (6-87)

$$\ln \frac{f^L}{P^*} = \int_{P^*}^{P'} \frac{v^V}{RT}\, dP + \int_{P'}^{P} \frac{v^L}{RT}\, dP \tag{6-113}$$

To circumvent problems that might arise in the limit of $P^* \longrightarrow 0$, we can subtract the equation

$$\ln \frac{P'}{P^*} = \int_{P^*}^{P'} \frac{1}{P}\, dP$$

to obtain the working equation

$$\ln \frac{f^L}{P'} = \int_{P^*}^{P'} \left(\frac{v^V}{RT} - \frac{1}{P}\right) dP + \int_{P'}^{P} \frac{v^L}{RT}\, dP \tag{6-114}$$

It should be clear, as well, that the same procedures can be applied to calculation of the fugacities of pure solids by noting that for a solid and gas in equilibrium (through sublimation of the solid), $f^S = f^V$. Equation (6-114) can be applied by changing all L superscripts to S and understanding P' to mean the vapor pressure of the pure solid at the temperature in question.

SAMPLE PROBLEM 6-8. To illustrate the procedures above, calculate the fugacity of liquid water at 400 psia and 300°F, using data from the steam tables in Appendix B.

Solution: The steam tables in Appendix B give data for the specific volume of water at 300°F as indicated in the first two columns of Table SP 6-8. The remaining columns were computed from the handbook data using values for

Table SP 6-8

Pressure, psia	v, ft^3/lb$_m$	$\frac{v}{RT}$, psia^{-1}	$\frac{v}{RT} - \frac{1}{P}$, psia^{-1}
1	452.3	0.9996	-4×10^{-4}
2	226.0	0.4995	-5×10^{-4}
5	90.25	0.1995	-5×10^{-4}
10	45.00	0.0995	-5×10^{-4}
20	22.36	0.0494	-6×10^{-4}
40	11.040	0.0244	-6×10^{-4}
60	7.259	0.0160	-6×10^{-4}
67.013	6.466 (vapor)	0.0143	-6×10^{-4}
67.013	0.01745 (liquid)	3.85×10^{-5}	—
200	0.01744	3.85×10^{-5}	—
400	0.01742	3.85×10^{-5}	—

R and T as follows:

$$R = 10.731 \frac{\text{psia ft}^3}{\text{lb-mole °R}} \frac{1}{18.016} \frac{\text{lb-mole}}{\text{lb}_m} = 0.5956 \text{ psia ft}^3/\text{lb}_m \text{ °R}$$

$$T = 459.7 + 300 = 759.7 \text{ °R}$$

Because the difference $v/RT - 1/P$ is so nearly constant, the first integral in equation (6-114) can be approxmiated as follows:

$$\int_0^{P'} \left(\frac{v^V}{RT} - \frac{1}{P}\right) dP = \left(\frac{v^V}{RT} - \frac{1}{P}\right)_{\max} (P' - 0)$$
$$= (-6 \times 10^{-4})(67.013)$$
$$= -0.0402$$

A more careful numerical procedure gives the value -0.0386.

The second integral of equation (6-114) also has a constant integrand, with the result

$$\int_{P'}^{P} \frac{v^L}{RT} dP = (3.85 \times 10^{-5})(400 - 67.013)$$
$$= 0.0128$$

Thus the fugacity of water at both the vapor pressure, P', and 400 psia, P, can now be calculated.

$$\ln \frac{f^{\text{sat vap}}}{P'} = -0.0402$$

$$f^{\text{sat vap}} = (67.013) \exp(-0.0402) = 64.4 \text{ psia}$$

and

$$\ln \frac{f^L}{P'} = -0.0402 + 0.0128$$

$$f^L = (67.013) \exp(-0.0274) = 65.1 \text{ psia}$$

The striking thing about this result is that the fugacity of the compressed liquid is not very different from that of the saturated vapor or saturated liquid. The reason for this is evident from an inspection of equation (6-114) in light of the data above. The volume of the liquid is so much less than the volume of the vapor that the second integral is much smaller than the first.

Sample Problem 6-8 has shown how we may compute the fugacity of a pure liquid from specific volume data and has revealed that the fugacity does not change greatly as the liquid is compressed beyond the saturation pressure. In fact, we can say, with reasonable accuracy, that

$$f^L(P, T) = f^{\text{sat liq}}(T) = f^{\text{sat vap}}(T) \tag{6-115}$$

In other words, *the fugacity of a compressed liquid is approximately equal to its fugacity at its vapor pressure at the same temperature*. This approximation will be enormously helpful in our later work, for we will not need to have liquid specific volume data to obtain values of the fugacity for compressed liquids.

The approximation in equation (6-115) will involve negligible error provided the second integral in equation (6-114) is small. In other words, it will be valid if $v^L \ll v^V$ and P is not $\gg P'$. The former is true except near the critical conditions and the latter is true provided the liquid is not compressed well beyond the vapor pressure.

6.12 Clausius–Clapeyron Equations

The pressure–temperature plot for a pure component introduced in Fig. 2-4 included three curves, which represented liquid–vapor, solid–vapor, and liquid–solid equilibrium. The Clausius–Clapeyron equation provides information on the slope of these curves, $(dP/dT)_{\text{equilibrium}}$, which is obtained by noting that at all points on these equilibrium curves $g^{\text{I}} = g^{\text{II}}$, where the superscripts denote different phases. Thus, for a change dT and dP along this curve

$$dg^{\text{I}} = dg^{\text{II}}$$

or

$$v^{\text{I}}\,dP - s^{\text{I}}\,dT = v^{\text{II}}\,dP - s^{\text{II}}\,dT \tag{6-116}$$

Solving this equation for dP/dT and making use of the fact that

$$g^{\text{I}} = h^{\text{I}} - Ts^{\text{I}} = h^{\text{II}} - Ts^{\text{II}} = g^{\text{II}} \tag{6-117}$$

or

$$s^{\text{II}} - s^{\text{I}} = \frac{h^{\text{II}} - h^{\text{I}}}{T} \tag{6-117a}$$

to eliminate entropy yields the result known as the *Clausius–Clapeyron equation*,

$$\left(\frac{dP}{dT}\right)_{\text{eq}} = \frac{s^{\text{II}} - s^{\text{I}}}{v^{\text{II}} - v^{\text{I}}} \tag{6-118}$$

$$= \frac{h^{\text{II}} - h^{\text{I}}}{(v^{\text{II}} - v^{\text{I}})T} \tag{6-118a}$$

The Clausius–Clapeyron equation as we have derived it is applicable to all materials and all examples of phase equilibria. Thus, for example, we may use equation (6-118a) to describe such diverse phase behavior as the effect of changing pressure on the freezing point of water, or the effect of temperature on the vapor pressure of liquid metals.

Sample Problem 6-9. An ice skate is able to glide over the ice because the skate blade exerts sufficient pressure on the ice that a thin layer of ice is melted. The skate blade then glides over this thin lubricating layer. Determine the pressure an ice skate blade must exert to allow smooth ice skating at $-10°C$. You may use $\Delta h_{\text{fusion}} = 1440$ cal/g-mol, $\rho_{\text{liq}} = 1.000$ g/cm³, and $\rho_{\text{ice}} = 0.917$ g/cm³ over the range of temperatures and pressures involved.

Solution: The Clausius–Clapeyron equation will be used:

$$\left(\frac{dP}{dT}\right)_{\text{eq}} = \frac{\Delta h}{\Delta v\, T}$$

Multiply by dT and integrate:

$$\int_{P_1}^{P_2} dP = \int_{T_1}^{T_2} \frac{\Delta h}{\Delta v}\frac{dT}{T}$$

or

$$P_2 - P_1 = \frac{\Delta h}{\Delta v} \ln \frac{T_2}{T_1} \tag{a}$$

But at $P = 1$ atm we know $T_1 = 0°C$. Since we desire the pressure, P_2, at which $T_2 = -10°C$, we may use equation (a) directly: Before substituting, however, we must convert into some consistent set of units. For this problem we use S.I. units:

$$P_1 = 1 \text{ atm} = 1.013 \times 10^5 \text{ N/m}^2$$

$$T_1 = 0°\text{C} = 273\ °\text{K}$$

$$T_2 = -10\ °\text{C} = 263\ °\text{K}$$

$$\Delta h = 1440 \text{ cal/g-mol} = 3.34 \times 10^5 \text{ J/kg} = 3.34 \times 10^5 \text{ Nm/kg}$$

$$v^{\text{liq}} = \frac{1}{\rho_{\text{L}}} = 1 \text{ cm}^3/\text{gm} = 1.000 \times 10^{-3} \text{ m}^3/\text{kg}$$

$$v^{\text{sol}} = \frac{1}{\rho_{\text{S}}} = 1.091 \text{ cm}^3/\text{g} = 1.091 \times 10^{-3} \text{ m}^3/\text{kg}$$

Therefore

$$P_2 - 1.013 \times 10^5 \text{ (N/m}^2) = \frac{3.34 \times 10^5 \text{ Nm/kg}}{(1.000 - 1.091) \times 10^{-3} \text{ m}^3/\text{kg}} \ln \frac{263}{273}$$

$$= \left[-\frac{3.34 \times 10^5}{9.1 \times 10^{-5}} \ln (0.963)\right] \text{N/m}^2$$

$$P_2 - 1.01 \times 10^5 \text{ N/m}^2 = 1.38 \times 10^8 \text{ N/m}^2$$

or

$$P_2 = 1.38 \times 10^8 \text{ N/m}^2 = 1.36 \times 10^3 \text{ atm} = 20.0 \times 10^3 \text{ psi}$$

This pressure is clearly quite high and is achieved with the ice skate blade by having only a small portion of the blade surface contact the ice at any given time. If the temperature drops much lower than about $-15°C$, it is not possible to generate sufficient pressure to melt the ice, and conventional ice skating is not possible.

If phase II in equation (6-118) is vapor and phase I is liquid, then $h^{\mathrm{II}} - h^{\mathrm{I}} = \Delta h_V$, the enthalpy change on vaporization, and $v^{\mathrm{II}} - v^{\mathrm{I}} = \Delta v_V$, the volume change on vaporization. An approximate form of this equation may be obtained by noting that usually $v^{\mathrm{L}} \ll v^{\mathrm{V}}$ and that

$$\Delta v_V = \frac{RT}{P} \tag{6-119}$$

if the vapor is an ideal gas. Substituting these simplifications into equation

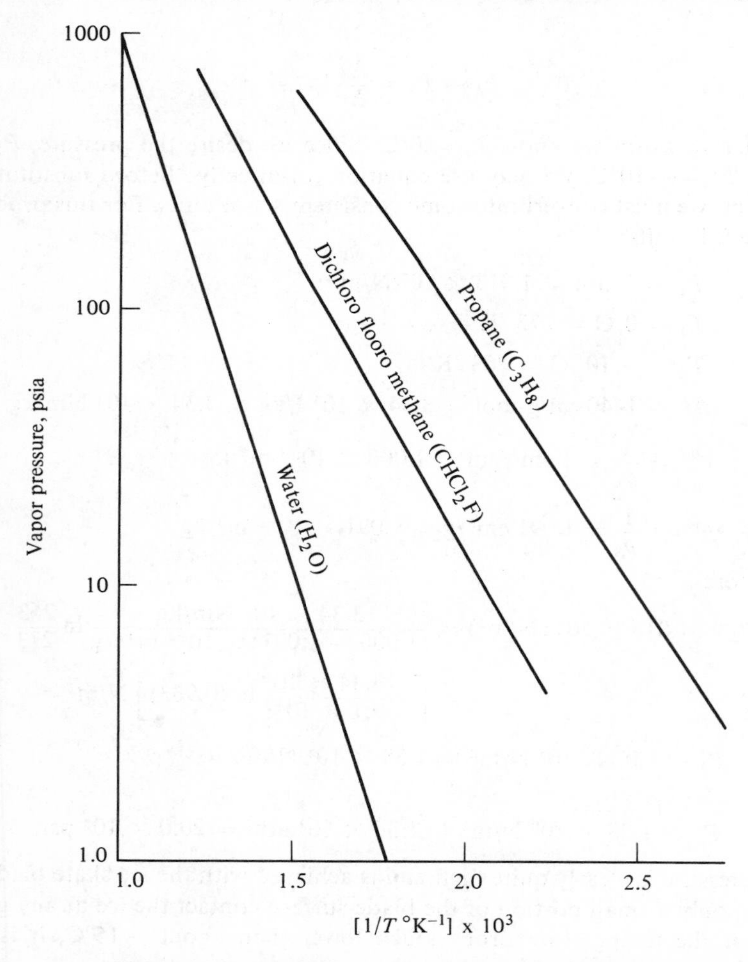

Figure 6-4. Vapor-pressure plots for selected substances.

(6-118a) reduces the Clausius–Clapeyron equation to

$$d \ln P' = -\frac{\Delta h_V}{R} d\left(\frac{1}{T}\right) \tag{6-120}$$

or

$$\frac{d \ln P'}{d(1/T)} = -\frac{\Delta h_V}{R} \tag{6-120a}$$

Equation (6-120a) shows that a plot of $\ln P'$ vs $1/T$ should yield a straight line with slope $-(\Delta h_V/R)$. Thus we may use such a plot to determine the enthalpy change on vaporization when the appropriate vapor-pressure–temperature data are available. An example of such plots for several substances is shown in Figs. 6-4 and 6-5. If we attempt to extend these curves into regions of higher pressures and temperatures, we will find some curvature occurring as the assumption of ideal-gas behavior, negligible liquid specific volume, and constant Δh_V begin to break down. Although the plots begin to curve, this curvature turns out to be much lower than we might anticipate since the three factors tend to cancel each other. Equation (6-120) in turn may be integrated over a range of temperatures in which Δh_V is constant to give

$$\ln \frac{P_2}{P_1} = -\frac{\Delta h_V}{R}\left(\frac{1}{T_2} - \frac{1}{T_1}\right) \tag{6-121}$$

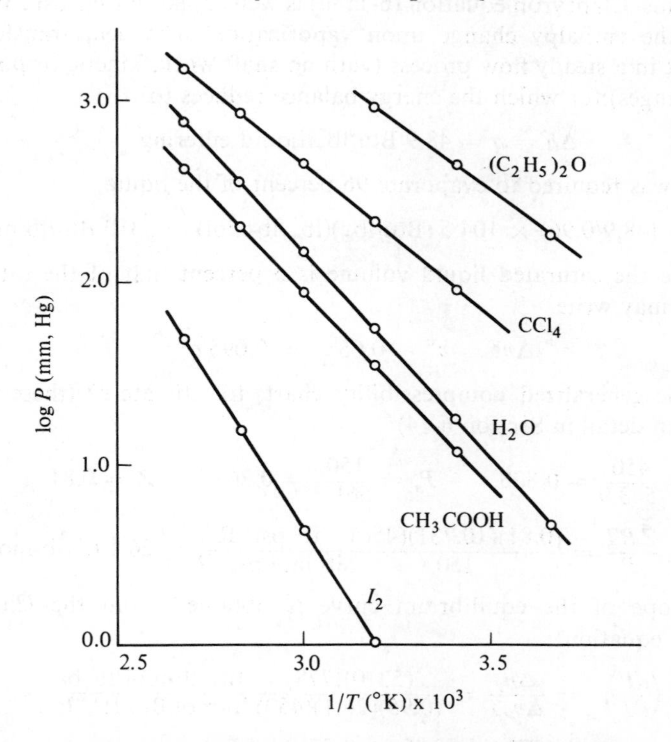

Figure 6-5. Vapor-pressure plots for selected substances.

Thus, for example, if we know the vapor pressure at any one temperature, and the heat of vaporization, we may calculate the vapor pressure at any other temperature by means of equation (6-121).

SAMPLE PROBLEM 6-10. Freon 13 is being used as a refrigerant in a commercial refrigerating system. At one point in the system saturated liquid Freon 13 enters an evaporator (a heat exchanger) under a pressure of 150 psia and a temperature of $-10°F$. The Freon 13 absorbs 48.9 Btu/lb_m of liquid entering and leaves the evaporator at the same temperature and pressure but with a quality of 96 percent. It has become necessary to lower the temperature of the Freon 13 in the evaporator. Would the pressure have to be increased or decreased to carry out the evaporation at $-20°F$? By how much?

The specific volume of the saturated liquid at 150 psia and $-10°F$ is estimated to be 5 percent that of the saturated vapor at the same temperature and pressure. Other constants of Freon 13 are as follows:

$$T_c = 83.9°F, \qquad P_c = 561.3 \text{ psia}, \qquad \text{mol wt} = 104.5$$

Solution: If the slope of the equilibrium curve at 150 psia and $-10°F$ can be estimated, then we can make a reasonable estimate of the pressure at $-20°F$. The Clausius–Clapeyron equation (6-118a) is well suited to this task. We first calculate the enthalpy change upon vaporization. The heat transfer was carried out in a steady flow process (with no shaft work, kinetic or potential energy changes) for which the energy balance reduces to

$$\Delta h = q = 48.9 \text{ Btu/lb}_m \text{ liquid entering}$$

Since this was required to evaporate 96 percent of the liquid,

$$\Delta h_V = (48.9/0.96) \times 104.5 \text{ (Btu/lb}_m\text{)(lb}_m\text{/lb-mol)} = 5310 \text{ Btu/lb-mol}$$

Because the saturated liquid volume is 5 percent that of the saturated vapor, we may write

$$\Delta v^V = v^V - 0.05\, v^V = 0.095\, v^V$$

and use the generalized compressibility charts to estimate v^V (these will be described in detail in Section 6-14):

$$T_r = \frac{450}{543.9} = 0.828, \qquad P_r = \frac{150}{561.3} = 0.267, \qquad Z = 0.81$$

$$v^V = \frac{ZRT}{P} = \frac{(0.81)(10.731)(450)}{150} \frac{\text{ft}^3 \text{ psi °R}}{\text{lb-mol psi °R}} = 26.1 \text{ ft}^3\text{/lb-mol}$$

The slope of the equilibrium curve is obtained from the Clausius–Clapeyron equation:

$$\left(\frac{dP}{dT}\right)_{eq} = \frac{\Delta h_V}{\Delta v_V T} = \frac{(5310)(778)}{(0.95)(26.1)(450)} \frac{\text{Btu lb-mol/ft-lb}_f}{\text{lb-mol Btu ft}^3 \text{ °R}}$$
$$= 370 \text{ lb}_f\text{/ft}^2 \text{ °R} = 2.57 \text{ psia/°R}$$

Note that the conversion factor 778 ft-lb$_f$/Btu has been employed here so as to yield familiar pressure units. The units Btu/ft^3 are pressure units but not convenient ones.

To answer the question originally posed, it is now evident that a pressure decrease of 25.7 psia will be necessary to decrease the temperature by 10°F. The approximate form of the Clausius–Clapeyron equation, (6-120a), is not sufficiently accurate here because the vapor is not an ideal gas ($Z = 0.81$).

6.13 Property Changes for Real Gases Using Equations of State

We now consider the calculation of property changes for real gases. As we shall see, many of the terms that conveniently dropped out when we were dealing with ideal gases no longer vanish. These terms are responsible for many of the difficulties encountered in calculations involving real gases.

The differential equations for *du*, *dh*, and *ds* may be written for any gas (real or ideal) as follows:

$$du = C_V\,dT + \left[T\left(\frac{\partial P}{\partial T}\right)_v - P\right]dv \tag{6-122}$$

$$dh = C_P\,dT - \left[T\left(\frac{\partial v}{\partial T}\right)_P - v\right]dP \tag{6-123}$$

$$ds = \frac{C_P}{T}\,dT - \left(\frac{\partial v}{\partial T}\right)_P dP \tag{6-124}$$

For real gases, the simplifications which result from the ideal-gas equation of state are frequently not valid. Rarely will the dP or dv terms in the dh or du expression vanish or be easily evaluated in the ds expression. Second, C_P and C_V are functions of T and P or v for real gases, so even evaluation of the integral $\int C_P\,dT$ is complicated. Third, we shall usually not know the exact path (that is, P–T or v–T relation) along which the process in question is proceeding, so we shall not be able to evaluate the integrals needed to find Δu, Δh, or Δs, even if the derivatives are expressible in terms of P, v, and T.

Therefore, we develop an alternative approach which takes advantage of the fact that u, h, and s are state functions. Thus any changes that occur in these properties depend only on the initial and final states of the substance and are not functions of the path between the states. Thus, if we evaluate Δu, Δh, and Δs for any specific path between the initial and final states of the system, the result must be the same for all other paths between the same initial and final states. However, we recognize that the change in a state property over any path may be evaluated as the sum of the changes of the property over all segments of the original path. That is, by evaluating $\Delta h_{1\text{-}2}$ and $\Delta h_{2\text{-}3}$ (where $\Delta h_{1\text{-}2} = h_2 - h_1$), the total change in enthalpy between state 1 and state 3 is obtained from the sum

$$\Delta h_{1\text{-}3} = \Delta h_{1\text{-}2} + \Delta h_{2\text{-}3}$$

Thus if we wish to evaluate a property change along any arbitrary path, we may break this path into a series of smaller paths (or subpaths) along which the change in the desired property can be evaluated.

Since in general we will know C_P or C_V only for very low pressures, where $C_P = C_P^*$ and $C_V = C_V^*$, it is frequently necessary to choose the integration paths in such a way that changes in temperature occur only at low pressures. Thus, if we wished to evaluate Δh or Δs between points 1 and 2 on a P–T diagram, we would use a three-part path as indicated in Fig. 6-6.

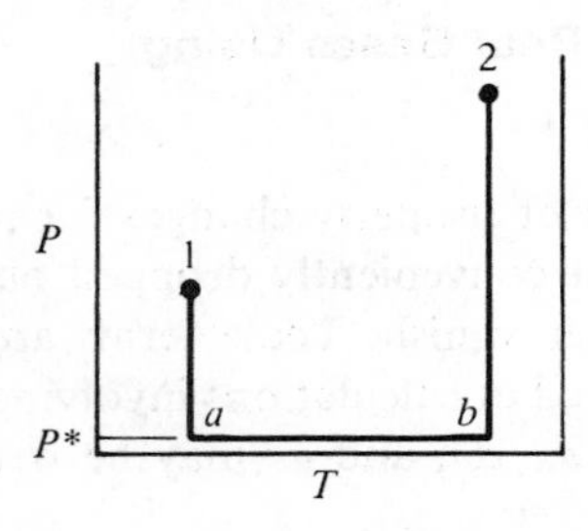

Figure 6-6. Three-step process.

On the first leg of the path from point 1 to point a, T is held constant at T_1, and the pressure is reduced from P_1 to P^* (where P^* is some very low pressure at which C_P is known). Since T is held constant, $dT = 0$ and $\Delta h_{1\text{-}a} = h_a - h_1$ may be evaluated by integrating equation (6-53):

$$\int_1^a dh = \Delta h_{1\text{-}a} = h_a - h_1 = \int_{P_1}^{P^*} \left[v - T_1\left(\frac{\partial v}{\partial T}\right)_P\right] dP \qquad (6\text{-}125)$$

Similarly

$$\Delta s_{1\text{-}a} = \int_{P_1}^{P^*} -\left(\frac{\partial v}{\partial T}\right)_P dP \qquad (6\text{-}126)$$

Along the second leg, from a to b, the pressure is held constant at P^* and the temperature changed from T_1 to T_2. Since P is constant, $dP = 0$, and $\Delta h_{a\text{-}b}$ and $\Delta s_{a\text{-}b}$ are calculated from

$$\Delta h_{a\text{-}b} = h_b - h_a = \int_{T_1}^{T_2} C_P^* \, dT \qquad (6\text{-}127)$$

$$\Delta s_{a\,b} = s_b - s_a = \int_{T_1}^{T_2} \frac{C_P^*}{T} dT \qquad (6\text{-}128)$$

where C_P^* indicates C_P at P^*.

Along the last leg of the path from b to point 2, the temperature is held constant at T_2 so $dT = 0$, and the pressure P is raised from P^* to P_2. Thus $\Delta h_{b\text{-}2}$ and $\Delta s_{b\text{-}2}$ become

$$\Delta h_{b\text{-}2} = h_2 - h_b = \int_{P^*}^{P} \left[v - T_2\left(\frac{\partial v}{\partial T}\right)_P\right] dP \qquad (6\text{-}129)$$

$$\Delta s_{b\text{-}2} = s_2 - s_b = \int_{P^*}^{P} -\left(\frac{\partial v}{\partial T}\right)_P dP \qquad (6\text{-}130)$$

The overall $\Delta h_{1\text{-}2}$ or $\Delta s_{1\text{-}2}$ is then simply the sum of the Δh's and Δs's of the three path segments:

$$\begin{aligned}\Delta h_{1\text{-}2} &= h_2 - h_1 = \Delta h_{1\text{-}a} + \Delta h_{a\text{-}b} + \Delta h_{b\text{-}2}\\ \Delta s_{1\text{-}2} &= s_2 - s_1 = \Delta s_{1\text{-}a} + \Delta s_{a\text{-}b} + \Delta s_{b\text{-}2}\end{aligned} \tag{6-131}$$

Substitution of the expressions for the three path-segment changes then gives us the total change in enthalpy or entropy. Similar expressions for Δu or Δs can be developed using T and v, rather than T and P, as the integration variables.

SAMPLE PROBLEM 6-11. Helium enters a reversible isothermal compressor at 540°R and 12 atm and is continuously compressed to 180 atm. Calculate the work per mole of helium needed to run the compressor and the amount of heat per mole of helium that must be removed from the compressor if

(a) Helium behaves as an ideal gas.

(b) Helium behaves according to the equations of state

$$Pv = RT - \frac{a}{T}P + bP$$

where $a = 11.13$°R ft³/lb-mol
$b = 0.2445$ ft³/lb-mol

Solution: We may picture the process as shown in Fig. SP 6-11a. We may write energy and entropy balances around the compressor and its contents as follows:

Energy:

$$\left(h + \frac{V^2}{2g_c} + \frac{gZ}{g_c}\right)_{\text{in}} \delta M_{\text{in}} - \left(h + \frac{V^2}{2g_c} + \frac{gZ}{g_c}\right)_{\text{out}} \delta M_{\text{out}} + \delta Q - \delta W = d(Me)_{\text{sys}}$$

Entropy:

$$(s\,\delta M)_{\text{in}} - (s\,\delta M)_{\text{out}} + \frac{\delta Q + \delta LW}{T_{\text{sys}}} = d(Ms)_{\text{sys}}$$

Since the compressor operates under steady-state conditions, $d(Me)_{\text{sys}} = d(Ms)_{\text{sys}} = 0$, and $\delta M_{\text{in}} = \delta M_{\text{out}}$. Kinetic and potential energy changes in the compressor will be assumed to be negligible. Since the compressor operates reversibly, $\delta LW = 0$. Therefore, the energy and entropy equations reduce to

Energy:

$$\Delta h = q - w$$

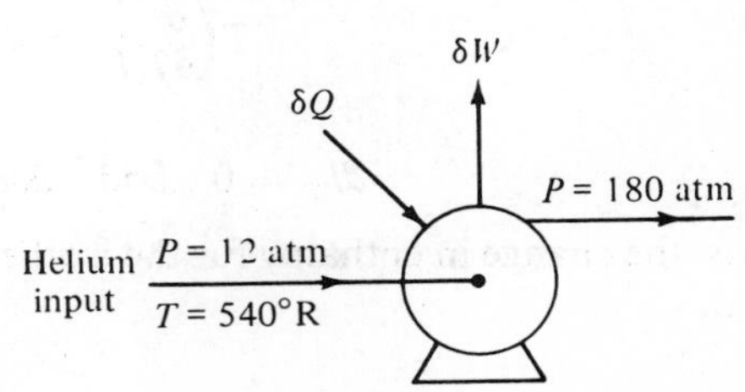

Figure SP6-11a

Entropy:

$$\Delta s = \frac{q}{T_{sys}}$$

or

$$q = T_{sys}(\Delta s)$$

But

$$T_{sys} = 540°\text{R}$$

Therefore,

$$q = 540°\text{R}\ (\Delta s)$$

Thus, if we can find Δs for the process, we can evaluate q. Then if we can find Δh, we can use q and the energy equation to determine w. On a P–T diagram the compression process is given as shown in Fig. SP 6-11b.

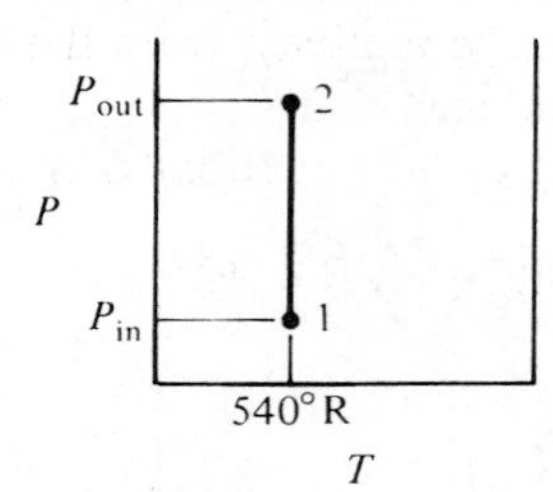

Figure SP6-11b

Since no temperature change occurs, it is not necessary to use a three-step process; a simple one-step integration from P_{in} to P_{out} at constant temperature $T = 540°\text{R}$ will be sufficient. For an isothermal process we have shown that Δh and Δs may be expressed by the relations

$$(\Delta h_{1\text{-}2})_T = \int_{P_1}^{P_2} \left[v - T\left(\frac{\partial v}{\partial T}\right)_P \right] dP$$

$$(\Delta s_{1\text{-}2})_T = -\int_{P_1}^{P_2} \left(\frac{\partial v}{\partial T}\right)_P dP$$

We must now evaluate these integrals for the desired equations of state.

(a) Ideal gas: $Pv = RT$.

$$\left(\frac{\partial v}{\partial T}\right)_P = \frac{R}{P} = \frac{v}{T}$$

Thus

$$T\left(\frac{\partial v}{\partial T}\right)_P = v$$

so

$$dh_T = 0 \quad \text{and} \quad \Delta h_{1\text{-}2} = 0$$

That is, the change in enthalpy for the isothermal compression of an ideal gas is zero.

$$\Delta s_{1\text{-}2} = -\int_{P_1}^{P_2} \frac{R}{P}\, dP = -R \ln \frac{P_2}{P_1}$$

$$= -1.987 \text{ Btu/lb-mol °R} \ln \frac{180}{12} = -5.36 \text{ Btu/lb-mol °R}$$

$$q = T\Delta s_{1\text{-}2} = -(540)(5.36) \text{ Btu/lb-mol}$$

$$= -2900 \text{ Btu/lb-mol}$$

But the energy balance gives

$$\Delta h = q - w = 0$$

Therefore,

$$q = w = -2900 \text{ Btu/lb-mol}$$

(b) Real gas: $Pv = RT + [(-a/T + b)]P$.

$$v = \frac{RT}{P} + \left(b - \frac{a}{T}\right)$$

So

$$\left(\frac{\partial v}{\partial T}\right)_P = \frac{R}{P} + \frac{a}{T^2}$$

Therefore,

$$\Delta s_{1\text{-}2} = -\int_{12}^{180} \left(\frac{R}{P} + \frac{a}{T^2}\right) dP$$

$$= R \ln \frac{180}{12} - \frac{a}{T^2}(180 - 12)$$

$$= -5.36 \frac{\text{Btu}}{\text{lb-mol °R}} - \frac{11.13}{(540)^2}$$

$$\times \frac{(180 - 12)(144)(14.7)}{778 \text{ lb-mol° R}^2} \frac{\text{°R ft}^3 \text{ lb}_f\text{-in.}^2 \text{ atm Btu}}{\text{lb-mol °R}^2 \text{ in.}^2 \text{ ft}^2 \text{ atm lb}_f\text{-ft}}$$

$$= -5.36 \frac{\text{Btu}}{\text{lb-mol °R}} - 0.0175 \frac{\text{Btu}}{\text{lb-mol °R}}$$

$$= -5.38 \frac{\text{Btu}}{\text{lb-mol °R}}$$

$$q = T\,\Delta s = -2910 \text{ Btu/lb-mol}$$

Enthalpy:

$$v - T\left(\frac{\partial v}{\partial T}\right)_P = -\frac{a}{T} + b - \frac{a}{T} + \frac{RT}{P} - \frac{RT}{P}$$

$$= b - \frac{2a}{T}$$

Therefore,

$$\Delta h_{1\text{-}2} = \int_{P_1}^{P_2} \left(b - \frac{2a}{T}\right) dP = \left(b - \frac{2a}{T}\right)(P_2 - P_1)$$

Therefore,

$$\Delta h_{1\text{-}2} = \left[0.2445 \frac{\text{ft}^3}{\text{lb-mol}} - \frac{(2)(11.13)\ {}^\circ\text{R ft}^3}{540\ {}^\circ\text{R lb-mol}}\right](180 - 12)\ \text{atm}$$

or

$$\Delta h_{1\text{-}2} = (0.2032)(180 - 12) \frac{\text{ft}^3\ \text{atm}}{\text{lb-mol}}$$

$$= 34.1 \frac{\text{ft}^3\ \text{atm}}{\text{lb-mol}}$$

$$= 34.1 \times \frac{(14.7)(144)}{778} \frac{\text{ft}^3\ \text{atm psi psf Btu}}{\text{lb-mol atm psi lb}_f\text{-ft}}$$

$$= 93.5\ \text{Btu/lb-mol}$$

But

$$\Delta h = q - w$$

or

$$w = q - \Delta h = -2910 - 93 = -3003\ \text{Btu/lb-mol}$$

Isentropic Processes with Real Gases

As we previously mentioned, many compressors, expanders, and other devices operate in a manner which may be treated as isentropic for design purposes. Since it is often necessary to calculate the work produced or required by these machines, we must devise methods for evaluating the changes in h and u which occur during isentropic processes. By and large, compressor or expander designs specify the inlet conditions (pressure and temperature) and the desired outlet pressure. Since the change in entropy of the fluid as it passes through an isentropic process is zero, the entropy of the fluid leaving an isentropic process must equal the entropy of the fluid entering the process. Thus if we know any two properties of the fluid entering the process, we can fix the fluid's state, and therefore determine its entropy. From the inlet entropy, we know the outlet entropy, which, together with the outlet pressure, fixes the state of the exiting fluid. Therefore, all other outlet properties must have fixed, but not necessarily known, values. As we are about to see, determination of these values is a straightforward, but often long and tedious, job.

We shall discuss the determination of the final temperature first. Once the final temperature is known, the three-step process may be used to evaluate changes in u and h, and thereby allow us to calculate work transferred to the surroundings during the process.

Since the equations that must be solved to determine the exit temperature are frequently extremely complicated, a trial-and-error solution for the final temperature is usually indicated. We may envision the logic of this trial-and-error solution as follows:

1. Estimate a final temperature T_2.
2. Using the three-step process and equations (6-126), (6-128), and (6-130) calculate the total change in entropy that would occur in a process between P_1, T_1 and P_2, T_2.
3. From the Δs of step 2, estimate a new value for T_2 to bring Δs closer to zero.
4. Repeat steps 2 and 3 until the change in T of step 3 is within the desired accuracy.

Another method of solution is to simply guess a whole series of T_2's and then calculate the Δs's for a process between P_1, T_1 and each P_2, T_2. A plot of Δs versus T_2 is then interpolated to $\Delta s = 0$, which gives T_2.

The second method of solutions has the advantage that it does not require us to devise a systematic method of correcting our last guess on T_2 (step 3 of the first method) and, therefore, for one-shot calculations is probably easier to use than the first method. However, for calculations that are likely to be repeated frequently (as in design of a whole series of similar compressors), the first method will require less calculations per case studied once the correction algorithm is devised and, therefore, is preferred. In addition, the first method can be programmed for computer solution; the second cannot.

SAMPLE PROBLEM 6-12. If the cooling water to the compressor of Sample Problem 6-11 fails so that the compressor now operates adiabatically (and reversibly) rather than isothermally, determine the outlet temperature of the gas and the amount of work per mole of helium that must be supplied to the compressor to maintain an outlet pressure of 180 atm if

(a) Helium behaves as an ideal gas.
(b) Helium behaves according to the equation of state

$$Pv = RT + \left(b - \frac{a}{T}\right)P$$

where $a = 0.385$ m³ °K/kg-mol
$b = 0.0153$ m³/kg-mol
C_P^* may be assumed to be constant:

$$C_P^* = 2.09 \times 10^4 \text{ J/kg-mol °K}$$

Solution: Application of the energy and entropy balances around the compressor yields the following relations:

Energy:

$$\left(h + \frac{V^2}{2g_c} + \frac{gZ}{g_c}\right)_{\text{in}} \delta M_{\text{in}} - \left(h + \frac{V^2}{2g_c} + \frac{gZ}{g_c}\right)_{\text{out}} \delta M_{\text{out}} + \delta Q - \delta W$$
$$= d(Me)_{\text{sys}}$$

Entropy:

$$(s\,\delta M)_{\text{in}} - (s\,\delta M)_{\text{out}} + \int_{\text{area}} \frac{\delta q\, dA}{T} + \int_{\text{volume}} \frac{\delta lw\, dV}{T} = d(Ms)_{\text{sys}}$$

Since the compressor operates at steady state, $d(Me)_{\text{sys}} = d(Ms)_{\text{sys}} = \delta M_{\text{in}} - \delta M_{\text{out}} = 0$. The compressor also operates adiabatically and reversibly, so $\delta q = \delta lw = 0$. Neglecting potential and kinetic energy changes in the compressor, the energy and entropy equations reduce to

Energy:

$$\Delta h_{1\text{-}2} = -w$$

Entropy:

$$\Delta s_{1\text{-}2} = 0$$

Therefore, we know that the entropy of the material leaving the compressor is identical to the entropy of the material entering the compressor. We may now use this condition to calculate the temperature of the material leaving the compressor, which may then be used to calculate Δh_{1-2} across the compressor, which in turn tells us the work that must be supplied.

(a) Ideal gas: $Pv = RT$. We have shown previously that the final T and P of an ideal gas (with constant C_P) undergoing an isentropic process are related according to the expression

$$C_P \ln \frac{T_2}{T_1} = R \ln \frac{P_2}{P_1}$$

(For the purpose of illustration, we will solve this problem using S.I. units exclusively.) That is,

$$\ln \frac{T_2}{T_1} = \frac{R}{C_P} \ln \frac{P_2}{P_1}$$

or

$$T_2 = T_1 \left(\frac{P_2}{P_1}\right)^{R/C_P}$$

But

$$T_1 = 540°\text{R} = 300\ °\text{K}$$

$$P_2 = 180 \text{ atm} = 182.4 \text{ Bar}$$

$$P_1 = 12 \text{ atm} = 12.2 \text{ Bar}$$

$$C_P = 2.09 \times 10^4 \text{ J/kg-mol °K}$$

$$R = 8.31 \times 10^3 \text{ J/kg-mol °K}$$

Therefore,

$$T_2 = 300\ °\text{K}\left(\frac{182.4}{12.2}\right)^{(8.31/20.9)} = 886\ °\text{K}$$

and the total change in enthalpy may be calculated as

$$\Delta h_{1-2} = \int_{T_1}^{T_2} C_P\, dT = (2.09)(886 - 300) \times 10^4 \text{ J/kg-mol}$$

$$= 1.22 \times 10^7 \text{ J/kg-mol} = 12.2 \text{ MJ/kg-mol}$$

(b) Real gas: $Pv = RT + [b - a/T)]P$. We must now attempt to find T_2 such that Δs_{1-2} between P_1, T_1 and P_2, T_2 is zero. According to the first procedure outlined in the text, the following trial-and-error solution will be attempted.

1. Guess an initial value of T_2.
2. Calculate Δs from P_1, T_1 to P_2, T_2.
3. Correct last guess for T_2.
4. Repeat steps 2 and 3.

For our initial guess of T_2, let us choose $T_2 = 860°\text{K}$. (Usually a wise initial choice for T_2 is something near the value determined for the ideal-gas case.)

We must now set up our three-step path for the evaluation of Δs. A P–T diagram for the path is given in Fig. SP 6-12a and the total entropy change

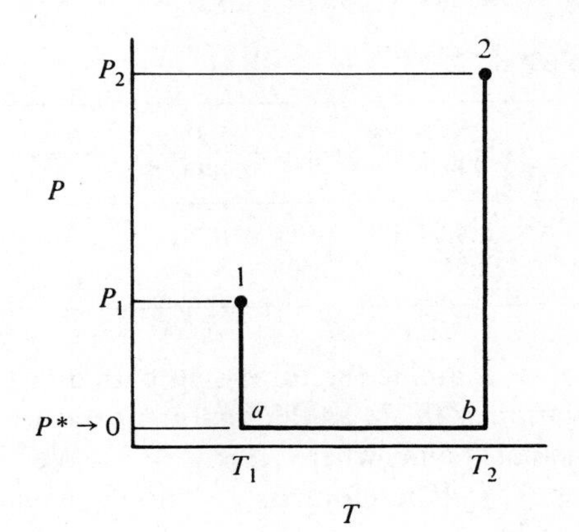

Figure SP6-12a

Δs_{1-2} is given by

$$\Delta s_{1-2} = \Delta s_{1-a} + \Delta s_{a-b} + \Delta s_{b-2}$$

where

0

$$\Delta s_{1-a} = \int_{P_1}^{P^*} -\left(\frac{\partial v}{\partial T}\right)_P dP = \int_{P_1}^{P^*} -\left(\frac{R}{P} + \frac{a}{T^2}\right) dP = -R \ln \frac{P^*}{P_1} - \frac{a}{T_1^2}(P^* - P_1)$$

$$\Delta s_{a-b} = \int_{T_1}^{T_2} \frac{C_P^*}{T}\, dT = C_P^* \ln \frac{T_2}{T_1}$$

0

$$\Delta s_{b-2} = \int_{P^*}^{P_2} -\left(\frac{\partial V}{\partial T}\right)_P dP = -R \ln \frac{P_2}{P^*} - \frac{a}{T_2^2}(P_2 - P^*)$$

where P^* is a very low pressure approaching zero. Therefore,

$$\Delta s_{1-2} = -R \ln \frac{P_2}{P_1} + C_P^* \ln \frac{T_2}{T_1} + \frac{aP_1}{T_1^2} - \frac{aP_2}{T_2^2}$$

but

$$-R \ln \frac{P_2}{P_1} = -8310 \text{ J/kg-mol °K} \ln \frac{182.4}{12.2} = -2.25 \times 10^4 \text{ J/kg-mol °K}$$

and

$$\frac{aP_1}{T_1^2} = \frac{(0.385)(1.22 \times 10^6)}{(300)^2} \frac{\text{Nm}}{\text{kg-mol °K}} = 5.21 \text{ J/kg-mol °K}$$

$$aP_2 = 0.385 \times 1.82 \times 10^7 \frac{\text{Nm °K}}{\text{kg-mol}} = 7.0 \times 10^6 \text{ J °K/kg-mol}$$

Therefore,

$$\Delta s_{1-2} = -22{,}500 \text{ J/kg-mol °K} + C_P^* \ln \frac{T_2}{300} - \frac{7.0 \times 10^6}{T_2^2} \text{ J °K/kg-mol}$$

We set up Table SP 6-12 in order to evaluate Δs_{1-2} as a function of T_2. We now see that our first guess of T_2 was too low and we must raise it—but by how much? We may frequently obtain a pretty good estimate for the next guess of T_2 as follows.

Table SP 6-12

T_2	$\frac{T_2}{300}$	$C_P^* \ln \frac{T_2}{300}, \frac{J}{\text{kg-mol °K}}$	$\frac{7.0 \times 10^6}{T_2^2}, \frac{J}{\text{kg-mol °K}}$	$\Delta \underline{S}_{1-2}, \frac{J}{\text{kg-mol °K}}$
860	2.87	2.195×10^4	9.5	−540
883	2.94	22500	9.4	~0

Let us examine the three-step path actually followed during the previous calculation with $T_{2'}$ as the final temperature (Fig. SP 6-12b). Let point 2 be the actual point where $\Delta s_{1-2} = 0$. We have evaluated $\Delta s_{1-2'} = -540$ J/kg-mol °K. Therefore $\Delta s_{2'-2}$ should equal $+540$ J/kg-mol °K. But $(\Delta s)_P$ is given by

$$(\Delta s)_P = \int_{T_{2'}}^{T_2} \frac{C_P}{T} \, dT$$

Now as an approximation, assume that $C_P = C_P^*$. Therefore,

$$\Delta s_{2'-2} = C_P^* \ln \frac{T_2}{T_{2'}}$$

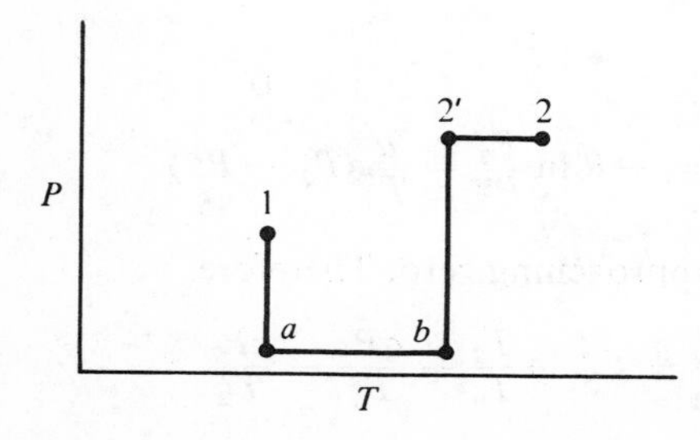

Figure SP6-12b

or

$$540 \text{ J/kg-mol °K} = 2.09 \times 10^4 \left[\ln\left(\frac{T_2}{860}\right)\right] \text{J/kg-mol °K}$$

Therefore,

$$\ln \frac{T_2}{860} = 0.0258$$

or

$$\frac{T_2}{860} = 1.026$$

so

$$T_2 = 883 \text{ °K}$$

which we now use as our second guess for T_2. Performing the calculations a second time, we find that our second guess, $T = 883$°K, satisfies the $\Delta s_{1-2} = 0$ condition to within the accuracy of our calculations and therefore is the final temperature.

With the final temperature, we may now calculate the change in enthalpy of the gas as it passes through the process from the familiar three-part path:

$$\Delta h_{1-2} = \Delta h_{1-a} + \Delta h_{a-b} + \Delta h_{b-2}$$

where

0

$$\Delta h_{1-a} = \int_{P_1}^{P^*} \left[v - T\left(\frac{\partial v}{\partial T}\right)_P\right] dP = \left(b - \frac{2a}{T_1}\right)(P^* - P_1)$$

$$\Delta h_{a-b} = \int_{T_1}^{T_2} C_p \, dT = C_p^*(883 - 300) \text{ °K}$$

0

$$\Delta h_{b-2} = \int_{P^*}^{P_2} \left[v - T\left(\frac{\partial v}{\partial T}\right)_P\right] dP = \left(b - \frac{2a}{T_2}\right)(P_2 - P^*)$$

Substitution of the values for a, b, P_1, P_2, T_1 and T_2 yields the value

$$\Delta h_{1-2} = 1.22 \times 10^7 \text{ J/kg-mol}$$

so

$$w = -1.22 \times 10^7 \text{ J/kg-mol}$$

6.14 Property Changes Using the Law of Corresponding States

Enthalpy Changes

In the evaluation of Δh_{1-2} using the three-step process previously described, we showed that evaluation of the integrals

$$\Delta h_{1-a} = h_a - h_1 = \int_{P_1}^{P^*} \left[v - T_1\left(\frac{\partial v}{\partial T}\right)_P\right] dP \qquad (6\text{-}125)$$

$$\Delta h_{b-2} = h_2 - h_b = \int_{P^*}^{P_2} \left[v - T_2 \left(\frac{\partial v}{\partial T} \right)_P \right] dP \qquad \text{(6-129)}$$

was necessary. Since these integrals depend solely on the P–v–T behavior of the substance in question, we might hope to express the integrands only in terms of compressibility, Z, and the reduced variables, P_r and T_r. We could then use the generalized compressibility charts to evaluate these integrals as functions of P_r and T_r. We shall now show how this evaluation may be performed.

In the generalized charts, the compressibility factor, $Z = Pv/RT$, is expressed as a function of P_r and T_r. Therefore, before we can use these charts for the evaluation of the equations for Δh_{1-a} and Δh_{b-2}, we must first express the integrals in terms of only reduced variables and compressibilities. This is accomplished as follows. Rearrangement of the definition of Z gives an expression for v:

$$v = \frac{RTZ}{P} \qquad \text{(6-132)}$$

which may be differentiated with respect to T at constant P to give

$$\left(\frac{\partial v}{\partial T} \right)_P = \frac{TR}{P} \left(\frac{\partial Z}{\partial T} \right)_P + \frac{RZ}{P} \qquad \text{(6-133)}$$

Combination of equations (6-132) and (6-133) yields

$$v - T \left(\frac{\partial v}{\partial T} \right)_P = \frac{RT}{P} \left[Z - Z - T \left(\frac{\partial Z}{\partial T} \right)_P \right] \qquad \text{(6-134)}$$

Simplification and multiplication by dP yields the expression

$$\left[v - T \left(\frac{\partial v}{\partial T} \right)_P \right] dP = \frac{RT^2}{P} \left(\frac{\partial Z}{\partial T} \right)_P dP \qquad \text{(6-135)}$$

The T's and P's of equation (6-135) can be converted to reduced variables by division by T_c and P_c. The reduced version of equation (6-135) is then substituted into equation (6-125) to give

$$h - h^* = \int_{P^*}^{P} \left[v - T \left(\frac{\partial v}{\partial T} \right)_P \right] dP = \int_{P_r^*}^{P_r} \frac{RT_c T_r^2}{P_r} \left(\frac{\partial Z}{\partial T_r} \right)_{P_r} dP_r \qquad \text{(6-136)}$$

Division by RT_c gives

$$\frac{h - h^*}{RT_c} = \int_{P_r^*}^{P_r} \frac{T_r^2}{P_r} \left(\frac{\partial Z}{\partial T_r} \right)_{P_r} dP_r \qquad \text{(6-137)}$$

At a given T_r and P_r we may evaluate $(\partial Z/\partial T_r)_{P_r}$ by graphically differentiating the data presented in the generalized compressibility chart. With the values of $(\partial Z/\partial T_r)_{P_r}$ so determined, we can graphically integrate equation (6-137) by plotting $(T_r^2/P_r)\,(\partial Z/\partial T_r)_{P_r}$ versus P_r from some very low pressure P_r^* to the desired pressure P_r. The value of the integral is then the area under the curve. In this manner we can determine the isothermal enthalpy change for a process between zero pressure and any desired reduced pressure P_r. Although the term $1/P_r$ in equation (6-137) would appear to cause trouble as $P_r \longrightarrow 0$, it is known that $(\partial Z/\partial T)_{P_r}$ approaches zero as $P_r \longrightarrow 0$. However, l'Hôspital's rule tells us that

$$\lim_{P_r \to 0} \left(\frac{\partial Z/\partial T_r}{P_r}\right) = \frac{\partial}{\partial P_r} \frac{(\partial Z/\partial T_r)_{P_r}}{1.0}\bigg|_{P_r \to 0} = \frac{\partial}{\partial P_r}\left(\frac{\partial Z}{\partial T_r}\right)_{P_r}\bigg|_{P_r \to 0} \tag{6-138}$$

Since

$$\frac{\partial}{\partial P_r}\left(\frac{\partial Z}{\partial T_r}\right)_{P_r}$$

is bounded as $P_r \longrightarrow 0$, the integrand is finite and the integral can be evaluated.

In order to eliminate the need for evaluating the integral of equation (6-137) whenever an enthalpy calculation must be performed, values of this integral have been tabulated as functions of T_r and P_r. These tables are found in many standard reference books. In addition to tables, graphs of $(h - h^*)/RT_c$, as functions of P_r and T_r, have also been assembled. A set of these graphs has been included in Appendix D.

In terms of the generalized $(h - h^*)$ charts, we can now express Δh for any process by means of our three-step path as

$$\begin{aligned}\Delta h_{1-2} &= h_2 - h_1 \\ &= -(h - h^*)_1 + (h - h^*)_2 + (h_2^* - h_1^*)\end{aligned} \tag{6-139}$$

where

$$h_2^* - h_1^* = \Delta h_{a-b} = \int_{T_1}^{T_2} C_P^* \, dT$$

is the change in enthalpy at zero pressure, while $(h^* - h)_1 = \Delta h_{1-a}$ and $(h^* - h)_2 = -\Delta h_{b-2}$.

The Fugacity Coefficient, ν

We have shown that the fugacity coefficient may be evaluated from

$$\ln \nu = \int_{P=0}^{P} \left(\frac{v}{RT} - \frac{1}{P}\right) dP \tag{6-94a}$$

Equation (6-94a) may be rearranged to give

$$\ln \nu = \int_{P=0}^{P} \left(\frac{Pv}{RT} - 1\right) d \ln P \tag{6-140}$$

or, if we introduce compressibilities and reduced pressures,

$$\ln \nu = \int_{P_r^*}^{P_r} (Z - 1)\, d \ln P_r \tag{6-141}$$

In the limit of $P \longrightarrow 0$, both $(Z - 1)$ and $P_r \longrightarrow 0$, so that the integrand of equation (6-141) becomes indeterminant. However, application of l'Hôspital's rule tell us that

$$\lim_{P \to 0} \frac{Z - 1}{P_r} \longrightarrow \left(\frac{\partial Z}{\partial P_r}\right)_{P_r = 0} \tag{6-142}$$

Examination of the compressibility charts indicates that as $P_r \longrightarrow 0$, $\partial Z/\partial P_r$ is finite, so the integrand is bounded and the integral is finite. Values of f/P, or ν, may then be graphically determined from the area under a plot of $(Z - 1)/P_r$ versus P_r.

As with equation (6-137), the integral on the right side of equation (6-142) may be evaluated once and for all as a function of P_r and T_r and then stored for future use either as a table or chart of f/P versus P_r and T_r. A chart of (f/P) versus P_r and T_r is also included in Appendix A.

Entropy Changes

In the evaluation of Δs_{1-2} using the three-step process, we found it necessary to evaluate the integrals

$$\Delta s_{1-a} = -\int_{P_1}^{P^*} \left(\frac{\partial v}{\partial T}\right)_P dP \tag{6-126}$$

$$\Delta s_{b-2} = -\int_{P^*}^{P_2} \left(\frac{\partial v}{\partial T}\right)_P dP \tag{6-130}$$

As with the enthalpy, we shall attempt to express these integrals in terms of reduced variables and compressibilities. However, it will be more to our advantage to take a slightly different approach in evaluating the two isothermal portions of the three-part Δs path. We begin by rearranging the enthalpy property relation

$$dh = T\,ds + v\,dP$$

to solve for ds:

$$ds = \frac{dh}{T} - \frac{v\,dP}{T} \tag{6-143}$$

At constant temperature, equation (6-143) may be integrated to yield

$$\Delta s_{1-a} = \frac{\Delta h_{1-a}}{T} - \frac{1}{T}\int_{P_1}^{P^*} v\,dP \tag{6-144}$$

where the path $1 - a$ represents one of the isothermal legs of the three-step process. However, we have shown in Section 6-12 how to evaluate Δh_{1-a}. Therefore, all we must do now is evaluate the integral $\int_{P_1}^{P^*} v\,dP$.

In Section 6-8 we have defined the fugacity such that

$$RT\left(\frac{\partial \ln f}{\partial P}\right)_T = v \tag{6-87}$$

Thus the second term in equation (6-144) may be expressed in terms of fugacities as

$$\frac{1}{T}\int_{P_1}^{P^*} v\,dP = R\int_{P_1}^{P^*} d\ln f \tag{6-145}$$

or

$$\frac{1}{T}\int_{P_1}^{P^*} v\,dP = R\ln\frac{f^*}{f_1} \tag{6-145a}$$

which may be substituted into equation (6-144) to give

$$\Delta s_{1-a} = \frac{\Delta h_{1-a}}{T} - R\ln\frac{f^*}{f_1} \tag{6-146}$$

where f^* is the fugacity evaluated at some low pressure, P^*. However, according to equation (6-88) $f^* \longrightarrow P^*$ as $P^* \longrightarrow 0$. Therefore, equation (6-146) becomes

$$\Delta s_{1-a} = \frac{\Delta h_{1-a}}{T_1} - R\ln\frac{P^*}{f_1} \tag{6-147}$$

Application of equation (6-147) for the second isothermal leg of the three-step path yields

$$\Delta s_{b-2} = \frac{\Delta h_{b-2}}{T_2} - R\ln\frac{f_2}{P^*} \tag{6-147a}$$

The change in entropy for the isobaric portion of the process is simply

$$\Delta s_{a-b} = \int_{T_1}^{T_2}\frac{C_P}{T}\,dT$$

which may be combined with equations (6-147) and (6-147a) to give the total change in entropy for the whole process as

$$\Delta s_{1-2} = \Delta s_{1-a} + \Delta s_{a-b} + \Delta s_{b-2} \tag{6-148}$$

$$\Delta s_{1-2} = \frac{\Delta h_{1-a}}{T_1} - R\ln\frac{P^*}{f_1} + \int_{T_1}^{T_2}\frac{C_P}{T}\,dT + \frac{\Delta h_{b-2}}{T_2} - R\ln\frac{f_2}{P^*} \tag{6-148a}$$

However, the two terms in $\ln(f/P^*)$ may be combined to yield

$$\Delta s_{1-2} = \frac{\Delta h_{1-a}}{T_1} + \frac{\Delta h_{b-2}}{T_2} + \int_{T_1}^{T_2}\frac{C_P}{T}\,dT - R\ln\frac{f_2}{f_1} \tag{6-149}$$

from which the overall entropy change may be evaluated.

SAMPLE PROBLEM 6-13. Ethylene is to be compressed adiabatically and reversibly from 200 psia and 140°F to 1500 psia. Compute the minimum work per pound of ethylene. $C_P^* = 5.34 + 0.0064T$ (°R), critical temperature = 508°R, and critical pressure = 645.0 psia.

Solution: (See Fig. SP 6-13a.) As illustrated in Sample Problem 6-12, the energy and entropy balances taken around the compressor and its contents as

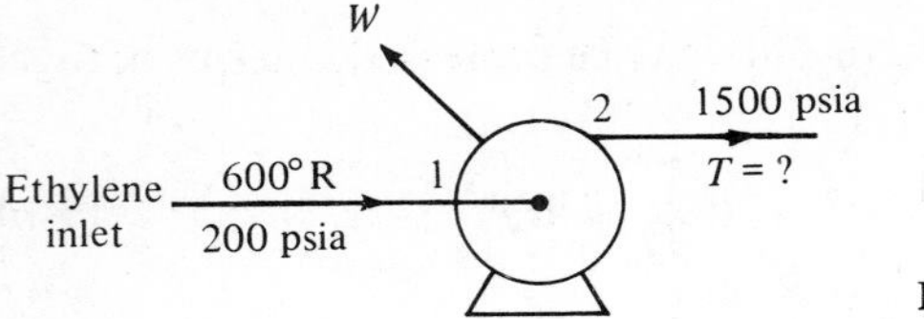

Figure SP6-13a

a system reduce to

Energy:

$$\Delta h = -w$$

Entropy:

$$\Delta s = 0$$

As in Sample Problem 6-12, we must use the constant entropy condition to determine the temperature of the material exiting from the compressor. From this temperature, we may then calculate the change in enthalpy and hence the work needed to drive the compressor. Since no information is given to us regarding the P–v–T behavior of ethylene, we are forced to use the generalized charts for the required volumetric behavior.

We may establish our three-step process for the evaluation of Δs and Δh across the compressor as indicated in Fig. SP 6-13b. The total entropy change

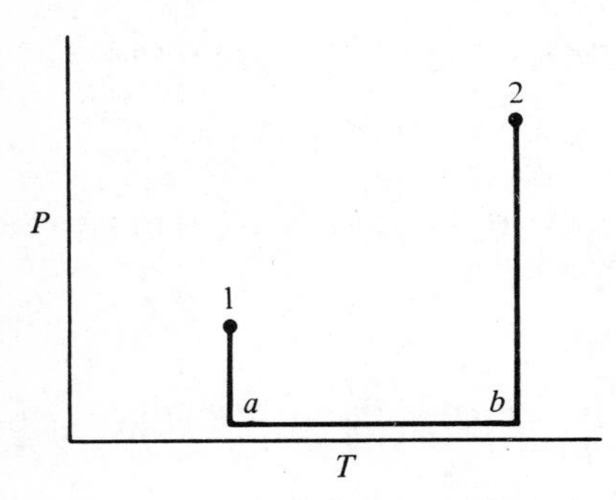

Figure SP6-13b

Δs_{1-2} is split into its three component parts:

$$\Delta s_{1-2} = \Delta s_{1-a} + \Delta s_{a-b} + \Delta s_{b-2}$$

where Δs_{1-a} is evaluated at constant $T = T_1$

Δs_{a-b} is evaluated at constant $P = P^* \longrightarrow 0$

Δs_{b-2} is evaluated at constant $T = T_2$

As given in equation (6-147), Δs_{1-a} and Δs_{b-2} may be expressed as

$$\Delta s_{1-a} = \frac{\Delta h_{1-a}}{T_1} - R \ln \frac{f^*}{f_1}$$

$$\Delta s_{b-2} = \frac{\Delta h_{b-2}}{T_2} - R \ln \frac{f_2}{f^*}$$

while Δs_{a-b} is given by

$$\Delta s_{a-b} = \int_{T_1}^{T_2} \frac{C_P}{T}\,dT = \int_{T_1}^{T_2} \frac{5.34 + 0.0064T}{T}\,dT$$

$$= 5.34 \ln \frac{T_2}{T_1} + 0.0064(T_2 - T_1) \frac{\text{Btu}}{\text{lb-mol °R}}$$

Therefore, the overall change in entropy is given by

$$\Delta s_{1-2} = \frac{\Delta h_{1-a}}{T_1} - R \ln \frac{f^*}{f_1} + \int_{T_1}^{T_2} \frac{C_P}{T}\,dT + \frac{\Delta h_{b-2}}{T_2} - R \ln \frac{f_2}{f^*}$$

or

$$\Delta s_{1-2} = \frac{\Delta h_{1-a}}{T_1} + \frac{\Delta h_{b-2}}{T_2} - R \ln \frac{f_2}{f_1} + \int_{T_1}^{T_2} \frac{C_P}{T}\,dT$$

Δh_{1-a} and f_1 may be calculated directly from the generalized charts and P_{1_r} and T_{1_r}. However, Δh_{b-2} and f_2 cannot be calculated before T_2 is known. Thus we must use a trial-and-error procedure to find the T_2 at which $\Delta s_{1-2} = 0$. Because of the complex nature of the relation between Δs_{1-2} and T_2, it seems that the second type of trial-and-error procedure suggested earlier will be more appropriate. Therefore, we now guess a series of T_2's and calculate the Δs's corresponding to these T_2's.

Inlet Data:

$$P_1 = 200 \text{ psi}, \qquad P_{1_r} = 0.268$$

$$T_1 = 140°\text{F} = 600°\text{R}, \qquad T_{1_r} = 1.18$$

Therefore, $(f/P)_{\text{inlet}} = 0.95$, and $f_1 = (0.95)(200) = 190$ psia.

$$\frac{h^* - h_1}{T_c} = 0.40 \text{ Btu/lb-mol °R} = \frac{h_a - h_1}{T_c} = \frac{\Delta h_{1-a}}{T_c}$$

But

$$\frac{\Delta h_{1-a}}{T_1} = \frac{\Delta h_{1-a}/T_c}{T_1/T_c} = \frac{\Delta h_{1-a}/T_c}{T_{1r}} = \frac{0.40}{1.18} \frac{\text{Btu}}{\text{lb-mol °R}} = 0.34 \frac{\text{Btu}}{\text{lb-mol °R}}$$

Outlet Data:

$$P_2 = 1500 \text{ psi} \qquad P_{2_r} = 2.01$$

Therefore, we calculate Δs_{1-2} as shown in Table SP 6-13.

We now plot Δs versus T (see Fig. SP 6-13c) and extrapolate to $\Delta s = 0$, giving $T = 918°$R. Now we must evaluate Δh_{1-2}. But

$$\Delta h_{1-2} = \Delta h_{1-a} + \Delta h_{a-b} + \Delta h_{b-2}$$

$$\Delta h_{1-a} = T_c \frac{h_a^* - h_1}{T_c} = (508)(0.40) \frac{\text{Btu °R}}{\text{lb-mol° R}} = 203 \text{ Btu/lb-mol}$$

$$\Delta h_{a-b} = \int_{600}^{918} C_P\,dT = (5.34)(918 - 600) + (0.0032)(918^2 - 600^2)$$

$$= 1670 + 1920 = 3590 \text{ Btu/lb-mol}$$

Table SP 6-13 Entropy Calculation

T_2, °R	T_{2r}	f_2, psia	$1.987 \ln \frac{f_2}{f_1}$, $\frac{\text{Btu}}{\text{lb-mol °R}}$	$5.34 \ln \frac{T_2}{T_1}$	$0.0064(T_2 - T_1)$	$\int_{T_1}^{T_2} \frac{C_P}{T} dT$	$\frac{\Delta h_{b-2}}{T_c}$	$\frac{\Delta h_{b-2}}{T_2}$	$\left(\frac{\Delta h_{b-1}}{T_1} + \frac{\Delta h_{b-2}}{T_2}\right)$	Δs_{1-2}, $\frac{\text{Btu}}{\text{lb-mol °R}}$
800	1.57	1320	3.87	1.55	1.28	2.83	−2.0	−1.27	−0.93	−1.97
900	1.78	1400	3.92	2.16	1.92	4.08	−1.45	−0.815	−0.475	−0.32
1000	1.97	1440	4.06	2.73	2.56	5.29	−1.20	−0.61	−0.27	0.96
1200	2.36	1500	4.1	3.17	3.84	7.01	−0.85	−0.36	−0.02	2.89

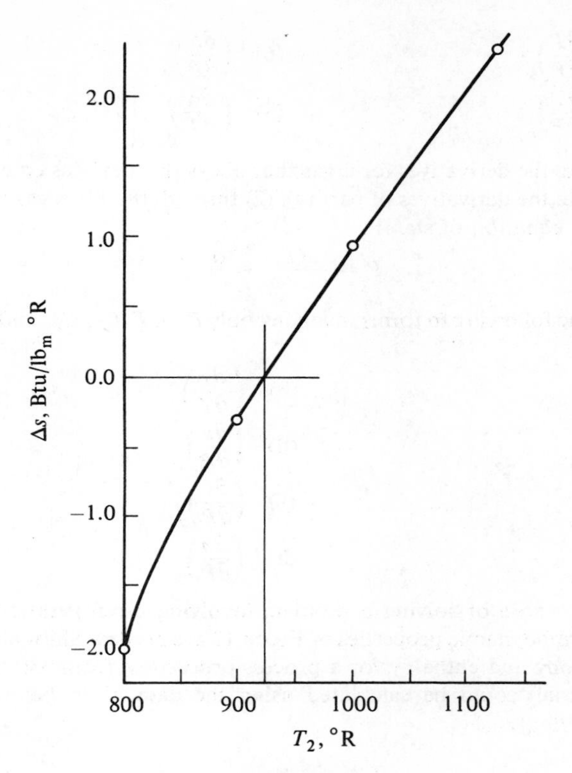

Figure SP6-13c

$$\Delta h_{b-2} = -T_c \frac{h_b^* - h_2}{T_c} = (508)(1.40)\,\frac{\text{Btu °R}}{\text{lb-mol °R}} = -711 \text{ Btu/lb-mol}$$

Therefore,

$$\Delta h_{1-2} = [203 + 3590 - 711] \text{ Btu/lb-mol} = 3080 \text{ Btu/lb-mol}$$

$$w = -\Delta h_{1-2} = -3080 \text{ Btu/lb-mol}$$

Problems

6-1. (a) Evaluate the following derivatives in terms of only P, v, T, C_P, C_V, and their derivatives.

(1) $\left(\frac{\partial^2 v}{\partial T^2}\right)_P$ (2) $\left(\frac{\partial s}{\partial P}\right)_T$

(3) $\left(\frac{\partial u}{\partial v}\right)_P$ (4) $\left(\frac{\partial P}{\partial v}\right)_s$

(5) $\left(\frac{\partial T}{\partial P}\right)_h$ (6) $\left(\frac{\partial h}{\partial P}\right)_T$

(7) $\left(\frac{\partial T}{\partial P}\right)_s$ (8) $\left(\frac{\partial v}{\partial T}\right)_u$

(b) Evaluate the derivatives for a gas that obeys the ideal-gas equation of state.

(c) Evaluate the derivatives of part (a), (2) through (8), for a gas that obeys the van der Waals equation of state:

$$P = \frac{RT}{v - b} - \frac{a}{v^2}$$

6-2. Reduce the following to forms involving only P, v, T, C_P, C_V, and their derivatives:

(a) $\left(\frac{\partial u}{\partial s}\right)_v$ (b) $\left(\frac{\partial v}{\partial T}\right)_u$

(c) $\left(\frac{\partial P}{\partial T}\right)_s$ (d) $\left(\frac{\partial u}{\partial P}\right)_T$

(e) $\left(\frac{\partial u}{\partial P}\right)_s$ (f) $\left(\frac{\partial s}{\partial P}\right)_T$

(g) $\left(\frac{\partial T}{\partial P}\right)_h$ (h) $\left(\frac{\partial h}{\partial P}\right)_T$

6-3. For the purpose of solving a problem involving a refrigeration cycle, some changes in thermodynamic properties of Freon 12 are needed. Show clearly how the change in entropy and enthalpy for a process proceeding from 330 to 350 psia at 400°F (isothermal) could be calculated using the data given below for specific volumes (in ft^3/lb_m):

Pressure, psia	380°F	390°F	400°F	410°F	420°F
300	0.22522	0.22893	0.23260	0.23625	0.23987
340	0.19589	0.19927	0.20261	0.20592	0.20921
380	0.17271	0.17583	0.17892	0.18197	0.18500

6-4. (a) During certain types of nozzle calculations it is necessary to know the rate at which the enthalpy of the nozzle fluid varies with density. Since the nozzle flow is usually approximated as isentropic, the derivative in question is

$$\left(\frac{\partial h}{\partial \rho}\right)_s$$

Express this derivative in terms of only P, v, T, C_P, C_V, and their derivatives. Evaluate for a gas that obeys the ideal-gas equation of state.

(b) It has been suggested that the temperature of an organic liquid might be reduced by adiabatically expanding the liquid through a reversible turbine. The coefficient of thermal expansion of the liquid $[\beta = (1/v)(\partial v/\partial T)_P]$ is

$$\beta = 4.0 \times 10^{-3}\ {}^\circ R^{-1}$$

and the liquid has a specific gravity of 0.8, $C_P = 0.09$ Btu/lb$_m$ °R, and a molecular weight of 100. Determine the isentropic temperature decrease with pressure, $(\partial T/\partial P)_s$, at $T = 200$°F.

6-5. (a) Evaluate

$$\left(\frac{\partial u}{\partial s}\right)_a$$

in terms of P, v, T, C_P, C_V, and their derivatives. Your answer may include absolute values of s if it is not associated with a derivative.

(b) Find an expression for the partial derivative

$$\left(\frac{\partial u}{\partial P}\right)_T$$

in terms of P, v, T, C_P, C_V, and their derivatives.

(c) Evaluate the above derivative for a gas that obeys the Redlich–Kwong equation of state.

6-6. Estimate the work required to produce droplets 0.001 ft in diameter per pound of water. Assume that the process is conducted adiabatically and reversibly and that the fluid is incompressible. [*Note:* The property relationship can be extended to include surface effects by adding a term, $\sigma\, da$, so that

$$du = T\,ds - P\,dv + \sigma\,da$$

where σ is surface tension (4.93×10^{-3} lb$_f$/ft) and a is surface area per pound mass.]

6-7. (a) Prove that

$$\left(\frac{\partial h}{\partial P}\right)_T = -\mu C_P$$

(b) The data below give μ(°C/atm) for air as a function of temperature and pressure:

Pressure, atm	0°C	25°C	50°C	75°C
1	0.2663	0.2269	0.1887	0.1581
20	0.2494	0.2116	0.1777	0.1490
60	0.2143	0.1815	0.1527	0.1275
100	0.1782	0.1517	0.1283	0.1073

Source: J. R. Roebuck, *Proc. Amer. Acad. Arts Sci.*, **60**, 535–596 (1925).

Estimate the change of heat capacity with pressure at 50°C. The observed value was 7.3×10^{-3} cal/g-mol °K atm. The molar heat capacity of air ($P = 100$ atm) is

$$C_P = \begin{cases} 7.91 \text{ cal/g-mol °K} & \text{at } 50\text{°C} \\ 7.79 \text{ cal/g-mol °K} & \text{at } 75\text{°C} \end{cases}$$

[*Hint:* Differentiate the equation of part (a) with respect to T at constant P.]

6-8. Calculate the final temperature after a Joule–Thomson expansion from 1000 bar and 300°K to 1 bar.

$$C_P = 7 \text{ cal/g-mol °K}$$

$$v = \frac{RT}{P} + (10^{-4})T^2$$

where v is in cm³/g-mol and P is in bar.

6-9. It is desired to develop a general-purpose computer program for constructing a pressure–enthalpy chart from a knowledge only of an equation of state for the gas, its C_P^* as a function of T, and the vapor–pressure curve. In particular, a method for determining the lines of constant entropy would be extremely useful.

(a) Prove that the slope of the isentrope is given by:

$$\left(\frac{\partial P}{\partial h}\right)_s = \frac{1}{v}$$

(b) Indicate clearly how the expression derived above may be integrated at constant entropy.

[*Hint:* How would one relate v to P at constant entropy knowing only the information given in the problem statement? Be specific in indicating what must be done, but do not attempt to carry out the calculations for a specific equation of state.]

6-10. A portable compressed air tank is frequently used to inflate flat tires as shown in Fig. P6-10. The tank is initially filled with compressed air at 150 psia and 70°F; the tire may be assumed to be fully evacuated (a *really* flat tire). After connecting the air hose to the tire, the valve is opened and air rushes into the tire. In the specific circumstances under study the valve is closed when the pressure in the tank has fallen to 75 psia. The volume of the tank is $\frac{1}{2}$ ft³; that of the tire is 1 ft³. Assume that the tire is rigid, so its walls do not move during the filling operation, and that the filling takes place rapidly enough that the process may be considered adiabatic; $k_{air} = 1.40$.

(a) Taking the gas that remains in the tank as a system, has this system undergone a reversible or irreversible expansion? Justify your answer.

(b) What are the final temperatures in the tire and tank?

(c) What is the final pressure in the tire (psig)?

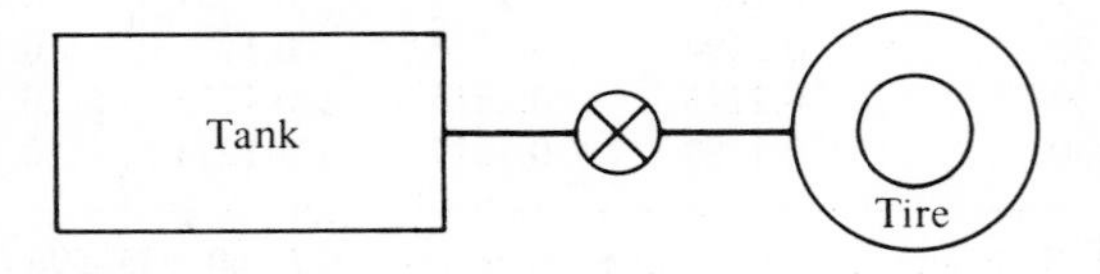

Figure P6-10

6-11. An air-actuated control device is designed as shown in Fig. P6-11. Chambers I and II are separated by a perfectly insulated frictionless piston which is connected to a movable control rod. Motion of the control rod is controlled by addition or removal of air from chamber I. Chamber II is completely airtight, and the whole

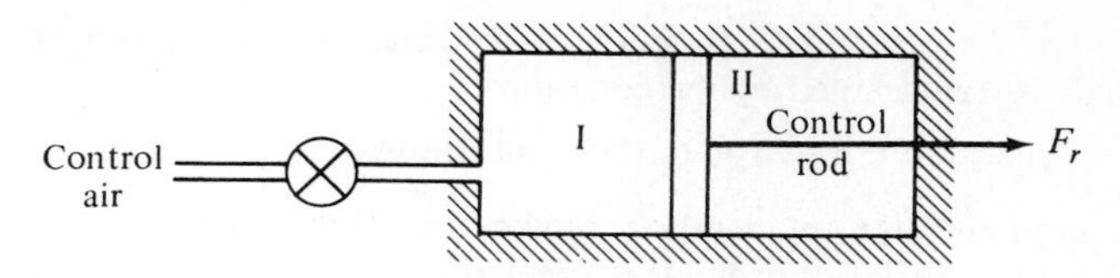

Figure P6-11

device is insulated from the surroundings. Chamber I initially occupies 10 percent of the total volume and contains air at 7 bar and 60°C; chamber II initially occupies 90 percent of the total volume and contains air at 1.2 bar and 20°C. If the resisting force on the control rod, F_r, almost counterbalances the pressure difference between chambers I and II so that the motion of the control rod can be considered reversible, determine the conditions that exist in the chambers when the control rod ceases moving. What fraction of the total volume is now occupied by chamber I? $(C_P^*)_{air} = 29.0$ kJ/kg-mol°K, and you may assume that air is an ideal gas under the problem conditions.

6-12. A diesel engine requires no spark plug when it is operating. Ignition of the fuel is carried out by adiabatically compressing the fuel–air mixture until its temperature exceeds the ignition temperature and the fuel burns. Assume that the fuel–air mixture is pumped into the cylinder at 5 bar and 10°C and is adiabatically and reversibly compressed in the closed cylinder until its volume is $\frac{1}{10}$ the initial volume. Assuming that no ignition has occurred at this point, determine the final T and P, as well as the work needed to compress each mole of air–fuel mixture. C_V^* for the mixture may be taken as 53 kJ/kg-mol°K, and the gas may be assumed to obey the equation of state

$$Pv = RT + AP$$

where A is a constant whose value is

$$A = 3 \text{ ft}^3/\text{lb-mol}$$

$$R = 0.73 \text{ atm ft}^3/\text{lb-mol °R}$$

Do not assume that C_V is independent of v.

6-13. A rigid container of negligible mass holds 18 lb_m of propane (C_3H_8) gas at 30 psia and 60°F. If the container is heated until the pressure becomes 45 psia, determine

(a) The final temperature.
(b) The amount of heat transferred.
(c) The change of internal energy of the propane.
(d) The change of entropy of the propane.
(e) The volume of the container.

Propane gas follows the P–v–T relation

$$\frac{PV}{nRT} = 1 - \frac{0.4T_cP}{P_cT}$$

where $P_c = 617.4$ psia and $T_c = 666°R$. The constant-volume heat capacity of propane may be represented by the equation

$$C_V^* = 4 + 0.021T(°R) \text{ Btu/lb-mol °R}$$

6-14. Two insulated tanks of identical volumes are connected by a pipe with a valve. One tank contains 11 lb_m of propane (C_3H_8) at 250 psia and 160°F. The other tank is completely evacuated. The valve is opened for a short time until the flow of propane ceases; the valve is then closed. Estimate, using reasonable assumptions, the mass of propane in each tank at the end of the process.

Propane gas follows the P–V–T relation

$$\frac{PV}{nRT} = 1 - \frac{0.4T_cP}{P_cT}$$

where $P_c = 617.4$ psia and $T_c = 666°R$. The constant-volume heat capacity of propane may be represented by the equation

$$C_V = 4 + 0.021T(°R) \text{ Btu/lb-mol °R}$$

6-15. A tank of oxygen (volume .05 m^3) is stored outdoors for a long time in a place where the temperature is 5°C. The pressure in the tank at this temperature is measured to be 100 bar. The tank is brought indoors and stored near a furnace, where its temperature rises to 80°C.

(a) What is the new pressure in the tank?

(b) How much heat was transferred to the contents of the tank to heat them from 5 to 80°C?

(c) With the tank at 80°C, suppose the valve is opened slightly and some gas allowed to flow into the room. Assuming the velocity of the stream to be small, what will be the temperature of the oxygen as it leaves the tank (the first increment)?

For oxygen:

$$P_c = 49.7 \text{ atm}$$
$$T_c = 154.3°K$$
$$v_c = 1.19 \text{ ft}^3\text{/lb-mol}$$
$$C_P^* = 29.3 \text{ kJ/kg-mol °K}$$
$$\text{Melting point} = 54.36°K$$
$$\text{Boiling point} = 90.19°K$$

6-16. A diesel engine operates without a spark plug by using the high-temperature gas generated during the compression stage to ignite the fuel. During a typical compression, pure air that is originally at 70°F and 0.95 atm is reversibly and adiabatically compressed to $\frac{1}{20}$ of its original volume by the piston, as shown in Fig. P6-16.

(a) If air is assumed to obey the ideal-gas equation of state, find the pressure and temperature in the cylinder at the end of the compression. $(C_P)_{air} = 7.0$ Btu/lb-mol. What is the work of compression?

(b) Using the generalized charts, determine a (hopefully) better approximation for the final temperature and pressure in the cylinder.

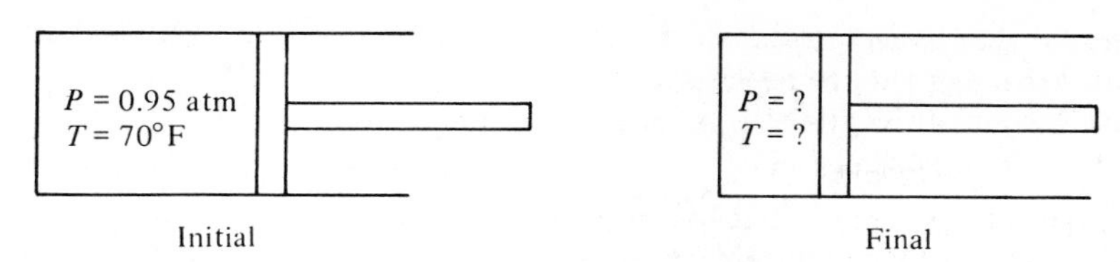

Figure P6-16

6-17. Ethylene is contained in a steel vessel at 100 bar and 20°C. A valve is opened and the tank pressure drops immediately to atmospheric pressure. If the temperature of the surroundings is 15°C, calculate the temperature of the gas remaining in the tank immediately after the expansion.

Critical properties for ethylene:

$$T_c = 282.5°\text{K}$$

$$P_c = 50 \text{ atm}$$

$$C_P^* = 2.71 + 16.20 \times 10^{-3}T(°\text{R}) - 2.80 \times 10^{-6}T^2 \text{ Btu/lb-mol °R}$$

6-18. Air is to be cooled before entering the distillation tower, where pure nitrogen and oxygen are produced. The cooling will take place in a gas-refrigeration process as follows. Air enters the refrigeration plant at 1 atm and 75°F. It is compressed to 1100 psia; water cooled to 50°F and then cooled to −150°F by heat exchange with the separated nitrogen and oxygen. In the final step, the compressed air is throttled to 1 atm and fed to the distillation tower. Assuming that the throttling operation is adiabatic, determine the temperature of the air that enters the tower. (You may assume negligible changes in potential and kinetic energy.)

The following data may prove useful:

$$C_P^* = 7.0 \text{ Btu/lb-mol °R in this temperature region}$$

$$P_c = 37.2 \text{ atm}$$

$$T_c = -140°\text{C}$$

$$v_c = 2.86 \text{ cm}^3/\text{g}$$

6-19. If the compressor in Problem 6-18 operates adiabatically and reversibly, what is the temperature of the outlet air stream? What is the work of compression? How much heat is removed in the water cooler, where the compressed oxygen is cooled to 50°F? Use the data of Problem 6-18.

6-20. Carbon dioxide is to be expanded isentropically through an insulated nozzle of proper design. The carbon dioxide will be supplied to the nozzle entrance at 800 psia and 100°F, and with negligible velocity. The discharge pressure will be 500 psia. At a pressure of 1 atm the heat capacity of carbon dioxide is given by the equation

$$C_P = 6.85 + 0.00474T(°\text{R}) - (7.64 \times 10^{-7})T^2 \text{ Btu/lb-mol °R}$$

Determine the velocity (ft/sec) at the nozzle outlet

(a) Assuming CO_2 to be an ideal gas.
(b) Assuming CO_2 to be a van der Waals gas.
(c) Using generalized plots.

6-21. Natural gas (assumed to be pure methane) flows from a pipeline into an initially empty insulated underground gas-storage reservoir. The gas flowing in the pipeline is at 25°C and 200 bar. Determine the temperature of the gas in the reservoir when the pressure reaches 200 bar. (You may assume that the filling operation is adiabatic.)

$$T_c = 191.1°\text{K}$$

$$P_c = 45.8 \text{ atm}$$

$$C_P^* = 5.52 + 0.00737T(°\text{R}) \text{ Btu/lb-mol °R}$$

6-22. The fugacity, f, has been defined in such a manner as to correct for the nonideal behavior of real gases. The compressibility factor, Z, has been defined for the same reason. Therefore, it has been suggested that the equation defining Z ($Pv = ZRT$) may be rewritten as

$$fv = RT$$

Comment in depth on the correctness of this suggestion.

6-23. Show that G is a minimum at equilibrium for the closed, isothermal, constant-pressure system shown in Fig. P6-23. Note that both pistons are free moving, massless, and perfectly conducting. In writing the energy balance, be sure to account for the work done in changing the total volume of the system as well as the work of the reversible engine. The equilibrium conditions correspond to the conditions where no work may be obtained *from the engine.*

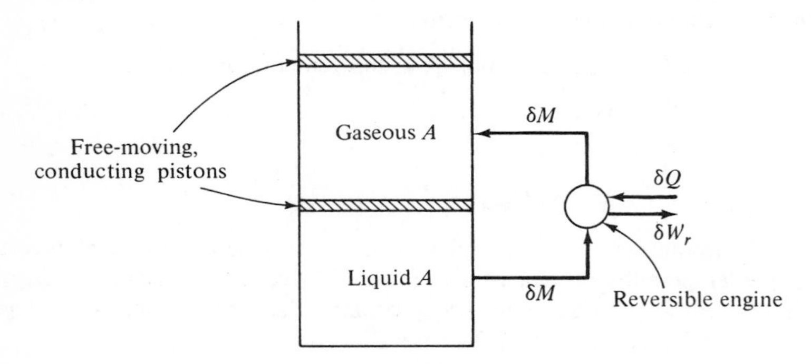

Figure P6-23

6-24. Show that A is a minimum at equilibrium for the closed, isothermal, constant-total-volume system that results if the upper piston in Fig. P6-29 is assumed to be fixed in position. Show also that the specific Gibbs free energies of the two phases are equal at equilibrium.

6-25. Listed below are properties of saturated carbon dioxide at temperatures below the triple point.

		Volume, ft^3/lb_m	
T, °F	P, psia	Solid	Vapor
−140	3.18	0.01008	24.320
−120	8.90	0.01018	9.179
−100	22.22	0.01032	3.804
−90	33.98	0.01040	2.525
−80	50.85	0.01048	1.700

Compute the Δh and Δu of sublimation of solid carbon dioxide at −100°F in Btu/lb_m.

6-26. Water and ice are in equilibrium at 0°C and 1 atm pressure. The latent heat of fusion of ice to water under these conditions is 1438 cal/g-mol. The specific volume of water is 18 cm^3/g-mol, of ice 19.7 cm^3/g-mol. The heat capacity of water is 18 cal/g-mol °C, and of ice 9 cal/g-mol °C.

(a) Compute Δg for the change of subcooled water to ice at −20°C and 1 atm.
(b) Compute the pressure at which ice and water are in equilibrium at −20°C.

6-27. Normal butane (C_4H_{10}) is stored as a compressed liquid at 190°F and 14 atm pressure. In order to use the butane in a low-pressure gas-phase process, it is throttled to 1.5 atm and passed to a vaporizer. From the vaporizer the butane emerges as a gas at 160°F and 1.5 atm. Using only the data below and any assumptions you believe are reasonable, calculate the heat that must be supplied the vaporizer per pound of butane passing through it. Describe quantitatively the condition of the butane entering the vaporizer.

For normal butane:

$$T_c = 765.3\text{°R}$$

$$P_c = 550.7 \text{ psia}$$

Heat capacity of the liquid at any constant pressure, $C_P = 0.555 + 0.0005T(\text{°F})$ Btu/lb_m °F.

Vapor pressure given by:

$$\ln P' \text{ (atm)} = \frac{-4840}{T(\text{°R})} + 9.92$$

Pressure has negligible effect on liquid enthalpy. Volume of saturated liquid = $0.025 + 0.0004T(\text{°F})$ ft^3/lb_m.

6-28. Liquid butane is pumped to a vaporizer as a saturated liquid under a pressure of 237 psia. The butane leaves the exchanger as a wet vapor of 90 percent quality and at practically the same pressure as it entered. From the following information, estimate the heat load on the vaporizer per pound of butane entering.

$$P_c = 37.48 \text{ atm}$$

$$T_c = 305.3\text{°F}$$

$$\ln P' = -4840/T + 9.92$$

where P is in atm and T is in °R. At 237 psia the specific volume of the saturated liquid is estimated as $\frac{1}{10}$ that of saturated vapor.

6-29. A certain cycle operates as shown by the path 1–2–3–4 in Fig. P6-29. Derive

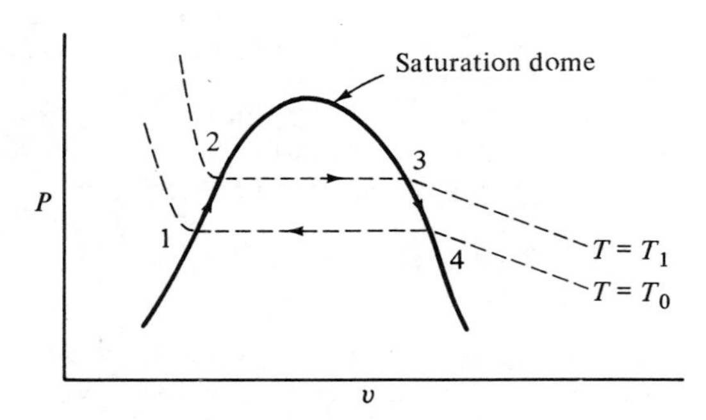

Figure P6-29

an expression involving only the heat of vaporization and the temperature T_1 and T_0 for the maximum work per cycle that can be removed per unit weight of working fluid. You may assume that the heat of vaporization is constant over the range T_0 to T_1. Note that for a closed cycle

$$\oint P\,dv = -\oint v\,dP$$

SEVEN

thermodynamics of fluid flow, compression, and expansion

Introduction

Our development thus far has stressed the basic differences between thermal and mechanical forms of energy, as well as their respective thermodynamic potentials—temperature, pressure, etc. We have rather casually alluded to the interconvertibility of mechanical forms, but have said relatively little about how the interconversion can be achieved in a manner which minimizes the degradation of mechanical potentials. In this and the following chapters we will consider systems which maximize the conversion of thermal and chemical energy to the mechanical form, as well as those which efficiently use mechanical energy to elevate thermal energy from one temperature to a higher one (for example, a refrigerator or air conditioner).

These applications often require the compression and expansion of fluids to achieve the desired conversions. Our ability to vary the pressure potential with a minimum of irreversibilities is essential if the interconversion of energy forms is to be accomplished efficiently. The pressure potential is the driving force for transporting fluids in pipelines and process equipment and for the conversion of internal energy or enthalpy into kinetic energy or work. In all of the above cases, it is the engineer's objective to either utilize or produce the pressure potential as reversibly as possible so as to minimize the work required to raise a fluid's pressure, maximize the work or thrust produced from an expanding gas, or to maximize the fluid flow produced for a given pressure drop.

Compression of fluids is typically achieved with pumps for liquid phases, and compressors for gases and vapors. Either positive displacement, or centrifugal devices can be used, with the choice depending on fluid properties and the desired compression ratio (P_{out}/P_{in}). Expanders, like pumps and compressors, can be of either the rotary (gas or liquid turbine) or positive displacement (reciprocating or piston) type. Expansion can also be achieved in nozzles if the objective is to produce *thrust*, as in the case of jet and rocket engines. In each of these applications, the expander is designed to minimize irreversibilities and maximize use of the pressure potential to produce work, or kinetic energy. Throttling devices, such as valves, are typically used for flow control and pressure let-down where the investment required for a work-producing expander cannot be justified in terms of the work recovered.

7.1 The Mechanical Energy Balance

In Chapter 6 we showed how the energy and entropy balances could be combined in such a way as to eliminate all path variables, and thereby derived the property relations. In this chapter we shall discuss a slightly different combination of the energy and entropy balances which produces the *mechanical energy balance*. This energy balance will then be applied to the study of fluid flow problems, including some interesting and useful applications for the study of compressible flows through pipelines and nozzles.

Let us examine the flow of a fluid through a piece of process equipment such as a pipeline, pump, or expander, as shown in Fig. 7-1. If we choose the piece of equipment and its contents at any instant as a system, the first-law energy balance can be written as

$$dE_{sys} = \delta Q - \delta W - \Delta[(h + ke + pe)\cdot\delta M] \tag{7-1}$$

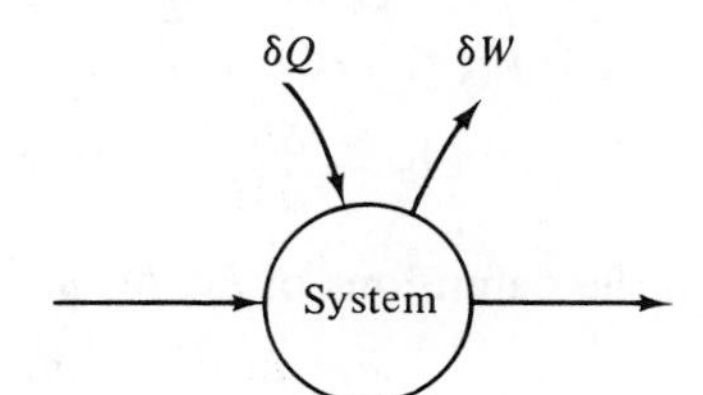

Figure 7-1. System.

The entropy balance for the system may be written as

$$dS_{sys} = \int_{volume} \frac{\delta lw}{T}\, dV + \int_{surface} \frac{\delta q}{T}\, dA - \Delta(s\,\delta M) \tag{7-2}$$

For a steady-state process $\Delta(\delta M) = dE_{sys} = dS_{sys} = 0$, so equations (7-1) and (7-2) yield, respectively,

$$Q - W - \Delta\left(h + \frac{V^2}{2g_c} + \frac{gZ}{g_c}\right)\delta M = 0 \tag{7-3}$$

$$\int_{volume} \frac{\delta lw}{T}\, dV + \int_{area} \frac{dq}{T}\, dA - (\Delta s)\,\delta M = 0 \tag{7-4}$$

If we now turn our attention to an infinitesimal portion of the system under investigation as shown in Fig. 7-2, the changes in h, s, V, and Z may be represented as differential quantities rather than as differences. (Later we shall

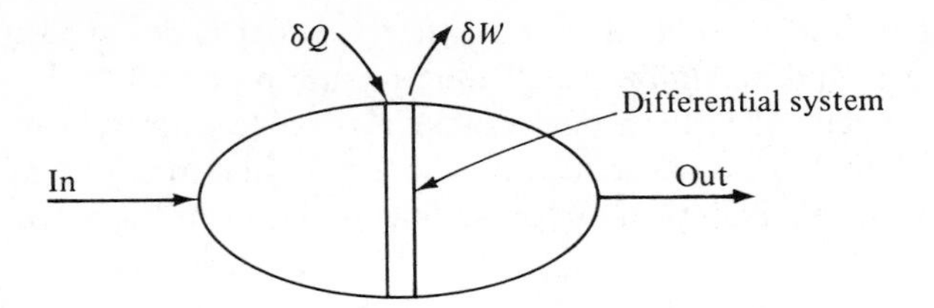

Figure 7-2. Differential system.

integrate the differential equations to obtain the desired difference forms.) The differential forms of equations (7-3) and (7-4) are given by

$$\delta Q - \delta W - d\left(h + \frac{\mathrm{V}^2}{2g_c} + \frac{gZ}{g_c}\right) \delta M = 0 \qquad (7\text{-}5)$$

$$\frac{\delta \mathbf{lw}}{T}\, dV + \frac{\delta \mathbf{q}}{T}\, dA = (ds)\, \delta M \qquad (7\text{-}6)$$

Since the temperature and δM are constant across the differential system, we may multiply equation (7-6) by $T/\delta M$ to obtain

$$\frac{\delta \mathbf{lw}\, dV}{\delta M} + \frac{\delta \mathbf{q}\, dA}{\delta M} = T\, ds \qquad (7\text{-}6a)$$

or

$$\mathbf{lw}\, dV + \mathbf{q}\, dA = T\, ds \qquad (7\text{-}7)$$

where

$$\mathbf{lw} = \frac{\delta \mathbf{lw}}{\delta M} \quad \text{and} \quad \mathbf{q} = \frac{\delta \mathbf{q}}{\delta M}$$

However. from the definitions of $\mathbf{lw}$, lw, $\mathbf{q}$, and q we see

$$\mathbf{lw}\, dV = \delta lw$$

$$\mathbf{q}\, dA = \delta q$$

so that equation (7-7) reduces to

$$\delta\, lw + \delta q = T\, ds \qquad (7\text{-}7a)$$

The heat term, δq, may now be eliminated between equations (7-5) and (7-7a) to give

$$-\delta\, lw - \delta w + T\, ds - d\left(h + \frac{\mathrm{V}^2}{2g_c} + \frac{gZ}{g_c}\right) = 0 \qquad (7\text{-}8)$$

or

$$T\, ds - dh - \delta lw - \delta w - d\left(\frac{\mathrm{V}^2}{2g_c} + \frac{gZ}{g_c}\right) = 0 \qquad (7\text{-}8a)$$

However, the enthalpy property relation states that

$$dh - T\, ds = -v\, dP$$

Therefore,

$$T\,ds - dh = -v\,dP \tag{7-9}$$

Equation (7-9) may then be substituted into equation (7-8a) to yield

$$-\delta lw - \delta w - v\,dP - d\left(\frac{V^2}{2g_c} - \frac{gZ}{g_c}\right) = 0 \tag{7-10}$$

Equation (7-10) is then integrated across the whole process under consideration (by summing the changes that occur across all the differential elements which comprise the original system) to give

$$-lw - w - \int_{P_1}^{P_2} v\,dP - \Delta\left(\frac{V^2}{2g_c} + \frac{gZ}{g_c}\right) = 0 \tag{7-11}$$

Equation (7-11) is the *mechanical energy balance*. The only assumption made in deriving this equation is that the process under consideration is at steady state! (We have also implicitly assumed that the process under consideration occurs through a series of equilibrium states so that the property relation may be used to eliminate $T\,ds - dh$.)

The work term in the mechanical energy balance is simply the sum of all the works transferred to the differential portions of the system, and, therefore, is the total work transferred from the system to the surroundings. The quantities $\int v\,dP$ and lw, on the other hand, are path functions and must be evaluated over the path followed through the actual process. Thus application of the mechanical energy balance is a very difficult task unless the state of the fluid is known at all times throughout the system.

7.2 Applications of the Mechanical Energy Balance to Fluid Flow

Let us now examine some of the simplifications that are commonly applied to reduce the mechanical energy balance to a more usable form.

Incompressible Fluid Flow: For an incompressible fluid, the specific volume v is constant, independent of T and P, so that it may be removed from under the integral sign. That is,

$$\int_{P_1}^{P_2} v\,dP = v\int_{P_1}^{P_2} dP = v\,\Delta P \tag{7-12}$$

Since $v = 1/\rho$, equation (7-12) becomes

$$\int_{P_1}^{P_2} v\,dP = \frac{\Delta P}{\rho} \tag{7-13}$$

Equation (7-13) may be substituted into (7-11) to give

$$-lw - w - \Delta\left(\frac{P}{\rho} + \frac{V^2}{2g_c} + \frac{gZ}{g_c}\right) = 0 \tag{7-14}$$

which is *Bernoulli's equation* for one-dimensional fluid flow with work, and lost work. If $w = lw = 0$, then equation (7-14) reduces to the more familiar form of Bernoulli's equation:

$$\Delta\left(\frac{P}{\rho} + \frac{V^2}{2g_c} + \frac{gZ}{g_c}\right) = 0 \tag{7-15}$$

or

$$\frac{P}{\rho} + \frac{V^2}{2g_c} + \frac{gZ}{g_c} = \text{const} \tag{7-16}$$

SAMPLE PROBLEM 7-1. A pump is taking 0.3 m³/min of water at 30°C from an open tank and delivering it to nozzles at the top of a spray tower that is 20 m high (see Fig. SP7-1). The pump motor produces 6 hp, and the pump has a mechanical efficiency of 75 percent. The water flows through a 0.08-m-i.d. pipe, and losses due to friction in the pipe are estimated to be 25 J/kg. If a pressure gauge was installed at the inlet of the nozzle, what would be the reading on the gauge? If there was a frictional loss across the nozzle of 15 J/kg of water entering the nozzle, what would be the velocity of the water at the nozzle outlet?

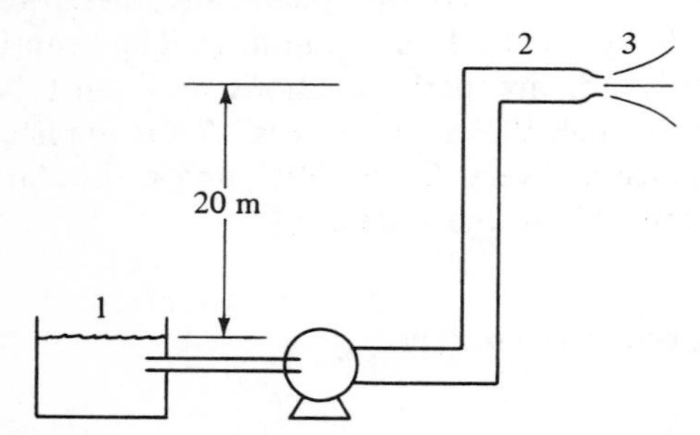

Figure SP7-1

Solution: For the first question, we take as a system the tank, the pump, and all the piping up to, but not including, the nozzle. That is everything between points 1 and 2 on the diagram. A mechanical energy balance may be written around this system as follows:

$$-\int_{P_1}^{P_2} v\,dP - \left(\Delta\frac{V^2}{2g_c} + \frac{g}{g_c}\Delta Z\right) - lw - w = 0$$

In the problem statement we are told that the pump motor delivers 6 hp to the pump. Therefore,

$$\dot{W} = -6\text{ hp} \times \frac{746\text{ J/sec}}{\text{hp}} = -4470\text{ J/sec}$$

is removed from the system. But

$$w = \frac{\dot{W}}{\dot{M}}$$

where

$$\dot{M} = \text{mass flow rate} = 0.3\frac{\text{m}^3}{\text{min}} \times 1000\,\frac{\text{kg}}{\text{m}^3} \times \frac{1\ \text{min}}{60\ \text{sec}}$$
$$= 5.0\ \text{kg/sec}$$

Therefore,

$$w = \frac{-4470}{5.0} = -894\ \text{J/kg}$$

However, of this, 25 percent, or 223 J/kg, is degraded to lost work in the pump. Another 25 J/kg is lost in the piping. Therefore,

$$lw = 25 + 223 = 248\ \text{J/kg}$$

At point 1,

$$P = \text{atmospheric pressure} = 0\ \text{(reference pressure)}$$
$$\text{V} = 0$$
$$Z = \text{reference height} = 0$$

at point 2,

$$P = \text{some pressure}\ P_2,\ \text{N/m}^2$$
$$Z = 20\ \text{m above point 1}$$
$$\text{V} = \frac{\text{volumetric flow rate}}{\text{area of 0.08-m-i.d. pipe}}$$

or

$$\text{V} = 0.3\frac{\text{m}^3}{\text{min}} \times \frac{1\ \text{min}}{60\ \text{sec}} \times \frac{1}{(0.08\ \text{m})^2 \cdot \pi/4}$$
$$= \frac{3.18}{\pi}\ \text{m/sec} = 1.01\ \text{m/sec}$$

and the mechanical energy balance reduces to

$$\int_{P_1}^{P_2} v\,dP = -\left[\frac{\text{V}_2^2 - \text{V}_1^2}{2g_c} + \frac{g}{g_c}(Z_2 - Z_1)\right] - lw - w$$
$$= -\left[\frac{(1.01)^2}{2g_c} + \frac{g}{g_c}(20 - 0)\right] - 248\ \text{J/kg} + 894\ \text{J/kg}$$

or, neglecting V_2^2,

$$\int_{P_1}^{P_2} v\,dP = (-196 - 248 + 894)\ \text{J/kg} = 450\ \text{J/kg}$$

where

$$g = 9.8\ \text{m/sec}^2$$
$$g_c = 1\ \text{kg m/N sec}^2$$

But liquid water is essentially incompressible; therefore, $v = \text{const}$ and

$$\int_{P_1}^{P_2} v\,dP = v\int_{P_1}^{P_2} dP = v(P_2 - P_1)$$

But $P_1 = 0$, therefore,

$$vP_2 = 450 \text{ J/kg} = 450 \text{ N m/kg}$$

$$v = 0.001 \text{ m}^3\text{/kg}$$

Therefore, $P_2 = 450{,}000 \text{ N/m}^2 = 4.50 \text{ Bar} \approx 4.5 \text{ atm}$.

Now choose the nozzle as the system:

$$w = 0, \qquad lw = 15 \text{ J/kg}, \qquad \int_{P_2}^{P_3} v\,dP = v(P_3 - P_2)$$

But $P_3 = 0$; therefore,

$$\int_{P_2}^{P_3} v\,dP = 450 \text{ J/kg}$$

Application of the mechanical energy balance around the nozzle then gives

$$\frac{\Delta V^2}{2g_c} = -\frac{\Delta P}{\rho} - lw = [-(-450) - 15] \text{ J/kg}$$

$$= +435 \text{ J/kg}$$

Therefore,

$$\Delta V^2 = V_3^2 - V_2^2 = 870 \text{ m}^2\text{/sec}^2$$

But $V_2^2 = 1.0 \text{ m}^2\text{/sec}^2$; therefore,

$$V_3 = 29.5 \text{ m/sec}$$

which is the velocity of the fluid leaving the nozzle tip.

Evaluation of the Lost Work for Flow Through Pipes and Fittings

For horizontal flow of an incompressible liquid through a constant area conduit, the mechanical energy balance reduces to

$$lw = \frac{-\Delta P}{\rho} \tag{7-17}$$

That is, the whole pressure drop is associated with the irreversibilities of friction. For flow through circular cylinders, the frictional pressure drop may be expressed in terms of the Fanning friction factor:

$$f = \frac{1}{2}\frac{Dg_c(-\Delta P)}{\rho V^2 L} \tag{7-18}$$

or

$$\frac{-\Delta P}{\rho} = \frac{2V^2 fL}{Dg_c} \tag{7-18a}$$

Thus the lost work is expressible in terms of the friction factor, fluid velocity, and the L/D ratio of the pipe:

$$lw = -\frac{\Delta P}{\rho} = \frac{2V^2 fL}{Dg_c} \tag{7-19}$$

By means of the technique of dimensional analysis,[1] it may be shown that the friction factor is a function of the Reynolds number ($N_{Re} = DV\rho/\mu$) of the flow, the ratio of pipe roughness to pipe diameter (ϵ/D), and pipe L/D ratio. That is,

$$f = \Phi\left[N_{Re}, \frac{\epsilon}{D}, \frac{L}{D}\right] \tag{7-20}$$

For most problems of engineering significance, the L/D ratio is not an important parameter (that is, the flow is fully developed), and the functional dependence of equation (7-20) reduces to

$$f = \Phi\left[N_{Re}, \frac{\epsilon}{D}\right] \tag{7-20a}$$

In Fig. 7-3 the Fanning friction factor is given as a function of the Reynolds number and ϵ/D ratio. Because of the difficulty in accurately describing the irreversibilities in a flowing fluid, the information in Fig. 7-3 has been determined experimentally.

Values of ϵ for several commonly used piping materials are listed in Table 7-1.

Table 7-1 Roughness of Commonly Used Piping Materials†

Material	Roughness, ft	Material	Roughness, ft
Riveted steel	0.003–0.03	Concrete	0.001–0.01
Cast iron	0.00085	Galvanized iron	0.0005
Commerical steel	0.00015	Wrought iron	0.00015
Drawn tubing	0.000005		

†From S. Whitaker, *Introduction to Fluid Mechanics*, Prentice-Hall, Inc., Englewood Cliffs, N. J., 1968, p. 299.

For noncircular ducts of constant cross section, equation (7-19) is still commonly used to relate the lost work to the friction factor. The pipe diameter, D, is replaced by the "hydraulic diameter" D_h:

$$D_h = \frac{4\text{ (conduit area)}}{\text{wetted perimeter}} \tag{7-21}$$

The friction factor is evaluated from Fig. 7-3 using a Reynolds number based on the hydraulic diameter. (Note that for a conduit of circular cross section the hydraulic and actual diameters are identical.)

Equation (7-19) is useful for relating the lost work to the fluid velocity, pipe L/D ratio, and the Fanning friction factor for flows where the fluid

[1]R. B. Bird, W. E. Stewart, and E. N. Lightfoot, *Transport Phenomena*, John Wiley & Sons, Inc., New York, 1960.

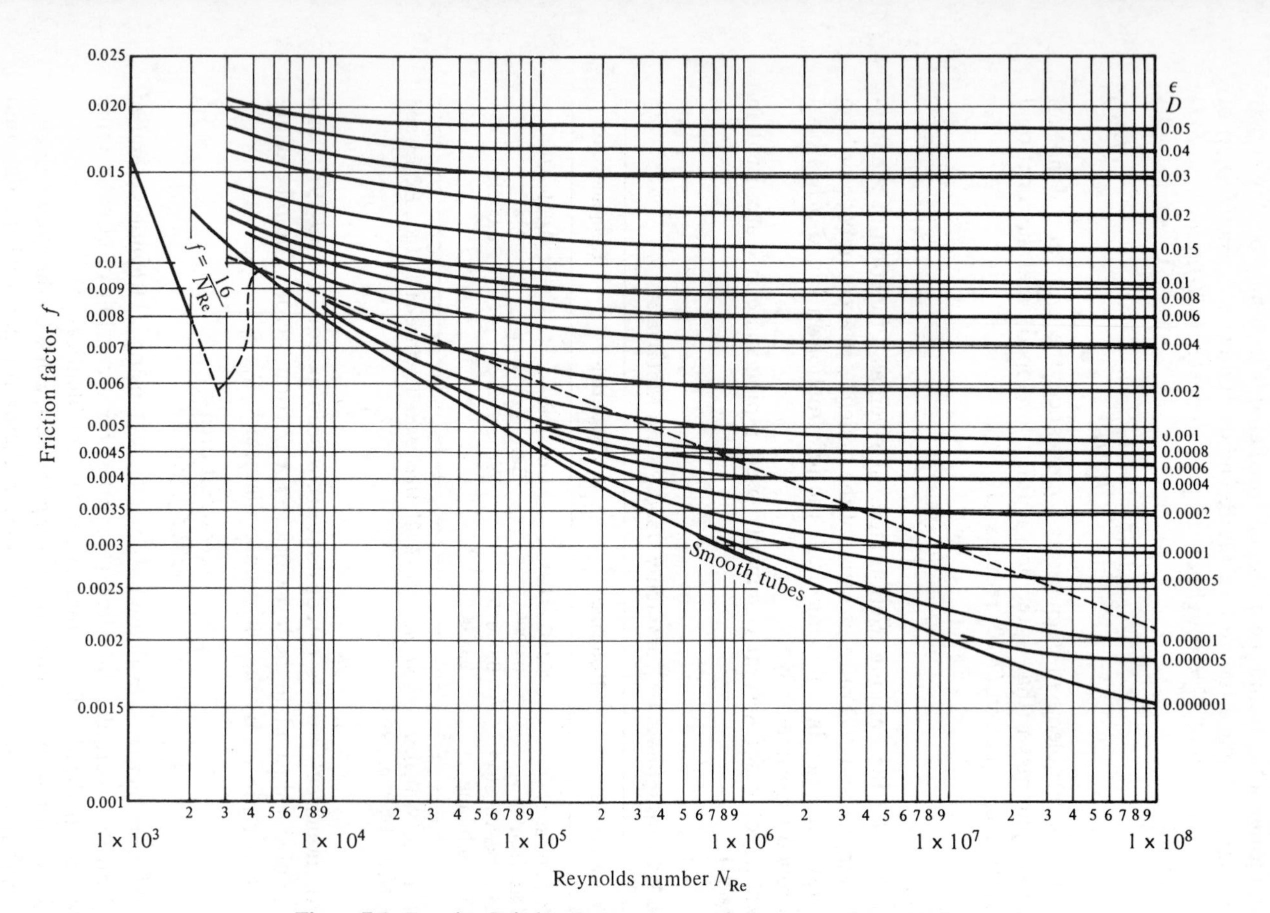

Figure 7-3. Fanning Friction factor vs. Reynolds number. (From Chemical Engineers' Handbook, R. H. Perry et al. eds. McGraw-Hill, Inc., New York, 1963, pg. 20.)

velocity remains constant. If the fluid velocity is changing with position down the pipe (either because the pipe area is changing or because of compressibility effects), we must use a differential form of equation (7-19):

$$\delta lw = \frac{2V^2 f}{Dg_c}\, dL \tag{7-22}$$

and integrate over the length of the pipeline:

$$lw = \int_0^L \frac{2V^2 f}{Dg_c}\, dL \tag{7-22a}$$

If the fluid velocity is changing gradually as the fluid moves down the pipeline, we can assume that the local velocity profiles are fully developed, so the friction factor-Reynolds number relation portrayed in Fig. 7-3 may be used to express the variation of f with the fluid velocity and fluid properties. For many problems of practical interest, the Reynolds numbers are quite high, say greater than 50,000. In this region Fig. 7-3 shows that the friction factor is reasonably insensitive to changes in the Reynolds number—particularly for commercial pipes with the larger roughness. Therefore, we may often remove the friction factor from under the integral sign in equation (7-22a) and simplify the treatment significantly.

The lost work associated with fittings and valves is usually expressed in terms of the "loss coefficient," k, of the valve or fitting:

$$lw = \frac{kV^2}{2g_c} = \left(\frac{-\Delta P}{\rho}\right)_{\text{valve or fitting}} \tag{7-23}$$

For many flow situations of interest (that is, Reynolds numbers greater than 20,000) it is observed that the loss coefficient defined by equation (7-23) is reasonably constant, independent of both diameter and Reynolds number. A list of commonly used loss coefficients for valves and fittings is presented as Table 7-2.

It is interesting to note that equation (7-19), which describes the frictional losses in straight pipe, may also be arranged in terms of a "pseudo loss coefficient":

$$lw = \frac{4fL}{D}\frac{V^2}{2g_c} \tag{7-19}$$

where the pseudo loss coefficient is

$$k_{\text{std pipe}} = \frac{4fL}{D} \tag{7-24}$$

In the preceding development we have tacitly assumed that frictional effects were the sole causes of any pressure changes that may have occurred. In this way we were able to express the lost work as $\Delta P/\rho$. The pressure drop was then expressed in terms of the Fanning friction factor or loss coefficient. In situations where pressure changes due to other than frictional effects are

Table 7-2 Loss Coefficients for Various Valves and Fittings†

Type of Fitting or Valve	Loss Coeffi-cient, *k*	Type of Fitting or Valve	Loss Coeffi-cient, *k*
45° ell		Union	0.04
standard	0.35	Gate valve	
long radius	0.20	open	0.20
90° ell		$\frac{3}{4}$ open	0.90
standard	0.75	$\frac{1}{2}$ open	4.5
long radius	0.45	$\frac{1}{4}$ open	24.0
square or miter	1.3	Diaphragm valve	
180° bend, close return	1.5	open	2.3
Tee, standard		$\frac{3}{4}$ open	2.6
along run, branch blanked off	0.4	$\frac{1}{2}$ open	4.3
used as ell, entering run	1.3	$\frac{1}{4}$ open	21.0
used as ell, entering branch	1.5	Globe valve	
branching flow	~1	open	6.4
Coupling	0.04	$\frac{1}{2}$ open	9.5

†From R. H. Perry et al., *Chemical Engineers' Handbook*, McGraw-Hill, Inc., New York, 4th ed., 1963, Sec. 5. p. 33.

present, we shall assume that the lost-work term can still be evaluated as if no other effects were present. That is, we assume that equations (7-19) and (7-23) still apply and that the values of friction factors, or loss coefficients, are unchanged.

SAMPLE PROBLEM 7-2. A large chemical plant requires cooling water at a rate of 1500 gal/min for use in its reactors and condensers. The cooling water will be taken from the municipal water supply and is available at 35 psig. Upon entering the plant, the water is pumped to high pressure and then passed through the plant pipeline. At the end of the pipeline the water is fed to the boiler house. The boiler house requires that the water be available at a pressure greater than 10 psia (to avoid cavitation in the boiler feed pumps). The pipeline comprises two segments, each 2500 ft long, as shown in Fig. SP 7-2a. In addition to the straight section, the following valves and fittings are present: ten 6-in. and ten 8-in. elbows (90°); five 6-in. and five 8-in. gate valves; and two 6-in. and one 8-in. globe valves.

(a) Find the minimum-sized motor that can run the pump (hp).

(b) If pump and motor have an overall efficiency of 70 percent, what is the actual motor size (hp)? If 440 volt electric service is available, what current will be drawn by the motor?

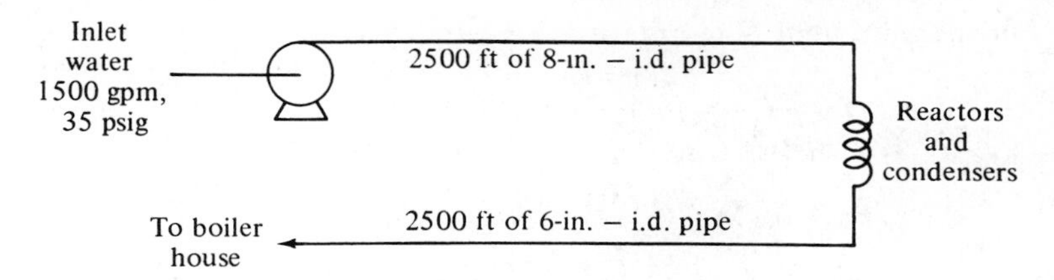

Figure SP7-2a

(c) What is the pressure at the outlet of the pump? At the reactors and condensers?

Solution: Let us begin by calculating the velocities and Reynolds numbers for both sections of pipe:

$$\dot{V} = \text{volumetric flow rate} = 1500 \text{ gal/min} = 200 \text{ ft}^3\text{/min}$$
$$= 3.33 \text{ ft}^3\text{/sec}$$

But

$$\dot{M} = \dot{V}\rho = (3.33 \text{ ft}^3\text{/sec})(62.4 \text{ lb}_m\text{/ft}^3) = 206 \text{ lb}_m\text{/sec}$$

For the 8-in. line:

$$A = \frac{\pi D^2}{4} = \frac{\pi}{4}\left(\frac{8}{12}\right)^2 = 0.35 \text{ ft}^2$$

but

$$\text{V} = \frac{\dot{V}}{A} = \frac{3.33 \text{ ft}^3\text{/sec}}{0.35 \text{ ft}^2} = 9.5 \text{ ft/sec}$$

The Reynolds number is defined by

$$N_{\text{Re}} = \frac{D\text{V}\rho}{\mu} = \frac{D\text{V}}{\nu}$$

But $\nu = 8.0 \times 10^{-5}$ ft^2/sec for H_2O at 100°F. Thus

$$N_{\text{Re}} = \frac{(8/12)(9.5)(\text{ft})(\text{ft/sec})}{8.0 \times 10^{-5} \text{ ft}^2\text{/sec}} = 7.8 \times 10^5$$

If we assume that the pipeline is made of commercial steel, Table 7-1 tells us that $\epsilon = 0.00015$ ft; so

$$\frac{\epsilon}{D} = \frac{(12)(0.00015)}{8} = 0.0002$$

From ϵ/D and N_{Re}, Fig. 7-3 gives us the Fanning friction factor:

$$f_{8\text{-in. line}} = 0.00375$$

Finally, the term $\text{V}^2/2g_c$ has the value

$$\frac{\text{V}^2}{2g_c} = \frac{9.5}{(2)(32.17)} \frac{(\text{ft/sec})^2}{(\text{lb}_m/\text{lb}_f)(\text{ft/sec}^2)} = 1.40 \text{ ft-lb}_f/\text{lb}_m$$

For the 6-in. line:

$$A = \frac{\pi D^2}{4} = \frac{\pi(\frac{1}{2})^2}{4} = 0.196 \text{ ft}^2$$

$$V = \frac{3.33}{0.196} = 17.0 \text{ ft/sec}$$

$$N_{Re} = \frac{DV}{\nu} = \frac{(\frac{1}{2})(17.0)}{8.0 \times 10^{-5}} \frac{(\text{ft})(\text{ft/sec})}{(\text{ft}^2/\text{sec})} = 1.06 \times 10^6$$

$$\frac{\epsilon}{D} = 0.0002, \qquad f = 0.00375$$

$$\frac{V^2}{2g_c} = \frac{(17.0)^2}{(2)(32.17)} \frac{(\text{ft/sec}^2)}{(\text{lb}_m/\text{lb}_f)(\text{ft/sec}^2)} = 4.5 \text{ ft-lb}_f/\text{lb}_m$$

Now let us add up the total lost work associated with the 8- and 6-in. sections of pipe.

$$lw = \frac{V^2}{2g_c} \sum k$$

where k = loss coefficient for various elements in pipeline.

For the 8-in. line:

For the straight pipe $k = 4fL/D$.

$$k_{8\text{-in. pipe}} = \frac{(4)(0.000375)(2500)}{8/12} = 56.4$$

k's for the valves and fittings:

10-90° elbows = (10)(0.75)	=	7.5
5 gate valves = (5)(0.2)	=	1.0
1 globe valve	=	6.4
total k for 8-in. pipeline	=	71.3

and the lost work in the 8-in. line is

$$lw = \frac{V^2}{2g_c} k = (1.40)(71.3) \text{ ft-lb}_f/\text{lb}_m$$

or

$$lw = 100 \frac{\text{ft-lb}_f}{\text{lb}_m}$$

For the 6-in. line:

$$k_{\text{pipe}} = \frac{(4)(0.00375)(2500)}{0.5} = 75.0$$

Valves and fittings:

$$\begin{aligned}
\text{10-90° elbows} &= (10)(0.75) &&= 7.5 \\
\text{5 gate valves} &= (5)(0.2) &&= 1.0 \\
\text{2 globe valves} &= (2)(6.4) &&= \underline{12.8} \\
\text{total } k & &&= 96.3
\end{aligned}$$

and the lost work in this section is

$$lw = (96.3)(4.5)\ \text{ft-lb}_f/\text{lb}_m = 433\ \text{ft-lb}_f/\text{lb}_m$$

The total lost work in the circulation system is the sum of the lost work in the 8- and 6-in. sections:

$$lw_{\text{total}} = (100 + 433)\ \text{ft-lb}_f/\text{lb}_m = 533\ \text{ft-lb}_f/\text{lb}_m$$

(a) We may now determine the work requirements of the pump by applying the mechanical energy balance around the whole pipeline, Fig. SP 7-2b:

$$-lw - w - \Delta\left(\frac{P}{\rho} + \frac{V^2}{2g_c} + \frac{gZ}{g_c}\right) = 0$$

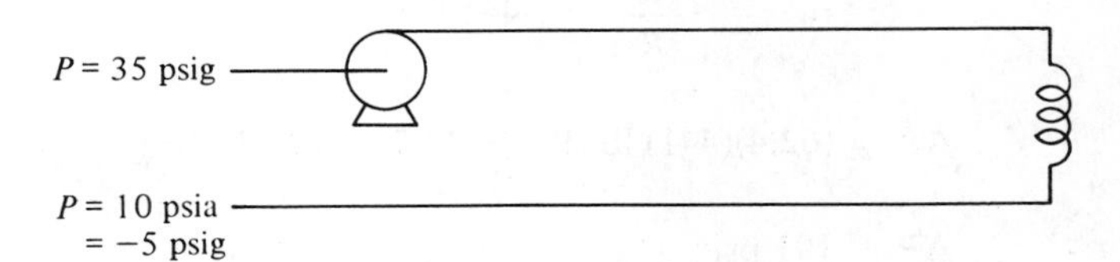

Figure SP7-2b

Assume that $\Delta gZ/g_c = 0$. We have seen that $\Delta V^2/2g_c$ is quite small, so neglect it. Thus the mechanical energy balance reduces to

$$w = -lw - \Delta\left(\frac{P}{\rho}\right)$$

But

$$lw = 533\ \text{ft-lb}_f/\text{lb}_m$$

$$\frac{\Delta P}{\rho} = \frac{P_2 - P_1}{\rho} = \frac{-5 - 35}{\rho}\ \text{psi} = -\frac{40}{\rho}\ \text{psi}$$

Since $\rho = 62.4\ \text{lb}_m/\text{ft}^3$,

$$\frac{\Delta P}{\rho} = -\frac{(40)(144)}{62.4}\ \frac{\text{lb}_f/\text{ft}^2}{\text{lb}_m/\text{ft}^3} = -92\ \text{ft-lb}_f/\text{lb}_m$$

$$w = -533\ \text{ft-lb}_f/\text{lb}_m - (-92\ \text{ft-lb}_f/\text{lb}_m) = -441\ \text{ft-lb}_f/\text{lb}_m$$

(The minus sign indicates that work must be supplied to the pump!)

The mass flow rate $\dot{M} = 206\ \text{lb}_m/\text{sec}$; so the rate of work supplied is

$$\dot{W} = w(\dot{M}) = (-441)(206)\ \text{ft-lb}_f/\text{sec}$$
$$= -9.1 \times 10^4\ \text{ft-lb}_f/\text{sec} = -165\ \text{hp}$$

for an ideal pump.

(b) If the efficiency of the pump is 70 percent, then the actual work *supplied* must be

$$\dot{W}_{\text{act}} = \frac{\dot{W}_{\text{ideal}}}{\text{eff}} = \frac{165}{0.7} = 235\ \text{hp} = 174\ \text{kW}$$

so the electrical consumption will be 174 kW. If 440-V electrical service is available, the amperage is

$$\text{current} = \frac{\dot{W}}{V} = \frac{174{,}000}{440} = 395\ \text{A!}$$

(c) We may determine the pressure at the pump outlet by applying the mechanical energy balance around the pump. We use the ideal work requirement, neglecting any lost work in the pump. Also assume that $\Delta gZ/g_c = \Delta V^2/2g_c = 0$, so the mechanical energy balance reduces to

$$w = -\frac{\Delta P}{\rho} = -441\ \text{ft-lb}_f/\text{lb}_m$$

Thus

$$\Delta P = (62.4)(441)\ \text{lb}_f/\text{ft}^2 = 2.75 \times 10^4\ \text{lb}_f/\text{ft}^2$$

or

$$\Delta P = 191\ \text{psi}$$

But $P_{\text{in}} = 35$ psig, so $P_{\text{outlet}} = 226$ psig.

Compressible Flows

In the study of compressible flows a relation between v and P is necessary before $\int v\, dP$ can be evaluated. An equation of state for the fluid provides a relation among pressure, temperature, and volume. Thus if the temperature can be related to the pressure, the desired relationship between P and $\hat{v}$ can be developed and the integral evaluated. For the major portion of the work in this book the ideal-gas equation of state will be used, although from time to time the effects of nonideal behavior will be discussed.

Two limiting examples of compressible flows will be studied. Most real flows fall some place between the two, but the limiting conditions serve to set extremes between which all real flows fall.

In certain limited cases, it is possible to have a compressible flow in which the temperature remains constant. For example, consider the flow of natural gas in an underwater section of pipeline where the surrounding water keeps the fluid at a more-or-less constant temperature. For such constant temperature flows v may be evaluated directly as a function of P from the equation of state. The velocity, V, may then be directly related to the density through the equation of conservation of mass:

$$\rho \text{V} A = \dot{M} = \text{const} \tag{7-25}$$

where A is the flow area, ρ the density of the flowing fluid, and $\dot{M}$ the mass flow rate.

In this manner, the velocity can be expressed as a function of pressure and/or area; so equation (7-10) can be integrated for many problems of interest.

SAMPLE PROBLEM 7-3. The differential form of the mechanical energy balance for the flow of a compressible fluid through a pipeline may be written as

$$\frac{\text{V}\,d\text{V}}{g_c} + \frac{dP}{\rho} + \delta lw = 0$$

The lost-work term is evaluated from the equation (7-22):

$$\delta lw = \frac{2f}{D}\frac{\text{V}^2}{g_c}\,dL$$

where ρ = fluid density
D = pipe diameter
dL = differential length of pipe
P = pressure
f = friction factor, assumed independent of N_{Re}
V = mean fluid velocity

(a) By proper manipulation and integration of the mechanical energy balance, derive the relation

$$\int_{P_1}^{P_2} \rho\,dP + \frac{G^2}{g_c}\ln\frac{\rho_1}{\rho_2} + \frac{2f}{g_c}\frac{L}{D}G^2 = 0$$

where $G = \rho\text{V}$ [=] $\text{lb}_\text{m}/\text{ft}^2$ hr.

(b) Ethylene is to be pumped along a 0.2-m-i.d. pipe a distance of 8 km at a mass flow rate of 1 kg/sec. The pressure at the end of the pipeline is to be 2 Bar, and the flow may be assumed to be isothermal with $T = 20°\text{C}$. If the friction factor may be assumed constant at $f = 0.003$ and ethylene behaves as an ideal gas, determine the required inlet pressure.

Solution: (a) Substitution of the friction-factor equation for the lost-work term in the differential mechanical energy balance yields the relation

$$\frac{dP}{\rho} + \frac{\text{V}\,d\text{V}}{g_c} + \frac{2f\text{V}^2}{Dg_c}\,dL = 0$$

Since the velocity, V, changes with position in the pipeline, evaluation of the integral of $(2f\mathrm{V}^2/D)\,dL$ requires prior knowledge of the relation between V and L. Therefore, it is to our advantage to eliminate V^2 from the $(2f\mathrm{V}^2/D)$ dL term before attempting the integration. This is accomplished by dividing the whole equation by V^2 to form

$$\frac{dP}{\rho \mathrm{V}^2} + \frac{d\mathrm{V}}{\mathrm{V}g_c} + \frac{2f}{Dg_c}\,dL = 0$$

We may relate V to ρ through the equation of continuity:

$$\rho \mathrm{V} = G = \text{const}$$

Therefore, $\mathrm{V} = G/\rho$ and

$$d\mathrm{V} = d\frac{G}{\rho} = G\left(d\frac{1}{\rho}\right) = \frac{-G}{\rho^2}\,d\rho$$

which is substituted above. Multiplication by G^2 and integration over the pipeline then yields the desired relation:

$$\int_{P_1}^{P_2} \rho\,dP + \frac{G^2}{g_c}\ln\frac{\rho_1}{\rho_2} + \frac{2fG^2L}{Dg_c} = 0$$

which is true for all gases and thermal boundary conditions.

(b) If the gas flowing through the pipeline is an ideal gas, the pressure, density, and temperature are related by the ideal-gas equation of state:

$$\frac{P}{\rho} = \frac{RT}{M} \quad \text{or} \quad \rho = \frac{PM}{RT}$$

where R = ideal-gas constant
M = molecular weight (M has been included here in anticipation of expressing ρ in terms of mass rather than moles)

If the flow through the pipeline is also isothermal, then $\rho_1/\rho_2 = P_1/P_2$ and $\int_{P_1}^{P_2} \rho\,dP$ may be integrated to yield

$$\int_{P_1}^{P_2} \rho\,dP = \int_{P_1}^{P_2} \frac{PM}{RT}\,dP = \frac{M}{2RT}(P_2^2 - P_1^2)$$

These relations may be substituted into the previously derived equation to give

$$\frac{g_c}{G^2}\frac{0.5M}{RT}(P_2^2 - P_1^2) + \ln\frac{P_1}{P_2} + \frac{2fL}{D} = 0$$

from which it is possible to relate P to L for any isothermal pipeline flow of an ideal gas. From the problem statement, we are told that

$$P_2 = 2\ \text{Bar} = 2.0 \times 10^5\ \text{N/m}^2$$

$$G = \frac{1\ \text{kg/sec}}{\text{pipe area}} = \frac{1\ \text{kg/sec}}{\pi D^2/4} = 31.8\ \text{kg/m}^2\ \text{sec}$$

$$f = 0.003, \qquad L = 8 \text{ km} = 8000 \text{ m}$$
$$D = 0.2 \text{ m}, \qquad T = 20°\text{C} = 293°\text{K}$$

The ideal-gas constant is

$$R' = \frac{R}{M} = \frac{8314}{M}\frac{\text{J}}{\text{kg °K}} = \frac{8314}{28}\frac{\text{J}}{\text{kg °K}} = 297 \text{ N m/kg °K}$$

Therefore,

$$\frac{g_c 0.5}{G^2 R' T} = \frac{(0.5)(1) \text{ kg m/sec}^2 \text{ N}}{(31.8)^2(297)(293)\ (\text{kg}^2/\text{m}^4 \text{ sec}^2)(\text{N m/kg °K})\ °\text{K}}$$
$$= 5.68 \times 10^{-9} \text{ m}^4/\text{N}^2$$

Thus the energy balance becomes

$$5.68 \times 10^{-9} \text{ m}^4/\text{N}^2 [(2.0 \times 10^5 \text{ N/m}^2)^2 - P_1^2] + \ln\left(\frac{P_1 \text{ m}^2}{2.0 \times 10^5 \text{ N}}\right)$$
$$+ \frac{(2.0)(0.003)(8000)}{0.2} = 0$$

or, solving for P_1^2,

$$P_1^2 = 8.2 \times 10^{10} \text{ N}^3/\text{m}^4 + 1.76 \times 10^8 \text{ N}^2/\text{m}^4 \ln\left(\frac{P_1}{2.0 \times 10^5 \text{ N/m}^2}\right)$$

Solution by trial and error for P_1 yields

$$P_1 = 2.88 \times 10^5 \text{ N/m}^2 = 2.88 \text{ Bar}$$

A careful examination of the equation relating G, P_2, and L for isothermal flow of an ideal gas,

$$\frac{1}{2}\frac{g_c}{G^2 R' T}(P_2^2 - P_1^2) - \ln\frac{P_2}{P_1} + \frac{2fL}{D} = 0 \tag{7-26}$$

derived in Sample Problem 7-3 can lead us to some rather startling conclusions. Rearrange equation (7-26) and solve for G, the mass flow rate per unit area of duct, to obtain

$$G^2 = \frac{g_c}{2R'T}\left[\frac{P_2^2 - P_1^2}{\ln(P_2/P_1) - (2fL/D)}\right] \tag{7-27}$$

or

$$G^2 = \frac{P_1^2 g_c}{2R'T}\left[\frac{1 - (P_2/P_1)^2}{(2fL/D) - \ln(P_2/P_1)}\right] \tag{7-27a}$$

Now suppose we fix the $2fL/D$ term in equation (2-27a) and examine the variation of G^2 with P_2. As P_2 approaches P_1, the numerator goes to zero, and so does G. This is not surprising—no pressure difference, no flow. But as P_2 approaches zero, we find that the $\ln(P_1/P_2)$ term in the denominator becomes infinite and G_2 approaches zero again. Thus according to equation

(7-27a), the G–P_2 curve should appear as shown in Fig. 7-4. The pressure corresponding to the maximum is termed the critical pressure, P_c. (Here and elsewhere in this chapter, P_c is to be distinguished from the pressure used in the evaluation of reduced properties.)

Our physical intuition tells us clearly something is amiss. Decreasing P_2 cannot lead to a decrease in the mass flow rate. As we will soon show (at least for the case of compressible flows through nozzles), the critical pressure is the minimum pressure that can exist *within* the pipeline. If the external pressure is less than P_c an expansion wave will develop at the exit of the pipeline. In the expansion wave the pressure drops almost discontinouusly from P_c to the external pressure. Thus the actual curve of mass flow rate, G, versus pressure ratio, P_2/P_1, is as shown in Fig. 7-5.

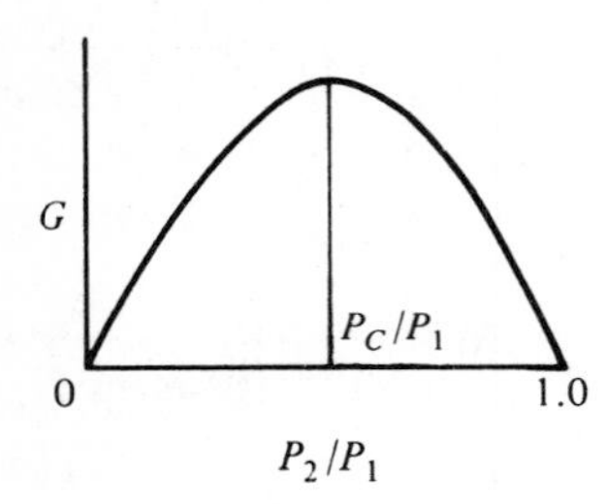

Figure 7-4. Mass flow rate vs. pressure ratio from equation (7-27a).

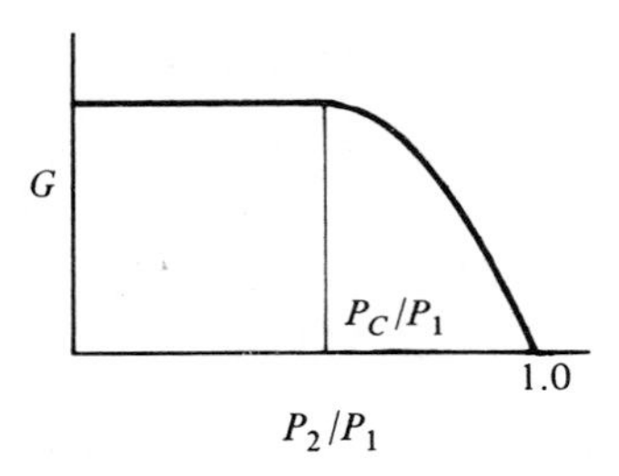

Figure 7-5. Mass flow rate vs. pressure ratio for isothermal duct flow.

We may determine the critical pressure ratio (P_c/P_1) by setting the derivative of G (or G^2) with respect to P_2/P_1 equal to zero:

$$\frac{d(G^2)}{d(P_2/P_1)} = \frac{d}{d(P_2/P_1)}\left[\left(\frac{g_c P_1^2}{2R'T}\right)\frac{1-(P_2/P_1)^2}{(2fL/D)-\ln(P_2/P_1)}\right] = 0 \qquad (7\text{-}28)$$

Performing the differentiation for a fixed value of P_1 yields

$$-\frac{2(P_c/P_1)}{(2fL/D)-\ln(P_c/P_1)} + \frac{1-(P_c/P_1)^2}{(P_c/P_1)[(2fL/D)-\ln(P_c/P_1)]^2} = 0 \qquad (7\text{-}29)$$

Simplification and collection of terms then yields

$$\frac{2fL}{D} + 0.5 = \ln\left(\frac{P_c}{P_1}\right) + 0.5\left(\frac{P_1}{P_c}\right)^2 \qquad (7\text{-}30)$$

For a particular pipeline and flow conditions the maximum flow rate may be determined as follows:

1. Evaluate $2fL/D$.
2. The critical pressure ratio P_c/P_1 is determined from equation (7-30).
3. The critical pressure ratio is substituted into equation (7-27a) to determine the maximum flow rate.

If it is necessary to pass a specified mass flow rate through the pipeline, the pipeline length must be less than L_{max}, where L_{max} is obtained as follows:

1. We assume that $P_2 = P_c$ in equation (7-27a) and use equation (7-30) to eliminate $2fL/D$ with the result

$$P_c^2 = \frac{G^2 R'T}{g_c} \tag{7-31}$$

2. P_c is then substituted into equation (7-30) to determine L_{max}, the length of pipeline that corresponds to critical conditions at the end of the pipeline. If the pipeline is shorter than L_{max}, P_2 will be greater than P_c (assuming G is held constant). If the pipeline is of length L_{max}, the outlet pressure must be the critical pressure, if the desired mass flow rate is to be obtained. If the pipeline is longer than L_{max}, there is no outlet pressure which satisfies the equations for the specified mass flow rate. That is, the specified mass flow rate cannot be obtained, and the actual flow rate will be less than the desired one. This phenomenon, in which the desired mass flow rate cannot be obtained no matter what the outlet pressure, is termed *choking* and is a phenomenon we shall encounter again in our discussion of flow through nozzles.

Adiabatic Flows

In adiabatic flows it is not possible to establish a priori the relationship between pressure and density, because the temperature is not known as a function of pressure or density. Thus the relationship between P and ρ must be determined before the mechanical energy balance can be integrated. We begin by examining the energy balance around a differential segment of pipeline:

$$d\left(h + \frac{V^2}{2g_c} + \frac{gZ}{g_c}\right)\delta M + \delta Q - \delta W = d\left[M\left(u + \frac{V^2}{2g_c} + \frac{gZ}{g_c}\right)\right]_{sys} \tag{7-32}$$

Since the pipeline will be assumed to operate at steady state, the right-hand side of equation (7-32) equals zero. For adiabatic flows, $\delta Q = 0$, and by assumption $\delta W = 0$ and $g\,dZ/g_c = 0$. Thus the energy balance reduces to

$$d\left(h + \frac{V^2}{2g_c}\right) = 0 \tag{7-33}$$

or

$$dh = -\frac{V\,dV}{g_c} \tag{7-34}$$

The enthalpy may in turn be expressed in terms of P, v, T, and C_P as

$$dh = C_P\,dT + \left[v - T\left(\frac{\partial v}{\partial T}\right)_P\right]dP \tag{7-35}$$

so the energy balance reduces to

$$C_P\, dT + \left[v - T\left(\frac{\partial v}{\partial T}\right)_P\right] dP = -\frac{\mathrm{V}\, d\mathrm{V}}{g_c} \tag{7-36}$$

The velocity term may be eliminated by application of the continuity equation:

$$\rho \mathrm{V} = G = \frac{\dot{M}}{A} = \text{const} \tag{7-37}$$

or

$$-\mathrm{V}\, d\mathrm{V} = \frac{G^2}{\rho^3}\, d\rho \tag{7-38}$$

so

$$C_P\, dT + \left[v - T\left(\frac{\partial v}{\partial T}\right)_P\right] dP = \frac{G^2}{\rho^3 g_c}\, d\rho \tag{7-39}$$

Equation (7-39) and the equation of state for the fluid, $f(P, v, T) = 0$, then allow determination of the relationship between P and ρ (or v). For a gas which obeys the ideal-gas equation of state, this relationship can be expressed in a reasonably simple closed form. For this case it is possible to integrate the mechanical energy balance exactly to obtain pressure, temperature, density, and velocity as functions of position within the pipeline. For nonideal equations of state, the relationship between pressure and density becomes extremely complex, and it is usually necessary to resort to numerical integration of the mechanical energy balance.

7.3 Compression Processes

Most commerical processes involve continuous-flow systems. For such systems the minimum work required to compress a fluid is obtained for a reversible process. For the reversible case $lw = 0$ in the mechanical energy balance. If we also neglect kinetic and potential energy changes within the system, the work of compression reduces to

$$w_{\text{rev}} = -\int_{P_1}^{P_2} v\, dP \tag{7-40}$$

The integral, $\int v\, dP$, is path-dependent and requires knowledge of how v varies with pressure. If we are to minimize the reversible work requirements for compressing between two fixed pressures, then clearly v must be minimized.

Liquids

The specific volume of most liquids is nearly independent of pressure and temperature until the critical temperature is approached. Thus, the specific volume in equation (7-40) may, in most cases, be removed from under the integral sign so that the reversible work of compression (or expansion) reduces to:

$$w_{\text{rev}} = -v(\Delta P) \tag{7-41}$$

Thus the work required is directly proportional to the desired pressure increase.

Since the specific volumes of liquids (again except near their critical temperatures) are much smaller than for the corresponding vapor, the work required to raise the pressure of a liquid is considerably smaller than for its corresponding vapor. In many applications it is possible to use this difference to increase the overall performance of a system in which the pressure of either a vapor or a liquid phase must be increased. This point will be discussed briefly in our description of the Rankine cycle in Chapter 8. Sample Problem 7-1 includes a calculation of the pressure increase for an incompressible fluid resulting from a given work input.

Gases and Vapors

Because of the changes in intermolecular forces which occur as the pressure is increased, the compression of gaseous fluids is more difficult to analyze than the compression of liquids. The specific volume of a gas varies with both T and P. Thus the changing temperature during the compression of a gas can have a marked influence on the work required to achieve a given pressure increase. At any given pressure the specific volume of a gas decreases with decreasing temperature. Thus, the reversible work required to compress a gas between two fixed pressures can be minimized by performing the compression at the lowest possible temperature. In practice, the inlet temperature and pressure of the gas are generally fixed by other process considerations. Thus, our primary concern generally relates to minimizing the work needed to compress from these fixed inlet conditions, to some specified outlet pressure.

The compression of a gas, even if performed reversibly, will always produce an increase in temperature unless heat is removed. The path for a typical adiabatic and reversible compression is shown on a P–v diagram in Fig. 7-6 and a P–h diagram in Fig. 7-7. Since the compression is isentropic,

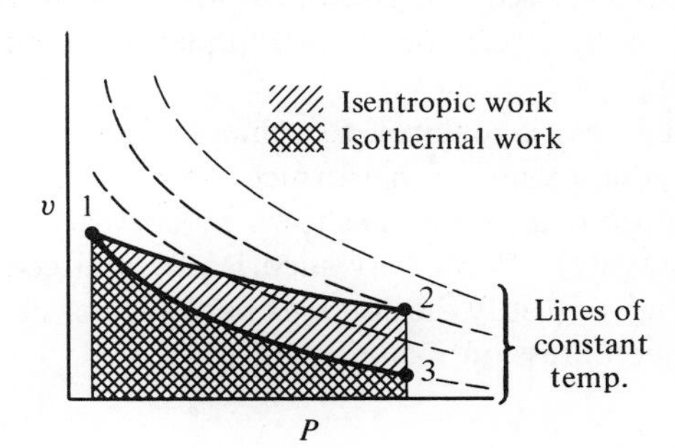

Figure 7-6

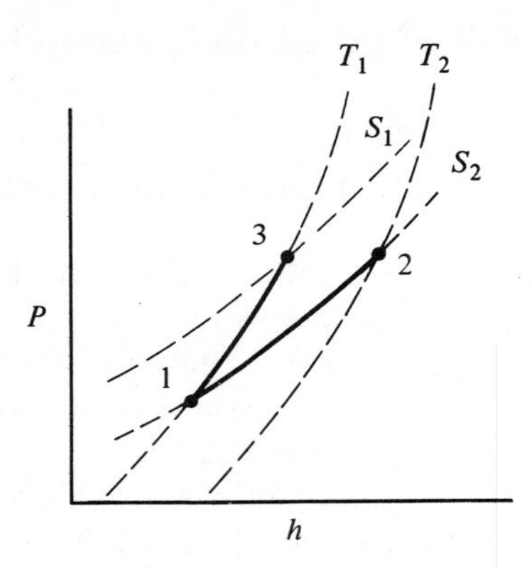

Figure 7-7

the process follows an isentrope from the inlet conditions to P_2. As mentioned previously, the fluid temperature increases as the pressure does. As we have shown, the work required for the reversible compression is:

$$w = -\int v\, dP$$

However, for the isentropic process, the integral is also given by:

$$\int v\, dP = (\Delta h)_S$$

Thus, the adiabatic, reversible work requirement can be determined from either Δh, or $\int v\, dP$. Typically the functional relationship $v = \phi(P, s)$ is not available and we generally use enthalpy tabulations or plots to determine the reversible work requirements. For systems where enthalpy information is not available but heat capacity data are, the generalized charts and procedures developed in Chapter 6 can be used to calculate Δh and thereby w for any initial and final states. Sample Problems 6-11, 6-12 and 6-13 illustrate the calculation of work requirements for compressors using equations of state and the generalized charts.

We may gain valuable insight into the work requirements for compression processes by observing the effect which different paths have on the integral, $\int v\, dP$, between two given pressures. The work required for the compression is given by $\int v\, dP$, or the area under a v–P curve as shown by cross-hatching in Fig. 7-6.

The work required to achieve a given pressure increase is minimized if the compression is performed isothermally (assuming no decrease in temperature is possible) as shown by curves 1–3 in Figs. 7-6 and 7-7. However, the savings afforded by such isothermal compression are usually not large enough to justify the complex hardware or decreased throughput that isothermal compression would require.

7.4 Staged Compression

Isothermal operation can be approximated to some degree by *staging* the compression as shown in Fig. 7-8.

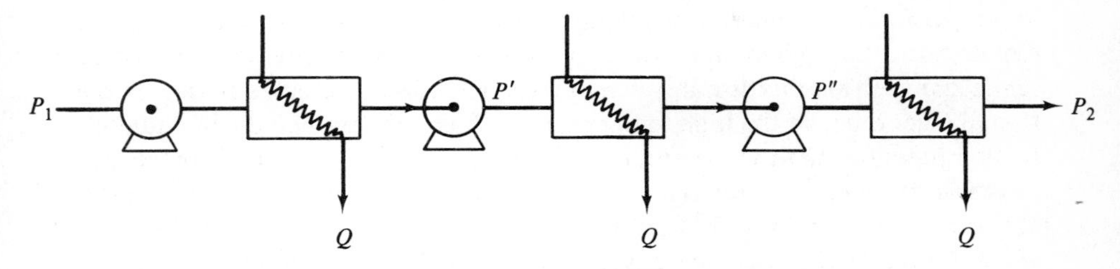

Figure 7-8

At the end of each compression stage the compressed gas is cooled in an *intercooler*. The work requirements for such a staged compression with intercooling are illustrated in Fig. 7-9. All paths are assumed to be reversible, but as can be seen, the area beneath the curves (and hence the work required) decreases as the number of stages increases. In the limit of an infinite number of stages and intercoolers, the compression path and work requirements approach those for the isothermal case.

Clearly, in practice the design engineer considers cost as his primary design criteria, not energy consumption per se. Since energy has contributed relatively little to the cost of most products as contrasted with labor or equipment costs, the use of intercoolers has been justified more often in terms of process considerations other than reduced energy consumption. Increased energy costs, if substantial, would clearly result in greater use of intercooling, particularly as compact, efficient, and relatively low-cost heat exchangers are developed for such applications.

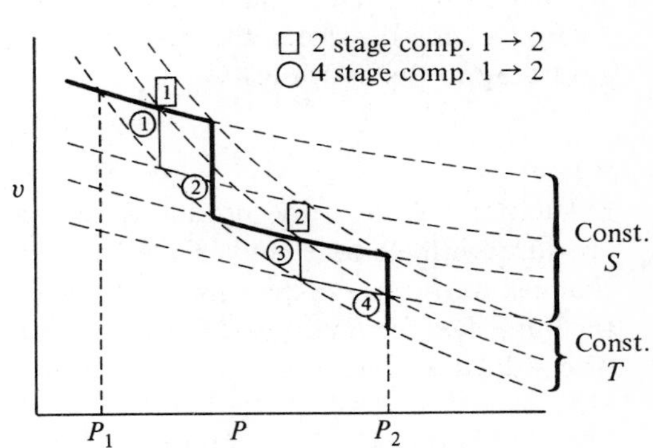

Figure 7-9. The staged compression.

If a process requires a substantial increase in pressure, it is usually necessary to stage the compression in any case because of the inherent limits on compression ratios (defined as P_{out}/P_{in}) which a single compressor can provide. Large compression ratios can produce outlet temperatures so high that equipment failure may result. In such cases, intercooling as described above may be used to limit the outlet gas temperatures.

In the following chapter on energy conversion we will learn that the diesel engine uses the high temperatures provided by the compression stroke to ignite the fuel–air mixture in the cylinder. Thus in some cases it is possible to make good use of the high temperatures which are attained during adiabatic compression. On the other hand, these high temperatures are not always desirable in internal-combustion engines. The compression ratio of a standard automobile engine is limited by the ignition characteristics of the fuel–air mixture. Higher-octance gasoline is required to avoid pre-ignition or "knocking" in engines which use higher compression ratios to improve engine performance.

7.5 Expansion Processes

The industrial revolution moved into high gear with the development of the steam engine, the first practical device for deriving work from a fuel via an expanding gas. Since that time more sophisticated engines and fuels have been developed, but in virtually every case an expanding gas plays a principal role in converting raw energy to useful work. From the wood- or coal-fired steam engine, technology has produced in rapid succession gasoline-fueled, piston-driven automobiles and airplanes, diesel-powered trucks and locomotives, jet-powered aircraft, and most recently chemically propelled rockets. Paralleling these developments was the electrification of much of the world as the result of improved steam-turbine technology capable of accommodating increasingly higher steam temperatures and pressures.

For a closed system the reversible expansion of a gas produces work according to the relationship

$$w_{rev} = \int_{V_1}^{V_2} P\,dv \tag{7-42}$$

Evaluation of the work produced requires knowledge of the variation of pressure with volume and, since both depend on the temperature which changes during the expansion, the evaluation of the integral is not easily accomplished for a real gas. Since the work produced for such an expansion (if conducted adiabatically) can be given by the expression:

$$w_s = \int_{V_1}^{V_2} P\,dv_s = \Delta u_s \tag{7-43}$$

it is possible to calculate work knowing only the initial and final states of the gas and the corresponding values of internal energy, u. Property tabulations or generalized charts, discussed in earlier chapters, can be used to evaluate Δu. It is frequently necessary to use the relationship $\Delta u = \Delta h - \Delta(Pv)$ since tabulations of enthalpy are more common than tabulations of internal energy.

The actual work delivered by the system is typically less than the reversible work which the gas transfers to the piston because of frictional effects and other system irreversibilities. This topic was discussed in considerable detail in Chapter 4.

For open systems the reversible work derived from an expanding gas is given by the expression:

$$w_{\text{rev}} = -\int_{P_1}^{P_2} v\,dP \tag{7-44}$$

Again the integral is path-dependent and requires information as to how p, v, and T vary during the process. If the process is carried out both adiabatically and reversibly (and hence isentropically) the work performed is given by

$$w_s = -\Delta h_s \tag{7-45}$$

For many gases and vapors property tabulations can be used to evaluate Δh from the initial and final state conditions. Generalized property charts can be used to estimate the change for most other cases, as was illustrated in Chapter 6.

While the above expressions permit us to calculate the work derivable from an isentropic expansion (or any expansion for which the gases initial and final states are known), they provide little insight into the conditions necessary to achieve an isentropic expansion. For such a process in which entropy remains constant as the pressure is decreased, all other fluid properties, including T, v, and h, are fixed for a single-component fluid at any given pressure. However the continuity equation must also be satisfied at all times. It requires that

$$\dot{M} = \frac{\text{V}A}{v}$$

and thus, for any value of the mass velocity, $\dot{M}$, and v, it requires that the product of the fluid velocity, V, and the cross-sectional area, A, for fluid flow must be equal to the product $\dot{M}v$. Only if these conditions are met can the flow be isentropic. Furthermore, all expanders which produce work or thrust, except closed-system piston expanders, rely on a conversion of enthalpy to kinetic energy before work or thrust can be derived. Thus we can also apply the energy balance to determine the intermediate velocities in such an open-system expansion

$$-\Delta h = \frac{\Delta \text{V}^2}{2g_c}$$

It is therefore possible to relate the cross-sectional area which the expansion channel must have for a given $\dot{M}$ at any pressure level P since the v and h corresponding to that P are fixed. Such conditions would lead to the maximum gas velocity, V, which could be produced by the expansion. The gas's kinetic energy can be used to provide thrust, as in a jet, or to produce work by impinging on a turbine blade and transferring its momentum. It becomes clear that the proper design of the flow channel requires a careful matching of pressure drop and channel cross section at all points in the expansion if the expansion is to occur reversibly.

These conditions are most easily achieved in a simple nozzle in which no attempt is made to extract work during the expansion, as is done in a turbine. The thermodynamics of nozzle flow will be discussed in Section 7.7 and can provide considerable insight into the design conditions which apply to a turbine expander in which expanding gases are used to produce shaft work.

7.6 Expanders

In conventional piston or reciprocating engines the expansion produces an axial force which is converted to torque (or rotary motion) with appropriate linkages. Rotary piston engines have been less common over the years but began receiving increased attention as a means of reducing engine size and hopefully improving mileage and emission performance.

Nozzles and turbines rely on a near isentropic conversion of the gases' enthalpy to kinetic energy. They derive work either by ejecting the gases directly at high velocity, thus utilizing the principle of Newton's second law to produce thrust (as in a jet plane or a rocket), or by directing the high-velocity gases on a turbine blade at such an angle that a torque is created and rotary motion results.

Piston Expanders

The expansion of a gas against a confining piston represents a straightforward conversion of the internal energy to work. If the gas's pressure potential is sufficiently large to produce a force on the piston which exceeds the resisting force, motion will result and work is performed (as long as the resisting force is finite and the force imbalance exists). We have made considerable use of piston–cylinder examples in our earlier discussions. The development of the mechanical linkages and valving required to make use of this simple expansion process on a continuous basis to produce either linear or rotating shaft work was an engineering contribution of immense significance to society.

In steam engines combustion takes place outside of the engine and steam is generated in a boiler. The steam at a high pressure and temperature is admitted to the cylinder and expands against the piston. In a double-acting system the piston is returned to its original position by admitting steam on the opposite side of the piston and forcing it back to its original position. Work can be extracted from both strokes and the expanded steam is simply exhausted to the atmosphere on the return stroke to minimize any back pressure. Such an engine requires not only fuel, but also water to operate; thus it found applications primarily in industry and railroads where locomotives could carry adequate water as well as fuel. Such an engine had obvious disadvantages in automotive applications, although many of them have been partially overcome, and today there are zealous advocates of steam engines for automobiles.

Early advocates of steam engines found the internal-combustion engine difficult to compete with because it used the combustion gases themselves as the working fluid and offered compact engines and light vehicles. Only the fuel need be transported, and the ready availability of oil (and fuels derived therefrom) led to the rapid switch to gasoline- and diesel-powered transportation systems.

Expansion of a gas in a piston-cylinder has inherent irreversibilities due to the frictional effects between the piston and cylinder walls. However, engineering advances in piston rings and lubricants has reduced these losses considerably, and today the expansion step is a reasonably efficient method for converting a gas's internal energy to work. The relatively small leakage of high-pressure gases past the piston to the low-pressure side makes it a particularly good system for high-torque applications. Such systems thus find application in heavy industry as well as in systems which involve frequent starting that requires sizable torques.

However, these systems, because of the mechanical linkages required to generate the torque, have additional frictional losses which reduce their conversion efficiencies. They also typically discharge the hot spent steam or exhaust gases to the surroundings, thus further degrading their thermodynamic performance.

7.7 Nozzles

Let us now examine the applications of the thermodynamic principles which apply to the study of compressible flow within nozzles and diffusers. A nozzle is a device whose function is to increase the velocity of a fluid by decreasing its pressure; a diffuser is a device that increases the pressure within a fluid by decreasing its kinetic energy. Through the years the term "nozzle" has come to be used for both nozzles and diffusers and will be so used in the following pages.

Since most flows through nozzles occur rapidly, there is little time for heat to be transferred into or out of the working fluid. Therefore, in many practical cases it can be assumed that the flow is adiabatic. In practice, it is found that nozzles may be designed to operate in an almost reversible fashion. That is, practically all the pressure gradient within the nozzle is used to increase the kinetic energy of the working fluid, whereas only a small portion is needed to overcome viscous effects. Therefore, we shall assume that all nozzles studied herein operate in a fully reversible fashion—that is, $lw = 0$.

Since the flow in the nozzle is adiabatic and reversible, it is isentropic, that is, the entropy is constant. Thus the pressure, temperature, density, and enthalpy of the fluid at any point within the nozzle are related by the isentropic conditions. For an ideal gas these relations may be expressed in relatively simple closed form, whereas for nonideal gases these relations are most simply obtained from a suitable property chart, such as a Mollier diagram. (If a suitable property chart is not available, the relationships must be established using the techniques developed in Chapter 6.)

Let us now apply the energy balance and continuity equations around a portion of the simple converging–diverging nozzle shown in Fig. 7-10. If the nozzle is operating at steady state, with no heat, no work, and negligible potential-energy effects, the energy balance around the indicated section reduces to

$$(h_2 - h_1) + \frac{V_2^2 - V_1^2}{2g_c} = 0 \tag{7-46}$$

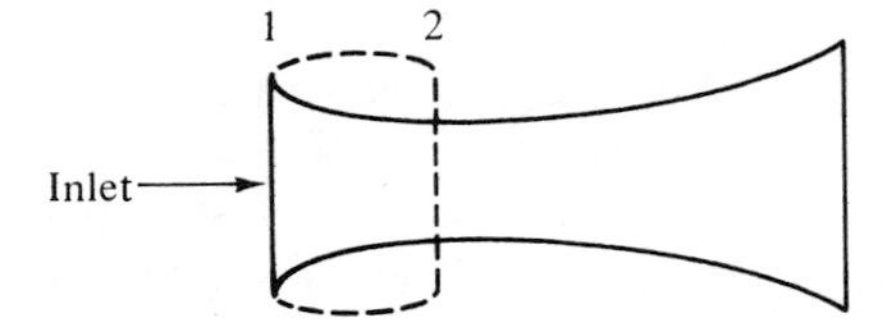

Figure 7-10. Converging-diverging nozzle.

(We shall work with the energy balance directly rather than the mechanical energy balance, because the path integral $\int v\, dP$ would be extremely difficult to evaluate for an isentropic process involving nonideal gases.) If the inlet cross-sectional area of the nozzle is large compared to the cross-sectional areas elsewhere in the nozzle, then $V_1 < V_2$, so $V_1^2 \ll V_2^2$. In this case we may neglect V_1^2 and reduce the energy balance to

$$V_2^2 = 2g_c(h_1 - h_2) \tag{7-47}$$

(If the inlet velocity is not "negligible," as in the case of many diffusers, then the V_1^2 term must be retained, but the analysis will proceed almost exactly as follows.)

If the pressure at point 2 is known, the enthalpy and temperature, h_2 and T_2, may be found from a suitable property chart and the inlet conditions, P_1 and T_1. From equation (7-47) the velocity at point 2 may then be calculated. The density at point 2 may then be calculated from the equation of state and the known pressure and temperature. In addition, if the mass flow rate through the nozzle is specified, the area at point 2 may be determined from the continuity equation:

$$\dot{M} = \rho_2 V_2 A_2 = \text{const} \tag{7-48}$$

or

$$A_2 = \frac{\dot{M}}{\rho_2 V_2} \tag{7-49}$$

SAMPLE PROBLEM 7-4. Steam at 730 psia and 780°F is to be expanded in a converging–diverging nozzle to obtain a high-speed flow. Assume that the incoming steam has negligible velocity and that the flow is isentropic. For a mass flow rate of 10 lb_m/sec determine the velocity, area, and temperatures as functions of pressure throughout the nozzle.

Solution: We may picture the nozzle as shown in Fig. SP 7-4a. The energy balance may be written about any portion of the nozzle. In particular, let us write the balance around the entrance to the nozzle, and any other point within the nozzle. Assume that $\delta Q = \delta W = g\Delta Z/g_c = 0$ and that V_{in}^2 is negligible. For steady state $\delta M_{in} = \delta M_{out}$, and the energy balance reduces to equation (7-47),

$$V^2 = -2g_c(h - h_1)$$

or

$$V^2 = (h - h_1) \times 2.0 \times 32.17 \times 778 \frac{lb_m}{lb_f} \frac{ft}{sec^2} \frac{lb_f\text{-ft}}{Btu}$$

$$= (h - h_1)(5.0 \times 10^4) \text{ ft}^2\text{-lb}_m/\text{sec}^2 \text{ Btu}$$

where h is the enthalpy at the point under consideration.

We may now proceed to determine the pressure–area profile within the nozzle as follows:

1. Choose a pressure P.
2. Using the initial conditions and the isentropic condition, determine

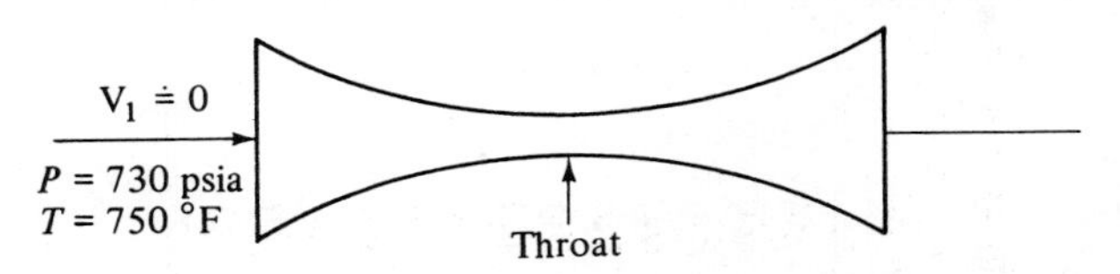

Figure SP7-4a

Table SP 7-4 Property Profiles for Converging–Diverging Nozzle

P, psia	T, °R	v, ft^3/lb_m	h, Btu/lb_m	$h - h_1$ Btu/lb_m	V^2, ft^2/sec^2	V, ft/sec	A, ft^2
730	780	0.9472	1390	0	0	0	∞
700	770	0.9806	1385	5	25×10^4	500	1.96×10^{-2}
650	750	1.039	1376	14	70×10^4	835	1.245×10^{-2}
600	725	1.099	1366	24	120×10^4	1090	1.00×10^{-2}
550	700	1.178	1355	35	175×10^4	1320	0.89×10^{-2}
500	680	1.276	1344	46	230×10^4	1512	0.845×10^{-2}
450	650	1.380	1332	58	280×10^4	1670	0.816×10^{-2}
400	620	1.513	1318	72	360×10^4	1895	0.799×10^{-2}
350	588	1.678	1304	86	430×10^4	2070	0.810×10^{-2}
300	550	1.890	1288	102	510×10^4	2258	0.838×10^{-2}
250	510	2.18	1268	122	610×10^4	2465	0.884×10^{-2}
200	460	2.59	1246	144	720×10^4	2680	0.967×10^{-2}
150	400	3.22	1220	170	850×10^4	2910	1.11×10^{-2}

the temperature and enthalpy corresponding to P from the Mollier diagram (Appendix A).

3. Using the temperature and pressure, determine the density (specific volume) from the steam tables (Appendix B).

4. Using the energy balance and the enthalpy, determine the velocity at the point under consideration.

5. From the known mass flow rate, the density, and the velocity, determine the area corresponding to P ($A = \dot{M}/\rho V = \dot{M}v/V$).

The calculations are summarized in Table SP 7-4.

We may now plot the velocity, area, and temperature, versus pressure as shown in Fig. SP 7-4b. An examination of the area–pressure curve indicates

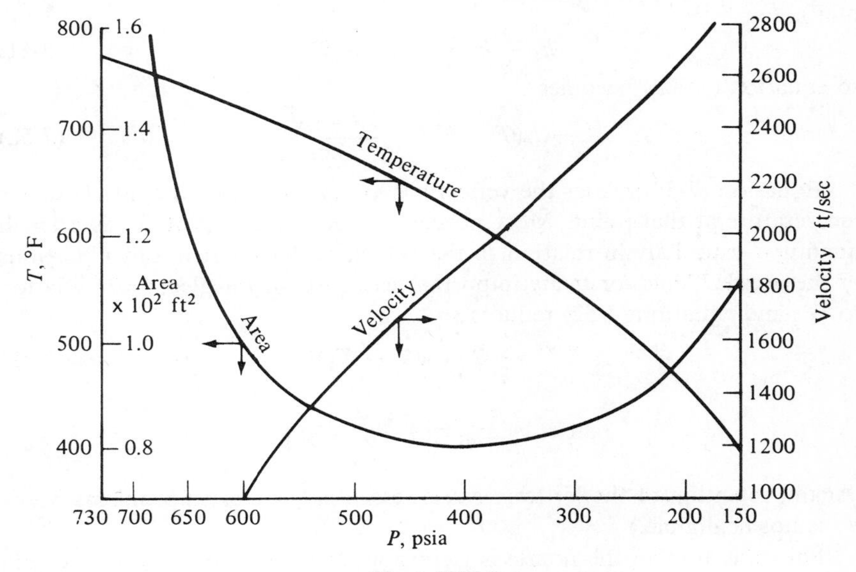

Figure SP7-4b

that the area of the nozzle passes through a minimum at $P = 395$ psia. As we will show later, this minimum area point, or "throat," is of major importance in determining the operating characteristics of a nozzle. The temperature and velocity at the throat also possess special significance; for the problem at hand they are

$$V_{\text{throat}} = 1860 \text{ ft/sec}, \qquad T_{\text{throat}} = 623°\text{F}$$

For a gas that obeys the ideal-gas equation of state the pressure, temperature, and density may be simply related during an isentropic process. Therefore, it is possible to develop closed-form expressions for the relationships among $P, T, \rho, \mathrm{V}, \dot{M}$, and A during the isentropic flow of an ideal gas through a nozzle. We shall now establish these relationships and study some of the phenomena associated with them.

We begin, as in the nonideal gas studies, by examining the energy balance written around some general portion of the nozzle as shown in Fig. 7-10:

$$(h_2 - h_1) + \frac{\mathrm{V}_2^2 - \mathrm{V}_1^2}{2g_c} = 0 \tag{7-50}$$

or, if we drop the subscript 2 for convenience,

$$h - h_1 = \frac{\mathrm{V}^2 - \mathrm{V}_1^2}{2g_c} \tag{7-50a}$$

For a gas that obeys the ideal-gas equation of state (and has C_P independent of T),

$$h_1 - h = C_P(T_1 - T) \tag{7-51}$$

so equation (7-50a) becomes

$$-C_P(T - T_1) = \frac{\mathrm{V}^2 - \mathrm{V}_1^2}{2g_c} \tag{7-52}$$

Equation (7-52) relates the velocity at any point within the nozzle to the temperature at that point. Most nozzles are designed so that V_1 is a small quantity—especially in relation to the velocities that exist in other portions of the nozzle. Thus for many important cases we may neglect V_1^2 in relation to V^2, and equation (7-52) reduces to

$$\mathrm{V}^2 = -2g_cC_P(T - T_1) \tag{7-53}$$

or

$$\mathrm{V}^2 = -2g_cC_PT_1\left(\frac{T}{T_1} - 1\right)$$

(Again, retention of the V_1^2 term causes essentially no extra problems where V_1 is not negligible.)

Since the flow in the nozzle is isentropic, the pressure, temperature, and density are related by the equations derived in Chapter 6 for isentropic processes of gases that obey the ideal gas equation of state:

$$\frac{T}{T_1} = \left(\frac{P}{P_1}\right)^{(k-1)/k} \tag{7-54}$$

and

$$\frac{P}{P_1} = \left(\frac{\rho}{\rho_1}\right)^k = \left(\frac{v_1}{v}\right)^k \tag{7-55}$$

Equation (7-54) is substituted into (7-53a) and yields

$$\mathrm{V}^2 = -2g_cC_PT_1\left[\left(\frac{P}{P_1}\right)^{(k-1)/k} - 1\right] \tag{7-56}$$

But

$$C_pT_1 = \frac{k}{k-1}\frac{P_1}{\rho_1} \tag{7-57}$$

and thus the velocity, V, at any point can be expressed solely as a function of the inlet conditions, the ratio of heat capacities, and the pressure:

$$V^2 = -2g_c\frac{k}{k-1}\frac{P_1}{\rho_1}\left[\left(\frac{P}{P_1}\right)^{(k-1)/k} - 1\right] \tag{7-58}$$

We may now relate P to the cross-sectional area of the nozzle, A, by means of the continuity equation:

$$\dot{M} = \rho\,VA = \text{const} \tag{7-59}$$

(where $\dot{M}$ is the flow rate through the nozzle). Equation (7-59) is rearranged to the form

$$A = \frac{\dot{M}}{\rho V} \tag{7-60}$$

Equation (7-58) is now substituted for V in equation (7-60) to form

$$A = \frac{\dot{M}}{\rho\sqrt{2g_c\frac{k}{k-1}\frac{P_1}{\rho_1}\left[1 - \left(\frac{P}{P_1}\right)^{(k-1)/k}\right]}} \tag{7-61}$$

The ρ in equation (7-61) may be expressed in terms of the initial conditions and P by means of equation (7-55):

$$\rho = \rho_1\left(\frac{P}{P_1}\right)^{1/k} \tag{7-55}$$

Therefore,

$$A = \frac{\dot{M}}{\sqrt{\frac{2kg_c}{k-1}P_1\rho_1\left[\left(\frac{P}{P_1}\right)^{2/k} - \left(\frac{P}{P_1}\right)^{(k+1)/k}\right]}} \tag{7-62}$$

Thus we have derived the desired relation between the pressure at any point within the nozzle and the area at that point. Note, however, that the pressure within the nozzle is a function of both the area at that point and the rate of mass flow through the nozzle.

Some very interesting behavior may be observed if we examine equation (7-62). Since the denominator of this equation is a function only of the initial conditions (P_1 and ρ_1) and the pressure within the nozzle, for any given initial conditions equation (7-62) may be rewritten in the form

$$A = \frac{\dot{M}}{\phi(P)} \tag{7-63}$$

where

$$\phi(P) = \sqrt{\frac{2kg_c}{k-1}P_1\rho_1\left[\left(\frac{P}{P_1}\right)^{2/k} - \left(\frac{P}{P_1}\right)^{(k+1)/k}\right]} \tag{7-64}$$

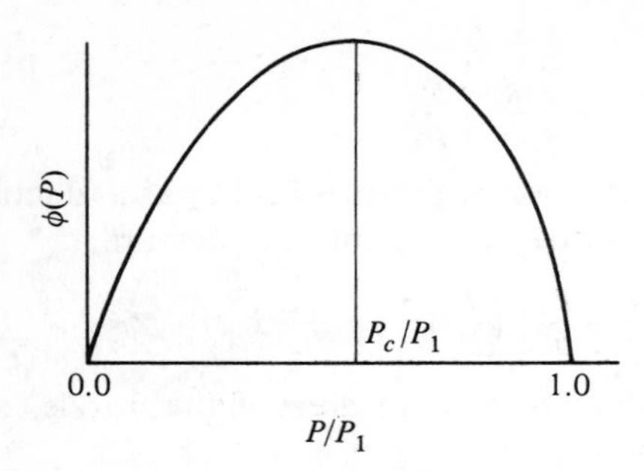

Figure 7-11. Function $\phi(P)$.

Examination of $\phi(P)$ indicates that $\phi(P) = 0$ for $P = 0$, or $P = P_1$ and is > 0 for $1 > P/P_1 > 0$. Thus it is apparent that $\phi(P)$ must have the form shown in Fig. 7-11. That is, $\phi(P)$ passes through a maximum at some $0 < P < P_1$. We define the pressure at the point where $\phi(P)$ is a maximum as the critical pressure, P_c, of the nozzle. It is a function only of the specific heat ratio and the initial conditions of the fluid entering the nozzle. We may now plot the relation between the area and the pressure within a nozzle, for some specified mass flow rate, $\dot{M}$, as shown in Fig. 7-12. Since $\phi(P)$ goes through a maximum at $P = P_c$, the nozzle area must go through a minimum at this pressure, as shown in the figure. Thus we see that if we wish to expand a gas to a pressure less than P_c, we must have a converging-diverging nozzle, that is, a nozzle in which the area first decreases and then increases with distance from the nozzle entrance. The point at which the nozzle area goes through a minimum is known as the *throat* of the nozzle, and the area at this point is appropriately called the *throat area*. We shall now examine the operating characteristics of two different types of nozzles.

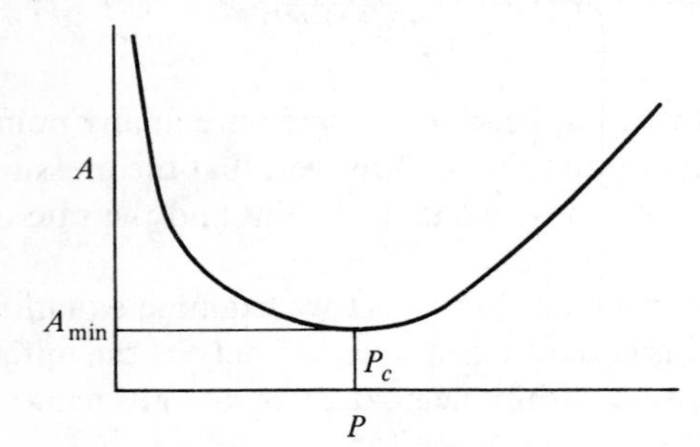

Figure 7-12. Nozzle area as a function of pressure.

Simple Converging Nozzles

A simple converging nozzle is a nozzle whose area continually decreases with increasing distance from the nozzle entrance (Fig. 7-13). The exit of the nozzle is the throat.

Let us examine the behavior of a simple converging nozzle as the exit

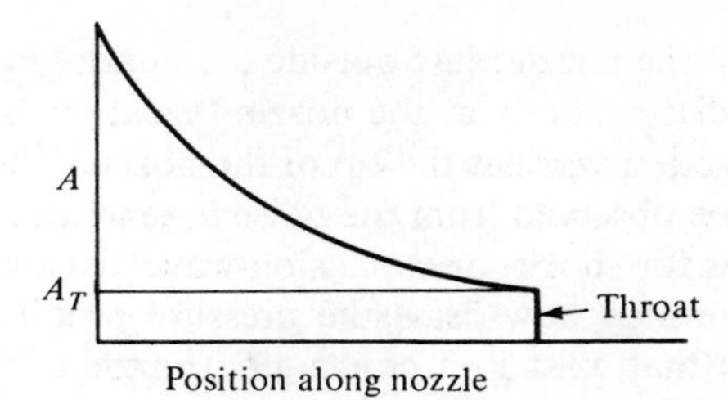

Figure 7-13. Converging nozzle.

pressure is dropped below the inlet pressure. Equation (7-63) is rearranged to the form

$$\dot{M} = A\phi(P) \tag{7-65}$$

and equation (7-65) is applied to the point just before the exit of the nozzle. Therefore,

$$\dot{M} = A_{\min}\phi(P_2) \tag{7-66}$$

where P_2 = the exit pressure.

Examination of Fig. 7-11 shows that for $P_2 = P_1, \phi(P_2) = 0$, and no flow occurs. This is equivalent to saying: No pressure drop, no flow! However, as P_2 is reduced, $\phi(P_2)$ increases and, therefore, $\dot{M}$ increases. As P_2 approaches P_c, $\phi(P_2)$ approaches its maximum and so does $\dot{M}$. As P_2 is further decreased, equations (7-63) and (7-64) predict that $\phi(P_2)$ and $\dot{M}$ should decrease again. However, Fig. 7-12 shows that for *any* specific mass flow rate, the lowest pressure that can be attained *within* a simple converging nozzle (where the flow area continually decreases) is the critical pressure P_c. Therefore, for external pressures less than P_c, the pressure just within the nozzle must still be P_c, and both $\phi(P_2)$ and $\dot{M}$ will have their maximum levels. Thus for a particular nozzle we may plot the mass flow rate as a function of the external pressure, as shown in Fig. 7-14. [The dashed line in Fig. 7-14 is the portion of the $\phi(P_2)$ curve that can never be attained in a simple converging nozzle.]

For P_2 less than P_c we find that a sharp discontinuity in the pressure must occur at the nozzle exit, since P_c is the lowest pressure that can exist just

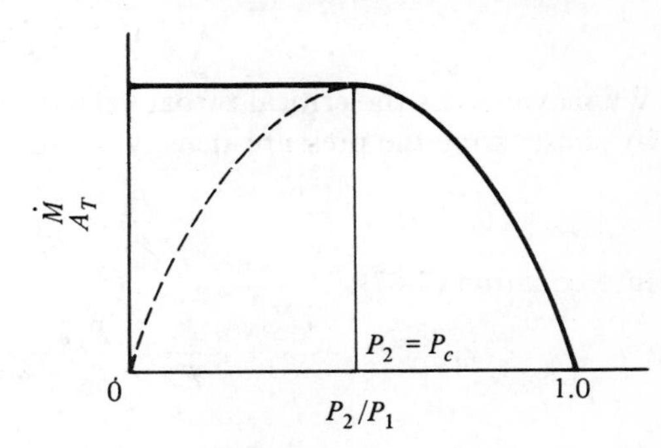

Figure 7-14. Mass flow rate ve. pressure ratio in a converging nozzle.

within the nozzle; just outside the nozzle, P_2 is less than P_c. This sharp pressure discontinuity at the nozzle throat leads to the formation of a standing rarefaction wave at the tip of the nozzle. The presence of this standing wave may be observed from the noise it generates, or by any of a number of techniques for shock- or expansion-wave visualization.

We may now draw the pressure profiles (that is, the pressure–position plots) that exist in a converging nozzle such as illustrated in Fig. 7-13 for three different exit pressures, $P_2 > P_c$, $P_2 = P_c$, and $P_2 < P_c$, as shown in Fig. 7-15.

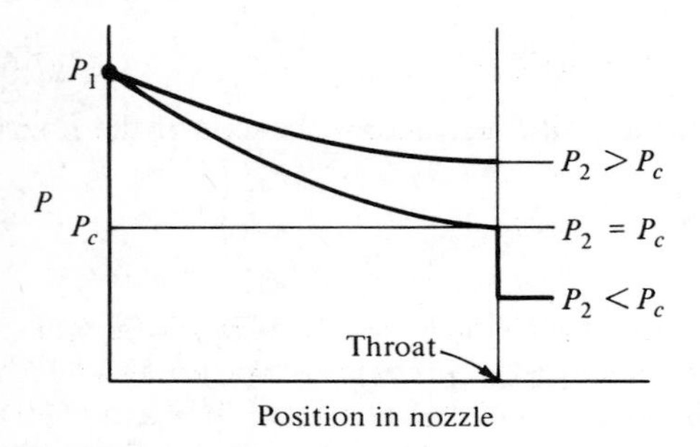

Figure 7-15. Pressure plots for a converging nozzle.

Let us now determine the maximum mass flow rate that may pass through a simple converging nozzle. For a given value of the throat area, the maximum flow rate will occur where $\phi(P_2)$ is a maximum, that is, at $P_2 = P_c$. The maximum in the $\phi(P_2)$ may be determined by setting the derivative of $\phi(P_2)$ with respect to $P_2 = 0$. Performing the required differentiation followed by algebraic simplification eventually gives the result:

$$P_c = P_1\left(\frac{2}{k+1}\right)^{k/(k-1)} \tag{7-67}$$

Thus P_c depends only on the inlet pressure and the heat capacity ratio.

The critical throat velocity may be calculated by substituting equation (7-67) into (7-58) to yield

$$\mathrm{V}_c = \sqrt{\frac{2k}{k+1}\frac{P_1}{\rho_1}g_c} \tag{7-68}$$

We may express the critical throat velocity in terms of the throat conditions by substituting the pressure–density relation

$$\frac{P_1}{\rho_1^k} = \frac{P_c}{\rho_c^k} \tag{7-55}$$

into equation (7-67):

$$\frac{P_1}{\rho_1} = \frac{P_c}{\rho_c}\frac{k+1}{2} \tag{7-69}$$

Equation (7-69) is then substituted into equation (7-68) to give

$$V_c = \sqrt{k\frac{P_c}{\rho_c}} = \sqrt{kRT_c g_c} \tag{7-70}$$

This value, however, corresponds to the speed of sound, C, at the throat temperature T_c:

$$V_c = C = \sqrt{kRT_c g_c} \tag{7-71}$$

Thus, the maximum velocity of an ideal gas in the throat of a simple converging nozzle is identical to the speed of sound at the throat conditions.

We may express the maximum mass flow rate that will pass through a converging nozzle by substituting equation (7-67) for P_c into equation (7-62) to give

$$\dot{M}_c = A_T\left(\frac{2}{k+1}\right)^{1/(k-1)} \sqrt{\frac{2kg_c}{k+1}P_1\rho_1} \tag{7-72}$$

Thus the critical mass flow rate can be calculated if we know the inlet conditions and the throat area. For exit pressures higher than P_c, the mass flow rate will be less than the critical value; for external pressures less than P_c, the throat pressure will be the critical pressure, and the mass flow rate will equal $\dot{M}_c$. On the other hand, equation (7-72) may be used to determine the minimum throat area that will pass a specified mass flow rate. If the area is greater than the minimum, P_2 will be greater than the critical pressure. If the area is less than the minimum, the flow will "choke" and less than the specified mass flow rate will be passed.

SAMPLE PROBLEM 7-5. Air at 20 Bar and 40°C enters a converging nozzle with negligible initial velocity. The nozzle exit has a cross-sectional area of 5×10^{-4} m². Assuming the flow to be isentropic and air to be an ideal gas at these conditions, calculate the mass flow rate (kg/sec) and the linear velocity (m/sec) at the discharge end

(a) If the discharge pressure is 13.3 Bar.

(b) If the discharge pressure is 10 Bar.

For air,

$$C_P = 29.4 \text{ kJ/(kg-mol °K)}$$

$$C_V = 21.0 \text{ kJ/(kg-mol °K)}$$

$$M = 29.0 \text{ kg/kg-mol}$$

Solution: Before we can proceed we must determine the critical pressure for the nozzle:

$$P_c = P_1\left(\frac{2}{k+1}\right)^{k/(k-1)}$$

But we are told that for air $C_P = 29.4$ and $C_V = 21.0$. Therefore,

$$k = \frac{29.4}{21.0} = 1.40$$

The critical pressure is then calculated from

$$P_c = 20\left(\frac{2}{2.4}\right)^{1.4/0.4} = 10.7 \text{ Bar}$$

For part (a), the exit pressure is greater than P_c; the flow is subsonic and equations (7-62) and (7-58) may be used directly to determine $\dot{M}$ and V. For part (b), the exit pressure is less than the critical value; so the flow is sonic at the throat and equations (7-68) and (7-72) must be used to calculate V and $\dot{M}$.

(a)
$$\dot{M} = A_T\sqrt{\frac{2kg_c}{k-1}P_1\rho_1\left[\left(\frac{P_2}{P_1}\right)^{2/k} - \left(\frac{P_2}{P_1}\right)^{(k+1)/k}\right]}$$

But $P_2 = 13.3$ Bar and $P_1 = 20$ Bar; therefore,

$$\frac{P_2}{P_1} = \frac{13.3}{20} = 0.666$$

We now use the ideal-gas equation to determine ρ_1:

$$\rho_1 = \frac{MP_1}{RT_1} = \frac{2.9 \times 2.0 \times 10^6 \text{ (kg/kg-mol) N/m}^2}{(8314)(313)\text{ (Nm/kg-mol °K) °K}} = 22.2 \text{ kg/m}^3$$

Therefore,

$$P_1\rho_1 = (2.0 \times 10^6)(22.2)(\text{N/m}^2)(\text{kg/m}^3)$$
$$= 4.44 \times 10^7 \text{ N kg/m}^5$$
$$A_T = 5 \times 10^{-4} \text{ m}^2$$

Therefore, $\dot{M}$ is calculated as

$$\dot{M} = 5 \times 10^{-4} \text{ m}^2$$
$$\cdot\sqrt{\frac{(2.0)(1.4)}{0.4}(4.44 \times 10^7)\left[\left(\frac{2}{3}\right)^{2.0/1.4} - \left(\frac{2}{3}\right)^{2.4/1.4}\right]1\frac{\text{kg m}}{\text{N sec}^2}\cdot\frac{\text{N kg}}{\text{m}^5}}$$

Therefore,

$$\dot{M} = 5 \times 10^{-4}\sqrt{18.7 \times 10^6 \text{ kg}^2/\text{sec}^2}$$

or

$$\dot{M} = 2.16 \text{ kg/sec}$$

We may now calculate V from the relation

$$\text{V} = \sqrt{\frac{2g_ck}{k-1}\frac{P_1}{\rho_1}\left[1 - \left(\frac{P_2}{P_1}\right)^{(k-1)/k}\right]}$$

Therefore,

$$\text{V} = \sqrt{\frac{(2)(1.4)}{0.4}\frac{1(2.0 \times 10^6)}{22.2}\left[1 - \left(\frac{2}{3}\right)^{0.4/1.4}\right]\frac{\text{kg m}}{\text{N sec}^2}\frac{\text{N}}{\text{m}^2}\frac{\text{m}^3}{\text{kg}}}$$

or

$$V = \sqrt{6.8 \times 10^4 \text{ m}^2/\text{sec}^2} = 260 \text{ m/sec}$$

(b) Since the exit pressure is below the critical pressure, we must now use the formulas derived for critical flows. Therefore,

$$\dot{M}_c = A\left(\frac{2}{k+1}\right)^{1/(k-1)} \sqrt{\frac{2kg_c}{k+1} P_1 \rho_1}$$

or

$$\dot{M}_c = 5.0 \times 10^{-4} \text{ m}^2 \left(\frac{2}{2.4}\right)^{1/0.4} \sqrt{\frac{2.8}{2.4}(4.44 \times 10^7)(1) \frac{\text{kg m}}{\text{N sec}^2} \frac{\text{N kg}}{\text{m}^5}}$$

Therefore,

$$\dot{M}_c = 3.17 \times 10^{-4} \sqrt{51.8 \times 10^6 \text{ kg}^2/\text{sec}^2} = 2.28 \text{ kg/sec}$$

and is the maximum amount of gas this nozzle can pass at an inlet pressure of 20 Bar. The throat velocity is sonic and is evaluated from

$$V_c = \sqrt{\frac{2kg_c}{k+1} \frac{P_1}{\rho_1}} = \sqrt{\frac{2(1.4)(1)}{2.4} \frac{2.0 \times 10^6}{22.2} \frac{\text{kg m}}{\text{N sec}^2} \frac{\text{N}}{\text{m}^2} \frac{\text{m}^3}{\text{kg}}}$$

Therefore,

$$V_c = \sqrt{10.5 \times 10^4 \text{ m}^2/\text{sec}^2} = 324 \text{ m/sec}$$

7.8 The Converging–Diverging Nozzle

We have shown earlier that the maximum exhaust velocity that may be achieved with a simple converging nozzle is the velocity of sound at the throat conditions. However, many nozzle applications, such as thrust nozzles for jet and rocket engines, require exhaust velocities many times the speed of sound. Therefore, the pressures within the nozzles at their exits must be considerably lower than the critical pressure, and the exit areas of these nozzles must be greater than the throat area. That is, the nozzle must contain a converging section that ends at the throat, followed by a diverging section as shown in Fig. 7-16. This type of nozzle is termed a converging–diverging nozzle.

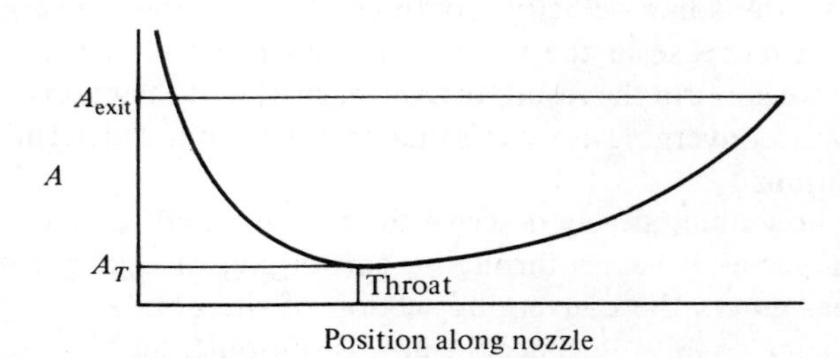

Figure 7-16. Converging-diverging nozzle.

We may obtain some valuable information about the flows that exist in each of the portions of the converging–diverging nozzle by examining the rate at which P (and thus V) varies with area in the nozzle:

$$A = \frac{\dot{M}}{\rho\sqrt{[2g_c k/(k-1)](P_1/\rho_1)[1-(P/P_1)^{(k-1)/k}]}} \tag{7-62}$$

Equation (7-62) has been derived as an expression between P and A. Differentiation of this expression by P at constant S, followed by by some agebraic manipulation and simplification eventually gives the following result:

$$\left(\frac{\partial A}{\partial P}\right)_S = \frac{A}{\rho V^2}\left(1 - \frac{V^2}{C^2}\right) = \frac{A}{\rho V^2}(1 - M^2) \tag{7-73}$$

where the Mach number, M, has been defined as

$$M = \frac{V}{C} \tag{7-74}$$

Since A, ρ_1, and V^2 are always positive, we may now make some observations about the way pressure will change with area. In particular, we note that a decrease in nozzle area may cause either an increase or a decrease in the nozzle pressure (and hence a decrease or increase in the velocity), depending on the Mach number, M. The effect of area changes on P and V, for constant $\dot{M}$, may be summarized as shown in Table 7-3. For example, if a gas

Table 7-3 Effect of Area Changes on *P* and V

Flow Type	Mach No.	Change of P with A	Change of V with A
Subsonic	$M < 1$	$\partial P/\partial A > 0$	$\partial V/\partial A < 0$
Supersonic	$M > 1$	$\partial P/\partial A < 0$	$\partial V/\partial A > 0$
Transonic	$M = 1$	$\partial P/\partial A = \infty$	$\partial V/\partial A = 0$

enters a converging nozzle at subsonic speed, the maximum velocity this gas can ever achieve is sonic, because if the gas at any point reached a speed slightly above the sonic velocity, continued flow in the converging nozzle would cause a decrease in the velocity and an increase in the pressure. In this manner, we confirm the result, derived earlier, that the maximum velocity attainable in a converging nozzle is the speed of sound determined at the throat conditions.

We may now qualitatively describe the pressure and velocity versus area relation of a gas as it passes through a converging–diverging nozzle as follows. The gas enters the converging section of the nozzle at high pressure and low velocity, so $M < 1$. Since the flow is subsonic, the pressure decreases and the velocity increases as the area decreases. This continues until the throat is reached. At the throat, the flow may be sonic or subsonic (but never supersonic), depending on the exit pressure. If the flow is subsonic, the pressure will increase, and the velocity will decrease in the diverging portion of the nozzle as the area increases. However, if the flow is sonic at the throat, two distinct flow regimes are possible in the diverging portion of the nozzle. If the exit pressure is low enough, the pressure will continue to drop along the nozzle and the flow will become supersonic, with increasing velocity. However, if the exit pressure is above a certain value, the pressure within the nozzle will have to increase in the diverging section of the nozzle, and the flow will be subsonic.

Let us now quantitatively examine the behavior of the converging–diverging nozzle. It is assumed that the throat and exit areas, A_T and A_2, respectively, of the nozzle under consideration are known (or are to be determined). Since the converging section of the converging–diverging nozzle behaves in a manner exactly analogous to a simple converging nozzle, there is a maximum flow rate that may pass through the throat. This maximum occurs when the throat pressure is P_c and the throat velocity is sonic.

Once the maximum flow rate of the nozzle is determined, one may determine the exit pressure P_2 such that $\dot{M}_c/A_2$ satisfies equation (7-62):

$$\frac{\dot{M}}{A} = \sqrt{\frac{2kg_c}{k-1}P_1\rho_1\left[\left(\frac{P}{P_1}\right)^{2/k} - \left(\frac{P}{P_1}\right)^{(k+1)/k}\right]} \tag{7-62}$$

It is found that two different pressures, known as the *design pressures* of the nozzle, may satisfy equation (7-62). At the upper design pressure it is found that $P_2 > P_c$; at the lower design pressure it is found that $P_2 < P_c$. The flow conditions in the diverging portion of the nozzle for the two design pressures may be summarized as follows:

1. Upper design pressure: Since $P_2 > P_c$, the pressure in the nozzle must increase with the increasing area. Thus the velocity decreases and the flow is subsonic.
2. Lower design pressure: Since $P_c > P_2$, the pressure within the nozzle must decrease with increasing distance from the throat. Therefore, the velocity continues to increase beyond sonic velocity, and the flow is supersonic.

The pressure profiles (pressure–position plots) for cases 1 and 2 are presented as curves A and B, respectively, in Fig. 7-17.

If the exit pressure, P_2, is greater than the upper design pressure but still less than P_1, it will not be possible for sonic flow to occur in throat, and subsonic flow will be encountered throughout the nozzle. This case is quite similar to flow through a converging nozzle with $P_2 > P_c$. The discharge rate of the nozzle will be less than $\dot{M}_c$ and can be calculated directly from equation (7-62). The pressure profile for the case of P_2 greater than the upper design pressure is illustrated by curve C in Fig. 7-17.

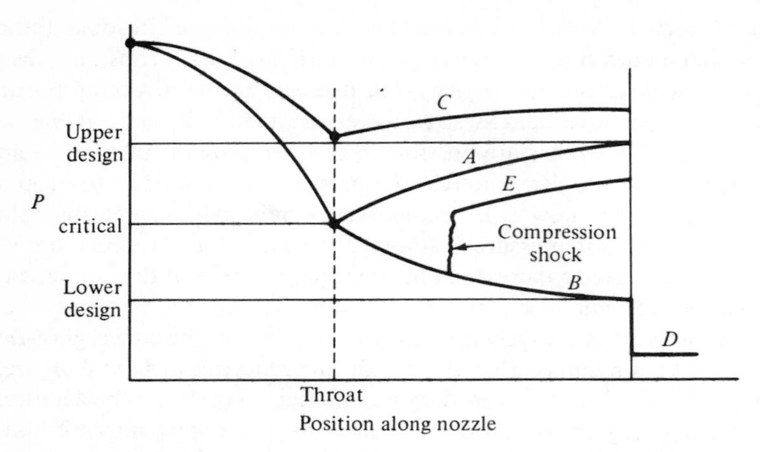

Figure 7-17. Pressure profiles for a converging-diverging nozzle.

If the external pressure, P_2, is less than the lower design pressure, a situation develops which is analogous to that for a simple converging nozzle operating with $P_2 < P_c$. We find that critical flow develops in the throat. Since the pressure decreases as the area increases, supersonic flow develops in the diverging portion of the nozzle. However, the lower design pressure is the lowest pressure that may be attained within the nozzle; the surroundings are at a still lower pressure. Thus we find that a pressure discontinuity (an expansion wave) develops at the nozzle exit. This case is illustrated by curve *D* of Fig. 7-17.

If the exit pressure is between the upper and lower design pressure, sonic flow develops at the throat. Since the external pressure is less than the upper design pressure, the flow in the diverging portion of the nozzle begins to expand and become supersonic, as if the exit pressure was at the lower design pressure. However, as the exit of the nozzle is approached, the gas within the nozzle begins to feel the pressure of the surroundings, which is greater than the lower design pressure. Since the gas cannot leave the nozzle at a pressure less than that of the surroundings, there will be a sharp pressure increase some place within the nozzle. This pressure increase manifests itself in the form of a "normal compression shock wave." As the gas passes through the compression shock wave, it is decelerated from supersonic to subsonic flow and experiences a sharp pressure increase. Since the flow leaving the shock is subsonic, its velocity continues to fall while the pressure rises in the remainder of the nozzle. The pressure profile corresponding to operation with an exit pressure between the two design pressures is represented in curve *E* of Fig. 7-17.

When a nozzle is operated so that the exit pressure is between the design pressures, the working fluid is said to be overexpanded. This condition is usually to be avoided in any nozzle designed for long-term operation, because the violent forces developed in the compression shock wave may be strong enough to damage the nozzle. The simplest solution to overexpansion is usually to decrease the exit area until the exit pressure equals one of the design pressures.

SAMPLE PROBLEM 7-6. Liquid hydrogen is to be burned in the combustion chamber of a liquid-fuel rocket. Liquid oxygen, at the stoichiometric rate, will be used as the oxidizing medium. The combustion is designed to operate at 600 psig, and initial calculations indicate that the combustion gases will enter the thrust nozzle at 5000°R. The thrust nozzle is to be designed so that there are equal pressure drops for equal length increments, and must be capable of exhausting to the atmosphere. If the hydrogen flow rate to the combustion chamber is 50 lb_m/sec and the total length of the nozzle is expected to be 5 ft,

(a) Plot the cross-sectional area of the nozzle and the steam velocity as a function of position.

(b) What is the throat area for this nozzle?

(c) It has been suggested that extra thrust may be obtained from the rocket in outer space by expanding the exhaust gases below 14.7 psia. What must the exit area be for an exit pressure of 1 psia? What is the exit velocity? Does this seem to be a reasonable method of obtaining extra thrust?

You may assume that steam is an ideal gas with $C_V = 6.64$ Btu/lb-mol°R under the conditions expected in the nozzle.

Solution: (a) We may picture the combustion chamber and thrust nozzle as shown in Fig. SP 7-6a. The pressure within the nozzle will vary linearly with position from 614.7 psia at the thrust chamber to 14.7 psia, as shown in Fig. SP 7-6b. The hydrogen flow rate to the combustion chamber is 50 lb_m/sec = 25 lb-mol/sec. Therefore, the steam flow rate into the thrust nozzle is 25 lb-mol/sec = $\dot{M}$.

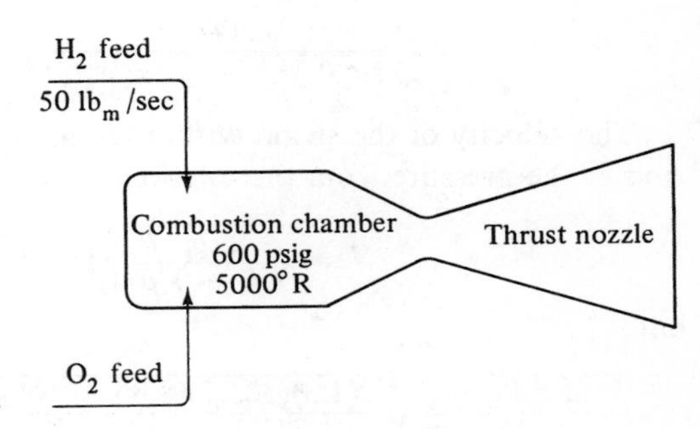

Figure SP7-6a

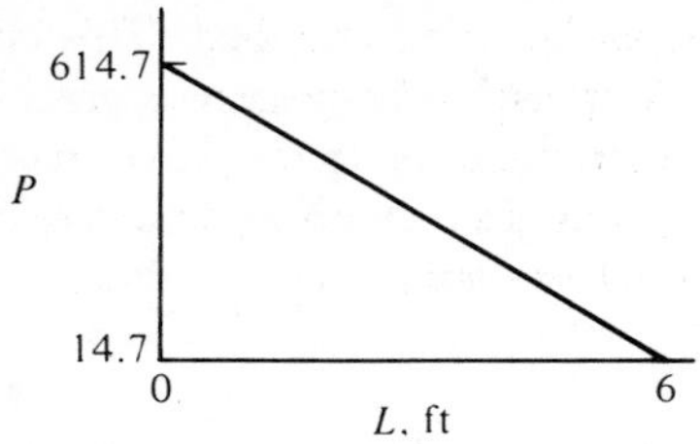

614.7

P

14.7

0 6

L, ft

Figure SP7-6b

The area at any point within the nozzle is related to the pressure through the relation

$$A = \frac{\dot{M}}{\sqrt{\dfrac{2kg_c}{k-1} P_1 \rho_1 \left[\left(\dfrac{P}{P_1}\right)^{2/k} - \left(\dfrac{P}{P_1}\right)^{(k+1)/k}\right]}}$$

where

$$P_1 = 614.7 \text{ psia} = 8.85 \times 10^4 \text{ lb}_f/\text{ft}^2$$

$$\rho_1 = \frac{P_1}{RT_1} = \frac{614.7 \text{ psia}}{(10.75)(5000)} \frac{\text{lb-mol}}{\text{psia ft}^3}$$

$$= 1.11 \times 10^{-2} \text{ lb-mol/ft}^3$$

$$k = \frac{8.63}{6.64} = 1.300; \quad \frac{2}{k} = 1.54; \quad \frac{k+1}{k} = 1.77; \quad \frac{k-1}{k} = 0.231$$

$$\dot{M} = 25 \text{ lb-mol/sec}$$

but

$$\sqrt{\frac{2kg_c P_1 \rho_1}{k-1}} = \sqrt{\frac{2(1.3)32.17(8.85 \times 10^4)(1.11 \times 10^{-2})}{0.3(18) \text{ lb}_m/\text{lb-mol}} \frac{\text{lb}_m\text{-ft}}{\text{lb}_f \text{ sec}^2} \frac{\text{lb}_f}{\text{ft}^2} \frac{\text{lb-mol}}{\text{ft}^3}}$$

$$= \sqrt{1.5 \times 10^4 \text{ lb-mol}^2/\text{sec}^2 \text{ ft}^4} = 123 \text{ lb-mol/sec ft}^2$$

Therefore,

$$A = \frac{25 \text{ lb-mol/sec}}{123 \text{ (lb-mol/sec ft}^2) \sqrt{(P/P_1)^{2/k} - (P/P_1)^{(k+1)/k}}}$$

$$= \frac{0.204 \text{ ft}^2}{\sqrt{(P/P_1)^{2/k} - (P/P_1)^{(k+1)/k}}}$$

The velocity of the steam within the nozzle may be calculated as a function of the pressure from the relation

$$V = \sqrt{\frac{2kg_c}{k-1} \frac{P_1}{\rho_1} \left[1 - \left(\frac{P}{P_1}\right)^{(k-1)/k}\right]}$$

but

$$\sqrt{\frac{2kg_c P_1}{(k-1)\rho_1}} = \sqrt{\frac{2(1.3)(32.17)(8.85 \times 10^4)}{(0.3)(1.11 \times 10^{-2})(18)} \frac{(\text{lb}_m\text{-ft/lb}_f \text{ sec}^2)}{(\text{lb-mol/ft}^3)} \frac{(\text{lb}_f)/\text{ft}^2}{(\text{lb}_m/\text{lb-mol})}}$$

Table SP 7-6 Values of Pressure, Area, and Velocity as a Function of Position

X	P, psia	$\frac{P}{P_1}$	$\left(\frac{P}{P_1}\right)^{2/k}$	$\left(\frac{P}{P_1}\right)^{(k+1)/k}$	$\left(\frac{P}{P_1}\right)^{2/k} - \left(\frac{P}{P_1}\right)^{(k+1)/k}$	A, ft^2	$\left(\frac{P}{P_1}\right)^{(k-1)/k}$	$-\left(\frac{P}{P_1}\right)^{(k-1)/k}$	V, ft/sec
0	614.7	1	1	1	0	∞	1	0	0
1	514.7	0.838	0.763	0.731	0.032	1.14	0.96	0.04	2240
2	414.7	0.674	0.545	0.497	0.048	0.930	0.913	0.087	3310
3	314.7	0.512	0.358	0.305	0.053	0.888	0.853	0.147	4330
4	214.7	0.349	0.197	0.155	0.042	0.994	0.784	0.216	5190
5	114.7	0.189	0.077	0.052	0.025	1.29	0.680	0.320	6330
6	14.7	0.0240	0.0032	0.0014	0.0018	4.80	0.422	0.578	8520

Therefore,

$$\sqrt{\frac{2kg_cP_1}{(k-1)\rho_1}} = \sqrt{1.24 \times 10^8 \text{ ft}^2/\text{sec}^2} = 11{,}100 \text{ ft/sec}$$

or

$$V = 11{,}100\sqrt{1 - \left(\frac{P}{P_1}\right)^{(k-1)/k}} \text{ ft/sec}$$

We may now evaluate both A and V as functions of pressure and position as shown in Table SP 7-6.

We may also plot the cross-sectional area and steam velocity as a function of position as shown in Fig. SP 7-6c.

(b) From the plot of A versus X, we may read the throat area as

$$A_T = 0.885 \text{ ft}^2$$

(c) For an exit pressure of 1.0 psia, $P_2/P_1 = 1.63 \times 10^{-3}$. Therefore,

$$A = \frac{0.204 \text{ ft}^2}{\sqrt{(0.00163)^{1.54} - (0.00163)^{1.74}}} = 8.2 \text{ ft}^2$$

$$V = 11{,}100\sqrt{1 - (0.00163)^{0.231}} \text{ ft/sec}$$

Therefore,

$$V = 9800 \text{ ft/sec}$$

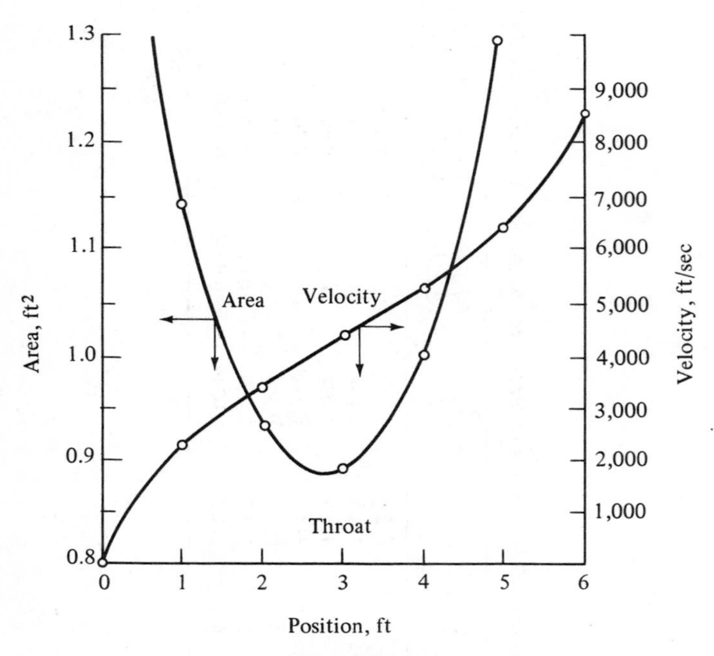

Figure SP7-6c

Thus we find that we must increase the size, and hence weight, of the nozzle tremendously in order to expand the gas from 14.7 to 1.0 psia, while the exit velocity increases only slightly. This is *not* a practical method of obtaining more thrust from a nozzle, because the added weight of the nozzle will more than likely balance the slight increase in exit velocity.

7.9 The Nozzle Equations in Terms of Mach Numbers

The nozzle equations just derived can all be expressed in a particularly simple form when the Mach number is used as the independent variable. The development of these relationships proceeds as follows:

We begin with equation (7-53):

$$V^2 = 2g_c C_P(T_1 - T) \tag{7-53}$$

but

$$C_P = \frac{kR}{k-1} \tag{7-75}$$

So that

$$V^2 = \frac{2g_c(kRT)}{k-1}\left(\frac{T_1}{T} - 1\right) \tag{7-76}$$

but the local speed of sound, C, is given by:

$$C = \sqrt{g_c kRT} \tag{7-71}$$

Thus equation (7-76) reduces to

$$V^2 = C^2\frac{2}{k-1}\left(\frac{T_1}{T} - 1\right) \tag{7-77}$$

or

$$\frac{V^2}{C^2} \equiv M^2 = \frac{2}{k-1}\left(\frac{T_1}{T} - 1\right) \tag{7-78}$$

Solving for T/T_1 in terms of the Mach number then gives:

$$\frac{T}{T_1} = \left[1 + \left(\frac{k-1}{2}\right)M^2\right]^{-1} \tag{7-79}$$

For the isentropic nozzle flow the pressures and density are related to the temperatures through:

$$\left(\frac{T}{T_1}\right) = \left(\frac{P}{P_1}\right)^{-(k-1/k)} = \left(\frac{\rho}{\rho_1}\right)^{-(k-1)} \tag{7-80}$$

Thus the pressure and density are related to the local Mach numbers by:

$$\frac{P}{P_1} = \left[1 + \frac{k-1}{2}M^2\right]^{-k/(k-1)} \tag{7-81}$$

$$\frac{\rho}{\rho_1} = \left[1 + \frac{k-1}{2}M^2\right]^{-1/(k-1)} \tag{7-82}$$

Thus for any given Mach number and inlet conditions, the local T, P, and ρ can be calculated from equations (7-79) to (7-82). The throat conditions can be obtained directly by setting $M = 1$:

$$\frac{T_c}{T_1} = \frac{2}{k+1}$$
$$\frac{P_c}{P_1} = \left(\frac{2}{k+1}\right)^{k/(k-1)} \tag{7-83}$$
$$\frac{\rho_c}{\rho} = \left(\frac{2}{k+1}\right)^{1/(k-1)}$$

These results agree with those obtained in our previous discussion based the maximum in $\phi(P)$.

The area of a nozzle may also be expressed in terms of the Mach number. We begin with the continuity equation:

$$\dot{M} = \rho \mathrm{V} A \tag{7-59}$$

but $\rho = P/RT$. Thus

$$\dot{M} = \frac{PA\mathrm{V}}{RT} \tag{7-84}$$

Introducing $C = \sqrt{g_c kRT}$ and $M = \mathrm{V}/C$ gives:

$$\dot{M} = MPA\sqrt{\frac{g_c k}{RT}} = \frac{MP}{\sqrt{T_1}}\sqrt{\frac{kg_c}{R}\left(\frac{T_1}{T}\right)} \tag{7-85}$$

Substituting equation (7-79) for (T_1/T) gives:

$$\dot{M} = \frac{AMP}{\sqrt{T_1}}\sqrt{\frac{kg_c}{R}\left[1 + \frac{k-1}{2}M^2\right]} \tag{7-86}$$

Likewise, substituting equation (7-80) for P/P_1 and solving for A gives:

$$A = \frac{\dot{M}}{P_1}\sqrt{\frac{RT_1}{g_c k}}\left[\frac{\left(1 + \frac{k-1}{2}M^2\right)^{(k+1)/(2k-2)}}{M}\right] \tag{7-87}$$

The throat area, A_T, can again be obtained by setting $M = 1$:

$$A_T = \frac{\dot{M}}{P_1}\sqrt{\frac{RT_1}{g_c k}}\left[\frac{k+1}{2}\right]^{(k+1)/(2k-2)} \tag{7-88}$$

The ratio of nozzle area to throat area can be obtained by dividing equation (7-87) by (7-88):

$$\frac{A}{A_T} = \frac{1}{M}\left[\left(\frac{2}{k+1}\right)\left(1 + \frac{k-1}{2}M^2\right)\right]^{(k+1)/(2k-2)} \tag{7-89}$$

A plot of A/A_T vs M for a gas with $k = 1.4$ is shown in Fig. 7-18.

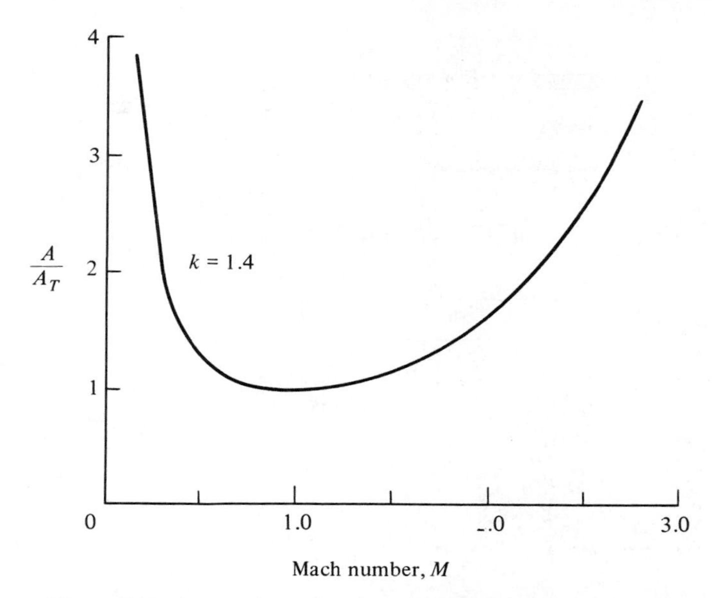

Figure 7-18. Area ratio as function of Mach number for converging-diverging nozzle.

7.10 Turbines

A turbine converts enthalpy to shaft work by first expanding the gas (or vapor) through a nozzle and then directing the high-velocity gas on a work surface at such an angle that a reversible transfer of momentum from the gas to the blade occurs at constant pressure. Each such combination of nozzle and blade is termed a *stage*. Each stage consists of many stationary nozzles mounted around the periphery of the flow channel and oriented in such a way that the exhaust gases impinge on the turbine blades (or buckets) which are mounted at the periphery of a disk attached to a shaft as shown in Fig. 7-19.

The gas impinges on the turbine blade as shown in Fig. 7-20, creating tangential and axial forces on the blade. The former causes tangential torque and motion while the latter is counterbalanced by the thrust bearings. The gas transfers part of its momentum to the blade. It emerges from the first stage with less kinetic energy and its momentum reduced and redirected as illustrated in Fig. 7-20. Subsequent stages of blading can be used to extract the remaining kinetic energy in the gas by simply reorienting the flow between each stage prior to introducing it to the next turbine stage. The use of several rows of blading to extract the energy from a single row of nozzles is referred to as *velocity compounding* and is used frequently in the first stage of turbines

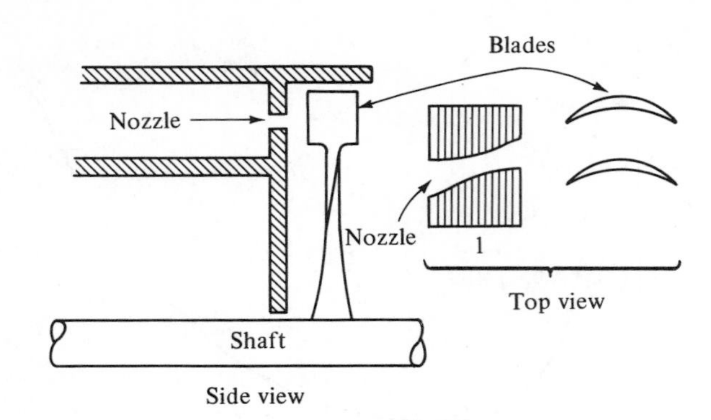

Figure 7-19

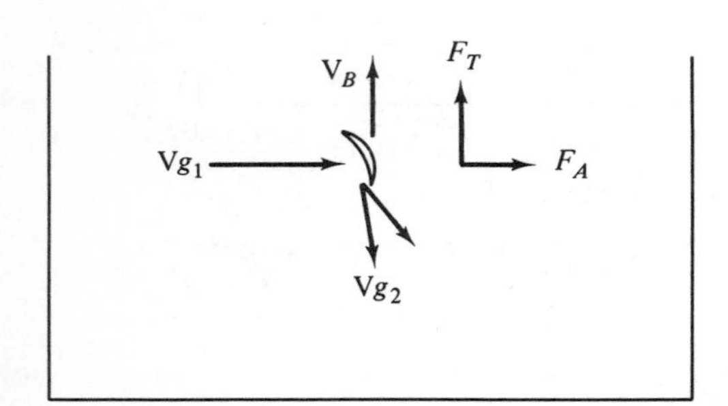

Figure 7-20

so that the gas can be expanded enough to reduce its temperature before it contacts the first blades. Figure 7-21 illustrates the path the expanding gas would follow after it leaves the nozzle and passes through moving blade and stationary divertor passages alternately. Up to three moving rows are used with a single impulse stage and alternating rows of flow divertors. The use of several rows to extract the momentum permits the momentum transfer to occur over several stages and thus avoids high rotational speed which severely stresses the rotating components.

Ideally the flow through each passage will be isentropic with the pressure drop occurring primarily in the nozzle. Both the rotating blades and the stationary divertors (stators) are designed to minimize flow irreversibilities as the gas passes through each stage.

Expansion of the gas in a stationary nozzle followed by a constant-pressure momentum transfer to the rotating blade is known as an *impulse stage*. It is also possible to design the rotating blade passages as nozzles so that expansion occurs between the moving blades. This is known as a *reaction stage* and is illustrated in Fig. 7-22.

Turbine designs frequently use both impulse and reaction staging. Steam turbine efficiencies of 95 percent have been realized in advanced designs.

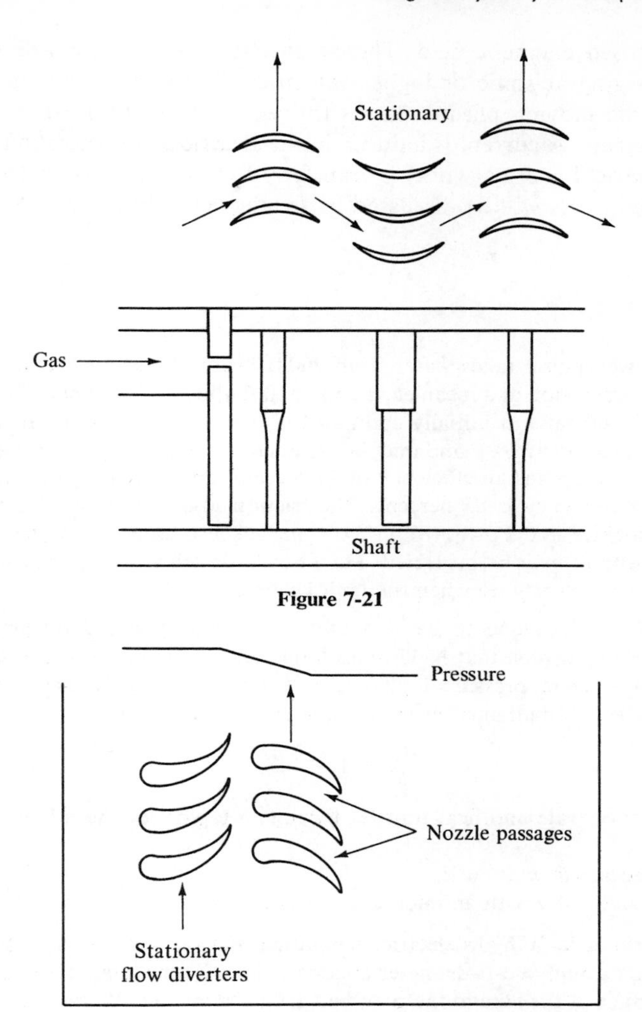

Figure 7-21

Figure 7-22

Aircraft gas turbines typically operate with lower efficiencies because of the need to keep the weight per hp down.

7.11 Magnetohydrodynamics (MHD)

An expanding gas, if electrically conducting, can generate electrical (mechanical) energy directly if passed through a magnetic field. The process termed *magnetohydrodynamics* (or MHD) utilizes no moving parts and relies on the interaction of the electric field produced by the flowing ionized gas (or plasma) and an imposed magnetic field. The expansion occurs in a nozzle with the plasma emerging at sonic or higher velocities. Enthalpy is extracted in the diffuser as the moving plasma passes through a magnetic field induced by external magnets. A current is induced in the electrical collectors in the channel and electrical energy (work) is transferred directly to the surroundings. MHD power cycles will be discussed in greater detail in Chapter 8.

Problems

7-1. A deep-well pump takes water from 200 ft below the ground and delivers it to a closed storage tank at a mean elevation of 10 ft above the ground. The tank has a volume of 5000 gal and initially contains only air at atmospheric pressure. The air and the water are at 70°F and may be assumed to remain at that temperature at all times. The pump has an efficiency of 75 percent, and it is driven by an electric motor whose efficiency is 85 percent. The friction loss in the pipe for the usual pumping velocities is 0.2 ft-$lb_f/(lb_m)$/(ft of pipe). The presence of water vapor in the air in the tank may be neglected. How much electrical energy will have been drawn from the power lines when the tank becomes $\frac{7}{8}$ full?

7-2. Natural gas (assumed to be all methane) is being pumped through a 1-ft-diameter horizontal pipe that is 35 miles long. The upstream pressure is 750 psia and the down-stream pressure is 400 psia. A Fanning friction factor of 0.0035 may be assumed. If methane obeys the equation of state

$$\frac{Pv}{RT} = 1 - \frac{0.4P_r}{T_r}$$

calculate the flow rate and heat transfer through the pipe for the following conditions:

(a) Isothermal flow at 60°F.

(b) Adiabatic flow with an inlet temperature of 60°F.

7-3. The turbine in a hydroelectric generating station receives water from an artificial lake through a 3-ft-diameter concrete pipe 0.5 mile long. The pipe extends from the surface of the lake to the river bed 250 ft below. At all practical flow rates the flow is sufficiently turbulent that a constant Fanning friction factor of $f = 0.005$ may be assumed. In an effort to increase the total work produced by the hydroelectric generators, more water will be allowed to flow through the turbines.

(a) Will an increase in the water flow rate (ft^3/min) always lead to an increased power production? Neglect all losses except friction in the concrete pipe. Justify your answer.

(b) If the answer to (a) is *no*, then at what flow rate will the power production of the generating station be maximized?

7-4. One hundred million standard cubic feet (60°F, 1 atm) per day of radioactive waste gas at 1000°F must be released at a height of 400 ft above the ground to avoid contamination of the surrounding area. A circular stack of uniform diameter is to be used. A draft at the base of the stack of 1 in. of water is available. (The pressure inside the stack base is 1 in. of water less than barometric pressure.) The barometric pressure at the base of the stack is 740 mm Hg and the ambient temperature 60°F. The gas has a molecular weight of 32 and may be considered an ideal gas. Assume that the gas passes through the stack isothermally. What diameter will be required? You can use a Fanning friction factor, $f = 0.016$, and you may assume that V^2 is constant throughout the tower.

7-5. A high-pressure reactor is vented to the surroundings by means of a 3-in.-i.d. pipe that is 4 ft long, as shown in Fig. P7-5. During normal operation a rupture disc seals the vent tube. However, if an upset occurs so that the reactor pressure exceeds 100 psig, the rupture disc disintegrates and the reactor is emptied of its contents.

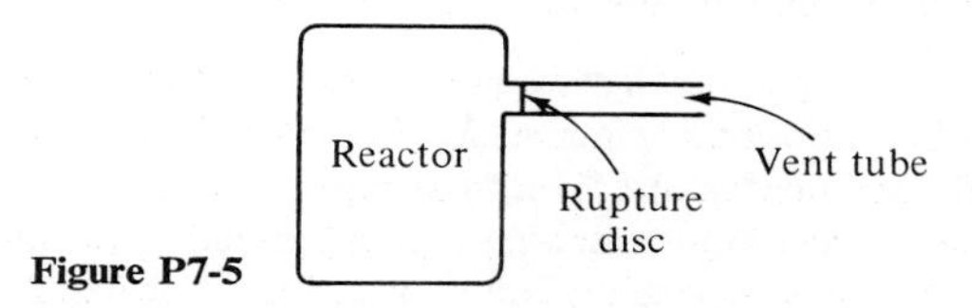

Figure P7-5

(a) During a test of the vent system the reactor is filled with air at 100 psig when the rupture disc fails. If the flow in the vent line is assumed to be isothermal and of very high Reynolds number, so the friction factor, f, is constant, derive the relation between the rate (lb_m/sec) at which air leaves the reactor (assuming the reactor pressure is 100 psig) and the exit pressure at the vent outlet. (You need *not* substitute numbers.) Use a Fanning friction factor $f = 0.003$.

(b) The vent line would normally be vented to the atmosphere. However, it has been suggested that placing vacuum pumps on the end of the vent line (to decrease the pressure at the end of the vent line) can be used to increase the initial rate at which the reactor empties. If the reactor is filled with air at 100 psig, would these vacuum pumps cause an increase, decrease, or no change at all in the initial rate at which the reactor is emptied? *Justify your answer carefully*!

(c) What is the initial rate at which air at 100 psig would exit from the vent line if the vent leads to the atmosphere? (Assume that the flow is fully developed in the pipeline, so flow field development can be neglected.)

7-6. A high-pressure air line contains a small bleed line whose diameter is $\frac{1}{4}$ in. i.d. and has an effective length of 3 ft. The air in the line is at 800 psia and 80°F. If the valve on the bleed line is accidentally opened, what is the maximum mass flow rate (express in lb_m/hr) of air through the bleed line? What is the exit pressure for the maximum flow rate? What do you suspect will happen if the atmospheric pressure is less than the critical pressure? Calculate the outlet velocity at maximum flow rate, and compare this with sonic velocity for isothermal flow. You may assume that air is an ideal gas at these conditions, the flow in the bleed line may be assumed to be isothermal, and a Fanning friction factor $f = 0.002$ may be assumed.

7-7. An ideal gas whose constant-volume heat capacity is 9.0 Btu/lb-mol °R is passed through two different converging nozzles. In each case the diameter of the discharge throat of the nozzle is 0.4 in., and in each case the gas enters the nozzle at 150 psia and 80°F and is discharged at 120 psia. The nozzles operate differently, however, according to the following description:

(a) This nozzle is well insulated and the friction effect or turbulence is negligible.

(b) This nozzle operates isothermally and the friction effect is negligible.

Show by numerical calculations which nozzle will pass the greatest amount of gas per unit time.

7-8. Gas enters an adiabatic isentropic nozzle at 40 psia and 600°F and with negligible velocity, and is continuously expanded to 1 atm.

(a) At what velocity (ft/sec) will the gas leave the nozzle?

(b) For a gas flow rate of 10 lb_m/hr, what cross-sectional area (ft^2) will be required at the nozzle exit?

The equation of state is $P(v - b) = RT$, where $b = 0.5$ ft^3/lb-mol; C_P at 1 atm = 7.0 Btu/lb-mol °F; mol wt of gas = 30.

7-9. A special research project requires that 9 g-mol of air be fed to a reaction vessel (*A*), which is initially evacuated (Fig. P7-9). The entire quantity of air is to be fed over a 60-sec interval, and the rate of supply must be constant over that interval. It is proposed to accomplish this by feeding the air from tank *B* through a quick-opening valve and a converging isentropic critical flow nozzle. Tank *B* is to be maintained at constant pressure by a freely floating (assume frictionless) piston. The nozzle will have a cross-sectional area of 0.025 cm^2 at its throat, and the tanks, line, valve, and nozzle will be well insulated. What is the minimum permissible volume for tank *A* if the air in *B* is initially at 38°C and air behaves ideally at all times?

$$C_P = 7 \text{ cal/g-mol °K}$$

$$C_V = 5 \text{ cal/g-mol °K}$$

$$\frac{P_c}{P_1} = 0.5275 \text{ for air}$$

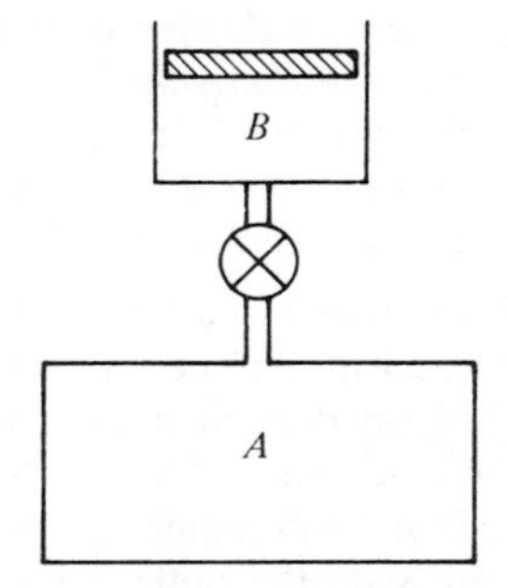

Figure P7-9

7-10. A "critical flow prover" is commonly used to measure flow rates accurately in order to calibrate other devices for flow measurement. The instrument consists of a carefully machined orifice stamped with the "air time," defined as the number of seconds required to pass 1 ft^3 of air measured at the inlet (upstream) pressure and an inlet temperature of 60°F. The definition thus automatically provides for changes in inlet pressure, while the existence of critical flow ensures that the air time will be independent of discharge pressure. Assuming isentropic flow and ideal gas behavior,

(a) Calculate the air flow (lb_m/hr) through an orifice having an air time of 11.4 sec when the inlet temperature is 60°F, the inlet pressure is 40 psia, and the discharge pressure is 15 psia.

(b) Calculate the cross-sectional area (ft^2) of an orifice having an air time of 8.5 sec.

(c) Calculate the air flow through the orifice of (a) if the inlet temperature is 80°F, all other conditions remaining the same as in (a).

$$k = 1.4$$

$$\text{mol wt} = 29$$

$$\frac{P_c}{P_1} = 0.5275$$

7-11. Develop the expression for flow through an adiabatic, frictionless nozzle for a gas that has the following properties:

$$\frac{Pv}{RT} = 1 - 0.4\frac{P_r}{T_r} \qquad C_P = \text{const with } T$$

Find the critical pressure ratio.

7-12. A rocket exhaust gas composed of equal amounts of H_2O, CO_2, CO, and N_2 is to be ejected at a rate of 22.7 kg/sec from an ICBM. The combustion chamber will operate at 2800°K and 13.6 bar and the gases may be assumed to have a heat-capacity ratio, k, equal to 1.2. *Estimate* the cross-sectional throat area of a nozzle to achieve this performance. If an efficiently designed nozzle has a length approximately 10 times its throat diameter, could this nozzle be used on a rocket that was equipped to use nozzles up to 60 cm in length? If not, suggest an engineering solution to the problem which permits the use of the existing rocket with the same mass ejection rate. Indicate any assumptions made in your solution.

7-13. A nozzle is being designed for application with its exit pressure either very near the inlet pressure or much below it. The exit area of the nozzle is 15 times the throat area, and the nozzle will use air as the working fluid. The air will enter the nozzle at 500 psia and 100°F. Determine the range of pressures over which it is *not safe* to operate this nozzle.

7-14. The air intake of a jet airliner is a diffuser that has the opposite purpose of a nozzle: It increases the pressure of the air flowing through it by performing an adiabatic, reversible deceleration of the air. An intake is to be designed for a plane that is expected to fly at 600 mph at 30,000 ft, where the pressure is 4.6 psia and the temperature is −50°F. The diffuser must compress the air to 7.5 psia before the air enters a centrifugal compressor for further compression.

(a) What will be the velocity of the air at the exit from the diffuser?

(b) How large must the cross-sectional area of the inlet and discharge of the diffuser be if it is to compress air at the rate of 15 lb_m/sec?

7-15. Natural gas (essentially pure methane) is being compressed prior to transmission by pipeline. The gas enters the compressor at 70°F and 200 psia and is isothermally compressed to 3000 psia. Evaluate Δh, q, and w for the compression if

(a) The natural gas behaves as an ideal gas.

(b) The natural gas behaves according to the van der Waals equation of state. (*Hint:* Because of the nature of the equation of state, you may find it simpler to let $\Delta h = \Delta u + \Delta Pv$, and then calculate Δu and ΔPv.)

7-16. Gas enters an isothermal gas turbine at 70 bar and 27°C and discharges at 20 bar. Part of the work from the turbine is used to drive an adiabatic isentropic compressor in which all the gas discharged from the turbine is partially recompressed and leaves the compressor at 93°C.

(a) At what pressure does the gas leave the compressor?

(b) What fraction of the total work delivered by the turbine must be supplied to the compressor?

Equation of state: $Pv = RT + av$, where $a = 5.6$ bar. $C_V^* = 5.0$ cal/g-mol °K. Assume that $q_{\text{turb}} = T(s_0 - s_i)$.

7-17. Gas enters an adiabatic, isentropic nozzle at 40 atm and 500°F and with negligible velocity and is continuously expanded to 1 atm.

(a) At what velocity (ft/sec) will the gas leave the nozzle?

(b) For a gas flow rate of 10 lb_m/hr, what cross-sectional area (ft^2) will be required at the nozzle exit?

Equation of state: $P(v - b) = RT$, where $b = 0.5$ ft^3/lb-mol. C_P at 1 atm = 7.0 Btu lb-mol °F. Molecular weight of gas = 30.

7-18. Natural gas as received from gas wells usually contains moderate amounts of valuable heavy hydrocarbons (C_2 through C_{10}) which are removed from the gas before it is sold as fuel. The lean natural gas is then compressed to high pressures so that it can be economically transported through gas pipelines to the major fuel markets. A lean natural gas is available at 10°C and 68 bar. The gas must be compressed to 340 bar for economical pipeline transportation. Determine the work that must be supplied to the compressor per kilogram of gas under the following assumptions:

(a) The compression is isothermal and reversible and the natural gas behaves according to the following equations of state: (1) The ideal gas law, and (2)

$$1 - Z = \frac{P_r}{T_r}\left(-0.115 + \frac{0.132}{T_r} + \frac{0.356}{T_r^2}\right)$$

where $Z = Pv/RT$.

(b) The compression is adiabatic and reversible, and the natural gas behaves according to the same two equations of state as given in (a).

You may assume that the natural gas behaves as if it is pure methane with the following properties:

$$P_c = 45.8 \text{ bar}$$

$$T_c = 191.1\text{°K}$$

$$C_P^* = 4.52 + 0.00737T(\text{°R}) \text{ cal/g-mol °K}$$

7-19. A compressor takes in CO_2 at a rate of 1000 ft^3/hr at 70°F and atmospheric pressure. The discharge conditions are 115 psia and 85°F. The compressor is cooled by a jacket that removes heat to cooling water. By noting the rise in temperature of a measured quantity of the water, it has been found that 7500 Btu of heat is removed every hour. The motor that drives the compressor is drawing 2.8 kw of electrical power from the line. Determine the efficiency of the electric motor:

$$(C_P)_{CO_2} = 9.3 \text{ Btu/lb-mol °F}$$

7-20. A single-acting, two-stage compressor is handling hydrogen at 1 atm and 60°F. Exhaust is at 250 psia. A heat exchanger cools the gas between the stages to 60°F; otherwise operation is assumed to be adiabatic. For hydrogen k (heat capacity ratio) = 1.42.

(a) What pressure between the two stages minimizes the reversible work for the compressor?

(b) Calculate the minimum work per 100 lb-mol of gas.

Suppose now that the compressor is equipped with jacketed cylinders so that the operation is essentially isothermal at all times.

(c) What pressure between the two stages minimizes the reversible work?

(d) Compare the amount of heat removed from the jacketed cylinders in the second case with the amount of heat removed from the intercooler (heat exchanger) in the first case.

(e) Compare the work per 100 lb-mol for the two cases.

7-21. Chloropentafluoroethane (C_2ClF_5) is used in certain refrigeration cycles. At one point the gas at 80 psia and 160°F is compressed to 320 psia. Per 1000 ft^3 of gas at the suction side, calculate the heat and work effects for the following cases:

(a) The compressor operates reversibly and isothermally.

(b) The compressor operates reversibly and adiabatically.

If the compressed gas from case (a) is expanded adiabatically through a throttle valve to a pressure of 80 psia, what will be the exit temperature?

The critical properties of C_2ClF_5 are as follows:

$$P_c = 453 \text{ psia}$$

$$T_c = 175.9\text{°F}$$

$$v_c = 0.02687 \text{ ft}^3/\text{lb}_m$$

The heat capacity of the gas at low pressure is

$$C_P^* = [0.044976 + 0.00033T(\text{°R}) - 1.52 \times 10^{-7}T^2] \text{ Btu/lb}_m \text{ °R}$$

EIGHT

thermodynamics of energy conversion

Introduction

Most processes with which the engineer is associated involve energy, either as a product or as a means of producing a product. In the preceding chapters the basic thermodynamic relationships that govern energy conversion have been developed. In addition, the thermodynamic potentials that serve as the driving forces for energy flows, as well as the interrelationships among these potentials, have been developed. In this chapter these relationships are examined in order to study the methods and systems devised by engineers for the utilization of our energy resources in meeting our everyday needs.

The various forms of energy, such as thermal, mechanical, electrical, nuclear, and chemical, have been cited in the earlier chapters. Energy flows caused solely by temperature gradients were categorized as heat, while energy flows resulting from thermodynamic potentials other than temperature (for example, pressure, electrical potential, etc.) were classified as work. As we have previously indicated, the energy flows we call work all have one important common trait: They are, in theory, completely interconvertible. To be sure, these interconversions are not easily achieved without irreversibilities which lead to degradation of some of the mechanical energy to the thermal form as was discussed earlier. Frequently mechanical forms are intentionally converted to thermal forms, as in the case of an electrical heating unit.

The efficient conversion of thermal energy into mechanical energy has challenged scientists and engineers for several hundred years. This challenge continues today because a large portion of our energy reserves exists in forms such that the energy is extracted in a thermal form by first burning (or fissioning) the fuel or extracting steam or hot water from the earth. A portion of the thermal energy so obtained is then frequently converted to a mechanical form, in which it has far greater utility. In spite of our best efforts we have been unable to devise a procedure by which the thermal energy obtained in any of these forms can be completely converted to a mechanical form. In

Chapter 5 we explained the reason for such a limitation by analyzing the Carnot cycle. In this Chapter we will study the various systems that have been devised to perform this conversion in such a way that the maximum possible work is obtained from a unit of thermal energy.

8.1 Energy Conversion Efficiencies

Efficiencies of conversion devices and systems are shown in Fig. 8-1. Values range from less than 5% for the incandescent lamp to 99% for the most efficient electric generators. With the exception of hydroenergy our primary energy sources, fossil, nuclear, solar, and geothermal all either occur or are converted to thermal energy before work is derived. Less than half of the thermal energy so produced is used for space and process heating, the rest is used to produce mechanical (e.g. electrical) energy for final consumption.

As Figure 8-1 shows, there are means by which chemical, nuclear and solar forms of energy can be converted directly to mechanical or electrical energy. In some cases the theoretical efficiencies of these conversions processes are impressively large, as for magnetohydrodynamics and the fuel cell, but the technology required to achieve these performances economically has not yet been developed. Needless to say the heightened awareness of the finite nature of our energy resources has led to the intensified development of many of the options which promise to yield improved efficiencies.

Fossil fuels currently supply close to 96% of our energy needs. While nuclear energy supplied about 2% of the nation's energy in 1974, it is rapidly growing in importance and by the year 2000 is expected to provide a much as 50% of the nation's electrical energy. Both of these sources (fossil and fissionable fuels) when used for central station power production produce thermal energy first which then must be converted to a mechanical or electrical form by some sort of a heat engine.

A heat engine is defined as one which exchanges only heat and work with the surroundings. These engines use a working fluid, such as steam, and operate in a closed cycle to effect the desired energy transformation. While heat engines have been used primarily for central station energy conversion applications in the past, they have received increasingly serious consideration for use in vehicles because of the improved emissions performance they offer. The overall efficiency of energy utilization depends on each of the conversion processes needed to transform the fuel into a Btu or joule at the point of consumption.

The conversion of chemical to thermal energy can be accomplished with efficiencies which vary from about 60% in home oil fired furnaces to close to 88% in a large steam boiler, where the principal loss is represented by the sensible heat leaving through the stack. The Carnot engine, discussed in

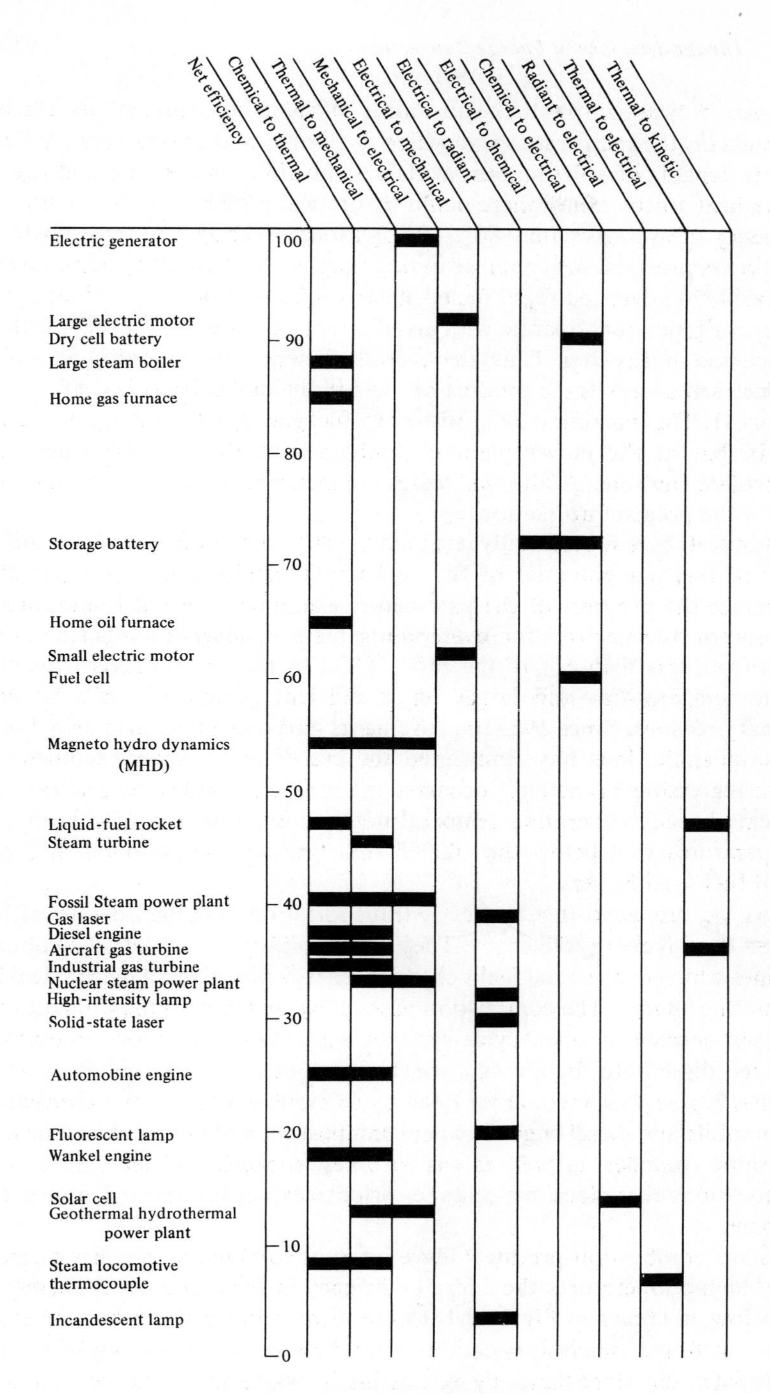

Figure 8-1. Energy conversion efficiencies.

Chapter 5, was shown to be the most efficient heat engine. Its efficiency depends on the temperatures at which it absorbs and rejects heat. A Carnot engine capable of utilizing heat at the combustion temperature and rejecting waste heat to the atmosphere could theoretically operate with a conversion efficiency of approximately 80%. For reasons which will become obvious as we discuss practical heat engines in this chapter, the best efficiencies currently realizable in advanced fossil fueled steam cycles is about 47%. Since modern electrical generators operate with an efficiency of close to 99%, very little loss is incurred in this step. Thus, the overall efficiency of converting a fossil fuel to electrical energy is the product of these individual efficiencies, $.88 \times .47 \times .99$ or .41. The remaining 59% of the raw fuel energy which is not delivered to the busbar at the power plant is discharged to the air and water in the immediate environs. Additional losses are incurred in transmitting the power and at the point of utilization.

While 41% is substantially less than the 80% that a Carnot engine utilizing the full thermal potential of the fuel might produce, it is appropriate to recognize the progress of the past before examining current limitations and options on the horizon for overcoming them. Conversion efficiencies have risen from less than 5% in the early 1900's to the present level with higher steam temperatures and larger, more efficient generating units being the primary reasons. Since 1960 improvements have been few, primarily because material limitations have prevented the use of higher steam temperatures. Some regression has actually occurred in the case of light water nuclear power systems because operating temperature limitations in the core keep steam temperatures well below the 1000°F that are commonly found in modern fossil fuel fired boilers.

As we discussed in Chapter I, transportation systems operate with the lowest overall energy efficiency. They have typically used internal combustion engines which convert the fuel's chemical energy directly to a mechanical form within the engine. The combustion gases serve as the working fluid and these engines operate as open cycle systems with the combustion products discharged directly to the atmosphere. In the context of this treatment *internal combustion engines* include all open cycle systems such as the conventional automobile and diesel engines, where combustion and expansion occur within the same chamber, as well as gas turbines, rockets, and jet engines where combustion takes place in a chamber prior to expansion through a nozzle or a turbine.

Since combustion products leave internal combustion engines at moderately high temperatures the overall efficiency in these engines is correspondingly low, as shown in Figure 8-1. Consequently, internal combustion engines have not found much application in the large-scale generation of electrical energy. On the other hand, by exhausting its waste heat with the combustion products, the internal combustion engine avoids the need for an external source of cooling and is ideally suited for nonstationary propulsion applications such as automobiles and airplanes.

8.2 Practical Considerations in Heat Engines

In our previous discussion of the Carnot cycle it was assumed that all steps occurred reversibly. Such operation will produce the maximum conversion of thermal to mechancial energy for a given source and sink temperature. However, reversible operation is both generally impossible and impractical to obtain and thus represents only a theoretical maximum against which a real system's performance can be compared. For example, if one attempted to operate the heat exchangers reversibly, only an infinitesimal temperature difference would exist between the hot and cold sinks and the respective portions of the working fluid. Thus infinitely large heat exchangers (boiler and condenser), which cost an infinite amount of money, would be required. Since this is clearly intolerable, it is necessary to provide a finite temperature difference between the working fluid and the hot and cold sinks. This permits the heat exchanger to be reasonably scaled and priced. Therefore, the high temperature at which the cyclic fluid operates is less than the temperature of the hot sink, and its low temperature is greater than the temperature of the cold sink by an amount necessary to achieve reasonable heat-transfer rates. The overall temperature difference between the fluid temperatures at the two extremes of the cycle is thus less than that between the hot and cold sinks, and the overall efficiency of any heat engine (not just the Carnot engine) is reduced commensurately. The crosshatched area in Fig. 8-2 represents work that could have been obtained from Carnot engines operating around the cycles 1-2-2′-1′ and 4′-3′-3-4 had it not been necessary to provide the temperature driving forces in both exchangers.

In addition, it must be remembered that even the best expanders and compressors available today do not operate reversibly. Consequently some

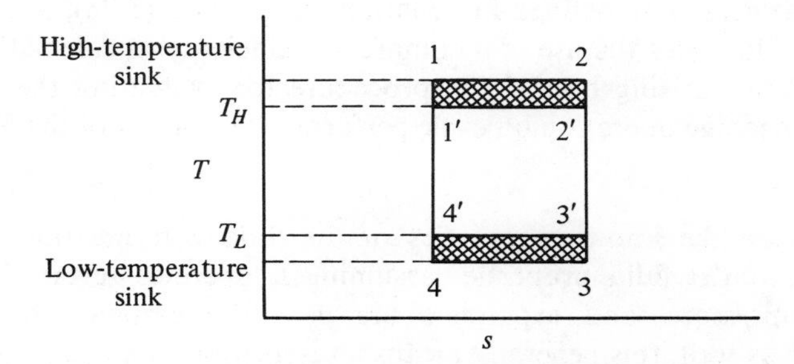

Figure 8-2. Carnot engine with finite temperature differences.

dissipation of mechanical to thermal energy will occur in the expander and the compressor; the compressor requires more work than the reversible minimum, and the expander produces less work than the reversible maximum. Since irreversible processes result in the fluid experiencing an entropy increase (if conducted adiabatically), the two work paths must both display increasing entropies, as seen in Fig. 8-3 for cycle 1-2-3′-4′. At first glance one is tempted to conclude by considering enclosed areas that w_{net} for cycle 1-2-3′-4′ is greater than for the Carnot cycle, 1-2-3-4, but, as cautioned earlier, *for cycles with irreversible work steps, the enclosed area does not represent* w_{net}. Instead, we recall that for any cyclic heat engine,

$$w_{net} = q_H + q_L$$

and that $q_H = T_H \, \Delta s_H$ = area under curve 1-2. Now contrast the area under curves 3-4 and 3′-4′, both of which are negative. Clearly the area under 3′-4′ is the larger and hence w_{net} for cycle 1-2-3′-4′ is appreciably less than for the Carnot cycle 1-2-3-4, even though both cycles receive exactly the same amount of heat at T_H.

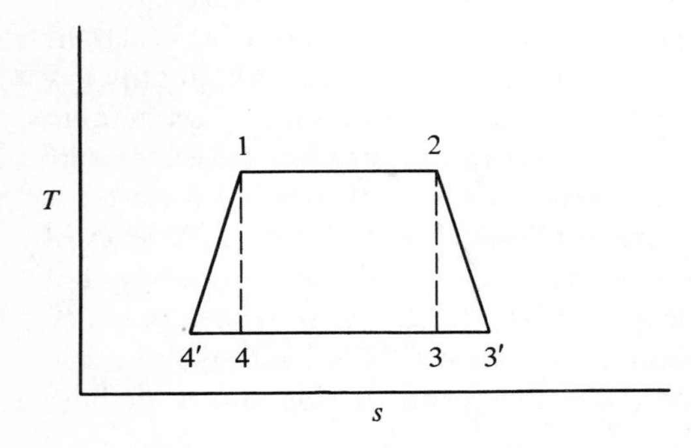

Figure 8-3. The effect of irreversible pumps and expanders on the Carnot cycle.

In evaluating the effect that compressor and expander irreversibilities have on cycle performance we typically utilize machine efficiencies which were discussed earlier in Chapter 5. Efficiencies for work-producing and work-consuming devices were defined in equations (5-27) and (5-28), and Sample Problem 5-5 illustrates the use of a compressor efficiency in calculating cycle properties. Summarizing briefly, the procedure for evaluating the effect of irreversible machine operation on cycle performance consists of the following steps:

1. Based on the known properties of the fluid entering the machine, calculate the outlet fluid properties assuming it operates reversibly. Since pumps, compressors, and expanders are generally assumed to operate adiabatically as well, this generally means an isentropic change.

2. Cycle conditions normally fix one other condition at the exit of the machine (frequently T or P) so that, given the entropy, all other properties can be calculated including the enthalpy, h.

3. Since $w = -\Delta h$ for an adiabatic process, the efficiency which relates the actual work, w_{act}, to the reversible work, w_{rev}, can also be used to calculate the Δh_{act} to the Δh_{rev} which can be calculated according to steps 1 and 2 above.

4. Knowing Δh_{act} one can calculate the actual h at the machine exhaust and then use this along with the other fixed conditions to define the fluid state and calculate all other properties.

This procedure applies to all of the cycles to be discussed in the following chapters. Machine efficiencies vary considerably depending on type (reciprocating versus centrifugal), size, and application. Larger rotating machinery is generally more efficient than smaller reciprocating devices. Nozzles can be designed to operate in a nearly reversible fashion, but when incorporated in turbine design there are numerous other irreversibilities which the fluid experiences as it expands through the passages, as well as the mechanical inefficiencies of the machine itself. Reciprocating expanders have substantial losses in terms of piston friction and mechanical friction in the linkages. These factors are significant in the relatively poor overall performance of internal combustion engines such as the Otto and Diesel cycles to be discussed later in this chapter. Representative values for machine efficiences are listed in Table 8-1.

Table 8-1 Typical Efficiencies for Rotating and Reciprocating Machines

Machine	Efficiency
Nozzles	95–99%
Steam turbines	85–95%
Gas turbines	80–90%
Rotating pumps and compressors	80–90%
Reciprocating machines	60–80%

SAMPLE PROBLEM 8-1. Steam is expanded in the final stage of a steam turbine from 830°F and 90 psia to a pressure of 3.7 psia. If the expander efficiency is 90 percent, calculate the work produced and the outlet condition of the steam.

Solution: From the steam tables for $T_1 = 830°\text{F}$ and $P_1 = 90$ psia:

$$h_1 = 1443 \text{ Btu/lb}_m \quad \text{and} \quad s_1 = 1.868 \text{ Btu/lb}_m{}°\text{R}$$

For a reversible (and adiabatic) expansion $\Delta s = 0$; therefore

$$s_2 = s_1 = 1.868 \text{ Btu/lb}_m{}°\text{R}$$

Using s_2 and $P_2 = 3.7$ psia we can find from the steam tables:

$$h_2 = 1125 \text{ Btu/lb}_m \quad \text{and} \quad T_2 = 150°\text{F}.$$

For a turbine

$$\text{efficiency} = \frac{w_{act}}{w_{rev}} = \frac{\Delta h_{act}}{\Delta h_{rev}} = 0.90$$

$$\Delta h_{rev} = h_2 - h_1 = 1125 - 1443 = -318 \text{ Btu/lb}_m$$

$$\Delta h_{act} = (0.90)(-318) = -286.2 \text{ Btu/lb}_m$$

$$w_{act} = -\Delta h_{act} = 286.2 \text{ Btu/lb}_m$$

$$h_2 = h_1 + \Delta h = 1443 + (-286.2) = 1156.8 \text{ Btu/lb}_m$$

Using $P_2 = 3.7$ psia and $h_2 = 1156.8$ Btu/lb$_m$, we obtain from the steam tables:

$$T_2 = 218°\text{F}, \qquad s_2 = 1.914 \text{ Btu/lb}_m°\text{R}$$

The steam is seen to be in a superheated condition.

8.3 The Rankine Cycle

In previous discussions the Carnot engine has always been sketched on a T–s diagram within the liquid–vapor region. This practice has been followed because an isothermal absorption or rejection of a finite amount of heat would result in violation of the isothermal requirement if a single phase (gas or liquid) was passed through the exchanger. In addition to the highly idealized nature of Carnot cycles, other considerations arise if one attempts to operate a heat engine with the fluid path in the two-phase region.

The work requirements for reversibly compressing (or pumping) a fluid between two pressure levels is directly related to its specific volume as shown in Chapter 7:

$$w = -\int_{P_L}^{P_H} v\, dP$$

Thus if one condenses the fluid completely before it enters the pump, the work required to elevate its pressure is considerably less because of the much lower value of v^L than v^V (for fluids far removed from their critical points). In addition to this consideration, the pumping of two-phase gas–liquid mixtures poses serious mechanical difficulty with most pumps or compressors.

As shown in Fig. 8-4, the entry point to the compressor is shifted to the saturated liquid curve in an effort to improve performance of the cycle. An adiabatic reversible compression of the liquid from the condenser pressure to the boiler pressure takes the fluid along the isentropic path 5-6. Since the tem-

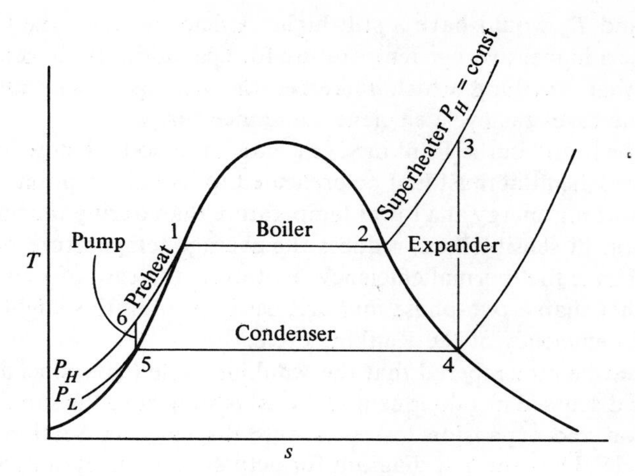

Figure 8-4. Rankine cycle.

perature drops sharply along lines of constant pressure as one moves away from the saturated liquid curve into the liquid region, the fluid emerges at point 6 in a subcooled state and its temperature is considerably below that corresponding to saturation at P_H. Consequently, the fluid must first experience a sensible heat addition (path 6-1) in the boiler before vaporization commences.

The discharge (from the expander) of the Carnot cycle discussed earlier also occurs in the two-phase region. This, too, is a highly undesirable situation, because the liquid droplets in the two-phase mixture cause serious erosion and vibration problems in the high-velocity portions of the expander. One would thus prefer that the expanding fluid remain in the vapor phase until the expansion is completed. For an isentropic expansion this may be achieved by superheating (at constant pressure) the working fluid before it enters the expander (path 2-3). By controlling the amount of superheat it is possible to predetermine the state of the fluid in the expander exhaust, because both its entropy and pressure are then fixed. For our purposes we shall assume that enough superheat is added so that the fluid exits from the expander as a saturated vapor. The modified cycle shown in Fig. 8-4 is called a *Rankine cycle* and, although still subject to the consideration of irreversibilities discussed earlier, represents a far more practical (although slightly less efficient) cycle than the Carnot cycle.

In discussing the efficiencies of heat engines it is useful to think in terms of the average temperatures at which heat is received and rejected. Since the cycle with superheat has a higher average temperature of heat addition than the comparable cycle without superheat, its efficiency would be higher. On the other hand, a Carnot cycle operating between the maximum superheat temperature and T_L would have a still higher efficiency, since the Carnot cycle would have a higher average temperature for heat addition. In general we may conclude that anything which increases the average temperature of heat addition increases the cycle efficiency, and vice versa.

Since the liquid in the Rankine cycle enters the boiler below its saturation temperature, the fluid must first be preheated to its boiling point. Because the fluid is absorbing energy at a lower temperature than during the phase change, this addition of sensible heat reduces the average temperature of heat addition, and hence the thermal efficiency. However, the ease of pumping a pure liquid, rather than a two-phase mixture, easily justifies this slight decrease in the thermal efficiency of the Rankine cycle.

It should be remembered that the Rankine cycle (as well as all modifications to be discussed in subsequent sections) is subject to the same irreversible compression and expansion losses as were discussed in Section 8.2 for the Carnot cycle. Thus the T–s diagram for actual Rankine cycle would deviate from Fig. 8-4 in that the isentropic work steps would both be replaced by segments along which the entropy increased as shown in Fig. 8-5. Similarly, a small amount of liquid subcooling is generally provided in the condenser to assure that no vapor will pass into the pump. This, too, can be observed at point 6 in Fig. 8-5.

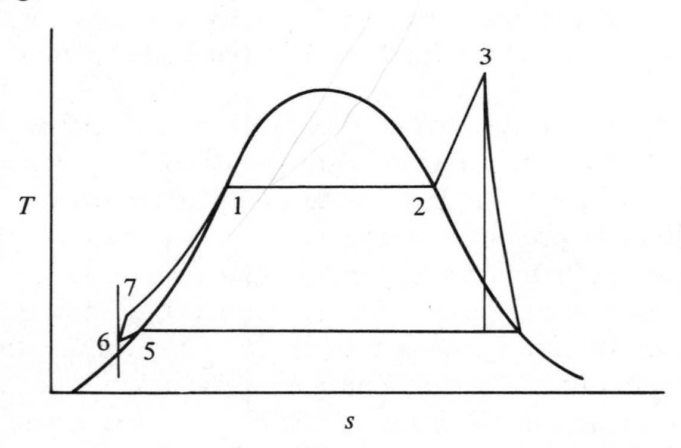

Figure 8-5. Rankine cycle with irreversibilities.

The Rankine cycle with slight modifications (which will be discussed in the next section) forms the basis for most of the thermal power generated in this country. The most recently designed facilities are capable of producing 1 kw hr of electricity per 8000 Btu of thermal energy supplied, for an overall efficiency of 42 percent. Older installations usually require about 10,000 Btu of thermal energy per kw hr, for an overall efficiency of 34 percent. These figures illustrate that even our best performance still results in a loss to our rivers and lakes of slightly more than 1 Btu for every Btu of electrical energy produced.

SAMPLE PROBLEM 8-2. A steam-generating station operates on the Rankine cycle with superheat. The boiler tubes are restricted to a maximum pressure of 7 Bar, and cooling water is available at 50°C. The condenser is designed for a 15°C temperature difference, so the low temperature in the cycle is 65°C. The superheater and condenser operate so that the expander and pump handle only single-phase materials. The pump and expander may be considered to be adiabatic and reversible.

(a) Determine the superheat (°C) needed, and the thermal efficiency of the cycle.

(b) Compare this efficiency with that of a Carnot cycle operating between 65°C and (1) the boiler temperature, and (2) the maximum superheater temperature.

(c) If new boiler tubes that can withstand a pressure of 15 Bar are installed, determine the new thermal efficiency of the cycle.

Solution: (a) We may picture the cycle as shown in Fig. SP 8-2a or, in terms of the actual equipment used, Fig. SP 8-2b.

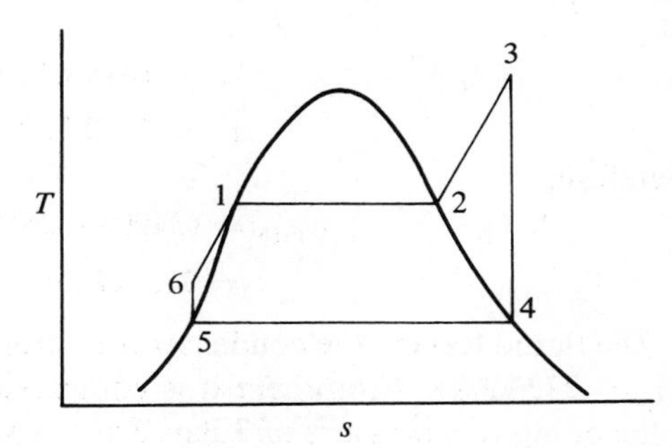

Figure SP8-2a

The steam leaving the boiler section (point 2) is a saturated vapor at 7 Bar, so $T = 165°C$. It is then superheated to a point where it can be adiabatically and reversibly expanded in the turbine to the point where it is a saturated vapor at 65°C and 0.25 Bar. Since the expansion in the turbine is adiabatic and reversible, the entropy balance tells us that it is also isentropic. Thus we can find the temperature that the superheater must achieve by following the isentrope from the outlet of the turbine to a pressure of 7 Bar. This gives us a superheater temperature of 462°C, so the steam is (462 − 165)° = 297°C superheated.

Since we now know the conditions of the steam entering and leaving the turbine we can determine the amount of work produced during the expansion. If we neglect potential and kinetic energy changes across the turbine and assume steady-state, adiabatic operation, the energy balance reduces to

$$-w_{\text{turb}} = \Delta h_{3\text{-}4} = h_4 - h_3$$

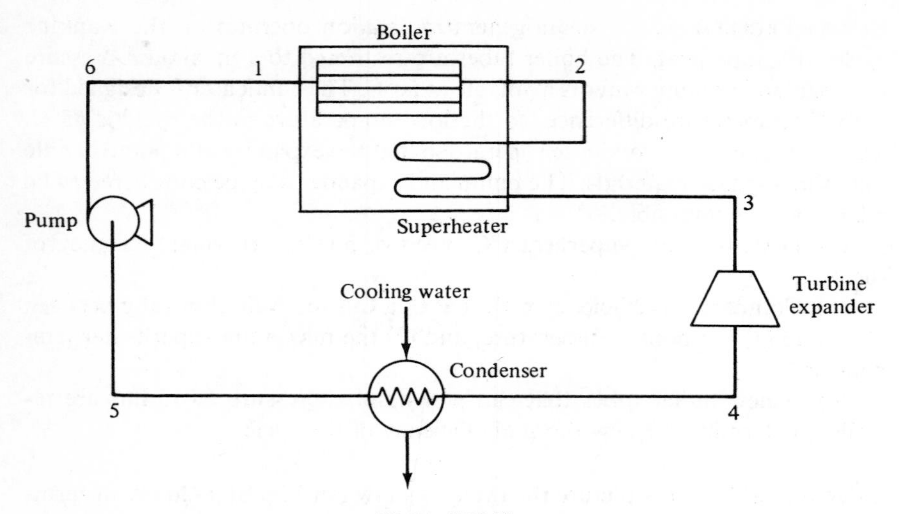

Figure SP8-2b

but

$$h_4 = 2618 \text{ kJ/kg}$$
$$h_3 = 3400 \text{ kJ/kg}$$

Therefore,

$$\begin{aligned} w_{\text{turb}} &= (3400 - 2618) \text{ kJ/kg} \\ &= 782 \text{ kJ/kg} \end{aligned}$$

The liquid leaving the condenser is a saturated liquid at 65°C. Its enthalpy is $h_5 = 272$ kJ/kg. From here it is compressed adiabatically and reversibly by the pump to a pressure of 7 Bar. The work that the pump consumes can be found from the expression

$$w_{\text{pump}} = -\int_{P_5}^{P_6} v\, dP$$

However, since the liquid is essentially incompressible, v is not a function of P and can be removed from under the integral sign to yield

$$\begin{aligned} w_{\text{pump}} &= -v \int_{P_5}^{P_6} dP = -v(P_6 - P_5) \\ &= -(10 \times 10^{-3})(7.0 - 0.25) \times 10^5 \text{ m}^3/\text{kg N/m}^2 \\ &= -0.675 \times 10^3 \text{ Nm/kg} = 0.675 \text{ kJ/kg} \end{aligned}$$

We now determine the enthalpy of the liquid entering the boiler as follows. An energy balance around the pump, assuming negligible changes in potential and kinetic energy and steady-state, adiabatic operation, gives

$$-w_{\text{pump}} = \Delta h_{5\text{-}6} = h_6 - h_5$$

or

$$h_6 = h_5 - w_{\text{pump}}$$

Since $h_5 = 272$ kJ/kg, $h_6 = 273$ kJ/kg.

We now evaluate the heat input in the boiler–superheater from an energy balance. Assuming negligible changes in potential and kinetic energy and steady-state operation, the energy balance around the boiler–superheater gives

$$\Delta h_{6\text{-}3} = q - w$$

But $w = 0$, so

$$\begin{aligned} q = \Delta h_{6\text{-}3} &= h_3 - h_6 \\ &= (3400 - 273) \text{ kJ/kg} \\ &= 3117 \text{ kJ/kg} \end{aligned}$$

The net work produced by the cycle is the sum of the work produced in the turbine and the pump. That is,

$$\begin{aligned} w_{\text{net}} = w_{\text{turb}} + w_{\text{pump}} &= (782 - 1) \text{ kJ/kg} \\ w_{\text{net}} &= 781 \text{ kJ/kg} \end{aligned}$$

and the thermal efficiency of the cycle is given by

$$\eta = \frac{w_{\text{net}}}{q_{\text{boiler}}} = \frac{781}{3117} = 0.250 \quad \text{or} \quad 25.0\%$$

(b) For the Carnot cycle $\eta = (T_H - T_L)/T_H$. (1) Therefore, for a Carnot cycle operating between the boiler temperature $T = 165°\text{C} = 438°\text{K}$ and the condenser temperature $T = 65°\text{C} = 338°\text{K}$, the thermal efficiency is

$$\eta = \frac{438 - 338}{438} = 0.228$$

(2) The maximum temperature in the superheater is $462°\text{C} = 635°\text{K}$. The Carnot efficiency of a cycle with this maximum temperature is then

$$\eta = \frac{625 - 338}{635} = 0.468$$

Thus we see that the Rankine cycle with superheat is slightly more efficient than the Carnot cycle operating at the boiler temperature. However, a Carnot cycle operating at the maximum superheater temperature is considerably more efficient than the Rankine cycle.

(c) If the old 7-Bar boiler tubes are replaced by tubes capable of withstanding 15 Bar, we must repeat the calculation of part (a), with the new boiler pressure. The *T–s* diagram for the new cycle is given in Fig. SP 8-2c. We may now fill in the property values at points 3, 4, and 5 as given in Fig. SP 8-2c. The work produced by the turbine is then

$$\begin{aligned} -w_{\text{turb}} &= \Delta h_{3\text{-}4} = (3678 - 2618) \text{ kJ/kg} \\ w_{\text{turb}} &= 1060 \text{ kJ/kg} \end{aligned}$$

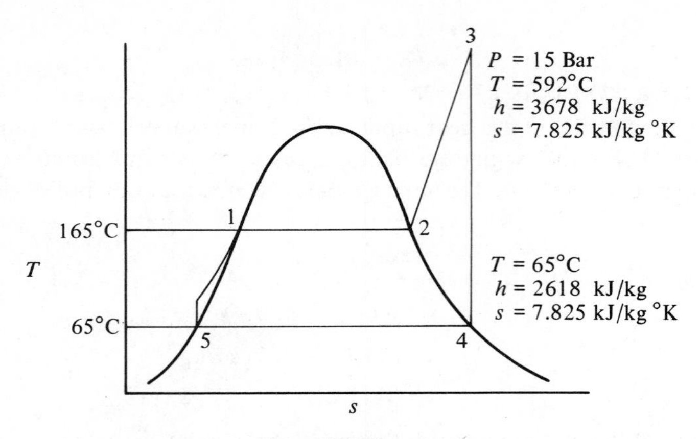

Figure SP8-2c

The work consumed by the pump is

$$-w_{\text{pump}} = v(P_6 - P_5) = (1.0 \times 10^{-3})(15 - 0.25) \times 10^5(\text{m}^3/\text{kg}) \times \text{N/m}^2$$
$$= -1.475 \times 10^3 \text{ Nm/kg} = -1.5 \text{ kJ/kg}$$

The enthalpy of the liquid entering the boiler is

$$h_6 = h_5 - w_{\text{pump}} = 274 \text{ kJ/kg}$$

The heat addition in the boiler–superheater is given by

$$q_{\text{boiler}} = \Delta h_{6\text{-}3} = (3678 - 274) \text{ kJ/kg}$$
$$= 3404 \text{ kJ/kg}$$

The net work produced by the cycle is

$$w_{\text{net}} = w_{\text{turb}} + w_{\text{pump}} = (1060 - 2) \text{ kJ/kg}$$
$$= 1058 \text{ kJ/kg}$$

Thus the thermal efficiency is

$$\eta = \frac{w_{\text{net}}}{q_{\text{boiler}}} = \frac{1058}{3404} = 0.312$$

which is a significant improvement over the previous values.

8.4 Improvements in the Rankine Cycle

Rankine with Reheat

As we saw in Sample Problem 8-2, the thermal efficiency of the Rankine cycle can be significantly increased by using higher boiler pressures. Unfortunately, this requires ever-increasing superheats. [In part (c) of Sample Problem 8-2 a superheat of approximately 427°C (759°F) was needed for a boiler pressure of only 15 Bar.] Since the maximum temperature in the superheater is limited by the temperature the tubes can stand, superheater temperatures are usually restricted to less than about 1100°F. Since the major fraction of the heat supplied to Rankine cycle is supplied in the boiler, not the superheater, we must increase the boiler temperatures (and hence pressures) if significant efficiency improvements are to be obtained. From Sample Problem 8-2 we saw that boiler pressures much above 200 psia (13 Bar) are not possible in a simple Rankine cycle if the expanding steam is to remain as a single phase throughout the expansion and a maximum temperature of 1100°F is allowed in the superheater.

The problem of excessive superheater temperatures may be solved while still avoiding two-phase mixtures in the expansion by "reheating" the expanding steam part way through the expansion as shown in Fig. 8-6. Thus the steam leaving the boiler section (point 2) is superheated to an acceptable temperature and then expanded (and work removed) until it intersects the saturation curve. The steam is then "reheated" in a second superheater section and expanded in a second turbine (with more work removed) until the saturation curve is again encountered. The steam is then condensed and pumped back into the boiler.

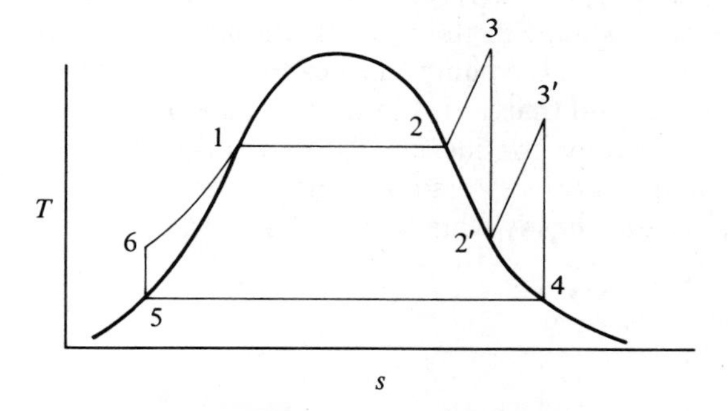

Figure 8-6. Rankine cycle with reheat.

Although the preceding example considered only one reheat operation, clearly we may use more if desired. In this fashion it is possible to use higher boiler pressures (thereby obtaining increased efficiencies) without having to increase the maximum superheater temperature above the working limits of the superheater tubes.

SAMPLE PROBLEM 8-3. The tubes in the boiler of Sample Problem 8-2 are replaced by tubes that can withstand pressures up to 60 Bar but no temperatures above 462°C. The condenser is still assumed to operate at 65°C.

(a) If a standard Rankine cycle with no reheat is used, what is the maximum boiler pressure and the cycle efficiency?

(b) If the boiler pressure is raised to 60 Bar and reheat added to allow single-phase expansions: (1) How many reheat cycles are needed? (2) What is the maximum temperature in the last superheater? (3) What is the overall cycle efficiency? (*Note:* You may assume that the pump and turbine are adiabatic and reversible.)

Solution: (a) Since 462°C is the maximum superheater temperature of Sample Problem 8-2 when the boiler pressure is 7 Bar, that is still the maximum boiler pressure that may now be used. If a boiler pressure above this value is used, it will not be possible to superheat the vapor enough to avoid condensation in the turbine. The cycle efficiency is thus the same as part (a) of the Sample Problem 8-2: $\eta = 25.0$ percent.

(b) We may trace out the reheat cycles needed with a boiler pressure of 60 Bar as shown in Fig. SP 8-3. The steam leaves the boiler at point 2 as a saturated vapor at 60 Bar and is superheated to 482°C at constant P (point 3). The steam is now adiabatically and reversibly (that is, isentropically) expanded until it hits the saturation curve which occurs at a pressure of 4.5 Bar. It is then reheated at 4.5 Bar until it hits either the 482°C isotherm or the isentrope that intersects the saturation curve at 65°C. It hits the isentrope first at a temperature of 395°C. From this point the steam is expanded in a second turbine until it intersects the saturation curve at a temperature of 65°C. Thus one reheat is necessary, and the overall cycle will be as shown in Fig. SP 8-3.

The net work produced in the cycle is the sum of the works of the two turbines, plus the work of the pump; the heat input is the sum of that added in the preheater–boiler and that added in the two superheaters. Let us begin the analysis of this cycle by considering the two turbines. An energy balance around each of the turbines (assuming no change of potential or kinetic energy, and adiabatic, steady-state flow) gives

$$-w_T = \Delta h$$

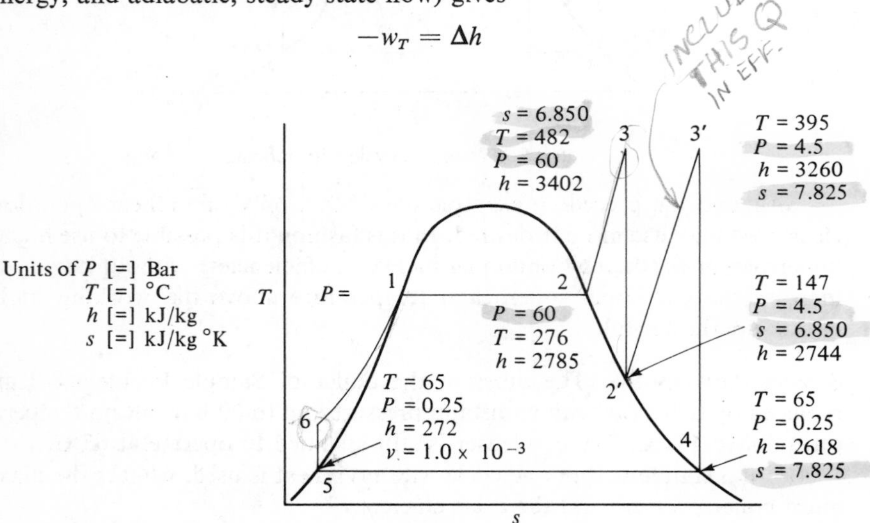

Figure SP8-3

Thus the work produced by the first turbine is

$$w_{T_1} = h_3 - h_{2'} = (3402 - 2744)\ \text{kJ/kg} = 658\ \text{kJ/kg}$$

while that produced in the second is

$$w_{T_2} = h_{3'} - h_4 = (3260 - 2618)\ \text{kJ/kg} = 642\ \text{kJ/kg}$$

The work produced in the pump is given by

$$w_P = -v\Delta P = -1.0 \times 10^{-3}(60 - 0.25) \times 10^5\ (\text{m}^3/\text{kg})(\text{N/m}^2)$$
$$= -5.98 \times 10^3\ \text{Nm/kg} = 6\ \text{kJ/kg}$$

Therefore, the net work produced by the cycle is

$$w_{\text{net}} = (658 + 642 - 6)\ \text{kJ/kg} = 1294\ \text{kJ/kg}$$

We may obtain the enthalpy of the liquid water entering the boiler from an energy balance around the pump:

$$-w_P = \Delta h = h_6 - h_5$$

or

$$h_6 = h_5 - w_P = (272 + 6)\ \text{kJ/kg} = 278\ \text{kJ/kg}$$

The heat added to the boiler and first superheater is then obtained from an energy balance around this section. Assuming negligible potential and kinetic energy changes and steady-state flow with no work, the energy balance becomes

$$q_{B\text{-}S} = \Delta h_{B\text{-}S} = h_3 - h_6 = (3402 - 278)\ \text{kJ/kg}$$

or

$$q_{B\text{-}S} = 3124\ \text{kJ/kg}$$

Similarly, the heat added in the reheater is given by

$$q_R = \Delta h_R = h_{3'} - h_{2'} = (3260 - 2744)\ \text{kJ/kg}$$
$$= 516\ \text{kJ/kg}$$

The total heat addition is $q_H = q_{B\text{-}S} + q_R$, or

$$q_H = (3124 + 516)\ \text{kJ/kg} = 3640\ \text{kJ/kg}$$

and the thermal efficiency is given by

$$\eta = \frac{w_{\text{net}}}{q_H} = \frac{1294}{3640} = 0.355 \quad \text{or} \quad 35.5\%$$

an increase of almost 50 percent above the efficiency of the simple Rankine cycle with no reheat operating at the same maximum temperature. Even greater increases can be obtained by operating at still higher boiler pressures, but the same superheater temperature, by using two, three, or even four reheat stages.

Rankine with Regereration

The thermal efficiency of the Rankine and reheat cycles can be further increased by the use of "regenerative" heat exchange, as shown in Fig. 8-7. In the regenerative cycle, a portion of the partially expanded steam is drawn off, between the high- and low-pressure turbines (or an intermediate tap if only a single turbine is used). This steam is used to preheat the condensed liquid before it is returned to the boiler–superheater. In this way we can reduce the amount of heat added at the low temperatures, so we increase the average temperature at which heat is added to the cycle and thereby increase the thermal efficiency. Although in theory any number of preheaters may be used, the gain in efficiency drops off fairly quickly with the number of preheaters, so it is extremely rare to find more than five preheaters, three being a more typical value.

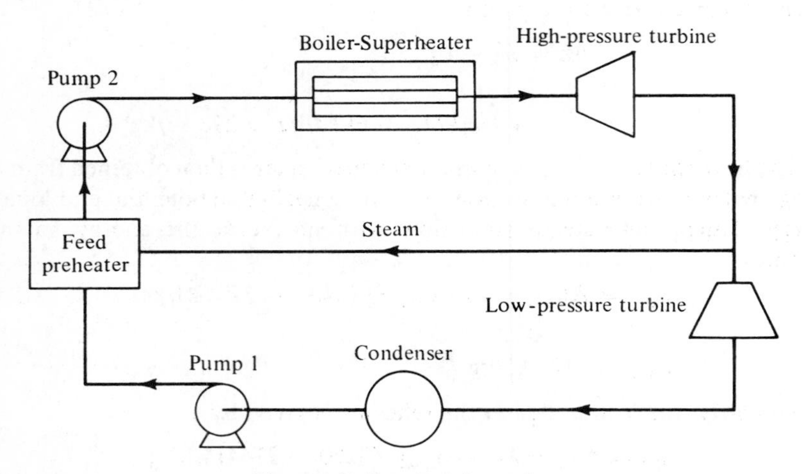

Figure 8-7. Regenerative cycle.

Two types of feed preheaters are commonly used: the open preheater and the closed preheater. In the open preheaters shown in Fig. 8-7, the steam is simply mixed with the feedwater in a tank. The steam condenses and heats the water. The mixture is then pumped to the next stage. In the closed preheater, on the other hand, the steam and feedwater are kept separate. The steam condenses in tubes that pass through the preheater. The condensate is removed through traps and allowed to enter the feed stream at a lower pressure, where it may do so without the need for additional pumps. A two-stage closed preheat scheme is shown in Fig. 8-8.

The advantages of the open preheaters are excellent heat-transfer characteristics, ease of operation and design, and inexpensive construction. They have, however, the disadvantage of requiring an extra feed pump for each preheater stage.

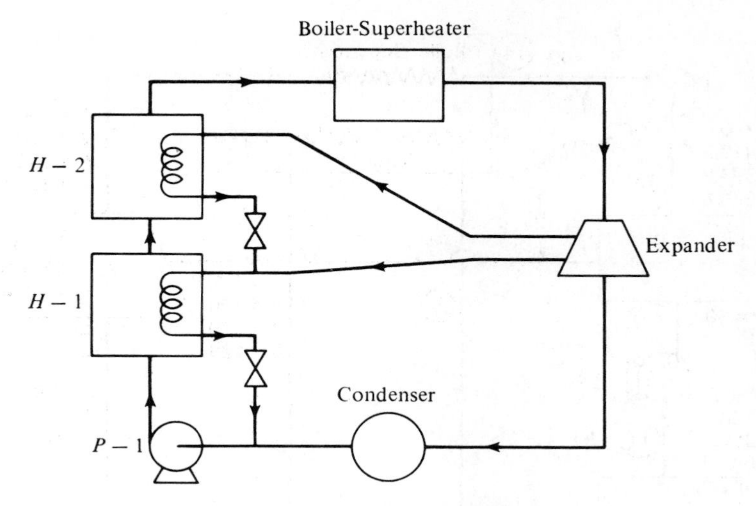

Figure 8-8. Closed preheaters.

SAMPLE PROBLEM 8-4. The reheat cycle of Sample Problem 8-3 is to be augmented by three stages of feed preheat as shown in Fig. SP 8-4. The boiler pressure is 900 psia, and a maximum temperature of 860°F is allowable. The steam for the feed preheaters is available at pressures of 300, 90, and 30 psia, and the liquids leaving the preheaters are to be saturated liquids at the pressure of the preheater steam. The condenser will operate as a total condenser discharging saturated liquid at a temperature of 150°F. Compare the thermal efficiency of this reheat cycle with and without regenerative preheating of the feed. You may assume that the pumps and turbines are adiabatic and reversible.

Solution: In Sample Problem 8-3 we learned that the efficiency of the cycle without feed preheating was 35 percent. Let us now calculate the efficiency with feed preheat. We may now determine the values of the properties of many of the streams from the steam tables. These streams and their properties are shown in Fig. SP 8-4.

With these property values we may now begin to analyze the cycle. Because of the number of streams leaving the turbines, and the number of pumps, we will calculate W^*_{net} from an overall energy balance around the cycle rather than from the sum of the individual works. Thus

$$W^*_{\text{net}} = \sum Q^* = Q^*_{\text{boiler}} + Q^*_{\text{reheater}} + Q^*_{\text{cond}}$$

The thermal efficiency is then

$$\eta = \frac{Q^*_{\text{boiler}} + Q^*_{\text{reheater}} + Q^*_{\text{cond}}}{Q^*_{\text{boiler}} + Q^*_{\text{reheater}}}$$

An energy balance around the condenser gives

$$Q^*_C = q_C = \Delta h_C = (117.9 - 1125) \text{ Btu/lb}_{\text{m}} = 1007.1 \text{ Btu/lb}_{\text{m}}$$

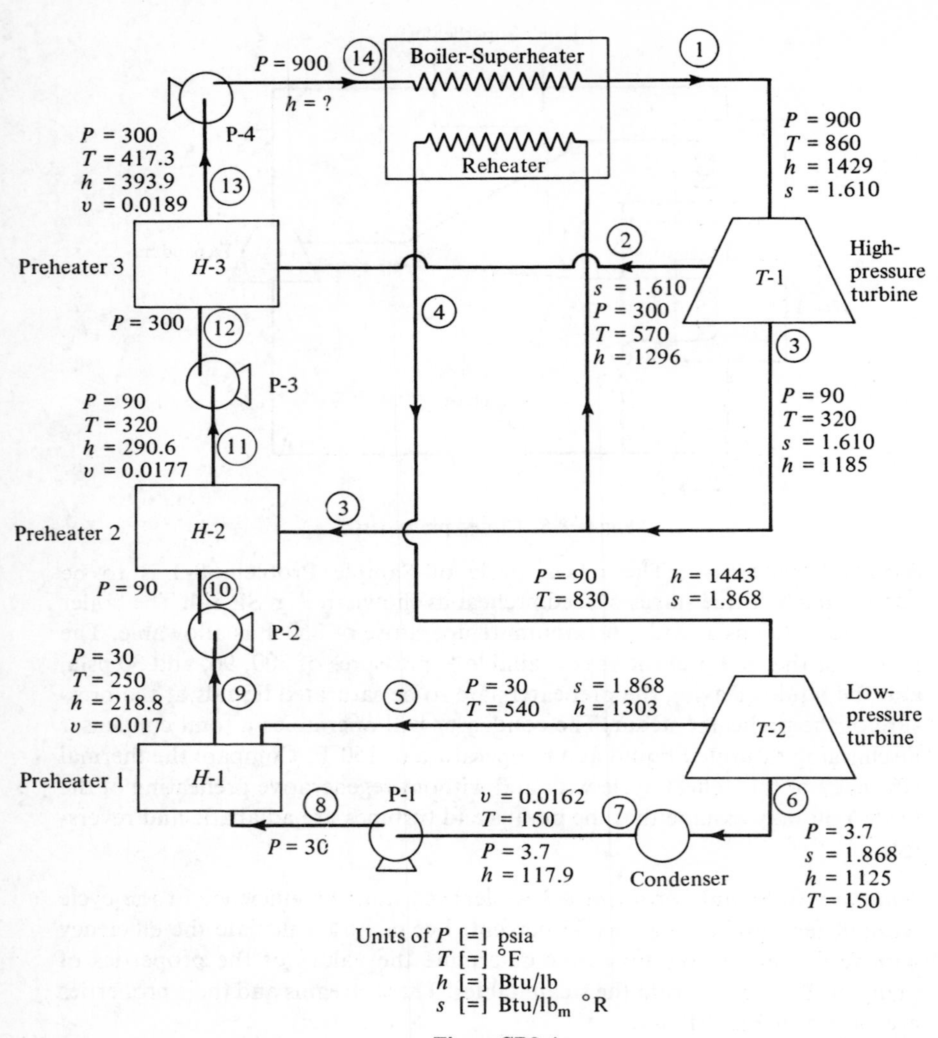

Figure SP8-4

(*Note:* For this problem the lower case refers to 1 lb_m through any particular piece of equipment. The superscript asterisk will be used to refer to 1 lb_m through the condenser. The two quantities are related as follows:

$$Q^* = M^* q$$

where $M^* = \dfrac{lb_m \text{ through particular equipment}}{lb_m \text{ through the condenser}}$.)

We may now determine the enthalpy of the liquid entering H–1 from an energy balance around P–1:

$$\Delta h = h_8 - h_7 = -w_{P\text{-}1} = \int v\,dP = v\,\Delta P$$

or, substituting values,

$$\Delta h = (0.0162)(30 - 3.7)\left(\frac{144}{778}\right)\frac{\text{ft}^3}{\text{lb}_\text{m}}\frac{\text{lb}_\text{f}}{\text{in.}^2}\frac{\text{in.}^2}{\text{ft}^2}\frac{\text{Btu}}{\text{lb}_\text{f}\text{-ft}} = 0.08\ \text{Btu/lb}_\text{m}$$

Therefore, $h_8 = 118.0\ \text{Btu/lb}_\text{m}$.

We may now determine the flow rate of steam to H–1 from an energy balance around preheater 1. Assuming negligible changes in potential and kinetic energy and steady, adiabatic operation we get

$$\sum M^*h = 0 \quad \text{and} \quad M_8^*h_8 + M_5^*h_5 - M_9^*h_9 = 0$$

A mass balance around the preheater gives

$$M_9^* = M_8^* + M_5^*$$

which is substituted in the energy balance. The energy balance is then solved for M_5^* to give

$$M_5^* = M_8^*\frac{h_9 - h_8}{h_5 - h_9}$$

But $M_8^* = 1$, so M_5^* becomes

$$M_5^* = \frac{218.8 - 118.0}{1303 - 218.8} = 0.093\ \text{lb}_\text{m}/\text{lb}_\text{m}\ \text{through condenser}$$

$$M_9^* = 1.093$$

We now proceed in an identical manner to determine the flow rates leaving the other two preheaters. The enthalpy of the fluid leaving P–2 is obtained from

$$h_{10} - h_9 = \Delta h = -w_{P\text{-}2} = v\,\Delta P = (0.017)(90 - 30)\left(\frac{144}{778}\right)\ \text{Btu/lb}_\text{m}$$

$$= 0.2\ \text{Btu/lb}_\text{m}$$

Thus

$$h_{10} = (218.8 + 0.2)\ \text{Btu/lb}_\text{m} = 219\ \text{Btu/lb}_\text{m}$$

The mass flow rate of steam into preheater H–2 is obtained from

$$M_3^* = M_{10}^*\frac{h_{11} - h_{10}}{h_3 - h_{11}}$$

But

$$M_{10}^* = M_9^* = 1.093$$

Therefore,

$$M_3^* = 1.093\frac{(290.6 - 218.8)\ \text{Btu/lb}_\text{m}}{(1185 - 290.6)\ \text{Btu/lb}_\text{m}} = 0.087$$

$$M_{11}^* = M_3^* + M_{10}^* = 1.180$$

Continue through P–3:

$$h_{12} - h_{11} = \Delta h_{P\text{-}3} = -w_{P\text{-}3} = v\,\Delta P$$

$$= (0.0177)(300 - 90)\left(\frac{144}{778}\right)\ \text{Btu/lb}_m = 0.7\ \text{Btu/lb}_m$$

$$h_{12} = (290.6 + 0.7)\ \text{Btu/lb}_m = 291.3\ \text{Btu/lb}_m$$

The steam flow rate into H–3 is given by

$$M_2^* = M_{12}^* \frac{h_{13} - h_{12}}{h_2 - h_{13}} \quad \text{but} \quad M_{12}^* = M_{11}^* = 1.180$$

therefore,

$$M_2^* = 1.180\frac{(393.9 - 291.3)\ \text{Btu/lb}_m}{(1296 - 393.9)\ \text{Btu/lb}_m} = 0.135$$

$$M_{13}^* = M_2^* + M_{12}^* = 1.315\ \text{lb}_m/\text{lb}_m \text{ through condenser}$$

Since $M_{14}^* = M_{13}^*$, 1.315 lb_m of water flows through the boiler per lb_m of fluid through the condenser.

We can determine the enthalpy of the liquid entering the boiler–superheater from the relation

$$h_{14} - h_{13} = \Delta h_{P\text{-}4} = -w_{P\text{-}4} = +v\,\Delta P$$

$$= (0.0189)(900 - 300)\left(\frac{144}{778}\right)\text{Btu/lb}_m = 2.1\ \text{Btu/lb}_m$$

therefore,

$$h_{14} = (393.9 + 2.1)\ \text{Btu/lb}_m = 396\ \text{Btu/lb}_m$$

The heat load in the boiler-superheater is obtained from the energy balance around the boiler–superheater:

$$q_B = \Delta h_B = h_1 - h_{14} = 1429 - 396.0\ \text{Btu/lb}_m$$

or

$$q_B = 1033\ \text{Btu/lb}_m$$

and

$$q_B^* = M_{14}^* q_B = (1.315)(1033) = 1362\ \text{Btu/lb}_m \text{ through condenser}$$

The heat absorbed in the reheater is obtained from an energy balance around the reheater:

$$q_R = \Delta h_R = h_4 - h_3 = 1443 - 1185\ \text{Btu/lb}_m$$

or

$$q_R = 258\ \text{Btu/lb}_m$$

The mass flow rate through the reheater is obtained from a mass balance around T – 2:

$$M_4^* = M_5^* + M_6^* = 0.093 + 1.0 = 1.093\frac{\text{lb}_m \text{ through reheater}}{\text{lb}_m \text{ through condenser}}$$

Therefore, the total heat absorbed in the reheater is given by

$$Q_R^* = M_4^* q_R = (1.093)(258)$$
$$= 281 \text{ Btu/lb}_m \text{ through condenser}$$

The heat load from the condenser is obtained from an energy balance around the condenser:

$$q_C = \Delta h_C = h_7 - h_6 = (117.9 - 1125) \text{ Btu/lb}_m = -1007.1 \text{ Btu/lb}_m$$

Since $M_C^* = 1$,

$$Q_C^* = q_C = -1007.1 \text{ Btu/lb}_m$$

The net work produced by the cycle is given by

$$W_{net}^* = Q_B^* + Q_R^* + Q_C^*$$
$$= 636 \text{ Btu/lb}_m \text{ through condenser}$$

The thermal efficiency of the whole cycle is then

$$\eta = \frac{W_{net}^*}{Q_B^* + Q_R^*} = \frac{636}{1643} = 0.387$$

This compares with an efficiency of 35 percent for the cycle without feed preheating. This is not nearly as impressive an increase in the efficiency as was obtained by using reheat. However, it must be remembered that in large generating plants which produce millions of kilowatts of electricity, an efficiency increase of a single percent represents annual fuel savings of hundreds of thousands, or even millions, of dollars. Thus no increase is too small to be considered.

Supercritical Rankine Cycle

The largest, most advanced generating plants utilize a supercritical Rankine cycle, shown in Fig. 8-9. Standard pressures for these cycles are 3500 psia (the critical pressure for water is 3206 psia) and temperatures have generally been limited to 1000°F. Higher pressures and temperatures were

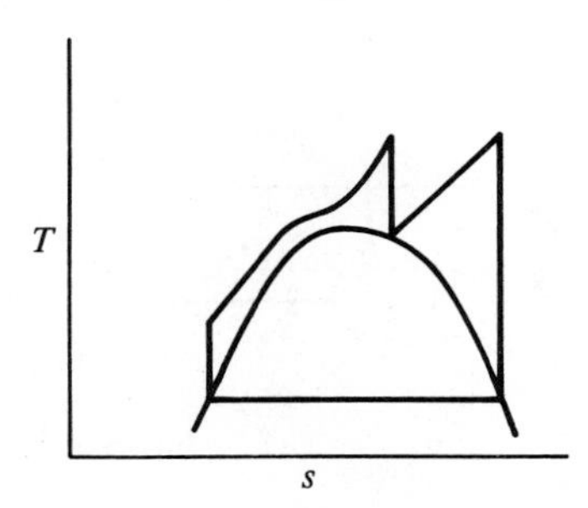

Figure 8-9

tried in commercial units, but operating difficulties as the limit of materials technology was pushed caused utilities to adopt the 3500 psia and 1000°F as practical operating limits in the mid 1960s.

As shown in Fig. 8-9, reheat is essential with the supercritical cycle if excessive condensation is to be avoided during the expansion step. "Regenerative" heating of the fluid with bleed steam from the expansion prior to its entering the boiler can be used to further increase the average temperature at which heat is supplied to the cycle. A single reheat to 1000°F is typically used. Such plants are capable of overall thermal efficiencies of 43 percent.

8.5 The Brayton Cycle

If one wants to take full advantage of the temperature potential available from the combustion of fossil fuels, high-temperature nuclear processes, or the sun to achieve higher thermal efficiencies in the conversion of heat to work, it is necessary to consider working fluids other than steam. The supercritical steam cycle has pushed steam conditions just about as far as metallurgical conditions are likely to permit. If temperatures above 2000°F are to be achieved in working fluids, gaseous working fluids such as helium, CO_2, or air are likely candidates. These fluids would remain in the gaseous state, and in the case of air (or combustion gases) could operate in an open cycle, much as they do in aircraft turbines. The discharge gases, after having been cooled by expansion, can be used as a heat source for a steam Rankine cycle, and thus make more effective use of the temperature potential in the conversion process. Helium or CO_2 would operate in a closed cycle, but could give up heat to a steam Rankine cycle in a like manner.

Gas-turbine technology has been rapidly advanced by the aircraft industry so that a high-temperature cycle using a gaseous working fluid is a

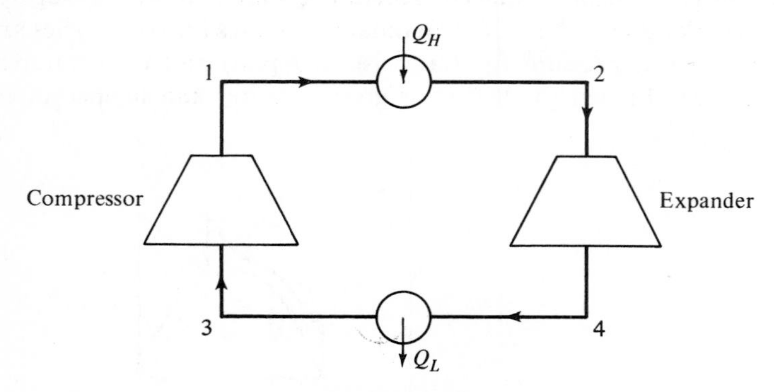

Figure 8-10

practical means of energy conversion for many applications, including not only aircraft propulsion, but also more efficient power plants for automotive and central station use. The idealized cycle for a gaseous working fluid which receives and discharges heat at a constant pressure is the Brayton cycle.

The Brayton cycle, unlike the Rankine cycle, utilizes a single-phase gaseous working fluid throughout the cycle. As shown in Fig. 8-10, the components of such a cycle are similar to those of cycles discussed earlier. The principal difference relates to use of a compressor rather than a pump to supply the work needed to return the fluid to the boiler pressure.

The *T–s* path followed by the working fluid in a Brayton cycle is shown in Fig. 8-11. The ideal Brayton cycle consists of isentropic expansion and compression steps and constant-pressure heat addition and removal. For an ideal gas,

$$q_H = C_P(T_2 - T_1) = \text{area } 1\text{–}2\text{–}b\text{–}a\text{–}1 \tag{8-1}$$

$$q_L = C_P(T_4 - T_3) = \text{area } 4\text{–}3\text{–}b\text{–}a\text{–}4 \tag{8-2}$$

and $w_{\text{rev}} = q_H + q_L =$ crosshatched area.

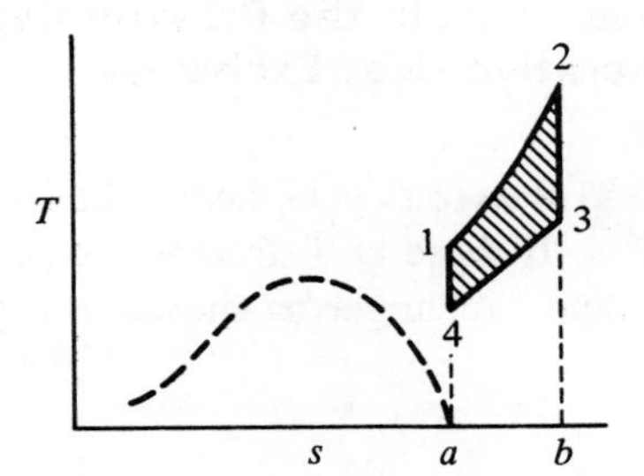

Figure 8-11

The thermal efficiency of the cycle can be expressed as

$$\eta = \frac{w_{\text{rev}}}{q_H} = \frac{q_H + q_L}{q_H} = \frac{C_P[(T_2 - T_1) + (T_4 - T_3)]}{C_P(T_2 - T_1)} = \left[1 - \frac{T_3 - T_4}{T_2 - T_1}\right]$$

$$= \left[1 - \frac{T_4(T_3/T_4 - 1)}{T_1(T_2/T_1 - 1)}\right] \tag{8-3}$$

Since $P_1 = P_2$ and $P_3 = P_4$,

$$\frac{P_1}{P_4} = \frac{P_2}{P_3} \quad \text{or} \quad \frac{P_1}{P_2} = \frac{P_4}{P_3}$$

For isentropic processes with an ideal gas,

$$\frac{P_1}{P_4} = \left(\frac{T_1}{T_4}\right)^{k/(k-1)} = \frac{P_2}{P_3} = \left(\frac{T_2}{T_3}\right)^{k/(k-1)}$$

Thus

$$\frac{T_1}{T_4} = \frac{T_2}{T_3} \quad \text{and} \quad \frac{T_2}{T_1} = \frac{T_3}{T_4}$$

therefore

$$\eta = 1 - \frac{T_4}{T_1} = 1 - \frac{1}{(P_1/P_4)^{(k-1)/k}} \tag{8-4}$$

As equation (8-4) shows, the thermal efficiency of the cycle can be expressed in terms of the pressure ratio or the temperature ratio across the expander and compressor. A larger pressure ratio P_1/P_4 increases the efficiency; in terms of temperatures, efficiencies increase with higher heat-addition temperatures and lower heat-rejection temperatures. It is left as an exercise for the student to show that the Brayton-cycle efficiency is less than that of a Carnot cycle operating between T_2 and T_4.

The performance of the Brayton cycle is severely limited by the relatively large work requirement to compress the gas (Step 4-1 in Fig. 8-11). Since compressor work equals $\int v\,dP$, the much larger values of v for gases as compared to liquids exacts a sizable penalty which the Rankine cycle avoids by operating within the liquid-phase region.

8.6 Improvements in the Brayton Cycle Regenerative Heat Exchange

In the idealized Brayton cycle, heat is added along path 1–2 and is removed along path 3–4. If these two streams are passed countercurrently through a regenerative heat exchanger as shown in Fig. 8-12, energy at relatively high

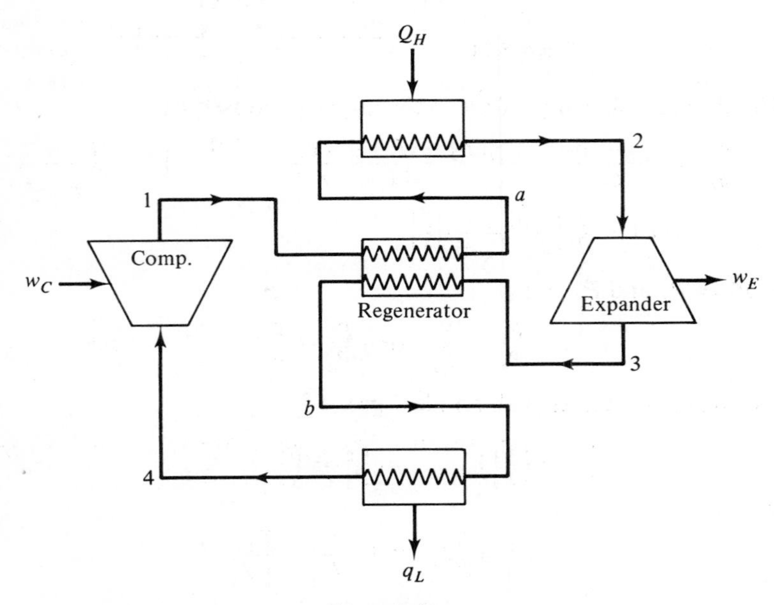

Figure 8-12

temperature, which would otherwise be discarded, can be used to supply much of the energy needed at lower temperatures. Figure 8-13 shows the regenerative Brayton cycle on a T–s plot.

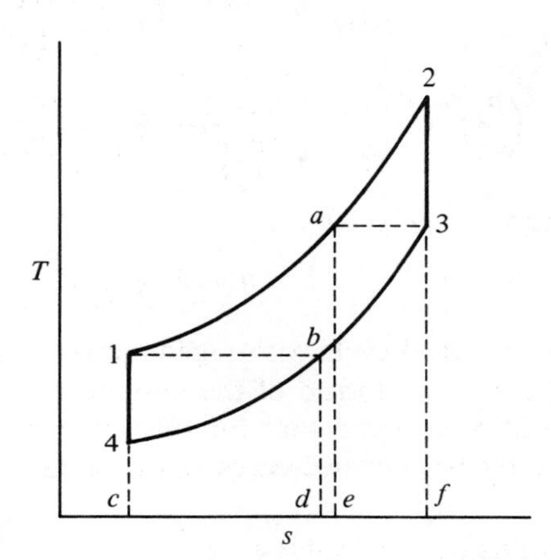

Figure 8-13

The fluid being heated actually goes from T_1 to T_a by picking up heat from the fluid being cooled as it goes from T_3 to T_b in the regenerator. Thus, heat to the cycle from the heat source occurs only from T_a up to T_2, and heat is discharged to the surroundings from T_b down to T_4, resulting in a considerable increase in cycle thermal efficiency.

The efficiency of the Brayton cycle with regeneration can be calculated (assuming ideal-gas conditions for the working fluid) if we assume that the exit temperature $T_{a'}$ of the fluid being heated in the regenerator equals the turbine exhaust temperature, T_3. (This would be an *ideal regenerator* and $T_a = T_3$ only if the heat-transfer area were very large.)

$$\eta = \frac{w_{\text{rev}}}{q_H} = \frac{w_{\text{exp}} + w_{\text{comp}}}{q_H}$$

$$w_{\text{exp}} = -\Delta h_{\text{exp}} = -C_P(T_3 - T_2)$$

$$q_H = C_P(T_2 - T_a)$$

but for $T_a = T_3$,

$$q_H = C_P(T_2 - T_3) = w_{\text{exp}}$$

Thus

$$w_{\text{comp}} = -\Delta h_{\text{comp}} = C_P(T_4 - T_1)$$

$$\eta = \frac{w_{\text{exp}} + w_{\text{comp}}}{w_{\text{exp}}} = 1 + \frac{w_{\text{comp}}}{w_{\text{exp}}} = 1 - \frac{C_P(T_1 - T_4)}{C_P(T_2 - T_3)}$$

$$\eta = 1 - \frac{T_4\left(\dfrac{T_1}{T_4} - 1\right)}{T_2\left(1 - \dfrac{T_3}{T_2}\right)} = 1 - \frac{T_4}{T_2}\left[\frac{\left(\dfrac{P_1}{P_4}\right)^{(k-1)/k} - 1}{1 - \left(\dfrac{P_3}{P_2}\right)^{(k-1)/k}}\right]$$

but $P_3 = P_4$ and $P_2 = P_1$. Thus

$$\eta = 1 - \frac{T_4}{T_2}\left[\frac{\left(\frac{P_1}{P_4}\right)^{(k-1)/k} - 1}{1 - \left(\frac{P_4}{P_1}\right)^{(k-1)/k}}\right]$$

$$\left(\frac{P_1}{P_4}\right)^{(k-1)/k}\left[1 - \frac{1}{\left(\frac{P_1}{P_4}\right)^{(k-1)/k}}\right] = \left(\frac{P_1}{P_4}\right)^{(k-1)/k}\left[1 - \left(\frac{P_4}{P_1}\right)^{(k-1)/k}\right]$$

Therefore

$$\eta = 1 - \frac{T_4}{T_2}\left(\frac{P_1}{P_4}\right)^{(k-1)/k} \tag{8-5}$$

For the ideal cycle with regeneration, the efficiency depends on both the pressure ratio and ratio of the minimum to maximum cycle temperature. It should also be noted that for this cycle (in contrast to the simple Brayton cycle) the efficiency decreases as the pressure ratio increases.

Regenerator Efficiency

The above calculation assumed that $T_a = T_3$ or that we had an ideal regenerator. The regenerator efficiency is defined as

$$\eta_{\text{reg}} = \frac{h_a - h_1}{h_3 - h_1} \tag{8-6}$$

For constant heat capacity, the efficiency can be expressed as

$$\eta_{\text{reg}} = \frac{T_a - T_1}{T_3 - T_1} \tag{8-7}$$

The gains in efficiency from regenerative exchange in the Brayton cycle are analogous to improvements realized in the Rankine cycle by bleeding off fluid from the turbine to supply sensible heat to the boiler feedwater. This regenerative heating in the Brayton cycle is easier to achieve because both fluids are

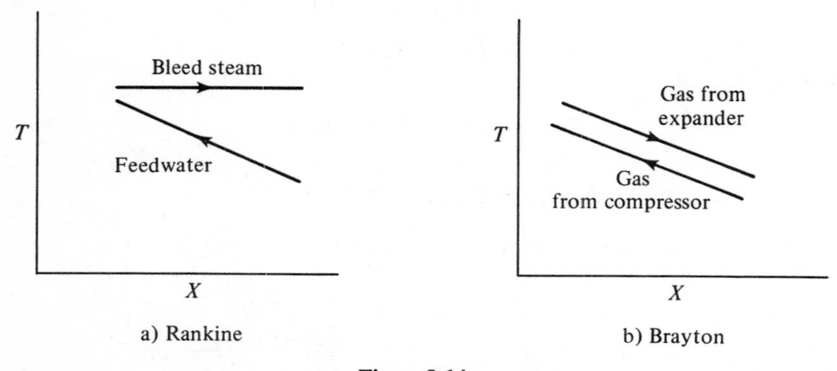

Figure 8-14

gases with comparable heat capacities. In the Rankine cycle the bleedsteam contains much of its energy as latent heat which it gives up at a constant temperature as it condenses. The two processes are contrasted in Fig. 8-14. The heat exchange in the Brayton regenerator clearly can be made to more nearly approach a reversible exchange than can the latter.

Brayton with Regeneration, Reheat, and Intercooling—Ericcson Cycle

Incorporating reheat along with regenerative exchange and intercooling in the compression yields the most efficient form of the Brayton cycle for a given set of conditions. Such a cycle is shown in Fig. 8-15. By increasing the number of reheats and intercooling steps, the cycle can be made to approximate the Ericcson cycle shown in Fig. 8-16, which consists of two isothermal

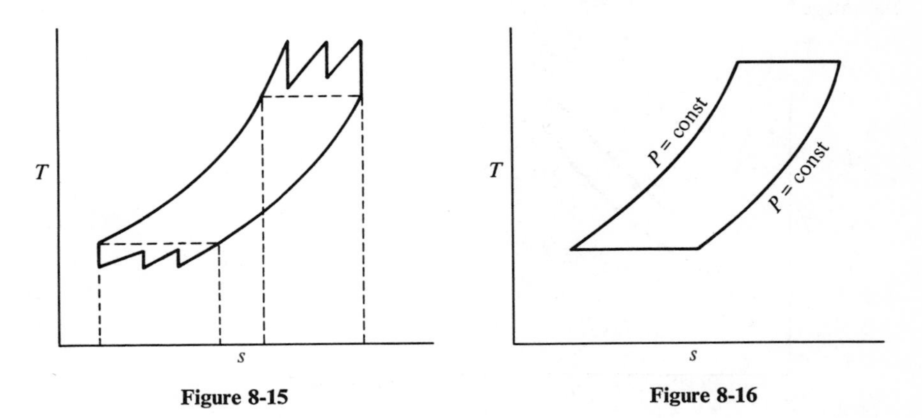

Figure 8-15 **Figure 8-16**

and two isobaric legs. The Ericcson cycle has an efficiency equal to the Carnot cycle inasmuch as they both exchange their heats with the *surroundings* under constant-temperature conditions. The regenerative Brayton cycle could approach this performance if the following conditions can be met:

1. Heat could be added continuously to the expander to maintain isothermal conditions.
2. Heat could be removed continuously in the compressor.
3. The regenerator operates reversibly, that is, with only infinitesimal temperature differences between the two streams.
4. Piping losses (pressure drop) are negligible, and expansion and compression are reversible.

Conditions 1 and 2 are extremely difficult to achieve at large throughputs. The extent to which they are approximated by reheat and intercooling in practice is purely a matter of economics. They require increased capital expenditures to incorporate and thus can be justified only if the fuel savings offset these costs. As fuel costs rise due to tightening supply and environmental concerns, industry will look with renewed interest at options which use the fuels more effectively.

Applications of the Brayton Cycle

In practice there are irreversibilities in each of the steps. Turbines and compressors can be designed with efficiencies of 88–94 percent based on reversible processes. Piping losses are impossible to avoid, particularly at large flow rates. As a result of irreversibilities, the simple Brayton cycle is slightly modified as shown in Fig. 8-17 (1–2–3′–4′–1) with corresponding reductions in efficiency. The same behavior illustrated for the simple cycle would also apply to the cycle with reheat, intercooling, and regenerative exchange.

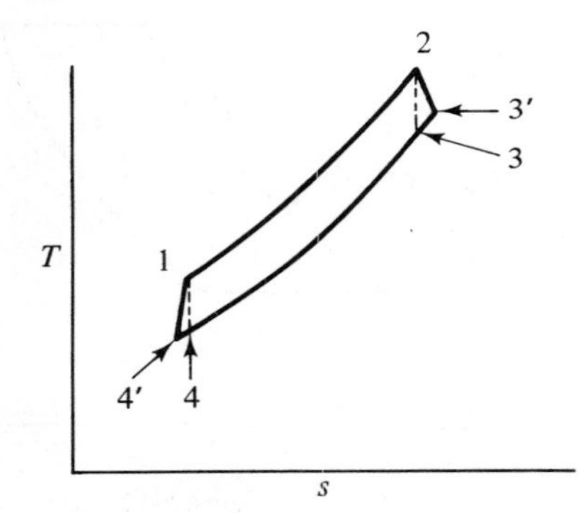

Figure 8-17

The Brayton cycle appears particularly attractive in conjunction with the high-temperature, gas-cooled nuclear reactor which uses helium as a reactor coolant. Figure 8-18 contrasts a typical Rankine cycle used today with a helium Brayton cycle. Efficiencies for the two cycles are comparable, but the Brayton cycle discharges its energy over a temperature range of 370°F–130°F as contrasted to the Rankine cycle which typically use condensers at about 100°F. The former can discharge heat directly to the atmosphere and thus avoid local heating of lakes and rivers.

8.7 Open-Cycle Brayton Systems

The rapid development of gas-turbine technology for the aircraft industry has made the gas turbine a relatively attractive prime mover. Open-cycle gas- and oil-fired turbines have been used with increasing frequency for meeting the peak power needs of the utilities because of their relatively low capital cost

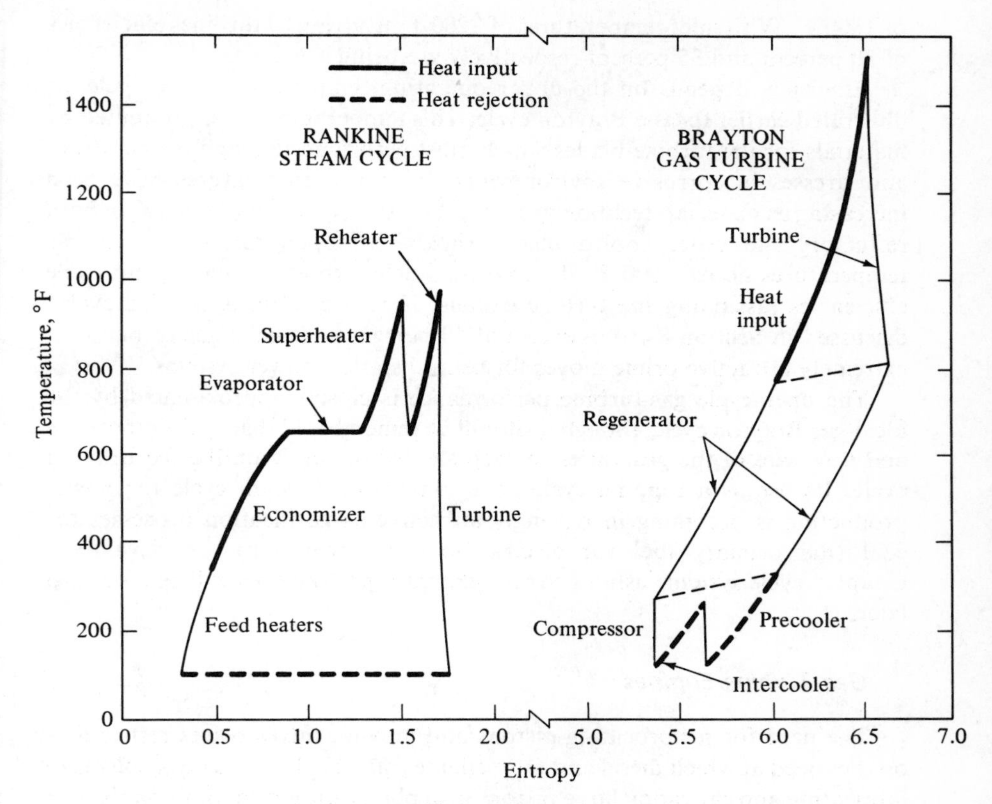

Figure 8-18

and rapid start-up capability. Figure 8-19 illustrates the open-cycle system with and without a regenerator. Compressed air, with or without preheat in the regenerator, is fed to a combustor along with the fuel. The hot combustion gases expand through the turbine which converts enthalpy to shaft work. These power plants typically operate with efficiencies of 26 percent and 32 percent with and without regenerators, respectively, for gas inlet temperatures

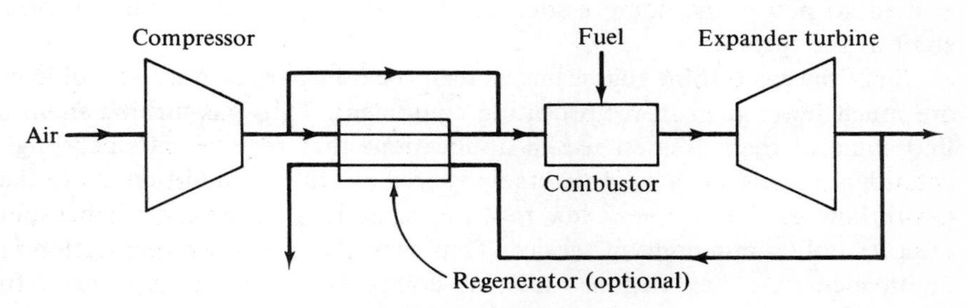

Figure 8-19

of 1800°F. With inlet temperatures of 2200°F in advanced turbines efficiencies of 30 percent and 35 percent respectively are being achieved.

Efficiency depends on the upper operating temperature of the cycle, as illustrated earlier for the Brayton cycle. This temperature has been limited by materials for the turbine blades which must withstand the high temperatures and stresses in corrosive environments. Inlet gas temperatures have been increasing as materials technology improves. Efforts continue to develop both refractory and water cooled blades capable of operation with inlet gas temperatures above 3000°F. If successful such turbines would permit cycle efficiencies (assuming the turbine exhaust heat is used in a coupled cycle–discussed in Section 8-10) in excess of 50 percent and make gas turbines an extremely attractive prime mover for central station power systems.

The open-cycle gas-turbine performance is closely approximated by the ideal-gas Brayton cycle, though it should be remembered that the composition and flow rate of the gas varies in each step of the cycle unlike the Brayton cycle. Its use as a topping cycle to a conventional steam cycle for power production is becoming increasingly attractive as gasification processes for coal (the primary fuel for electrical-energy generation) are developed. Coupled-cycle systems using several different topping cycles will be discussed later.

Gas-Turbine Engines

The need for reciprocating pistons and moving valves places restrictions on the speed at which diesel and Otto engines may be driven. The problems in fabricating and operating large pistons also places a practical limit on the size of these engines. By replacing the reciprocating-piston expansion device with a turbine, these restrictions are greatly reduced. Since the piston is no longer available to perform the precombustion compression, a rotary compressor is added to the engine.

The gas-turbine engine is shown schematically in Fig. 8-20. The inlet air is compressed in the compressor and mixed with fuel in the combustion chamber. The combustion products are then expanded in the turbine and finally exhausted to the atmosphere. A portion of the work produced in the turbine is used to power the compressor, which is usually mounted on the same shaft as the turbine.

Since the gas-turbine engine has no reciprocating parts, vibration problems are much fewer than in reciprocating equipment. Thus gas-turbine engines find some of their greatest use in applications that require extremely high rotational speeds, such as in turbine-powered aircraft. In addition, since the gas-turbine engine has very few moving parts, it is extremely reliable and requires only a minimum of service. Thus they also find great application in unattended field applications, such as compressors on gas pipelines. (In

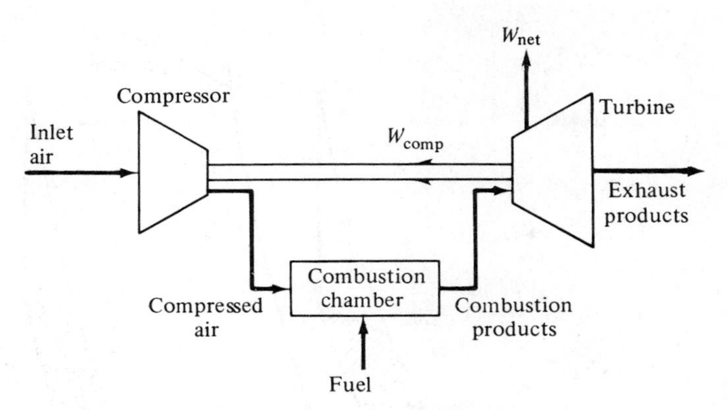

Figure 8-20. Gas-turbine engine.

commercial aircraft where failure must be kept as low as possible, turbine engines are completely overhauled after 10,000 hours of flying, as opposed to 200 to 2000 hours for piston-powered engines.)

The open-cycle gas turbine operates on a modified Brayton cycle as discussed earlier. Thermal efficiencies for simple open-cycle turbines depend on the pressure ratio, turbine inlet temperature, and the turbine exhaust temperature. The efficiency can be improved significantly if regenerative heat exchange can be used. The availability of compact and efficient regenerators has made turbine engines far more attractive for smaller mobile power systems, such as the automobile.

8.8 Feher Cycle

The Feher cycle is a single-phase cycle operating entirely in the supercritical regime. It combines the attractive features of the Rankine and Brayton cycles as shown in Fig. 8-21, which contrasts the three cycles. By operating above the critical pressure, where v of the fluid in the compressor is much lower, the Feher cycle incorporates the efficient pumping of the Rankine cycle with the regenerative heating features of the Brayton cycle to achieve higher theoretical efficiencies.

Carbon dioxide with a critical pressure of 1072 psia and a critical temperature of 88°F appears to be the best working fluid for such a cycle. A proposed carbon-dioxide cycle is shown in Fig. 8-22 operating between 4000 psia and 1300°F and 1950 psia and 100°F. The components of the cycle are identical to those of the Brayton cycle except that the higher pressures permit a substantial reduction in the size of all components, making it ideally suited for mobile power sources as well as stationary power plants.

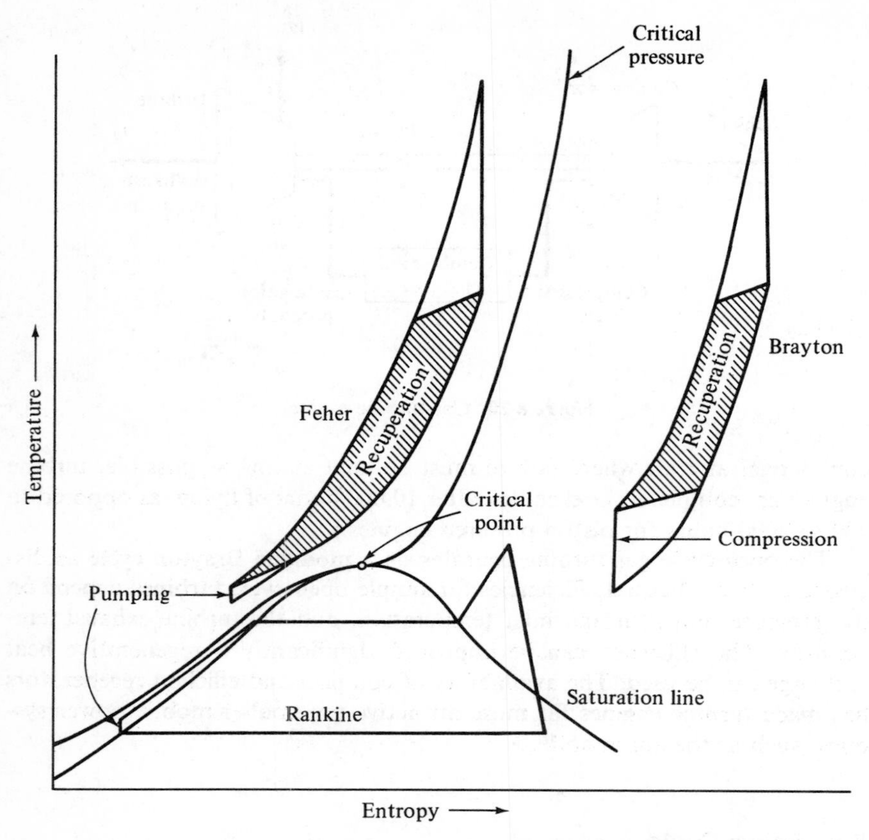

Figure 8-21. Comparison of closed heat engine cycles.

Figure 8-22 shows the *T–s* path that the Feher cycle would follow in actual operation assuming efficiencies of 88 percent and 85 percent for the turbine and pump, respectively. Assuming a boiler efficiency of 85 percent and reasonable recuperator performance, this cycle would have an overall efficiency of 41 percent, which is comparable to today's best fossil-fueled plants.

Figure 8-23 compares the theoretical efficiencies of Rankine, Brayton, and Feher cycles over the temperature range where Rankine and Brayton cycles have typically operated in the past. For comparison, the Carnot-cycle performance over the same temperature range is also included. Thermodynamically there is little difference between the Rankine and Feher cycles at lower temperatures and between the Brayton and Feher cycles at very high temperatures (as is suggested by extrapolating the curves to temperatures in excess of 3000°F).

Although the Feher cycle appears thermodynamically attractive in the intermediate temperature range, the commercial attractiveness of any cycle is

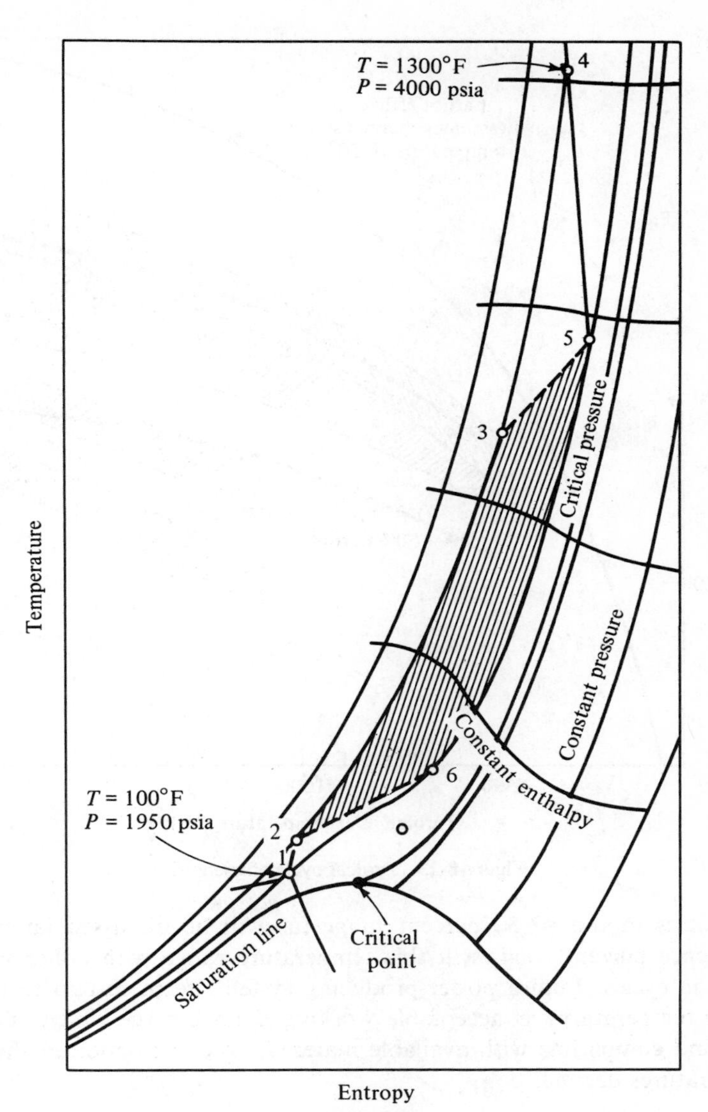

Figure 8-22. Feher cycle.

determined by the overall system cost and reliability as well as its thermodynamic performance. The high operating pressure of the CO_2 Feher cycle poses engineering design difficulties which offset some of the theoretical performance advantages. The steam Rankine cycle has amassed an enormous number of operating hours in power generation and will not be easily displaced in temperature ranges below 1200°F. At higher temperatures metallic working fluids could be used in a Rankine cycle to achieve thermodynamic

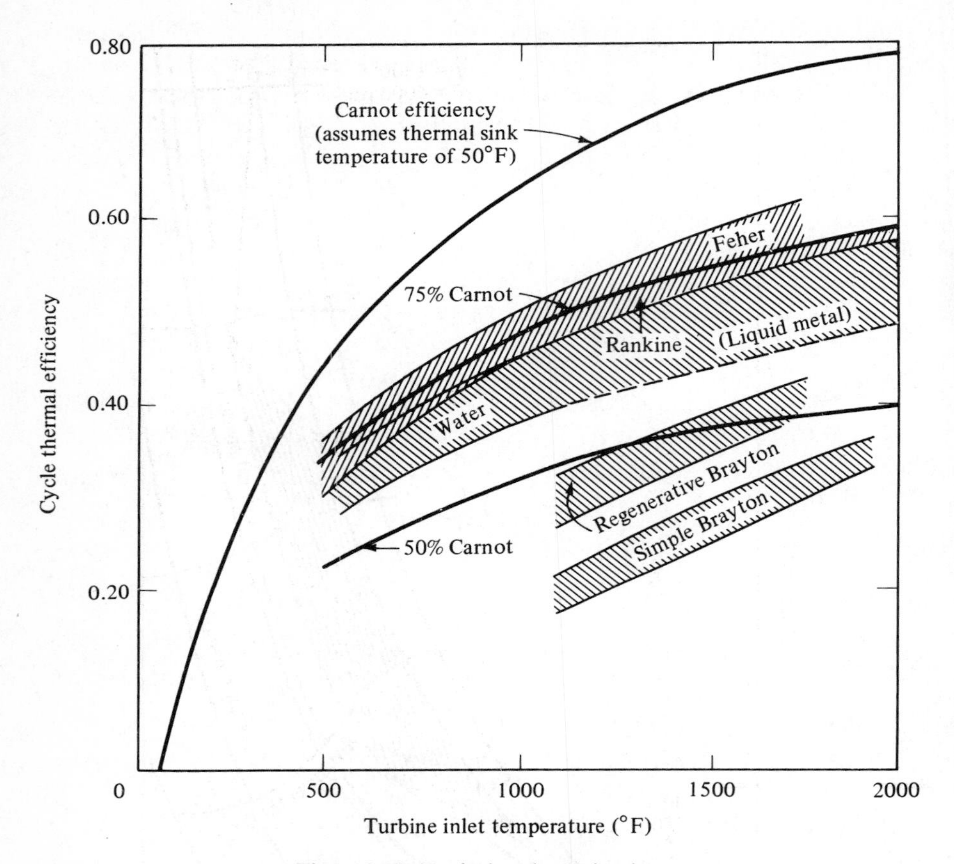

Figure 8-23. Typical cycle efficiencies.

efficiencies in the 40–50 percent range (as will be discussed later). Little experience currently exists in this temperature range with either Feher or Rankine cycles. Future power-producing systems seem certain to move to higher temperatures as acceptable working fluids emerge which are stable, safe, and compatible with available materials of construction at the higher temperatures desired.

8.9 Magnetohydrodynamics

Magnetohydrodynamic (MHD) conversion produces electrical power directly from conducting fluids as they move at high velocity through a magnetic field. Often referred to as an "electrical turbine," it accomplishes in a single channel without rotating components what conventional rotating machinery requires two devices to achieve—the first a turbine producing the necessary torque (or shaft work) to drive the second, a generator which produces electrical power from the interaction of the rotating electrical field and the magnetic field. In MHD the conducting fluid is first accelerated by the expansion of the driving fluid (a gas or plasma) in a nozzle. The conversion of enthalpy to work (or electrical energy) occurs in the diverging portion of the channel where the high-velocity conducting fluid interacts with the magnetic field supplied by magnets which are external to the channel.

The interaction of the flowing conductor (or current) and the magnetic field produces an electric field in a direction perpendicular to both. The integral of the electric field across the channel provides the external voltage of the generator and causes a current to flow in the external circuit (see Fig. 8-24). The internal current through the generator combines with the magnetic field to form a retarding force that decelerates the fluid and results in enthalpy extraction (or reduction).

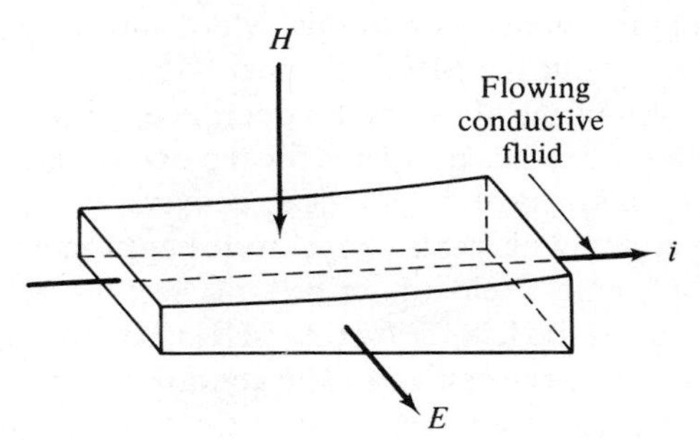

Figure 8-24. MHD generator.

MHD cycles require a good conducting fluid so that the internal resistance of the generator is low. This can be achieved by using a high-temperature ionized gas (or plasma) or a conducting liquid, such as a metal. In the case of the plasma, it is necessary to heat the gases to extremely high temperatures to ionize the gas sufficiently so that MHD conversion is possible. If the gas is seeded with components which ionize more readily, such as potassium or cesium, the necessary conductivity can be achieved at a somewhat lower temperature than would otherwise be possible with pure combustion product gases. However, the cost of the seed material requires recycling of the seed if the cycle is to be economically attractive. Cesium is more easily ionized than potassium but is considerably more expensive. Its use would require a closed-cycle system in which the combustion gases are used to heat the working fluid, which would be an inert gas doped with cesium. Such a working fluid can attain good conductivity at lower temperatures and appears capable of reasonable generator performance if heated to 3000°F–3500°F.

Open-cycle MHD uses combustion gases seeded with potassium to increase the fluid conductivity, but temperatures of 4000°F–5000°F are required. Such temperature levels are well above the adiabatic flame temperature of most conventional fuels in air. Thus it is necessary to either use oxygen for combustion or preheat the combustion air before injecting the fuel. Either adds to the overall system cost and complexity. Materials of construction for electrodes become the major problem when operating at these temperatures where even most ceramics begin to deteriorate. Thus channel design and cooling pose significant engineering challenges.

A third approach to MHD which avoids the high temperatures required for plasmas uses a liquid metal as the conductor. The liquid itself cannot be expanded and accelerated since its v is relatively independent of P and T. Thus it must either be partially vaporized or an inert gas added to act as a driver fluid. Promising results have been obtained in laboratories in the U. S. and the U. S. S. R. with both alkali-metal/inert gas and tin/steam working fluids. The ability to operate at lower temperatures eases the material problems but introduces the complexities of expanding a two-phase liquid–gas mixture under conditions which must retain a continuous liquid phase throughout the MHD channel.

All of the above cycles operate on either an open or closed Brayton cycle. If the materials problems for the open-cycle system can be solved, such a system may have a practical conversion efficiency of 55–60 percent when operated with peak temperatures approaching 5000°F in conjunction with a bottoming cycle (described in Section 8.10). Since the gas conductivity becomes too low for further MHD expansion at temperatures around 4000°F, the MHD exhaust gases still contain much of their enthalpy and must be used in lower temperature cycles if the overall system performance is to be attractive. Combined cycles using MHD topping systems will be discussed later.

The closed-cycle plasma and metal MHD systems operate at lower temperatures as closed-cycle Brayton systems. The plasma system, with cesium seed, would likely operate between 3000°F and 3500°F in the channel as shown in Fig. 8-25. As with the open cycle, only a fraction of the total enthalpy is extracted in the MHD cycle, and it is essential that the discharged enthalpy be used in a lower-temperature bottoming cycle.

Closed-cycle liquid metal MHD systems offer a compromise between the high cycle efficiencies of the gas systems and the material problems posed by such a system. Liquid metals possess a conductivity five orders of magnitude greater than gaseous fluids and can operate at considerably lower temperatures—1500°F to 2000°F—without a reduction in conductivity. Drag effects reduce the velocity achievable with liquid-metal systems by a factor of 100. Since power density varies linearly with conductivity and with the square of the velocity, the power density with liquid metals may be a factor of 10 times greater than that with plasmas. This permits liquid-metal MHD systems to operate with lower magnetic fields which can be provided by ordinary elec-

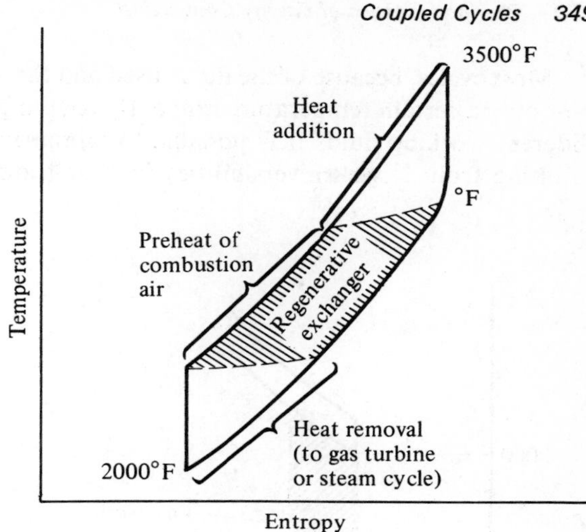

Figure 8-25

tromagnets as opposed to superconducting magnets which will probably be required by the plasma systems.

Liquid-metal systems could use either fossil fuel or nuclear heat sources and appear capable of achieving overall thermal efficiencies of up to 45 percent with operating temperatures of 2000°F if coupled with a conventional steam-generating system.

Alkali-metal technology has advanced appreciably as a result of energy conversion efforts in the space and liquid-metal breeder-reactor programs. While the expansion process in two-phase MHD systems is more complex than the expansion of a metallic vapor used in the Rankine topping cycle (to be discussed later), the flow configuration is much simpler in the MHD system. Thus liquid-metal MHD systems offer a potentially attractive means of using existing technology to take advantage of the higher temperatures at which metallic systems can operate.

8.10 Coupled Cycles

Carnot's principle tells us that the conversion of thermal to mechanical energy is increased at higher heat-addition temperatures for any given sink temperature. It is equally important to recognize that for maximum conversion efficiency it is not only necessary to utilize reversible expanders and pumps and compressors, but also to assure that all heat transfer between fluids and cycles is conducted as reversibly as possible. Thus if one utilizes an MHD cycle at a peak temperature of 5000°F, but then produces steam at a 1000°F with hot exhaust gases at 3000°F, a significant fraction of the thermal potential has been sacrificed and considerable lost work results.

Most cycles, because of the fluids used and the materials required, operate best over a certain temperature range. By coupling different cycles and using different working fluids it is possible to minimize the amount of lost work resulting from large irreversibilities in heat transfer, as illustrated in Fig. 8-26.

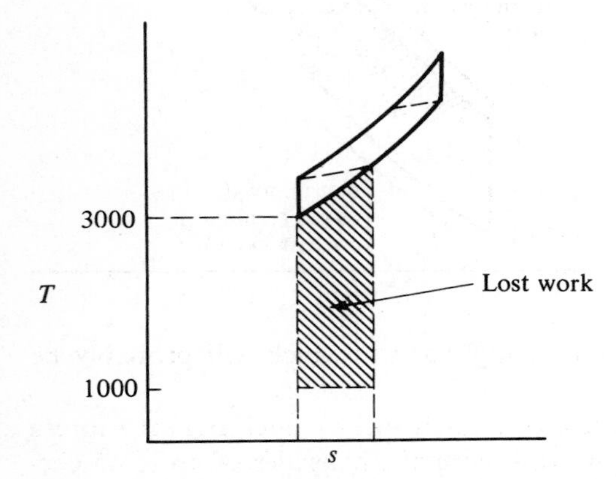

Figure 8-26

Before discussing possible cycle coupling, there are several general points regarding heat sources, fluid properties, materials of construction, and economics which must be considered in assessing the feasibility of coupling cycles to improve conversion efficiencies. First, one should recognize that materials limitations have provided the major barrier to higher conversion temperatures to date. Fossil fuels are capable of producing combustion temperatures in excess of 2000°F, but efforts to push the steam Rankine cycle (the work horse of power production) to temperatures above about 1050°F have not as yet been successful because of materials reliability. Gas-turbine technology has been advanced rapidly for aviation systems, and gas inlet temperatures in the 2000–2500°F seem achievable for industrial gas turbines in the near future. At these levels and above, it becomes necessary to utilize regenerative heat exchange to preheat combustion air sufficiently to realize the desired peak temperatures in the combustor with conventional fossil fuels. As was pointed out earlier, MHD systems operating above 3000°F require appreciable preheat or use of oxygen to achieve the necessary temperature levels.

Materials pose not only a major technical barrier to higher temperature operation, but also an economic one since high-temperature materials are seldom inexpensive. In addition, whether one uses regenerators or oxygen to achieve higher combustion and hence cycle temperatures, increased capital costs in conversion equipment are required. Industrial decisions have been based on economics, not conversion efficiencies. If one has cheap fuel available, the incentive to use that energy more efficiently is less, and large capital investments to increase temperatures may not result in lower kilowatt-hour costs.

Nuclear power has actually resulted in lower steam-plant temperatures than had been common in advanced fossil-fueled plants because of the temperature limitations in the core of the reactor. Nevertheless, the low nuclear fuel costs have justified large capital costs for water-cooled reactors in spite of their lower thermal efficiencies (about 32 percent) than conventional fossil-fuel-fired Rankine cycles. Actually, nuclear fission can produce much higher temperatures than currently used, and as cladding and coolant technology progresses it should be possible to considerably improve the performance of nuclear systems. The helium-cooled, high-temperature reactor discussed earlier and the sodium-cooled breeder reactor are examples of advanced systems. Fusion which occurs at temperatures in excess of 100,000,000° poses severe materials problems, but also offers exciting possibilities for new conversion processes.

Finally, it is important to recognize that the basic cycles discussed have general applicability over wide temperature ranges. As was shown in Fig. 8-23, the relative attractiveness of different cycles in terms of theoretical performance varies with temperature. Brayton cycles look considerably more attractive at temperatures above 2000°F than at lower temperatures if thermal efficiency is a prime criterion. However, the more important consideration in determining a cycle's viability in any given application is the availability of a working fluid whose physical properties match the cycle needs and which is compatible with available materials of construction. In addition, its cost, safety, and availability are also important. Mercury—although it looked good thermodynamically as a working fluid for a high-temperature Rankine cycle —was abandoned in part because of its safety problems.

While the Feher cycle appears to be the most attractive cycle between 1000°F and 2000°F, it requires a fluid with very special physical properties for its operation. Only CO_2 looks like it can come close to matching the primary cycle needs, but the pressure levels are so high that materials and design problems are encountered. When used at higher temperatures in the Rankine cycle, steam also causes difficulties because of its high vapor pressure. The specific volume of steam increases rather rapidly at the discharge conditions of the conventional steam plants ($\sim$5 inches of Hg) and thus rather large turbomachinery is required to fully expand steam.

These basic limitations on the steam cycle have led to efforts to find other fluids and/or cycles which can operate efficiently above 1000°F and reject their heat to a conventional steam cycle. As fuel costs increase and thermal discharge requirements become more stringent, attention is again being given to lower-temperature (bottoming) cycles that could utilize the heat rejected by the steam cycle as input to another cycle which uses a fluid with a higher vapor pressure that is capable of operating more efficiently at lower temperatures.

The thermal efficiency of coupled cycles can be expressed in terms of the individual cycle efficiencies as follows:

$$\eta_H = \frac{Q_H + Q'_L}{Q_H} = \frac{W_H}{Q_H} \tag{8-8}$$

and

$$\eta_L = \frac{Q'_L + Q_L}{Q'_L} = \frac{W_L}{Q'_L} \tag{8-9}$$

where Q'_L represents the heat flow between cycles (Q'_L is negative for the high-temperature cycle and positive for the lower-temperature cycle).

$$\eta_{\text{coupled}} = \frac{W_H + W_L}{Q_H} = \frac{\eta_H Q_H + \eta_L Q'_L}{Q_H} \tag{8-10}$$

$$\eta_{\text{coupled}} = \eta_H + \eta_L \frac{Q'_L}{Q_H} \tag{8-11}$$

but

$$\eta_H = 1 - \frac{Q'_L}{Q_H} \quad \text{or} \quad \frac{Q'_L}{Q_H} = 1 - \eta_H$$

then

$$\eta_{\text{coupled}} = \eta_H + (1 - \eta_H)\eta_L = \eta_H + \eta_L - \eta_H \eta_L \tag{8-12}$$

The remainder of this section deals with the coupling of cycles to achieve increased thermal efficiencies.

Metallic-Fluid Rankine Topping Cycles

If the thermal efficiency for Rankine cycles shown in Fig. 8-23 is extrapolated to higher temperatures, its theoretical performance remains quite attractive relative to other alternatives. The problems which develop with higher temperature and pressure steam cycles can be partially avoided by using a metallic fluid with a lower vapor pressure. Earlier attempts to utilize binary cycles with a mercury topping cycle were abandoned because of mercury's cost and toxicity. More recently the use of alkali metals for high-temperature-space Rankine cycles and for reactor cooling has led to an increased interest in binary cycles.

Potassium appears to be the most attractive candidate for a topping cycle in conjunction with the steam cycle. Steam-cycle peak temperatures have stabilized at 1000°F in recent years, and at this temperature level potassium has a vapor pressure of about 1.5 psia. At 2000°F the vapor pressure of potassium is only 152.4 psia, which poses no containment problem from a pressure point of view. However corrosion problems may arise with stainless steels if operating temperatures are pushed much above 1600°F. More expensive materials are available which can be used up to 2200°F if economics justify their use. The potassium–water reaction is a matter of some concern, but the problems to be overcome are similar to those which must be solved with the sodium–steam generator portion of the liquid-metal breeder-reactor and are not considered insurmountable.

The potassium topping cycle would probably operate with a high side temperature between 1500°F and 1800°F. The condensing temperature would be selected to optimize overall system performance, and would probably be between 1000°F and 1100°F. Figure 8-27 illustrates a potassium topping cycle with single reheat, rejecting its heat to a supercritical steam cycle with reheat. Expansion of the potassium vapor must not exceed the point where the moisture content causes erosion problems on the turbine blades. If the turbine and pump have efficiencies of 75 percent and 70 percent, respectively, the topping-cycle efficiency will be about 22 percent. At peak potassium temperatures of 2200°F, efficiencies approaching 30 percent for the topping cycle alone could

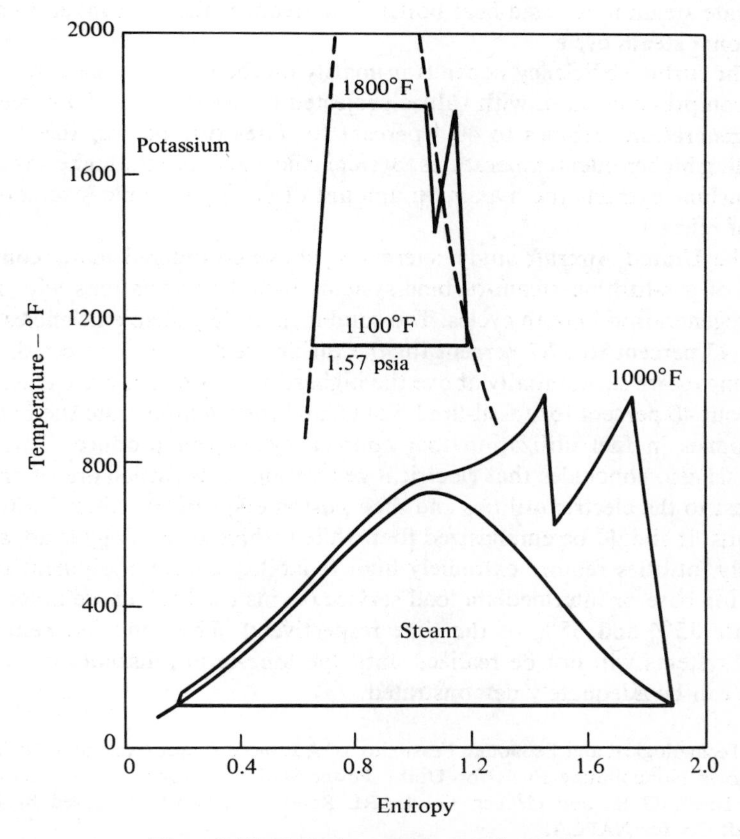

Figure 8-27. Typical potassium topping cycle.

be achieved. Binary cycle efficiencies ranging from 48–62 percent are theoretically achievable under these conditions.

The cycle illustrated in Fig. 8-27 would have an overall efficiency of about 53 percent.

Gas-Turbine Topping Cycles

The rapid development of gas turbines for the aircraft industry has accelerated their use in central station power plants. Initial use has been primarily for meeting the utility's peak power needs, but improved reliability and performance has led to their use in gas-turbine–steam-turbine systems, commonly called *combined cycles.*

Such systems can be fired with gas, oil, or even coal if the coal is gasified or liquefied first. In the most desirable system of the many possible combinations considered,[1] the combustion gases pass through the turbine where they are fully expanded. Thermal energy from the exhaust gases is used to generate steam in a waste-heat boiler. The steam is then expanded in a conventional steam cycle.

The turbine efficiency depends primarily on the inlet turbine temperature and compression ratio, with values projected to range from 27.3 percent for first-generation turbines to 40.4 percent for later turbines capable of withstanding higher inlet temperatures. System efficiencies are the highest when the gas turbine extracts the maximum amount of energy possible (excluding the use of reheat).

The United Aircraft study referenced above considered many combinations of gas-turbine–steam-turbine systems including variations with reheat and regeneration in both cycles. The combined-cycle system efficiencies range from 47 percent to 57.7 percent for the turbine technology projected. These efficiencies are substantially above the highest present steam-cycle efficiencies of about 40 percent for fossil-fired plants, and they demonstrate the potential economies in fuel utilization that coupled cycles can produce. This same analysis also concludes that electrical generating costs (which are of primary interest to the electric utilities and their customers) can be reduced with such systems. It should be emphasized that while turbine technology is advancing rapidly, utilities require extremely high reliability before equipment can be used for base or intermediate load service (terms used to denote units which operate 35% and 65% of the time respectively). Thus, the full benefits of these systems will not be realized until the long-term reliability of gas turbines can be adequately demonstrated.

[1]"Technological and Economic Feasibility of Advanced Power Cycle and Methods of Producing Nonpolluting Fuels for Utility Power Stations," Robson, F. L., Giramonti, A. J., Lewis, G. P., and Gruber, G., UARL Report J-970855-B. Prepared by United Aircraft Co. for NAPCA.

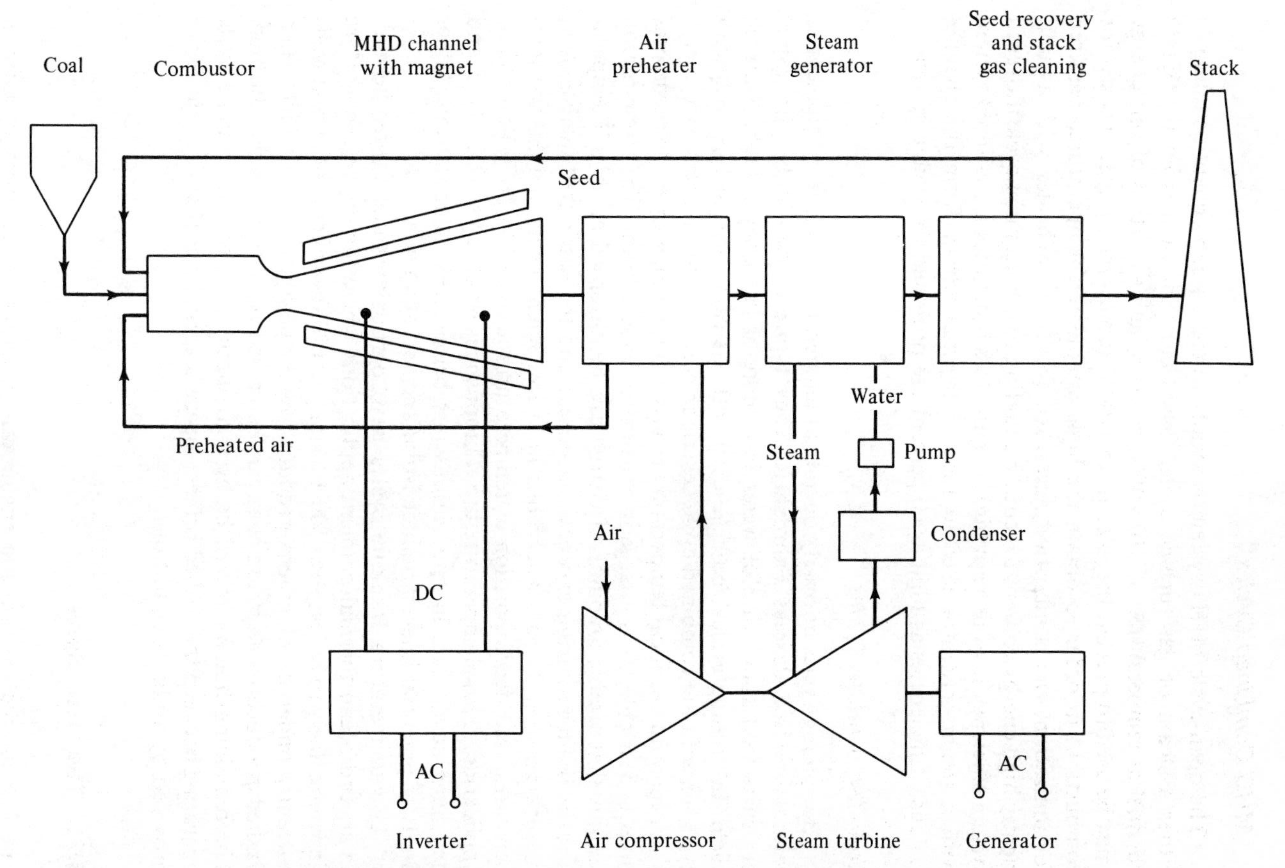

Figure 8-28. Schematic diagram of open-cycle MHD steam power system.

MHD Combined Cycles

The open-cycle MHD system is ideally suited for subsequent coupling to either a steam or gas turbine cycle. Gases are exhausted from the MHD channel at temperatures of 3100–4000°F. Although a portion of this energy must be used to preheat the inlet air to 3100°F (unless pure oxygen is used for combustion) in order to obtain the high combustion temperatures required for ionization, the combustion gases exit from the preheater with a sufficiently high energy content to require coupling to a bottoming cycle if overall system efficiency is to be kept high. Figure 8-28 illustrates a coal-fired system with a steam generator used to recover thermal energy from the channel exhaust. Efficiencies as high as 60 percent are projected for such a system.

Low-Temperature Cycles

Bottoming cycles normally operate at temperature levels from ambient to several hundred degrees Fahrenheit. They typically use fluids with vapor pressures well above that of water in a more-or-less conventional Rankine cycle. Bottoming cycles could be used with a steam cycle to form a binary cycle, or with the combined cycle described earlier to form a ternary cycle.

A high fluid-critical temperature relative to the upper operating temperatures is desirable, as is a large latent heat of vaporization. The combination assures that sizable amounts of energy can be absorbed at "high" temperatures without requiring excessive superheat and the associated inefficiencies of superheaters. The fluorocarbons have low latent heats and critical temperatures, and decomposition at temperatures above 325–400°F limits their application to low-temperature use. Thiophene is stable up to 600–700°F and has some advantages in applications just below these temperatures. Other fluids of potential interest include isobutane and ammonia.

Low-temperature Rankine cycles have particular value in geothermal, solar, and ocean-thermal-gradient applications where the upper temperature limits are likely to be less than 500°F. Many of the hot-water geothermal wells currently under development produce low-temperature brines which when flashed produce some steam that can be passed directly through a turbine. The remaining hot water will be used to supply heat to a bottoming cycle operating between the flashed brine temperatures and a heat-sink temperature provided by ambient conditions.

8.11 The Otto Cycle

The first internal-combustion engine was produced in Germany by Otto in the mid 1860s. This engine with only minor modifications is still found in almost every automobile in use today. The cycle shown in Fig. 8-29 consists

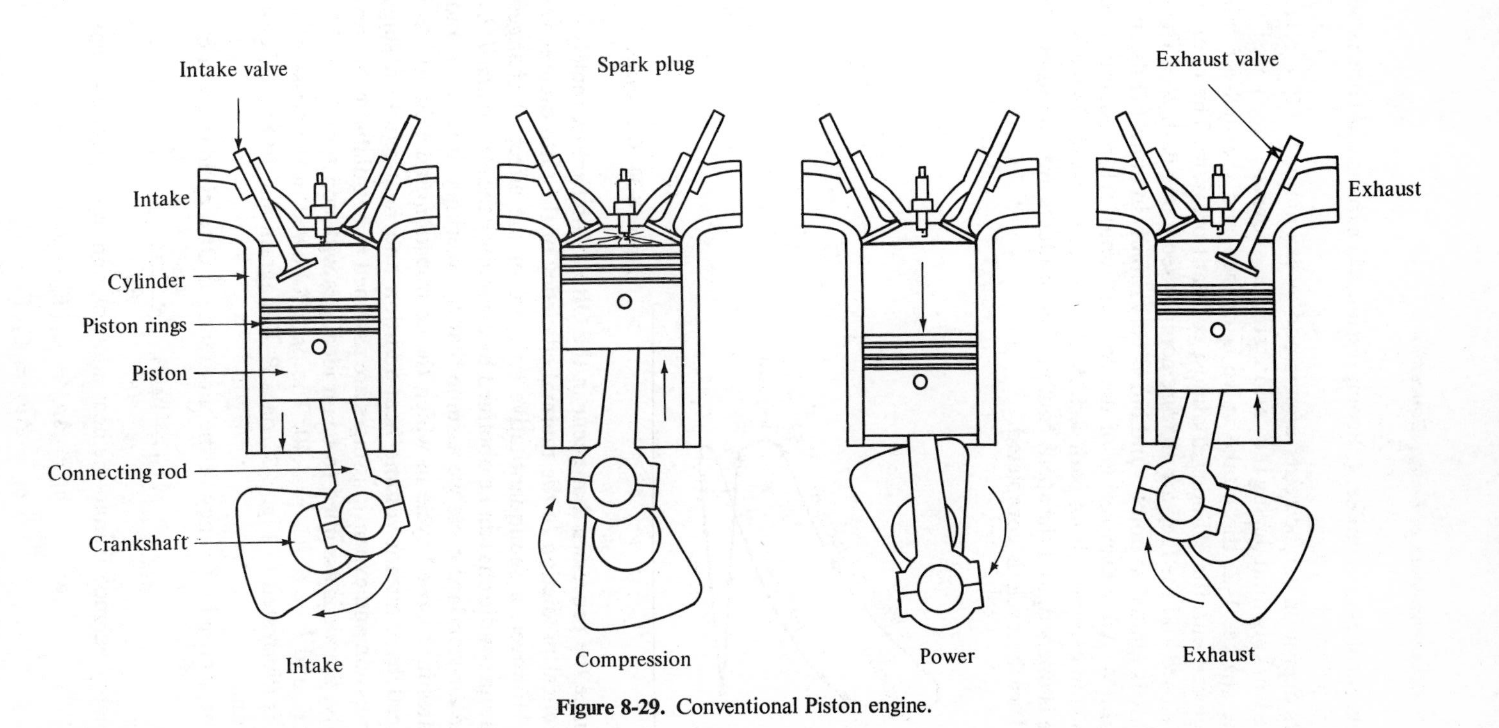

Figure 8-29. Conventional Piston engine.

of four strokes: (1) intake, (2) compression, (3) expansion/work production, and (4) exhaust.

Its operation can be better understood by examining the P–V relationship of the working fluid during the various strokes shown in Fig. 8-30. During the intake stroke, 1–2, the intake valve is opened and the fuel–air mixture is introduced into the cylinder at almost constant pressure. The intake valve is then closed and the fuel–air mixture compressed along path 2–3. The fuel–air mixture is ignited at point 3 and burns very rapidly at almost constant volume to point 4. After completion of the combustion, work production occurs as the piston expands along path 4–5. At the end of the work stroke the exhaust valve is opened, and the spent combustion products are exhausted along line 5–6–1 as the cycle is completed.

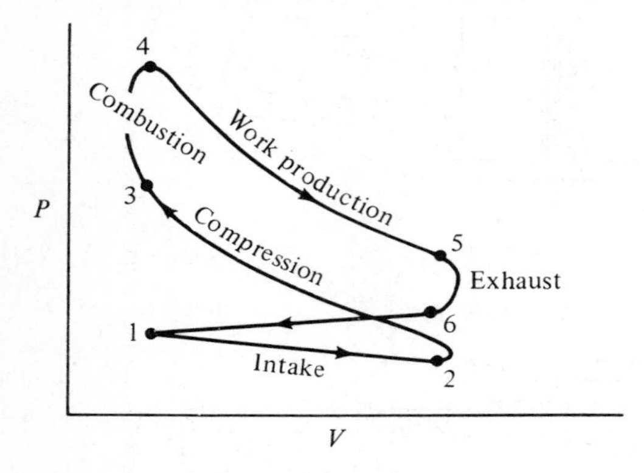

Figure 8-30. Otto cycle.

Since the processes that occur in the Otto engine are extremely complex, theoretical prediction of the thermal efficiency of the Otto cycle is very difficult. However, a semiquantitative estimate of the effect of changes in the operating conditions can be obtained by examining the *air-standard Otto cycle*. In the air-standard cycle we assume that the working fluid is air and that it operates in a closed cycle in which the combustion and exhaust strokes are replaced by constant-volume heat-addition and heat-rejection stages. The compression and expansion stages are assumed to be adiabatic and reversible. Finally, the air is assumed to be an ideal gas with constant $C_V = 5.0$ Btu/lb-mol °F, and $k = C_P/C_V = 1.40$. The path followed in the air-standard Otto cycle is illustrated in Fig. 8-31 on a P–V diagram and in Fig. 8-32 on a T–S diagram.

The thermal efficiency of the air-standard Otto cycle is given by

$$\eta = \frac{w_{\text{net}}}{q_H} = \frac{q_H + q_L}{q_H} \tag{8-13}$$

An energy balance around the heat-addition and heat-rejection steps gives

$$q_H = \Delta u_H = C_V(T_4 - T_3)$$
$$q_L = \Delta u_L = C_V(T_2 - T_5)$$

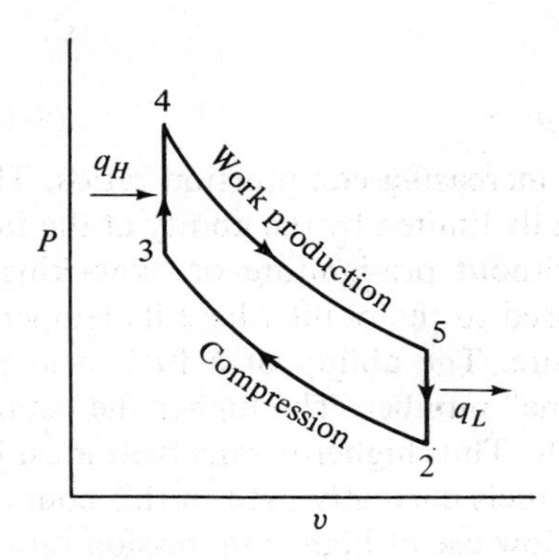

Figure 8-31. Air-standard Otto cycle.

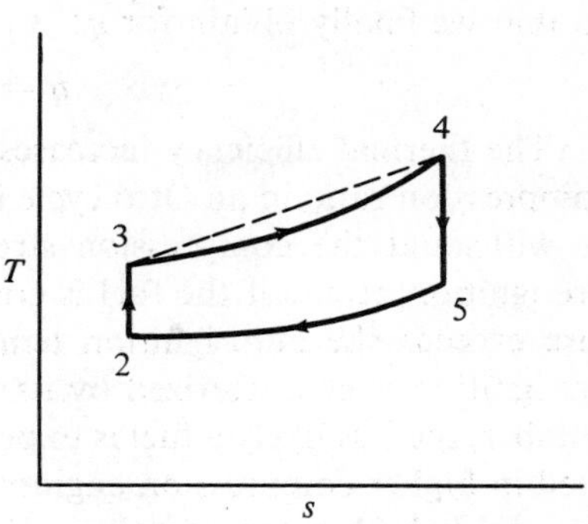

Figure 8-32. Air-standard Otto cycle.

since no work is involved in these nonflow operations. The thermal efficiency then becomes

$$\eta = \frac{q_H + q_L}{q_H} = \frac{C_V[(T_4 - T_3) + (T_2 - T_5)]}{C_V(T_4 - T_3)} = 1 + \frac{T_2 - T_5}{T_4 - T_3} \qquad (8\text{-}14)$$

Since the air is assumed to be an ideal gas, $T = PV/nR$, which can be substituted into the thermal efficiency to give

$$\eta = 1 + \frac{P_2V_2 - P_5V_5}{P_4V_4 - P_3V_3} \qquad (8\text{-}15)$$

The *compression ratio*, ρ, is defined as the ratio of the uncompressed gas volume to that of the compressed gas, V_2/V_3. Since $V_3 = V_4$ and $V_2 = V_5$,

$$\eta = 1 + \frac{V_2}{V_3}\left[\frac{P_2 - P_5}{P_4 - P_3}\right] = 1 + \rho\left[\frac{P_2 - P_5}{P_4 - P_3}\right] = 1 + \rho\frac{P_5(P_2/P_5 - 1)}{P_3(P_4/P_3 - 1)} \qquad (8\text{-}16)$$

Since the compression and expansion are assumed to be adiabatic and reversible, they are isentropic, and the pressure and volume are related by the expression $PV^k = \text{const}$, so

$$P_4V_4^k = P_5V_5^k; \qquad P_2V_2^k = P_3V_3^k$$

However, since $V_3 = V_4$ and $V_2 = V_5$, this gives

$$P_4V_3^k = P_5V_2^k; \qquad P_2V_2^k = P_3V_3^k$$

which gives

$$\frac{P_4}{P_3} = \frac{P_5}{P_2}$$

Thus the thermal efficiency is given by

$$\eta = 1 + \rho\frac{P_5(P_2/P_5 - 1)}{P_3(P_4/P_3 - 1)} = 1 - \frac{P_5}{P_4}\rho \qquad (8\text{-}17)$$

But from the relation between P and V,

$$\frac{P_5}{P_4} = \frac{P_2}{P_3} = \left(\frac{V_3}{V_2}\right)^k = \rho^{-k}$$

so that we finally obtain for η:

$$\eta = 1 - \rho^{1-k} \tag{8-18}$$

The thermal efficiency increases with increasing compression ratios. The compression ratio in an Otto cycle is usually limited by the ability of the fuel to withstand the compression stroke without pre-igniting or "knocking." Pre-ignition occurs if the fuel is compressed to the point where its temperature exceeds the auto-ignition temperature. The ability of a fuel to avoid pre-ignition is characterized by its "octane" number. The higher the octane number, the less likely a fuel is to pre-ignite. Thus higher-octane fuels must be used in higher-compression engines. The fuels currently used in this country have fairly high octane numbers which allow use of high compression ratios. Automobile engines that run on "regular" gasoline typically have compression ratios of about 8.5/1; engines that use "high test" or "ethyl" gasoline have compression ratios of about 10.5/1.

SAMPLE PROBLEM 8-5. The compression ratio in an air-standard Otto cycle is 8. If the air before compression is at 60°F and 1 atm and 800 Btu/lb_m is added to the gas per cycle, calculate using the air tables (Appendix F):

(a) Temperature and pressure at each point of the cycle.
(b) The heat that must be removed.
(c) The thermal efficiency of the cycle.

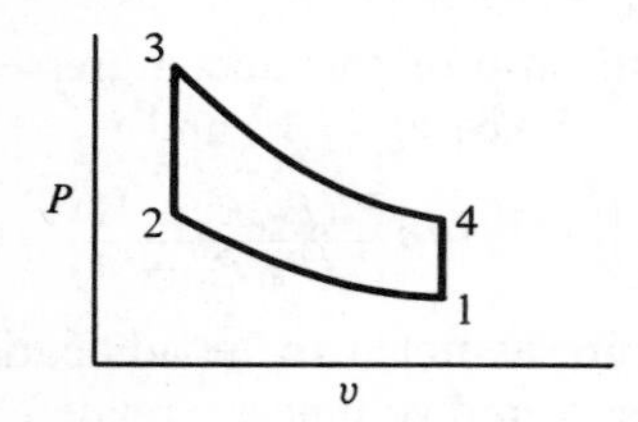

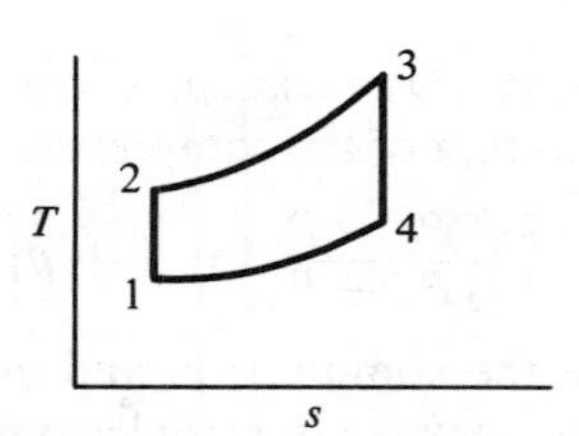

Figure SP8-5

Solution: (a) For $P_1 = 14.7$ psi and $T_1 = 520°R$

$$v_1 = \frac{53.34 \times 520}{14.7 \times 144} = 13.0 \text{ ft}^3/\text{lb}_m$$

From the air tables at point 1

$$h_1 = 124.27 \frac{\text{Btu}}{\text{lb}_m}, \quad p_{r1} = 1.2147, \quad v_{r1} = 158.58, \quad u_1 = 88.62 \text{ Btu/lb}_m$$

For an isentropic process

$$\frac{v_1}{v_2} = \frac{v_{r1}}{v_{r2}} = 8$$

therefore

$$v_{r2} = \frac{v_{r1}}{8} = \frac{158.58}{8} = 19.82$$

From the air tables at point 2 for $v_{r2} = 19.82$

$$T_2 = 1170°\text{R}$$

$$p_{r2} = 21.37, \qquad P_2 = P_1 \times \frac{p_{r2}}{p_{r1}} = (14.7)\frac{(21.37)}{1.21} = 260 \text{ psi}$$

$$u_2 = 203.3 \text{ Btu/lb}_m, \qquad v_2 = 1.64 \text{ ft}^3/\text{lb}_m$$

Going from 2 to 3 involves a constant-volume heat addition of 800 Btu/lb$_m$; therefore

$$q = \Delta u = 800 = u_3 - u_2$$

$$u_3 = 800 + 203.3 = 1003.3 \text{ Btu/lb}_m$$

From air tables:

$$\left.\begin{array}{ll} 5000°\text{R}: & u = 1050.12 \\ 4500°\text{R}: & u = 931.39 \end{array}\right] \Delta u = 118.73$$

$$1003.3 - 931.4 = 71.9$$

$$T_3 = 4500 + \frac{71.9}{118.7}(500) = \underline{4803°\text{R}}$$

$$v_3 = v_2 = 1.64 \text{ ft}^3/\text{lb}_m \qquad v_{r3} = 0.246$$

$$P_3 = \frac{(53.34)(4803)}{(1.64)(144)} = \underline{1085} \text{ psi}, \qquad p_{r3} = 5521 + \frac{71.9}{118.7}(3316) = 7530$$

An isentropic path is followed from 3–4; therefore

$$\frac{P_3}{P_4} = \frac{p_{r3}}{p_{r4}} \quad \text{and} \quad \frac{v_3}{v_4} = \frac{v_{r3}}{v_{r4}} = \frac{1}{8}$$

$$v_{r4} = 8v_{r3} = 8(0.246) = 1.968$$

$$\left.\begin{array}{ll} \text{at } 2550°\text{R}, & v_r = 1.996 \\ \text{at } 2600°\text{R}, & v_r = 1.876 \end{array}\right] \Delta = 0.120$$

$$1.996 - 1.968 = 0.028$$

$$T_4 = 2550 + \frac{0.028}{0.120}(50) = \underline{2562°\text{R}}$$

$$P_4 = \frac{p_{r4}}{p_{r3}} P_3$$

$$p_{r4} = 473.3 + \frac{0.028}{0.120}(41) = 482.9$$

$$P_4 = \frac{482.9}{7530}(1085) = 69.6 \text{ psi}$$

(b) The final leg from 4 to 1 requires heat removal at constant volume:

$$q = \Delta u = u_1 - u_4$$

From the air tables

$$u_4 = 485.3 + \frac{0.028}{0.120}(11)$$

$$u_4 = 485.3 + 2.6 = 487.9 \text{ Btu/lb}_\text{m}$$

$$u_1 = 88.6 \text{ Btu/lb}_\text{m}$$

$$\Delta u = -389.3 \text{ Btu/lb}_\text{m}$$

$$q = -389.3 \text{ Btu/lb}_\text{m}$$

(c) The thermal efficiency of the cycle is:

$$\eta = \frac{w_{net}}{q_H} = \frac{q_H + q_L}{q_H} = \frac{800 - 389.3}{800} = \frac{410.7}{800} = .51$$

$$\eta = 51\%$$

It is instructive to contrast the results using the air tables with those obtained using ideal-gas relationships and the assumption of constant heat capacity.

	Air Tables		Ideal Gas–Const. C_v	
	T (°R)	P (psi)	T (°R)	P (psi)
1	520	14.7	520	14.7
2	1170	260	1197	270
3	4803	1085	5887	1331
4	2562	69.6	2558	73
q_L	−389.3 Btu/lb$_\text{m}$		−348 Btu/lb$_\text{m}$	
η	51%		56.5%	

Since the heat capacity increases significantly with temperature, the temperature rise from step 2 to step 3 is appreciably less than that calculated assuming a constant value of 0.171 Btu/lb$_\text{m}$ air. The thermal efficiency taking this into account is about 10 percent lower than equation (8-18) indicates.

Although the theoretical efficiency of an air-standard Otto cycle with a 10: 1 compression ratio is about 62 percent, actual engines seldom operate much above 25 percent and generally at 20 percent or below. As was mentioned earlier, combustion products leaving at 300–400°F represent a major energy loss to the system. The expansion and compression steps both involve considerable irreversibilities because of the friction inherent in a piston engine. The need to maintain cylinder and valve temperatures low enough to avoid excessive engine wear and valve leakage results in heat losses from the hot combustion gases to the engine coolant. In addition, the valve restrictions result in a significant pressure drop to get air–fuel mixtures in and exhaust out of the cylinder. Finally, the combustion process itself is incomplete as considered by the unburned hydrocarbons and carbon monoxide in the exhaust. Leaner fuel mixtures to remedy the latter would reduce the peak combustion temperature and thus the efficiency.

Rotary Engines: The Wankel

Virtually all automotive engines mass-produced to date have utilized the piston-cylinder configuration. In all of these engines the linear movement of the piston is translated into rotary motion through a crankshaft. In 1920 Felix Wankel started work on a rotary engine in his laboratory in Germany. Following major modifications of Wankel's original design through the collaborative efforts of engine manufacturers in Germany and the U.S., the rotary engine was first introduced on a significant scale in 1965 by a German manufacturer, and several years later by a Japanese firm. The engine utilizes a triangularly shaped rotor mounted off-center in a stationary housing as shown in Fig. 8-33. Its operation is identical to the four-cycle Otto engine discussed previously, except that the expanding gases act against the rotor, producing rotational action directly. The engine uses no valves and thus avoids the pressure losses associated with the intake and exhaust steps of conventional piston engines. Unlike the four-stroke IC engine for which the crankshaft must make two revolutions for each power stroke, the Wankel crankshaft makes one revolution for each power stroke delivered. Consequently the engine displacement is more effectively used which permits a significant size reduction for a given horsepower rating. Its reduced size is one of the most attractive features to automotive engineers. The engine operates

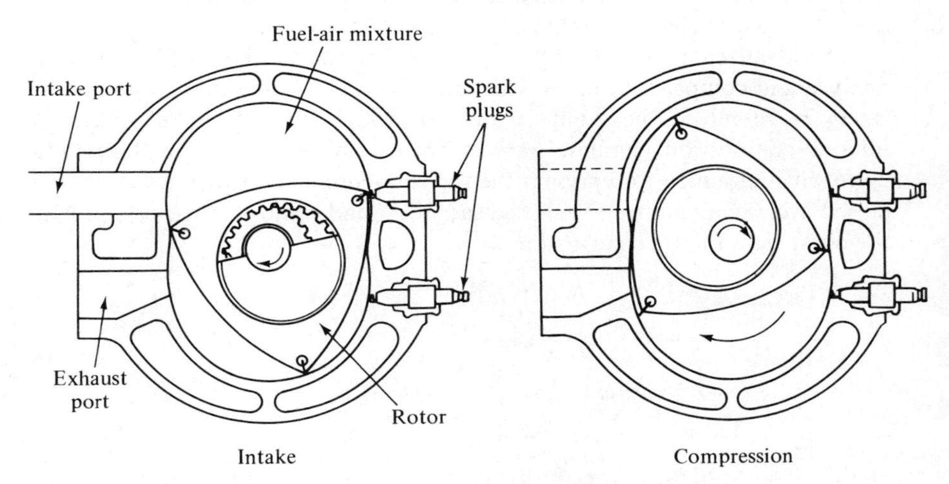

Figure 8-33. Wankel, or rotary, engine has three small chambers of variable size each of which undergoes a complete four-stroke Otto cycle in one revolution of the rotor.

at larger compression ratios on low-octane fuel and can use unleaded gasoline because the engine has no valves (which are lubricated by lead in conventional piston engines). Although the efficiency is lower than in conventional engines, its smaller size leads to a weight reduction which should ultimately produce overall fuel economies. This is an important consideration in view of the tightening energy supply picture. Unburned hydrocarbon and carbon monoxide are reportedly higher in the exhaust, but nitrogen oxides are significantly lower. Complete combustion of unburned hydrocarbons and carbon monoxide can be handled with relative ease, particularly with the additional room that the compact engine provides under the automobile hood. Nitrogen oxides have proved particularly difficult to eliminate and thus improved performance in this regard could be an important advantage for the Wankel engine.

8.12 The Diesel Cycle

Whereas pre-ignition is harmful and usually avoided in the Otto cycle, the diesel cycle uses the high temperatures that are attained in the compression step to avoid the need for an external combustion initiator (typically a spark plug in the Otto engine). In the diesel cycle the fuel is not mixed with the air until the end of the compression stage. The fuel is then added slowly, so that the combustion occurs, ideally, at constant pressure. The expansion actually commences before combustion is complete and continues isentropically until the piston reaches its lower dead-center position. The combustion products are exhausted in the same manner as in the Otto engine.

The idealized air-standard diesel cycle, illustrated in Fig. 8-34, is used to analyze and compare the cycle with others because of the difficulty in accurately representing the complex process which actually takes place in internal-combustion engines. As was the case with the air-standard Otto cycle, air is assumed to represent the fuel–air–combustion-gases mixture in a closed-cycle configuration with heat addition and removal used to simulate the combustion and exhaust steps.

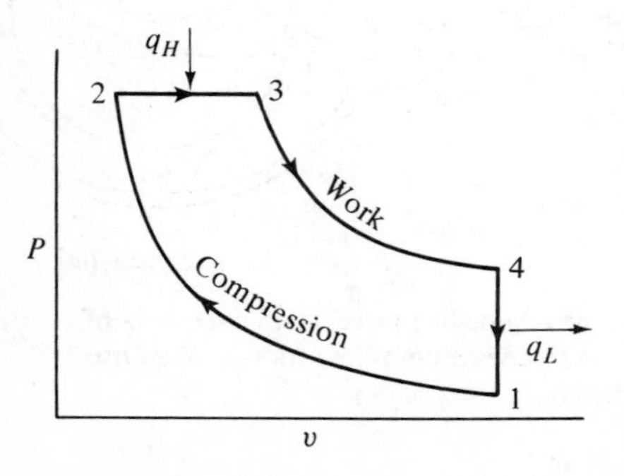

Figure 8-34. Air-standard diesel cycle.

Path 2–3 represents the constant-pressure expansion during combustion of the fuel. In the idealized cycle this is represented by a heat addition, q_H, which increases the temperature of the fluid as the fluid is expanding. When combustion is completed, the fluid reaches its peak temperature, T_3, after which the expansion continues isentropically along path 3–4. Work is actually produced by the engine over both legs 2–3 and 3–4.

Heat removal, q_L, occurs from 4–1 at constant volume once the expansion is completed. The fluid is then compressed isentropically back to its original pressure, P_2, at which point heat addition again commences.

The thermal efficiency of the air-standard diesel cycle can be calculated as was done for the Otto cycle, except that the heat addition occurs at constant pressure. The air is assumed to behave as an ideal gas with a constant heat capacity. (The significance of the latter assumption was illustrated in Sample Problem 8-5.)

$$\eta = \frac{w_{\text{net}}}{q_H} = \frac{q_H + q_L}{q_H} = 1 - \frac{|q_L|}{q_H} \tag{8-19}$$

$$q_H = C_P(T_3 - T_2) \quad \text{and} \quad q_L = C_V(T_1 - T_4)$$

Since $C_P/C_V = k$

$$\eta = 1 - \frac{T_4\left(\dfrac{T_1}{T_4} - 1\right)}{kT_2\left(\dfrac{T_3}{T_2} - 1\right)} \tag{8-20}$$

If v_3/v_2 is defined as the cutoff pressure ratio, ρ_c, then the efficiency can be expressed in terms of the compression ratio, $\rho = v_1/v_2$, and ρ_c as

$$\eta = 1 - \rho^{1-k}\left[\frac{\rho_c^k - 1}{k(\rho_c - 1)}\right] \tag{8-21}$$

For $\rho_c = 1$, the above relationship corresponds to that of the Otto cycle. For values of $\rho_c > 1$ the diesel efficiencies fall below those of the Otto cycle at corresponding compression ratios. Examination of the diesel cycle on a T–s diagram (Fig. 8-35) reveals this difference clearly. The Otto cycle, 1–2–3′–4–1,

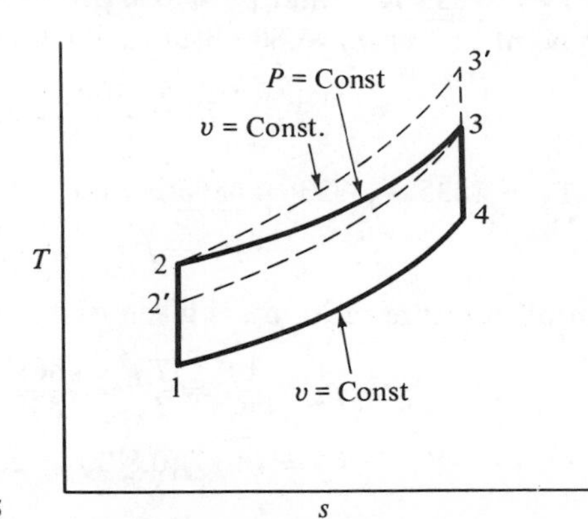

Figure 8-35

and the diesel cycle, 1–2–3–4–1, have identical compression ratios, but as can be seen, heat addition along path 2–3′ is at a higher temperature than along path 2–3 in the diesel cycle. However the diesel cycle can operate with higher compression ratios since fuel injection does not occur until compression is completed. At the same maximum temperature and pressure for the two cycles (as illustrated by path 1–2′–3–4–1 for the Otto cycle and 1–2–3–4–1 for the diesel cycle) the diesel has the higher thermal efficiency. The diesel engine, because of the delayed fuel injection, does not require high-octane fuel. It typically operates at pressure ratios of about 15 and achieves overall efficiencies of close to 40 percent.

SAMPLE PROBLEM 8-6. For the same initial conditions and heat input used for the Otto cycle in Sample Problem 8-5 and for a compression ratio of 15, calculate cycle temperatures and pressures, the cutoff pressure ratio, and the cycle thermal efficiency.

Solution: Since air-standard cycles are so highly idealized, comparisons are generally made neglecting the effect of temperature on heat capacities which was shown to have an appreciable effect in Sample Problem 8-5. Calculations here will assume constant values of the heat capacities, $C_P = 0.24$ Btu/°R lb$_m$ and $C_V = 0.171$ Btu/°R lb$_m$, for air.

At point 1: $T_1 = 520°R$ and $P_1 = 14.7$ psi. From the ideal gas law, $v_1 = 13.08$ ft^3/lb$_m$.

At point 2: For a compression ratio of 15, $v_1/v_2 = 15$.

$$v_2 = \frac{v_1}{15} = 0.872\,\frac{\text{ft}^3}{\text{lb}_m}$$

For an ideal gas undergoing an isentropic process,

$$\frac{T_2}{T_1} = \left(\frac{v_1}{v_2}\right)^{k-1} \quad \text{and} \quad \frac{P_2}{P_1} = \left(\frac{v_1}{v_2}\right)^{k} \quad \text{where } k = \frac{C_P}{C_V} = 1.4$$

Thus, $T_2 = \underline{1535°R} =$ and $P_2 = \underline{650}$ psi.

At point 3: For $q_H = 800$ Btu/lb$_m$ the temperature increase is given by

$$T_3 - T_2 = \frac{800}{0.24} = 3333°R$$

Thus, $T_3 = 1535 + 3333 = \underline{4868°R}$ and

$$P_3 = P_2 = \underline{650}\text{ psi}$$

The cutoff pressure ratio, ρ_c, is given by v_3/v_2.

$$\rho_c = \frac{v_3}{v_2} = \frac{T_3}{T_2} = \frac{4868}{1535} = 3.17$$

$$v_3 = (3.17)(0.872) = 2.76\text{ ft}^3/\text{lb}_m$$

At point 4: In going from 3–4 work is extracted as the expansion continues isentropically, until $v_4 = v_1 = 13.08\ \text{ft}^3/\text{lb}_\text{m}$

$$\frac{T_4}{T_3} = \left(\frac{v_3}{v_4}\right)^{k-1} \quad \text{and} \quad \frac{P_4}{P_3} = \left(\frac{v_3}{v_4}\right)^{k}$$

$$\frac{v_3}{v_4} = \frac{2.76}{13.08} = 0.211$$

Thus,

$$T_4 = T_3(0.211)^{0.4} = 4868(0.537) = \underline{2614°\text{R}}$$

$$P_4 = P_3(0.211)^{1.4} = 650(0.113) = \underline{73.45}\ \text{psi}$$

$$q_L = C_V(T_1 - T_4) = 0.171(520 - 2614) = (0.171)(-2094)$$

$$= -358\ \text{Btu/lb}_\text{m}$$

$$\eta = \frac{q_H + q_L}{q_H} = \frac{800 - 358}{800} = \frac{442}{800} = 0.55$$

$$= 55\%$$

Contrasting the performance of the Otto and diesel cycles from Sample Problems 8-5 and 8-6, under comparable initial conditions and heat input but for $\rho = 8$ for the Otto and $\rho = 15$ for the diesel, we observe thermal efficiencies of 56.5 percent and 55 percent, respectively. The two cycles are contrasted on a P–v diagram and a T–s diagram in Fig. SP 8-6. We can see

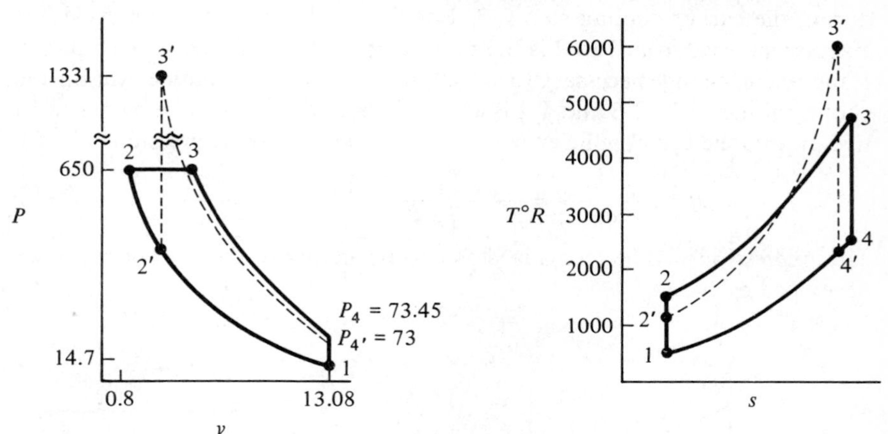

Figure SP8-6

that the average temperature at which heat addition occurs is comparable for the two cycles, and thus the efficiencies are approximately the same. However, peak pressure and temperature are appreciably lower for the diesel cycle, despite the much higher compression rate.

8.13 Stirling Cycle

The Stirling cycle first proposed by Robert Stirling (1840–1878) has received considerable interest in recent years for automotive applications, because of its high theoretical efficiency (equal to the Carnot cycle) and its potential for reducing exhaust emissions associated with internal combustion engines. The latter advantage arises from the fact that external combustion is used to supply heat to a gaseous cyclic fluid, and the relative ease of reducing carbon monoxide, unburned hydrocarbons, and nitrogen oxide with external combustion.

The ideal Stirling cycle consists of heat addition and removal at constant temperature, as is the case in the Carnot engine, with two constant-volume paths of changing pressure. Theoretically a heat regenerator operating reversibly would be used to exchange heat between the fluid as it moves between the high- and low-temperature parts of the cycle. Such a device is easy to specify, but much harder to design.

The working fluid, as illustrated on the P–v and T–s diagrams in Fig. 8-36, expands as it is heated at constant temperature along path 1–2, thus delivering work. It must then be cooled at constant volume along path 2–3, to a lower temperature. Heat is then rejected as the fluid is compressed along path 3–4 to its initial volume, after which it is heated back to its original temperature regeneratively, by exchanging heat reversibly with energy stored during the earlier cooling step 2–3. For the idealized air Stirling cycle, the entropy increase from 1 to 2 is the same as the entropy decrease from 3 to 4. If the heat exchange necessary to accomplish the constant volume heating and cooling along paths 2–3 and 4–1 is accomplished within the cycle by regenerative means, the cycle's efficiency is given by the familiar expression:

$$\eta = \frac{q_H + q_L}{q_H} = \frac{T_H \, \Delta s_H + T_L \, \Delta s_L}{T_H \, \Delta s_H} = \frac{T_H - T_L}{T_H} \tag{8-22}$$

This expression will be recognized as corresponding to the Carnot efficiency developed earlier.

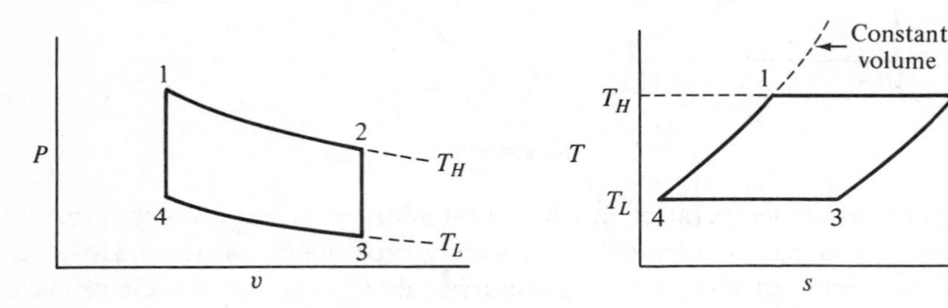

Figure 8-36

While seemingly simple in principle, the development of hardware to perform each of the steps in the cycle is quite challenging, and thus the cycle has not had widespread application to date. Engines have been developed with efficiencies in excess of 40 percent, slightly better than the diesel, but considerably above the typical Otto cycles used for automotive applications.

SAMPLE PROBLEM 8-7. Calculate the maximum theoretical efficiency of a Stirling cycle engine operated with a heat source temperature (presumably a flame) maintained below 900°C to minimize NO_X formation. Assume the engine must discharge to ambient air at a temperature of 40°C. Contrast the ideal engine with one which allows a 100°C temperature driving force for heat exchange between the same heat source and ambient conditions.

Solution:

$$\eta = \frac{T_H - T_L}{T_H}$$

For

$$T_H = 900 + 273 = 1173°\text{K} \quad \text{and} \quad T_L = 40 + 273 = 313°\text{K}$$

$$\eta = \frac{1173 - 313}{1173} = \frac{860}{1173} = 0.733 \quad \text{or} \quad 73.3\%$$

For

$$T_H = 800 + 273 = 1073°\text{K}, \qquad T_L = 140 + 273 = 413°\text{K}$$

$$\eta = \frac{1073 - 413}{1073} = \frac{660}{1073} = 0.615 \quad \text{or} \quad 61.5\%$$

This sample problem illustrates the effect that irreversibilities in heat transfer external to the cycle have on its performance. It should be recognized that further degradation in performance can be expected as a result of regenerator inefficiency and irreversibilities in the expander and compressor. If one is limited to the heat source temperature specified in this problem, attainment of the 40 percent efficiency achieved with some Stirling cycles to date would be most difficult because of the inherent difficulties in providing reversible expanders, compressors, and heat exchangers at reasonable costs.

It is worth noting that the Stirling cycle is theoretically capable of achieving thermal efficiencies approximately the same as those of the Otto and diesel cycles in Sample Problems 8-5 and 8-6 with a peak cycle temperature of only 1860°R as contrasted to 5887°R for the Otto, and 4868°R for the diesel, despite the fact that the lowest temperature in the Stirling cycle is 560°R as contrasted with 520°R for both the Otto and diesel examples. In the latter two cases the average temperature of heat rejection is much higher which more than compensates for the lower average temperature at which heat addition occurs in the Stirling. The deceptive feature of the Stirling cycle relates to the regenerative heat exchange between two fluids under conditions of constant volume. Without the regenerative feature, the average temperature of heat addition and heat rejection can rapidly approach one another, and the efficiency thus falls quickly.

8.14 Jet and Rocket Engines

A nozzle is perhaps the most simple and efficient device for converting thermal energy to a mechanical form, in this case kinetic energy. Gases are typically exhausted from nozzles at supersonic velocities and produce a thrust in accord with Newton's second law. Nozzles provide the thrust for jet and rocket engines. Jet engines typically utilize a turbocompressor as shown in Fig. 8-37 to compress the incoming air prior to injection and combustion of

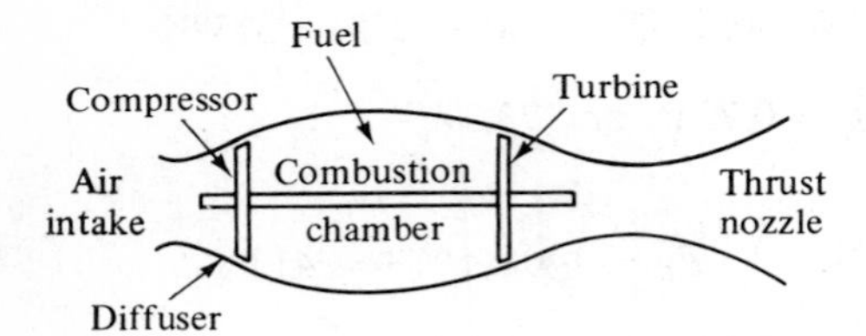

Figure 8-37. Jet engine.

the fuel. A diffuser is used at the engine inlet to reduce the compression ratio of the turbocompressors. Rocket engines, unlike jets, use an oxidizer, such as LOX (liquid oxygen), which is carried aboard the vehicle along with the fuel. The fuel and oxidizer are injected into the combustion chamber where they react and sustain a sufficiently high temperature and pressure to power the rocket by means of expanding the combustion products through the exhaust nozzle.

Problems

8-1. A steam locomotive is powered by a simple steam engine illustrated in Fig. P8-1. The boiler operates at 10 bar and discharges a saturated vapor. The steam is expanded to 1.4 bar in the piston and discharged to the atmosphere. If feedwater is available at 1 bar and 70°C and the engine is operating at steady state, determine the thermal efficiency of the engine. What is the quality of the steam exhausted from the piston?

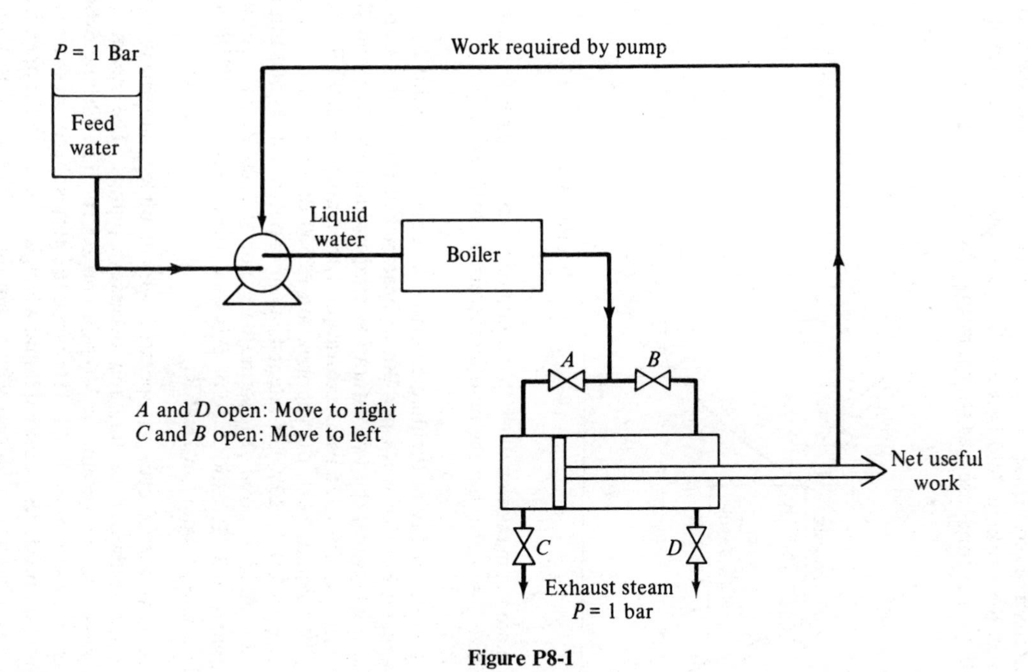
P = 1 Bar
Feed water
Work required by pump
Liquid water
Boiler
A
B
A and D open: Move to right
C and B open: Move to left
Net useful work
C
D
Exhaust steam
P = 1 bar

Figure P8-1

8-2. A space program requires a "portable" power-generating unit that can produce electrical energy for extended periods of time during orbital flight. A modified Rankine cycle (see Fig. P8-2) that utilizes a metallic fluid has been proposed for this purpose. Heat will be extracted at very high temperatures directly from a nuclear reactor. Heat from the condenser will be discharged by radiation to space. If sodium is circulated as the working fluid between the pressures of 126 psia (in the boiler) and 14.7 psia (in the condenser) as shown in Fig. P8-2,

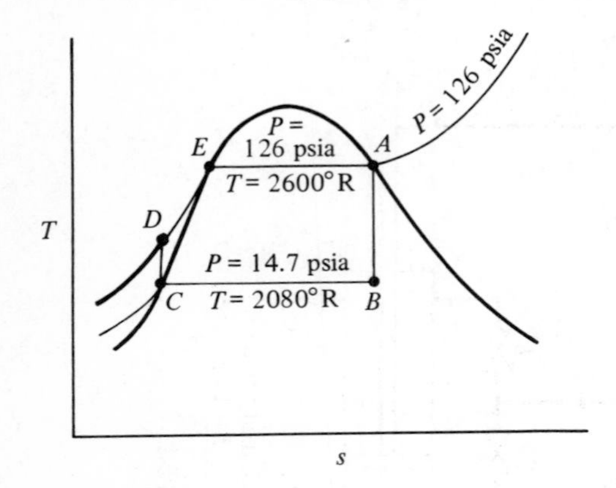

Figure P8-2

(a) Calculate the maximum work that could be produced per Btu of heat picked up in the reactor. Assume that the fluid temperature at the exit of the pump (point *D*) is 2100°R.

(b) What is the quality of the fluid entering the condenser (point *B*)?

(c) How many Btu's of work per Btu of heat picked up could a Carnot cycle deliver operating between the two saturation temperatures (2600°R and 2080°R)?

For sodium: At $P = 126$ psia, saturation temperature = 2600°R; $(C_P)_{\text{liq}} =$ 0.28 Btu/lb$_m$ °R; latent heat of vaporization = 500 Btu/lb$_m$. At $P = 14.7$ psia, saturation temperature = 2080°R; latent heat of vaporization = 560 Btu/lb$_m$.

In Fig. P8-2, fluid at point *A* is saturated vapor at 2600°R, fluid at point *C* is saturated liquid at 2080°R, fluid at point *D* is subcooled liquid at 2100°R, and fluid at point *E* is saturated liquid at 2600°R.

8-3. A steam heat engine cycle is operated with a boiler, superheater, turbine, condenser, open feedwater heater, and necessary pumps. The steam from the superheater passes to the turbine at 310 psia and 600°F. The turbine exhausts to the condenser at a vacuum of 28.4 in. Hg when the atmospheric pressure is 29.9 in. Hg. The condenser produces saturated liquid which is pumped to the open feedwater heater, which is operating at 50 psia with steam that is bled from an intermediate stage of the turbine. Also, some of the 50 psia steam bled from the turbine is sent out to a chemical process area for use in an evaporator. This steam is returned to the heat-engine cycle as saturated liquid at 50 psia, where it is combined with the feedwater heater product and pumped into the boiler. The turbine is 80 percent efficient compared to a reversible turbine, and all pumps are 70 percent efficient on the same basis. All equipment is well insulated and pressure drops are negligible in the piping, condenser, boiler, super-heater, and feedwater heater.

(a) For the power cycle alone, without considering the amount of process steam withdrawn for the evaporator, determine the net work delivered per Btu of heat supplied to the boiler and superheater.

(b) Repeat the calculation of (a), assuming that the feedwater heater is not used.

(c) If steam delivered to the turbine is valued at 55 cents per 1000 lb_m what would be a fair charge per 1000 lb_m of steam used by the process department for its evaporator?

8-4. A standard Rankine cycle with superheat (so no moisture condenses in the turbine) is to be operated with two closed feed preheaters as shown in Fig. P8-4.

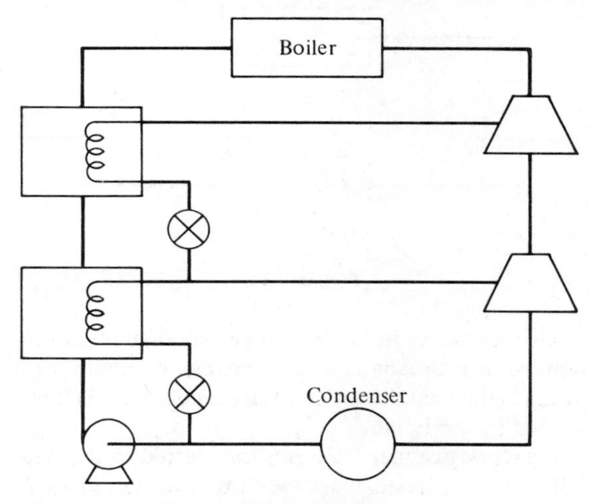

Figure P8-4

The boiler will operate 13.6 bar, and the turbine takeoffs will occur at 4.8 and 1 bar. The condenser will operate at 38°C and will discharge saturated liquid. The feed preheaters discharge liquid at the same temperature as the heating steam is condensing, and the condensed steam leaves each preheater as a saturated liquid. If the pumps and compressors operate adiabatically and all process equipment is reversible, determine the thermal efficiency of this cycle.

8-5. The boiler of a Rankine-cycle steam plant furnishes steam at 350 psia and 800°F to the turbine at the rate of 1000 lb_m/min. From the turbine the steam passes to the condenser, which operates at a pressure of 4 in. Hg using cooling water at 70°F. Assume that the turbine is adiabatic and reversible, that the condensate leaving the condenser is saturated liquid, and that the pump returning the condensate to the boiler is adiabatic and reversible. Calculate

(a) The work produced by the turbine (Btu/lb_m of steam).

(b) The pump work.

(c) The heat transferred in the boiler (Btu/lb_m of steam).

(d) The maximum (Carnot) work per Btu transferred in the boiler for a cycle receiving all its heat at the temperature of the boiler, 2000°F, and discharging at 70°F.

(e) The turbine work (Btu/lb_m of steam) if 500 lb_m/min of process steam is removed from the turbine at the 40-psia stage.

8-6. A Rankine-cycle steam plant (Fig. P8-6) is operating with a steam pressure of 275.3 psig, a steam temperature of 600°F, and a condenser vacuum of 28.4 in. Hg. Barometer indicates 29.9 in. Hg. Saturated liquid leaves the condenser.

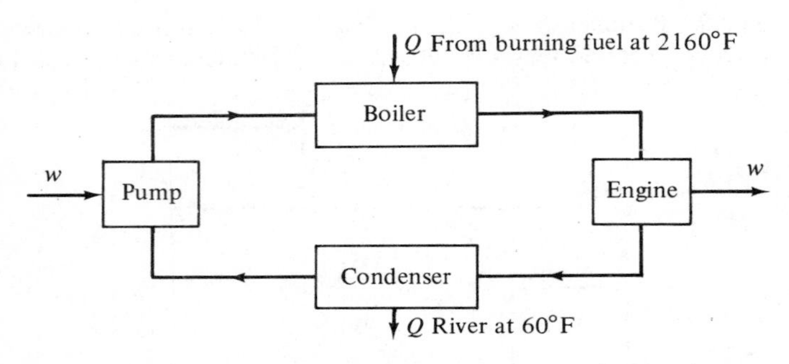

Figure P8-6

(a) Assuming that all steam lines are well insulated and that the work of expansion and of compression is adiabatic and reversible, compute for 1 lb_m of steam: (1) the engine work, (2) the heat discarded in the condenser, (3) the pump work, and (4) the heat absorbed in the boiler.

(b) Calculate the work produced per Btu transferred to the steam in the boiler. Compare this with the calculated work per Btu transferred in the boiler for a reversible Carnot engine receiving all of its heat at 2100°F and discharging its heat at 60°F.

(c) How many pounds of steam would be required per horsepower-hour of plant output and what would be the fuel requirement of the plant if it were delivering 10,000 hp while using fuel oil with a heating value of 19,000 Btu/lb_m and a furnace and boiler having an efficiency of 75 percent? What power would be required to drive the pump?

8-7. Steam enters a bleeder turbine at 21 bar and 371°C and exhausts at .05 bar absolute. The saturated condensate from the condenser is heated in an open feedwater heater before being returned to the boiler. Heating is accomplished with steam extracted from the 3.4 bar stage of the turbine. The pressure in the heater is 3.4 bar and the liquid leaving the heater is saturated at this pressure. Assume adiabatic reversible expansion of the steam in the turbine and negligible work to pump any liquid into the boiler. Determine the work produced by the turbine per joule of heat transferred to the steam in the boiler

(a) When using the feedwater heater as described above.

(b) When operating with no feedwater heater (that is, the condensate from the condenser is pumped directly to the boiler.)

How do you account for the difference in work for the two cases?

8-8. Figure P8-8 illustrates the HTGR (High Temperature Gas Reactor)[1] operating with the regenerative Brayton cycle and dry cooling (heat discharge is to the atmosphere). 1) Calculate the cycle efficiency for the conditions shown. 2) Calculate the maximum efficiency of a simple Brayton cycle operating between the same top temperature, 1500°F, and bottom temperature, 130°F without regenerative heating. 3) Calculate the ratio of compressor work required to power turbine work delivered for the two cycles.

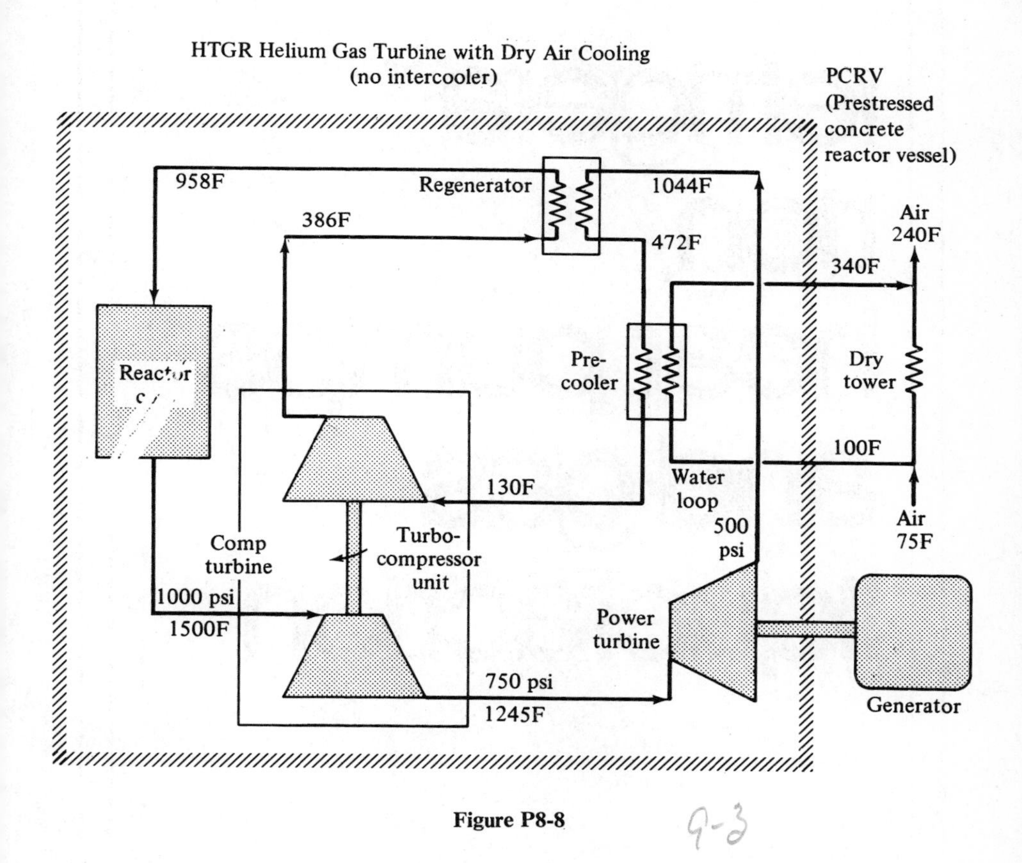

Figure P8-8

9-3

8-9. Compare the thermal efficiency of a Brayton cycle with intercooled compression with that of a simple Brayton cycle assuming no regenerative heating. How do you explain the lower efficiency for the intercooled cycle in spite of the reduced compressor work required for the intercooled cycle? If regenerative heating is utilized would you expect the results to differ? Justify your answer.

[1]The HTGR is a helium cooled nuclear reactor which can utilize the heated helium to generate steam for a conventional Rankine cycle or use the helium coolant as a working fluid in a Brayton cycle.

NINE

refrigeration, heat pumps, and gas liquefaction

Introduction

Emphasis in our discussion thus far has been placed on the conversion of thermal to mechanical energy and the thermodynamic constraints inherent in the process. While we have observed that it is always possible to convert mechanical energy to thermal energy completely, we also observed in our treatment of the second law of thermodynamics that thermal energy never flows spontaneously from any temperature level to a higher one. Such a transfer, which is required for any refrigeration application, requires the expenditure of work (or mechanical energy) in some manner if it is to occur.

Although the amount of energy required to meet society's refrigeration needs is still relatively small, it is among the fastest growing. Refrigeration for food processing and storage is now regarded as a basic need in modern societies. Air-conditioning is rapidly becoming a part of our space-conditioning needs. In recent years cryogenic processes have led to liquefied fuels and oxidizers for the space program and the basic oxygen steel-making process and the introduction of gas liquefaction for the transport and storage of methane. In the future, the potential of superconductivity as it applies to the transmission of electrical energy and the development of superconducting magnets for electrical generators, MHD, and fusion reactors will be realized only if efficient and economic cryogenic systems capable of sustaining temperatures approaching absolute zero are available.

In this chapter we discuss the thermodynamics of refrigeration systems and their applications to many of the above processes. We will also consider application of these cycles as heat pumps for multiplying the heating effect that one can accomplish with a given amount of electrical energy.

9.1 The Carnot Refrigeration Cycle

As indicated earlier, the Carnot cycle can be operated in the reverse direction from the cycles we have discussed earlier. By supplying work to the cycle, heat can be absorbed from a low-temperature sink and discharged to a high–temperature sink. In this fashion the Carnot cycle may be used to produce a refrigeration effect. The term *heat pump* is generally used when the heat effect of interest occurs at the high-temperature level (as in heating applications), and the term *refrigerator* is used when the heat effect of interest is at the low-temperature level (as in refrigeration or air-conditioning). Heat pumps have become of considerable interest as a means of increasing the efficiency of electrical heating in residential applications and will be discussed later in this chapter.

As with the heat engines, the Carnot refrigeration cycle represents the most efficient (Btu-of-cooling/Btu-of-work supplied) cycle and therefore is used as a standard against which we compare the more practical cycles that are actually used. The Carnot refrigeration cycle (or heat pump) consists of two isothermal and two isentropic processes and uses the same components as the heat engine, but operates in reverse sequence, as shown in Fig. 9-1.

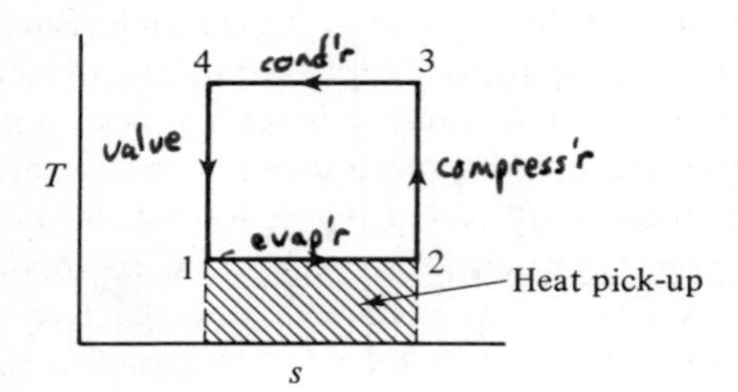

Figure 9-1. Carnot cycle.

The fluid picks up heat isothermally in a low-temperature heat exchanger (here assumed to be an evaporator), during which time its entropy is increased (step 1–2). The fluid's temperature is then increased by an isentropic compression (step 2–3). The fluid then discharges heat isothermally to a high-temperature sink and experiences an entropy reduction (step 3–4) that precisely offsets its increase at the lower temperature. The fluid is then expanded isentropically to the pressure and temperature at which it began the cycle (step 4–1). During this step reversible expansion work is extracted.

Application of the energy balance to the cycle, a closed, steady-state system, yields

$$W_{\text{net}} = Q_H + Q_L$$

or, on the basis of a unit mass of refrigerant flowing,

$$w_{\text{net}} = q_H + q_L \tag{9-1}$$

Similarly application of the entropy balance to the cycle yields for a unit mass of refrigerant:

$$\frac{q_H}{T_H} + \frac{q_L}{T_L} = 0 \tag{9-2}$$

q_H can be shown to be represented by the area under the curve corresponding to the fluid path through the high-temperature heat exchanger (step 3–4) by simply applying the entropy balance to that piece of equipment. Since step 3–4 occurs isothermally and reversibly,

$$\frac{q_H}{T_H} + s_{\text{in}} - s_{\text{out}} = 0 \tag{9-3}$$

or

$$q_H = T_H(s_{\text{out}} - s_{\text{in}}) = T_H \, \Delta s_H \tag{9-4}$$

Δs_{3-4} is negative, as is q_H, for a process where heat leaves the system at the high temperature. Similarly, $q_L = T_L \, \Delta s_L$ and is represented by the area beneath curve 1–2 in Fig. 9-1. q_L is positive, because heat is absorbed in the evaporator. Equation (9-1) indicates that the net work, w_{net}, required is the sum of q_H and q_L. Graphically this is represented by the area enclosed by the cycle in Fig. 9-1, because the areas beneath curve 1–2 exactly cancel. This result is precisely the same as observed for the Carnot heat engine with the *important exception* that the net area is *negative* ($|q_H| > |q_L|$ and $q_H < 0$) in the case of the heat pump. Whereas a heat engine's performance is improved by maintaining T_H as high as possible and T_L as low as possible, precisely the reverse is seen to be true for the Carnot refrigeration cycle or heat pump. Thus the most efficient refrigeration cycle would contract the area by raising T_L and decreasing T_H. However, T_L cannot exceed the temperature of the sink from which heat is to be extracted (or heat will not flow into the fluid), and T_H cannot be less than the sink to which heat must be discarded. Thus the ideal refrigeration cycle operates with T_H infinitesimally greater than the hot sink and T_L infinitesimally lower than the cold sink.

The *coefficient of performance* (COP) of a refrigeration cycle is defined as the amount of cooling produced per unit of work supplied to the cycle. We may develop an expression for the COP of a Carnot refrigerator by eliminating q_H between equations 9-1 and 9-2 to give

$$w_{\text{net}} = q_L\left(1 - \frac{T_H}{T_L}\right) \tag{9-5}$$

Rearrangement yields the coefficient of performance

$$\text{COP} = \frac{-q_L}{w_{\text{net}}} = \frac{-1}{1 - T_H/T_L} = \frac{T_L}{T_H - T_L} \tag{9-6}$$

Thus the coefficient of performance of a Carnot refrigerator can be expressed as a function of only the hot and cold sink temperatures. (It should be emphasized that these are *absolute* temperatures.)

9.2 Vapor Compression Refrigeration Cycles

Although it is possible in theory to build a gas-phase Carnot cycle, it would be possible to transfer finite amounts of heat to a fluid at constant temperatures only if the fluid flowed at an infinite rate or an isothermal work transfer occurred. We may avoid this difficulty by simply operating the cycle in a two-phase (gas–liquid) region as shown in Fig. 9-2. The working fluid, or refrigerant, boils (evaporates) at constant temperature and pressure as heat is absorbed. Heat is rejected in the condenser as the refrigerant condenses—again at constant temperature and pressure. Since boiling and condensing are excellent heat transfer mechanisms, these steps cause little

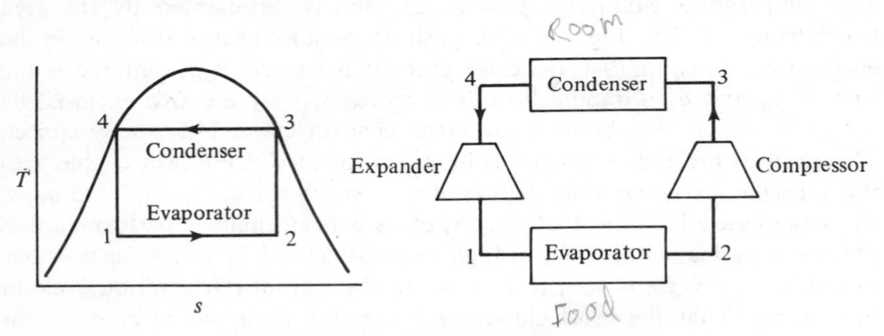

Figure 9-2. Two-phase Carnot refrigerator.

Examination of Fig. 9-2 indicates that both the compression and expansion steps in this Carnot refrigerator occur within the two-phase region. As was discussed for heat engines, the compression of a gas–liquid two-phase mixture is generally avoided. The problem can be eliminated by simply allowing the refrigerant to evaporate completely in the evaporator, producing a saturated vapor as shown in Fig. 9-3. Isentropic compression of the refrigerant to the condenser pressure then produces a superheated vapor at point 3 in Fig. 9-3. This superheating is necessary in order that the path from point 3 to point 3′ be a simple constant-pressure cooling step involving no work. This constant-pressure cooling occurs in the first portion of the condenser. Since this heat removal involves single-phase gas-heat exchange, it is a rather inefficient process and thus should be kept to a minimum.

Since the fluid passing through the expander is mostly a liquid, its specific volume is relatively low, so the amount of work produced by the expander is not appreciable. Thus little would be lost, in the way of efficiency, if the reversible expander were replaced by a simple isenthalpic throttling expansion. More importantly, the throttling expansion devices are much less

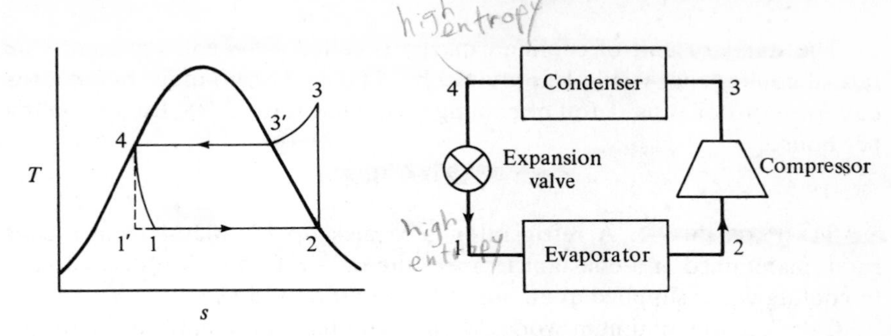

Figure 9-3. Vapor-compression refrigeration.

expensive than their work-producing counterparts and are essentially maintenance free. Consequently, most refrigeration cycles use a throttling expansion rather than the slightly more efficient work-producing expansion. Since the throttling device, either a fine capillary tube or an expansion valve, involves no moving parts, it is possible to operate in the two-phase region without experiencing serious operating problems.

The throttling process is irreversible and causes an increase in the fluid's entropy (path 4–1), as shown in Fig. 9-3. Less refrigeration is obtained in cycle 1–2–3–4 than in 1′–2–3–4 because of the reduced q_L. (The area under curve 1–2 is less than under 1′–2, but the net work required is still greater.) However, this sacrifice is compensated for by the practical considerations mentioned above, and the resulting cycle, the vapor-compression refrigeration cycle, is used in essentially all electrically driven home air conditioners, freezers, and refrigerators.

Since enthalpy values at the various points in the cycle are frequently needed in order to perform calculations on a cycle under consideration, it is common to prepare refrigeration-cycle plots on pressure-enthalpy, or *P–h* plots, as well as on *T–s* diagrams. A typical *P–h* diagram for the vapor-compression refrigeration cycle is illustrated in Fig. 9-4.

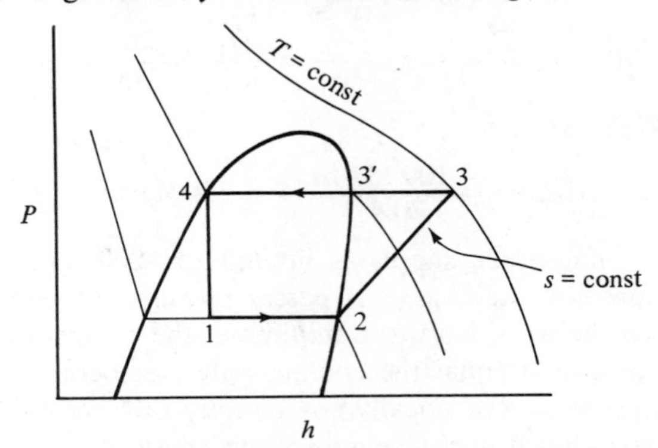

Figure 9-4. Vapor compression refrigeration cycle.

The standard unit of cooling capacity is called a *ton* and represents the rate of cooling that would be provided by 2000 lb_m (1 ton) of ice melting per day. In terms of Btus, 1 ton of cooling is equivalent to 12,000 Btu of cooling per hour:

$$1 \text{ ton} = 12{,}000 \text{ Btu/hr}$$

SAMPLE PROBLEM 9-1. A refrigerator is to pick up 100 MJ/hr from a cold room maintained at a constant temperature of −8°C, and will discard heat to cooling water supplied at an initial temperature of 5°C.

Calculate the minimum work (MJ/hr) required to accomplish the refrigeration for each of the following cases:

(a) If the supply of cooling water is unlimited.

(b) If the cooling-water supply is limited to a maximum of 1000 kg/hr.

(c) Same as (b), with additional specification that the refrigerant enter the condenser as a saturated vapor and leave as a saturated liquid. Assume that the pressure drop across the condenser is negligible.

(d) Same as (c), with the additional specification that a minimum "driving force" of 12°C be maintained in both the cold room and the condenser.

(e) Same as (d), with the additional specification that the adiabatic, reversible expander in the cycle be replaced with an adiabatic throttling valve. (*Note:* The outlet water temperature will be 50°C for this case.)

Solution: Since we are attempting to calculate the minimum work requirements of the refrigeration scheme, we shall assume that the cycle is reversible in all ways except as it would violate the statement of the problem.

(a) If the cooling-water supply is unlimited, we may assume that the high-temperature sink is at a constant temperature of 5°C = 278°K. The cold room is maintained at a constant temperature of −8°C = 265°K. The absolute minimum work requirements for the desired cooling would be that for a Carnot refrigerator working between the temperatures 265 and 278°K. The COP for a Carnot refrigerator between these temperatures would be

$$\text{COP} = \frac{T_L}{T_H - T_L} = -\frac{q_L}{w_{\text{net}}} = \frac{\dot{Q}_L}{\dot{W}_{\text{net}}} = \frac{265}{13} = 20.4$$

Since $\dot{Q}_L = 100$ MJ/hr

$$\dot{W}_{\text{net}} = \frac{-100 \text{ MJ/hr}}{20.4} = -5.4 \text{ MJ/hr.}$$

(b) If the cooling-water supply is limited to 1000 kg/hr, the cooling-water temperature will increase as it passes through the condenser. Thus, if we are to keep the work load to a minimum, the refrigerant temperature in the condenser should equal the cooling-water temperature at all points within the condenser. (Realistically, of course, this would be extremely difficult to achieve, but it might be approached by using a large number of small compressors and condensers, as shown in Fig. SP 9-1.)

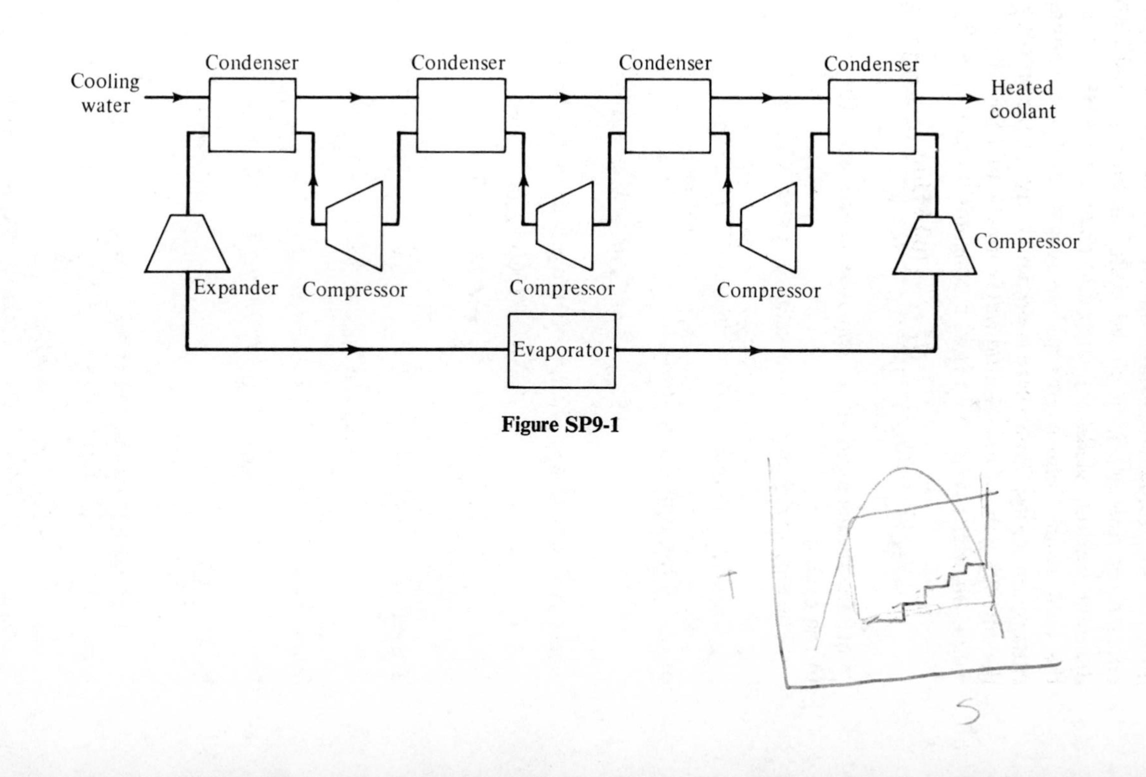

Cooling water
Condenser
Condenser
Condenser
Condenser
Heated coolant
Expander
Compressor
Compressor
Compressor
Compressor
Evaporator

Figure SP9-1

If the cycle is to be completely reversible, we know that the entropy of the universe will remain constant during the refrigeration process. Since the refrigerator operates in a closed cycle, its entropy does not change. Thus the only portions of the universe that change because of the refrigerator are the cold room, whose entropy decreases, and the cooling water, whose entropy increases. Since there is no net change in the entropy of the universe, the entropy gain of the cooling water must equal the entropy loss in the cold room. Since the cold room is at a constant temperature, T_L,

$$\Delta\dot{S}_{CR} = -\Delta\dot{S}_L = \frac{-\dot{Q}_L}{T_L} = \frac{-100 \text{ MJ/hr}}{265°\text{K}} = -0.377 \text{ MJ/hr°K}$$

(Note the minus sign, since heat is removed from the cold room but added to the cycle.)

The entropy change in the cooling water is given by

$$d\dot{S}_{H_2O} = \frac{\dot{M}C_P\,dT}{T}$$

or, assuming that $C_P = \text{const}$,

$$\Delta\dot{S}_{H_2O} = \dot{M}C_P \ln\frac{T_{out}}{T_{in}} = -\Delta\dot{S}_L$$

where $\Delta\dot{S}_L$ = entropy change of refrigerant in condenser. But

$$\dot{M} = 1000 \text{ kg/hr}$$
$$C_P = 4.182 \text{ kJ/°K kg}$$
$$T_{in} = 278°\text{K}$$

Therefore,

$$-\Delta\dot{S}_L = \Delta\dot{S}_{H_2O} = 4.18\left(\ln\frac{T_{out}}{278}\right) \text{ MJ/hr°K}$$

Since $\Delta\dot{S}_H + \Delta\dot{S}_L = 0$,

$$\Delta\dot{S}_H = -\Delta\dot{S}_L = 0.377 \text{ MJ/hr°K} = 4.18\left(\ln\frac{T_{out}}{278}\right) \text{ MJ/hr°K}$$

Solving for T_{out} gives

$$T_{out} = 304.4°\text{K}$$

The heat rejected to the cooling water is then given by an energy balance around the condenser:

$$\dot{Q}_H = -\dot{M}(\Delta h)_{H_2O}$$
$$= -\dot{M}C_P(304.4 - 278) = -110.4 \text{ MJ/hr}$$

The work required by the cycle is obtained from an energy balance around the whole cycle:

$$\dot{W}_{net} = \dot{Q}_H + \dot{Q}_L = -110.4 \text{ MJ/hr} + 100 \text{ MJ/hr}$$

or

$$\dot{W}_{\text{net}} = -10.4 \text{ MJ/hr}$$

so 10.4 MJ/hr of work must be supplied to the cycle.

(c) If we add the additional restriction that only one condenser be used, the high temperature in the refrigerator cycle is a constant and must be *no lower* than the maximum temperature of the cooling water. Thus, if T_H is the maximum temperature of the cooling water, it is also the condenser temperature of the refrigeration cycle. Since the refrigeration cycle can still operate as a Carnot cycle,

$$\frac{\dot{Q}_H}{T_H} = \frac{-\dot{Q}_L}{T_L} = -0.377 \text{ MJ/hr°K}$$

Therefore,

$$\dot{Q}_H = -T_H(0.377) \text{ MJ/hr}$$

But, from an energy balance around the condenser,

$$\dot{Q}_H = -\dot{Q}_{H_2O} = -\dot{M}C_P(T_H - 278) = -4.18(T_H - 278) \text{ MJ/hr}$$

Equating the two expressions for $\dot{Q}_H$ gives

$$-T_H(0.377) = -4.18(T_H - 278)$$

or, solving for T_H and $\dot{Q}_H$,

$$T_H = 305.5\text{°K}, \qquad \dot{Q}_H = -115.2 \text{ MJ/hr}$$

The net work is again obtained from an overall energy balance:

$$\dot{W}_{\text{net}} = \dot{Q}_H + \dot{Q}_L = (-115.2 + 100) \text{ MJ/hr} = -15.2 \text{ MJ/hr}$$

(d) We now add the requirement of at least a 12°C temperature difference everywhere in both the evaporator and the condenser. Thus the evaporator temperature is 12°C less than the cold room, or $-20\text{°C} = 253\text{°K}$.

The condenser temperature will be 12°C above the maximum temperature of the cooling water. Let the maximum temperature in the cooling water be T_W, the outlet temperature. The condenser temperature, T_H, will then be $T_W + 20$. We now proceed in the same way as in part (c):

$$\dot{Q}_H = -T_H\frac{\dot{Q}_L}{T_L} = -T_H\left(\frac{100}{253}\right) = -0.395\, T_H \text{ MJ/hr}$$

$$-\dot{Q}_{H_2O} = \dot{Q}_H = -\dot{M}C_P(T_W - 278) = -4.18(T_H - 290) \text{ MJ/hr}$$

Eliminating $\dot{Q}_H$ gives

$$-4.18(T_H - 290) \text{ MJ/hr} = -.395T_H \text{ MJ/hr}$$

Solving for T_H and Q_H then yields

$$T_H = 320.2\text{°K}, \qquad T_W = 308.2\text{°K}, \qquad \dot{Q}_H = -126.5 \text{ MJ/hr}$$

The work produced is given by

$$\dot{W}_{\text{net}} = \dot{Q}_H + \dot{Q}_L = (-126.5 + 100) \text{ MJ/hr} = -26.5 \text{ MJ/hr}$$

(e) If we are told the outlet water temperature, we can get $\dot{Q}_H$ from the relation

$$-\dot{Q}_{H_2O} = \dot{Q}_H = -4.18(T_W - 290°K)\ \text{MJ/hr}$$

independent of the cycle workings. Thus

$$\dot{Q}_H = -4.18(323 - 290)\ \text{MJ/hr} = -137.9\ \text{MJ/hr}$$

The work production is then given by

$$\dot{W}_{net} = (-137.9 + 100)\ \text{MJ/hr} = -37.9\ \text{MJ/hr}$$

so 37.9 MJ/hr of work must be provided to the cycle.

We can summarize the results of the five calculations as shown in Table SP 9-1.

Table SP 9-1

Case	Restrictions	Work Requirements, MJ/hr	C.O.P.
(a)	Pure Carnot cycle, $T_H = 278°K$, $T_L = 265°K$	5.40	18.5
(b)	Cooling water limited to 1000 kg/hr, $T_L = 265°K$	10.4	9.6
(c)	Limited cooling H_2O, single condenser, $T_L = 265°K$	15.2	6.4
(d)	Same as (c) but with 12°C ΔT in both condenser and evaporator	26.5	3.8
(e)	Same as (d) except use throttle instead of reversible expansion	37.9	2.6

A quick examination of the results of Sample Problem 9-1 serves to illustrate the extreme differences that can exist between the absolute minimum work requirements for a particular cooling job and the minimum requirements when more realistic operating conditions are assumed. In any real refrigeration scheme more work than suggested in part (e) would be required because of the mechanical inefficiencies associated with the compressor. With this factor taken into account the actual work requirements for this cooling load would be closer to 60 or 70 MJ/hr. Indeed, it is effects of just this type that cause the household air-conditioner (which, after all, pumps heat across only a 10 to 20°C temperature difference) to be so expensive to operate, when a simple analysis such as that of part (a) indicates a very small work requirement.

SAMPLE PROBLEM 9-2. A central home air-conditioner is being designed to provide 36,000 Btu/hr (or 3 tons) of cooling capacity. The air-conditioner will operate on a standard vapor-compression refrigeration cycle with throttling expansion. Freon 22 will be used as the working fluid. The evaporator temperature will be held at 50°F, and the condenser will operate at 140°F. Determine the flow rate of the working fluid, the compressor horsepower, and the total heat load of the condenser. Assume reversible compression.

Solution: If we assume that the compression is adiabatic and reversible, it is isentropic. If the throttling expansion is also adiabatic, then it is isenthalpic. If, in addition, we assume that the evaporator and condenser operate at constant pressure we can plot the proposed refrigeration cycle on a *P–h* diagram for Freon 22 as shown in Fig. SP 9-2.

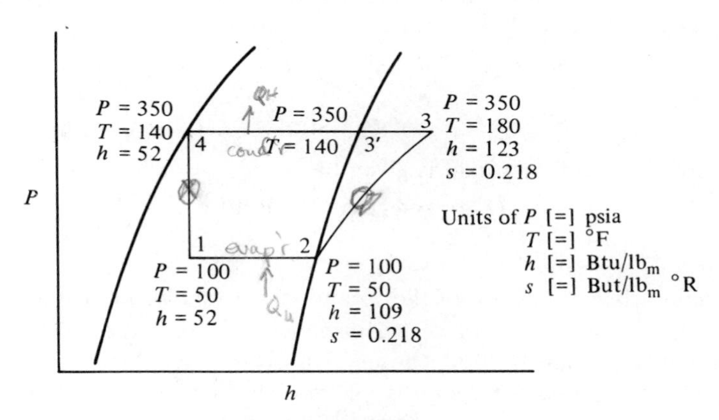

Figure SP9-2

The values of the various properties have been taken from the *P–h* diagram for Freon 22 which appears in Appendix D. With these values for the properties we are now in a position to evaluate the behavior of the various cycle components.

Evaporator: The heat absorbed by the working fluid in the evaporator can be obtained from an energy balance around the evaporator. Assuming negligible changes in potential and kinetic energy and steady-state operation, the energy balance reduces to

$$q_L = \Delta h_{\text{evap}} = h_2 - h_1 = 109 - 52 \text{ Btu/lb}_m$$

or

$$q_L = 57 \text{ Btu/lb}_m$$

Since the conditioner is to absorb 36,000 Btu/hr, the required flow rate of refrigerant is given by

$$\dot{M} = \frac{\dot{Q}_L}{q_L} = \frac{36{,}000 \text{ Btu/hr}}{57.0 \text{ Btu/lb}_m} = 632 \text{ lb}_m\text{/hr}$$

Compressor: The work production of the compressor can be obtained from an energy balance around the compressor. Assuming no changes in potential and kinetic energy and adiabatic steady-state flow, the energy balance reduces to

$$w = -\Delta h = -(h_3 - h_2) = -(123 - 109)\ \text{Btu/lb}_\text{m}$$

or

$$w = -14\ \text{Btu/lb}_\text{m}$$

The total work production is given by

$$\dot{W} = \dot{M}w = (632)(-14)\ \text{Btu/hr} = -8.85 \times 10^3\ \text{Btu/hr}$$

Since 1 hp = 2545 Btu/hr, the compressor motor will consume 3.47 hp.

Condenser: The total heat load of the condenser can be directly evaluated from an overall energy balance around the refrigerator:

$$\dot{W} = \dot{Q}_H + \dot{Q}_L$$

or

$$\dot{Q}_H = \dot{W} - \dot{Q}_L = -(8.85 \times 10^3 + 36 \times 10^3)\ \text{Btu/hr}$$

Therefore, the condenser load is given by

$$\dot{Q}_H = -44.9 \times 10^3\ \text{Btu/hr}$$

In our earlier discussion of the second law we showed that the efficiency (or, similarly, the coefficient of performance) of a Carnot cycle was independent of the working fluid chosen. Unfortunately, this statement is no longer true for the vapor-compression refrigeration cycle. Thus the choice of refrigerant can significantly affect the performance of a particular cycle. In addition to favorable thermodynamic properties, a good refrigerant should possess several other characteristics. It should be (1) nonflammable, (2) nontoxic, (3) noncorrosive, (4) nonexplosive, (5) not too expensive, and (6) have suitable vapor pressures at the condenser and evaporator temperatures so that neither extremely high nor extremely low pressures are involved.

When vapor-compression refrigeration was first introduced, sulfur dioxide, carbon dioxide, and ammonia were the commonly used refrigerants. However, none of these fluids proved entirely satisfactory, and they have been almost completely replaced by the halogenated (chlorine and fluorine) hydrocarbons–commonly known by the duPont trade name Freon®. The Freons® have proved to be nearly ideal refrigerants for many applications.

9.3 Heating with Heat Pumps

In our discussion thus far in this chapter our primary concern has related to using mechanical energy to remove heat from a low-temperature source and discard it to a high-temperature sink. Attention has been focused on maintaining the low-temperature condition for refrigeration or comfort purposes. In the wintertime when comfort conditions require heat input to a building, the heat pump could be used to elevate energy from the highest available temperature to a sufficiently high temperature level to heat a home. Such heat sources might include the air, the earth, or a solar-heated water storage system.

Typically, heating needs have been met by fossil fuels, primarily oil and gas in recent decades. However, electrical heating is becoming increasingly popular and will probably experience an even more rapid growth if adequate supplies of natural gas and oil are not available to meet the residential demand in the future. Since electricity can be produced from coal and nuclear energy, both of which are abundant domestically, electricity will likely supply an increasingly large fraction of our residential needs.

In electrically heated homes today the total heat load is supplied by the conversion of electrical energy (the most versatile and costly form of energy) to heat. If the electrical energy is used to operate a compressor in a heat pump cycle, the same amount of electrical energy consumption can deliver several times as much energy as heat at the temperature level required for comfort control.

In the case of the heat pump, the coefficient of performance is defined in terms of the heat delivered by the cycle at the high-temperature level.

$$\text{COP} = \frac{Q_H}{W} \tag{9-7}$$

For the Carnot cycle, equation (9-7) becomes

$$\text{COP} = \frac{Q_H}{W_{\text{rev}}}$$

therefore

$$\text{COP} = \frac{T_H}{T_H - T_L} \tag{9-8}$$

Equation (9-8) illustrates that the COP improves as T_L increases for a given T_H. Since $|W_{\text{rev}}| < |W|$, realistic cycles will always operate below the reversible COP. The actual performance of the cycle depends on the amount of heat-exchanger surface available at both the high- and low-temperature levels. The greater the area, the closer T_H and T_L can approach one another and the higher the COP. Cycle optimization becomes a matter of balancing the initial cost of heat-exchanger surface with the higher operating costs for larger differences in temperature levels.

Sample Problem 9-1 (part e) illustrates that 1 unit of electrical energy can be used to pick up about 2.7 units of thermal energy at a temperature of about −8°C and discharge 3.7 units as heat at an average temperature of 27°C. Such a system would have a COP of 3.7 and provide almost four times better efficiency in the use of electrical energy than if it were dissipated directly as heat. Such schemes have been considered in the past, but the cost to install the heat pump and heat-source coils made it less attractive than other alternatives on an initial-cost basis.

The heat pump can also be used to air-condition a residence in the summer as shown in Fig. 9-5. The air entering the dwelling is cooled by supplying the energy necessary to vaporize the working fluid, presumatly from the surroundings, and the compressor work requirements. In this mode the heat exchanger which heats air going to the house replaces the condenser.

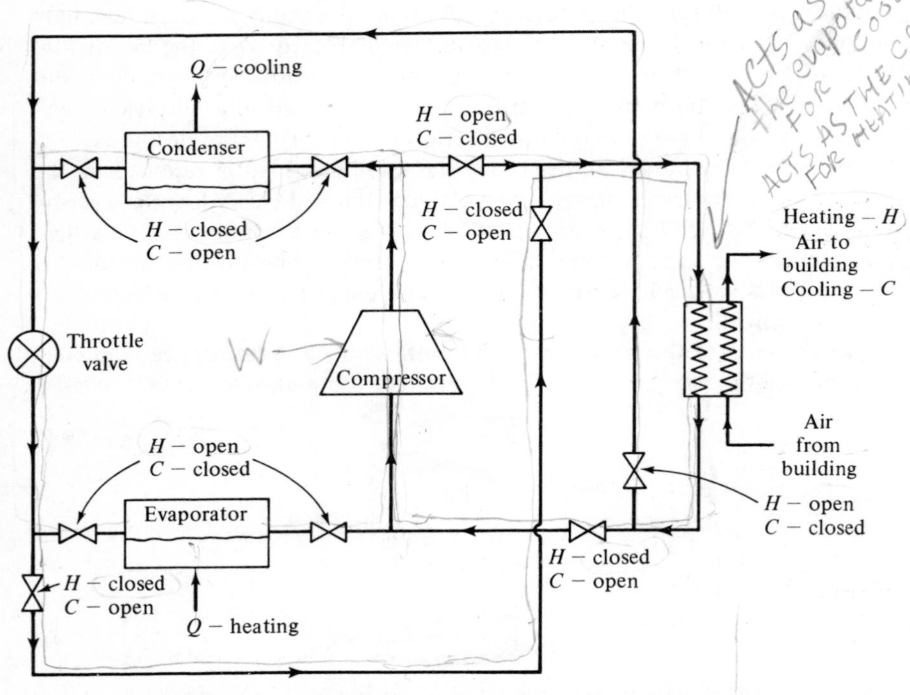

Figure 9-5. Schematic diagram of heat pump for heating (H) and cooling (C).

The cycle discharges heat to the atmosphere in the high-temperature heat exchanger (condenser) in summer operation and the heat exchanger which cools the circulating air replaces the evaporator. Thus the investment in the heat pump can provide year-round comfort control.

With air-conditioning rapidly increasing in regions where heating needs are appreciable during the winter months and with energy costs which have remained relatively low in the past increasing significantly, heat pumps should become more economically attractive. Clearly, from an energy conservation point-of-view they are already attractive compared with straight electrical heating.

9.4 Absorption Refrigeration Cycles

The compressor work requirement for a vapor compression refrigeration cycle is given by the expression

$$w = -\int v \, dP$$

Since v is relatively large for a vapor phase, the compressor energy consumption is substantial. Absorption systems using a refrigerant, such as ammonia, circumvent the large work requirement by first absorbing the refrigerant vapor at low pressure in a liquid solvent, and then pumping the liquid phase to the high-pressure side of the cycle. Since the liquid specific volume is typically several orders of magnitude lower than that of refrigerant vapors, the work requirements for pumping the liquid phase are much less, as was discussed in Chapter 7.

A schematic diagram of the ammonia absorption refrigeration system is shown in Fig. 9-6. As can be seen, the reduced work requirement is offset in part by the need for heat addition to the stripper, Q_S, where the ammonia is driven out of solution and becomes a vapor again, this time at high pressure and temperature. The important consideration from an energy point-of-view is that thermal energy with a much lower temperature potential can be used to supply the stripper heat requirements. It is more easily met than work requirements and with a considerably lower overall energy expenditure when one considers the efficiency of converting chemical or thermal energy to the mechanical form required by the pump or compressor. For example, low-pressure steam or solar-derived energy, which would otherwise be uneconomical for producing power to drive a compressor, could meet the stripper needs.

While energy savings can be realized, particularly if thermal energy is readily available at temperature levels of 100–200°C (for NH_3–H_2O systems), the absorption system is more complex and requires an absorber, stripper, and heat exchanger in addition to the components required for the vapor-compression system. Consequently, the size and cost of these systems are larger than comparably sized vapor-compression systems. They are typically used for industrial applications, as opposed to meeting residential refrigeration or air-conditioning needs. However, the interest in solar heating and cooling of buildings has generated considerable interest in smaller, more reliable absorption refrigeration equipment. It should be remembered that early refrigerators used gas-fired absorption refrigeration technology, but the heat was supplied at a higher temperature level.

The coefficient of performance, COP, defined previously is not an appropriate measure of the performance of an absorption system. The energy

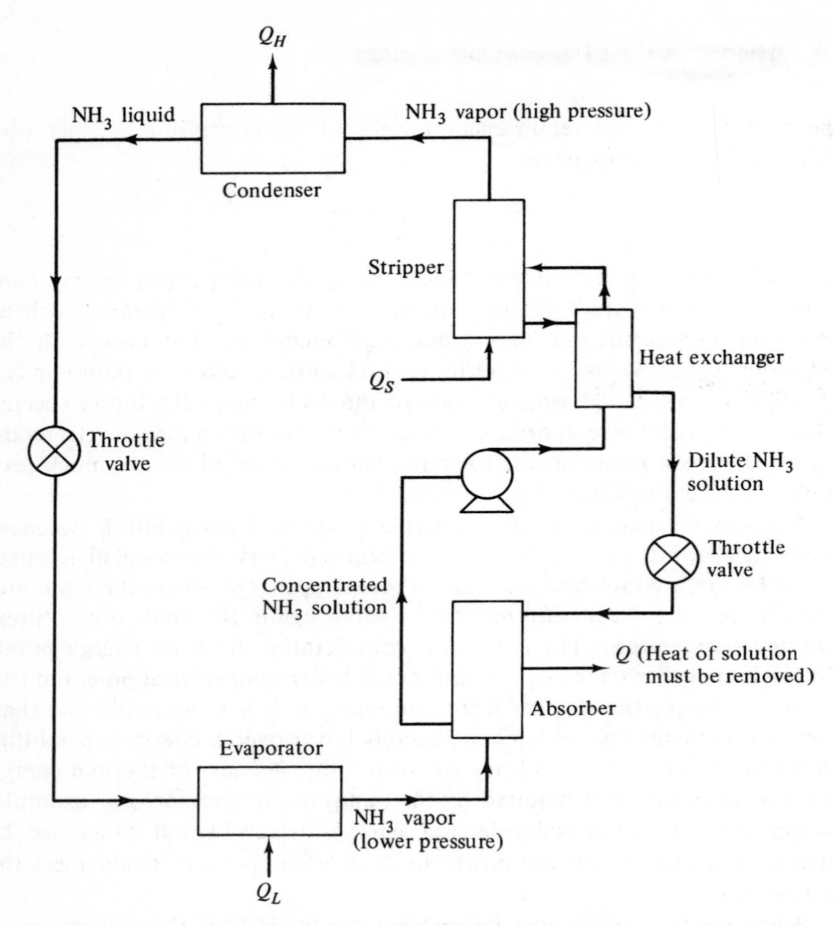

Figure 9-6. Absorption refrigeration system.

requirements as work are almost trivial compared to the heat requirements. One might thus define a modified coefficient of performance as

$$\text{COP}^* = \frac{q_L}{q_S} \tag{9-9}$$

where q_L represents the energy absorbed by 1 lb of the refrigerant (primarily NH_3 in the cycle discussed) in the evaporator. q_S represents the heat addition required in the stripper to liberate 1 lb of NH_3 vapor from the stripper. q_L is essentially the latent heat of vaporization of the NH_3 at the evaporator pressure, and we can determine q_S by applying the energy balance to the stripper:

$$q_S = (h)_{\text{HN}_3} + (mh)_{\text{dil NH}_3} - (mh)_{\text{conc NH}_3} \tag{9-10}$$

q_S represents the stripper heat requirements per unit mass of NH_3 released, h_{NH_3} is the specific enthalpy of NH_3 at the exit conditions of the stripper, $(mh)_{dil\ NH_3}$ is the enthalpy of the dilute ammonia solution at its exit conditions per unit mass of NH_3 vaporized, and $(mh)_{conc\ NH_3}$ is the enthalpy of the concentrated solution entering the stripper per unit mass of NH_3 vaporized.

The treatment of multicomponent systems in Chapter 10 should aid the student in making such calculations. Physical property information is presented and discussed and sample problems on absorption refrigeration (SP 10–7 and 8) are included.

9.5 Cascade Cycles

The recent development of the cryogenics industry has spurred great interest in the attainment of very low temperatures. Examination of the thermodynamic properties of the commonly used refrigerants quickly shows that it is not possible to achieve these extremely cold temperatures (less than −300°F) in a single vapor-compression cycle (practically any refrigerant that did not freeze in the evaporator would be above its critical point in the condenser). These problems can be avoided by means of a *cascade cycle*, as shown in Fig. 9-7. Although cascade cycles can theoretically be built with any number of steps, three or four appear to be the maximum practical limit. Note the similarity of the cascade cycle to the binary power cycle we discussed previously.

Cycle III would use a refrigerant with a normal boiling point around the lowest temperature to be achieved. The heat is absorbed in the evaporator of cycle III. The condenser of cycle III rejects its heat to the evaporator of

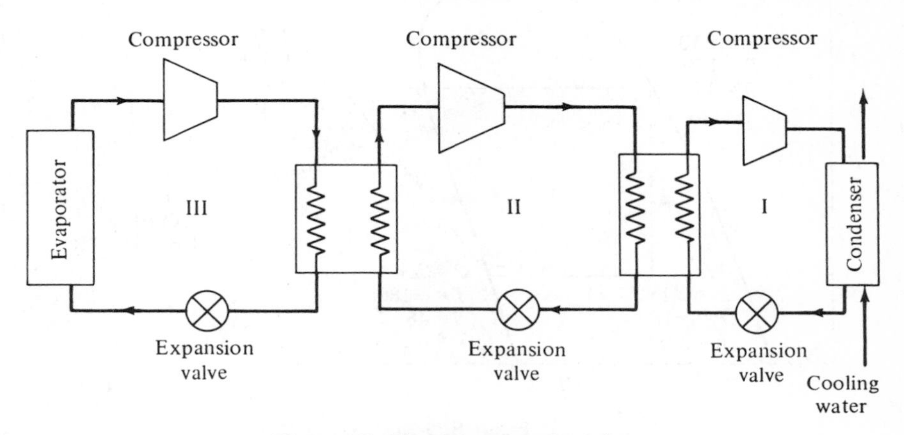

Figure 9-7. Three-step cascade cycle.

cycle II. The refrigerant in cycle II would be one with a normal boiling point about two thirds of the way from ambient to the evaporator temperature in cycle III. The condenser for cycle II discharges its heat into the evaporator of cycle I, which finally rejects the heat to cooling water or some other convenient thermal sink. The refrigerant for cycle I would be one with a normal boiling point about halfway between the cooling-water temperature and evaporator temperature in cycle II. In a three-step cascade cycle designed to provide refrigeration for the liquefaction of natural gas the refrigerants would commonly be: cycle I, propane; cycle II, ethylene; and cycle III, methane.

SAMPLE PROBLEM 9-3. A plant for processing quick-frozen poultry requires the freezing chambers to be held at a temperature of −160°F. A two-stage cascade refrigeration unit is being designed to provide the required cooling. The low-temperature evaporator is to operate at −180°F; the high-temperature condenser will be at 100°F. The low-temperature cycle will use Freon 503 as its working fluid and will discharge its heat of condensation at −30°F. The high-temperature cycle will use Freon 22 as the refrigerant and will absorb heat at −50°F. If the refrigerator is to absorb 20,000 Btu/hr from the freezing chamber, determine the compressor horsepower for both cycles. You may assume that the compressions are adiabatic and reversible and that the expansions are isenthalpic.

Solution: We may plot the low-temperature cycle on a *P–h* diagram (assuming isentropic compression and isenthalpic expansion) for Freon 503 as shown in Fig. SP 9-3a. The heat absorbed in the evaporator per lb_m of

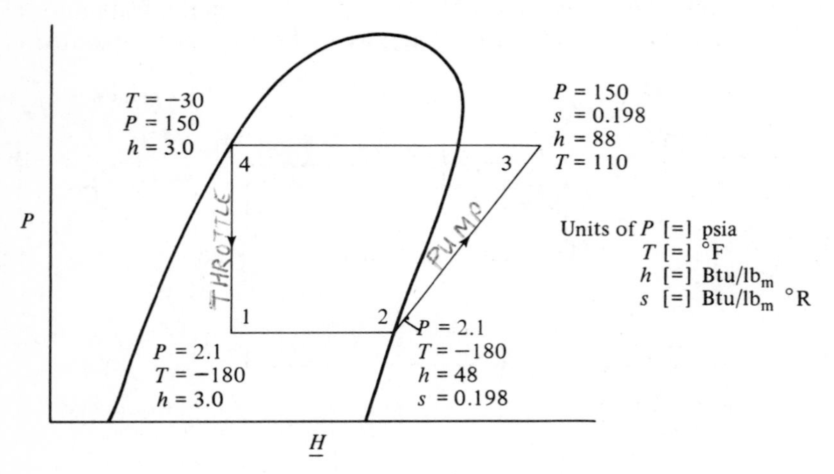

Figure SP9-3a

refrigerant is then given by

$$q_L = \Delta h_L = (48 - 3.0) \text{ Btu/lb}_m = 45 \text{ Btu/lb}_m$$

Since it is desired to absorb 20,000 Btu/hr the refrigerant rate in this cycle is given by

$$\dot{M} = \frac{\dot{Q}_L}{q_L} = \frac{20{,}000 \text{ Btu/hr}}{45 \text{ Btu/lb}_m} = 445 \text{ lb}_m\text{/hr}$$

The compressor load in this cycle is given by

$$-w_C = \Delta h_L = (88 - 48) \text{ Btu/lb}_m = 40 \text{ Btu/lb}_m$$

The total compressor load is then

$$-\dot{W}_C = -w_C\dot{M} = (40)(445) \text{ Btu/hr} = 17{,}800 \text{ Btu/hr}$$

or, converting to horsepower,

$$-\dot{W}_C = 7.0 \text{ hp}$$

The heat rejected in the low-temperature condenser is then given by

$$-q_H = -\Delta h_H = (88 - 3.0) \text{ Btu/lb}_m = 85 \text{ Btu/lb}_m$$

and

$$-\dot{Q}_H = \dot{M}q_H = (85)(445) \text{ Btu/hr} = 37{,}800 \text{ Btu/lb}_m$$

We can now attack the high-temperature Freon 22 cycle. The P–h diagram for the cycle (again assuming isentropic compression, and isenthalpic expansion) is as shown in Fig. SP 9-3b. The heat absorbed in the evaporator is given by

$$q_L = \Delta h_L = (99 - 39) \text{ Btu/lb}_m = 60 \text{ Btu/lb}_m$$

However, the total heat absorbed in this evaporator equals the heat discharged by the low-temperature condenser-37,800 Btu/hr. Thus the flow rate

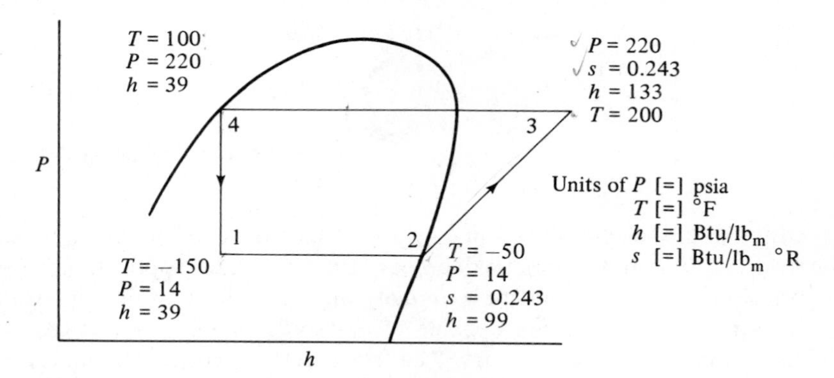

Figure SP9-3b

of refrigerant in the Freon 22 cycle is given by

$$\dot{M} = \frac{\dot{Q}_L}{q_L} = \frac{37{,}800 \text{ Btu/hr}}{60 \text{ Btu/lb}_\text{m}} = 630 \text{ lb}_\text{m}\text{/hr}$$

The work supplied to the compressor is given by

$$-w_C = \Delta h_C = (133 - 99) \text{ Btu/lb}_\text{m} = 34 \text{ Btu/lb}_\text{m}$$

Thus the total work requirement of the compressor is given by

$$\dot{W}_C = w_C\dot{M} = (-34)(630) \text{ Btu/hr} = -21{,}400 \text{ Btu/hr}$$

Therefore, the Freon 22 compressor horsepower is

$$-\dot{W}_C = 8.4 \text{ hp}$$

The total compressor power necessary is the sum of the two compressor requirements and equals 15.4 hp.

9.6 Liquefaction of Gases—Cryogenic Temperatures

The first successful liquefaction of a "permanent gas" was achieved by a Frenchman, Georges Claude, who liquefied air around the turn of the century. Claude's cycle used air as the working fluid and consisted of an isothermal compression of the gas to a high pressure, followed by isentropic expansion to a low pressure. During the isentropic expansion the gas temperature decreases and finally the gas condenses, as shown in Fig. 9-8.

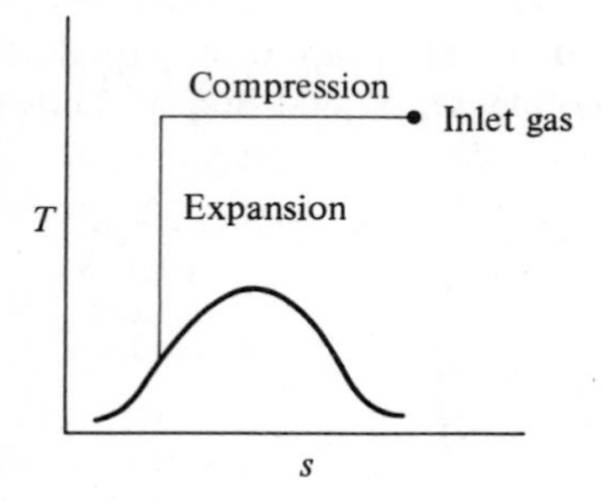

Figure 9-8. Claude cycle for air liquefaction.

Although the scheme shown in Fig. 9-8 is the most efficient gas-liquefaction process, it is not a practical process for large-scale applications. The compression pressures would be extremely high. In addition, no expansion devices are available that are capable of efficiently handling the degree of expansion that would be necessary. The first practical scheme for liquefying gases was developed in the early 1900s in Germany by Carl von Linde. The basic Linde process is shown in Fig. 9-9. The incoming gas is mixed with the

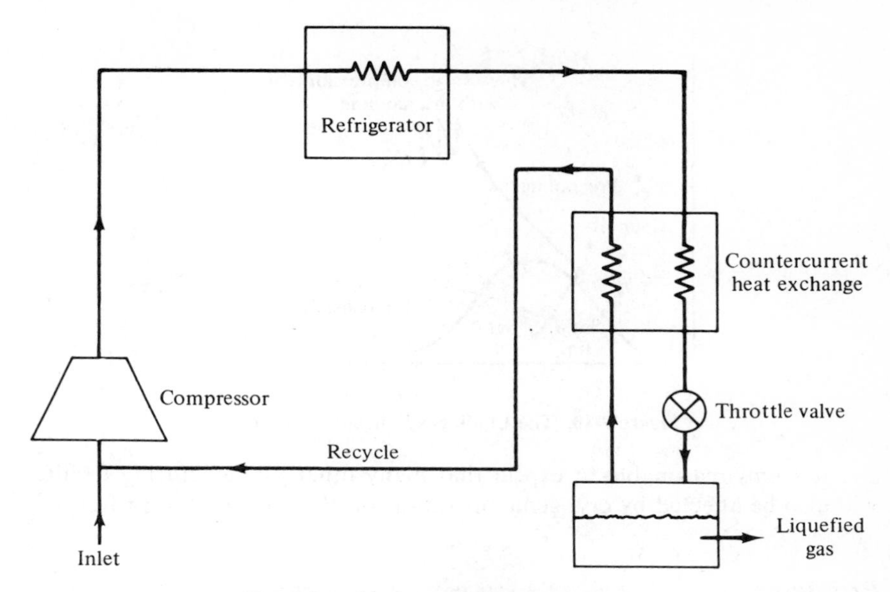

Figure 9-9. Linde gas-liquefaction process.

returning unliquefied gas, and both are compressed to about 1000 psia. The compressed gas is precooled by an external refrigeration scheme. The precooled gas is further cooled by countercurrent heat exchange with the unliquefied gas. After the heat exchanger the gas passes through a throttling Joule–Thomson expansion, where the gas temperature is reduced and the gas partially liquefied. The liquid is removed and the gas returned through the heat exchanger. The Joule–Thomson expansion requires no moving parts and hence can easily be performed over large pressure ratios. However, the Joule–Thomson expansion will provide the desired temperature decrease only if the gas is first cooled to near its critical temperature, thus the need for the precooling and countercurrent heat exchange. A T–s plot of the Linde process is shown in Fig. 9-10.

Gas liquefaction has become of increasing commercial importance since the 1950s, owing partially to its use in the space program but even more to the natural gas industry. Large quantities of natural gas are now being transported in the liquefied state by tankers from gas fields to areas of high consumption. Storage facilities, both above and below ground, are being used by suppliers and users of liquefied natural gas to store the fuel. To many users, liquefied natural gas supplies large amounts of refrigeration capacity as it is returned to the gaseous state prior to use as a fuel or a raw material.

The recently commercialized basic oxygen steelmaking process makes use of large quantities of purified oxygen obtained from the distillation of liquefied

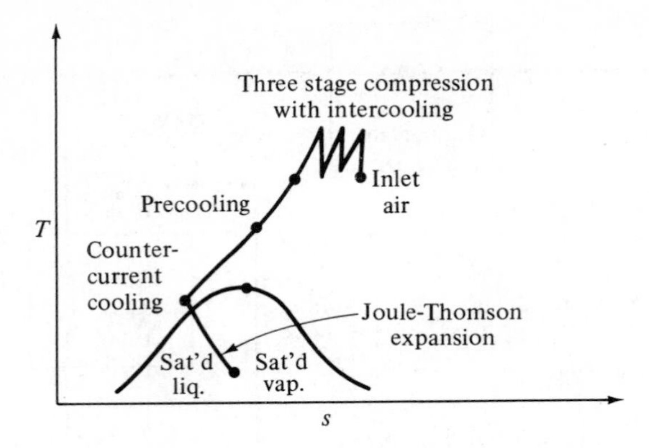

Figure 9-10. The Linde gas liquefaction process.

air. It seems reasonable to expect that many other manufacturing facilities will also be affected by cryogenic operations in the not-too-distant future.

Problems

9-1. (a) In a certain deep-freeze unit the heat transferred into the cold chest from the surroundings is 3000 Btu/day. The chest is maintained at 0°F while the surroundings are at 80°F. What is the power rating (in horsepower) of the smallest motor that can be used to drive an ideal refrigeration cycle to hold the chest at 0°F?

(b) Suppose the cycle uses Freon 12 as the working fluid and has a condenser pressure of 172.35 psia and an evaporator pressure of 19.189 psia. What is the horsepower rating of the smallest motor that can be used to drive this cycle?

(c) The actual refrigeration cycle shown in Fig. P9-1 is to be used to maintain

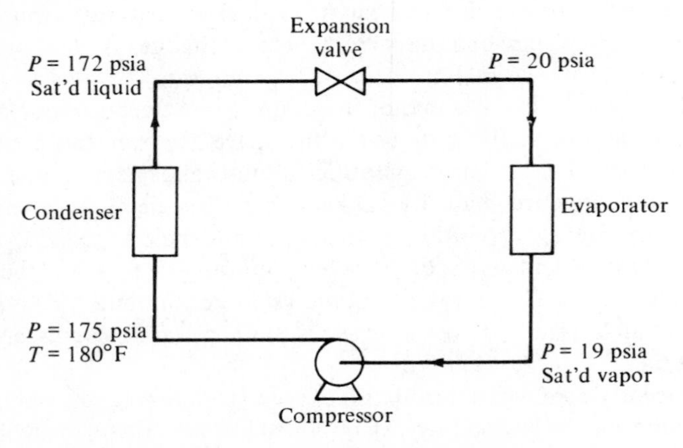

Figure P9-1

the chest at 0°F. (1) What circulation rate (lb_m/hr) of Freon 12 is required? (2) What is the smallest motor that could be used to drive the pump? (Give the horsepower rating.)

9-2. Water is supplied at 60°F for cooling and for ice manufacture. Ice is to be delivered at 20°F. The cooling water is heated from 60 to 80°F.

(a) Compute the minimum theoretical horsepower required for an ideal reversible operation producing 5 tons of ice per hour under the above conditions.

(b) Draw on the *T–s* diagram the path followed by the water in forming ice and by the cooling water under the above conditions.

In an actual vapor compression plant using ammonia as a refrigerant the conditions are as given in Fig. P9-2.

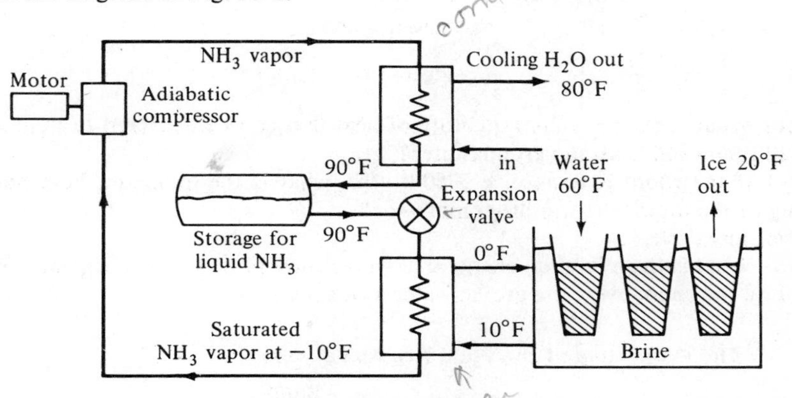

Figure P9-2

(c) Compute the horsepower required by the ammonia compressor for producing 5 tons of ice per hour. Assume the mechanical efficiency of the compressor to be 80 percent and adiabatic compression of the ammonia vapor.

The Saturated Properties of Ammonia Are As Follows:

		Saturated Liquid		Vapor	
P, psia	*T*, °F	*h*, Btu/lb_m	*s*, Btu/lb_m °R	*h*, Btu/lb_m	*s*, Btu/lb_m °R
23	−10	32.11	0.0738	608.5	1.3558
180	90	143.3	0.2954	432	1.1850
180	248			735.3	1.3558

$(C_P)_{brine} = 0.8$ Btu/lb_m°F, $(C_P)_{water} = 1.0$ Btu/lb_m°F, $(C_P)_{ice} = 0.5$ Btu/lb_m°F

Latent heat (Δh) of melting ice at 32°F = 143.8 Btu/lb_m.

9-3. A heat pump is being used to maintain a room at 70°F by removing heat from ground water and discharging heat to the room. For the cycle shown in Fig. P9-3,

(a) What is the quality of the fluid leaving the evaporator?

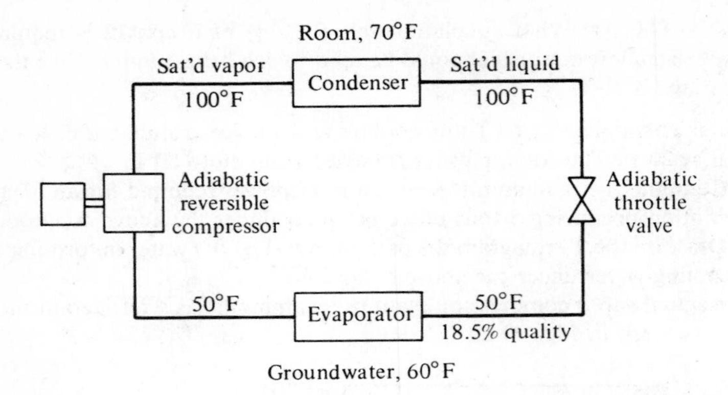

Figure P9-3

(b) What is the maximum quantity of heat that can be delivered to the room per Btu removed from the groundwater?

(c) If the room is to receive 5750 Btu/hr, what is the minimum horsepower rating of the motor driving the compressor?

For any cycle,

(d) What is the minimum energy that would have to be supplied to pump 5750 Btu into the room using the groundwater as a source?

The Properties of the Fluid Are As Follows:

T, °F	P_{sat}, psia	$s_{sat\ liq}$ (Btu/lb$_m$ °R)	Δs_{vap} (Btu/lb$_m$ °R)	$s_{sat\ vap}$ (Btu/lb$_m$ °R)
50	81.4	0.0413	0.1266	0.1679
98	128.0	0.0623	0.1036	0.1659
100	131.6	0.0632	0.1026	0.1658
102	135.3	0.0640	0.1018	0.1658

9-4. In a preliminary cooling step of an air liquefaction plant a stream of air at atmospheric pressure and -15°C enters a well-insulated compressor. The compressor discharges the air at 8.5 bar to a heat exchanger cooled by the ambient air surroundings. The heat exchanger operates with negligible pressure drop so that the compressed air leaves at 8.5 bar and 40°C. The cooled compressed air is then passed to a well-insulated turbo expander which discharges at atmospheric pressure. The expanded air goes through a second heat exchanger (also isobaric), where it picks up heat from a low-temperature stream in the liquefaction process. The expanded air leaving the second heat exchanger is now at -15°C, in which condition it goes into the compressor to go through the cycle again. The compressor and expander are 80 percent efficient (based on reversible machines operating over the same pressure ranges). Air may be assumed to behave as an ideal gas with a constant-pressure heat capacity of 6.97 cal/gm-mol °K. If the work derived from the turbo expander is used to help drive the compressor, determine the net work for the above cycle per unit of heat picked up in the low-temperature heat exchanger. Sketch this cycle on a T–s diagram and compare it with a completely reversible cycle performing the same job.

9-5. A conventional refrigerator uses Freon 12, dichlorodifluoromethane, as its working fluid. In the cycle of operations saturated vapor at $-10°F$ enters an insulated compressor whose compression ratio is 9 : 1 (that is, the outlet pressure is 9 times the inlet pressure). The compressor is 70 percent efficient based on an adiabatic reversible compression over the same range of pressure. The compressed gas from the compressor is cooled and condensed isobaricly to a saturated liquid. The saturated liquid is passed through an insulated throttle valve whose downstream pressure corresponds to a saturation temperature of $-10°F$. The mixture from the throttle valve goes to the evaporator, where it absorbs just enough heat from the ice trays and the interior of the refrigerator box to become saturated vapor, which enters the compressor and repeats the cycle. Under summertime conditions (greatest load) it is expected that 110 Btu/min will have to be absorbed in the evaporator.

(a) What is the rate of circulation of Freon 12 in lb_m/min?

(b) How large a motor (rated in terms of its output) should be specified for the compressor?

(c) If the throttle valve were replaced with a turbo expander whose efficiency is 60 percent (based on adiabatic reversible expansion over same pressure range), and the work from this expander used to help drive the compressor, how large a motor would be needed to supply the remaining work to the compressor?

(d) If a perfect heat pump was operating between the same evaporating and condensing temperatures, how large a motor would it require?

9-6. The refrigeration cycle shown at the left in Fig. P9-6 has been proposed to replace the more conventional cycle at the right. If the circulating fluid is CO_2 and

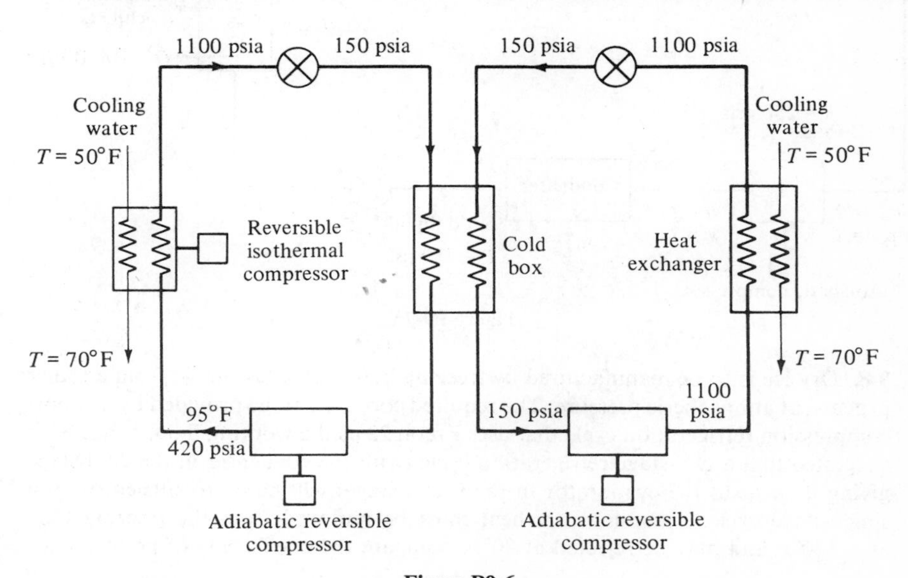

Figure P9-6

the temperatures entering and leaving the cold box are to be the same for both systems,

(a) Calculate the heat removed from the cold box per pound of CO_2 circulated through the proposed cycle (conventional cycle not operating).

(b) What is the total work input (Btu per pound of CO_2 circulated) to the proposed cycle?

(c) What is the work input (Btu per pound of CO_2 circulated) to the conventional cycle?

(d) Sketch the shape of each cycle on a T–s diagram.

(e) Why is one cycle more efficient than the other?

(f) What is the absolute minimum work of refrigeration for the given conditions entering and leaving the cold box and the given inlet and outlet cooling water temperatures (Btu per pound of CO_2 circulated)?

9-7. For the refrigeration flow diagram of Fig. P9-7, using ammonia, compute the values indicated using

(a) A chart of ammonia properties.

(b) The generalized charts. The compressor may be assumed to be adiabatic and reversible, although an efficiency of 70 to 80 percent would be reasonably expected for real operating equipment.

$$C_P^* = 6.70 + 0.0035T(°\text{R})\ \text{Btu/lb-mol °R}$$

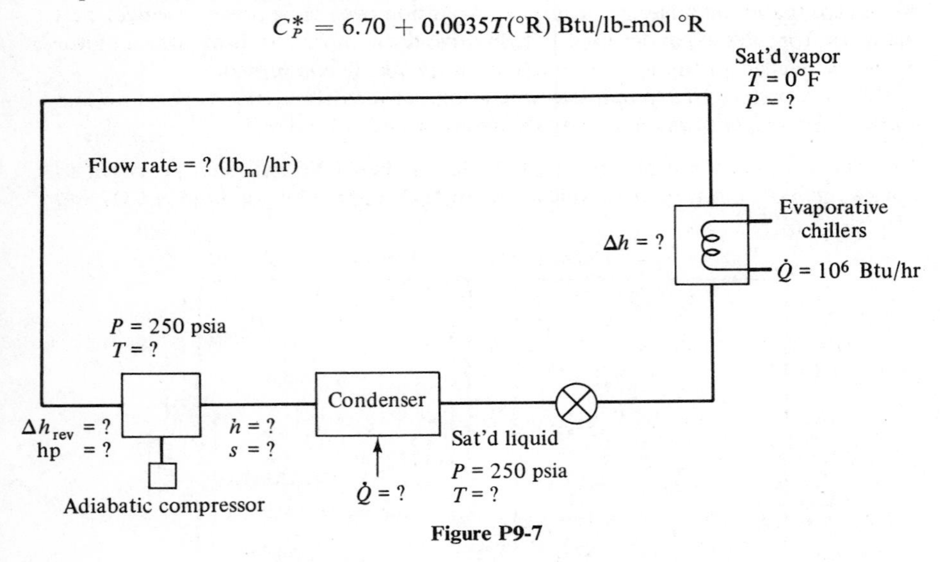

Figure P9-7

9-8. Dry ice is to be manufactured by freezing liquid CO_2 (available from another process) at atmospheric pressure. The required cooling is to be provided by a vapor-compression refrigeration cycle that uses Freon 22 as the working fluid. It has been suggested that a two-stage refrigeration cycle (with the condenser in the cold stage giving its heat to the evaporator in the warm stage) will be more efficient than a single-stage cycle. Assuming that heat must be removed from the freezing CO_2 at -130°F and may be rejected at 80°F, compare the coefficients of performance

(Btu absorbed from cold source/Btu of work supplied to refrigerator) for a two-stage and a one-stage refrigeration cycle.

(*Note:* Assume isenthalpic expansions for both cycles. You may base your two-stage calculations on a temperature of $-20°F$ in the coupled evaporator-condenser.)

9-9. A refrigerator operates with carbon dioxide as the working fluid. Saturated vapor from the evaporator coils is compressed adiabatically and reversibly from 150 to 800 psia. From the compressor the vapor passes to a condenser, where it becomes a saturated liquid at 800 psia. The liquid is then expanded through a throttle valve to 150 psia, in which condition it enters the evaporator. Use the information below to calculate the work required per Btu of heat picked up at the low temperature in the evaporator.

1. For CO_2,

$$P = \frac{RT}{v-b} + \frac{A_2 + B_2T + C_2 \exp(-kT)}{(v-b)^2} + \frac{A_3 + B_3T + C_3 \exp(-kT)}{(v-b)^3} + \frac{A_4}{(v-b)^4} + \frac{A_5 + B_5T + C_5 \exp(-kT)}{(v-b)^5}$$

where for P in psia, T in °R, and ρ in lb_m/ft^3, the constants are

$$\begin{aligned} R &= 0.24381, & C_3 &= 4.705805 \\ b &= 0.007495, & A_4 &= -2.112459 \times 10^{-3} \\ A_2 &= -8.9273631, & A_5 &= 7.017835 \times 10^{-6} \\ B_2 &= 5.262476 \times 10^{-3}, & B_5 &= 1.023511 \times 10^{-8} \\ C_2 &= -150.97587, & C_5 &= -4.55437 \times 10^{-4} \\ A_3 &= 0.17948621, & k &= 0.01 \\ B_3 &= -5.770542 \times 10^{-5}, & & \end{aligned}$$

2.

$$C_V^* = \alpha + \frac{\beta}{T} + \frac{\gamma}{T^2}$$

where for C_V^* in Btu/lb-mol °R and T in °R,

$$\alpha = 14.214, \qquad \beta = -6.53 \times 10^3, \qquad \gamma = 1.14 \times 10^6$$

3. The vapor pressure and saturated liquid volume data may be taken from R. H. Perry et al., *Chemical Engineers' Handbook* (McGraw-Hill Book Company, Inc., New York, 4th ed., 1963).

Check your answer by using the tables of thermodynamic properties for CO_2.

9-10. Carbon dioxide is available at 60°F and 1200 psia. It is proposed to make solid carbon dioxide from it by flash evaporation to 15 psia. If the CO_2 is passed through a valve and the gas allowed to escape freely into the surroundings, and if the solid particles are caught in a bag,

(a) What will be temperature of the solid?

(b) What fraction of CO_2 will be recoverable as solid?

9-11. Suppose the Linde process shown in Fig. P9-11 is used in Problem 9-10.

(a) What is the advantage, if any, of circulating the gas from the snow chamber over the coil?

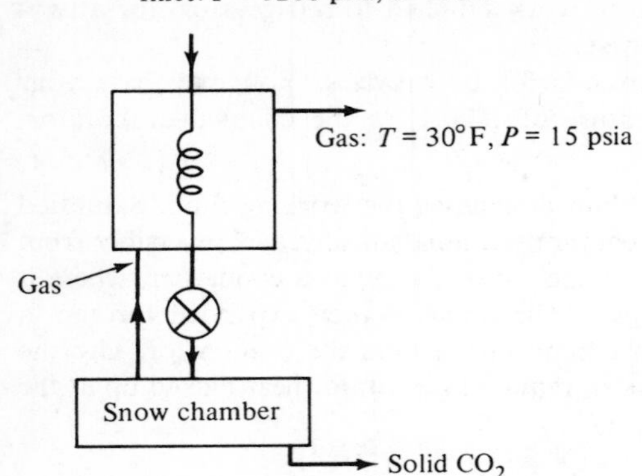

Figure P9-11

(b) What fraction of the liquid entering the coil shows up as a solid? Assume no extraneous heat losses.

9-12. Calculate the compressor size (horsepower) required for a heat pump (vapor compression cycle) to maintain a temperature of 22°C year-round in a climate where winter temperatures fall to −20°C and summer temperatures reach 35°C. Use cycle temperatures which allow for temperature differences of at least 10°C in all heat exchangers and a compressor efficiency of 80%. Assume a peak heat load of 250,000 Btu/hr. in the summer and peak load of 500,000 Btu/hr in the winter. If electrical energy costs 30 cents per kilowatt hour, compare the operating cost of such a system in the winter with a heating system (assume 70% efficient) using fuel at a cost of $3.00/10^6 Btu.

TEN

multi component systems

Introduction

Up to this point our thermodynamic relationships have been limited in application to systems with uniform chemical composition and have dealt only briefly with equilibrium between materials in two or more distinct phases. For instance, we have discussed the properties of pure materials (such as water) and of mixtures with constant composition (such as air). But if we add alcohol or salt to the water or hydrogen to the air, thus changing the composition, our present relationships are inadequate to describe the change in the properties of the system.

We will go well beyond the question of simply determining the properties of multicomponent systems and will discuss the ways and means by which these properties may be used to predict and describe phase-equilibrium problems. For example, what is the minimum pressures needed to recover freshwater from seawater by reverse osmosis, or what will be the composition of a vapor in equilibrium with a multicomponent liquid, and how does pressure or temperature affect this vapor composition? We shall develop specific methods for treating such problems.

This chapter will be divided into two major areas. In the first area we will extend the thermodynamic relationships developed in Chapter 6 to cover systems with two or more chemical components. In the second area we will show how these properties may be used to describe multicomponent phase equilibrium. Chemically reacting mixtures are specifically excluded in this chapter but will be considered in the next.

To describe multicomponent systems we must first expand our thermodynamic relations to include the effects of composition. This task requires careful definition and development of several novel concepts.

10.1 Partial Molar Properties

To insure that we understand the meaning of thermodynamic terms for a multicomponent system, we shall begin with very basic ideas, recalling first our conclusions for pure components and extending these in a rational manner to multicomponent systems.

We have seen that the total internal energy, U, of a one-phase, pure-component system is a function of two state variables and the size of the system (or number of moles n). For instance,

$$U = U(P, T, n) \tag{10-1}$$

or

$$dU = \left(\frac{\partial U}{\partial P}\right)_{T,n} dP + \left(\frac{\partial U}{\partial T}\right)_{P,n} dT + \left(\frac{\partial U}{\partial n}\right)_{P,T} dn \tag{10-2}$$

The internal energy of a one-phase system of C components must then be given by

$$U = U(P, T, n_1, n_2, n_3, \ldots, n_C) \tag{10-1a}$$

etc., that is, it is a function of pressure, temperature, and the number of moles of each component present. For any change of state of the system, the change in internal energy will be given by the total differential,

$$dU = \left(\frac{\partial U}{\partial P}\right)_{T,n_i} dP + \left(\frac{\partial U}{\partial T}\right)_{P,n_i} dT + \sum_{i=1}^{C} \left(\frac{\partial U}{\partial n_i}\right)_{P,T,n_j} dn_i \tag{10-3}$$

The subscript n_i in the first two terms denotes that all compositions are held unchanged; the subscript n_j in the third term denotes that the amounts of all species except the ith component are held fixed.

In equation (10-3) we identify the quantity

$$\bar{U}_i \equiv \left(\frac{\partial U}{\partial n_i}\right)_{P,T,n_j} \tag{10-4}$$

as the *partial molar internal energy*. The quantity

$$\bar{U}_i \, dn_i = \left(\frac{\partial U}{\partial n_i}\right)_{P,T,n_j} dn_i \tag{10-5}$$

represents the change in internal energy of the system brought about by adding an increment, dn_i, of component i to the system while holding the pressure, temperature, and all other component amounts constant. The partial molar internal energies, $\bar{U}_i$, of each component in a mixture will depend on the pressure, temperature, and composition of the system but will be independent of the total size of the system. We will use a capital italic symbol with an overbar to represent partial molar properties.

Special attention should be paid to the conditions placed on the differentiation to obtain a partial property since the procedure will be applied to all extensive properties and will be used frequently in the remaining text. A partial molar quantity is *always* obtained by differentiation of an extensive property with respect to the number of moles of one of the components *under conditions of constant pressure, temperature, and number of moles of all other components.*

A useful relationship between the partial molar internal energy, $\bar{U}_i$, and the total internal energy, U, can be found in the following manner. Suppose a system of n moles with internal energy U is increased in size by adding to it dn moles of additional material of the same composition. If the total system volume is allowed to increase proportionally, this may be done without changing the temperature and pressure. The change in internal energy of the system is found from equation (10-3):

$$dU = \sum_{i=1}^{C} \bar{U}_i \, dn_i \tag{10-6}$$

Since U is an extensive property and we have not changed the temperature, pressure, or relative mole numbers but only the total system size, the increase in internal energy must be proportional to the change in total system size,

$$dU = \frac{U}{n} \, dn \tag{10-7}$$

The increase in each of the mole numbers is proportional to the relative amounts of each component originally present,

$$dn_i = \frac{n_i}{n} \, dn \tag{10-8}$$

Substitution of equations (10-7) and (10-8) in (10-6) and elimination of the factor dn/n on both sides yields the general result

$$U = \sum_{i=1}^{C} \bar{U}_i n_i \tag{10-9}$$

In short, the total internal energy is just the weighted sum of the partial molar internal energies. This result also suggests why the adjective "partial" is employed.

Similar results can be obtained for all other extensive properties of the mixture and their corresponding partial molar properties. For instance,

$$dH = \left(\frac{\partial H}{\partial P}\right)_{T,n_i} dP + \left(\frac{\partial H}{\partial T}\right)_{P,n_i} dT + \sum_{i=1}^{C} \bar{H}_i \, dn_i$$

$$\bar{H}_i = \left(\frac{\partial H}{\partial n_i}\right)_{P,T,n_j} \quad \text{and} \quad H = \sum_{i=1}^{C} \bar{H}_i n_i \tag{10-10}$$

Exactly equivalent results can be written for S, V, A, and G! Moreover, the relations between these propeties,

$$\begin{aligned} H &= U + PV \\ A &= U - TS \\ G &= H - TS \end{aligned} \tag{10-11}$$

can be differentiated with respect to n_i at constant temperature, pressure, and n_j to produce corresponding relations for the partial molar properties:

$$\begin{aligned}\bar{H}_i &= \bar{U}_i + P\bar{V}_i \\ \bar{A}_i &= \bar{U}_i - T\bar{S}_i \\ \bar{G}_i &= \bar{H}_i - T\bar{S}_i\end{aligned} \tag{10-12}$$

It is frequently useful to refer to the change on mixing of an extensive property such as volume. The total volume of the pure components before mixing is given by

$$V_{\text{components}} = \sum_{i=1}^{C} n_i v_i \tag{10-13}$$

Hence the volume change upon mixing is

$$\begin{aligned}\Delta V_{\text{mixing}} &= V_{\text{mixture}} - V_{\text{components}} \\ &= \sum_{i=1}^{C} n_i \bar{V}_i - \sum_{i=1}^{C} n_i v_i \\ &= \sum_{i=1}^{C} n_i(\bar{V}_i - v_i)\end{aligned} \tag{10-14}$$

Similar results are obtained for all other extensive properties.

In the ensuing discussion of mixtures we shall often find it necessary to refer to the property of a pure component at the same pressure, temperature, and phase as the mixture. This will be done by using the symbol for the pure-component property with a subscript denoting the component. Thus v_i denotes the specific volume of pure component i at the same temperature, pressure, and phase as the mixture.

SAMPLE PROBLEM 10-1. Using Table SP 10-1a,

(a) Calculate $\bar{V}_{CH_3CH_2OH}$ for a solution consisting of 40.04 percent ethyl alcohol by weight.

(b) Calculate $\bar{V}_{CH_3CH_2OH}$ for a solution consisting of 100 percent ethyl alcohol by weight.

(c) Calculate v for the solution consisting of 40.04 percent ethyl alcohol by weight.

(d) Calculate v for the solution consisting of 100 percent ethyl alcohol by weight.

Solution: (a) Several techniques may be used to solve this part of the problem. A few of these methods are discussed below:

1. The brute-force technique: By definition,

$$\bar{V}_i = \left(\frac{\partial V}{\partial n_i}\right)_{P,T,n_j} = \lim_{\Delta n_i \to 0} \left(\frac{\Delta V}{\Delta n_i}\right)_{P,T,n_j}$$

Table SP 10-1a Specific Gravity of Mixtures of Ethyl Alcohol and Water by Volume and by Weight†,‡

Specific Gravity	Percent Alcohol by Volume	Percent Alcohol by Weight	Grams Alcohol per 100 cm³
1.00000	0.00	0.00	0.00
0.98391	12.40	10.00	9.84
0.97149	24.50	20.01	19.44
0.95745	36.20	30.00	28.73
0.94008	47.00	39.67	37.29
0.93990	47.10	39.76	37.37
0.93971	47.20	39.85	37.45
0.93953	47.30	39.95	37.53
0.93934	47.40	40.04	37.61
0.93916	47.50	40.13	37.69
0.93898	47.60	40.22	37.77
0.93879	47.70	40.32	37.85
0.93861	47.80	40.41	37.93
0.8773	75.00	—	—
0.7939	100.00	100.00	—

†From *Handbook of Chemistry and Physics*, Chemical Rubber Co., Cleveland, 36th ed., 1954–1955, pp. 1932–1938. Used by permission of the Chemical Rubber Co.

‡Giving the specific gravity at 15.56°C, referred to water at the same temperature. To reduce to specific gravity referred to water at 4°C, multiply by 0.99908.

That is, $\bar{V}_i$ may be obtained by plotting $(\Delta V/\Delta n_i)_{P,T,n_j}$ versus Δn_i and extrapolating to $\Delta n_i \longrightarrow 0$. Thus $\bar{V}_{\text{alc}}$ in a 40.04 percent (by weight) alcohol–water solution may be found as follows.

We begin by considering 100 cm³ of solution with 40.04 wt. percent alcohol, 59.96 wt. percent water. From the data given in the problem, this solution has a density $\rho =$ (specific gravity) (0.99908) $= 0.93848$ g/cm³. Thus the 100 cm³ of solution weighs 93.848 g. Of this, 40.04 percent, or 37.576 g, is alcohol, and 56.272 g is water. To this original solution, let us add $\Delta n = X$ grams of alcohol, so the solution concentration increases to 40.13 percent (the next entry on the data table). The density of the new solution is $\rho = (0.93916)(0.99908) = 0.93829$ g/cm³. The total weight of the new solution is

$$M = (93.848 + X)$$

The amount of water in this solution is $(0.5987)(93.848 + X)$ and is equal to the amount in the original 40.04 percent solution. Thus

$$(0.5987)(93.848 + X) = 56.272$$

which may be solved for X to give

$$X = 0.1423 \text{ g of alcohol added}$$

The volume of the new solution is then $V = M/\rho = 93.990/0.93829 = 100.172\ \text{cm}^3$, so $\Delta V = 0.172\ \text{cm}^3$ and

$$\frac{\Delta V}{\Delta n} = \frac{0.172}{0.143} = 1.20\ \text{cm}^3/\text{g}$$

We perform the same calculations for the next few entries in Table SP 10-1b.

Table SP 10-1b

x_{alc}	Specific Gravity	ρ	x_{H_2O}	$56.272/x_{H_2O}$	$X = [(\ \ ) - 93.848]$	$V = \dfrac{93.848 + X}{\rho}$	ΔV	$\Delta V/V\Delta n$
40.04	0.93934	0.93848	0.5996	93.848	0	100.00		
40.13	0.93916	0.93829	0.5987	93.990	0.1423	100.172	0.172	1.20
40.22	0.93898	0.93812	0.5978	94.132	0.2838	100.341	0.341	1.20
40.32	0.93879	0.93793	0.5968	94.290	0.4415	100.529	0.529	1.20

Thus we see that for all intents and purposes the finite difference $\Delta V/\Delta n$ approximates the required derivative and $\bar{V}_{alc} = 1.20\ \text{cm}^3/\text{g}$.

2. A second technique to obtain $\bar{V}$ would be as follows. Take 100 g of water as a base. Then determine the number of grams of alcohol that must be added to this base to form a 10 percent, 20 percent, . . . alcohol (by weight) solution. Determine volume of these solutions as functions of concentration. Then plot volume of solution/100 g of water as a function of the number of grams of alcohol (Fig. SP 10-1). The slope of this curve at any concentration is

$$\text{slope} = \left(\frac{\partial V}{\partial n_{alc}}\right)_{P,T,n_{water}}$$

and hence is the partial molar volume of alcohol at the concentration in question. Thus all one needs to do is determime the slope of this curve at $x_{alc} = 40.04$ percent.

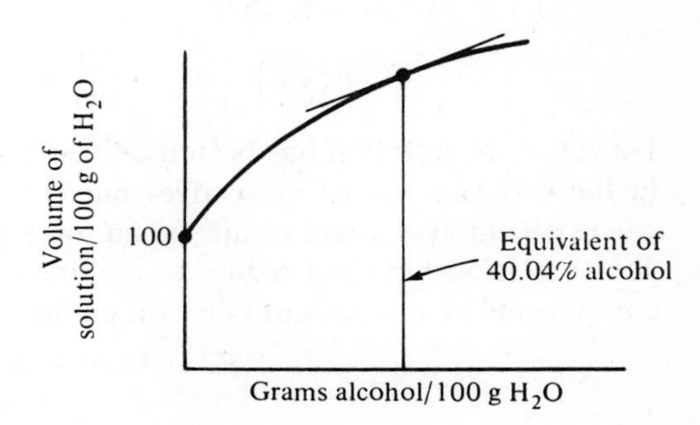

Figure SP10-1

(b) $\bar{V}_{alc}$ for the pure alcohol is simply v_{alc} and is given by

$$\bar{V}_{alc} = \frac{1}{\rho_{alc}} = \frac{1}{(0.7939)(0.99908)} = 1.261 \text{ cm}^3/\text{g}$$

(c) $v_{mix} = 1/\rho_{mix} = 1/[(0.93934)(0.99908)] = 1.0656 \text{ cm}^3/\text{g}$.
(d) Same as (b): $v_{alc} = 1.261 \text{ cm}^3/\text{g}$.

A brief comment on the units obtained in the foregoing problem is in order. The derivations should have led the careful reader to expect that the proper units for $\bar{V}_i$ are cm³/g-mol, not cm³/g. Conversion requires only multiplication by the molecular weight, but this operation was not carried out because the given units are quite satisfactory. The case is much the same as that in earlier chapters where specific volume v was given as often on a mass basis as on a molar basis. The reader can demonstrate for himself that, with suitable redefinition, all results of this and ensuing sections can be interpreted on either a mass or molar basis. Since a great deal of engineering data is given on a mass basis, those will often be the more convenient units. Not until Chapter 11, where we deal with chemical reactions in which mass is conserved, but not always moles, will it be necessary to separate the bases. This apparent ambiguity, then, reduces to but another example of the importance of clearly specifying the units of all numerical answers.

10.2 Partial Molar Gibbs Free Energy— The Chemical Potential

One of the partial molar properties, the partial molar Gibbs free energy, is particularly useful in evaluating the behavior of mixtures. It will be seen later that this quantity plays a central role in the criteria for both chemical and physical equilibria, and hence it is appropriate here to examine the mathematical connection between this and the other thermodynamic properties of mixtures.

We have already seen that

$$dG = \left(\frac{\partial G}{\partial P}\right)_{T,n} dP + \left(\frac{\partial G}{\partial T}\right)_{P,n} dT + \sum_{i=1}^{C} \bar{G}_i \, dn_i \tag{10-15}$$

Hereafter the notation has been modified slightly; the index i is still implied in the first two partial derivatives but is omitted to prevent undue clutter where no ambiguity will result. When there is no change in composition, the above relationship must reduce to the one we have used previously for pure components or mixtures at constant composition:

$$dG = V\,dP - S\,dT \tag{10-16}$$

In other words, we want to preserve the relations established previously,

$$\left(\frac{\partial G}{\partial P}\right)_{T,n} = V \quad \text{and} \quad \left(\frac{\partial G}{\partial T}\right)_{P,n} = -S \tag{10-17}$$

Thus we now write equation (10-15) as

$$dG = V\,dP - S\,dT + \sum_{i=1}^{C} \bar{G}_i\,dn_i \tag{10-18}$$

We recall that the Maxwell relationship,

$$\left(\frac{\partial V}{\partial T}\right)_{P,n} = -\left(\frac{\partial S}{\partial P}\right)_{T,n} \tag{10-19}$$

was obtained by equating the derivatives,

$$\frac{\partial^2 G}{\partial P\,\partial T} = \frac{\partial^2 G}{\partial T\,\partial P} \tag{10-20}$$

Inlike manner, by equating the derivatives obtained from cross differentiation of two other terms of equation (10-18), we obtain

$$\left(\frac{\partial \bar{G}_i}{\partial P}\right)_{T,n} = \left(\frac{\partial V}{\partial n_i}\right)_{P,T,n_j} \tag{10-21}$$

so

$$\left(\frac{\partial \bar{G}_i}{\partial P}\right)_{T,n} = \bar{V}_i \tag{10-22}$$

where the subscript n stands for *all components* and the subscript n_j for *all components but i*. Similarly,

$$\left(\frac{\partial \bar{G}_i}{\partial T}\right)_{P,n} = -\left(\frac{\partial S}{\partial n_i}\right)_{P,T,n_j} = -\bar{S}_i \tag{10-23}$$

Even more useful than the latter is the derivative with respect to temperature of the ratio $\bar{G}_i/T$. Making use of equations (10-23) and (10-12) we find that

$$\begin{aligned}\left(\frac{\partial(\bar{G}_i/T)}{\partial T}\right)_{P,n} &= -\frac{\bar{G}_i}{T^2} + \frac{1}{T}\left(\frac{\partial \bar{G}_i}{\partial T}\right)_{P,n} \\ &= -\frac{1}{T^2}(\bar{H}_i - T\bar{S}_i) + \frac{1}{T}(-\bar{S}_i) \\ &= -\frac{\bar{H}_i}{T^2}\end{aligned} \tag{10-24}$$

Although the central role of the partial molar Gibbs free energy, $\bar{G}_i$, in the description of physical equilibrium has yet to unfold, equations (10-22) and (10-24) already suggest that the partial molar volume, $\bar{V}_i$, and the partial molar enthaply, $\bar{H}_i$, have important roles in conveying information about the pressure and temperature dependence of $\bar{G}_i$. Although the temperature dependence of $\bar{G}_i$ can also be learned from the partial molar entropy, $\bar{S}_i$, by equation (10-23), it is generally easier to obtain values for $\bar{H}_i$ than for $\bar{S}_i$. Thus we shall use equation (10-24) more frequently than equation (10-23).

It was shown in equation (10-10) that the total Gibbs free energy, G, is related to the partial molar Gibbs free energies by

$$G = \sum_{i=1}^{C} \bar{G}_i n_i \tag{10-25}$$

From this it follows that for any small change,

$$dG = \sum_{i=1}^{C} n_i d\bar{G}_i + \sum_{i=1}^{C} \bar{G}_i \, dn_i \tag{10-26}$$

By comparison to equation (10-18), it is clear that

$$\sum_{i=1}^{C} n_i \, d\bar{G}_i = V \, dP - S \, dT \tag{10-27}$$

and that for any change *at constant temperature and pressure*,

$$\sum_{i=1}^{C} n_i \, d\bar{G}_i = 0 \tag{10-28}$$

This result, known as the *Gibbs–Duhem equation*, will be used frequently in the ensuing material. Exactly similar results can be obtained for the other partial molar extensive properties by repeating the procedure of equations (10-25) to (10-28) for any other property. Thus for any change *at constant temperature and pressure* we may write $\sum_{i=1}^{C} n_i \, d\bar{H}_i = 0$, $\sum_{i=1}^{C} n_i \, d\bar{S}_i = 0$, and so on.

In many situations where the partial molar Gibbs free energy is employed, it is given another name, the chemical potential, and another symbol, μ_i. Indeed we may write

$$\mu_i \equiv \bar{G}_i \tag{10-29}$$

and, if we chose to do so, all the above equations could be rewritten inserting μ_i wherever $\bar{G}_i$ appears. Some authors prefer one symbol, some the other, and the student should be prepared to recognize either symbol. The name "chemical potential" was introduced into the thermodynamic literature because μ_i (or $\bar{G}_i$) is a measure of the chemical energy potential of a component. Thus, for example, one can say that pressure is a potential that determines the internal-energy change associated with a volume change, and temperature is a potential that determines the internal-energy change associated with an entropy change. By the same token, μ_i (or $\bar{G}_i$) is a potential that determines the internal energy change associated with changes in composition. Although such an explanation does not justify the use of the quantity in thermodynamic relationships, it does give some appreciation for the nature of this quantity.

10.3 Fugacity

Although we might use a method similar to that discussed earlier for $\bar{V}_i$ to determine the partial molar Gibbs free energy (or chemical potential), it is a cumbersome approach and would require that we have data on the entropy change on mixing as well as the heat of mixing. As we shall see, it is possible to circumvent this difficulty by making use of the fugacity concept introduced in Chapter 6. In Chapter 6 fugacity was defined as an intensive thermodynamic property which under isothermal conditions was related to differential changes in the Gibbs free energy by equation (6-87). The fugacity of a pure gaseous component, denoted by $f(P, T)$, satisfies the pair of equations

$$RT\left(\frac{\partial \ln f}{\partial P}\right)_T = v$$
$$\lim_{P\to 0} \frac{f(P, T)}{P} = 1 \tag{10-30}$$

which constitute the definition of fugacity.

Fugacity is an artificial property in that it is not directly measurable, but the same can be said of enthaply and the free energies. All are introduced out of convenience and are justified because they help us to express certain thermodynamic relationships. For mixtures we introduce another such property, the partial fugacity of a component *i*. It will be denoted by the symbol $\bar{f}_i(P, T)$ and is defined by the pair of equations

$$RT\left(\frac{\partial \ln \bar{f}_i}{\partial P}\right)_{T,n} = \bar{V}_i$$
$$\lim_{P\to 0} \frac{\bar{f}_i(P, T)}{y_i P} = 1 \tag{10-31}$$

where y_i is the mole fraction of *i* in the mixture. Note that partial fugacity is an intensive property and thus is not a "partial property" in the same sense as discussed for the extensive properties earlier. Rather, as we shall see shortly, it is more like a partial pressure.

The convenience of this property in describing the behavior of mixtures will be discussed in later sections, but first we shall summarize the behavior of partial fugacity with changes in pressure at constant composition. This information will be needed in our later discussions of mixture properties.

The defining equations (10-30) and (10-31) give the dependence of the fugacities on pressure at constant temperature and composition. The dependence of fugacity on temperature may be determined as follows. In light of equation (10-22) we can write

$$RT\left(\frac{\partial \ln f}{\partial P}\right)_T = v = \left(\frac{\partial g}{\partial P}\right)_T \tag{10-32}$$

$$RT\left(\frac{\partial \ln \bar{f}_i}{\partial P}\right)_{T,n} = \bar{V}_i = \left(\frac{\partial \bar{G}_i}{\partial P}\right)_{T,n} \tag{10-33}$$

These equations may be integrated at constant temperature over the range P^* to P, where the asterisk denotes a reference state at a pressure low enough that $f^* = P^*$ and $\bar{f}^* = y_i P^*$. (The use of an asterisk here has a meaning very similar to its use with specific heats, where C_P^* denoted the specific heat at a pressure low enough that the gas could be considered ideal.) The results of integrating equations (10-32) and (10-33) between these limits are

$$\ln \frac{f}{P^*} = \frac{g - g^*}{RT} \tag{10-34}$$

$$\ln \frac{\bar{f}_i}{y_i P^*} = \frac{\bar{G}_i - \bar{G}_i^*}{RT} \tag{10-35}$$

10.4 An Ideal-Gas Mixture

A mixture whose components are ideal gases and which itself behaves as an ideal gas is quite simple to describe. If the mixture is an ideal gas, its total volume obeys the ideal gas equation of state:

$$V = \frac{nRT}{P} = (n_1 + n_2 + \cdots)\frac{RT}{P} \tag{10-36}$$

The partial molar volumes are given by

$$\bar{V}_i = \left(\frac{\partial V}{\partial n_i}\right)_{P,T,n_j} = \frac{RT}{P} = v_i \tag{10-37}$$

and are equal to the pure-component specific volumes at the same temperature and pressure. Thus, the left-hand sides of equations (10-32) and (10-33) may be equated:

$$\left(\frac{\partial \ln f_i}{\partial P}\right)_T = \left(\frac{\partial \ln \bar{f}_i}{\partial P}\right)_T \tag{10-38}$$

which may be integrated at constant T from $P^* \longrightarrow 0$ to any desired pressure P:

$$\ln\left(\frac{f_i}{f_i^*}\right) = \ln \frac{\bar{f}_i}{\bar{f}_i^*} \tag{10-39}$$

or

$$\frac{f_i}{f_i^*} = \frac{\bar{f}_i}{\bar{f}_i^*} \tag{10-39a}$$

But from the definitions of f_i and $\bar{f}_i$, $f_i^* \longrightarrow P^*$ and $\bar{f}_i^* \longrightarrow y_i P^*$. Equation (10-39a) then reduces to:

$$\frac{f_i}{P^*} = \frac{\bar{f}_i}{y_i P^*} \tag{10-40}$$

or

$$\bar{f}_i = y_i f_i \tag{10-41}$$

Equation (10-41) is the *Lewis–Randall rule* and in general is used to define an *ideal solution*. Clearly, mixtures of ideal gases are ideal solutions. However, they are not the only class of ideal solutions: Most gaseous and vapor mixtures at low to moderate pressures obey the Lewis–Randall rule even if their volumetric behavior is not accurately described by the ideal-gas equation of state. In addition, many liquid mixtures—particularly series of compounds with similar chemical structure—also obey the Lewis–Randall rule. We shall deal quantitatively only with such mixtures. A brief discussion of techniques for dealing with nonideal mixtures is included later in this chapter for the interested reader.

For a gas which obeys the ideal-gas equation of state equation (10-32) can be integrated directly from P^* to P at constant T to give:

$$f_i = P_i \tag{10-42}$$

which in turn can be substituted into (10-41) to give

$$\bar{f}_i = y_i P_i \tag{10-43}$$

Thus the fugacities are equal to the pressures and partial pressures, respectively.

The absence of intermolecular forces in any gaseous system that exhibits ideal gas behavior permits us to conclude that there would be no heat effect in the isothermal mixing of ideal gases. Therefore, the internal energy of such mixtures would be equal to the sum of the internal energies of the individual components for isothermal mixing. Thus the partial molar internal energy must equal the specific internal energy at the same temperature and pressure:

$$\bar{U}_i = u_i \tag{10-44}$$

Since $\bar{U}_i = u_i$ and $\bar{V}_i = v_i$ for ideal-gas mixtures, we conclude that

$$\bar{H}_i = \bar{U}_i + P\bar{V}_i = u_i + Pv_i$$

or

$$\bar{H}_i = h_i \tag{10-45}$$

and thus, upon mixing, there is no change in either internal energy or enthalpy.

However, the mixing of two ideal gases to form an ideal-gas mixture is clearly an irreversible process, in that the separation of the mixture into the two pure components would never occur spontaneously. Thus we would expect an entropy increase to result from such a mixing process. To compute this entropy change, let us first recall that one characteristic of an ideal-gas mixture is that each component in such a mixture behaves as if it occupied the same total volume as the mixture at the temperature of the mixture, but at the partial pressure of the component. Thus the mixing process is identical to that in which given amounts of pure crystalline solids are mixed at constant temperature and pressure to form a mixed crystal whose volume is equal to the sum of the volumes of the original crystals.

In the mixing of both the ideal gases and ideal crystals, the entropy change on mixing is due to the additional number of particle arrangements, or microstates, that arise from the additional volume elements available to the individual molecules. We show in Appendix H that the entropy change that occurs during such a process is given by:

$$\Delta S_m = -R\left[\sum_{i=1}^{C} n_i \ln\left(\frac{n_i}{\sum_{i=1}^{C} n_i}\right)\right] = \sum_{i=1}^{C} n_i(\bar{S}_i - s_i) \tag{10-46}$$

or

$$\sum_{i=1}^{C} n_i(\bar{S}_i - s_i) = -R\left[\sum_{i=1}^{C} n_i \ln n_i + \sum_{i=1}^{C} n_i \ln\left(\sum_{i=1}^{C} n_i\right)\right] \tag{10-46a}$$

Differentiation with respect to n_k at constant pressure, temperature, and n_j gives:

$$(\bar{S}_k - s_k) + \sum_{i=1}^{C} n_i\, d\bar{S}_i = -R\left[\ln n_k + \frac{n_k}{n_k} - \sum_{i=1}^{C}\left(\frac{n_i}{\sum_{i=1}^{C} n_i}\right) - \ln \sum_{i=1}^{C} n_i\right] \tag{10-47}$$

Elimination of the second term in the left-hand side by means of the Gibbs–Duhem equation and simplification yields:

$$\bar{S}_k - s_k = -R \ln \frac{n_k}{\sum_{i=1}^{C} n_i} \tag{10-48}$$

Since $n_k/\sum_{i=1}^{C} n_i$ is simply y_k, the mole fraction of species k, equation (10-48) reduces to

$$\bar{S}_k - s_k = -R \ln y_k \tag{10-49}$$

We can now summarize the mixing behavior for ideal-gas systems:

$$\begin{aligned}
\bar{V}_i - v_i &= 0\\
\bar{U}_i - u_i &= 0\\
\bar{H}_i - h_i &= 0\\
\bar{S}_i - s_i &= -R \ln y_i\\
\bar{G}_i - g_i &= \bar{H}_i - h_i - T(\bar{S}_i - s_i) = RT \ln y_i\\
\bar{A}_i - a_i &= (\bar{U}_i - u_i) - T(\bar{S}_i - s_i) = RT \ln y_i
\end{aligned} \tag{10-50}$$

Thus the properties of an ideal-gas mixture can be completely described if the properties of each pure component are known.

We can obtain another useful result by subtracting equation (10-34) from equation (10-35):

$$RT \ln \frac{\bar{f}_i}{y_i f_i} = (\bar{G}_i - g_i) - (\bar{G}_i^* - g_i^*) \tag{10-51}$$

However, at vanishing small pressures $(\bar{G}_i^* - g_i^*)$ can be expressed according to equation (10-50):

$$(\bar{G}_i^* - g_i^*) = RT \ln y_i \tag{10-52}$$

Thus equation (10-51) reduces to

$$RT \ln \frac{\bar{f}_i}{f_i} = (\bar{G}_i - g_i) \tag{10-53}$$

or

$$\ln \frac{\bar{f}_i}{f_i} = \frac{\bar{G}_i - g_i}{RT} \tag{10-54}$$

a result which is applicable *in all cases.* We shall make repeated use of this relationship in Chapter 11.

10.5 Criteria for Phase Equilibrium

In the previous section we have developed methods for evaluating the properties of single-phase, multicomponent ideal-gas mixtures. Although these relations also hold for ideal-liquid mixtures, we have said little about the applications in which these properties may be useful. In the remainder of this chapter and all of the next we shall examine the use of these mixture properties in the description of phase and chemical equilibria. The remainder of this chapter will deal primarily with problems in phase equilibrium. This material is essential for the description of such diverse and widely encountered processes as oil–gas reservoir behavior, humidification, fractional distillation, zone melting, and reverse osmosis. Accordingly, we now reopen the discussion of phase equilibrium, but this time for multicomponent systems. We shall make further use of the criterion for equilibrium first discussed in Chapter 6, as well as the relations just developed to describe the properties of multicomponent mixtures.

One can choose to think of equilibrium as represented by a minimization of internal energy, U; the enthalpy, H, the Helmholtz free energy, A; the Gibbs free energy, G; or the maximization of entropy, S, depending on the constraints placed on the system. In developing the entropy concept we demonstrated that entropy was maximized at equilibrium for systems with fixed U and V. It can also be readily shown that if S and V are fixed, U will be minimized at equilibrium; if V and T are fixed, A is minimized, and if P and T are fixed, G is minimized. Since P and T are easily measured and controlled parameters in actual processes, the latter criterion has proved most useful in engineering applications.

In Chapter 6 we used the relationship $-W_{\text{rev}} = \Delta G_T$ to relate work production and the Gibbs free energy to our concept of equilibrium. At this point we choose (merely for brevity) to begin with minimization of the Gibbs free energy at constant temperature and pressure as the criterion from which we develop the working relationships we need for multicomponent systems. It should be observed that a development paralleling that in Chapter 6 for single-component phase equilibria could be included. The interested reader can refer to this treatment for additional justification for the condition $dG = 0$ at equilibrium.

The development that follows considers two phases I and II, but generalization to any number of phases is easily shown. If we consider a two-phase system where mass is being transferred from one phase to the other, we observe that the free-energy change for the system is the sum of the changes occurring in each phase:

$$dG = dG^{\text{I}} + dG^{\text{II}} \tag{10-55}$$

For a multicomponent system at constant T and P, the free-energy change for each phase is given by

$$dG = \sum_{i=1}^{C} \bar{G}_i \, dn_i \tag{10-56}$$

Substitution of equation (10-56) into (10-55) yields

$$dG = \sum_{i=1}^{C} \bar{G}_i^{\text{I}} \, dn_i^{\text{I}} + \sum_{i=1}^{C} \bar{G}_i^{\text{II}} \, dn_i^{\text{II}} \tag{10-57}$$

Since the summations cover the same range and, by conservation of mass, $dn_i^{\text{II}} = -dn_i^{\text{I}}$ for each component present, we can combine the summations:

$$dG = \sum_{i=1}^{C} (\bar{G}_i^{\text{II}} - \bar{G}_i^{\text{I}}) \, dn_i^{\text{II}} \tag{10-58}$$

Our criterion for equilibrium should be generally valid, that is, it should not depend on any particular choice of the dn_i. Thus the total Gibbs free energy will be a minimum ($dG = 0$) for an arbitrary choice of the dn_i only if

$$\bar{G}_i^{\text{II}} = \bar{G}_i^{\text{I}} \tag{10-59}$$

for all C components. Thus, at equilibrium the partial molar Gibbs free energy (or chemical potential) of each component must be the same in each phase present in the system. Equation (10-59) can be expressed in terms of partial fugacities:

$$\bar{f}_i^{\text{I}} = \bar{f}_i^{\text{II}} = \bar{f}_i^{\text{III}} = \cdots \tag{10-60}$$

As we shall see in the ensuing discussion, this relationship is the starting point for all multicomponent phase equilibrium computations.

Although the discussion above has assumed that both phases are at the same temperature and pressure (typically the case for phase equilibrium), it can be shown that the equilibrium criterion [equation (10-60)] is also valid even if the two phases are at different pressures, provided only that these pressures are constant. The development of the equilibrium requirement $dG = 0$ in Chapter 6 required only that the temperatures of the two phases be identical. Thus even if $P^{\text{I}} \neq P^{\text{II}}$, equation (10-60) can be obtained as long as $dP^{\text{I}} = dP^{\text{II}} = 0$, and $T^{\text{I}} = T^{\text{II}}$. Let us consider for example the diffusion of water across the semipermeable membrane of a reverse osmosis cell as shown in Fig. 10-1. If P^{I} is sufficiently greater than P^{II}, then water will diffuse from the salt solution into the pure water. If we adjust the pressure such that the two phases are in equilibrium, what conditions must be met?

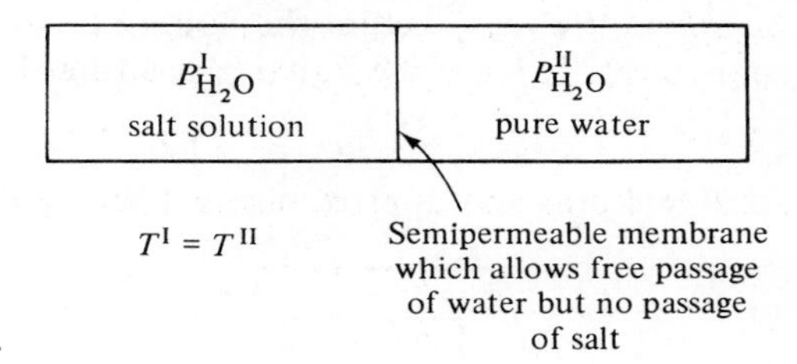

Figure 10-1. The reverse osmosis cell.

For the equilibrium conditions the Gibbs free energy must be at a minimum so that the dG of equation (10-58) vanishes. Thus

$$\sum_{i=1}^{2} (\bar{G}_i^{\text{II}} - \bar{G}_i^{\text{I}})\, dn_i = 0$$

However, the semipermeable membrane prevents the salt from diffusing, and thus $dn_{\text{salt}} = 0$. The equilibrium requirement thus reduces to equation (10-60) applied only to the water component. As we will show in our later discussion, the concentration difference between phases I and II will require $P^{\text{I}} > P^{\text{II}}$ if $\bar{G}^{\text{I}}_{H_2O} = \bar{G}^{\text{II}}_{H_2O}$. If we remove the semipermeable membrane, then salt is also permitted to diffuse and the new equilibrium conditions would require equality of the partial molar Gibbs free energies of the salt as well as the water. The only way in which both the salt and water partial molar Gibbs free energies can be constant across an open boundary would be for the salt concentration and the pressure to be uniform. This example illustrates that the conditions for equilibrium are greatly dependent on the constraints placed upon the process.

In Section 10.4 we discussed the Lewis–Randall rule and the concept of ideal mixtures. Although the Lewis–Randall rule is by no means applicable to all systems, it does adequately describe a large number of commercially important systems and, except where specifically noted, will be assumed in all remaining discussions in this text. Likewise, we will also assume that all gaseous phases and mixtures obey the ideal-gas equation of state.

Sample Problem 10-2. In the evaporative desalinization of seawater, the saltwater is first heated to its boiling point and then partially vaporized. The vapor, which is essentially pure water, is then condensed and collected to give the product stream. If the entering seawater is initially 1.5 mol percent NaCl, estimate the boiling temperature of the solution at 1 atm. (1.0135 Bar) $\Delta h_V = 2257$ kJ/kg and may be assumed constant. The vapor may be assumed to be ideal and to have negligible density with respect to the liquid. Carefully list any assumptions you make; you should not need any data other than those supplied.

Solution: We may picture the boiling as shown in Fig. SP 10-2. Water will spontaneously leave the liquid phase (that is, the solution will boil) when

$$\bar{f}^{L}_{H_2O} > \bar{f}^{V}_{H_2O}$$

At $P = 1$ atm and approximately 100°C, $(f/P)^V = 1$, so $\bar{f}^{V}_{H_2O} = 1.013$ Bar.

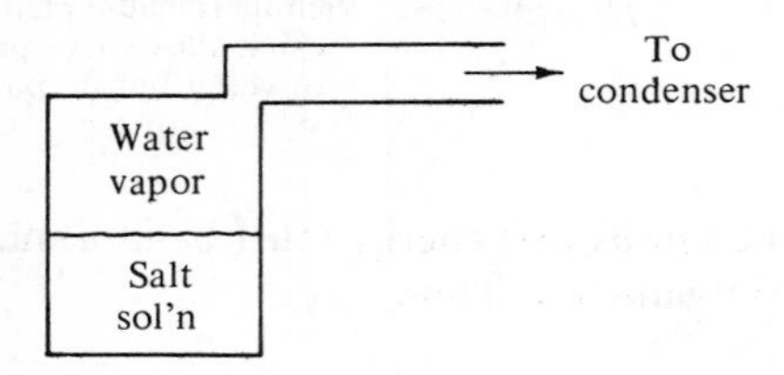

Figure SP10-2

Thus the solution will boil when $\bar{f}^{L}_{H_2O} > 1.013$ Bar. Since the mole fraction of water in the solution is very close to unity, we may assume that the solution is ideal—at least with regard to water. Thus we may use the Lewis–Randall rule for $\bar{f}^{L}_{H_2O}$:

$$\bar{f}^{L}_{H_2O} = x_{H_2O} f_{H_2O}$$

At this pressure and temperature $f_{H_2O} = P'_{H_2O}$ (the vapor pressure at the temperature of the solution). Thus we find that

$$\bar{f}^{L}_{H_2O} = x_{H_2O} P'_{H_2O} > 1.013 \text{ Bar}$$

Since $x_{H_2O} = 0.985$, we find that

$$P'_{H_2O} > \frac{1.013}{0.985} = 1.030 \text{ Bar}$$

Thus the minimum solution boiling temperature will be the temperature at which the vapor pressure of water is 1.030 Bar. We may determine the variation of the vapor pressure of water by means of the Clausius–Clapeyron equation. However, at the conditions involved, the assumptions of the simplified form of this equation are justifiable, and we shall use it:

$$\frac{d(\ln P'_{H_2O})}{d(1/T)} = -\frac{\Delta h_V}{R}$$

which may be integrated to give

$$\ln \frac{P'_2}{P'_1} = -\frac{\Delta h_V}{R}\left(\frac{1}{T_2} - \frac{1}{T_1}\right) = \frac{\Delta h_V}{R}\left(\frac{T_2 - T_1}{T_2 T_1}\right)$$

Let us take condition 1 as the normal boiling point of water:

$$P'_1 = 1.013 \text{ Bar}, \qquad T_1 = 100° = 373°\text{K}$$

Condition 2 is the unknown temperature for which

$$P'_2 = 1.030 \text{ Bar}, \qquad \Delta h_V = 2257 \text{ kJ/kg} = 40.6 \text{ MJ/kg-mol}$$

Thus T_2 is found from

$$\ln \frac{1.030}{1.013} = \left[\frac{40.6 \times 10^3 \text{ kJ/kg-mol}}{(8.310) \text{ kJ/kg-mol°K}}\right]\left[\frac{T_2 - 373}{T_2(373)} \,°\text{K}^{-1}\right]$$

which is solved for T_2 to give

$$T_2 = 373.43°\text{K} = 100.43°\text{C}$$

for a boiling-point elevation of 0.43°C

10.6 Humidity and the Air–Water System

Vapor–liquid equilibrium in the air–water system is particularly simple to describe because we may "uncouple" the vapor and liquid portions by observing that air is only sparingly soluble in water under most conditions of interest. Thus we may treat the liquid phase as if it were pure water. The equilibrium water concentration in the vapor phase is then obtained from the requirements of phase equilibrium:

$$\bar{f}_i^L \equiv \bar{f}_i^V \tag{10-61}$$

and the Lewis–Randall rule $\bar{f}_i = y_i f_i$:

$$y_i^L f_i^L = y_i^V f_i^V \tag{10-62}$$

but $y_i^L = 1$; $f_i^L = P'$, the vapor pressure of water at the temperature of interest; and $f_i^V = P_T$, the total pressure on the system. Thus,

$$P'_i = y_i^V P_T \tag{10-63}$$

or

$$y_i^V = \frac{P'_i}{P_T} \tag{10-63a}$$

The *humidity*, **H**, of a vapor-phase air—water mixture is defined as the mass of water vapor per pound of water-free (bone-dry) air:

$$\mathbf{H} = \frac{M_{\text{water vapor}}}{M_{\text{dry air}}} \tag{10-64}$$

The *saturation humidity* $\mathbf{H}^{\text{sat}}$ is the humidity of a vapor-phase air–water mixture which is (or may be) in equilibrium with liquid water — that is, when the vapor phase is *saturated* with water vapor. If the humidity is made higher than $\mathbf{H}^{\text{sat}}$ the mixture is *supersaturated* and will spontaneously deposit liquid water droplets, or *dew*, on any solid surface which the vapor contacts. The saturation partial pressure, P_w^{sat} of water in the vapor phase is simply its vapor pressure, P', at the temperature of the mixture.

The humidity of any given air–water mixture may be related to the vapor-phase *mole* fraction as follows:

$$\mathbf{H} = \frac{M_{\text{water}}}{M_{\text{air}}} = \frac{\bar{V}_{\text{air}}}{\bar{V}_{\text{water}}} \tag{10-65}$$

where the partial volumes are expressed on a *mass* basis. Assuming ideal-gas behavior, the partial volumes may be determined from the ideal-gas equation of state:

$$\bar{V}_i = \frac{R_i T}{P_i} \tag{10-66}$$

where $R_i = R/MW_i$, the ideal-gas constant divided by the molecular weight of specie i. The humidity is then:

$$\mathbf{H} = \frac{R_{\text{air}} T P_{\text{water}}}{R_{\text{water}} T P_{\text{air}}} = \frac{R_{\text{air}} P_{\text{water}}}{R_{\text{water}} P_{\text{air}}} \tag{10-67}$$

but the partial pressures may be replaced by Py_i

$$\mathbf{H} = \frac{R_{\text{air}} y_{\text{water}} P}{R_{\text{water}} y_{\text{air}} P} = \frac{R_{\text{air}} y_{\text{water}}}{R_{\text{water}} y_{\text{air}}} \tag{10-68}$$

The ratio of ideal-gas constants is simply the inverse ratio of molecular weights, so that

$$\mathbf{H} = \frac{y_{\text{water}}}{y_{\text{air}}}\left(\frac{MW_{\text{water}}}{MW_{\text{air}}}\right) = \frac{18}{29}\left(\frac{y_{\text{water}}}{y_{\text{air}}}\right) \tag{10-69}$$

A final simplification results if we observe that

$$y_{\text{air}} + y_{\text{water}} = 1.0$$

or (10-70)

$$y_{\text{air}} = 1.0 - y_{\text{water}}$$

The humidity can then be expressed directly in terms of the mole fraction of water:

$$\mathbf{H} = \frac{18}{29}\left(\frac{y_{\text{water}}}{1 - y_{\text{water}}}\right) \tag{10-71}$$

The saturation humidity is then

$$\mathbf{H}^{\text{sat}} = \frac{18}{29}\left(\frac{y_{\text{water}}^{\text{sat}}}{1 - y_{\text{water}}^{\text{sat}}}\right) \tag{10-72}$$

where

$$y_{\text{water}}^{\text{sat}} = \frac{P'_{\text{water}}}{P}$$

Therefore the saturation humidity can be expressed solely in terms of the vapor pressure of the water and the total system pressure:

$$\mathbf{H}^{\text{sat}} = \frac{18}{29}\left(\frac{P'_{\text{water}}}{P - P'_{\text{water}}}\right) \tag{10-73}$$

Since the vapor pressure of water is a function of the system temperature, the saturation humidity is a function of both the system temperature and pressure. However, for the remainder of this discussion we will restrict ourselves to atmospheric pressures only.

The *relative humidity*, *RH*, is defined as the ratio of the humidity to the saturation humidity at that temperature (and pressure):

$$RH = \frac{\mathbf{H}}{\mathbf{H}^{\text{sat}}} = \frac{[y_{\text{water}}/(1 - y_{\text{water}})]}{[P'_{\text{water}}/(P - P'_{\text{water}})]} \tag{10-74}$$

The relative humidity may be considered as a fractional saturation and thus is commonly reported in percent saturation, or simply percent.

The enthalpy of the vapor-phase air–water mixture may be determined as a function of temperature and humidity as follows:

$$h_{\text{mix}} = y_{\text{air}}\bar{H}_{\text{air}} + y_{\text{water}}\bar{H}_{\text{water}} \tag{10-75}$$

Under most conditions of interest the mixtures are treated as ideal so the partial enthalpies can be replaced by the pure-component specific enthalpies. Thus:

$$h_{\text{mix}} = y_{\text{air}}h_{\text{air}} + y_{\text{water}}h_{\text{water}} \tag{10-75a}$$

The pure-component enthalpies are obtained from standard sources such as the gas and steam tables.

Thus, the vapor-phase mole fractions, humidity, relative humidity, and mixture enthalpies can all be determined provided that the system temperature and any one of the properties (excluding pressure) are known.

Sample Problem 10-3. 2.0×10^{-2} lb_m of water are evaporated into 1 lb_m of bone-dry air. If the resulting mixture is at 120°F, determine

(a) The absolute and relative humidities of the mixture.
(b) The saturation temperature of the mixture.

Solution: (a) The absolute humidity is simply the ratio of the weights of water and bone-dry air.

$$\mathbf{H} = \frac{M_{\text{water}}}{M_{\text{air}}} = \frac{2.0 \times 10^{-2}}{1.0} \frac{\text{lb}_\text{m}\ \text{water}}{\text{lb}_\text{m}\ \text{air}}$$

To determine the relative humidity we must first determine the saturation humidity at this temperatune. The saturation humidity is expressed in terms of the vapor pressure of water according to equation (10-73)

$$\mathbf{H}^{\text{sat}} = \frac{18}{29}\left(\frac{P'_{\text{water}}}{P - P'_{\text{water}}}\right)$$

The vapor pressure of water at 120°F is obtained from the steam tables

$$P'_{\text{H}_2\text{O}} = 1.69 \text{ psia}$$

also,

$$P = 1 \text{ atm} = 14.7 \text{ psia}$$

Therefore:

$$\mathbf{H}^{\text{sat}} = \frac{18}{29}\left(\frac{1.69}{14.7 - 1.69}\right) = 0.081 \frac{\text{lb}_{\text{m}} \text{ water}}{\text{lb}_{\text{m}} \text{ air}}$$

The relative humidity is then:

$$RH = \frac{\mathbf{H}}{\mathbf{H}^{\text{sat}}} = \frac{0.020}{0.081} \times 100\% = 24.8\%$$

(b) To determine the saturation temperature we must solve equation (10-73) for the vapor pressure of water such that the saturation humidity is 0.02 lb_{m} water/lb_{m} air. Once the vapor pressure of water is so determined, the saturation temperature is obtained from the steam tables:

Solving (10-73) for P'_{water} gives

$$P'_{\text{water}} = \frac{29\mathbf{H}^{\text{sat}}P}{18 + 29\mathbf{H}^{\text{sat}}}$$

or

$$P'_{\text{water}} = \left[\frac{29(0.02)(14.7)}{18 + 29(0.02)}\right] \text{psia} = 4.58 \times 10^{-1} \text{ psia}$$

The saturation temperature is then ~77°F.

The Adiabatic Humidification Curve

Consider the process of Fig. 10-2. Air at T_1 and $\mathbf{H}_1$ is mixed with water which enters the process at T_2. During passage through the process, moisture is transferred between the liquid and vapor stages. If the mixing of the phases is intimate, then the water and the outlet air will leave at the same temperature. The liquid water is recirculated and make-up water (at T_3) is added so that the total amount of liquid remains constant. The whole process is adiabatic since no heat is transferred across the system boundaries. If the process proceeds for a sufficient length of time, eventually the outlet temperatures will reach a steady state.

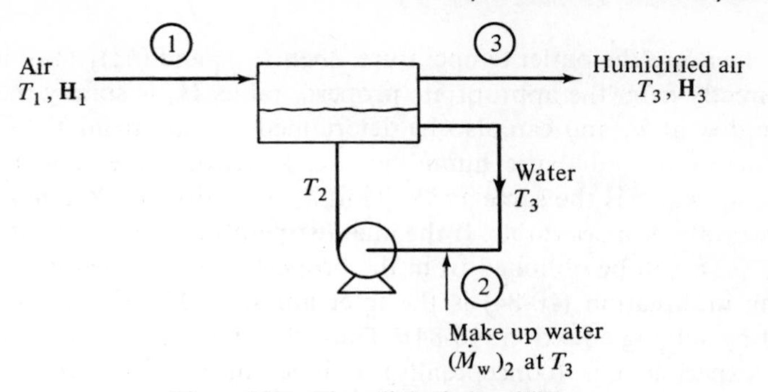

Figure 10-2. The adiabatic humidifier.

The process we have just described is termed *adiabatic humidification.* If the humidification chamber is long enough, then the outlet air stream will also be saturated with water. We may relate the inlet and outlet conditions for the adiabatic humidification as follows:

Mass balance on air:

$$(\dot{M}_a)_1 = (\dot{M}_a)_3 \tag{10-76}$$

Mass balance on water:

$$(\dot{M}_w)_1 + (\dot{M}_w)_2 = (\dot{M}_w)_3 \tag{10-77}$$

Energy balance:

$$[(\dot{M}_a)_1 + (\dot{M}_w)_1][h_1] + (\dot{M}_w)_2(h_w)_2 = [(\dot{M}_w)_3 + (\dot{M}_a)_3][h_3] \tag{10-78}$$

but

$$\dot{M}_a + \dot{M}_w = \dot{M}_a(1 + \mathbf{H})$$

So that (10-78) reduces to:

$$[\dot{M}_{a_1}(1 + \mathbf{H}_1)]h_1 + \dot{M}_{w_2}h_{w_2} = [\dot{M}_{a_3}(1 + \mathbf{H}_3)]h_3 \tag{10-79}$$

but $\dot{M}_{a_1} = \dot{M}_{a_3}$ so that

$$(1 + \mathbf{H}_1)h_1 + \left(\frac{\dot{M}_{w_2}}{\dot{M}_{a_1}}\right)(h_{w_2}) = (1 + \mathbf{H}_3)h_3 \tag{10-80}$$

From the water balance

$$\frac{\dot{M}_{w_2}}{\dot{M}_{a_1}} = \frac{\dot{M}_{w_3}}{\dot{M}_{a_1}} - \frac{\dot{M}_{w_1}}{\dot{M}_{a_1}} \tag{10-81}$$

so that equation (10-80) reduces to

$$(1 + \mathbf{H}_1)h_1 + (\mathbf{H}_3 - \mathbf{H}_1)(h_{w_2}) = (1 + \mathbf{H}_3)h_3 \tag{10-82}$$

Equation (10-82) can be simplified somewhat if we observe that:

$$(1 + \mathbf{H})h \equiv h_{\text{air}} + h_w^V\mathbf{H} \tag{10-83}$$

Thus:

$$[h_{\text{air}} + \mathbf{H}h_w^V]_1 + (\mathbf{H}_3 - \mathbf{H}_1)h_{w_2}^L = [h_{\text{air}} + \mathbf{H}h_w^V]_3 \tag{10-84}$$

If we measure the outlet temperature, then $(h_a)_3$ and $(h_w^V)_3$ may be determined directly from the appropriate property table. $\mathbf{H}_3$ is simply the saturation humidity at T_3 and can also be determined directly from T_3. From the description of the adiabatic humidification statement, the temperature of the make-up water is the same as T_3. Thus h_{w_2} can also be determined from the appropriate property table. If the inlet temperature is also measured, then $(h_a)_1$ and $(h_w^V)_1$ can be obtained from the property tables. The only unknown remaining in equation (10-84) is the inlet humidity $\mathbf{H}_1$ which can then be obtained by solving equation (10-84)! Thus, the adiabatic-saturation humidification experiment is (conceptually) a direct means for determining the humidity of a given air–water stream.

SAMPLE PROBLEM 10-4. Air originally at 120°F is passed through an adiabatic-saturation humidification. The outlet temperature is found to be 80°F. Determine the absolute and relative humidities of the incoming air. Assume C_P^* for air under these conditions is constant at $C_P^* = 7.0$ Btu/lb-mol°R.

Solution: We begin by expressing h_a (T). For convenience we set h_a (32°F) $\equiv 0$ (that is, the reference state). Since we are only interested in enthalpies at $P = 1$ atm, we may express

$$h_a(T) - h(T_0) = \int_{T_0}^{T} C_P(T)\, dT = C_P(T - T_0)$$

but

$$T_0 = 32°\text{F},$$

$$h_a(T_0) = 0$$

$$C_p = 7.0 \text{ Btu/lb-mol °R} = \left(\frac{7.0}{29}\right) \frac{\text{Btu}}{\text{lb}_\text{m}\text{°R}} = 0.241 \frac{\text{Btu}}{\text{lb}_\text{m}\text{°R}}$$

Thus:

$$h_a(T) = [0.241(T) - 7.72] \frac{\text{Btu}}{\text{lb}_\text{m}\text{°R}}$$

where T is °F.

The enthalpy of saturated water vapor is listed as a function of temperature in the steam tables.

We begin the calculation by determining the humidity of the outlet stream. We know the outlet stream is saturated with moisture from the requirements of the adiabatic saturation. Thus,

$$\mathbf{H}_3 = \mathbf{H}^{\text{sat}}(T_3) \equiv \mathbf{H}^{\text{sat}}(80°\text{F})$$

but

$$\mathbf{H}^{\text{sat}}(80°\text{F}) = \frac{18}{29}\left[\frac{P'_{\text{water}}(80°\text{F})}{P - P'_{\text{water}}(80°\text{F})}\right]$$

From the steam tables, P'_{water} (80°F) = 0.51 psia,

$$\mathbf{H}_3 = \frac{18}{29}\left(\frac{0.51}{14.7 - 0.51}\right) \frac{\text{lb}_\text{m}\text{ water}}{\text{lb}_\text{m}\text{ air}} = 0.0223 \frac{\text{lb}_\text{m}\text{ water}}{\text{lb}_\text{m}\text{ air}}$$

The enthalpy of the make-up water is obtained from the steam tables:

$$h_{w_2} = h^L_{\text{water}}(80°\text{F}) = 48.0\,\frac{\text{Btu}}{\text{lb}_m}$$

The enthalpies of water vapor at 80°F and 120°F are also obtained from the steam tables:

$$(h^V_w)_3 = h^V_w(80°\text{F}) = 1096.4\ \text{Btu/lb}_m$$

$$(h^V_w)_1 = h^V_w(120°\text{F}) = 1113.6\ \text{Btu/lb}_m$$

The enthalpy of air is obtained from equation:

$$(h_a)_3 = h_a(80°\text{F}) = 11.6\ \text{Btu/lb}_m$$

$$(h_a)_1 = h_a(120°\text{F}) = 21.1\ \text{Btu/lb}_m$$

Solving equation (10-84) for $\mathbf{H}_1$ gives:

$$\mathbf{H}_1 = \frac{(h_a + \mathbf{H}h^V_w)_3 - (h_a)_1 - \mathbf{H}_3 h_{w_2}}{h^V_{w_1} - h^L_{w_2}}$$

Substitution of the known values then allows determination of $\mathbf{H}_1$:

$$\mathbf{H}_1 = \frac{\left\{[11.6 + 0.0223)1096.4)] - 21.1 - 0.0223(48)\right\}\left(\dfrac{\text{Btu}}{\text{lb}_m\ \text{air}}\right)}{1113.6 - 48\left(\dfrac{\text{Btu}}{\text{lb}_m\ \text{water}}\right)}$$

or

$$\mathbf{H}_1 = 14 \times 10^{-2}\,\frac{\text{lb}_m\ \text{water}}{\text{lb}_m\ \text{air}}$$

Although the calculations illustrated in Sample Problem (10-4) are relatviely simple, they are a bit tedious. For this reason solutions have been generated for a series of inlet and outlet temperatures. The results of these calculations are presented in a convenient graphical form known as the *psychrometric chart* (see Fig. 10-3). On this chart the outlet (adiabatic-saturation) temperature is plotted along the 100% relative humidity line. The adiabatic saturation line which goes through this point is then followed to the inlet temperature where the inlet humidity and relative humidity are obtained.

In theory, the adiabatic-saturation experiment may always be used to determine the humidity of an air stream; in practice it is of course quite impractical. Rather, the *wet-bulb thermometer* is commonly used as practical means to determine humidities. The wet-bulb thermometer shown in Fig. 10-4 consists of three major parts: water reservoir, wick, and thermometer. The wick brings water from the reservoir to the bulb of the thermometer where it vaporizes into the surrounding air stream. During this vaporization the water absorbs energy from the surrounding air stream and thereby reduces

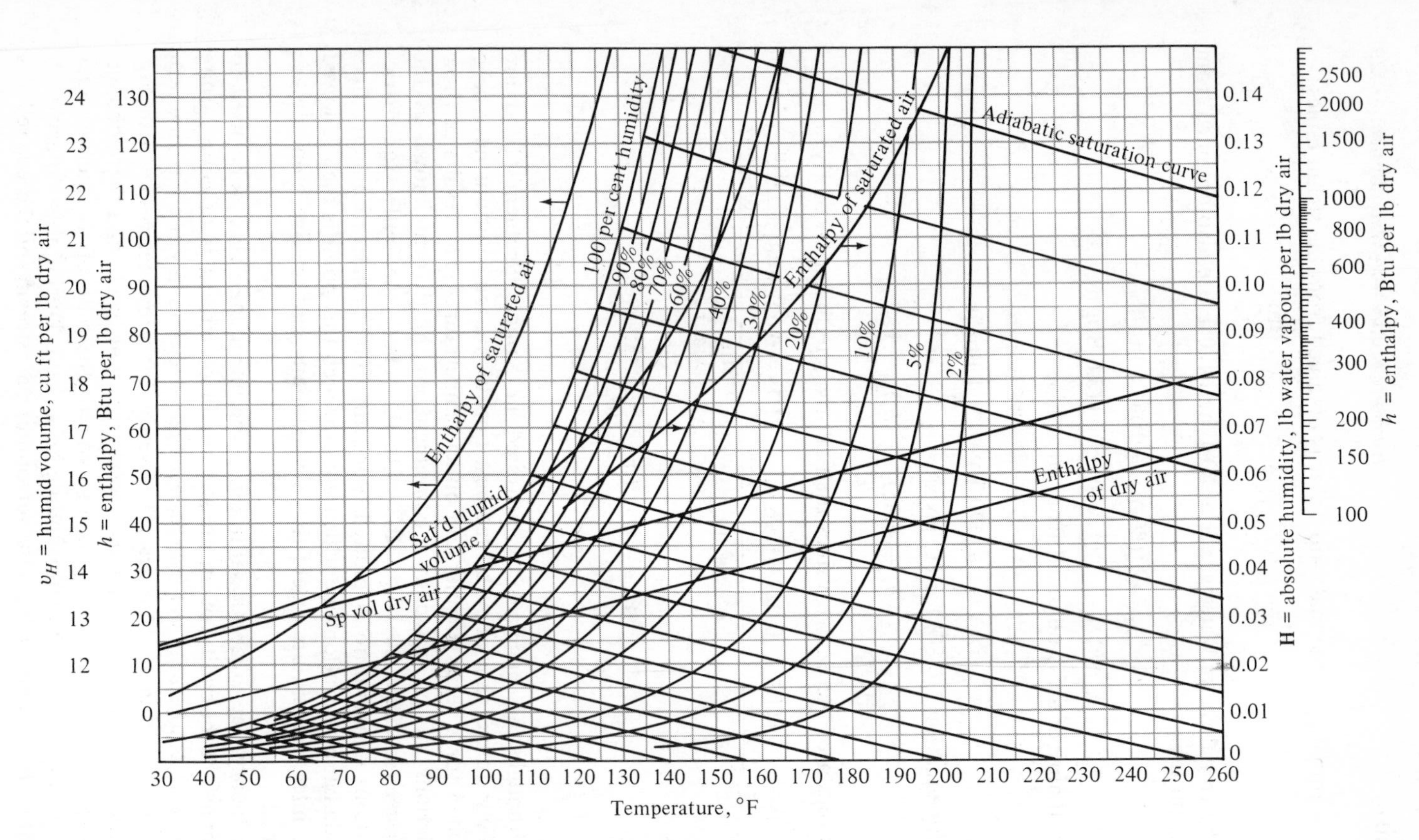

Figure 10-3. Psychrometric chart for air-water vapor, 1 atm. abd. (From: R.E. Treybal, *Mass Transfer Operations*, McGraw-Hill Book Co., New York, 1955.)

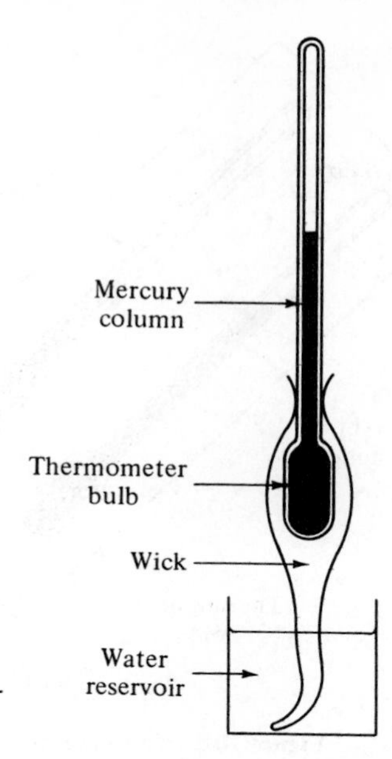

Figure 10-4. The wet-bulb thermometer.

its temperature. It is this reduced temperature which is measured by the wet-bulb thermometer. The wet-bulb temperature is a strong function of the local (or *dry-bulb*) temperature and humidity; it is also a function of the bulb and wick geometry and the local heat-transfer coefficient. However, if the air velocity around the thermometer exceeds $\sim$20 mph, the effect of the external physical parameters becomes minimized and the wet-bulb temperature is only a function of the dry-bulb temperature and humidity. Although this functional relationship may be derived relatively simply, it requires introduction of mass and heat-transfer concepts which are beyond the scope of this text. However, as shown by Treybal,[1] the form of this functional relationship is essentially equivalent to that of equation (10-84) where the wet-bulb conditions are those of state 3 and the dry-bulb conditions are those of state 1. Thus, the adiabatic-saturation lines shown on the psychrometric chart also serve as *wet-bulb lines*.

As we indicated earlier, the wet-bulb temperature should be measured in a $>$20-mph air stream. Although it is obviously inconvenient to require a 20-mph fan whenever the wet-bulb temperature is measured, this difficulty may be overcome by the use of a *sling psychrometer*, illustrated in Fig. 10-5. To use the psychrometer, the wick around the wet bulb is wet with water and the whole unit is swung in a circular path around the handle. By swinging the unit rapidly, the 20-mph air speed is easily exceeded, and after 2–3 minutes,

[1]See for example *Mass Transfer Operations* by R. E. Treybal, McGraw-Hill Book Co., New York, 1955.

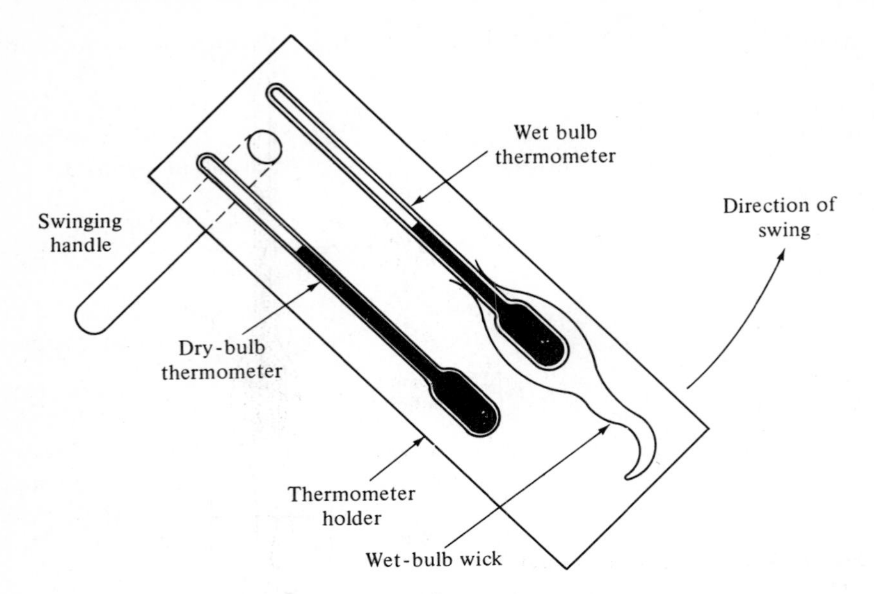

Figure 10-5. The sling psychrometer.

both the wet- and dry-bulb temperatures are read. The humidity and relative humidities are then obtained directly from the psychrometric chart.

SAMPLE PROBLEM 10-5. The wet- and dry-bulb temperatures during a humid afternoon are found to be 80°F and 95°F, respectively. Determine the relative and absolute humidities.

Solution: The wet-bulb temperature is plotted on the saturation line of the psychrometric chart. The adiabatic saturation line through the 80°F wet-bulb temperature is then followed back to $T = 95°F$. The absolute humidity is found to be 1.9×10^{-2} lb_m water/lb_m air, while the relative humidity is ~53 percent. Most people would be quite uncomfortable under these conditions.

Dehumidification

Frequently it is necessary to reduce the absolute and relative humidities of an air stream. If it is necessary to reduce only the relative humidity, with no restriction on temperature, then the temperature may be increased. Although this does not affect the absolute humidity, it does increase the saturation humidity and thereby reduces the relative humidity. This technique is applied, for example, in the household clothes dryer whose inside temperature is raised to ~170°F in order to reduce the relative humidity and increase the driving force for drying of the clothes.

On the other hand, it is frequently necessary to decrease both the relative and absolute humidity. In this case moisture must clearly be removed from the air stream under question. Thus let us consider the following simple process:

1. The air stream is cooled at constant absolute humidity until the saturation temperature is reached.
2. The air stream is further cooled and sufficient moisture collected as liquid water so that the air stream always remains saturated with water vapor.

These steps are shown as paths *a–b* and *b–c* on the psychrometric chart in Fig. 10-6. At point *C* the absolute humidity has been lowered, but the relative humidity is now 100 percent. To reduce the relative humidity we need simply add a third step, *c–d*, in which the air is reheated at constant moisture content to the initial temperature. Both the relative and absolute humidities of the outlet stream have then been reduced.

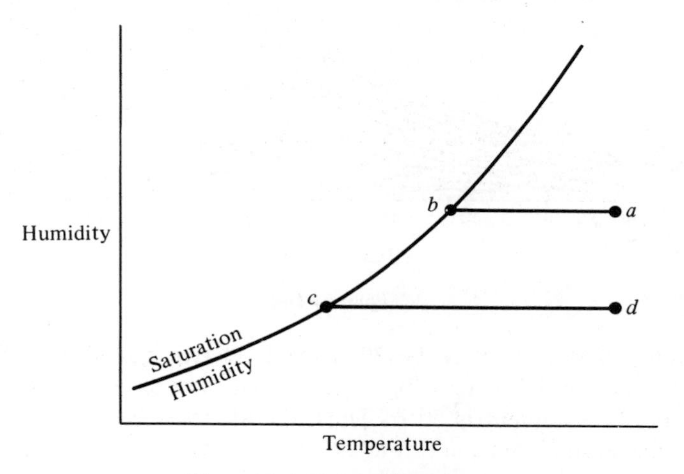

Figure 10-6. Dehumidification.

SAMPLE PROBLEM 10-6. An air stream originally has 75/85°F wet/dry bulb temperatures. To what temperature must this air stream be cooled before reheating to 85°F if the outlet relative humidity is to be less than 30 percent? How much liquid water must be removed per lb_m of bone-dry air dehumidified in this manner?

Solution: The dehumidification process described in the problem statement is outlined on the psychrometric chart shown in Fig. SP 10-6. The input air is found to have 63 percent relative humidity and 1.6×10^{-2} lb_m water/lb_m air.

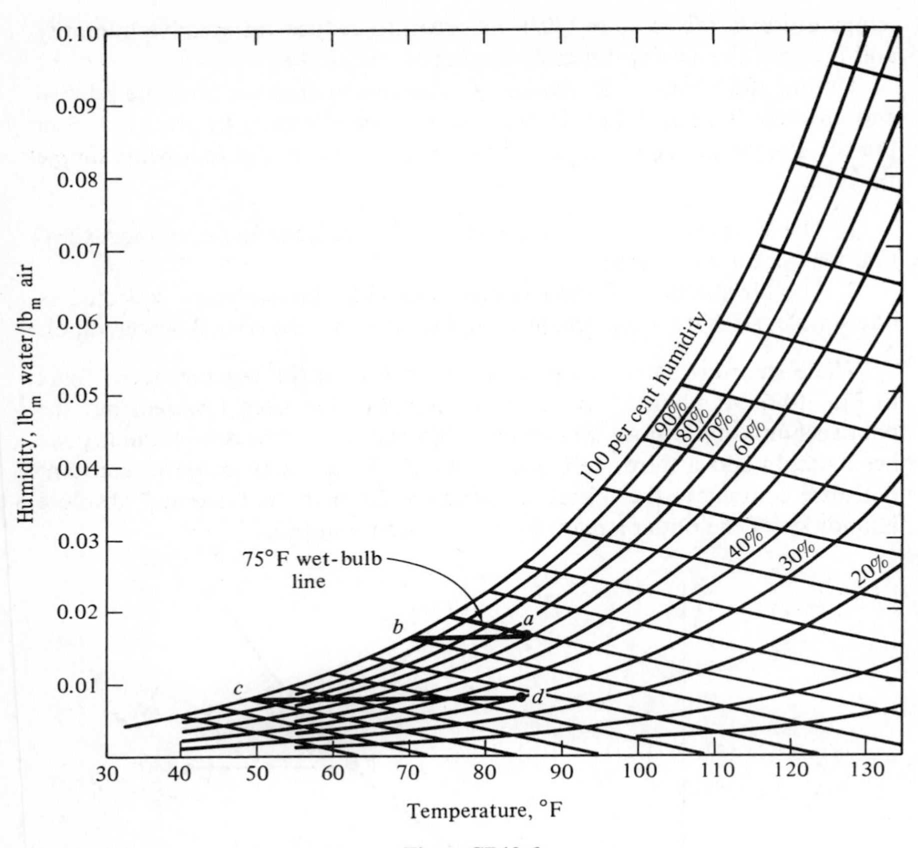

Figure SP10-6

The desired outlet air is to be 30 percent relative humidity at 85°C and thus is to contain 4.3×10^{-3} lb$_m$ water/lb$_m$ air. The saturation temperature at this moisture content is 50°F. Thus, the air must be cooled to 50°F to provide the desired moisture content. The, amount of water extracted is given by

$$\text{water removed} = \mathbf{H}_a - \mathbf{H}_d = \frac{1.2 \times 10^{-2}\ \text{lb}_\text{m}\ \text{water}}{\text{lb}_\text{m}\ \text{dry air}}$$

10.7 Enthalpy–Concentration Diagrams

One of the more useful property charts to the engineer concerned with mixing or fractionation processes is the enthalpy–concentration diagram, an example of which is presented in Fig. 10–7 for the NH_3—H_2O system. The *reference*

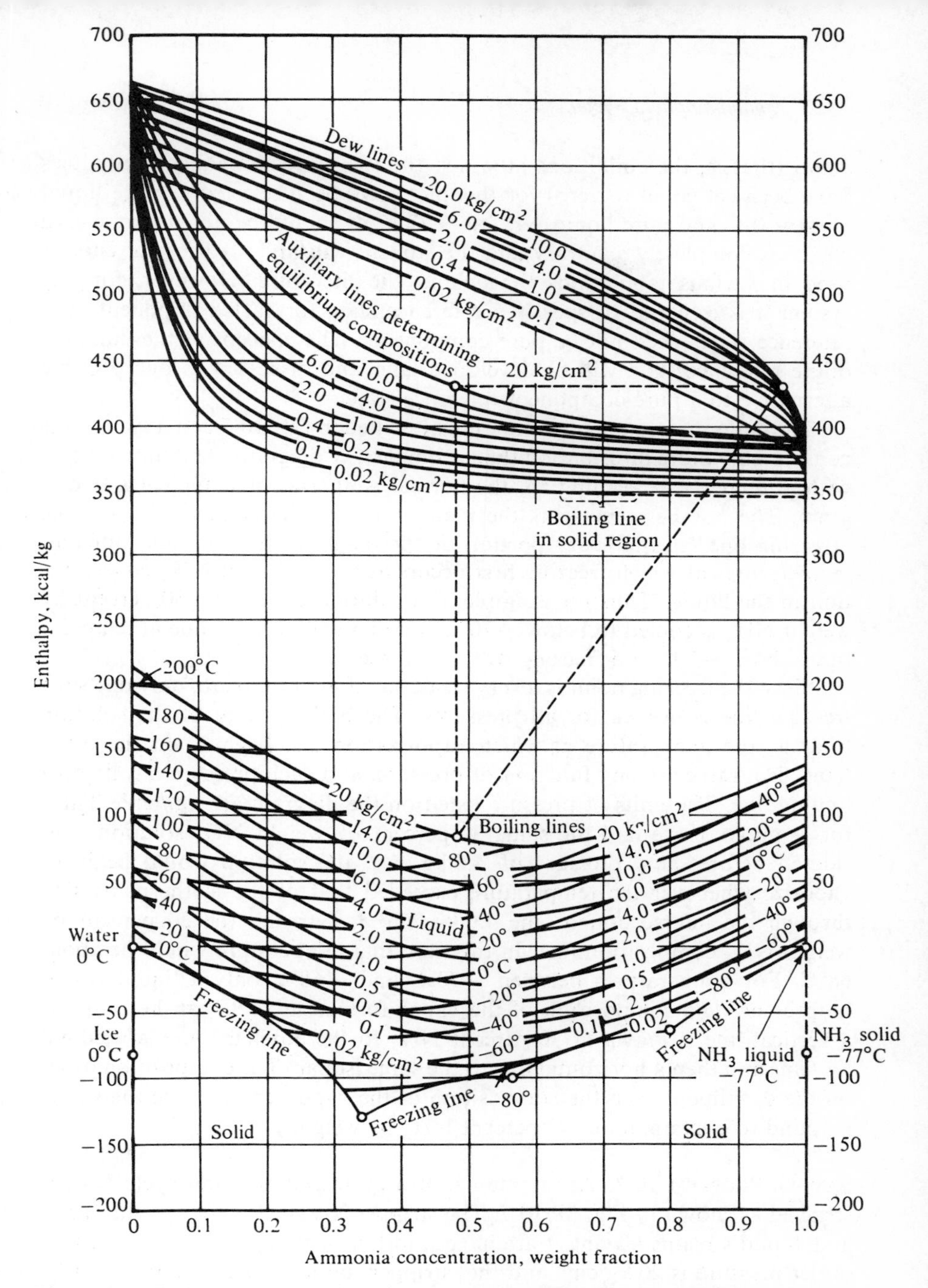

Figure 10-7. Enthalpy-concentration diagram for aqueous ammonia. Reference states: Enthalpies of liquid water at 0°C and liquid ammonia at −77°C are zero. *Note*: To determine equilibrium compositions, a vertical may be erected from any liquid composition on any boiling line and its intersection with the appropriate auxiliary line determined. A horizontal from this intersection will establish the equilibrium vapor composition on the appropriate dew line: At 48 percent ammonia and 20 kg/cm² is indicated. [From R.E. Perry et al., Chemical Engineers Handbook, McGraw-Hill, Inc., New York, 4th ed., (1963) pp 3-154.]

states (that is, the conditions at which the two pure-component enthalpies have been set equal to zero) for this plot have been selected as pure liquid H_2O at 0°C and pure liquid NH_3 at −77°C. It should be noted that such a choice is completely arbitrary and the student will find a variety of choices used in various tabulations throughout the literature. However, for any system it is essential to specify a reference state for each component. The reference states need not be pure components. The concept of the infinitely dilute reference state will be developed later and discussed as one possible alternative to a pure-component reference state.

The diagram for aqueous ammonia presented in Fig. 10-7 actually presents considerably more information than a simple h–x diagram. The lines labeled with a temperature reading are the normal isotherms of a typical h–x diagram. The number represents the temperature of the isotherm in °C. The "freezing line" in the lower portion of the diagram indicates the different temperatures at which freezing first occurs as a function of NH_3 concentration in the liquid. Thus for example, if a solution containing 30 percent by weight NH_3 is cooled to below −70°C freezing will occur, while at temperatures above −70°C no freezing will take place.

Since the freezing point is a very weak function of pressure, only one such freezing line is needed for all pressures. The boiling point of the solution (that is, the temperature at which vaporization just begins), on the other hand, is a rather strong function of pressure, and each pressure has its own boiling line. The units of pressure listed on the diagram are kg_f/cm^2. Thus, for example, if we had the same 30 percent by weight NH_3 solution considered before, at a pressure of 4.0 kg_f/cm^2 the solution would begin to vaporize whenever the temperature exceeded that of the isotherm passing through the intersection of the 4.0 kg_f/cm^2 isobar and the 30 percent by weight NH_3 composition line. In this case that temperature would be about 68°C. For temperatures between −70°C and +68°C only a liquid would be present. The composition of the vapor formed at 68°C can be found if a vertical line is drawn at 30 percent NH_3 to the auxiliary line labeled 4.0 kg_f/cm^2 and then a horizontal line to the same isobar. The composition listed on the dew line is then the composition of the vapor. In this case the vapor is found to contain about 95 percent NH_3 by weight.

SAMPLE PROBLEM 10-7. An ammonia-absorption refrigeration cycle is constructed as shown in Fig. SP 10-7. The stripper is a single stage, so the vapor and liquid streams leaving it are in equilibrium with each other. The pump outlet pressure is 20 kg/cm^2 and the stripper outlet temperature is 100°C. If the evaporator operates at 20°C determine:

(a) The composition of the vapor and liquid streams leaving the stripper.

(b) The minimum pressure at which the evaporator and absorber operate.

(c) The maximum temperature of the liquid leaving the condenser.

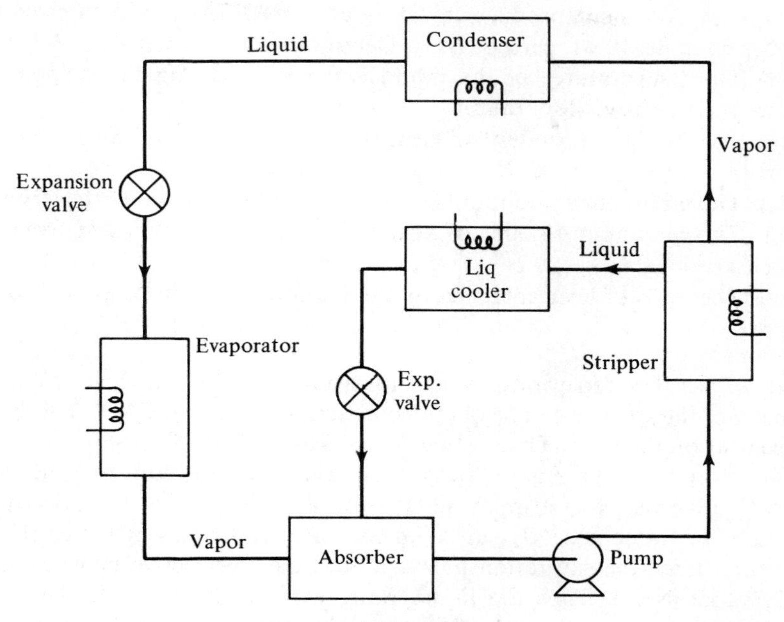

Figure SP10-7

Solution: Since we are told both the pressure and outlet temperatures of the stripper, we can read the liquid and vapor compositions directly from Fig. 10-7. At 20 kg/cm² and 100°C the liquid composition is ~45 percent NH_3 and the corresponding vapor is ~96 percent NH_3 (both by weight).

If the condenser condenses all of the vapor which enters it, then the liquid leaving the condenser must be the same 96 percent NH_3 as the vapor which entered it. The maximum temperature of this liquid is simply its boiling point at 20 kg/cm². From Fig. 10-7 this is ~50°C.

Likewise, if the evaporator is to vaporize all material entering it, the vapor leaving the evaporator will be 96 percent NH_3, and the minimum pressure here will be the dew-point pressure for a gas with 96 percent NH_3. We must therefore find the dew-line pressure whose corresponding liquid temperature is 20°C. By trial and error we find a dew-point pressure of ~0.05 kg/cm². The corresponding dew composition is ~25 percent NH_3.

Sample Problem 10-8. The single-stage stripper in the ammonia-absorption refrigerator of Sample Problem 10-7 is replaced by a multistage stripper which yields essentially pure gaseous NH_3 and pure liquid H_2O. The stripper pressure is still 20 kg/cm², and the condenser gives a saturated liquid. The liquid water is cooled to 30°C before entering the absorber. The evaporator operates at 10°C and produces a saturated vapor. The liquid entering the stripper contains 10 wt. percent NH_3. Determine:

(a) The temperatures of the liquid and vapor leaving the stripper, as well as their relative flow rates.

(b) The minimum amount of heat which must be withdrawn from the absorber.

(c) The minimum amount of heat which must be supplied to the stripper.

(d) The minimum amount of heat which must be withdrawn from the condenser.

(e) The ratio of heat absorbed by the evaporator to that supplied to the stripper.

Solution: (a) The refrigerator is as shown in Fig. SP 10-8a. The properties of many of the streams can be obtained directly from the NH_3–H_2O enthalpy concentration diagram. These values are shown on Fig. SP 10-8b.

The evaporator pressure is simply the saturation pressure of liquid NH_3 at 10°C. Likewise, the stripper outlet temperatures are the saturation temperatures of water and NH_3 at 20 kg/cm². The enthalpies of these streams are found from the saturation pressures and their respective phases (liquid H_2O, vapor NH_3). Thus the liquid water exits at 210°C while the vapor NH_3 leaves at 50°C. Since the liquid entering the stripper is 10 wt. percent NH_3, a mass balance around the stripper shows that 9 kg of liquid water are produced for each 1 kg of vapor NH_3. For convenience we will base all remaining calculations on 10 kg of liquid entering the stripper.

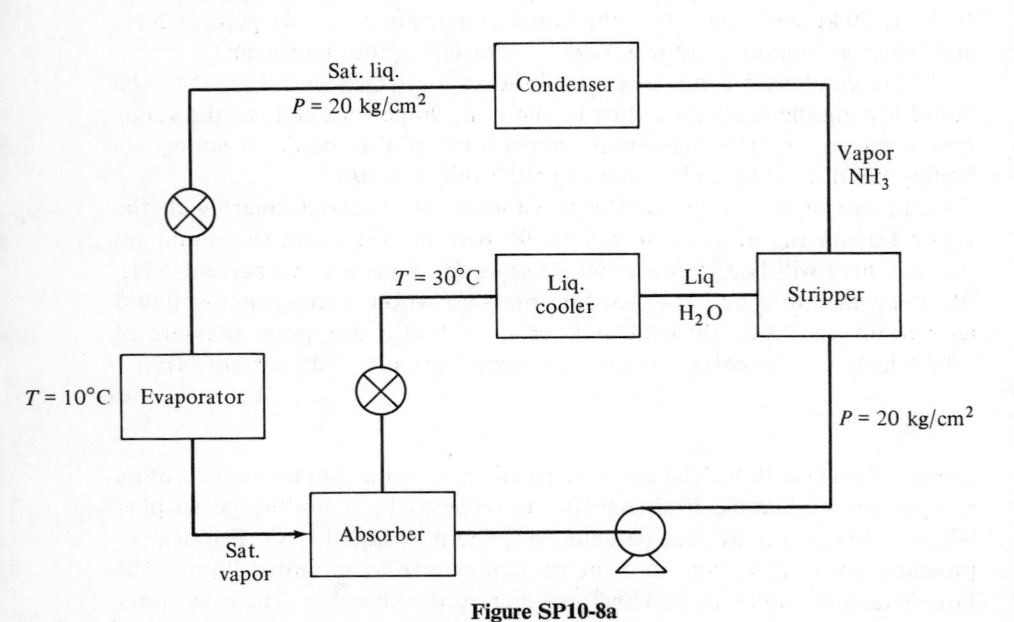

Figure SP10-8a

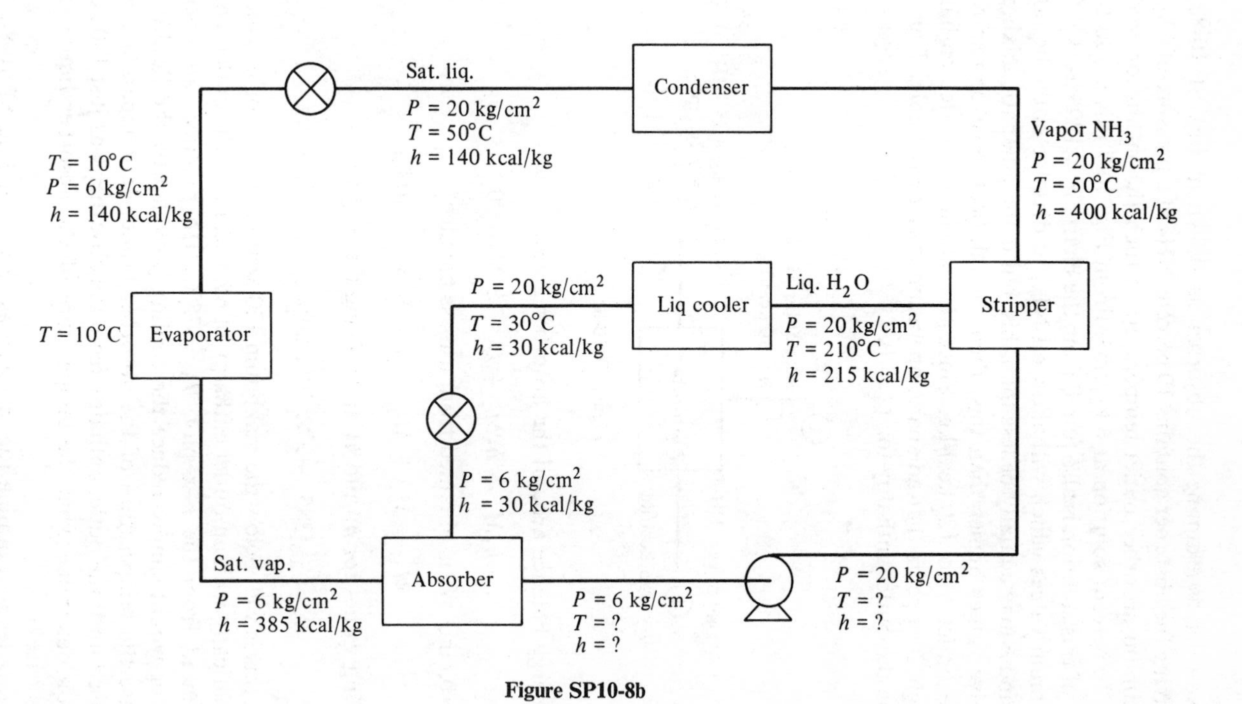

Figure SP10-8b

(b) Now let us examine the absorber as shown in Fig. SP 10-8c. The liquid leaving the absorber contains 10 percent NH_3 at a pressure of 6 kg/cm². The maximum absorber outlet temperature is then the saturation temperature of a 10 percent solution at 6 kg/cm². From Fig. 10-7 this is 130°C. The enthalpy of this solution is also found from the chart: $h = 120$ kcal/kg. These are maximum values which will allow all NH_3 to dissolve in the liquid. The outlet T and h may actually be less than this, but if the adiabatic mixing conditions give values higher than this, then enough heat must be removed to bring the outlet h to 120 kcal/kg. Since the maximum outlet enthalpy is pretty high, it appears that adiabatic operation may be possible. Thus let us determine the outlet enthalpy for $Q = 0$.

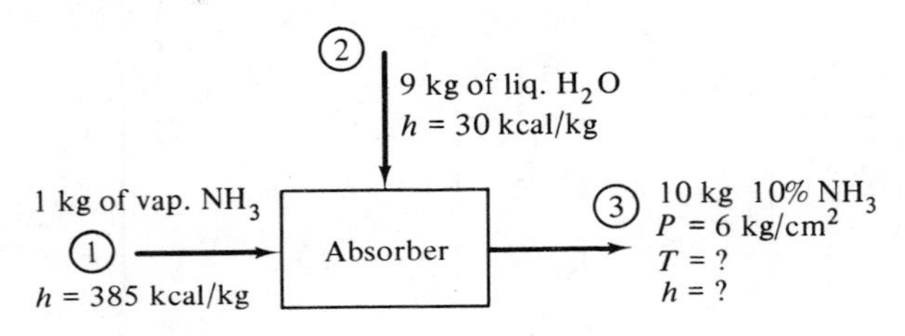

Figure SP10-8c

An energy balance around the absorber gives:

$$h_1 M_1 + h_2 M_2 + Q - h_3 M_3 = 0$$

but $Q = 0$, and M_3 is obtained from a mass balance:

$$M_3 = M_1 + M_2 = (1 + 9) \text{ kg} = 10 \text{ kg}$$

Substituting values for h_1 and h_2 and solving for h_3 gives:

$$h_3 = \tfrac{1}{10}[385 + 9 \times 30] \text{ kcal/kg} = 65.5 \text{ kcal/kg}$$

Since this is well below the maximum 130 kcal/kg, the absorber can operate adiabatically with an outlet enthalpy of 65.5 kcal/kg. The outlet temperature is found from the h–x plot, $T_3 = 75$°C. (In practice, an operating refrigerator would simply reduce the cooling provided by the liquid cooler to increase the temperature of the absorber liquid. Its temperature would be increased until the outlet enthalpy from the absorber was just 130 kcal/kg —Can you determine what the temperature of the absorber liquid should be in this case?)

(c) Now let us examine the stripper as shown in Fig. SP 10-8d. Note that we have assumed that the enthalpy of the liquid entering the stripper is the same as that leaving the absorber. Normally we would expect a small enthalpy increase across the pump, but this will be very small for the liquid

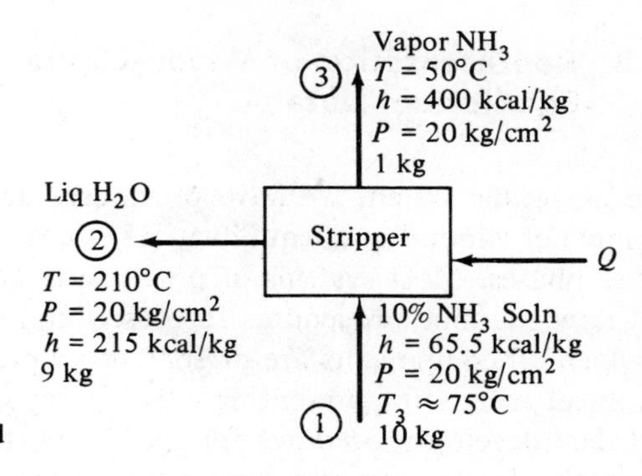

Figure SP10-8d

solution and has been neglected here. The stripper heat load can be calculated from an energy balance around the stripper.

$$h_1M_1 - [h_2M_2 + h_3M_3] + Q = 0$$

or

$$Q = h_2M_2 + h_3M_3 - h_1M_1$$

Substituting values gives

$$Q = [(9)(215) + 1(400) - 10(65.5)] \text{ kcal/10 kg fed}$$

or:

$$Q = 1680 \text{ kcal/10 kg fed to stripper}$$

(d) The heat load of the condenser is obtained from a simple energy balance around the condenser:

$$\begin{aligned} Q &= (Mh)_{\text{out}} - (Mh)_{\text{in}} \\ &= 1 \text{ kg } (140 - 400) \text{ kcal/kg} \\ &= -260 \text{ kcal} \quad \text{or} \quad -260 \text{ kcal/10 kg fed to stripper} \end{aligned}$$

with the minus sign indicating that heat is actually withdrawn.

(e) The heat absorbed by the evaporator is also obtained from a simple energy balance:

$$\begin{aligned} Q &= (Mh)_{\text{out}} - (Mh)_{\text{in}} \\ &= 1[385 - 140] \text{ kcal} \\ &= 245 \text{ kcal} \quad \text{or} \quad 245 \text{ kcal/10 kg fed to stripper} \end{aligned}$$

The ratio of heat absorbed in the evaporator to heat supplied to the stripper is then

$$\frac{Q_{\text{evap}}}{Q_{\text{stripper}}} = \frac{245 \text{ kcal/10 kg fed to stripper}}{1680 \text{ kcal/10 kg fed to stripper}} = 0.146$$

10.8 Representation of Vapor–Liquid Equilibrium Data

The air–water system we have previously described is a relatively simple example of vapor–liquid equilibrium since we can "uncouple" the vapor and liquid phases. Most systems of practical interest are not so simple, and we will now examine procedures for describing the phase behavior of systems in which all components are present in all phases. We begin by considering graphical means for presenting existing vapor–liquid equilibrium data; we will then develop procedures for predicting this behavior in simple systems where the data do not exist or are not readily available.

When we consider the graphical presentation of vapor–liquid data, the first question is what are we to plot, and against what variables. Here the choice is quite wide—pressure, temperature, composition, partial fugactiy, and chemical potential are just a few of the variables we could plot. Clearly, as we have shown in the previous sections, many of these are redundant and it would be wasted effort to prepare any more than the minimum number of such plots necessary to completely specify the remaining properties. In practice it turns out that the properties of greatest interest are usually the compositions of the equilibrium phases as functions of pressure and temperature. The remaining properties are of interest primarily because they would allow us to calculate the phase compositions if they were not known. In addition, since it is not possible to plot the phase compositions as functions of *both* temperature and pressure on a single two-dimensional graph, it is most common to find vapor–liquid (or any other type of phase equilibrium) equilibrium data presented in the form of temperature–composition (fixed-pressure) or pressure–composition (fixed-temperature) plots. These plots present in a highly useful form the great bulk of the information collected in a vapor–liquid equilibrium measurement. Example temperature–composition and pressure–composition plots for a typical binary mixture are presented in Fig. 10-8.

Let us consider for the moment the temperature–composition (or T–x) plot. Analysis of the pressure–composition (or P–x) curve is analogous. The vertical ordinate is temperature, while the horizontal coordinate is composition—in this case, mole fraction. The T–x region is split into three distinct regions by the curves labeled "dew-point curve" and "bubble-point curve." The two curves are obtained from a series of vapor–liquid equilibrium measurements, each performed at a different temperature, but all at the pressure used in developing the T–x diagram. Each experiment typically consists of placing a liquid sample into some form of boiling chamber or "still." The sample is then heated until boiling occurs and vapor is evolved. If the still is properly designed and operated (no small experimental task), the resulting

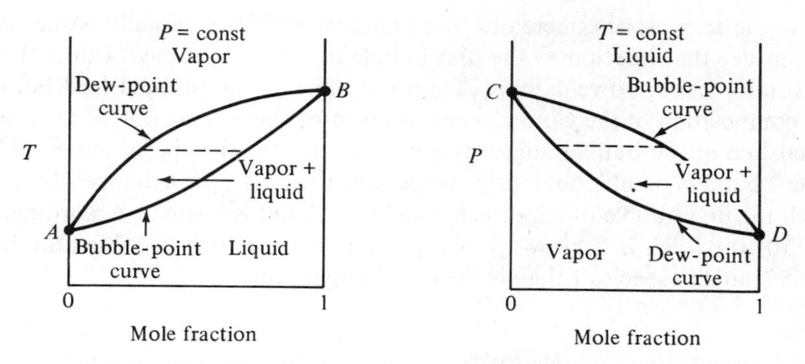

Figure 10-8. Temperature-composition and pressure-composition diagrams for binary mixtures.

vapor and liquid are in equilibrium. The temperatures of the mixture and both vapor and liquid compositions are then measured. (The composition measurement is also a rather difficult experimental task which must always be considered in great detail before real experiments are performed. In this discussion we'll assume our experimenter is careful and accurate, and we'll concern ourselves more with what the experiments tell us rather than how they would actually be performed.) The compositions of the liquid and vapor are plotted at the measured temperature to yield a single point each on the bubble- and dew-point curves, respectively. The liquid sample is then changed and the whole procedure repeated with the new liquid. After many such experiments, the entire dew-point and bubble-point curves are established.

The region below the bubble-point curve on the T–x diagram corresponds to a region of subcooled liquid. That is, the liquid at this specific temperature is below its first boiling (or bubble-point) temperature and hence no vapor can coexist with the liquid. Suppose we enclose this sample in a constant-pressure chamber as shown in Fig. 10-9 and slowly increase the temperature of the liquid.

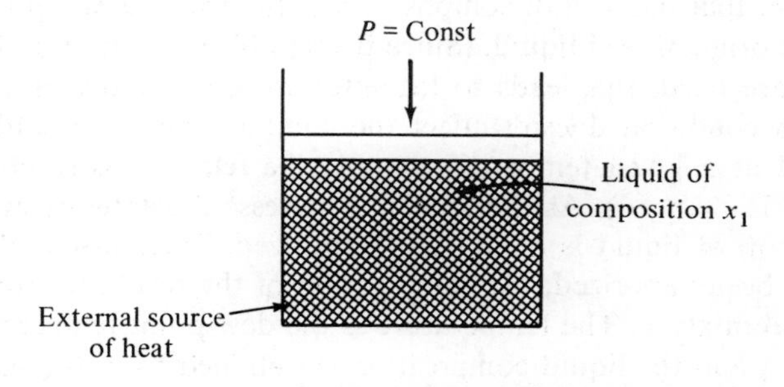

Figure 10-9. Initial boiling of a mixture.

As the temperature increases, the first boiling will eventually ensue. Let us consider the situation as the first bubble of vapor is formed. Under these conditions the liquid remaining is still the same as the original feed liquid. The composition of the vapor formed is given by the intersection of the horizontal line at the bubble-point temperature with the dew-point curve. (The name "bubble point" obviously arises because this is the temperature at which the first bubble of vapor is formed as a liquid is heated.) According to Fig. 10-10 (which is a blow-up of Fig. 10-8) this vapor is considerably less concentrated in species 1 than is the remaining liquid.

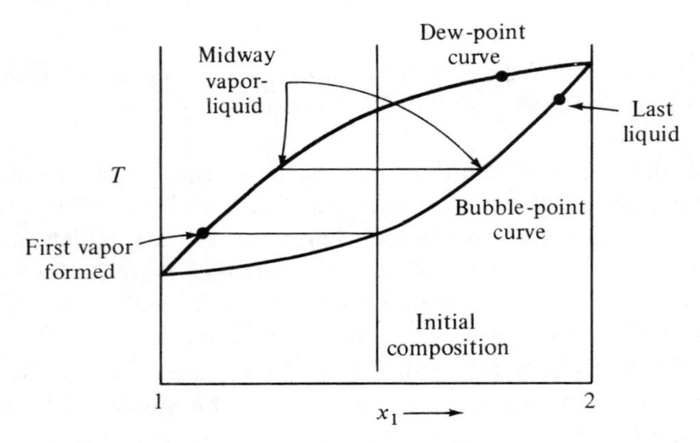

Figure 10-10. The $T - x$ diagram.

As the heating is continued, more vapor forms and the temperature of the vapor–liquid mixture increases. Since the vapor is always less concentrated in species 1 than is the liquid, the liquid composition of species 1 increases as the vaporization continues. The compositions of the equilibrium vapor and liquid at any specific temperature are given by the intersection of the horizontal line at the given temperature with the dew- and bubble-point curves. The relative quantities of vapor and liquid are determined by the requirement that the overall composition of the vapor and liquid must equal that of the original feed liquid. (Since the equilibrium vapor and liquid compositions are fixed, this leads to the interesting result that a change in the initial feed condition doesn't affect the compositions of equilibrium vapor and liquid at a given temperature—only the relative distribution of vapor and liquid is changed). As the heating progresses, eventually everything but the last drop of liquid is completely vaporized. Since essentially all of the liquid has been vaporized, the composition of the resulting vapor is that of the original mixture. The temperature is the dew-point temperature at composition x_1, and the liquid composition is obtained from the intersection of the horizontal line at the dew-point temperature with the bubble-point curve. (The name "dew point" arises since this is the first temperature at which liquid, or *dew*, is formed as a vapor is cooled). At temperatures above the dew-point temperature, we have superheated vapor.

SAMPLE PROBLEM 10-9. A mixture initially containing 35 mol percent benzene in heptane is at a temperature of 80°C and a pressure of 760 mm Hg.

(a) What is the state of the mixture at these conditions?

(b) Heat is applied at constant pressure until vapor just begins to form. What is the temperature at this point, and what is the composition of the first vapor that forms?

(c) If the heating is continued until $T = 89°C$, determine the relative amounts of vapor and liquid as well as the compositions of the two phases.

(d) At what temperature will 50 mol percent of the original mixture be vaporized?

(e) At what temperature is the last of the liquid vaporized, and what is the composition of this liquid?

A *T–x* diagram for the benzene-heptane system constructed from the experimental data of Meyers[2] is shown in Fig. SP 10-9a.

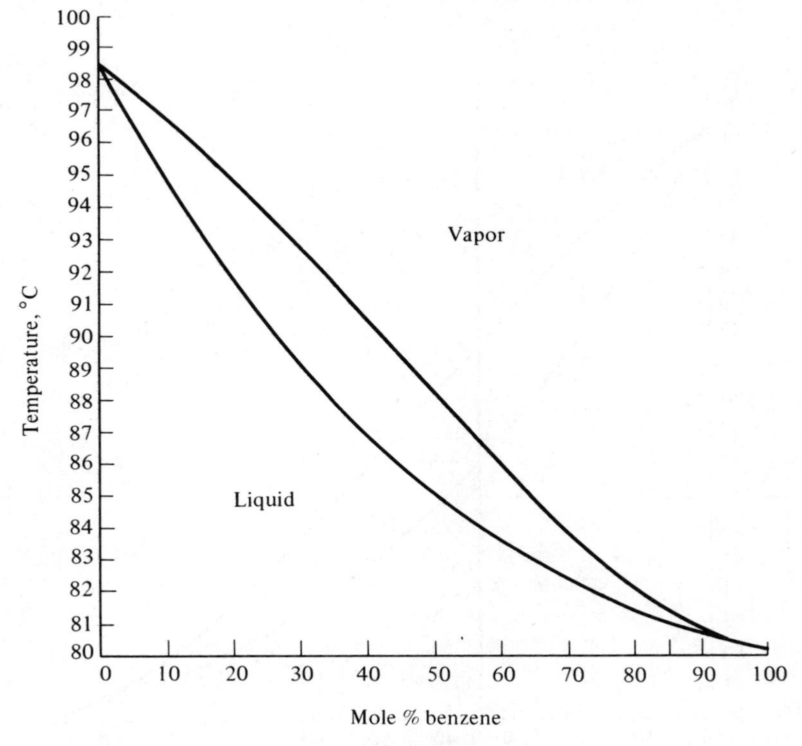

Figure SP10-9a

[2]H.S. Meyers, *Ind. & Eng. Chem.*, **47**, 10, 2215 (Oct., 1955). Reprinted by permission, Copyright American Chemical Society.

Solution: (a) At 80°C the 35 mol percent benzene mixture is well within the *subcooled liquid range*, and so we have a single phase (liquid) under these conditions.

(b) By drawing the vertical line at 35 mol percent benzene and following it up, we find that the bubble point is at 88.1°C. This is the place where vapor is first formed. This first vapor has a composition of 50 mol percent benzene.

(c) At 89°C we find that the liquid composition will be 31 mol percent benzene while the vapor composition is 47 mol percent benzene. The relative amounts of vapor and liquid are now obtained from a component mass balance on either the benzene or the heptane. We will work with benzene and base our calculations on 1 mol of initial feed. The total amount of liquid plus vapor at any stage is then equal to 1 mol (from an overall balance)

$$L + V = 1$$

where L = number of moles of liquid
V = number of moles of vapor

The component balance on benzene then gives:

$$x_b L + y_b V = (x_b)_F(1)$$

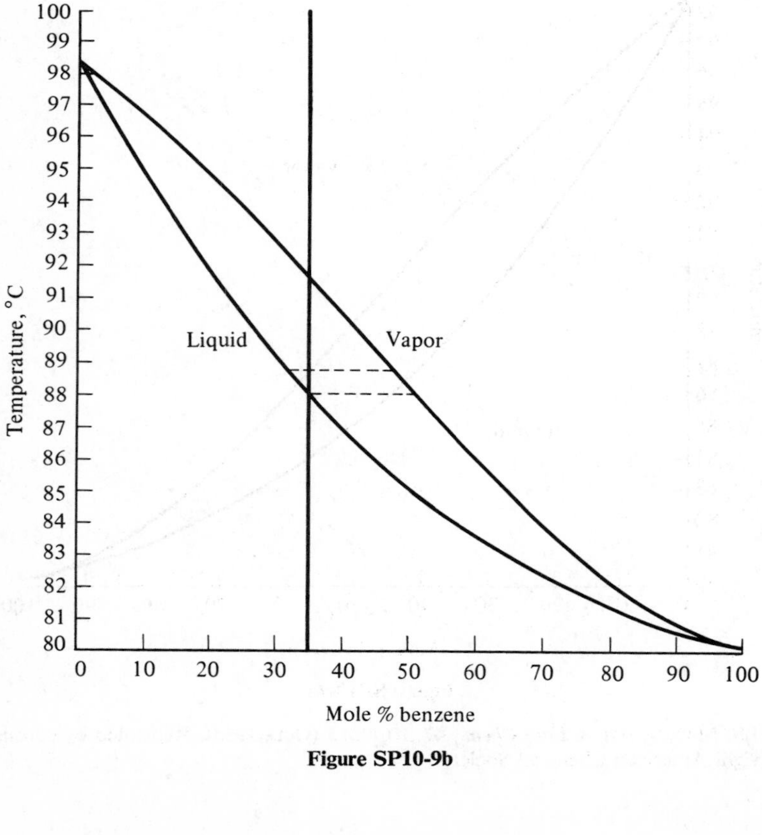

Figure SP10-9b

where x_b = liquid composition
y_b = vapor composition
$(x_b)_F$ = feed composition

Elimination of (say) V from the two equations gives:

$$x_b L + y_b(1 - L) = (x_b)_F$$

or

$$L(x_b - y_b) = (x_b)_F - (y_b)$$

which may be solved for L to give:

$$L = \frac{[y_b - (x_b)_F]}{[y_b - x_b]}$$

To interpret this result graphically we observe that $[y_b - (x_b)_F]$ is the horizontal distance between the dew-point curve and the feed composition at the temperature in question, while $[y_b - x_b]$ is the total distance between the dew- and bubble-point curves at this temperature as shown in Fig. SP10-9c:

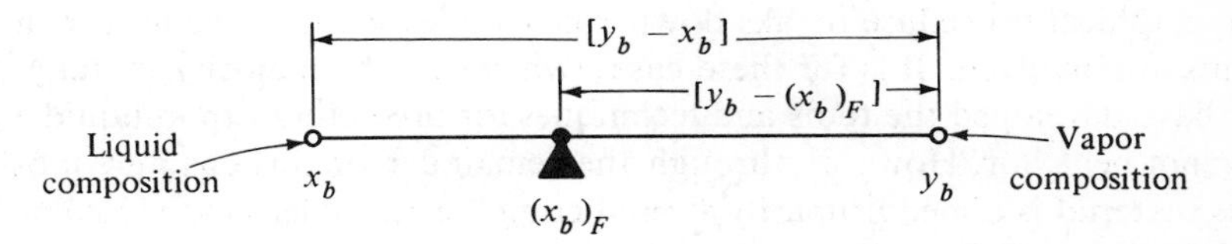

Figure SP10-9c

The result

$$L = \frac{[y_b - (x_b)_F]}{[y_b - x_b]}$$

when interpreted graphically in this manner is known as the *inverse-lever rule* since the fraction of the mixture that is liquid is given by the inverse-lever arm of an imaginary scale whose balance point is the feed composition. Substitution of the known bubble point, dew point, and feed compositions then gives

$$L = \frac{[y_b - (x_b)_F]}{[y_b - x_b]} = \frac{47 - 35}{47 - 31} = 0.75$$

The vapor is obtained from the overall balance

$$V = 1 - L = 0.25$$

Thus at these conditions 25 percent of the feed liquid has been vaporized.

(d) The point where 50 percent of the feed is vaporized can be found most simply by means of inverse-lever rule we have just derived. If 50 percent of the feed is to be liquid, then

$$L = \frac{[y_b - (x_b)_F]}{[y_b - x_b]} = 0.50$$

That is, the feed composition should fall exactly halfway between the liquid and vapor compositions. A couple of trial-and-error calculations quickly locate the temperature at which the feed composition is halfway between the liquid and vapor compositions: $T = 89.8°C$ and the liquid and vapor compositions are $(x_b) = 0.27$ mol fraction and $(y_b) = 0.43$ mol fraction.

(e) As the heating continues, the dew point is eventually reached at $T = 91.6°C$. The liquid composition at this temperature is found to be 0.31 mol fraction benzene. For temperatures above 91.6°C, only vapor is present and its composition is of course 0.35 mol fraction benzene.

Thus the *T–x* or *P–x* diagram portrays a great deal of information about the vapor–liquid equilibrium behavior of a given binary mixture. For those cases where these diagrams are available for the pressures and temperatures of interest, we can solve essentially any vapor–liquid equilibrium problem with only the information given on the diagram. Unfortunately, these diagrams are available for only a limited number of mixtures, at an even more limited number of pressures and/or temperatures. In addition, the graphical procedure breaks down when three or more chemical components are involved. It is for these cases (which are the majority by far.) that we have developed the tools and techniques for predicting vapor–liquid equilibrium behavior. However, through the remainder of this chapter, most of this material is aimed primarily at predicting the equivalent individual points on the *T–x* (or *P–x*) diagram for those cases where the diagram is either not available or where a simple two-dimensional diagram is insufficient.

The *T–x* diagram given in Sample Problem 10-9 is a relatively simple diagram that shows no particularly unusual behavior. It is possible to have considerably more complex diagrams than this. In addition, we have considered the *T–x* diagram as it applies to vapor–liquid equilibrium. Clearly, we may apply the same general principles in the determination of solid–liquid or even liquid–liquid equilibrium. The *T–x* diagram is exceedingly useful in describing the phase behavior of these systems as well.

Prediction of Vapor–Liquid Equilibrium

A comprehensive discussion of the prediction of vapor–liquid equilibrium is not feasible within this text because of the great variety of combinations possible.[3] Rather, we shall restrict this discussion to low-pressure ideal solutions.

At pressures less than 10 atm it is normally reasonable to assume that gases form ideal (or Lewis–Randall) mixtures. Thus under these conditions

[3]The reader wishing to explore these topics more comprehensively can find elaboration on nearly every aspect of this chapter in the excellent monograph by J. M. Prausnitz, *Molecular Thermodynamics of Fluid Phase Equilibria*, Prentice-Hall, Inc., Englewood Cliffs, N.J., 1969.

the general criterion for phase equilibrium, $\bar{f}_i^L = \bar{f}_i^V$, (Eqn. 10-60) reduces to:

$$x_i f_i^L = y_i f_i^V \tag{10-85}$$

At pressures less than 10 atm, it is also possible to equate the fugacities and the pressures without serious error. Under these circumstances $f_i^V = P_T$, the total pressure on the system, while $f_i^L = P'_i$, the vapor pressure of component i at the temperature of the system. Thus equation (10-85) reduces to:

$$x_i P'_i = y_i P_T = P_i \tag{10-86}$$

where P_i is the partial pressure of species i in the vapor phase. P_i is defined by $P_i = y_i P_T$. Equation (10-86) is generally referred to as Raoult's Law. It applies to systems for which ideal solution behavior characterizes both phases and ideal gas assumptions permit equating fugacitits and pressures. This is the simplest quantitative rule for phase equilibrium and is reliable for a limited number of real mixtures. However, as indicated in Fig. 10-11, it does provide a useful standard against which to compare vapor–liquid equilibrium. The dashed lines represent Raoult's-law predictions and the solid line, a typical nonideal solution. The total pressure on the system, P_{total}, is simply the sum of the partial pressures of the various components. The liquid and vapor compositions for the system shown in Figure 10-11 are plotted against each other in Figure 10-12. The vapor phase compositions were found by dividing the species partial pressures by the total system pressure.

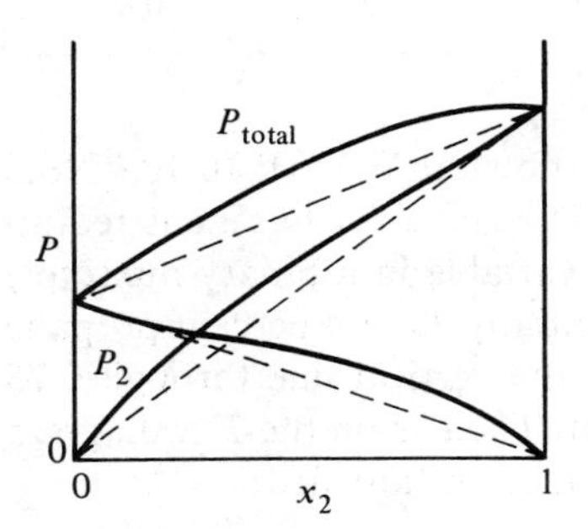

Figure 10-11. Comparison of Raoult's law with a typical nonideal binary mixture.

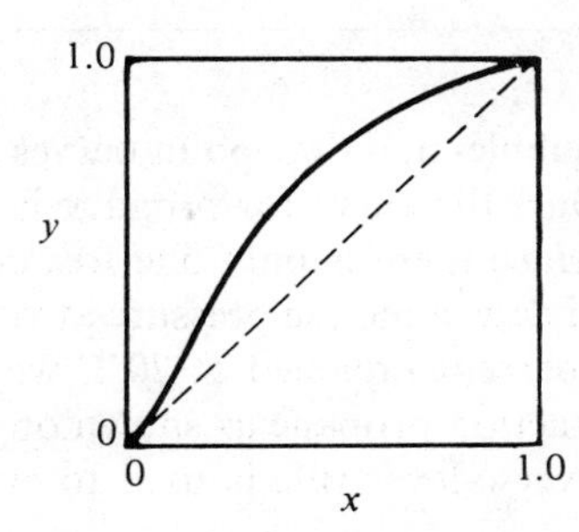

Figure 10-12. Vapor composition versus liquid composition for a binary mixture at either constant P or T.

SAMPLE PROBLEM 10-10. At 70°F mixtures of propane and n-butane may be assumed to obey Raoult's law for pressures below 10 atm. The vapor pressures of the pure components at this temperature are

propane: $P'_{70°\text{F}} = 124$ psia

n-butane: $P'_{70°\text{F}} = 32$ psia

Determine and plot the dew- and bubble-point curves for mixtures of C_3 and n-C_4 at 70°F. To what pressure must a gas mixture of composition 75 mol percent C_3 be compressed before 60 percent of the mixture has condensed. What is the composition of the resulting liquid?

Solution: Since the mixture is known to obey Raoult's law, determination of the total pressure and vapor composition is relatively simple once the liquid-phase composition is known. Thus to plot the bubble-point and dew-point curves let us choose a series of liquid-phase compositions. For each of these the total pressure and vapor–phase compositions are evaluated. The P–x plot may then be obtained directly if we plot x_i and y_i against P:

x_3	x_4	$P_3 = x_3P'_3$	$P_4 = x_4P'_4$	P_T	$y_3 = P_3/P_T$	$y_4 = P_4/P_T$
0.0	1.0	0 psia	32 psia	32 psia	0	1.0
0.1	0.9	12.4	28.8	41.2	0.301	0.699
0.2	0.8	24.8	25.6	50.4	0.493	0.507
0.3	0.7	37.2	22.4	59.6	0.624	0.376
0.4	0.6	49.6	19.2	68.8	0.721	0.279
0.5	0.5	62.0	16.0	78.0	0.795	0.205
0.6	0.4	74.4	12.8	87.2	0.853	0.147
0.7	0.3	86.8	9.6	96.4	0.900	0.100
0.8	0.2	99.2	6.4	105.6	0.94	0.060
0.9	0.1	111.6	3.2	114.8	0.975	0.025
1.0	0.0	124.0	0	124	1.0	0.0

The bubble- and dew-point curves are then as shown in Fig. SP 10-10. (Note that only the curve for propane is presented. The curve for butane is redundant since there is only one free composition variable in a binary mixture!)

To determine the pressure at which a gas initially 75 mol percent propane is 60 percent liquefied at 70°F we must draw the vertical line through 0.75 mol fraction propane as shown on the P–x plot. Then as on the T–x diagram the inverse-lever rule is used to evaluate the fraction liquefied:

$$L = \frac{y_{C_3} - (x_{C_3})_F}{y_{C_3} - x_{C_3}}$$

That is, we must find the pressure where the 0.75 mol fraction C_3 line lies 60 percent of the distance between the vapor and liquid lines, away from the vapor line. A quick trial-and-error calculation then yields $P = 92.8$ psia. The vapor and liquid compositions as read from the diagram are

$$x_{C_3} = 0.662, \qquad y_{C_3} = 0.883$$

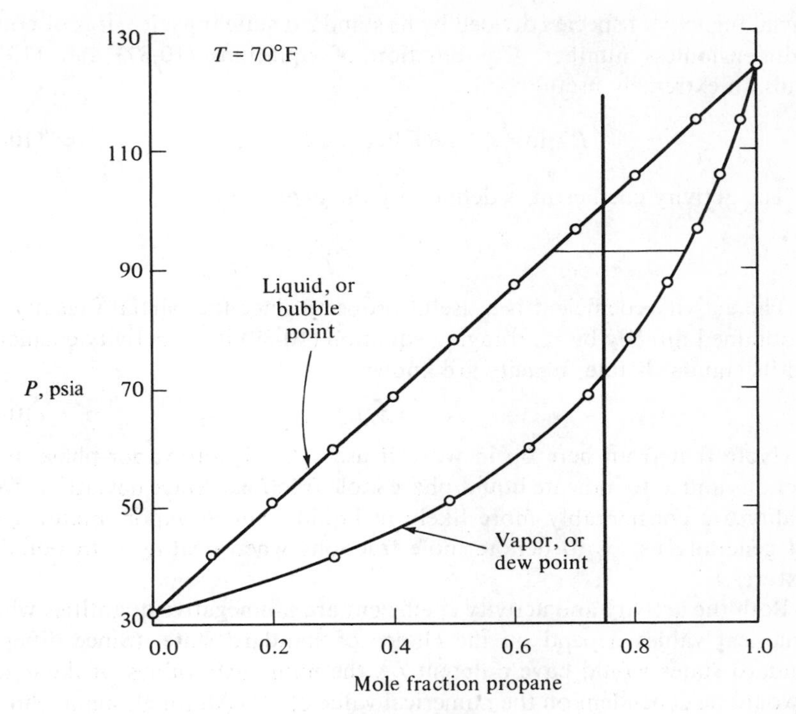

Figure SP10-10

10.9 Nonideal Solutions—The Activity and Activity Coefficient

Although the Lewis–Randall rule is quite useful for evaluating partial fugacities (or chemical potentials) for many gaseous mixtures and some liquid mixtures, it is not widely applicable to liquid mixtures in general. Since many liquid mixtures do not obey the Lewis–Randall rule, means for dealing with these mixtures are clearly needed. We thus introduce two new thermodynamic variables—the *activity*, a_i, and the *activity coefficient*, γ_i°. The activity of component i is defined to be

$$a_i = \bar{f}_i / f_i^\circ \tag{10-87}$$

where f_i° is the fugacity of component i in some *arbitrarily chosen reference* (or *standard*) state, denoted by the superscript °. The choice of the reference state is essentially arbitrary with the exception that the temperature of the reference state must be the same as the temperature of the mixture under consideration. The activity so defined is seen to be simply the ratio of the partial fugacity of specie i divided by its standard state fugacity; it is of course a dimensionless number. Combination of equations (10-87) and (10-53) yields an extremely useful result:

$$RT \ln \frac{\bar{f}_i}{f_i^\circ} = RT \ln a_i \equiv \bar{G}_i - g_i^\circ \tag{10-88}$$

The activity coefficient is defined by the expression

$$\gamma_i^\circ = \frac{a_i}{x_i} = \frac{\bar{f}_i}{x_i f_i^\circ} \tag{10-89}$$

The activity coefficient is a useful property since the partial fugacity can be obtained directly by rearranging equation (10-89) if the activity coefficient and its standard-state fugacity are known:

$$\bar{f}_i = x_i \gamma_i^\circ f_i^\circ \tag{10-90}$$

(Note that from here on in we will use y_i to denote vapor-phase mole fractions and x_i to indicate liquid-phase mole fractions. Since deviations from ideality are considerably more likely in liquid than in vapor mixtures, we will generally use x_i to denote mole fractions when dealing with nonideal mixtures.)

Both the activity and activity coefficient are nonnegative quantities whose numerical values depend on the choice of standard state. (Since different standard states would have different f_i°, the numerical values of the a_i and γ_i° would be dependent on the numerical value of f_i°.) Although many choices of standard state are possible, we will consider only two in our discussion.

1. The most common and generally useful standard state is the *pure component* (gas, liquid, or solid) *at the same temperature and pressure as the phase in the mixture*. Examining this choice of reference state, we observe that for the activity

$$a_i = \frac{\bar{f}_i}{f_i^\circ} = \frac{\bar{f}_i}{f_i} \tag{10-91}$$

The activity coefficient for this choice of standard state is given by

$$\gamma_i = \frac{a_i}{x_i} = \frac{\bar{f}_i}{x_i f_i} \tag{10-92}$$

where we have dropped the superscript $^\circ$ from γ_i to indicate this particular choice of standard state in the evaluation of γ_i. For an ideal solution and this choice of standard state, $a_i = x_i$ and $\gamma_i = 1$. The deviation of γ_i from unity then provides a measure of the nonideality of the mixture, as seen in Fig. 10-13. It should be noted that even if the mixture is nonideal, γ_i approaches unity in the limit as the mixture becomes pure i.

2. For some systems it is not convenient to use the pure component as the reference condition. For example, suppose we were describing the solu-

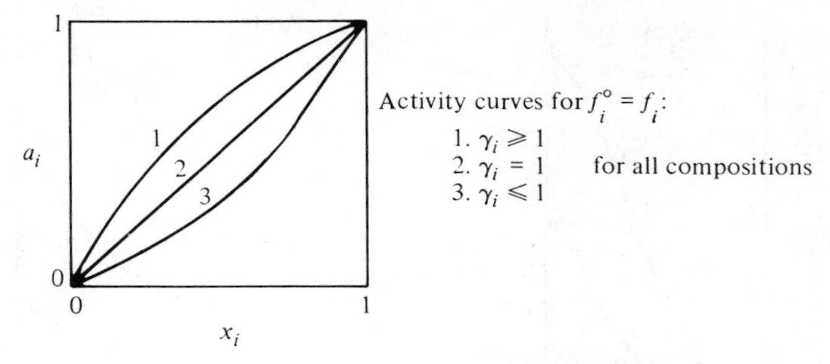

Figure 10-13. Plot of activity vs concentration.

bility of gaseous oxygen in water at 25°C. Clearly, pure liquid oxygen simply does not exist under these conditions and another standard state must be used. Such a standard is offered by the *infinite-dilution* state; the meaning of such a choice can be better understood by considering Fig. 10-14 in which the partial fugacity of a component in a mixture is plotted versus its concentration. If the limiting slope of the curve as $x_i \longrightarrow 0$ is used to draw a line (A–B) from the origin ($x_i = 0, \bar{f}_i = 0$) across the entire composition range, its intersection with the $x_i = 1$ line defines a hypothetical fugacity at point B, which is used as $\bar{f}_i^\circ$, the standard-state fugacity.

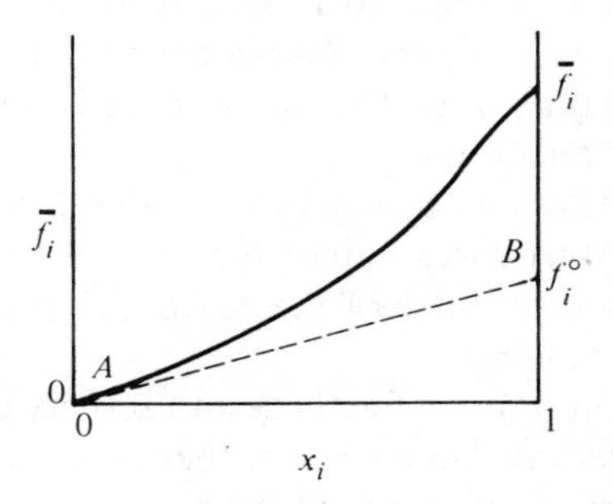

Figure 10-14. Partial fugacity vs concentration plot illustrating infinitely dilute standard state.

A plot of a_i–x_i using the infinitely dilute standard state is obtained by dividing each value of $\bar{f}_i$ along the curve by the value for f_i° and replotting it versus x_i as shown in Fig. 10-15.

This choice of standard state is particularly advantageous when one is interested in systems where solutes at low concentrations are involved, because in the dilute region $a_i = x_i$. Many metallurgical and electrochemical systems involve low solute concentrations in alloys, and thus this standard state is often employed. As Fig. 10-15 demonstrates, at high concentrations of i the activity may take on values greater than unity as a result of the standard state chosen. Also shown in Fig. 10-16 is the variation of γ_i° with concentration for the system illustrated in Fig. 10-15 when the infinitely dilute

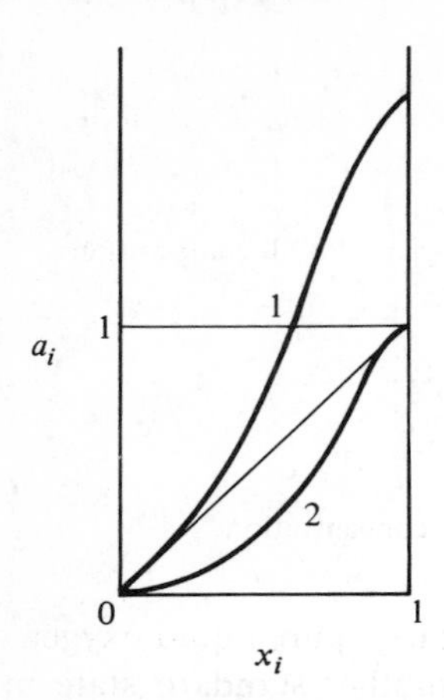

Figure 10-15. Effect of standard state on activities.

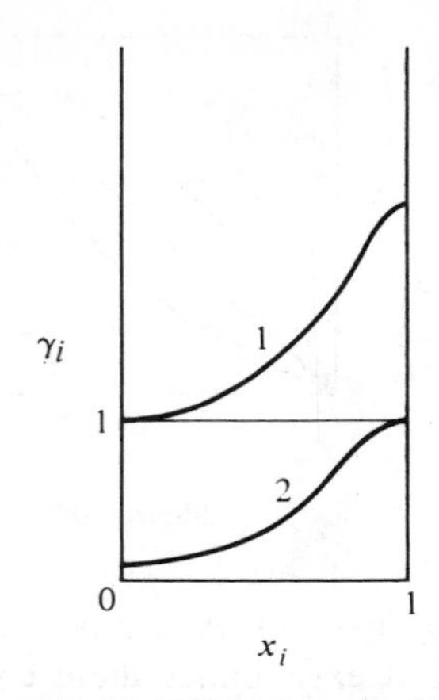

Figure 10-16. Effect of standard state upon activity coefficients. Curve 1, infinitely dilute standard states; curve 2, pure i at P and T of mixture as standard state.

standard state is used. Note that γ_i° has a value of unity in the dilute region and increases at higher values of x_i.

Curves 2 in Figs. 10-15 and 10-16 are included to illustrate a_i versus x_i and γ_i versus x_i if pure component i at the pressure, temperature, and phase of the mixture is used as the standard state. Values for activity then range between 0 and 1 and $\gamma < 1$.

It is left as an exercise to show that the behavior illustrated by curve 1 in Fig. 10-16 produces values for the activity and activity coefficient greater than unity over much of the composition range if the infinitely dilute standard states is used.

Since the values of activity and activity coefficient of a component depend on the reference state chosen, the reader should always be alert to ascertain what reference or standard state is to be associated with each component of a mixture.

As we have previously indicated, knowledge of the activity coefficients and standard-state fugacity give us the ability to calculate partial fugacities even in nonideal solutions. Usually the evaluation of f_i° presents only minor difficulty. On the other hand, evaluation of γ_i° is another story entirely. In general, it is *not* possible to predict γ_i° from first principles directly. Rather, the more common approach is to measure partial fugacities and compositions and from these evaluate γ_i°. By performing such a procedure on many different kinds of mixtures, it has been possible to separate many mixtures into different classes of solutions. By extrapolating the behavior of known mixtures within a class to untested mixtures in the same class, it is often possible to obtain reasonable estimates of the activity coefficients in the untested mixture. Such procedures are well beyond the scope of this text. For a detailed review of work in this area the reader is referred to the excellent text of Prausnitz.[4]

Nonideal Solutions at Low Pressures

For pressures below 10 atm or so, it is still reasonable to assume that the vapor phase is an ideal mixture even though the liquid phase may not be. In addition, at these low pressures the replacement of fugacities by pressures is often a justifiable simplification. Under these conditions the requirements of phase equilibrium reduce to:

$$x_i^{\circ}\gamma_i^{\circ L} f_i^{\circ L} = y_i P = P_i \tag{10-93}$$

or

$$x_i \gamma_i P'_L = y_i P = P_i \tag{10-94}$$

depending on the choice of reference state for the liquid-phase activity coefficient. If the common choice of pure liquid at the pressure and temperature of the system is made, then equation (10-94) would be the preferred form; for the more general choice of standard state, equation (10-93) would be applicable. We find form (10-93) most useful when one of the species involved is a gas well above its critical point—for example, if we were considering the solubility of oxygen or nitrogen in water at room temperatures. At these temperatures both oxygen and nitrogen are well above their critical temperatures, so that use of a pure-component liquid for either of these species would clearly be meaningless. Rather, we find in cases such as this that the use of an infinite-dilution standard state is quite convenient.

It is an experimentally observed fact that the solubility of many gases in various liquids is directly proportional to the partial pressure of the gas above the liquid—especially when the concentration of the dissolved gas is below 10 mol percent in the liquid. That is

$$y_i P = P_i = x_i H_i \tag{10-95}$$

where H_i is the proportionality constant between liquid-phase composition and vapor-phase partial pressure.

Equation (10-95) is known as *Henry's law*; the constant H_i is termed the *Henry's law constant* and is a function of the chemical make-up of both the solvent and the solute (the "dissolved" specie), as well as the temperature of the mixture. Although Henry's law, like Raoult's law, is a great simplification of real solution behavior, it frequently does correctly predict the behavior of solutes in dilute liquid solutions. Although Henry's law is rigorously

[4]J. M. Prausnitz, *Molecular Thermodynamics of Fluid-Phase Equilibrium*, Prentice-Hall, Inc., Englewood Cliffs, N.J., 1969.

correct in the limit of infinite dilution, the actual range of applicability may be quite small.

In view of our discussion of reference and standard states, it is pertinent to pause here and ask what reference states are commonly presumed when Henry's law is used. We begin with equation (10-93):

$$x_i \gamma_i^{\circ L} f_i^{\circ L} = P_i \tag{10-93}$$

We can select for the reference state either the pure liquid or the solute at infinite dilution, both at the temperature and pressure of the mixture. Chooing pure liquid as the standard state for the solute and assuming that $f_i^L = P_i'$, we obtain

$$x_i \gamma_i^{\circ L} f_i^{\circ L} = x_i \gamma_i P_i' = P_i = x_i H_i \tag{10-96}$$

or

$$H_i = \gamma_i^{\circ L} f_i^{\circ L} = \gamma_i P_i' \tag{10-97}$$

We see that with this choice of standard state, Henry's law implies a constant activity coefficient, although not necessarily unity. If $\gamma_i = 1$, then Henry's law reduces to Raoult's law directly.

For the infinite-dilution standard state we obtain

$$\gamma_i^{\circ L} f_i^{\circ L} = H_i \tag{10-98}$$

If we recall our requirement that the activity coefficient approach the value of unity in the standard state and note that Henry's-law coefficients are experimentally determined by fitting data in the limit of infinite dilution, we conclude that $\gamma_i^{\circ L} = 1.0$ and that $f_i^{\circ L} = H_i$ for the infinite-dilution standard state. We normally consider Henry's law to be applicable over whatever concentration range the vapor-phase partial pressure (fugacity) can be predicted by linear extrapolation of dilute-solution behavior. The slope of the partial fugacity versus the concentration curve in the dilute region equals H_i or f_i°. Beyond this range $\gamma_i^{\circ L}$ will differ from unity and Henry's law will cease to be valid.

10.10 Phase Equilibrium Involving Other Than Vapor–Liquid Systems

The engineer often encounters separation problems involving other than the vapor–liquid systems just discussed. For example, liquid–liquid extraction between two immiscible phases, adsorption on a solid surface from either a liquid or vapor phase, and leaching or crystallization involving liquid–solid phase equilibria are all examples familiar to the engineer. The same criteria discussed earlier for phase equilibria apply to these applications as well:

$$\bar{f}_i^L = \bar{f}_i^S, \bar{f}_i^V = \bar{f}_i^S \quad \text{or} \quad \bar{f}_i^{L_1} = \bar{f}_i^{L_2} \tag{10-99}$$

That is, the chemical potentials or partial fugacities of each specie must be constant throughout all phases which are in equilibrium. Thus, the prediction of the phase behavior in any multiphase system is dependent upon our ability to predict the properties of multicomponent mixtures in the various phases under consideration. As we have discussed, prediction of the pertinent vapor-phase properties is reasonably straight forward and readily accomplished. For liquid mixtures the picture is less optimistic—we can accurately predict the behavior for many systems, but for many others the prediction tools are simply not available. For the solid state the picture is considerably bleaker. The simple concepts—such as ideal solutions—which at least worked sometimes for liquid mixtures are simply insufficient to describe the exceedingly complex behavior of solid solutions. The state of development of more complex mixing rules for solid mixtures is truly primitive, with little hope for major breakthroughs in the near future. Thus it is not surprising that the main thrust in phase-equilibrium studies involving solid materials is still primarily experimental, that is, determination of the equilibrium phase compositions as functions of temperature.

Phase behavior in liquid–solid systems is normally mapped on a T–x diagram, much as described earlier for vapor–liquid systems. Whereas vapor–liquid systems show strong pressure dependence, the same is not true for liquid–solid regions of the phase diagram. Thus Fig. 10-17, which shows the nickel–copper system in which both components are soluble across the entire composition range for both phases, is an adequate representation of the phase behavior for a large range of pressures.

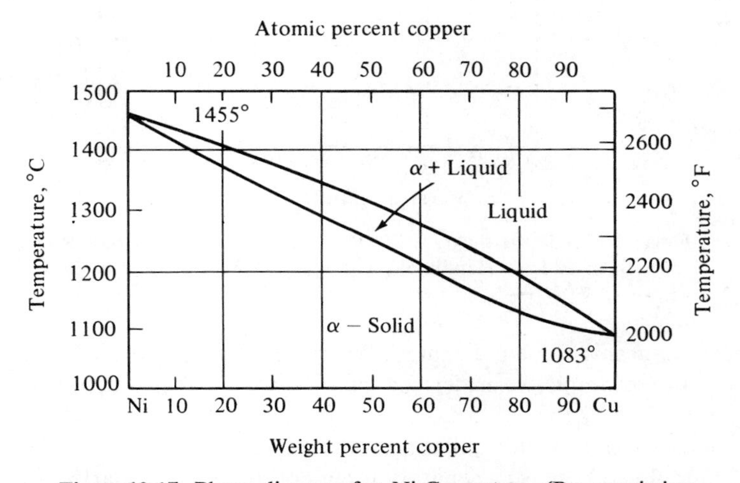

Figure 10-17. Phase diagram for Ni-Cu system. (By permission, from *Metals Handbook*, vol. 8, © American Society for Metals (1973).)

Although solid–liquid systems are by far the most commonly encountered solid-containing systems, solid–vapor systems are occasionally encountered. Since the solubility of most gases is quite low in solid materials, the solubility of gases in solids (at low pressures) is often described by a Henry's-law type of expression.

Liquid–solid systems can exhibit a wide variety of behavior other than the simple solution behavior shown in Fig. 10-17. For example, consider the lead–tin system illustrated in Fig. 10-18. Notice that the solid and liquid compositions converge at a composition of 61.9 percent. This point is known as the *eutectic point*. When the eutectic mixture freezes or melts it does so at constant composition and temperature. The central portion of the solidified region indicates a two-phase mixture. One of the phases is of course tin-rich, the other lead-rich. Although the solid mixture of lead and tin shows significant phase separation, there appears to be no liquid-phase immiscibility.

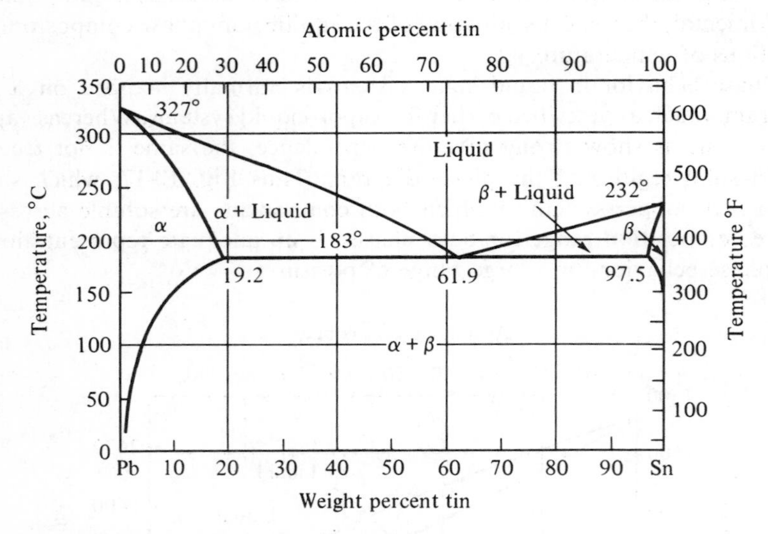

Figure 10-18. Phase diagram for the lead-tin system. (By permission, from Metals Handbook, vol 8, © American Society for Metals (1973).)

Phase diagrams for solid materials can on occasion become extremely complex. For example, consider the iron–carbon system illustrated in Fig. 10-19.

In much the same fashion that most processes which involve vapor–liquid equilibrium do so because a separation may be achieved by partial vaporization or condensation, a separation may also be affected by partial freezing or melting. Although a complete separation cannot usually be accomplished in

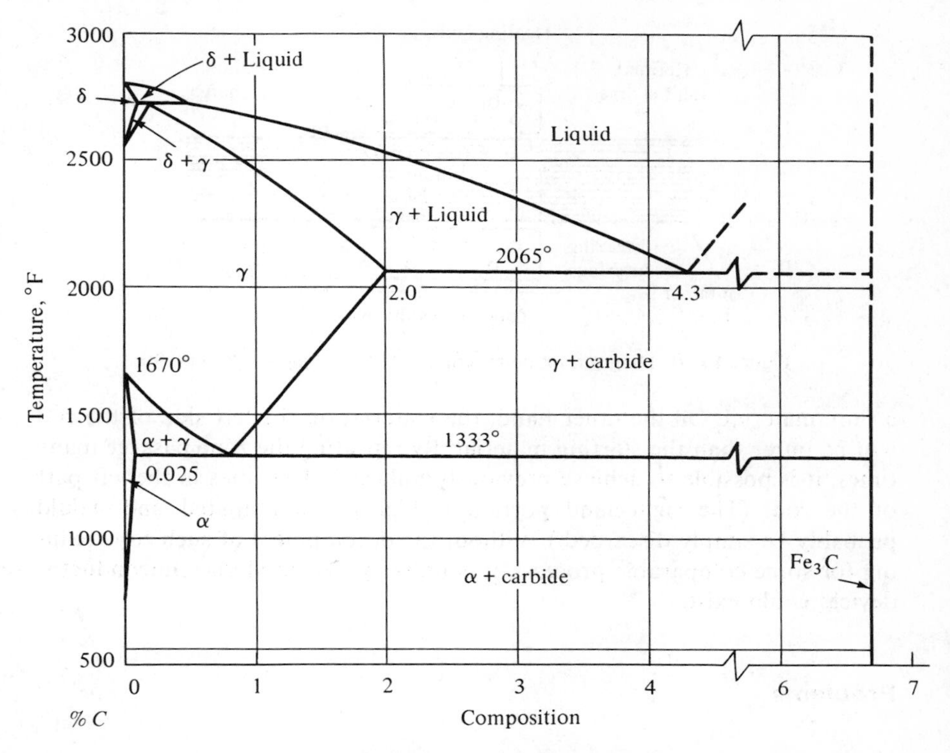

Figure 10-19. Phase diagram for the iron-carbon system.

a single freezing or melting, it is possible to repeatedly freeze and melt a mixture and thereby bring about rather spectacular separations. Probably the most successful industrial process for performing this repeated freezing and melting is the *zone-refining* process developed by Bell Laboratories for the production of semiconductor-grade silicon and germanium. The silicon and germanium produced by the zone-refining process commonly contains contaminants in concentrations less than 1 *part per billion*. The zone-refining process is illustrated in Fig. 10-20. The process proceeds as follows: A rod of the material to be purified is prepared. The rod is positioned such that one end is situated within a donutshaped heater known as the zone heater. The heater is turned on and forms a liquid zone (in this case at the left end of the rod). The rod is now slowly moved to the left. As the melted zone advances out from under the heater, it slowly freezes. The frozen material is "purer" than the original material because of the solid–liquid equilibrium behavior. The rod is continually moved to the left until eventually the molten zone is at the right-hand end of rod. This molten zone contains a very high concentration of contaminant and, when frozen, will be useless as

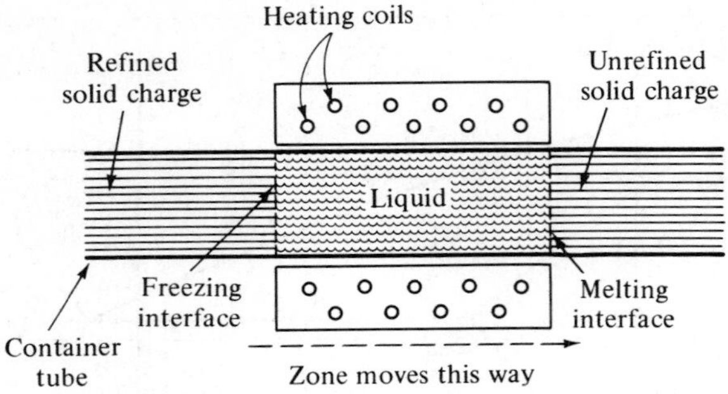

Figure 10-20. Diagram on zone refining. [*Chem. Eng.* 66 (8), (1959)]

a pure material. On the other hand, the material on the left side of the rod will be purer than the starting material. By repeating the zone passage many times, it is possible to achieve previously unheard of purities in the left part of the rod. (The right-hand portion is highly contaminated and would probably be simply discarded.) Without the development of such zone refining (or some comparable process), it is unlikely that modern semiconductor devices could exist.

Problems

10-1. The partial molar enthalpies for aqueous solutions of sulfuric acid at 25°C are as follows:

Moles of H_2O/mole of H_2SO_4	$\bar{H}_{H_2O}$, cal/g-mol	$\bar{H}_{H_2SO_4}$, cal/g-mol
∞	0	0
1,600	−0.48	4,811
200	−2.16	5,842
10	−233	9,632
2	−2,315	18,216
1	−4,731	21,451
0	—	23,540

(a) What are the reference states for the above data?

(b) Sketch the 25°C isotherm on an enthalpy–concentration diagram for the above reference states.

(c) Sketch the 25°C isotherm using as reference states pure H_2O and H_2SO_4 at 25°C. Indicate how the points on the curve would actually be calculated.

10-2. In a water–ethyl alcohol mixture with a water mole fraction of 0.4, the partial volume of ethyl alcohol is observed to be 57.5 cm^3. If the mixture density is 0.8494, calculate the partial volume of water.

10-3. An ammonia-absorption refrigeration cycle is pictured in Fig. P10-3. The solution in the absorber and stripper is ammonia and water. An enthalpy–concentration diagram for ammonia and water is given in Fig. 10-7.

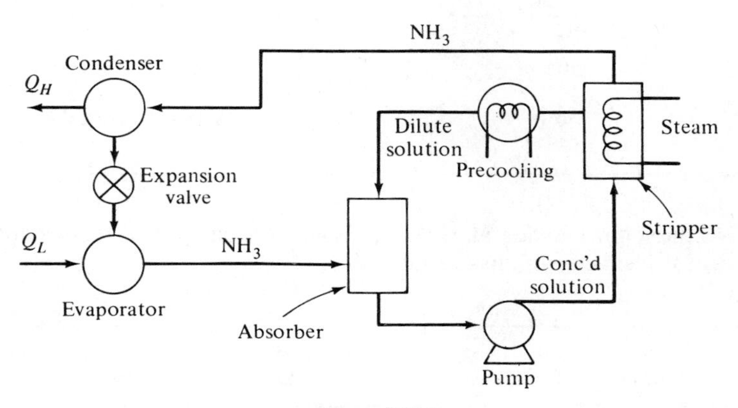

Figure P10-3

(a) Steam heating is used to remove the absorbed ammonia from its water solution in the stripper. If the stripper operates at 180°C and a pressure of 20 kg/cm^2, what is the composition of the dilute liquid leaving the stripper?

(b) Eight pounds of dilute solution enter the absorber for every pound of ammonia vapor. The dilute solution enters at 40°C and the ammonia arrives as saturated vapor from the evaporator, which operates at 1 kg/cm^2. The absorber produces concentrated solution at 20°C. How much heat is removed from the absorber per pound of concentrated solution?

(c) Could all the ammonia vapor be absorbed adiabatically? *If so*, what would be the temperature of the concentrated solution leaving the absorber? *If not*, what is the minimum amount of heat that must be removed?

Pressure, kg/cm^2	Enthalpy of Saturated Ammonia Vapor, kcal/kg
0.1	357
1.0	374
10	390
20	392

10-4. Calculate the activity coefficient of NH_3 in a 10-percent NH_3 solution using the following data at 32°F for a H_2O–NH_3 system. Use conventional standard states for liquids. Does this system follow Henry's law at 10 percent?

NH_3 in liquid, mol %	Partial Pressure of NH_3 in Equilibrium Vapor, psia
5	0.26
10	0.52
15	0.90
50	19.36
100	62.29

10-5. For the following data at 69.94°C, assume that the partial pressure of each component is identical with its fugacity.

x_1 (toluene)	x_2 (acetic acid)	P_1 (toluene)	P_2 (acetic acid)
0.0000	1.0000	0	136
0.1250	0.8750	54.8	120.5
0.2310	0.7690	84.8	110.8
0.3121	0.6879	101.9	103.0
0.4019	0.5981	117.8	95.7
0.4860	0.5140	130.7	88.2
0.5349	0.4651	137.6	83.7
0.5912	0.4088	154.2†	78.2
0.6620	0.3380	155.7	69.3
0.7597	0.2403	167.3	57.8
0.8289	0.1711	176.2	46.5
0.9058	0.0942	186.1	30.5
0.9565	0.0435	193.5	17.2
1.0000	0.0000	202	0

Source: International Critical Tables, 1st ed., Vol. III, pp. 217, 223, and 288, McGraw-Hill Book Company, New York, 1928.

†This is evidently a typographical error. The correct figure is near 145.2.

(a) Draw a graph of $\bar{f}_1$ versus x_1. Indicate Raoult's law by a dashed line.

(b) Draw a graph of $\bar{f}_2$ versus x_2. Indicate Raoult's law and Henry's law, each by a dashed line.

(c) Find the constant in Henry's law for acetic acid in toluene solutions by extrapolating a graph of $\bar{f}_2/x_2$ versus x_2 to infinite dilution.

(d) Calculate the activities and activity coefficients of acetic acid on the basis of an $f_2°$ established from Henry's law. Plot these values versus x_2.

(e) Calculate the activities and activity coefficients of acetic acid when the pure liquid is taken as the standard state. Plot these values on the same graph as in (d).

(f) Calculate the activities and activity coefficients of toluene, the solvent in these solutions. What standard states did you use?

10-6. Suppose we have a chamber separated into two halves by means of a membrane that is permeable to water vapor molecules but not to nitrogen molecules. At a given instant of time the temperature, pressures, and concentrations in the two halves are found to be the values in Fig. P10-6. Assuming ideal solutions where necessary, answer the following questions:

(a) What are the partial pressures of water in each half?

(b) Is the system at equilibrium subject to the constraints of the problem?

(c) If the system is not at equilibrium, calculate the driving force for attainment of equilibrium.

(*Note:* At these conditions both nitrogen and water may be assumed to follow the generalized compressibility chart.)

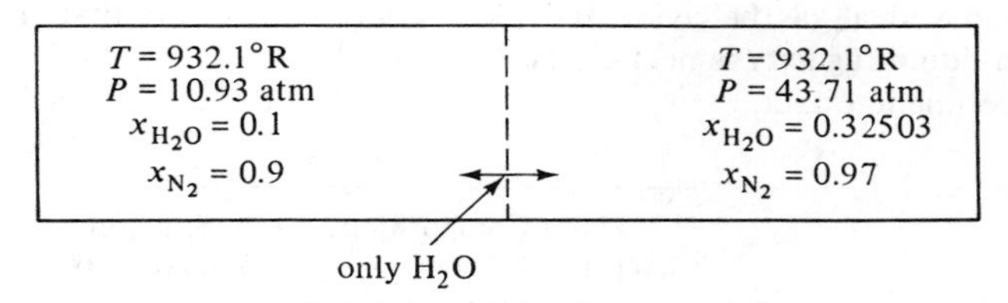

Figure P10-6

10-7. A vessel containing a sodium chloride solution at 25°C is connected by an overhead tube to one containing pure water. The temperature of the latter vessel is adjusted until no transfer of water takes place through the tube; that is, equilibrium conditions are attained. This temperature is found to be 22°C. Using only the fact that the heat of vaporization of water is 585 cal/g in this temperature range, determine the activity of water in the sodium chloride solution.

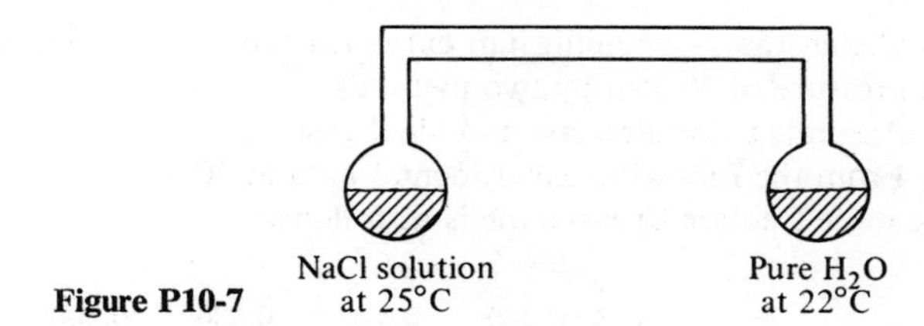

Figure P10-7

10-8. A salt solution (brine) containing 4.1 mol percent salt is slowly heated until boiling occurs. At a pressure of 1.0 atm, it is found that the brine boils at 216.2°F. The evolved vapor is essentially pure water, and may be assumed to be in thermodynamic equilibrium with the boiling brine. Calculate the liquid-phase activity coefficient for the water from the above information.

10-9. The following data were obtained experimentally at 1000 psia and 100°F.

Component	Equilibrium-Phase Compositions, Mole Fraction	
	Vapor	Liquid
Methane	0.657	0.284
Ethylene	0.187	0.1705
Isobutane	0.156	0.5455

Assuming ideal-solution behavior in both phases, compute the fugacity of pure methane in the liquid phase. Note that this is one method for obtaining such information above the critical point of a compound.

10-10. Compute and compare vapor–liquid distribution coefficients, $K = y/x$ for the methane–ethylene–isobutane system at 500 psia and 100°F by the following methods. Plot the results as log K versus log P.

(a) Assume ideal-gas behavior and ideal solutions (assume that the effect of pressure on liquid fugacity is negligible).

(b) Experimental data.

	Phase Compositions, Mole Fraction			
	500 psi, 100°F		1000 psi, 100°F	
	y	x	y	x
Methane	0.3355	0.069	0.657	0.284
Ethylene	0.4815	0.279	0.187	0.1705
Isobutane	0.1830	0.652	0.156	0.5455

Source: M. Benedict, E. Solomon, and L. C. Rubin, *Ind. Eng. Chem.*, **37**, 55 (1945).

10-11. Sketch the y–x equilibrium curve for the system n-butane and n-hexane at a total pressure of 20 atm by two methods.

(a) Assuming Raoult's law and ideal gases.

(b) From the following experimental data at 20 atm.

The mole fraction of n-butane is as follows:

x_4	0.200	0.400	0.600	0.800
y_4	0.349	0.619	0.799	0.912

It is believed that five select points should be adequate to represent the curve. As (0, 0) and (1, 1) are two excellent points, only three need be calculated for part (a).

10-12. A refinery by-product gas whose composition is 20 mol percent methane, 30 mol percent *n*-butane, and 50 mol percent *n*-pentane is available at 120°C and 10 atm. To what pressure must the mixture be compressed isothermally to liquify 50 mol percent of it? You may assume ideal solutions of ideal components.

10-13. A tank containing equimolal quantities of propane, *n*-butane, and *n*-pentane is under an absolute pressure of 15 atm. The temperature in the tank is 80°C. What is the condition of the mixture in the tank?

10-14. A mixture containing 30 mol percent propane, 50 mol percent butane, and 20 mol percent hexane is originally at 1 atm and 120°C. If this mixture is compressed isothermally to a point where condensation begins to occur, what is the pressure of the mixture? What is the composition of the first drop condensing?

10-15. What is the composition of the vapor in equilibrium with a liquid mixture of 60.6 mol percent *n*-propane and 39.4 mol percent *n*-butane at 80°F? Assume the vapor pressures of *n*-propane and *n*-butane at 80°F are not known, but the following data are available at 70°F.

	P' psia	Δh_V Btu/lb$_m$	T_c, °R	P_c, psia	Mol Wt.
n-propane	124	145	665	617	44.09
n-butane	32	156	767	530	58.12

For small ranges in temperature, Δh_V and the compressibility factor of the gas may be assumed to remain constant. The specific volume of the liquid may be assumed to be very small compared to the specific volume of the gas.

10-16. In a petroleum refining plant, an equimolar mixture of ethane, propane, and normal butane is fed to a flash separator in which the pressure is 200 psia and the temperature 100°F, as shown in Fig. P10-16. Assuming ideal-solution behavior, calculate the liquid to vapor ratio and the composition of each phase.

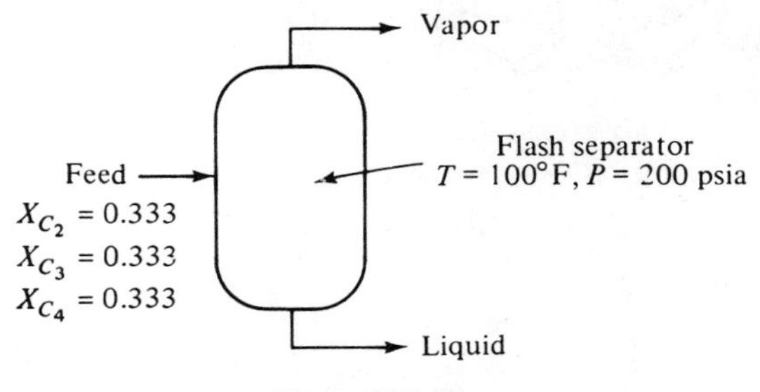

Figure P10-16

ELEVEN

thermodynamics of chemically reacting systems

Introduction

In the previous chapters we have considered the thermodynamic properties of nonreacting systems. We began by discussing the properties of single-phase, single-component systems. This discussion was extended in Chapter 10 to include multicomponent, multiphase systems in physical equilibrium. In this chapter we will describe the thermodynamics of reacting systems.

When substances react chemically, one of the results of the chemical reaction is frequently a substantial rearrangement of the types of energies the reacting species possess. The exact form this rearrangement takes is a function of the way in which the reaction is allowed to occur. For example, if hydrogen is allowed to react with oxygen at 25°C and energy is removed so that the reaction proceeds isothermally as shown in Fig. 11-1, it is observed that some 68,300 cal of heat will have to be removed from the reactor per mole of hydrogen reacted. On the other hand, if the reactor is insulated so that little heat is lost to the surroundings, water vapor leaves the reactor at temperatures as high as 2800°K. As we will soon show, the outlet temperature is a function of many things, including the quantity of oxygen supplied to the reactor and the presence or absence of inert materials such as nitrogen or argon. In the following section we will discuss the application of the energy

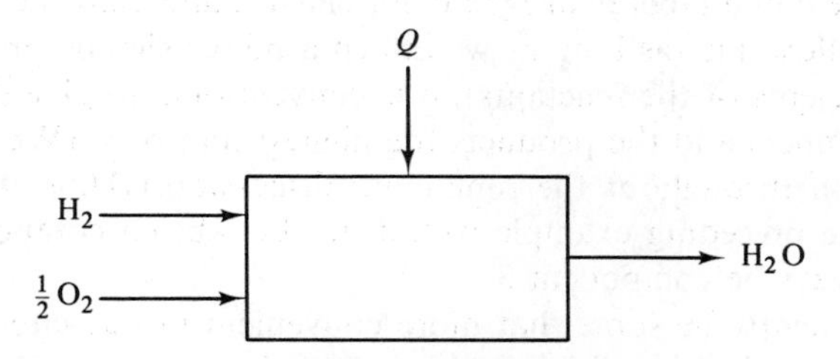

Figure 11-1. The isothermal chemical reactor.

balance to chemically reacting systems. In subsequent sections we will consider the production of work in chemically reacting systems, and we will finally develop and discuss the conditions for equilibrium in a chemically reacting system.

11.1 Heat Liberation During Isothermal Reactions

We begin by choosing a system in which a chemical reaction is occurring. For the purposes of discussion, let us examine the steady-flow isothermal chemical reactor shown in Fig. 11-2, where the components A_1, A_2, . . . react to form the products. . . , A_{C-1}, A_C according to the reaction

$$-\alpha_1 A_1 - \alpha_2 A_2 - \cdots \longrightarrow \cdots \alpha_{C-1} A_{C-1} + \alpha_C A_C \tag{11-1}$$

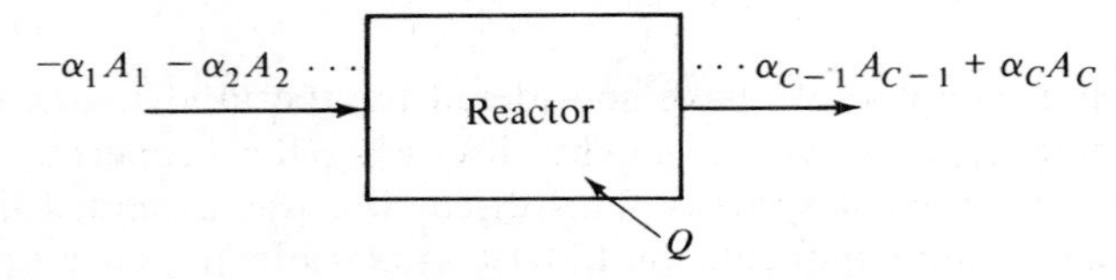

Figure 11-2. Steady-flow isothermal chemical reactor.

The stoichiometric coefficient of the ith species in the chemical reaction is α_i, and is negative or positive depending on whether the ith species is a reactant $(-)$ or product $(+)$. For example, the reaction between hydrogen and chlorine to form hydrogen chloride would be written

$$H_2 + Cl_2 = 2\ HCl \tag{11-2}$$

or

$$-\alpha_1 A_1 - \alpha_2 A_2 = \alpha_3 A_3 \tag{11-3}$$

where

$$\begin{aligned} \alpha_1 &= -1, \qquad & A_1 &= H_2 \\ \alpha_2 &= -1, \qquad & A_2 &= Cl_2 \\ \alpha_3 &= 2, \qquad & A_3 &= HCl \end{aligned} \tag{11-4}$$

Clearly, we could label hydrogen component 2 and chlorine 1. Although any order is allowable (as long as we attach a minus sign before the stoichiometric coefficients of the reactants), it is conventional to give the reactants the lowest numbers and the products the highest numbers. (We shall follow this convention throughout the remaining discussion.) Thus the hydrogen chloride of the preceding example would not be labeled component 1 or 2 but would always be component 3.

It will frequently be somewhat more convenient in our later work if we rearrange the chemical reactions so that all species appear on the same side:

$$\alpha_1 A_1 + \alpha_2 A_2 + \cdots + \alpha_{C-1} A_{C-1} + \alpha_C A_C = 0 \tag{11-5}$$

or

$$\sum_{i=1}^{C} \alpha_i A_i = 0 \tag{11-5a}$$

where the α's and A's have exactly the same meaning as before. Once again, a negative α_i indicates a reactant and a positive α_i represents a product. Using this scheme, the hydrogen–chlorine reaction would be written as

$$-H_2 - Cl_2 + 2\,HCl = 0$$

Although we have written the reaction between hydrogen and chlorine with the coefficients given in equation (11-4), we might just as correctly have written it as

$$-\lambda\, H_2 - \lambda\, Cl_2 + 2\lambda\, HCl = 0 \tag{11-6}$$

where λ is any number other than zero. Thus we find there is no unique set of α's that describe a particular reaction. For the sake of convenience and clarity we arbitrarily choose λ such that $-\alpha_1 = 1$. This means that we may get a different set of α's if we rearrange the order of the chemical species in a particular reaction, and therefore we must carefully specify which component is chosen number 1.

Let us now return to the isothermal chemical reactor of Fig. 11-1. We assume that reactants in their stoichiometric ratio (that is, the molar rates of reactants bear the same ratios to each other as their stoichiometric coefficients in the reaction equation) enter the reactor and are completely converted into the stoichiometric amounts of products. (We shall extend this discussion shortly to include other cases, but for the moment let's stick with the simplest case.) If we assume no work is recovered from the chemical reactor, potential and kinetic energy effects are negligible, and the reactor is operating at steady state, then the energy balance written around the reactor reduces to:

$$Q = \Delta H = (H)_{\text{prod}} - (H)_{\text{react}} \tag{11-7}$$

If we restrict ourselves to gaseous reactions at low pressures and/or those in which only ideal liquid or solid solutions are involved, then the enthalpies of the product and reactant streams can be expressed as the sums of the enthalpies of the individual reacting species, that is,

$$\begin{aligned} (H)_{\text{prod}} &= \sum_{\text{prod}} (n_i)_{\text{out}}(h_i)_{\text{out}} \\ (H)_{\text{react}} &= \sum_{\text{react}} (n_i)_{\text{in}}(h_i)_{\text{in}} \end{aligned} \tag{11-8}$$

In evaluating $(H)_{\text{prod}}$ we consider the quantity of each specie in the *outlet stream* and its enthalpy in the outlet stream; we consider the quantity of each specie and its enthalpy in the inlet stream in evaluating $(H)_{\text{react}}$. (This distinction is not terribly important for isothermal reactions, but it will be of crucial importance when we consider nonisothermal reactions. Thus we might as well get accustomed to keeping our notation straight right from the start.)

We may put equation (11-7) on a molar basis by dividing the whole equation by a given number of moles. Although there is no unique value that *must* be chosen, we are frequently interested in the heat liberation per mole of some specific fuel fed to the reactor. Let us label this component species number 1 and then divide the energy balance equation by the number of moles of species 1, $(n_1)_{\text{in}}$, that enter the reactor:

$$\frac{Q}{(n_1)_{\text{in}}} = \frac{\Delta H}{(n_1)_{\text{in}}} = \frac{(H)_{\text{prod}}}{(n_1)_{\text{in}}} - \frac{(H)_{\text{react}}}{(n_1)_{\text{in}}} \tag{11-9}$$

or

$$\mathscr{q} = (\Delta \mathscr{h})_T = \mathscr{h}_{\text{prod}} - \mathscr{h}_{\text{react}} \tag{11-9a}$$

where the lower case script character is read "per mole of species 1 entering the reactor." The distinction between a normal specific property such as h and the property $\mathscr{h}$ we have just defined is an extremely important one. A specific property is simply equal to the total property divided by the total mass (or moles); the script symbol represents the total property divided by the mass (or moles) of *a particular component*, in this case 1. (This distinction is not critical when specific properties are expressed on a mass basis, because total mass is conserved during any process. However, the number of moles is not necessarily constant during a chemical reaction, and thus when specific properties are expressed in terms of moles, extreme care must be exercised to insure consistency.)

Equation (11-8) for the enthalpies of the product and reactant streams may now be substituted into equation (11-9a) to give:

$$\mathscr{q} = \sum_{\text{prod}} \frac{(n_i)_{\text{out}}}{(n_1)_{\text{in}}}(h_i)_{\text{out}} - \sum_{\text{react}} \frac{(n_i)_{\text{in}}}{(n_1)_{\text{in}}}(h_i)_{\text{in}} \tag{11-10}$$

Since we have assumed a stoichiometric feed and complete conversion, the ratios $(n_i)_{\text{out}}/(n_1)_{\text{in}}$ and $(n_i)_{\text{in}}/(n_1)_{\text{in}}$ may be expressed directly in terms of the stoichiometric coefficients:

$$\begin{aligned} \frac{(n_i)_{\text{out}}}{(n_1)_{\text{in}}} &= \alpha_i \quad \text{for a product} \\ \frac{(n_i)_{\text{in}}}{(n_1)_{\text{in}}} &= -\alpha_i \quad \text{for a reactant} \end{aligned} \tag{11-11}$$

So that equation (11-10) reduces to:

$$\mathscr{q} = \sum_{\text{prod}} \alpha_i (h_i)_{\text{out}} + \sum_{\text{react}} \alpha_i (h_i)_{\text{in}} \tag{11-12}$$

Since both the products and reactants are at the temperature of the reactor (and we are *tacitly assuming* the enthalpies are *not a function of composition*), we may now drop the "in" and "out" subscripts and replace them with T, the reaction temperature.

$$\mathscr{q} = \sum_{\text{prod}} \alpha_i (h_i)_T + \sum_{\text{react}} \alpha_i (h_i)_T \tag{11-13}$$

The summations may now be combined into a single summation over *all* reacting species:

$$q = \sum_{i=1}^{C} \alpha_i (h_i)_T = \Delta h_R \tag{11-14}$$

This final summation is now called the *isothermal enthalpy change on reaction* or simply the *isothermal heat of reaction* and is clearly equal to the quantity of energy that is liberated when 1 mol of species 1 undergoes an isothermal chemical reaction with a stoichiometric quantity of each other reacting species.

Now let us consider the case in which one of the reactants other than species 1 is in excess of its stoichiometric amount, or when one or more of the product streams is present in the reactor feed stream. In this case we may, as before, write the energy equation as

$$q = \Delta h \tag{11-15}$$

Now, however, let us picture the chemical reactor as made up of three different sections as shown in Fig. 11-3.

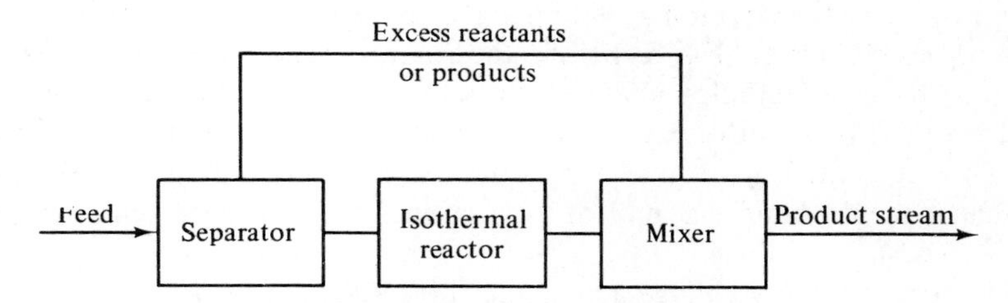

Figure 11-3. Nonstoichiometric isothermal reactor.

In the first stage of the model reactor the excess reactants and/or products are isothermally separated from the remaining stoichiometric mix. Since the process is isothermal and we are neglecting concentration effects on enthalpy, this separation occurs without a change in enthalpy. In the second stage the remaining stoichiometric feed stream is isothermally reacted. The enthalpy change here is simply Δh_R, exactly as in the previous analysis since the two reactors are identical. In the third stage the reaction products are isothermally mixed with the second stream from the separation. Again, if we neglect concentration effects on enthalpy, this isothermal mixing occurs without enthalpy change. The total enthalpy change across the three-step-model reaction is simply the sum of the three individual changes, which reduces to

$$\Delta h = (\Delta h)_R \tag{11-16}$$

That is, the material which passes through the isothermal reactor unreacted has no effect whatsoever on the overall enthalpy change.

Since the enthalpy is a state function its value must be independent of the way in which it is evaluated. Thus the enthalpy change across the actual chemical reactor will be identical to the enthalpy change calculated from our model three-step reactor provided only that the inlet and outlet conditions are the same. Since we have carefully chosen our model so that this is the case, the enthalpy change across the actual isothermal reactor will be

$$\Delta h = \Delta h_R \tag{11-16}$$

and, as before, the energy balance reduces to:

$$q = \Delta h_R \tag{11-17}$$

In the case where species 1 is not completely exhausted (either because one of the other reactants is used up before species 1 or because the reaction does not go to completion), the problem is handled in a manner quite similar to that above. We first assume that X fraction of specie 1 which enters the reactor is actually reacted, and $(1 - X)$ fraction passes through unreacted. The energy balance around the reactor still yields

$$q = \Delta h = (h)_{\text{prod}} - (h)_{\text{react}} \tag{11-18}$$

Now picture the reaction as taking place according to the three isothermal steps suggested in Fig. 11-4. As before, the separator and mixer are isothermal and thus have no enthalpy change across them. The reactor section is supplied with $(-X\alpha_i)$ mol of each reactant and produces $X\alpha_i$ mol of each product. Thus the enthalpy change across the reactor section is just X fraction of what it would be if $-\alpha_i$ mol of each species entered the reactor, that is, $\Delta h = X(\Delta h)_R$.

The total enthalpy change across the three stages is then

$$\Delta h = X(\Delta h)_R \tag{11-19}$$

so that

$$q = X(\Delta h)_R \tag{11-20}$$

Actually equation (11-20) is the most general form of the energy balance for

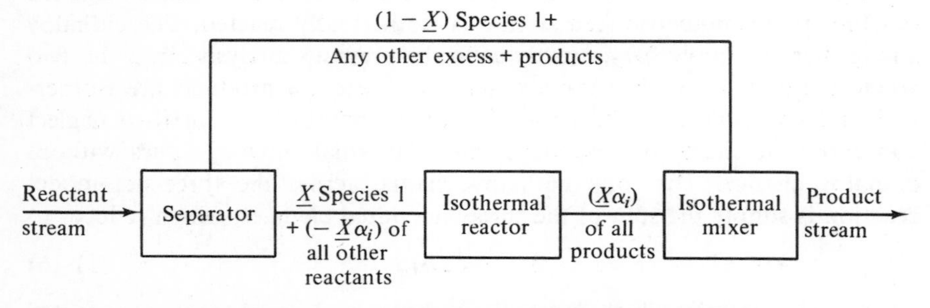

Figure 11-4. Nonstoichiometric isothermal chemical reactor.

isothermal reacting systems in the sense that all other cases may be handled by it, provided the appropriate value of X is used. Thus, if species 1 is totally reacted, $X = 1.0$ and equation (11-20) reduces to equations (11-14) and (11-17).

SAMPLE PROBLEM 11-1. Carbon monoxide is fed to gas burners where it is mixed with air (molecular weight 29, mole fraction $O_2 = 0.21$) and the resulting mixture burned according to the chemical reaction

$$CO + \tfrac{1}{2} O_2 = CO_2$$

The reaction is performed isothermally at 25°C. Determine the quantity of heat absorbed (joules) by the chemical reaction per kg of CO fed to the burners if the air rate is adjusted so that (a) 2.0, (b) 2.46, and (c) 3.0 kg of air is mixed with each kg_m of CO before burning. The isothermal enthalpy change on reaction for the CO oxidation is −67.6 kcal/g-mol CO.

Solution: Although the problem is presented in terms of mass flow rates, we will need the molar flow rates to check whether the CO or O_2 is in excess. Thus we convert the mass flow rates into molar flow rates. For the basis of the calculations use 28 kg or 1 kg mol of CO:

Case	CO (kg)	$\frac{\text{kg Air}}{\text{kg CO}}$	Air (kg)	Air (mol)	O_2 (mol)
(a)	28	2.0	56.0	1.93	0.405
(b)	28	2.46	68.9	2.38	0.500
(c)	28	3.0	84.0	2.90	0.608

Since the stoichiometric amount of oxygen is 0.50 mol of oxygen per mole of CO, there is sufficient oxygen for complete reaction in cases (b) and (c), and the CO will be completely converted to CO_2 giving $X = 1.0$. In case (a) there is only 0.405 mol of oxygen. When all the oxygen is consumed only $0.405/(-\alpha_{O_2}) = 0.405/(0.5)$ or 0.81 mol of CO has been reacted. Thus, for each mole of CO fed, 0.81 mol reacts and $X = 0.81$. The fractional conversions of carbon monoxide are summarized below:

Case	X	$X(\Delta h)_R$
(a)	0.81	−54,700 cal/g-mol
(b)	1.00	−67,600 cal/g-mol
(c)	1.00	−67,600 cal/g-mol

Converting the cal/g-mol to MJ/kg-mol and remembering that $q = X(\Delta h)_{\text{reaction}}$, we obtain:

Case	q (MJ/kg-mol) CO Fed to Reactor	q MJ/kg CO Fed to Reactor
(a)	−229.2	−8.19
(b)	−283.2	−10.1
(c)	−283.2	−10.1

The *minus sign* indicates, of course, that the isothermal reaction *transfers heat to the surroundings*

11.2 Evaluation of $(\Delta h)_R$ from Δh_f

The *enthalpy change on formation*, or Δh_f, of a chemical compound is defined as the enthalpy change which occurs during the isothermal formation of the compound from its constituent elements. For example, if 1 mol of H_2 and $\frac{1}{2}$ mol of oxygen are isothermally reacted at 25°C to form liquid water, 68,317 cal of heat will be lost to the surroundings. Thus the isothermal enthalpy change on reaction at 25°C is—68,317 cal. Since this reaction also leads to the formation of 1 mol of liquid water from its constituent elements (in this case oxygen and hydrogen), the isothermal enthalpy change on reaction is identical to the enthalpy change on formation for 1 mol of liquid water, that is, $(\Delta h_f)_{H_2O} = -68{,}317$ cal/g-mol.

Rather large and extensive tabulations of $(\Delta h_f)_i$ exist and are readily available.[1] In general these values are available for a pressure of 1 atm and a temperature of 25°C (298.15°K). For convenience a table listing $(\Delta h_f)_i$ for approximately 280 compounds is given in Appendix E. Since the enthalpies of most species are only weakly dependent on pressure, the values given in Appendix E may be safely used for pressures below about 10 atm. Although the effects of temperature are somewhat stronger, we shall generally neglect the variation of (Δh_f) with temperature and will assume it is constant for at least moderate temperature ranges.

The entries in the enthalpy of formation tables in Appendix E carry a small single letter (g, s, or l) after each chemical species. This letter indicates

[1] See, for example, F. D. Rossini et al., *Selected Values of Properties of Hydrocarbons and Related Compounds*, API Research Project 44, Carnegie Institute of Technology, Pittsburgh, 1953; F. D. Rossini et al., *Selected Values of Chemical Thermodynamic Properties*, National Bureau of Standards Circular 500, 1952.

that the value listed for that substance is evaluated on the basis of the material being in the gaseous (g), solid (s) or liquid (l) state. Pure elements in their natural state at 25°C and 1 atm by convention have an enthalpy of formation of zero, since no heat effects are associated with their formation. Pure elements in any other state have an enthalpy of formation equivalent to the amount of heat necessary to convert them from their natural state to the listed state. Thus, for example, the enthalpy of formation of solid iron is listed as 0 cal/g-mol, while that for gaseous iron is given as +96,680 cal/g-mol and indicates that 96,680 calories are needed to completely vaporize 1 g-mol of iron at 25°C. In the case of compounds, such as water, the difference between heats of formation in different states is simply the heat that must be supplied to change the compound from one state to the other. Thus, for example, the enthalpy of formation of liquid water is −68,317 cal/g-mol, while that for gaseous water is −57,798 cal/g-mol. Thus (−57,798 + 68,317) = 10,519 calories are needed to vaporize 1 g-mol of water at 25°C. When using the isothermal enthalpy change on formation, it is imperative that the enthalpies of formation for each species be chosen to correspond to the state in which that species actually exists. As observed above, the errors can be quite large if the wrong state is used.

The isothermal enthalpy change on reaction $\sum_{i=1}^{C} \alpha_i h_i$ can now be expressed in terms of the enthalpy changes of formation (at the same temperature) by:

$$\Delta h_R = \sum_{i=1}^{C} (\alpha_i h_i) \equiv \sum_{i=1}^{C} \alpha_i (\Delta h_i)_f \tag{11-21}$$

For example, consider the oxidation of CO to CO_2:

$$CO + \tfrac{1}{2}\, O_2 = CO_2 \tag{11-22}$$

By definition, Δh_R for this reaction is given by:

$$(\Delta h_R) = h_{CO_2} - h_{CO} - \tfrac{1}{2}\, h_{O_2} \tag{11-23}$$

In terms of the enthalpy changes on formation this becomes:

$$(\Delta h_R) = (\Delta h_{CO_2})_f - (\Delta h_{CO})_f - \tfrac{1}{2}(\Delta h_{O_2})_f \tag{11-24}$$

To show that forms (11-24) and (11-23) are indeed identical we expand the enthalpy changes of formation:

$$\begin{aligned} (\Delta h_{CO_2})_f &= h_{CO_2} - h_C - h_{O_2} \\ (\Delta h_{CO})_f &= h_{CO} - h_C - \tfrac{1}{2} h_{O_2} \\ (\Delta h_{O_2})_f &= h_{O_2} - h_{O_2} \equiv 0 \end{aligned} \tag{11-25}$$

Substitution of these into equation (11-24) gives:

$$\begin{aligned} (\Delta h_R) = {} & (h_{CO_2} - h_C - h_{O_2}) \\ & - (h_{CO} - h_C - \tfrac{1}{2}\, h_{O_2}) - \tfrac{1}{2}(h_{O_2} - h_{O_2}) \end{aligned} \tag{11-26}$$

Regrouping the terms in equation (11-26), we obtain.

$$(\Delta h_R) = [h_{CO_2} - h_{CO} - \tfrac{1}{2} h_{O_2}] - [h_C + h_{O_2} - (h_C + \tfrac{1}{2} h_{O_2} + \tfrac{1}{2} h_{O_2})] \tag{11-26a}$$

The first term in square brackets on the right-hand side of equation (11-26a) is simply the isothermal enthalpy change of reaction, Δh_R. The second term in square brackets is seen to reduce identically to zero; so equations (11-24) and (11-23) are indeed equivalent expressions for Δh_R. Although this illustration has dealt with only one simple chemical reaction, it is relatively straightfoward to prove the equivalency of equations (11-23) and (11-24) for all chemical reactions.

SAMPLE PROBLEM 11-2. Methane is to be burned according to the reaction:

$$CH_4 + 2\,O_2 = CO_2 + 2H_2O$$

Evaluate $(\Delta h)_{\text{reaction}}$ assuming that (a) liquid water and (b) gaseous water are formed. At 1 atm and 25°C, which is the correct value?

Solution: We may summarize the stoichiometry and enthalpies of formation for the given reaction as shown below:

			$(\Delta h_i)_f \dfrac{\text{cal}}{\text{g-mol}}$	
i	Species	α_i	a	b
1	CH_4	−1	−17,889	−17,889
2	O_2	−2	0	0
3	CO_2	+1	−94,052	−94,052
4	H_2O	+2	−68,317	−57,798

Since

$$(\Delta h)_R = \sum_{i=1}^{4} \alpha_i (\Delta h_i)_f$$

we find:

Case (a):

$$(\Delta h)_R = -212{,}797 \text{ cal/g-mol } CH_4 \text{ reacted}$$

Case (b):

$$(\Delta h)_R = -191{,}656 \text{ cal/g-mol } CH_4 \text{ reacted}$$

At 25°C and 1 atm, water is normally a liquid so that the actual isothermal enthalpy change on reaction at 25°C and 1 atm would be given by case (a): −212,797 cal/g-mol CH_4 reacted.

11.3 Applications to Nonisothermal Systems

Many chemical reactions—particularly those referred to as *combustion reactions*—are performed in a nonisothermal manner. In some cases this may occur because practical heat-transfer limitations prevent isothermal operation. In many cases, however, the primary purpose of the chemical reaction is to provide hot gas. Examples of the latter are found in such common items as the laboratory Bunsen burner, the cylinders of a conventional gasoline engine, and the combustion chambers of jet and rocket engines. Clearly, in most of these applications we are primarily concerned with determining the temperature that is attained during the combustion reaction. In this section we will show how this question is answered.

We begin by writing an energy balance around the nonisothermal reactor illustrated in Fig. 11-5. If we assume negligible changes in potential and kinetic energies and steady-state operation, then the energy balance reduces to:

$$q = \Delta h = h_{\text{prod}} - h_{\text{react}} \tag{11-27}$$

where Δh is the enthalpy difference between the products and reactants. Since the reaction is not isothermal, Δh is *not* the isothermal enthalpy change on reaction. In general we are looking for the maximum temperature that can be obtained from the combustion reaction. This maximum temperature is attained when the reaction is performed adiabatically (assuming of course that it is not possible to add heat to the reaction). The temperature obtained in this case is known as the *adiabatic flame temperature* and is calculated from the general expression we develop by simply setting $q = 0$.

We must now relate Δh back to quantities that we can determine and/or to those things in which we are interested. We now model the nonisothermal reactor as the two-step sequence shown in Fig. 11-6. The reactants enter the

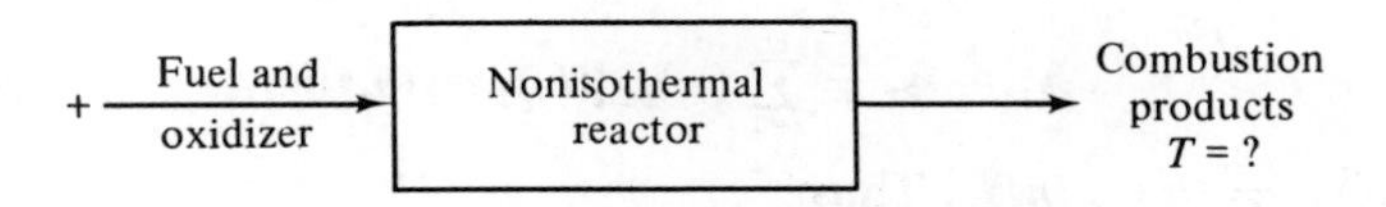

Figure 11-5. The nonisothermal chemical reactor.

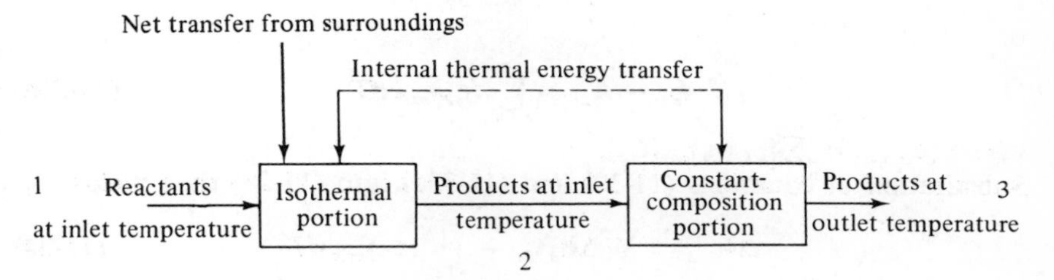

Figure 11-6. Two-stage nonisothermal reactor.

first section where they react isothermally to form the products. In the second section of the reactor the products absorb (or liberate) some fraction of the heat liberated (absorbed) in the first stage but do not change composition. (It is assumed that the material leaving the first stage is at the final product composition.)

The overall enthalpy change across the reactor, $\Delta \not{h}_{1-3}$, is expressed as the sum of the enthalpy changes across both sections of the model reactor:

$$\Delta \not{h}_{1-3} = \Delta \not{h}_{1-2} + \Delta \not{h}_{2-3} \tag{11-28}$$

where the enthalpy, $\not{h}$, of a stream is given by

$$\not{h} = \sum_{i=1}^{C} \frac{n_i h_i}{(n_1)_{\text{in}}} \equiv \sum_{i=1}^{C} \underline{n}_i h_i \tag{11-29}$$

If we assume that the enthalpies of the reacting species are not composition-dependent, then the only enthalpy changes that occur in the isothermal section of the model reactor are those due to the chemical reaction. Thus for every mole of species 1 that reacts in the first stage, the enthalpy change is, by definition, the enthalpy change on reaction, $\Delta \not{h}_R$, evaluated at the inlet temperature, T_1. Thus if each mole of species 1 that entered the reactor reacted, $\Delta \not{h}_{1-2}$ would simply equal $(\Delta \not{h}_R)_{T_1}$. If X fraction of species 1 that enters the reaction actually reacts, the total change in enthalpy due to reaction is given by

$$\Delta \not{h}_{1-2} = X(\Delta \not{h}_R)_{T_1} \tag{11-30}$$

Since the temperature change in section 2 of the model reactor occurs at constant pressure and composition, the enthalpy changes of the individual species are given by the integral of C_p (again assuming no composition-dependence).

$$(h_i)_3 - (h_i)_2 = \int_{T_2}^{T_3} (C_p)_i \, dT \tag{11-31}$$

Then the total enthalpy change in the second stage is given by

$$\not{h}_3 - \not{h}_2 = \sum_{i=1}^{C} (\underline{n}_i)_{\text{out}}[(h_i)_3 - (h_i)_2] \tag{11-32}$$

where $(\underline{n}_i)_{\text{out}} = (n_i)_{\text{out}}/(n_1)_{\text{in}}$. Thus

$$\not{h}_3 - \not{h}_2 = \sum_{i=1}^{C} (\underline{n}_i)_{\text{out}} \int_{T_1=T_2}^{T_3} (C_P)_i \, dT \tag{11-33}$$

or

$$\not{h}_3 - \not{h}_2 = \int_{T_1}^{T_3} (\mathcal{C}_P)_{\text{exit}} \, dT \tag{11-33a}$$

where $(\mathcal{C}_P)_{\text{exit}} = \sum_{i=1}^{C} (\underline{n}_i)_{\text{out}}(C_P)_i$.
Substitution of equations (11-30) and (11-33a) into (11-28) then yields

$$\Delta \not{h}_{1-3} = X(\Delta \not{h}_R)_{T_1} + \int_{T_1}^{T_3} (\mathcal{C}_P)_{\text{exit}} \, dT \tag{11-34}$$

Equation (11-34) is then substituted into the overall energy balance to give

$$q = X(\mathcal{h}_R)_{T_1} + \int_{T_1}^{T_3} (\mathcal{C}_P)_{\text{exit}}\, dT \tag{11-35}$$

Therefore, if we know $\Delta \mathcal{h}_R$ at T_1, the fractional conversion, X, the heat transfer q, and can express $(\mathcal{C}_P)_{\text{exit}}$ as a function of T, we may solve equation (11-35) for the outlet temperature T_3. Similarly, if we know T_1, $(\Delta \mathcal{h}_R)_{T_1}$, X, and T_3, we can determine q.

Sample Problem 11-3. Pure hydrogen is to be burned in a hydrogen torch using air as the oxidizing medium. Both the hydrogen and air enter the torch at 25°C. If the torch is adjusted to give 200 percent more air than the stoichiometric ratio and the combustion is adiabatic, what is the flame temperature (assume that air = 80 percent N_2, 20 percent O_2, molar). The data are given in Table P11-3a, where C_P = cal/g-mol°K and T = °K. Assume that the reaction goes to completion with respect to hydrogen.

Table SP 11-3a Heat Capacities, C_P

O_2	$6.14 + 3.102 \times 10^{-3}T$
N_2	$6.524 + 1.250 \times 10^{-3}T$
H_2	$6.947 - 0.120 \times 10^{-3}T$
H_2O	$7.256 + 2.290 \times 10^{-3}T$

Reaction $H_2 + \frac{1}{2}O_2 = H_2O$(vapor)

$(\Delta \mathcal{h}_R)_{25°C} = -57.8$ kcal/g-mol H_2

Solution: The basis of all calculations is 1 mol of H_2 entering the reaction. One mole of H_2 requires $\frac{1}{2}$ mol of O_2. But we are told that the burner is adjusted for a 200 percent excess of O_2. Therefore, the actual O_2 rate = 1.5 mol, which includes 6.0 mol of N_2. The inlet and outlet streams are then summarized as given in Table SP11-3b. The outlet stream assumes total combustion of H_2, so that $X = 1.0$.

Table SP 11-3b

	N_2	O_2	H_2	H_2O	Total
Inlet	6.00	1.50	1.0	0.0	8.50
Outlet	6.00	1.0	0.0	1.0	8.00

Since we have $\Delta \mathcal{h}_R$ at the temperature of the entering feed stream, we may substitute it directly into equation (11-35). (If this were not the case, we could calculate Δh_R at the feed temperature as shown in *Chemical Engineering Thermodynamics*[2]).

We now proceed to calculate $(\mathcal{C}_P)_{exit}$ as follows (basis, 1 mol of H_2 feed):

$$(\mathcal{C}_P)_{exit} = 6.00(C_P)_{N_2} + 1.0(C_P)_{O_2} + 1.0(C_P)_{H_2O}$$
$$= [(6.0)(6.524 + 1.250 \times 10^{-3}T) + (6.148 + 3.102 \times 10^{-3}T) + (7.256 + 2.30 \times 10^{-3}T)] \text{ cal/g-mol°K} \quad (T = \text{°K})$$
$$= (53.21 + 12.90 \times 10^{-3}T) \text{ cal/g-mol°K}$$

Substitution of $q = 0$, $X = 1.0$, $(\mathcal{C}_P)_{exit}$, and $(\Delta h_R)_{T_1}$ into equation (11-35) and integration between 298°K and the torch outlet temperature, T_2 yields

$$0 = -57{,}800 \frac{\text{cal}}{\text{g-mol}} + \int_{298}^{T_2} (53.21 + 12.90 \times 10^{-3}T)\, dT \text{ cal °K/g-mol°K}$$

or

$$53.21(T_2 - 298) + 6.45 \times 10^{-3}(T_2^2 - 298^2)] = 57{,}800$$

Collecting terms yields

$$6.45 \times 10^{-3}T_2^2 + 53.21T_2 - 74{,}300 = 0$$

or

$$T_2^2 + 8.25 \times 10^3 T_2 - 1.15 \times 10^7 = 0$$

We may now solve for T_2 by means of the quadratic equation

$$T_2 = \frac{-8.25 \times 10^3 + \sqrt{(8.25 \times 10^3)^2 + (4)(1.15 \times 10^7)}}{2.0}$$

Thus

$$T_2 = 1215\text{°K} \quad \text{or} \quad T_2 = 2187\text{°R} = 1727\text{°F}$$

Sample Problem 11-4. Metallurgical coke (essentially pure carbon) initially at 25°C is continuously burned in a fluidized bed reactor using 25 percent excess of air (21 mol percent O_2) also fed at 25°C according to the reaction

$$C + O_2 = CO_2$$

Cooling coils are used in the walls of the furnace to prevent excessive temperatures in the fire box and surrounding area. The temperature of the resulting combustion products is measured to be 1230°C. Determine the total heat losses during the reaction per long (metric) ton (1000 kg) of carbon burned. Assume the reaction goes to completion with respect to carbon.

Solution: The heat losses may be calculated from equation (10-35):

$$q = X(\Delta h_R)_{T_1} + \int_{T_1}^{T_2} (\mathcal{C}_P)_{exit}\, dT$$

[2]R. E. Balzhiser et. al., *Chemical Engineering Thermodynamics*, Prentice-Hall Inc., Englewood Cliffs, N.J., (1972).

Since the carbon is totally reacted, $X = 1$. The $\Delta h_{25°C}$ may be obtained from the enthalpy of formation of CO_2 in Appendix E:

$$(\Delta h_R)_{T_1} = (\Delta h_{CO_2})_f = -94{,}000 \text{ cal/g-mol carbon}$$

Each mole of carbon burned requires 1 mol of O_2 for the stoichiometric amount. With the 25-percent excess this means 1.25 mol O_2/mol of C. These 1.25 mol of O_2 also bring with them 1.25 (0.79/0.21) = 4.70 mol of N_2 from the air. Thus the inlet and exit streams are as summarized below (basis of calculation 1 g-mol of C):

	C	O_2	N_2	CO_2
Inlet	1	1.25	4.70	0.0
Outlet	0	0.25	4.70	1.0

$(\mathcal{C}_P)_{exit}$ is then obtained from

$$(\mathcal{C}_P)_{exit} = 0.25(C_P)_{O_2} + 4.70(C_P)_{N_2} + 1.0(C_P)_{CO_2}$$

The heat capacities of the individual species as functions of temperature are obtained from Appendix E:

$$(C_P)_{N_2} = 6.524 + 1.250 \times 10^{-3}T - 0.001 \times 10^{-6}T^2 \text{ cal/g-mol°K}$$
$$(C_P)_{O_2} = 6.148 + 3.102 \times 10^{-3}T - 0.923 \times 10^{-6}T^2 \text{ cal/g-mol°K}$$
$$(C_P)_{CO_2} = 6.214 + 10.396 \times 10^{-3}T - 3.545 \times 10^{-6}T^2 \text{ cal/g-mol°K}$$

when T = °K. Thus

$$(\mathcal{C}_P)_{exit} \equiv 38.4 + 17.0 \times 10^{-3}T - 3.77 \times 10^{-6}T^2$$

The outlet temperature from the combustion is found to be 1230°C = 1503°K. Thus q is obtained from:

$$\begin{aligned} q &= -94{,}000 \text{ cal/g-mol} + \int_{298}^{1503} [(38.4 + 17.0 \times 10^{-3}T \\ &\quad - 3.77 \times 10^{-6}T^2)\, dT] \text{ cal/g-mol} \\ &= \{-94{,}000 + [38.4(1503 - 298) + 8.5 \times 10^{-3}(1503^2 - 298^2) \\ &\quad - 1.26 \times 10^{-6}(1503^3 - 298^3)]\} \text{ cal/g-mol} \\ &= [-94{,}000 + 60{,}700] \text{ cal/g-mol} \\ q &= -33{,}300 \text{ cal/g-mol} \\ &= -139.5 \text{ MJ/kg-mol} \end{aligned}$$

So that 139.5 MJ are lost per kg-mol of carbon that is burned. Since there are $\frac{1000}{12}$ = 83.3 kg-mol in a long ton, the heat loss per long ton of carbon is then 83.3 × 139.5 MJ/ton = 11.6 × 10^9 joule/long ton.

11.4 Work Production from Chemically Reacting Systems

Consider the steady-flow isothermal chemical reactor shown in Fig. 11-7. As in our previous discussions the chemical reaction is written as

$$-\alpha_1 A_1 - \alpha_2 A_2 - \cdots \longrightarrow \cdots \alpha_{C-1} A_{C-1} + \alpha_C A_C \tag{11-36}$$

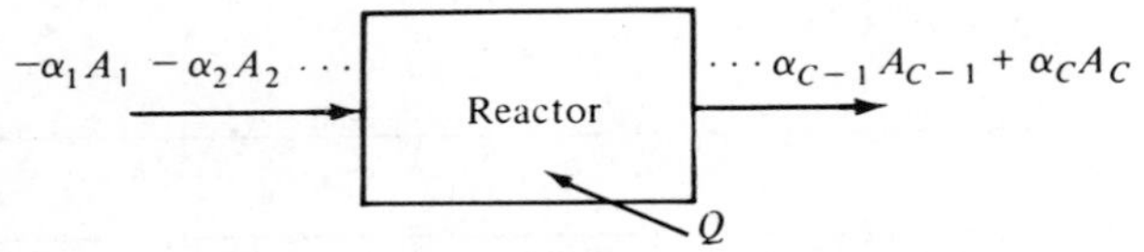

$-\alpha_1 A_1 - \alpha_2 A_2 \cdots$

Reactor

$\cdots \alpha_{C-1} A_{C-1} + \alpha_C A_C$

Q

Figure 11-7. Steady-flow isothermal chemical reactor.

We assume that reactants, in their stoichiometric ratio, enter the reactor and are completely converted into the stoichiometric amounts of products. It was shown in Chapter 6 that the maximum work which may be extracted from any isothermal flow process is the negative of the Gibbs-free-energy change. Thus for the chemical reactor we may write

$$(-W_R)_T = (\Delta G)_T = G_{\text{prod}} - G_{\text{react}} \tag{11-37}$$

where the work and Gibbs free energies are based on a given quantity of *mass* that enters and leaves the reactor. (The reactor is operating at steady state, so the *mass* flow rate out equals the *mass* flow rate in.) Equation (11-37) is put on a per mole basis by dividing the whole equation by a given number of moles. Although there is no unique value that *must* be chosen, we again choose the number of moles of species 1, n_1, that enter the reactor. Thus equation (11-37) becomes

$$\left[-\frac{W_R}{(n_1)_{\text{in}}}\right]_T = \left[\frac{\Delta G}{(n_1)_{\text{in}}}\right]_T = \frac{G_{\text{prod}}}{(n_1)_{\text{in}}} - \frac{G_{\text{react}}}{(n_1)_{\text{in}}}$$

or

$$(-w_R)_T = (\Delta \mathcal{G})_T = \mathcal{G}_{\text{prod}} - \mathcal{G}_{\text{react}} \tag{11-38}$$

The term $\mathcal{G}_{\text{prod}} - \mathcal{G}_{\text{react}}$ is now expressed in terms of the partial molar properties of the individual species as follows:

$$\mathcal{G}_{\text{prod}} - \mathcal{G}_{\text{react}} = \sum_{\text{prod}} \frac{(\bar{G}_i n_i)_{\text{out}}}{(n_1)_{\text{in}}} - \sum_{\text{react}} \frac{(\bar{G}_i n_i)_{\text{in}}}{(n_1)_{\text{in}}} \tag{11-39}$$

where $(n_1)_{\text{in}}$ = moles of species 1 which enter the reactor. If we also define

$$\underline{n}_i = \frac{n_i}{(n_1)_{\text{in}}}$$

equation (11-39) reduces to

$$\mathcal{G}_{\text{prod}} - \mathcal{G}_{\text{react}} = \sum_{\text{prod}} (\bar{G}_i \underline{n}_i)_{\text{out}} - \sum_{\text{react}} (\bar{G}_i \underline{n}_i)_{\text{in}} \tag{11-40}$$

Since it is assumed that all components are present in their stoichiometric ratio, $(\underline{n}_i)_{\text{in}} = -\alpha_i$ for a reactant and $(\underline{n}_i)_{\text{out}} = \alpha_i$ for a product. Thus equation (11-40) becomes

$$\mathscr{G}_{\text{prod}} - \mathscr{G}_{\text{react}} = \sum_{\text{prod}} (\alpha_i \bar{G}_i)_{\text{out}} + \sum_{\text{react}} (\alpha_i \bar{G}_i)_{\text{in}} \tag{11-41}$$

and equation (11-39) becomes

$$(-w_R)_T = \sum_{\text{prod}} (\alpha_i \bar{G}_i)_{\text{out}} + \sum_{\text{react}} (\alpha_i \bar{G}_i)_{\text{in}} \tag{11-42}$$

It is obviously impossible to tabulate partial molar Gibbs free energies for all possible combinations of temperature, pressure, and composition. However, we have developed (in Chapter 10) methods for predicting the variation of $\bar{G}_i$ with pressure, temperature, and composition. Thus if we tabulate values of the Gibbs free energies at a prespecified set of reference conditions, we may calculate the free energies at any other condition. Thus we now refer all partial molar Gibbs free energies to a set of arbitrary reference conditions. These conditions may be different for each component. The only requirement that we shall place upon each reference state is that it must be at the temperature of the system. Although a great variety of choices for the composition and pressure of the reference state exist, the reference state is most frequently (but not always) chosen as the pure component, at the temperature of the system and 1 atm pressure.

The partial molar Gibbs free energy of each specie is expressed in terms of the standard-state Gibbs free energy as shown below:

$$\bar{G}_i = (\bar{G}_i - g_i^\circ) + g_i^\circ \tag{11-43}$$

Equation (11-43) is then substituted into equation (11-42) for the various components to give

$$(-w_R)_T = \sum_{\text{prod}} \alpha_i[(\bar{G}_i)_{\text{out}} - g_i^\circ] + \sum_{\text{react}} \alpha_i[(\bar{G}_i)_{\text{in}} - g_i^\circ] + \sum_{i=1} \alpha_i g_i^\circ \tag{11-44}$$

Since the reference state has been chosen at the temperature of the system, we have shown (Chapter 10) that

$$\bar{G}_i - g_i^\circ = RT \ln \frac{\bar{f}_i}{f_i^\circ} \tag{11-45}$$

or

$$\bar{G}_i - g_i^\circ = RT \ln a_i \tag{11-46}$$

where a_i is termed the *activity* of the ith component and is defined as $a_i = \bar{f}_i/f_i^\circ$. Substitution of equation (11-46) into equation (11-44) then yields

$$(-w_R)_T = \sum_{\text{prod}} \alpha_i RT \ln a_i)_{\text{out}} + \sum_{\text{react}} \alpha_i RT \ln a_i)_{\text{in}} + \sum_{i=1} \alpha_i g_i^\circ \tag{11-47}$$

The term $\sum_{i=1}^{C} \alpha_i g_i^\circ$ is equal to the Gibbs-free-energy change that would occur if the reactants (based on 1 mol of species 1) in their standard states were completely converted to the stoichiometric amount of products in their standard state. This quantity is termed the *standard-state Gibbs-free-energy change of reaction* and is written as

$$\Delta \mathscr{G}^\circ = \sum_{i=1}^{C} \alpha_i g_i^\circ = \sum_{\text{prod}} \alpha_i g_i^\circ + \sum_{\text{react}} \alpha_i g_i^\circ \tag{11-48}$$

Thus equation (11-47) reduces to

$$(-w_R)_T = RT \sum_{\text{prod}} \alpha_i \ln (a_i)_{\text{out}} + RT \sum_{\text{react}} \alpha_i \ln (a_i)_{\text{in}} + \Delta \mathscr{G}^\circ \tag{11-49}$$

but

$$\alpha_i \ln a_i = \ln a_i^{\alpha_i} \tag{11-50}$$

$$\ln a_i^{\alpha_i} + \ln a_j^{\alpha_j} = \ln (a_i^{\alpha_i} \cdot a_j^{\alpha_j}) \tag{11-51}$$

so equation (11-49) becomes

$$(-w_R)_T = RT \ln [\prod_{\text{prod}} (a_i)_{\text{out}}^{\alpha_i}] + RT \ln [\prod_{\text{react}} (a_i)_{\text{in}}^{\alpha_i}] + \Delta \mathscr{G}^\circ \tag{11-52}$$

or

$$(-w_R)_T = RT \ln \frac{\prod_{\text{prod}} (a_i)_{\text{out}}^{\alpha_i}}{\prod_{\text{react}} (a_i)_{\text{in}}^{-\alpha_i}} + \Delta \mathscr{G}^\circ \tag{11-52a}$$

where the symbol $\prod_{\text{prod}}$ ($\prod_{\text{react}}$) represents "the repeated product" over all products (reactants). For example,

$$\prod_{i=1}^{4} B_i = B_1 \cdot B_2 \cdot B_3 \cdot B_4 \tag{11-53}$$

We now define the activity ratio, J_a, as

$$J_a = \frac{\prod_{\text{prod}} (a_i)_{\text{out}}^{\alpha_i}}{\prod_{\text{react}} (a_i)_{\text{in}}^{-\alpha_i}} \tag{11-54}$$

and equation (11-52a) becomes

$$(-w_R)_T = RT \ln J_a + \Delta \mathscr{G}^\circ \tag{11-55}$$

In the evaluation of the activity ratio, J_a, it must be remembered that *the activities of the various species in the products must be determined from the state of the products, while the activities of the species in the reactants must be evaluated from the state of the reactants.*

Equation (11-55) expresses the maximum (reversible) work that can be obtained from the complete conversion of reactants to products in a steady-flow isothermal chemical reactor. This work is expressed in terms of the standard-state change in Gibbs free energy between the reactants and products and the activities of the reactants and products. Although equation (11-55) has been derived specifically for the case where a stoichiometric quantity of reactants is completely converted to a stoichiomatic quantity of products, it may also be used to give (in most cases of interest) a reasonable estimate of the work production when these conditions are not completely satisfied—as for example when an inert or product material is in the feed stream, or when the reaction is not carried to completion. For these cases, however, we also need to recognize that the Gibbs-free-energy change and work production calculated above are based on 1 mol of species 1 being reacted, rather than on 1 mol of species 1 being fed to the reactor. Thus, if X fraction of species 1 is consumed in the reactor, the work produced per mole of species 1 *fed* to the reactor is given by:

$$w = -X(\Delta \mathcal{G}^\circ + RT \ln J_a) \tag{11-56}$$

This equation neglects the Gibbs-free-energy changes that the unreacted portions of the feed material undergo. These terms are generally small except for very low values of X, and hence this approximation is usually quite satisfactory. For a more detailed discussion of this question, the reader is referred to Balzhiser et al.[3]

In our later discussions we will have use for two quantities which look quite similar to J_a. For completeness, we now define these quantities:

$$(J_a)_{\text{out}} = \prod_{i=1}^{C} (a_i)_{\text{out}}^{\alpha_i} \tag{11-57}$$

$$(J_a)_{\text{in}} = \prod_{i=1}^{C} (a_i)_{\text{in}}^{\alpha_i} \tag{11-58}$$

$(J_a)_{\text{out}}$ is simply the activity ratio evaluated at the outlet conditions, while $(J_a)_{\text{in}}$ is the activity ratio evaluated at the inlet conditions. We will use these quantities when we discuss equilibrium in chemically reacting mixtures. During this discussion we will show that $(J_a)_{\text{out}}$ and $(J_a)_{\text{in}}$ are helpful in determining the direction in which a given reaction will proceed without outside interference.

SAMPLE PROBLEM 11-5. A hydrogen fuel cell which uses air as the oxidizing medium is being designed. The cell will be fed pure hydrogen and pure air. A 20-percent excess of air will be used, and the waste gas will be purged from the top of the cell. Pure liquid water is the main product. The cell will be operated at 25°C and 1 atm pressure. What is the maximum amount of work that can be recovered per mole of hydrogen (H_2) fed to the cell? Assume that all the hydrogen is reacted.

For the reaction $H_2 + \frac{1}{2} O_2 = H_2O$,

$$\Delta \mathcal{G}^\circ = -56{,}700 \text{ cal/g-mol}$$

where the standard states are pure liquid water and pure gaseous hydrogen and oxygen at 1 atm and 25°C.

Solution: We may picture the fuel-cell operation as shown in Fig. SP11-5. Since it is desired to know the work liberated per mole of H_2, let us call

[3]Balzhiser, Samuels, and Eliassen, *Chemical Engineering Thermodynamics*, Prentice-Hall, Inc., Englewood Cliffs, N.J., 1972.

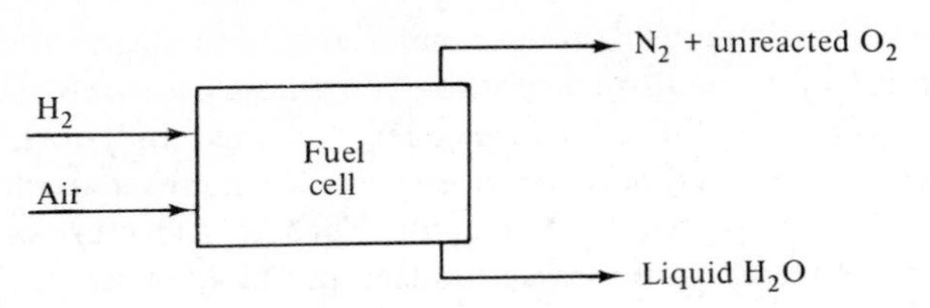

Figure SP11-5

hydrogen species 1, oxygen species 2, and water species 3. The stoichiometric coefficients for the reaction are then

$$\alpha_1 = -1 \quad \text{(hydrogen)}$$

$$\alpha_2 = -\tfrac{1}{2} \quad \text{(oxygen)}$$

$$\alpha_3 = 1 \quad \text{(water)}$$

and our basis of calculation is 1 mol of H_2 fed to the reactor. We use equation (11-55) to determine the reversible work. For each mole of hydrogen fed to the fuel cell, we have 0.6 mol of oxygen and 2.4 mol of N_2 (for a total of 3 mol of air). According to the problem statement, all hydrogen is consumed, so "1 mol of reaction" has occurred for each mole of H_2 fed to the cell. Thus $X = 1.0$ and w_R is given by

$$-w_R = \Delta\mathscr{G}^\circ + RT \ln J_a$$

But

$$\Delta\mathscr{G}^\circ = -56{,}700 \text{ cal/g-mol}$$

$$R = 1.987 \text{ cal/g-mol °K}$$

$$T = 25°\text{C} = 298°\text{K}$$

$$J_a = \frac{\prod_{\text{prod}} (a_i)_{\text{out}}^{\alpha_i}}{\prod_{\text{react}} (a_i)_{\text{in}}^{-\alpha_i}}$$

We now evaluate the various activities. Both the inlet hydrogen and outlet water are in their standard states, and thus $(a_{H_2})_{\text{in}} = (a_{H_2O})_{\text{out}} = 1.0$. We are at liberty to choose the standard state for nitrogen as the pure gas at 1 atm and 25°C and evaluate the activities of the oxygen and nitrogen in the inlet and outlet streams from the expression

$$a_i = \frac{\bar{f}_i}{f_i^\circ} = \frac{y_i f_i}{f_i^\circ} = \frac{y_i P}{f_i^\circ}$$

where we have assumed that the Lewis–Randall rule may be used to evaluate $\bar{f}_i$ in terms of y_i and f_i. We also assume ideal-gas behavior, so $f_i \equiv P$. For both the oxygen and nitrogen $f_i^\circ = 1$ atm. In addition, the pressure in the fuel cell is 1 atm, so $P = 1$ atm. The activities reduce to

$$a_i = y_i$$

or

$$(a_{O_2})_{in} = 0.20$$

$$(a_{O_2})_{out} = \frac{0.1}{2.5} = 0.0400$$

$$(a_{N_2})_{in} = 0.80$$

$$(a_{N_2})_{out} = \frac{2.4}{2.5} = 0.96$$

The reversible work production is then given by

$$w_R = -\left\{-56{,}700 + (1.987)(293)\left[\ln \frac{1.0}{(1.0)(0.2)^{1/2}}\right]\right\} \frac{\text{cal}}{\text{g-mol}}$$

or

$$w_R = (56{,}700 - 470)\ \text{cal/g-mol} = 56{,}230\ \text{cal/g-mol}$$

as opposed to a value of 56,070 cal/g-mol calculated by Balzhiser et al. using the completely rigorous equation for work evaluation when unreacted materials are present.

11.5 Equilibrium in Chemically Reacting Systems—the Equilibrium Constant

Let us now consider what happens to the total Gibbs free energy of a reacting mixture as the mixture passes through an isothermal, constant-pressure flow reactor as shown in Fig. 11-8. We assume that any reaction which can occur will occur spontaneously, that is, without the addition of work from the reactor surroundings.

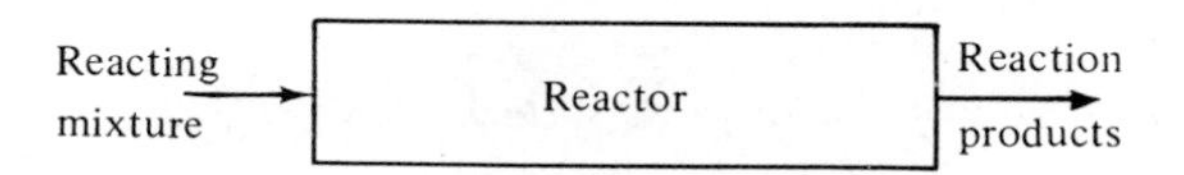

Figure 11-8. Chemical reactor.

We have shown in Chapter 6 that as long as the Gibbs free energy of a system undergoing change is decreasing, that change will spontaneously continue. Since the discussion in Chapter 6 was perfectly general (for any isothermal change), it also applies to processes in which chemical reactions are occurring. Thus an isothermal reacting mixture will continue to react as long as its total Gibbs free energy is decreasing.

However, when the Gibbs free energy of the reacting mixture reaches a minimum, the spontaneous reaction will proceed no further because the surroundings would have to provide work if the Gibbs free energy is to increase, and this violates our assumption of spontaneity. Similarly, the reaction cannot even proceed (spontaneously) in the reverse direction, because this would also cause an increase in the Gibbs free energy of the reacting mixture and hence require the surroundings to supply work. We may indicate these various situations on a plot of $\mathcal{G}_{mix}$ versus X as shown in Fig. 11-9. (Since $(n_1)_{in}$ is a constant, the minimum in $\mathcal{G}_{mix}$ is also the minimum in G_{mix}.) Once the reaction has proceeded to the minimum in the $\mathcal{G}$–X curve, it may go no further (in either direction) without the surroundings supplying work. Clearly, the reacting mixture at this point is at its equilibrium condition (subject to the imposed constraints of constant temperature and pressure). There is no way that this mixture can supply additional work to the surroundings or spontaneously change its chemical composition unless we alter the constraints.

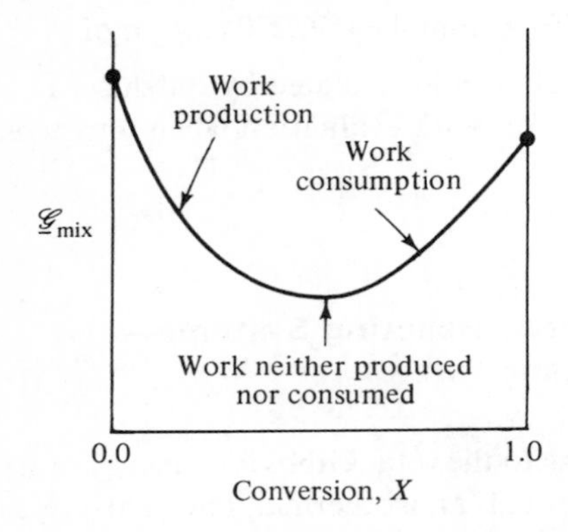

Figure 11-9. Gibbs free energy as a function of conversion.

We may determine the conditions that correspond to the minimum in the $\mathcal{G}$–X curve by setting $d\mathcal{G}/dX = 0$. (We know that this is a minimum rather than a maximum from the nature of the $\mathcal{G}$–X curve.) We begin by expressing the outlet Gibbs free energy as:

$$\mathcal{G}_{out} = \sum_{i=1}^{C} (\underline{n}_i)_{out}(\bar{G}_i)_{out} \tag{11-59}$$

The minimum in the outlet $\mathcal{G}$ is obtained by setting

$$\frac{d\mathcal{G}_{out}}{dX} = 0 = \frac{d}{dX}\left[\sum_{i=1}^{C} (\underline{n}_i)_{out}(\bar{G}_i)_{out}\right] \tag{11-60}$$

or

$$0 = \sum_{i=1}^{C} \frac{d}{dX}[(\underline{n}_i)_{out}(\bar{G}_i)_{out}] \tag{11-60a}$$

which upon differentiation becomes

$$0 = \sum_{i=1}^{C} \left[(\underline{n}_i)_{out}\frac{(d\bar{G}_i)_{out}}{dX} + (\bar{G}_i)_{out}\frac{(d\underline{n}_i)_{out}}{dX}\right] \tag{11-61}$$

However, the first term when summed over all components at constant T and

P gives zero via the Gibbs–Duhem equation; so equation (11-61) reduces to

$$0 = \sum_{i=1}^{C} (\bar{G}_i)_{\text{out}} \frac{(d\underline{n}_i)_{\text{out}}}{dX} \tag{11-62}$$

The outlet mole numbers are then expressed as:

$$(n_i)_{\text{out}} = (n_i)_{\text{in}} + X\alpha_i(n_1)_{\text{in}} \tag{11-63}$$

Equation (11-63) indicates that the amount of each species at the end of a reaction is equal to the amount of that species which entered the reactor plus that which is formed by the chemical reaction. (Since α_i is negative for reactants, equation (11-63) correctly predicts both the decreasing quantity of reactants and the increasing quantity of products.) Division of equation (11-63) by $(n_i)_{\text{in}}$ gives

$$(\underline{n}_i)_{\text{out}} = (\underline{n}_i)_{\text{in}} + X\alpha_i \tag{11-64}$$

where $(\underline{n}_i)_{\text{out}} = \dfrac{(n_i)_{\text{out}}}{(n_1)_{\text{in}}}$

$(\underline{n}_i)_{\text{in}} = \dfrac{(n_i)_{\text{in}}}{(n_1)_{\text{in}}}$

X = mol of reaction per mole of species 1 fed to the reactor

Since $\alpha_1 = -1$, this is also the fractional conversion of species 1. Thus the derivative in equation (11-62) becomes

$$\frac{d(\underline{n}_i)_{\text{out}}}{dX} = \alpha_i \tag{11-65}$$

where $\alpha_i = 0$ for an inert. Equation (11-62) reduces to

$$0 = \sum_{i=1}^{C} (\bar{G}_i)_{\text{out}}\alpha_i \tag{11-66}$$

Now, as before, $(\bar{G}_i)_{\text{out}}$ is referred to the reference state conditions

$$(\bar{G}_i)_{\text{out}} = RT \ln (a_i)_{\text{out}} + g^\circ \tag{11-67}$$

which is substituted into equation (11-66) to give

$$\sum_{i=1}^{C} \alpha_i \left[g_i^\circ + RT \ln (a_i)_{\text{out}}\right] = 0 \tag{11-68}$$

The first term in the summation is recognized as $\Delta\mathscr{G}^\circ$, the standard-state Gibbs-free-energy change, while the log terms are combined to yield the overall result:

$$\Delta\mathscr{G}^\circ + RT \ln \left[\frac{\prod_{\text{prod}} (a_i)_{\text{out}}^{\alpha_i}}{\prod_{\text{react}} (a_i)_{\text{out}}^{-\alpha_i}}\right] = 0 \tag{11-69}$$

Equation (11-69) *thus relates the standard-state Gibbs-free-energy change to the activities of the reacting species which coexist in a mixture that is at chemical equilibrium.* The term in brackets is called the *equilibrium constant,* K_a:

$$K_a = \frac{\prod_{\text{prod}} (a_i)_{\text{out}}^{\alpha_i}}{\prod_{\text{react}} (a_i)_{\text{out}}^{-\alpha_i}} = \prod_{i=1}^{C} (a_i)_{\text{out}}^{\alpha_i} \tag{11-70}$$

Equations (11-69) and (11-70) may then be solved for K_a to give

$$K_a = \exp\left(\frac{-\Delta\mathscr{G}^\circ}{RT}\right) \tag{11-71}$$

or

$$-\Delta\mathscr{G}^\circ = RT \ln K_a \tag{11-71a}$$

Thus if we can determine $\Delta\mathscr{G}^\circ$ for a given reaction, equation (11-71a) permits us to evaluate the equilibrium constant for the reaction.

Examination of equation (11-71a) indicates that the definition of the equilibrium constant greatly resembles our definition of $(J_a)_{\text{out}}$. Indeed, the definitions appear at first glance to be identical. However, there is a subtle, but major, distinction between the two that must be remembered: *The equilibrium constant applies only to the activities in a mixture that is at chemical equilibrium.* The outlet activity ratio, $(J_a)_{\text{out}}$, on the other hand, is simply a number that may always be calculated from the state of the stream exiting from the chemical reactor under study. *In the special case where the exit stream is at chemical equilibrium, $(J_a)_{\text{out}}$ and K_a are equal.* In the more general case, where the exit stream is not in chemical equilibrium, K_a is still calculated from equation (11-71) but $(J_a)_{\text{out}}$ will no longer be equal to the equilibrium constant so calculated.

We may determine much about the state of a chemically reacting mixture by examining the activity ratio calculated from its present compositions. For example, let us consider a stream entering a chemical reactor. If $(J_a)_{\text{in}}$ is equal to K_a, the mixture is at equilibrium, and no reaction will occur. If $(J_a)_{\text{in}}$ is less than K_a, the activities of the products are less than their equilibrium value, while the activities of the reactants are above their equilibrium value. If the reaction is allowed to proceed, the activities of the products will increase, while those of the reactants will decrease. At constant temperature and pressure, the only way to increase the activity of a species is to increase its mole fraction. Thus as the reaction proceeds, the mole fractions of the products will increase, while the mole fractions of the reactants will decrease. Therefore, reactants are consumed and products produced, so the reaction proceeds in the forward direction.

If $(J_a)_{\text{in}}$ is greater than K_a, the reverse is true. That is, the activities and mole fractions of the products are too high, and the reaction will proceed in the reverse direction—opposite to that assumed in the calculation of $(J_a)_{\text{in}}$ and K_a.

The term "equilibrium constant" is actually a misnomer because K_a is not a true constant but varies with temperature. However, K_a does not vary with pressure (unless the standard-state conditions change) or composition of the reacting species, and therefore has become referred to as a "constant."

As we shall soon demonstrate, the equilibrium constant is directly related to the equilibrium compositions in a reacting mixture. Therefore, knowledge of K_a will tell us much about the maximum extent to which any given reaction proceeds. From these observations we may begin to understand the great importance of knowing, or being able to calculate, K_a. The remainder of this chapter will deal with the determination and use of K_a under many different circumstances.

11.6 Evaluation of $\Delta \mathcal{g}^\circ$ from Gibbs Free Energies of Formation

In the same way that we have defined the enthalpy change on formation, we define the *Gibbs-free-energy change on formation* (Δg_f) as the change in Gibbs free energy which occurs when a chemical species is isothermally formed from its constituent elements. Unlike the enthalpy change on formation, however, the Gibbs-free-energy change on formation is decidedly a function of temperature *and pressure*. Thus it is critical that both T and P be specified whenever Δg_f is given. If the temperature and pressure are standard-state conditions, then the Gibbs free energy of formation is termed the *standard Gibbs free energy of formation*, (Δg_f°).

The Gibbs free energies of formation of several commonly encountered substances *at* 25°*C and* 1 *atm pressure* are given in Table 11-1. The free energies of all pure elements have been arbitrarily chosen as zero.

Table 11-1 Standard Gibbs Free Energies of Formation at 25°C (kcal/mol)†

Substance‡	Δg_f°	Substance	Δg_f°	Substance	Δg_f°
SO_3(g)	−88.52	NO(g)	20.66	CH_4(g)	−12.14
HCl(g)	−22.74	H_2S(g)	− 7.87	C_2H_6(g)	− 7.6
C_2H_4O(g) (ethylene oxide)	− 2.79	SO_2(g)	−71.7	C_3H_8(g)	− 5.61
CO (g)	−32.81	NH_3(g)	− 3.94	C_2H_4(g)	16.34
CO_2(g)	−94.26	H_2O(l)	−56.70	C_2H_2(g)	50.7
Methanol(l)	−40.0				
Benzene(l)	29.06				
Acetic Acid(l)	−94.5				
Ethanol(l)	−40.2				
Acetaldehyde(l)	−31.9				
Ethylene Glycol(l)	−77.12				

†Selected mainly from F. D. Rossini et al., *Selected Values of Hydrocarbons and Related compounds*, API Research Project 44, Carnegie Instit. of Tech. Pittsburgh (1953). A more extensive tabulation may be found in Appendix D.

‡The (l) or (g) after each component indicates the equilibrium phase at 1 atm.

In a manner exactly analogous to that derived in our earlier discussion of the enthalpy change on reaction, the Gibbs-free-energy change on reaction ($\Delta\mathcal{G}$) may be expressed in terms of the Gibbs-free-energy change on formation of the reacting species:

$$\Delta\mathcal{G} = \sum_{i=1}^{C} \alpha_i(\Delta g_f)_i \tag{11-72}$$

Thus, the standard-state Gibbs-free-energy change on reaction may be expressed in terms of the standard-state Gibbs-free-energy changes of formation as:

$$\Delta\mathcal{G}^\circ = \sum_{i=1}^{C} \alpha_i(\Delta g_f^\circ)_i \tag{11-73}$$

A similar result may be developed for any property change on reaction, in terms of the property changes of formation of the reacting species.

In cases where it is desired to use standard states other than those used in Table 11-1, it is usually a simple task to correct the values given in the table to whatever conditions are desired.

SAMPLE PROBLEM 11-6. Determine the free energy of formation of pure water vapor at 25°C and 1 atm.

Solution: From Table 11-1, (Δg_f) for liquid water is given by

$$\Delta g_f^{\text{liq}} = -56.70 \text{ kcal/g-mol} = g_{H_2O}^{\text{liq}} - \tfrac{1}{2}g_{O_2} - g_{H_2}$$

However, at 25°C the vapor pressure of liquid H_2O is only 23.76 mm Hg. Therefore, $f_{H_2O} = 23.76$ mm Hg (assuming negligible change of liquid fugacity with pressure and ideal-gas behavior). For water vapor at 1 atm and 25°C, $f_{H_2O}^{\text{vap}} = 760$ mm Hg. The change in g_{H_2O} from 23.76 mm Hg to 1 atm is given by

$$\begin{aligned}(g_{H_2O}^{\text{vap}}) - (g_{H_2O}^{\text{liq}}) &= RT \ln \frac{f^{\text{vap}}}{f^{\text{liq}}} \\ &= (1.987)(298) \text{ cal °K/g-mol °K} \cdot \ln \frac{760}{23.76} \\ &= 2.05 \text{ kcal/g-mol}\end{aligned}$$

Therefore,

$$\begin{aligned}(\Delta g_f^{\text{vap}}) &= g_{H_2O}^{\text{vap}} - \tfrac{1}{2}g_{O_2} - g_{H_2} \\ &= g_{H_2O}^{\text{liq}} - \tfrac{1}{2}g_{O_2} - g_{H_2} + (g_{H_2O}^{\text{vap}} - g_{H_2O}^{\text{liq}}) \\ &= -56.70 \text{ kcal/g-mol} + 2.05 \text{ kcal/g-mol}\end{aligned}$$

or

$$(\Delta g_f^{\text{vap}})\ 1 \text{ atm} = -54.65 \text{ kcal/g-mol}$$

which is the desired free energy of formation of water *vapor* at 1 atm and 25°C.

SAMPLE PROBLEM 11-7. Determine the equilibrium constant at 25°C for the reaction

$$CH_4 + H_2O = CO + 3\,H_2$$

Solution: Since both the standard-state free energies and the equilibrium constants are functions of the standard-state conditions, we must specify these quantities before we can calculate a meaningful value of the equilibrium constant. Since the free energy of formation data we have is for pure components at 1 atm pressure and 25°C, this is the most reasonable choice for standard-state conditions. The standard-state free-energy change may then be calculated from the free energies of formation as shown in equation (11-73).

$$\Delta \mathcal{g}^\circ = \sum_{i=1}^{4} \alpha_i (\Delta g_f^\circ)_i$$

Therefore,

$$\Delta \mathcal{g}^\circ = (\Delta g_f^\circ)_{CO} + 3(\Delta g_f^\circ)_{H_2} - (\Delta g_f^\circ)_{CH_4} - (\Delta g_f^\circ)_{H_2O}$$

But

$$(\Delta g_f^\circ)_{H_2} = 0$$

Therefore,

$$\Delta \mathcal{g}^\circ = -32.81 \text{ kcal/g-mol} - [-12.14 + (-56.70)] \text{ kcal/g-mol}$$

or

$$\Delta \mathcal{g}^\circ = 36.03 \text{ kcal/g-mol}$$

But, $RT \ln K_a = -\Delta \mathcal{g}^\circ$; therefore,

$$\ln K_a = \frac{-36.03 \text{ kcal/g-mol}}{(298.17)(0.00198) \text{ kcal/g-mol}} = -61.0$$

Therefore,

$$K_a = e^{-61.0} = 10^{-26.5}$$

or

$$K_a = 3.1 \times 10^{-27}$$

As we shall soon show, an equilibrium constant this small indicates that the reaction will not proceed to any appreciable extent at this temperature.

11.7 The Equilibrium Constant in Terms of Measurable Properties

Let us now show how the equilibrium constant may be expressed in terms of measurable, and easily calculable, properties. By the definition of K_a and a_i,

$$K_a = \prod_{i=1}^{C} a_i^{\alpha_i} = \prod_{i=1}^{C} \left(\frac{\bar{f}_i}{f_i^\circ}\right)^{\alpha_i} \tag{11-74}$$

Although the subscript "out" has not been written in equation (11-74) we should remember that the equilibrium constant only applies to outlet properties. We have dropped the "out" subscript for the sake of clarity. Equation (11-74) may be rewritten in the form

$$K_a = \frac{\prod_{i=1}^{C} \bar{f}_i^{\alpha_i}}{\prod_{i=1}^{C} f_i^{\circ\,\alpha_i}} \tag{11-75}$$

Therefore,

$$K_a = \frac{K_{\bar{f}}}{K_{f^\circ}} \tag{11-76}$$

where

$$K_{\bar{f}} = \prod_{i=1}^{C} \bar{f}_i^{\alpha_i}$$
$$K_{f^\circ} = \prod_{i=1}^{C} f_i^{\circ\,\alpha_i} \tag{11-77}$$

For single-phase gaseous reactions—that is, reactions which involve only gases—it is possible to relate K_a to the measurable properties y_i, P, and T in a relatively straightforward manner. For multiphase or liquid systems these simplifications are not directly applicable, and we will find it more convenient to handle these cases on an individual basis. However, for now, *we consider only single-phase gaseous systems*. In addition, *we will assume that all mixture properties can be evaluated on the basis of the Lewis–Randall rule* (that is, we restrict ourselves to *ideal solutions*). Thus, the ensuing discussion will be applicable primarily to gaseous mixtures at pressures less than 15 atm. (As we will see, this is not a particularly severe restriction.) According to the Lewis–Randall rule, the partial fugacity $\bar{f}_i$ of a specie can be related to the pure-component fugacity, f_i, by means of equation (10-41):

$$\bar{f}_i = y_i f_i$$

where f_i is the fugacity of pure i at the pressure and temperature of the reacting system. For simplicity we now restrict the remainder of our discussion to gaseous systems which also obey the ideal-gas equation of state; so that:

$$f_i = P \tag{11-78}$$

Substitution of equations (10-41) and (11-78) into equation (11-77) yields

$$K_{\bar{f}} = \prod_{i=1}^{C} (y_i P)^{\alpha_i} \tag{11-79}$$

Thus

$$K_{\bar{f}} = K_y K_P \tag{11-80}$$

where $K_y = \prod_{i=1}^{C} y_i^{\alpha_i}$

$K_P = \prod_{i=1}^{C} P^{\alpha_i} = P^{\sum_{i=1}^{C} \alpha_i}$

Thus K_a can be expressed as

$$K_a = \frac{K_y K_P}{K_{f^\circ}} \tag{11-81}$$

Under most circumstances we shall know P and T for the reaction under consideration, so K_P can be calculated directly. From the standard-state conditions, K_{f° can also be determined in a straightforward manner. Therefore, we see that K_a can then be directly related to the equilibrium compositions, as expressed by K_y (and vice versa). Thus measurement of one set of equilibrium conditions and compositions may be used to determine K_a (from which we can also obtain $\Delta\mathscr{G}^\circ$). Once K_a is determined for one set of conditions, we may then use this value to calculate K_y for many other sets of reaction conditions as long as T is constant. Because of the direct relation between K_y and the equilibrium compositions of a reaction, knowledge of K_a, P, and T along with the reaction feed composition is usually sufficient to completely determine the equilibrium compositions of the reaction. A calculation of this nature is presented below.

SAMPLE PROBLEM 11-8. Many metals are found in the form of sulfide ores. Copper, for example, is widely found as copper sulfide—CuS. To recover the metallic copper the ore is first "roasted" in hot air to form copper oxide and sulfur dioxide, SO_2. The copper oxide is recovered for further processing to produce the desired metallic copper. The sulfur dioxide formed has in the past been frequently vented to the atmosphere. However, in large amounts SO_2 is a serious air pollutant and efforts are now under way in many areas to recover the SO_2 in a form which has some economic value. In most cases this is accomplished by a further oxidation of the SO_2 to SO_3. The SO_3 may then be reacted with water to form sulfuric acid which is sold as a by-product of the SO_2 removal. Although the oxidation of SO_2 to SO_3 is highly favorable at low temperatures, the rates of reaction are extremely low and the oxidation simply doesn't "go." At higher temperatures the reaction rate increases but the equilibrium conversion begins to drop quickly with rising temperature. Experience with common catalysts (reaction-rate enhancers) indicates that the reaction rate begins to reach practical values at temperatures above 425°C and is extremely high for temperatures above 600°C.

At 610°C the equilibrium constant for the reaction

$$SO_2 + \tfrac{1}{2}\,O_2 = SO_3$$

is found to be $K_a = 8.5$ when pure gases at 1 atm pressure are used as the standard states.

The off gases from a typical ore-roasting oven contain 12 mol percent SO_2, 8 mol percent O_2, and 80 mol percent N_2. If this gas is passed into a catalytic reactor for conversion to SO_3 at a temperature of 610°C and 1-atm pressure, what is the equilibrium fractional conversion of SO_2 to SO_3?

Solution: The equilibrium constant K_a has been expressed in equation (11-81) as

$$K_a = \frac{K_y K_P}{K_{f^\circ}}$$

Since the standard states for all components have been chosen as the pure gases at 1 atm, the further assumption of ideal-gas behavior for all pure components leads to the results

$$f^\circ_{SO_2} = f^\circ_{SO_3} = f^\circ_{O_2} = 1 \text{ atm}$$

Consequently

$$K_{f^\circ} = 1 \text{ atm}^{-1/2}$$

Furthermore, because the reactor operates at $P = 1$ atm, $K_P = 1 \text{ atm}^{-1/2}$. Thus the equilibrium expression reduces to

$$K_a = K_y = \frac{y_{SO_3}}{y_{SO_2}(y_{O_2})^{1/2}} = 8.5$$

We may now express the *exit* compositions of all components in terms of one variable, say the amount of SO_2 converted to SO_3. These expressions are then substituted into the equilibrium expression and solved for the degree of conversion as follows.

For equilibrium calculations it is generally most convenient to base the calculations on a specified number of moles of total feed rather than on any one component. Thus let us base our calculations on 100 mol of feed; 100 mol of feed contains 0 mol of SO_3, 12 mol of SO_2, 8 mol of O_2, and 80 mol of N_2. Assume that X mol of reaction occurs, so X mol of SO_2 is converted to SO_3.

The outlet mole numbers for all species are obtained from

$$(n_i)_{\text{out}} = (n_i)_{\text{in}} + \alpha_i X$$

Thus the outlet stream contains X mol of SO_3, $12 - X$ mol of SO_2, $8 - 0.5X$ mol of O_2, and 80 mol of N_2, for a total of $100 - 0.5X$ mol. The mole fraction of each component in the exit stream is given by

$$y_{SO_3} = \frac{X}{100 - 0.5X}$$

$$y_{SO_2} = \frac{12 - X}{100 - 0.5X}$$

$$y_{O_2} = \frac{8 - 0.5X}{100 - 0.5X}$$

The mole fractions may now be substituted into the equilibrium expression to give

$$\frac{X/(100 - 0.5X)}{\dfrac{12 - X}{100 - 0.5X}\left(\dfrac{8 - 0.5X}{100 - 0.5X}\right)^{1/2}} = 8.5$$

or, upon simplification,

$$\frac{X}{12 - X}\sqrt{\frac{100 - 0.5X}{8 - 0.5X}} = 8.5$$

Solution of this equation for X then allows us to completely determine the equilibrium compositions.

Before attempting solution of equation for X it is wise to examine the equation to determine if it has any particular properties that will allow us to get a good approximate answer without going to the effort of obtaining an exact solution. The first thing to do is to check the range of possible values of X—easily seen to be $0 < X < 12$. Now let us check to see if the solution lies near either of the extremes of this range. As $X \longrightarrow 0$ the left-hand side of the equation approaches zero while the right-hand side is constant. Therefore, the solution will not lie at small X. On the other hand, as $X \longrightarrow 12$, the left-hand side of the equation gets much larger than the right side, so the solution is not in this region either. Thus the actual X must be someplace toward the middle of the range. Examination of the equation then shows that the term $\sqrt{(100 - 0.5X)/(8 - 0.5X)}$ will not vary greatly with moderate changes in X, in the range of X's we expect. Therefore, rearrange the equation to the form

$$\frac{X}{12 - X} = 8.5\sqrt{\frac{8 - 0.5X}{100 - 0.5X}} = C$$

and assume that over any one trial the right-hand side will remain constant. Solving for X we now obtain

$$X = (12 - X)C$$

$$X(1 + C) = 12C$$

or, in terms of an iteration scheme,

$$X^{n+1} = \frac{12C}{1 + C} \quad \text{where} \quad C = 8.5\sqrt{\frac{8 - 0.5X^n}{100 - 0.5X^n}}$$

Now we use a trial-and-error procedure to solve for X. As a first guess, choose $X = 6.0$ (see Table SP 11-8). The final solution will be extremely close to $X = 7.65$. Thus we see that by using a little analysis, we have developed a method of solution that will give a relatively accurate answer to a complex equation with a minimum of effort. (Although this particular equation could have been solved by analytical means—squaring both sides yields a cubic equation whose solution can be found—this is frequently not true for complex reactions, and therefore a trial-and-error solution is often used.) Since $X = 7.65$, $X/12$ or 63.5 percent of the incoming SO_2 is converted to SO_3 at equilibrium conditions! Under these conditions 36.5 percent of the inlet SO_2 would still remain unconverted at equilibrium. This value is quite high and in most cases would be unacceptable. In practice the conversion would more

Table SP 11-8

X	$0.5X$	$\sqrt{\dfrac{8-0.5X}{100-0.5X}}$	C	$12C$	$1+C$	X^{n+1}
6.0	3.0	0.227	1.93	23.2	2.93	7.95
7.95	3.98	0.205	1.74	20.9	2.74	7.62
7.62	3.81	0.210	1.77	21.3	2.77	7.68

likely be performed at a temperature closer to 500°C where the equilibrium conversion would be about 95 percent.

11-8 Equilibrium in Multiphase Reactions

Although our previous discussion in this chapter has dealt primarily with homogeneous gaseous reactions, the fundamental equations

$$\Delta\mathscr{G} = -w_R = X(\Delta\mathscr{G}^\circ + RT \ln J_a) \tag{11-56}$$

and

$$RT \ln K_a = -\Delta\mathscr{G}^\circ \tag{11-71a}$$

apply for all reacting mixtures, irrespective of the phase behavior of the products or reactants. Thus the problem of describing heterogeneous reactions simply reduces to a problem of choosing and specifying the proper standard states. As in the case of homogeneous reactions, the choices of standard states for heterogeneous reactions are completely arbitrary (except for the temperature—which must always be the temperature of the system). Therefore, we may choose different standard states for the different components. In particular, we shall see that vastly different standard states are frequently chosen for gaseous, liquid, or solid componenents. For example, let us consider the water–gas reaction

$$C_{\text{coke}} + H_2O_{\text{steam}} = CO + H_2$$

which plays an important role in the various coal-gasification processes which are under study in an effort to supplement the nation's dwindling natural-gas supply. The equilibrium activity ratio, K_a, is written as

$$K_a = \frac{a_{CO}a_{H_2}}{a_C a_{H_2O}} = \frac{(\bar{f}/f^\circ)_{CO}(\bar{f}/f^\circ)_{H_2}}{(\bar{f}/f^\circ)_C(\bar{f}/f^\circ)_{H_2O}} \tag{11-82}$$

Normally the standard states for the three gaseous components would be chosen as the pure gases at either 1 atm pressure or at the pressure of the system. The choice of a standard state for the carbon (coke) is still left. We could choose pure gaseous carbon at some specific pressure as the standard state. However, carbon at most normal temperatures is a solid with an extremely low, and hence difficult to measure, vapor pressure. Therefore, pure carbon in any gaseous state would be a rather inconvenient choice for the reference conditions. On the other hand, we may choose pure *solid* carbon at some pressure, usually 1 atm, and the temperature of the system as the standard state. When this choice is made, we find that the activity of the solid carbon becomes extremely simple to calculate. As long as any solid carbon remains (assuming negligible solubility of the gases in the solid carbon), this carbon is essentially in its pure form. Thus the partial fugacity of carbon is equal to its pure-component value independent of the extent of reaction.

The variation of the pure-component fugacity from the standard-state pressure to the pressure of the reaction is given by

$$\ln \frac{f_i}{f_i^\circ} = \int_{P^\circ}^{P} \frac{v_i}{RT}\, dP \tag{11-83}$$

However, v_i for most solids is *extremely* small with respect to RT, so f_i is essentially independent of pressure, except over extremely large pressure ranges. (Remember that we found this same result when considering the change in pure-component fugacity of a liquid with pressure.) Thus we find that the pure-component fugacity and partial fugacity of the solid carbon are essentially equal to the standard-state fugacity, except at very high pressures. Since the partial fugacities and standard-state fugacities are equal, the activity of the solid carbon is unity independent of the extent of reaction, provided only that some solid carbon is still present. Thus we see a very convenient choice of standard states for solid components is the pure solid component at some pressure and the temperature of the system. (You may assume that this choice of standard states has been made for *all* solid components for the remainder of this text, unless you are specifically told otherwise.)

Although examples of solid reactants appearing in their standard states are by far more numerous than examples of any other pure phases, they are by no means the only examples of pure phases that we may encounter. For example, the dissociation of pure liquid water into hydrogen and oxygen gas. If the standard state for water is chosen as pure liquid water at the temperature of the dissociation and some arbitrary pressure, then the liquid water is in its standard state and will have an activity of unity (provided, of course, that f° does not vary greatly with pressure). Thus we see that any time a pure phase is encountered, we should always consider the possibility that the material may be in its standard state and hence have an activity of unity.

Effect of Solid Phases Upon Equilibrium

As we have shown, the actual amount of a reacting solid phase does not affect the equilibrium composition, as long as there is enough of the solid component present to form a pure phase. This arises because the activity of the material in the solid phase is independent of the amount. In contrast, the activity of a component in a gas phase is quite dependent on the amount of the component present. Thus we can expect that the equilibrium behavior of a reacting mixture with solid phases may be quite different from that exhibited by a strictly gas-phase reaction. This difference can best be exhibited by looking at the respective equilibrium expressions for a simple gas-phase and a solid–gas reaction.

For purposes of illustration, let us examine the following reactions:

$$\begin{aligned} C(s) + \tfrac{1}{2}\,O_2 &= CO \qquad \text{(reaction 1)} \\ CO + \tfrac{1}{2}\,O_2 &= CO_2 \qquad \text{(reaction 2)} \end{aligned} \tag{11-84}$$

Assume that the reactions are occurring at 1 atm pressure, so all gas components may be assumed to behave in an ideal manner. If the standard states are pure components (either gases or solids) at the pressure and temperature of the system, the activity of the solid carbon is unity, and the activities of the gaseous components are equal to their mole fractions. Thus the equilibrium expressions for the two reactions may be written as

$$(K_a)^1 = \frac{y_{CO}}{(y_{O_2})^{1/2}} \tag{11-85}$$

$$(K_a)^2 = \frac{y_{CO_2}}{y_{CO}(y_{O_2})^{1/2}} \tag{11-86}$$

Now let us examine the following situation: Suppose carbon is mixed with a large excess of oxygen and reaction 1 allowed to proceed. Since K_a for reaction 1 is normally a very large number, y_{O_2} must be much greater than y_{O_2} at equilibrium. However, we are told that the reaction initially had a large excess of oxygen over carbon, and therefore a carbon balance will indicate that there is not sufficient carbon present to reduce the O_2 content to a point where equilibrium can exist. That is, the outlet activity ratio, $(J_a)_{out}$, will be less than K_a when all the carbon is *exhausted*. Therefore, this reaction cannot reach equilibrium until enough carbon is added so that the outlet activity ratio equals the equilibrium constant. Thus we see that it is possible to *completely consume* a solid-phase reactant (up to the point where a single crystal of solid remains) without reaching equilibrium conditions. In this manner the reaction may be driven to completion with respect to the solid component. Although this phenomenon is not overly important in the oxidation of carbon, it does form the cornerstone of many metallurgical operations, and hence is extremely important.

That this behavior is impossible to obtain in a gas-phase reaction can be shown as follows. Suppose we wish to burn CO in an excess of O_2. As the reaction proceeds and the amount of CO unreacted decreases, the mole fraction of CO also decreases, and the outlet activity ratio, $(J_a)_{out}$, increases. By reducing the CO content far enough, it is always possible to have the outlet activity ratio equal the equilibrium constant so that equation (11-86) is satisfied. Therefore, equilibrium can always be attained in a gas-phase reaction. By the same token we see that a gas-phase reaction can never be driven to completion (although if K_a is extremely high, the reaction may go *almost* to completion) in the way a reaction involving a solid phase can be.

As we have previously demonstrated, the direction a given reaction will proceed may be determined by comparing the inlet activity ratio with the equilibrium constant. If

$(J_a)_{in} < K_a$	reaction will proceed in direction written
$(J_a)_{in} = K_a$	reactants already in equilibrium with themselves; no reaction
$(J_a)_{in} > K_a$	reaction will proceed in reverse direction from that written

However, as we have seen, the tendency of a reaction to proceed toward equilibrium may be overridden by mass-balance considerations when pure solid (or pure liquid) phases are present.

11.9 Effect of Pressure on Equilibrium Conversions

We have shown that the equilibrium constant K_a is directly related to the standard-state free-energy difference between the reactants and the products by the relation

$$-RT \ln K_a = \Delta\mathscr{G}^\circ \tag{11-71a}$$

Since it is obviously inconvenient to tabulate K_a as a function of both T and P, we may wonder if it is not possible instead to calculate the changes that occur in K_a when the pressure and temperature of a system are changed. In addition, it is useful to have some feeling for how these variations affect the equilibrium compositions of the reacting mixture.

As shown in equation (11-71a) the equilibrium constant K_a is directly related to the change in standard-state Gibbs free energy by

$$\ln K_a = -\frac{\Delta\mathscr{G}^\circ}{RT} \tag{11-71b}$$

Since $\Delta\mathscr{G}^\circ$ is a function only of the choice of standard states for the products and reactants, it is not a function of the pressure of the system, unless of course the standard-state conditions are related to the system pressure. It will be assumed in everything that follows that the standard-state conditions are not related to the system pressure! Since the temperature of the system is also independent of the pressure, the equilibrium constant, K_a, is *independent* of the pressure of the system. However, since both $\Delta\mathscr{G}^\circ$ and T vary with the temperature, K_a will certainly be a function of T.

Although we have seen that the equilibrium constant, K_a, is independent of pressure, this in no way means that the equilibrium *composition* of a reacting mixture is necessarily independent of pressure. Let us now examine the ways in which changing pressure may vary the composition of a reacting mixture.

As shown in equation (11-81), the equilibrium constant for a *gas-phase reaction* can be split into three separate parts.[4]

$$K_a = \frac{K_{\hat{f}}}{K_{f^\circ}} = \frac{K_y K_P}{K_{f^\circ}} \tag{11-81}$$

where the K's have been defined in equations (11-77) and (11-80).

However, K_{f° is a function only of the standard states and, like K_a, is independent of pressure. Thus we find that

$$K_y K_P = K_a K_{f^\circ} = \text{function of temperature} \tag{11-87}$$

However, we are really interested in variations in K_y, since we may determine the equilibrium compositions directly once K_y is known. Qualitatively we observe: The greater the value of K_y, the further a reaction will proceed in the forward direction (reactants $\longrightarrow$ products). From equation (11-87) we may express K_y as

$$K_y = \frac{K_a K_{f^\circ}}{K_P} \tag{11-88}$$

At constant temperature K_y varies inversely as K_P; thus the greater K_P, the smaller will be the yield of the products. However,

$$K_P = P^{\sum_{i=1}^{C} \alpha_i} \tag{11-89}$$

Therefore, if $\sum_{\text{prod}} \alpha_i > -\sum_{\text{react}} \alpha_i$, K_P will increase with increasing pressure and K_y will decrease, thereby decreasing the equilibrium yield. For example, consider the reaction $\alpha_2 = 2$, $\alpha_1 = -1$, where the number of moles of products $>$ moles of reactants. We may expect the equilibrium yield of A_2 to decrease with increasing P. On the other hand, if $-\sum_{\text{react}} \alpha_i > \sum_{\text{prod}} \alpha_i$, K_P will decrease with increasing P and K_y will increase with P, thereby increasing the yield. For example, the equilibrium yield of NH_3 from the reaction

$$N_2 + 3\,H_2 = 2\,NH_3$$

may be increased by increasing P.

[4]For multiphase reactions, the following discussion is applicable if only the gaseous reactants and products are included in K_y K_P, and K_{f°.

If $-\sum_{\text{react}} \alpha_i = \sum_{\text{prod}} \alpha_i$, K_P is independent of pressure, and the equilibrium composition will be independent of pressure.

SAMPLE PROBLEM 11-9. It is desired to raise the equilibrium conversion of SO_2 to SO_3 in Sample Problem 11-8 from 63.5 percent at 610°C to 80 percent by changing the system pressure. At what pressure must the reactor now be operated? (You may still assume ideal-solution and ideal-gas behavior.)

Solution: We have shown that the equilibrium constant may be related to the pressure of the system and the equilibrium compositions by the expression

$$K_a = \frac{K_y K_P}{K_{f^\circ}}$$

Since the standard states for all components have been chosen as pure gases at 1 atm, the standard-state fugacities are all 1 atm, so $K_{f^\circ} = 1 \text{ atm}^{-1/2}$. Therefore, K_a reduces to

$$K_a = \frac{K_y K_P}{1 \text{ atm}^{-1/2}}$$

But, from sample Problem 11-8, $K_a = 8.5$; therefore,

$$K_y K_P = 8.5(1 \text{ atm})^{-1/2}$$

Determination of K_y will now allow us to solve for K_P, which is related to the reactor pressure by the expression

$$K_P = \frac{P}{PP^{1/2}} = P^{-1/2}$$

Therefore,

$$P^{-1/2} = \frac{8.5 \text{ atm}^{-1/2}}{K_y}$$

But

$$K_y = \frac{y_{SO_3}}{y_{SO_2}(y_{O_2})^{1/2}}$$

Base all future calculations on 100 mol of feed to reactor: Therefore, the feed contains 12 mol of SO_2, 8 mol of O_2, and 80 mol of N_2. For 80-percent conversion of SO_2 to SO_3, the product stream will contain

SO_3:	$(0.8)(12)$	$= 9.6$ mol of SO_3	$y_{SO_3} = 0.101$
SO_2:	$12 - (0.80)(12)$	$= 2.4$ mol of SO_2	$y_{SO_2} = 0.0252$
O_2:	$8 - (0.40)(12)$	$= 3.2$ mol of O_2	$y_{O_2} = 0.0337$
N_2:		$= 80.0$ mol of N_2	$y_{N_2} = 0.840$
		95.2	

Therefore,

$$K_y = \frac{0.101}{(0.0252)(0.0337)^{1/2}} = 21.8$$

The value for K_y is then substituted into the expression for the system pressure, to give

$$P^{-1/2} = \frac{8.5\ \text{atm}^{-1/2}}{21.8} = 0.39\ \text{atm}^{-1/2}$$

or

$$P = \left(\frac{21.8}{8.5}\right)^2 \text{atm} = 6.6\ \text{atm}$$

Thus, if the reactor is operated at 6.6 atm, an equilibrium conversion of 80 percent of the SO_2 to SO_3 may be expected.

11.10 Variation of K_a with Changes in Temperature

Although we have shown that K_a is not a function of the system pressure, it is well known that K_a may vary with system temperature. We shall now develop a means of calculating the variation of K_a with changes in temperature. Since K_a is related to $\Delta\mathscr{g}^\circ$ by the relation

$$R \ln K_a = -\frac{\Delta\mathscr{g}^\circ}{T} \tag{11-71b}$$

our problem reduces to one of determining the variation of $\Delta\mathscr{g}^\circ/T$ with respect to temperature. The derivative of K_a with T is given by

$$R\frac{d \ln K_a}{dT} = -\frac{d}{dT}\frac{\Delta\mathscr{g}^\circ}{T} \tag{11-90}$$

Since $\Delta\mathscr{g}^\circ$ refers to a specific pressure (that of the standard state), the derivative in equation (11-90) must be evaluated at constant pressure. This derivative can in turn be expressed as

$$\begin{aligned} \frac{\partial}{\partial T}\left(\frac{\Delta\mathscr{g}^\circ}{T}\right)_P &= \frac{\partial}{\partial T}\left(\frac{1}{T}\sum_{i=1}^{C}\alpha_i g_i^\circ\right)_P \\ &= \frac{\partial}{\partial T}\left(\sum_{i=1}^{C}\alpha_i\frac{g_i^\circ}{T}\right)_P = \sum_{i=1}^{C}\alpha_i\frac{\partial}{\partial T}\left(\frac{g_i^\circ}{T}\right)_P \end{aligned} \tag{11-91}$$

However, equation (10-24) for partial properties is also applicable to pure components. Thus

$$\frac{\partial}{\partial T}\left(\frac{g_i^\circ}{T}\right)_P = -\frac{h_i^\circ}{T^2}$$

which may be substituted into equation (11-91) to give

$$\frac{\partial}{\partial T}\left(\frac{\Delta\mathscr{g}_i^\circ}{T}\right)_P = -\sum_{i=1}^{C}\frac{\alpha_i h_i^\circ}{T^2} = -\frac{\Delta\mathscr{h}^\circ}{T^2} \tag{11-92}$$

where

$$\Delta\mathscr{h}^\circ = \sum_{i=1}^{C}\alpha_i h_i^\circ \tag{11-14}$$

is what we have termed the *standard-state enthalpy change* (or heat) *of reaction*. Equation (11-92) (which is commonly called the *Gibbs–Helmholtz equation*) may then be substituted into equation (11-90) to give the desired relation:

$$R\frac{d}{dT}(\ln K_a) = \frac{\Delta h^\circ}{T^2} \tag{11-93}$$

or

$$\frac{d}{dT}(\ln K_a) = \frac{\Delta h^\circ}{RT^2} \tag{11-93a}$$

Rearrangement of equation (11-93a) yields a frequently useful result:

$$d\ln K_a = \frac{\Delta h^\circ}{RT^2}\,dT = -\frac{\Delta h^\circ}{R}\,d\left(\frac{1}{T}\right) \tag{11-94}$$

Thus the slope of the ln (K_a) vs $1/T$ curve at any specific value of T will yield the local value of $-\Delta h^\circ/R$. Over moderate ranges of temperature, we may treat Δh° as independent of T, so that the ln K_a vs $1/T$ curve is a straight line, and the integrated form of equation (11-94) has a particularly simple form:

$$\ln\frac{(K_a)_{T_2}}{(K_a)_{T_1}} = -\frac{\Delta h^\circ}{R}\left(\frac{1}{T_2} - \frac{1}{T_1}\right) \tag{11-95}$$

SAMPLE PROBLEM 11-10. The oxide of tungsten may be reduced with hydrogen according to the reaction

$$WO_2(s) + 2\,H_2 \rightleftharpoons W(s) + 2\,H_2O$$

The reaction has been studied at atmospheric pressure, and the equilibrium ratio of the partial pressure of water to that of hydrogen has been measured at a number of temperatures and found to be as given in Table SP11-10a. From these data, determine the heat of reaction at 900°C.

Table SP 11-10a

T, °C	P_{H_2O}/P_{H_2}
734	0.45
800	0.53
828	0.65
865	0.75
900	0.82
941	1.00
1000	1.16
1036	1.29

Solution: We may determine the heat of reaction for a given reaction, if we know the variation of K_a with T, from the Gibbs–Helmholtz equation:

$$R\frac{d\ln K_a}{dT} = \frac{\Delta h^\circ}{T^2}$$

or

$$\frac{d(\ln K_a)}{d(1/T)} = -\frac{\Delta h^\circ}{R}$$

Thus, if we plot ln K_a versus $1/T$, the slope of the ensuing curve will be $-(\Delta h^\circ/R)$. Since the reaction is occurring at low pressure, we may assume that $(\partial h/\partial P)_T = 0$, so the heat of reaction is equal to the standard-state enthalpy change. Therefore,

$$\frac{d(\ln K_a)}{d(1/T)} = -\frac{(\Delta h)_R}{R}$$

We may now determine the equilibrium constant from the partial pressure ratio as follows:

$$K_a = \frac{a_W(a_{H_2O})^2}{a_{WO_2}(a_{H_2})^2}$$

Choosing the standard states as close to the reaction conditions as possible, we choose

W and WO_2: pure solids at 1 atm and T of system

H_2O and H_2: pure gases at 1 atm and T of system

With this choice of standard states $a_W = a_{WO_2} = 1.0$ (as long as they are present) and $f^\circ_{H_2O} = f^\circ_{H_2} = 1.0$ atm. The equilibrium expression thus reduces to

$$K_a = \left(\frac{\bar{f}_{H_2O}}{\bar{f}_{H_2}}\right)^2$$

At 1 atm we may assume ideal solutions of ideal gases, so the partial fugacities are equal to the partial pressures. Therefore,

$$K_a = \left(\frac{P_{H_2O}}{P_{H_2}}\right)^2$$

and we may relate K_a and T as shown in Table SP11-10b.

Table SP 11-10b

T, °C	T, °K	$1/T$, °K ($\times 10^{-3}$)	P_{H_2O}/P_{H_2}	K_a	ln K_a
734	1007	0.99	0.45	0.202	−1.600
800	1073	0.931	0.53	0.281	−1.268
828	1101	0.909	0.65	0.422	−0.815
865	1138	0.880	0.75	0.563	−0.574
900	1173	0.852	0.82	0.672	−0.397
941	1214	0.824	1.00	1.00	0.0
1000	1273	0.785	1.16	1.35	0.300
1036	1309	0.765	1.29	1.66	0.505

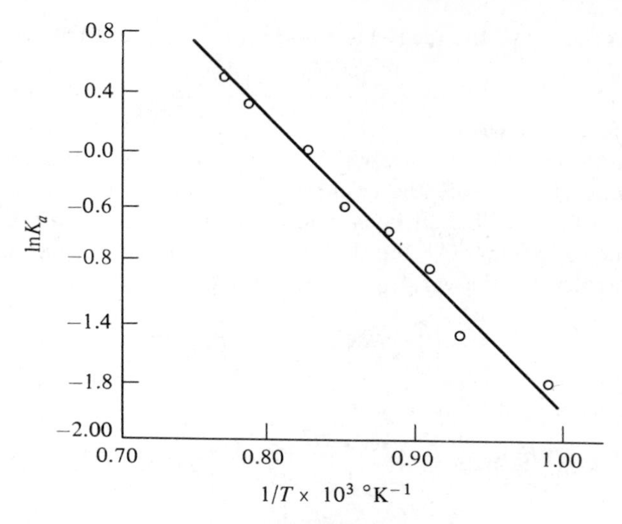

Figure SP11-10

We may now prepare a plot of ln K_a versus $1/T$ as shown in Fig. SP11-10. Although a fair amount of scatter is present in the data, we estimate the slope of the line to be about

$$\frac{-\Delta h^\circ}{R} = -1.19 \times 10^4 \text{ °K}$$

Therefore,

$$\Delta h^\circ = 1.19 \times 10^4 \times 1.987 \text{ (cal/g-mol)(°K/°K)}$$

or

$$(\Delta h)_R = 23.6 \text{ kcal/g-mol}$$

so that 23.6 kcal of heat must be added to isothermally reduce 1 g-mol of WO_2 with hydrogen at 900°C.

Sample Problem 11-11. A mixture of nitrogen and hydrogen, in the proportions 4 mol of hydrogen to 1 mol of nitrogen, is being fed continuously to a catalytic ammonia converter. The pressure in the converter is 200 atm, and the temperature is held constant at 450°F. Compute the maximum (equilibrium) conversion of nitrogen to ammonia from the reaction

$$N_2 + 3\,H_2 = 2\,NH_3$$

$K_a = 14.4 \times 10^{-6}$ at 500°C for this reaction. Standard states are pure components at 1-atm pressure. $\Delta h_R^\circ = -25{,}800$ cal/g-mol may be used over the temperature range of interest, and it may be assumed that the gas mixture behaves according to the ideal-gas equation of state and the Lewis–Randall rule.

Solution: Since we wish to calculate the equilibrium composition for this reaction at 450°F, our first task will be to determine the equilibrium constant at this temperature. From the equilibrium constant we shall determine K_y, from which the equilibrium conversion can be obtained. Since we do not have a value for K_a at 450°F but do have one at 500°C, we shall use equation (11-93a) to calculate the variation of K_a with T. Thus:

$$\ln \frac{(K_a)_{450^\circ F}}{(K_a)_{500^\circ C}} = \int_{500^\circ C}^{450^\circ F} \frac{\Delta h_R^\circ}{RT^2}\, dT$$

or after converting all temperatures to °K and substituting Δh°

$$R \ln \frac{(K_a)_{505^\circ K}}{(K_a)_{773^\circ K}} = \int_{773^\circ K}^{505^\circ K} \left[\frac{-25{,}800}{T^2}\right] dT \text{ cal/}^\circ\text{K}^2 \text{ g-mol}$$

Therefore,

$$1.987 \frac{\text{cal}}{\text{g-mol }^\circ\text{K}} \ln \frac{(K_a)_{505^\circ K}}{14.4 \times 10^{-6}} = \left[25{,}800\left(\frac{1}{505} - \frac{1}{773}\right)\right] \text{cal/g-mol }^\circ\text{K}$$

or

$$\ln \frac{(K_a)_{505^\circ K}}{14.4 \times 10^{-6}} = \frac{17.7}{1.987} = 8.85$$

Therefore,

$$(K_a)_{505^\circ K} = 14.4 \times 10^{-6} \cdot e^{8.85} = 0.0768 \text{ at } 450^\circ\text{F } (505^\circ\text{K})$$

K_a is related to the properties of the reacting mixture through the expression

$$K_a = \frac{K_y K_P}{K_f^\circ} \quad \text{or} \quad K_y = \frac{K_a K_{f^\circ}}{K_P}$$

The standard-state conditions are all pure components at 1 atm. Therefore, $f^\circ = 1$ atm, so $K_{f^\circ} = 1 \text{ atm}^{-2}$. K_P is related to the system pressure by the expression

$$K_P = \frac{P^2}{PP^3} = \frac{1}{P^2}$$

Therefore,

$$K_P = (200 \text{ atm})^{-2} = 2.5 \times 10^{-5} \text{ atm}^{-2}$$

Substitution of K_P and K_{f° into the above expression yields

$$K_y = \frac{K_a K_{f^\circ}}{K_P} = \frac{(0.0768)(1 \text{ atm}^{-2})}{2.5 \times 10^{-5} \text{ atm}^{-2}} = 3.07 \times 10^3$$

Let us now set up a table of property values (Table SP11-11a) for the reacting mixture based on 1 mol of nitrogen feed and X mol of N_2 reacting.

Table SP 11-11a

	N_2	H_2	NH_3	Total
Initial	1	4	0	5
Formed by reaction	$-X$	$-3X$	$2X$	$-2X$
Final	$1 - X$	$4 - 3X$	$2X$	$5 - 2X$
Final mole fraction	$(1-X)/(5-2X)$	$(4-3X)/(5-2X)$	$2X/(5-2X)$	1.0

But

$$K_y = \frac{(y_{NH_3})^2}{y_{N_2}(y_{H_2})^3}$$

Therefore,

$$3070 = \frac{\left(\dfrac{2X}{5-2X}\right)^2}{\dfrac{1-X}{5-2X}\left(\dfrac{4-3X}{5-2X}\right)^3}$$

or

$$3070 = \frac{(4X^2)(5-2X)^2}{(1-X)(4-3X)^3}$$

Solution of this equation for X then gives the equilibrium conversions of N_2 to NH_3. Since this is a fourth-order equation, a trial-and-error solution is appropriate. Examination of the range of possible value for X shows that $0 \leq X \leq 1$; a rapid check of the extremes $X \longrightarrow 0$ and $X \longrightarrow 1$ shows that the actual X must be fairly near 1.0. [How could

$$\frac{4X^2(5-2X)^2}{(1-X)(4-3X)^3}$$

be as large as 3000 without $1 - X$ being quite small?] Examination of the equation for X shows that the only term which will vary rapidly with X in the range $X \longrightarrow 1$ is the term $X - 1$. Therefore, as the trial-and-error scheme, rearrange the equation into the form

$$1 - X^{n+1} = \frac{4(X^n)^2(5-2X^n)^2}{(3070)(4-3X^n)^3}$$

and let $X = 1$ be a first trial variable (see Table SP11-12b). Thus the final solution is obtained on the second iteration.

Table SP 11-11b

X^n	$4(X^n)^2$	$(5-2X^n)^2$	$(4-3X^n)^3$	$1 - X^{n+1}$	X^{n+1}
1.0	4.0	9	1.0	0.012	0.988
0.996	3.90	9.14	1.11	0.010	0.990

The maximum conversion of N_2 to NH_3 under these conditions, with this feed, is then $X/1.0 = 0.99$, or 99-percent conversion.

Although calculation of the variation of K_a with T is reasonably simple if we assume Δh_R° is constant, this approximation is often not acceptable. For cases such as these it is necessary to develop an expression for $\Delta h_R^\circ(T)$ and then substitute this expression into equation (11-73a). The resulting calculations are not difficult, but can be quite tedious. Since these calculations are frequently encountered for a relatively small, but important, group of reactions, tables and graphs showing the variation of K_a with T for these select reactions have been prepared and are commonly available. Figures 11-10 and 11-11 present the results of such efforts for a series of reactions related to combustion and metallurgy. Although the number of reactions specifically presented in these figures is quite small, it is possible to use these results for a considerably wider variety of reactions. Consider, for example, the system of reactions:

$$H_2 + \tfrac{1}{2}O_2 = H_2O \qquad (1)$$

$$C + \tfrac{1}{2}O_2 = CO \qquad (2)$$

$$C + H_2O = CO + H_2 \qquad (3)$$

Equilibrium constants for reactions (1) and (2) are presented in Fig. 11-10, while that for reaction (3) is not. Let us see how we can get K_a for reaction (3) (which we will call $K_a^{(3)}$) in terms of $K_a^{(1)}$ and $K_a^{(2)}$: By definition

$$\begin{aligned} \Delta\mathcal{G}^{\circ(1)} &= \mathring{g}_{H_2O} - \mathring{g}_{H_2} - \tfrac{1}{2}\mathring{g}_{O_2} \\ \Delta\mathcal{G}^{\circ(2)} &= \mathring{g}_{CO} - \mathring{g}_{C} - \tfrac{1}{2}\mathring{g}_{O_2} \\ \Delta\mathcal{G}^{\circ(3)} &= \mathring{g}_{CO} + \mathring{g}_{H_2} - \mathring{g}_{C} - \mathring{g}_{H_2O} \end{aligned} \qquad (11\text{-}96)$$

Examination of these three expressions shows that

$$\Delta\mathcal{G}^{\circ(3)} = \Delta\mathcal{G}^{\circ(2)} - \Delta\mathcal{G}^{\circ(1)} \qquad (11\text{-}97)$$

[Likewise, reaction (3) can be obtained by substracting reaction (1) from reaction (2).] However, we know the equilibrium constants and $\Delta\mathcal{G}^\circ$'s are related by:

$$\begin{aligned} \Delta\mathcal{G}^{\circ(1)} &= -RT \ln K_a^{(1)} \\ \Delta\mathcal{G}^{\circ(2)} &= -RT \ln K_a^{(2)} \\ \Delta\mathcal{G}^{\circ(3)} &= -RT \ln K_a^{(3)} \end{aligned} \qquad (11\text{-}98)$$

Substitution ot these expressions into equation (11-97) and cancellation of $-RT$ then yields:

$$\ln K_a^{(3)} = \ln K_a^{(2)} - \ln K_a^{(1)} \qquad (11\text{-}99)$$

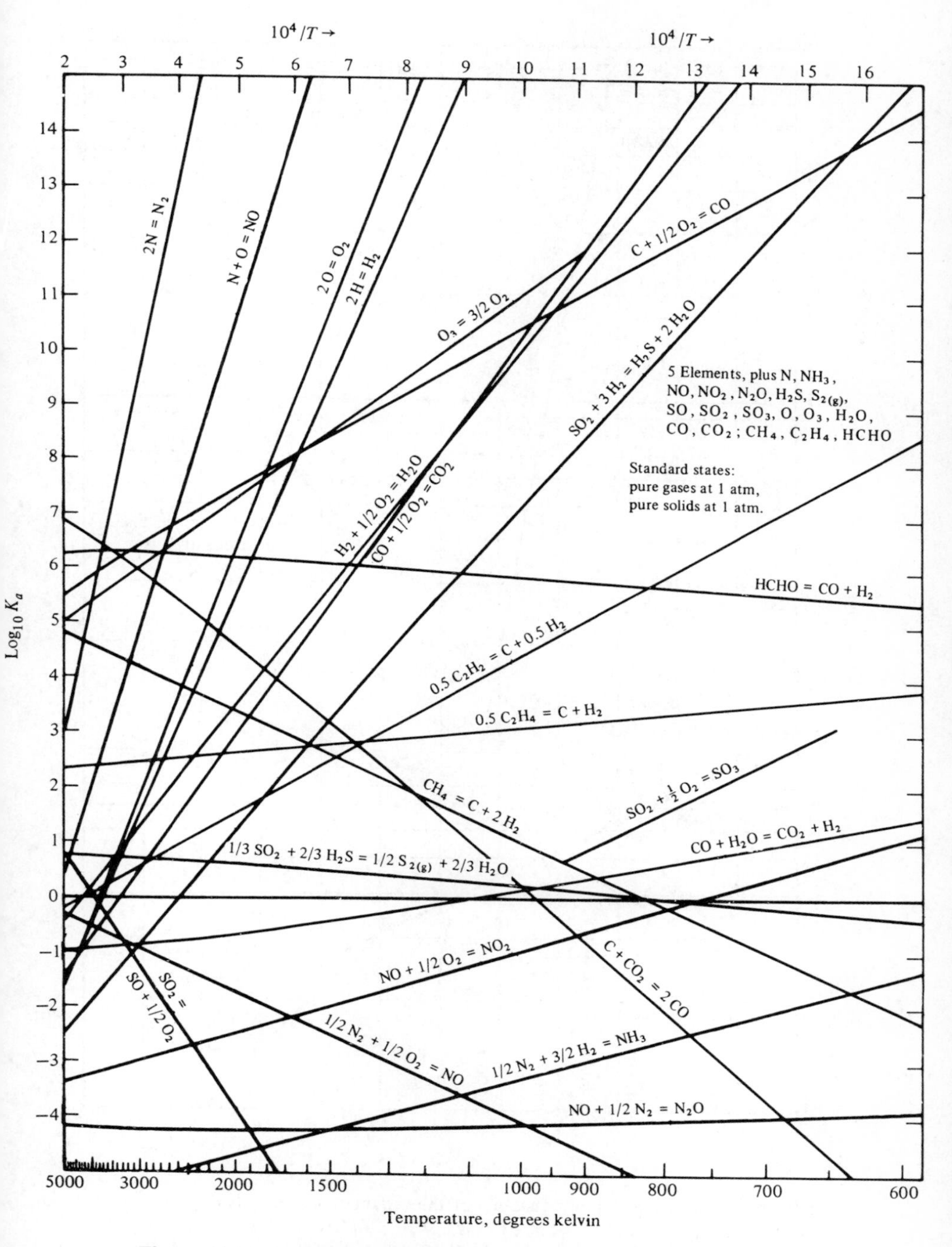

Figure 11-10. Combustion-related equilibrium constants. (Courtesy of Professor H. C. Hottel of the Massachusetts Institute of Technology.)

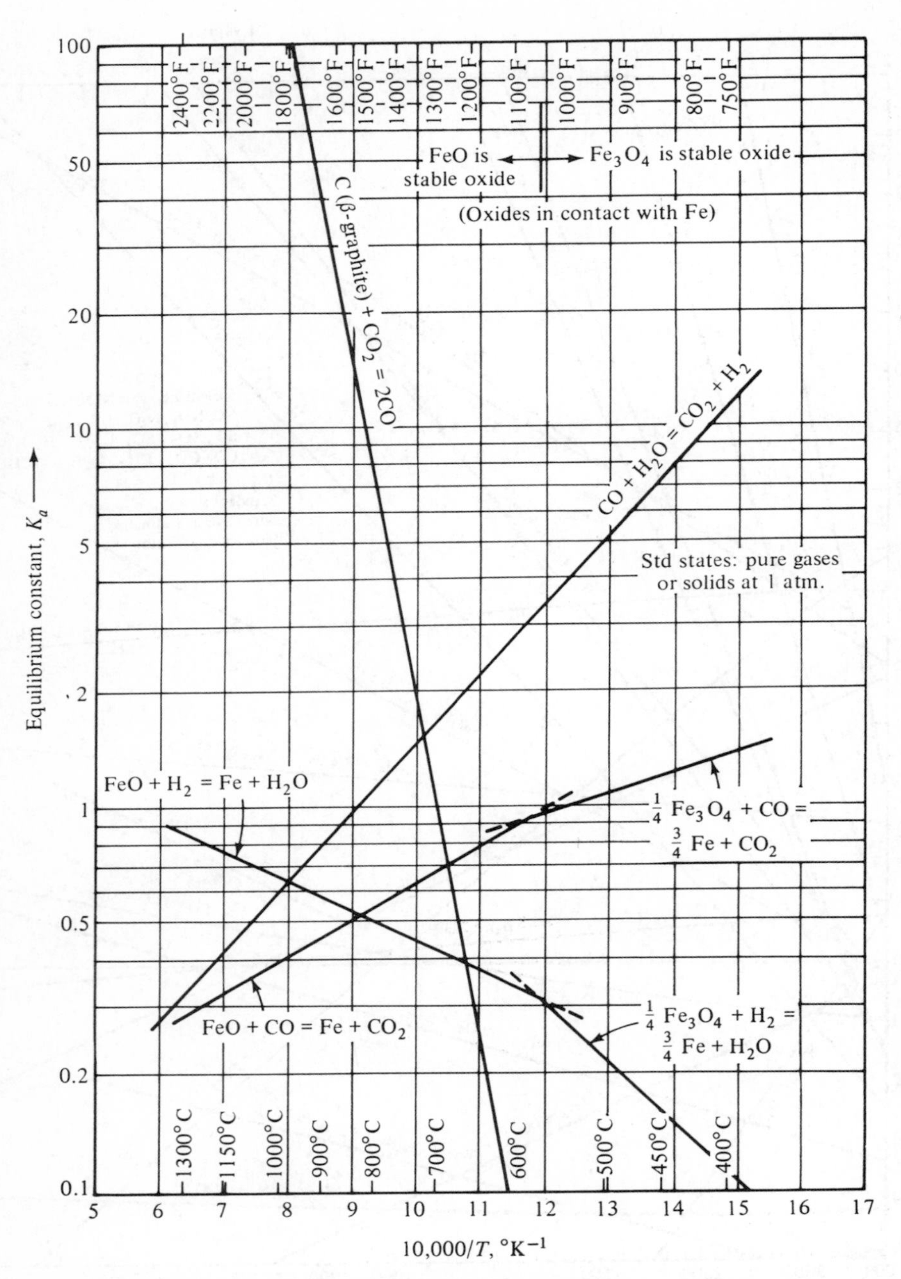

Figure 11-11. Equilibrium constants of the reduction of iron oxides and carbon dioxide. (From W. K. Lewis, et al., *Industrial Stoichiometry*, 2ed., McGraw-Hill, Inc., New York, 1954.)

or upon exponentiating

$$K_a^{(3)} = \frac{K_a^{(2)}}{K_a^{(1)}} \tag{11-99a}$$

Thus, we see the equilibrium constant for a complex reaction can be determined directly in terms of a sequence of simpler reactions from which the complex reaction can be formed. Although we have demonstrated this result for a reaction which is formed from only two simpler reactions, the same technique can be applied for any level of complexity.

SAMPLE PROBLEM 11-12. Current predictions suggest a severe shortage of natural gas (mainly methane) in the very near future. One means of producing synthetic natural gas is by the methanation of a hydrogen–carbon monoxide mixture (produced by reacting the carbon in coal with steam) according to the reaction

$$CO + 3\,H_2 = CH_4 + H_2O \tag{a}$$

Determine the equilibrium constant for this reaction at 850°K.

Solution: Examination of Fig. 11-10 indicates that the desired reaction is not included. However, the reacting species are all included in various different reactions, so it should be possible to decompose this reaction into a series of simpler ones which are listed in the figure. Although there are several ways in which this decomposition may be accomplished, the following reactions are applicable. (These have been found by trial and error—can you find another set that will work?)

$$CH_4 = C + 2H_2 \tag{b}$$

$$C + CO_2 = 2\,CO \tag{c}$$

$$CO + H_2O = CO_2 + H_2 \tag{d}$$

The desired reaction (a) can be formed from the negative of (b) + (c) + (d), that is,

$$\text{(a)} = -\text{(b)} - \text{(c)} - \text{(d)}$$

Therefore $\Delta\mathscr{G}^{\circ(a)} = -(\Delta\mathscr{G}^{\circ(b)} + \Delta\mathscr{G}^{\circ(c)} + \Delta\mathscr{G}^{\circ(d)})$ and $\ln K_a^{(a)} = -(\ln K_a^{(b)} + \ln K_a^{(c)} + \ln K_a^{(d)})$ or $\log K_a^{(a)} = -(\log K_a^{(b)} + \log K_a^{(c)} + \log K_a^{(d)})$, but from Fig. 11-10 at 850°K:

$$\log K^{(b)} = +0.2$$

$$\log K^{(c)} = -1.7$$

$$\log K^{(d)} = +0.5$$

Thus $\log K_a^{(a)} = -[0.2 - 1.7 + 0.5] = +1.0$ or $K_a^{(a)} = 10$, which is then the equilibrium constant for the desired methanation reaction at 850°K.

Problems

11-1. Hydrogen is to be formed by the steam cracking of methane according to the reaction

$$CH_4 + 2\,H_2O = CO_2 + 4\,H_2$$

The reaction will be performed at 600°C, where the equilibrium constant K_a is

$$\log_{10} K_a = 0.75$$

The standard states are pure gases at 1 atm. If the reaction pressure is 1 atm and a 50-percent excess of steam is used, what is the fractional conversion of CH_4 to H_2? What is the composition of the exhaust stream?

11-2. Methanol is manufactured commercially by reacting carbon monoxide and hydrogen according to the reaction

$$CO + 2\,H_2 = CH_3OH$$

Determine the equilibrium constant for this reaction at 25°C. What are your standard states? If a 20-percent excess of CO is used and the reaction is run at 15 atm and 25°C, what is the conversion of H_2 to alcohol? How much heat is liberated (or consumed) per mole of CO that reacts? (Values of the free energy and enthalpy of formation are given in Appendix E.)

11-3. The equilibrium constant for the gas-phase reaction $A = B + C$ at 400°K is $K_a = 1.0$ (standard states: pure gases at 1 atm and $T = 400$°K). It is proposed to obtain a conversion of A to B and C by passing pure A into a catalytic reactor operating at 400°K and with sufficiently long residence time that equilibrium is achieved.

(a) At what pressure should the reactor be operated to obtain 90-percent conversion of A to B and C?

(b) Instead of feeding pure A into the reactor, an equimolar mixture of A and an inert diluent I is used as the feed. If the pressure is maintained at 1 atm, what is the equilibrium conversion of A to B and C?

(c) Instead of using a constant-pressure flow reactor, the reaction will be run in an isothermal constant-volume reaction vessel. Initially (that is, before any reaction has occurred), the vessel is filled with pure A at 1 atm and 400°K. Find the equilibrium conversion and final pressure in the system.

11-4. Isomerization of a paraffin hydrocarbon is the conversion of a straight carbon chain to a branched chain. A petroleum plant is isomerizing normal butane to isobutane in a gas-phase reaction at 15 atm and 300°F. For this reaction, at 300°F the standard free-energy change, $\Delta\mathcal{G}^\circ = 97.5$ cal/g-mol nC_4. If the feed to the reactor is 93 mol percent n-butane, 5 mol percent isobutane, and 2 mol percent HCl (catalyst), what is the maximum conversion to isobutane? Assume ideal gas behavior.

11-5. Methanol can be synthesized in accordance with the reaction

$$CO + 2\,H_2 = CH_3OH(g)$$

K_a for this reaction is 1.377 at a temperature of 127°C.

(a) Calculate the equilibrium yield of methanol when 1 lb-mol of CO and 2 lb-mol of hydrogen react at 127°C and a pressure of 5 atm. (Assume that methanol is in the vapor phase.)

(b) How many lb-moles of CO should be contacted with 2 lb-mol of hydrogen if it is desired to obtain the maximum possible concentration of methanol in the equilibrium gas mixture at 127°C and 5 atm pressure?

11-6. Assume that 0.01 lb-mol of nitrogen is added to a constant-volume bomb that already contains 0.03 lb-mol of hydrogen. The bomb, which has a volume of 1.95 ft^3, is then placed in a constant-temperature bath at 910°R. When equilibrium has been established, the pressure in the bomb is 10 atm.

(a) Calculate $\Delta\mathcal{G}_T^\circ$ (Btu/lb-mol) for the reaction

$$N_2 + 3\,H_2 \rightleftharpoons 2\,NH_3$$

at 910°R and 10 atm. Refer to the usual standard states.

(b) How many lb-moles of ammonia would be present at equilibrium if 2 lb-mol of nitrogen are added to 3 lb-mol of hydrogen at a constant temperature of 910°R and a constant pressure of 10 atm? What is the concentration of the ammonia in the gas mixture?

(c) Repeat (b) for the addition of 1 lb-mol of nitrogen and 1 lb-mol of argon to 3 lb-mol of hydrogen.

It may be assumed that all three gases behave ideally at 910°R and 10 atm pressure.

11-7. Stoichiometric proportions of SO_2 and O_2 are placed in a constant-pressure bomb at 1 atm and allowed to come to equilibrium isothermally according to the reaction

$$SO_2 + \tfrac{1}{2}\,O_2 = SO_3$$

The resulting equilibrium gas analyses at two temperatures are as follows:

	1000°K	1200°K
SO_2	42.4	60.4
O_2	21.2	30.2
SO_3	36.4	9.4
	100.0	100.0

(a) If 2 g-mol of air and 1 g-mol of SO_2 were to be placed in the bomb and allowed to come to equilibrium isothermally at 1000°K, how many moles of SO_3 would be present at equilibrium?

(b) Calculate the isothermal enthalpy change on reaction (cal/g-mol of SO_3 formed), assuming it to be independent of temperature in the range 1000 to 1200°K.

(c) What is the change in entropy [cal/(°K)(g-mol of SO_3 formed)] accompanying the original reaction at a constant temperature of 1000°K and a constant pressure of 1 atm?

11-8. Gaseous hydrogen is to be manufactured by steam cracking of methane in a catalytic reactor according to the reaction

$$CH_4 + 2\,H_2O = CO_2 + 4\,H_2$$

at 900°F and 1 atm. Equilibrium data, K_a versus $1/T$, are provided in Fig. P11-8. The standard states for all gaseous components are pure components at 1 atm.

(a) If 5 mol of steam are fed to the reactor for every mole of methane, what is the equilibrium composition of methane in the product stream?

(b) Is this reaction exothermic or endothermic? How much heat must be added or removed from the reactor per mole of methane fed?

(c) Will an increase in pressure result in an increased or decreased yield?

(d) It has been suggested that the equilibrium yield of this reaction can be increased by diluting the feed stream with an inert gas. Do you agree? Justify your answer clearly. You may assume ideal gaseous mixtures.

$$\text{Yield} = \frac{\text{moles of } CH_4 \text{ reacted}}{\text{moles of } CH_4 \text{ fed}}$$

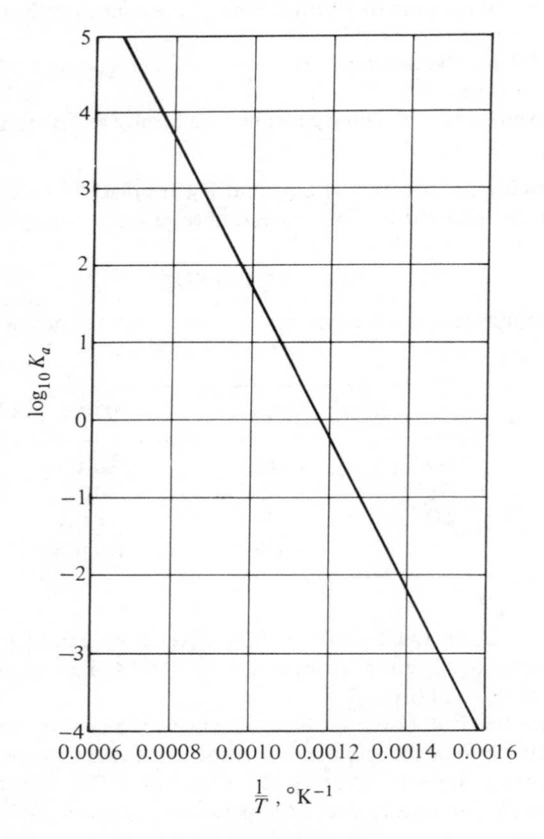

Figure P11-8

11-9. Methanol, CH_3OH, is decomposed by a catalyst into CO and H_2. At equilibrium conditions at 190°C and 1 atm pressure CH_3OH is 98 percent decomposed into CO and H_2. At 298°K the "enthalpy change of formation" (Δh_f) of gaseous H_2O from the elements with each at 1 atm pressure is −68,313 cal/g-mol. For the formation of CO_2 from CO and $\frac{1}{2}$ O_2 under the same conditions Δh_f is −67,623 cal. The Δh at 298°K for the following reaction is −182,580 cal/g-mol.

$$CH_3OH(g) + \tfrac{3}{2}\, O_2(g) = CO_2(g) + 2\, H_2O(g)$$

From these data derive an expression for ln K_a as a function of temperature for the formation of CH_3OH from CO and H_2.

11-10. The vapor-phase hydration of ethylene to ethanol is represented by the reaction

$$C_2H_4 + H_2O = C_2H_5OH$$

Experimental measurements have determined the equilibrium constant at two temperatures as

$$K_a = 6.8 \times 10^{-2} \text{ at } 145°\text{C}$$

$$K_a = 1.9 \times 10^{-3} \text{ at } 320°\text{C}$$

where the standard states are the usual ones—pure gases at 1 atm fugacity. If ethylene and water vapor are passed to a reactor at 400°C and 10 atm pressure, calculate the maximum percentage conversion of ethylene to ethanol if stoichiometric proportions of the reactants are used.

11-11. Nitric acid can be made by air oxidation of nitric oxide NO to NO_2 and absorption in water. It has been suggested that nitric oxide be made from air by heating the air to a very high temperature at atmospheric pressure. To make this process economically interesting, it is necessary to obtain 1 percent nitric oxide so that the recovery of the NO by absorption will not be too expensive. What is the minimum temperature that must be used to obtain this minimum concentration assuming the gases reach equilibrium at this temperature?

$$(\Delta g_f^\circ)_{298} = 20.72 \text{ kcal/mol NO}$$

$$(\Delta h_f^\circ)_{298} = 21.6 \text{ kcal/mol NO}$$

$$(\Delta s_f^\circ)_{298} = 50.34 \text{ kcal/mol NO °K}$$

where T = °K and the standard states are pure gases at 1 atm.

11-12. Ethyl acetate may be produced by the catalytic dehydrogenation of ethyl alcohol according to the reaction

$$2\, C_2H_5OH = CH_3COOC_2H_5 + 2\, H_2$$

Equilibrium constants for this reaction have been measured at 181 and 201.5°C as 1.075 and 1.705, respectively, with the standard states being pure gases at 1 atm pressure. Heat capacities of the several components are given in cal/g-mol°K (with T in °K) as

$$H_2: \quad C_P^* = 6.744 + 2.7 \times 10^{-4}T + 1.956 \times 10^{-7}T^2$$

$$C_2H_5OH: \quad C_P^* = 5.56 + 4.522 \times 10^{-2}T - 1.639 \times 10^{-5}T^2$$

$$CH_3COOC_2H_5: \quad C_P^* = 2.27 + 0.88T - 3.086 \times 10^{-5}T^2$$

It is proposed to design a reactor to be fed continuously with ethanol vapor at 150°C and 1 atm. The product stream is to be at 201.5°C and 1 atm. If equilibrium is attained in the product stream, how much heat must be supplied to the reactor per gram-mole of ethanol fed?

11-13. For the gaseous hydration of ethylene to ethanol,

$$C_2H_4 + H_2O = C_2H_5OH$$

Dodge[5] reports:

$$\Delta \mathcal{g}^\circ = (28.60T - 9740) \text{ cal/g-mol}$$

(T = °K) where the standard states are pure gases at 1 atm. If the hydration is performed at 1 atm and 300°C,

(a) Is this reaction endothermic or exothermic?

(b) How much heat is liberated (or consumed) during the formation of 1 g-mol of ethanol? How does this vary from Δh°?

11-14. Natural gas (essentially pure methane) is burned as a heating fuel with 25-percent excess air. Determine the flame temperature (assume adiabatic combustion) and the exit composition of the combustion products.

$$CH_4 + 2\,O_2 = CO_2 + 2\,H_2O$$

$$(C_P)_{O_2} = 6.148 + 3.10 \times 10^{-3}T - 9.23 \times 10^{-7}T^2$$

$$(C_P)_{CH_4} = 3.381 + 1.80 \times 10^{-2}T - 4.3 \times 10^{-6}T^2$$

$$(C_P)_{CO_2} = 6.214 + 1.04 \times 10^{-2}T - 3.55 \times 10^{-6}T^2$$

$$(C_P)_{H_2O} = 7.256 + 2.30 + 10^{-3}T + 2.83 \times 10^{-7}T^2$$

$$C_P = \text{cal/g-mol °K}, \qquad T = \text{°K}$$

11-15. Carbon monoxide is burned in a stoichiometric amount of air. Air and CO are fed to the burner at 25°C. If heat losses from the burner are estimated to be 47,500 Btu/lb-mol of CO feed, determine the composition and temperature of the gas leaving the burner.

$$(C_P)_{CO} = 6.42 + 1.0 \times 10^{-3}T$$

$$(C_P)_{CO_2} = 6.21 + 7 \times 10^{-3}T$$

$$(C_P)_{N_2} = 6.524 + 1.25 \times 10^{-3}T$$

$$(C_P)_{O_2} = 6.148 + 2.0 \times 10^{-3}T$$

$$C_P = \text{cal/g-mol °K}, \qquad T = \text{°K}$$

11-16. Rework Problem 11-15 assuming that only 3500 Btu/lb-mol is lost.

11-17. Ammonia has been proposed as a fuel for an experimental rocket motor (Fig. P11-17). Gaseous NH_3 and air (to supply oxygen) in the stoichiometric ratio are injected into the rocket motor at 150°F and 300 psia, where they react to form nitrogen and water according to the reaction

$$2\,NH_3 + \tfrac{3}{2}\,O_2 = N_2 + 3\,H_2O$$

[5]B. F. Dodge, *Chemical Engineering Thermodynamics*, McGraw-Hill Book Company, Inc., New York, 1944.

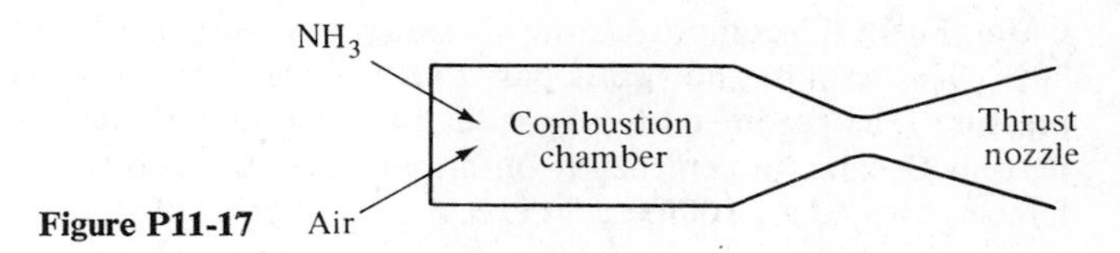

Figure P11-17

If the combustion chamber is insulated so that the reaction is essentially adiabatic, determine the temperature and approximate composition of the combustion gases as they enter the thrust nozzle. You may use the following average specific heats:

$$(C_P)_{N_2} = 8 \text{ Btu/lb-mol °R}$$

$$(C_P)_{H_2O} = 10.0 \text{ Btu/lb-mol °R}$$

$$(C_P)_{O_2} = 8.2 \text{ Btu/lb-mol °R}$$

$$(C_P)_{NH_3} = 13 \text{ Btu/lb-mol °R}$$

11-18. Gases from the burner in a contact sulfuric acid plant have the following composition (molar): 7.8 percent SO_2, 10.8 percent O_2, and 81.4 percent N_2. The gas mixture at 450°C and 1 atm is passed through a catalytic converter, where the SO_2 is oxidized to SO_3. If the converter operates adiabatically, what is the maximum percent of SO_2 converted to SO_3? How does this compare with the maximum conversion if the converter operates isothermally?

C_P in cal/g-mol °K (with T in °K)

$$SO_2 = 8.43 + 0.0024T$$

$$O_2 \text{ and } N_2 = 6.84 + 0.00034T$$

$$SO_3 = 10.6 + 0.002T$$

Δh° at 0°C for the above reaction is $-22{,}000$ cal/g-mol of SO_3 formed. The equilibrium constant K_a for the above reaction is

$$\log_{10} K_a = \frac{4950}{T(°K)} - 4.7$$

where the standard states are pure gases at atmospheric pressure.

11-19. A stationary gas-turbine plant is constructed according to the scheme shown in Fig. P11-19. Air at 70°F enters the compressor, where its pressure is raised to

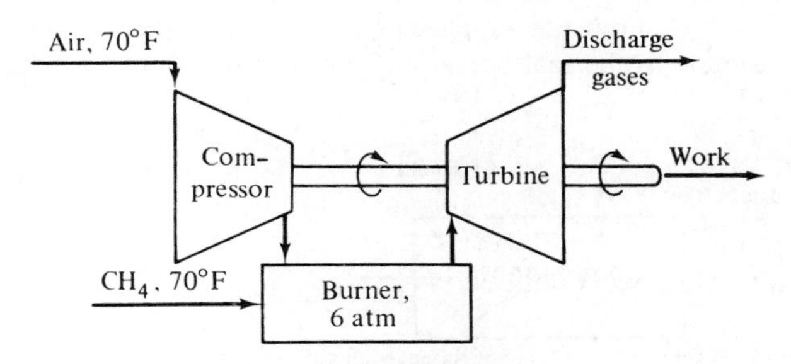

Figure P11-19

6 atm. From the compressor the air passes to a burner, where methane is added at 70°F. The resulting hot gases pass to a turbine, which discharges to atmospheric pressure. The barometer is normal at 14.7 psia, and air may be considered to be 21 percent O_2 and 79 percent N_2 on a mol basis. In order to keep the turbine inlet temperature down, 300-percent excess air will be used to burn the methane in the burner. Using the following data, calculate the work output of this plant per lb-mol of methane fed.

The compressor and turbine are both 80-percent efficient on an adiabatic reversible basis. There is negligible pressure drop in any line and negligible heat losses to the surroundings. All gases may be assumed to behave as ideal gases. The isothermal enthalpy change of combustion of methane at 70°F is −345,000 Btu/lb-mol. The average heat capacities of the various compounds in Btu/lb-mol °F are tabulated below:

Compound	C_P
CO_2	10.3
H_2	8.7
N_2	7.2
O_2	7.5

11-20. SO_2 is to be oxidized with a 50-percent excess of oxygen (from air) in a catalytic reactor (Fig. P11-20) according to the reaction

$$SO_2 + \tfrac{1}{2}\,O_2 = SO_3$$

For optimum yield from the reaction it is desirable to keep the reactor isothermal at a temperature of 425°C. This will be accomplished by preheating the reactants (available at 100°F) as shown in Fig. P11-20. The reactants are first passed through tubes in the catalyst bed, where the reactants are heated to a temperature of 425°C. The reactants are then passed through the catalyst bed, where they react at constant temperature. If the heat absorbed by the reactants does not equal that liberated during the reaction, steam or cooling water will be used to provide the necessary additional heating or cooling.

(a) If equilibrium is attained within the reactor, what is the exit composition?

(b) How much heat must be supplied or removed from the reactor per lb-mol of SO_2 fed to maintain isothermal operation? Must heat be added or removed from the reactor?

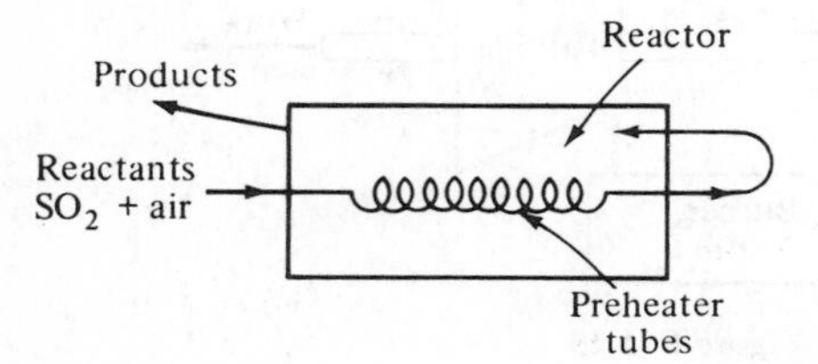

Figure P11-20

For the reaction

$$SO_2 + \tfrac{1}{2} O_2 = SO_3$$

$$\log_{10} K_a = \frac{5186.5}{T} + 0.611 \log_{10} T - 6.7497$$

(T = °K) where the standard states are pure gases at 1 atm pressure.[6] The heat capacities of the reactants may be taken as

$$(C_P)_{N_2} = 7.0 \text{ Btu/lb-mol °R}$$

$$(C_P)_{O_2} = 7.0 \text{ Btu/lb-mol °R}$$

$$(C_P)_{SO_2} = 12.0 \text{ Btu/lb-mol °R}$$

11-21. Hydrogen peroxide (H_2O_2) is used as a fuel for the small thruster rockets by which our astronauts turn and roll their space capsule. The hydrogen peroxide is stored as a pure liquid at 25°C and 1 atm pressure aboard the space capsule. When thrusting is needed, the peroxide is passed over a catalyst, where it decomposes according to the reaction

$$H_2O_2(l) = H_2O(g) + \tfrac{1}{2} O_2$$

The decomposition products are then fed to the thrust nozzles where they expand, and provide the desired force.

(a) Assuming that the decomposition goes to completion, what is the temperature of the steam and oxygen that enter the thruster?

(b) Obtain a reasonable estimate of the equilibrium constant for the reaction at this temperature. Is the assumption of part (a) reasonable?

At 25°C,

$$(\Delta g_f^\circ)_{H_2O_2(l)} = -28.2 \text{ kcal/g-mol}$$

$$(\Delta h_f^\circ)_{H_2O_2(l)} = -44.8 \text{ kcal/g-mol}$$

$$(C_P)_{O_2} = 8.0 \text{ Btu/lb-mol °R}$$

$$(C_P)_{H_2O} = 10.0 \text{ Btu/lb-mol °R}$$

The standard states are pure liquid H_2O_2 at 1 atm; pure vapors, H_2O, O_2 at 1 atm.

11-22. Carbon monoxide from a water gas plant is being burned with air in an adiabatic reactor. Both the carbon monoxide and air are fed to the reactor at 70°F and atmospheric pressure. Ten percent more air is used than is theoretically required by the stoichiometry of the reaction. The constant-pressure heat capacities in Btu/lb-mol °R are as follows:

$$CO: \quad 7.0 + 0.0005T$$

$$N_2: \quad 7.0 + 0.0005T$$

$$CO_2: \quad 9.0 + 0.0014T$$

$$H_2O: \quad 8.0 + 0.0012T$$

$$O_2: \quad 7.1 + 0.0006T$$

where T = °F. The ideal-gas law may be assumed.

[6]M. Bodenstein and W. Pohl, *Z. Electrochem.*, **11**, 373 (1905).

(a) If the products from the reactor are at atmospheric pressure, what is the maximum temperature they can attain?

(b) If the exact stoichiometric amount of air is used, show quantitatively what will happen to the temperature of the products.

11-23. Although the oxidation of carbon monoxide to carbon dioxide is highly favorable according to free-energy considerations, the kinetics are quite unfavorable at low temperatures and low CO concentrations. (This is the reason carbon monoxide is a major pollutant of the air.) In an effort to reduce the CO effluent from a chemical process, a catalytic converter is being designed as shown in Fig. P11-23.

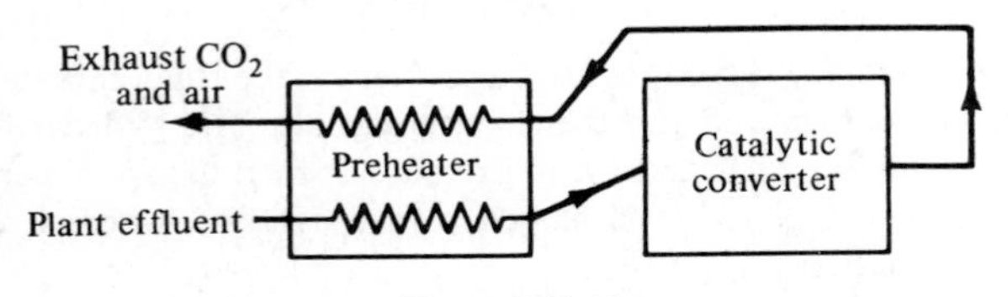

Figure P11-23

The plant effluent (25°C and 1 atm) contains 3 percent CO and 97 percent air (mole basis). The effluent is first passed through a preheater, where it is heated by the converter product stream. After the preheater the effluent is passed into the catalytic converter, which produces an exit stream that is essentially in equilibrium with itself. (We assume that no chemical reaction occurs in the preheater because no catalyst is present.) The exit stream from the converter is then used to preheat the entering gas in a countercurrent preheater. Assume that the preheater is of sufficient size that the plant effluent is preheated to a temperature that is 500°F cooler than the temperature of the gas leaving the catalytic converter.

(a) Determine the temperature of the streams entering and leaving the catalytic converter.

(b) Determine the composition of the gas leaving the converter.

You may assume that $(C_P)_{O_2} = (C_P)_{N_2} = 7.0$ Btu/lb-mol °R, while $(C_P)_{CO} =$ 8.0 Btu/lb-mol °R, and $(C_P)_{CO_2} = 13.5$ Btu/lb-mol °R over these temperatures.

11-24. A stream of monatomic hydrogen is available from an arc at 1000°K. These atoms are made to recombine on a suitable catalyst surface. What is the maximum temperature that may be generated by this reaction? (Remember to account for dissociation of H_2.) For $H_2 = 2\,H$,

$$\Delta\mathcal{G}^\circ = 81{,}000 - 3.5T \ln T + 4.5 \times 10^{-4}T^2 + 1.17T$$

where T is in °K, $\Delta\mathcal{G}^\circ$ is in cal/g-mol, standard states are pure gases at 1 atm. Assume as an approximation that

$$(C_P)_{H_2} = 7 \text{ cal/g-mol °K}$$

$$(C_P)_{H} = 5 \text{ cal/g-mol °K}$$

11-25. The reaction

$$H_2O + C(s) = CO + H_2$$

is frequently referred to as the *water–gas reaction.*

(a) From the data below calculate $\Delta\mathscr{G}^\circ$ for this reaction proceeding to the right at 298°K and 1 atm; also at 1000°K and 1 atm. Specify the standard states used.

(b) Would a mixture of H_2O and C if brought to equilibrium at either of these temperatures and 1 atm contain CO or H_2? Your answer must be justified.

(c) Express K_a in terms of (1) activities and (2) concentrations. Indicate any necessary assumptions.

Δg_f in kcal/g-atom of oxygen, at 1 atm:

T, °K	$H_2O(g)$	CO(g)
298	−54.6	−32.8
1000	−46.0	−47.9

11-26. The standard free-energy change for the formation of ethylene from the elements may be represented by the equation

$$\Delta\mathscr{G}^\circ = 5700 + 21.1T \text{ cal/g-mol}$$

where T is in °K and the reaction is given as

$$2\,C(\text{graphite}) + 2\,H_2(g) = C_2H_4(g)$$

On the basis of thermodynamic calculations, discuss the feasibility of the industrial preparation of ethylene by the above reaction.

11-27. At 828°F and 200 atm, the mole fraction of ammonia at equilibrium in the reaction

$$2\,FeN(s) + 3\,H_2(g) = 4\,Fe(s) + 2\,NH_3(g)$$

is 0.9142.

(a) Calculate K_a at the above pressure and temperature.

(b) Ammonia at 300 atm and 828°F is charged in a reactor with iron. The reactor is held at 250 atm and 828°F. A gas stream of hydrogen and ammonia containing 50 percent ammonia on a mole basis is drawn from the reactor at the temperature and pressure of the reactor. Calculate the maximum work extracted (or the minimum work required).

11-28. Hydrated sodium carbonate decomposes according to the reaction

$$Na_2CO_3 \cdot H_2O(s) = Na_2CO_3(s) + H_2O(g)$$

The equilibrium pressure in atmospheres of water vapor during this reaction is given by the relationship

$$\ln P' = 18.3 - \frac{6910}{T}$$

where T is in °K. Assuming that water vapor behaves ideally, derive an expression for $\Delta\mathscr{G}^\circ$ for the above reaction, as a function of temperature.

11-29. Ferrous oxide is reduced to metallic iron by passing a mixture of 20 percent CO and 80 percent N_2 over it at a temperature of 1000°C and a pressure of 1 atm:

$$FeO(s) + CO(g) = Fe(s) + CO_2(g)$$

At 1000°C, $\Delta\mathscr{G}^\circ$ for this reaction is 2300 cal/g-mol. How many pounds of metallic iron will be produced if 1000 ft^3 of CO (at 1000°C and 1 atm) is introduced and allowed to come to equilibrium?

11-30. The following partial pressure of CO and CO_2 are in equilibrium with austenite (a solid solution of iron and carbon) of the stated carbon content at 1000°C according to the equation

$$2\,CO(g) = C\text{ (in austenite)} + CO_2$$

C, wt. %	P_{CO}, atm	P_{CO_2}, atm
0.13	0.891	0.109
0.45	0.9660	0.9340
0.96	0.9862	0.0138
1.50	0.9928	0.0072

The solution of 1.50 percent is saturated with graphite at 1000°C.

(a) Find the activity of carbon in each case, taking pure graphite as the standard state.

(b) Plot the activities of carbon versus carbon concentration up to 2.0 percent carbon.

(c) What will be the ratio of pressures of CH_4 to H_2 (at 1 atm total pressure) in equilibrium with iron containing 0.6 percent carbon at 1000°C?

Δg_f for CH_4 at 1000°C and 1 atm is 11.78 kcal/mol.

11-31. It is desired to explore possibilities for decreasing the energy requirements necessary for decomposing Al_2O_3 to Al and O_2 according to the reaction

$$Al_2O_3(s) = 2\,Al(l) + \tfrac{3}{2}\,O_2(g)$$

$\Delta\mathscr{G}^\circ_{2500^\circ K} = 210$ kcal/mol of Al_2O_3. The standard states are pure solid Al_2O_3, pure liquid Al, and pure gaseous O_2 at 1 atm.

(a) If the oxygen pressure can be maintained at 10^{-6} atm, what is the maximum Al activity for which the reaction will proceed to the right?

(b) Can you think of any method by which this activity might realistically be achieved?

(c) Calculate the minimum work requirements to decompose 1 mole of Al_2O_3 to pure Al liquid and pure oxygen at a pressure of 0.1 atm.

11-32. It is proposed to "clean" some tungsten wires (that is, to remove all traces of the oxide WO_2) by heating the wires to 2000°K in a hydrogen atmosphere (1 atm total pressure).

(a) What is the maximum water impurity level that can be tolerated in the hydrogen?

(b) If the temperature of the wire is raised, would higher purity be required? Show your calculations.

For WO_2, $\Delta g_f = -131{,}600 + 36.6T$, at 1 atm; for H_2O, $\Delta g_f = -58{,}900 + 13.1T$, at 1 atm where Δg_f is in cal/g-mol of compound and T is in °K.

11-33. Determine the minimum pressure that must be exerted on solid graphite (carbon) to form diamond (also carbon). Assume that the diamonds and graphite form pure-solid phases that are incompressible and insoluble in each other. The temperature is 25°C. For the reaction

$$C_{\text{graphite}} = C_{\text{diamond}}$$

$$\Delta \mathscr{g}^\circ_{25°C} = 685 \text{ cal/g-mol}$$

(standard-state pure solids at 1 atm pressure). The density of graphite is 2.25 g/cm³; that of diamond is 3.51 g/cm³.

11-34. The following data were taken on the equilibrium pressure of CO_2 over $CaCO_3$ at various temperatures. In this reaction calcium carbonate decomposes to calcium oxide and carbon dioxide according to the equation $CaCO_3 = CaO + CO_2$. From the table of data estimate the standard heat of decomposition of $CaCO_3$ at 750°C.

T, °C	P, mm Hg
587	1.00
631	4
701	23
743	60
786	134
800	183
852	381
894	716

11-35. The following equilibrium data have been determined for the reaction

$$NiO(s) + CO(g) = Ni(s) + CO_2(g)$$

T, °C	663	716	754	793	852
$K_a \times 10^{-3}$	4.535	3.323	2.554	2.037	1.577

(a) Find $\Delta \mathscr{h}^\circ$, K_a, and $\Delta \mathscr{g}^\circ$ at 1000°K by use of graphical techniques.

(b) Would an atmosphere of 15 percent CO_2, 5 percent CO, and 80 percent N_2 oxidize nickel at 1000°K?

11-36. The heat of formation of silver oxide is −7300 cal/g-mol at 25°C. Calculate the approximate temperature at which silver oxide begins to decompose on heating (a) in pure oxygen and (b) in air (both at $P = 1$ atm). Use the following data:

	Entropy, cal/g-mol °K $s^{\circ}_{298°K, 1\ atm}$	C_P, cal/g-mol °K
Ag_2O	29.1	15.7
O_2	49.0	7.0
Ag	10.2	6.1

11-37. For the reaction

$$Si(s) + SiO_2(s) = 2\ SiO(g)$$

$$\Delta \mathscr{g}^{\circ} = 160{,}000 + 13.8T \log_{10} T - 121.55T \text{ cal/g-mol}$$

where $T = °K$,

(a) What is the pressure of SiO(g) above pure Si and pure SiO_2 at 1000°K?
(b) What is the $\Delta \mathscr{h}^{\circ}$ at 1000°K?
(c) What is the standard entropy of SiO(g) at 1000°K?

Substance	s° at 1000°K, cal/g-mol °K
Si	11.36
SiO_2	27.85

11-38. The standard free-energy change for the reduction of chromium oxide by hydrogen is given approximately as follows:

$$Cr_2O_3(s) + 3\ H_2(g) = 2\ Cr(s) + 3\ H_2O(g)$$

$$\Delta \mathscr{g}^{\circ} = 97{,}650 - 28.6T, \text{ cal/g-mol,} \qquad T = °K$$

(a) Find the maximum partial pressure of water vapor in otherwise pure hydrogen at 1 atm in which chromium can be heated without oxidation at 1500°K.

(b) Is the oxidation of chromium by water vapor exothermic or endothermic?

(c) What can you conclude from the above data about the heat effect at 1500°K for the oxidation of chromium by pure oxygen?

(d) Would the equilibrium in the equation above be affected by a change in pressure of the hydrogen–water vapor mixture from 1 to 2 atm? To 200 atm?

11-39. Limestone dissociates according to the reaction

$$CaCO_3(s) = CaO(s) + CO_2$$

with an equilibrium pressure (that is, the pressure of CO_2 in equilibrium with the mixture of solids) which is given, as a function of temperature, by the relationship

$$\log P = -\frac{11{,}355}{T} - 5.388 \log T + 26.238$$

where the units are atmospheres and °K. If limestone and coke are mixed at elevated temperatures, carbon monoxide is produced by the side reaction

$$C + CO_2 = 2\ CO$$

At 1000°C the equilibrium constant for the latter reaction is $K_a = 126$, based on standard states of pure solids at 1000°C and pure gases at 1000°C and 1 atm.

(a) Evaluate K_a for the decomposition of limestone at 1000°C.

(b) What will be the composition of the gas phase in equilibrium with an equimolar mixture of limestone and coke?

(c) What will be the pressure of the gas phase in (b)?

(d) If the pressure is decreased from the value found above, and maintained there, what effect will this have on the system?

11-40. It has been suggested that water is an extremely dangerous material to use for extinguishing hot fires where large amounts of iron (or other metals) are present. It is feared that the iron (or other metals that may be present) may decompose the water to form metal oxide and liberate elemental hydrogen. For iron the reaction would be

$$Fe + H_2O \rightleftharpoons FeO + H_2$$

(Equilibrium data for this reaction are in Appendix E).

(a) If the temperature in a certain fire is 1800°F, determine the ratio y_{H_2}/y_{H_2O} which would be in equilibrium with the iron.

(b) Does it appear that this reaction will proceed to any appreciable extent at 1800°F. (Justify your answer!)

(c) Determine the amount of heat liberated per mole of iron oxidized according to the above reaction.

11-41. At 800°C the equilibrium partial pressure of nitrogen for the reaction

$$Mg_3N_2(s) = 3\,Mg(l) + N_2$$

is 10^{-15} atm. If a liquid bismuth solution containing 2.5 mol percent magnesium is blanketed with an argon atmosphere at 2 atm pressure and 800°C, what is the maximum concentration of N_2 that can be tolerated in the gas without forming Mg_3N_2? Assume that N_2 and Mg_3N_2 are insoluble in bismuth and that the activity coefficient of the magnesium in the bismuth solution is $\gamma^{\circ}_{Mg} = 10^{-3}$, where the standard state of magnesium is pure liquid at 800°C and 1 atm pressure. Also assume that bismuth does not form a stable nitride under these conditions and that the vapor pressures of Mg and Bi are negligible.

appendices

A

mollier diagrams

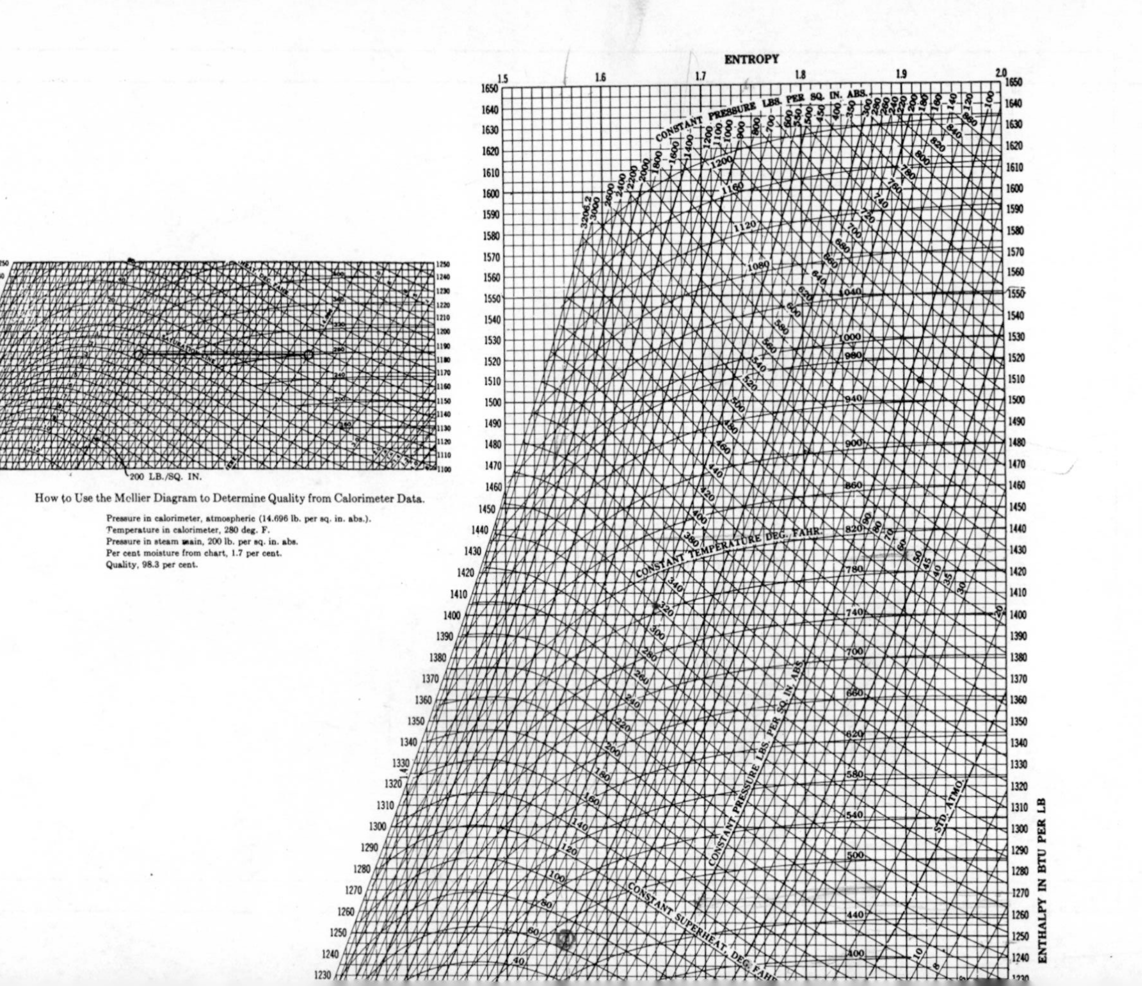

How to Use the Mollier Diagram to Determine Quality from Calorimeter Data.

Pressure in calorimeter, atmospheric (14.696 lb. per sq. in. abs.).
Temperature in calorimeter, 280 deg. F.
Pressure in steam main, 200 lb. per sq. in. abs.
Per cent moisture from chart, 1.7 per cent.
Quality, 98.3 per cent.

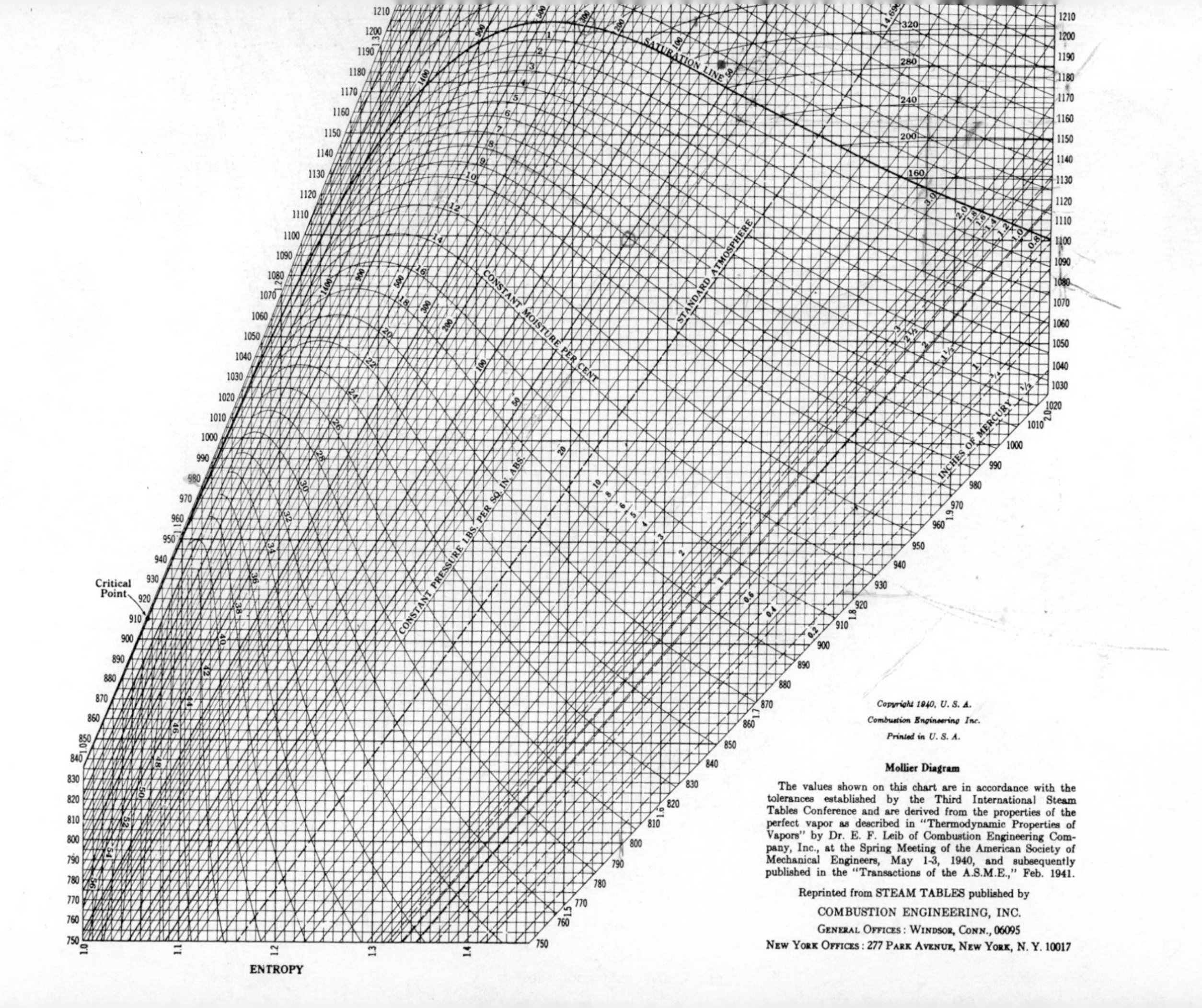
Copyright 1940, U. S. A.
Combustion Engineering Inc.
Printed in U. S. A.
Mollier Diagram
The values shown on this chart are in accordance with the tolerances established by the Third International Steam Tables Conference and are derived from the properties of the perfect vapor as described in "Thermodynamic Properties of Vapors" by Dr. E. F. Leib of Combustion Engineering Company, Inc., at the Spring Meeting of the American Society of Mechanical Engineers, May 1-3, 1940, and subsequently published in the "Transactions of the A.S.M.E.," Feb. 1941.
Reprinted from STEAM TABLES published by
COMBUSTION ENGINEERING, INC.
GENERAL OFFICES: WINDSOR, CONN., 06095
NEW YORK OFFICES: 277 PARK AVENUE, NEW YORK, N. Y. 10017
SATURATION LINE
STANDARD ATMOSPHERE
CONSTANT MOISTURE PER CENT
CONSTANT PRESSURE LBS. PER SQ. IN. ABS.
INCHES OF MERCURY
Critical Point
ENTROPY

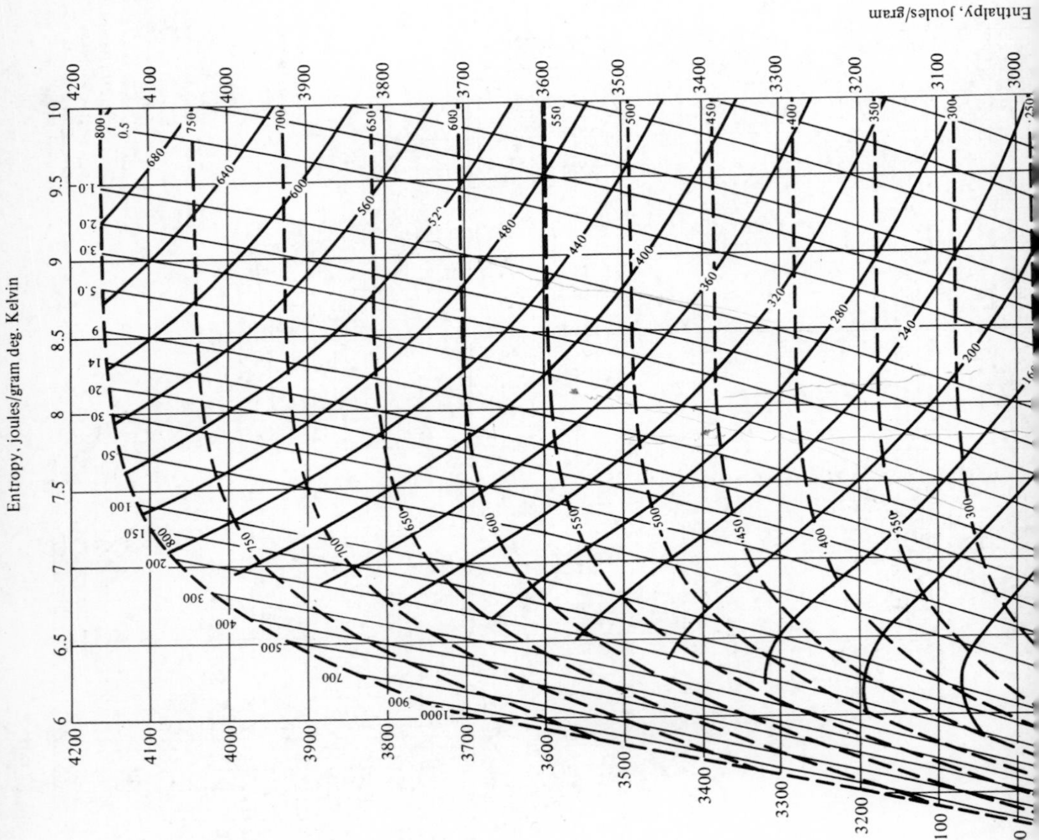

Entropy, joules/gram deg. Kelvin

Enthalpy, joules/gram

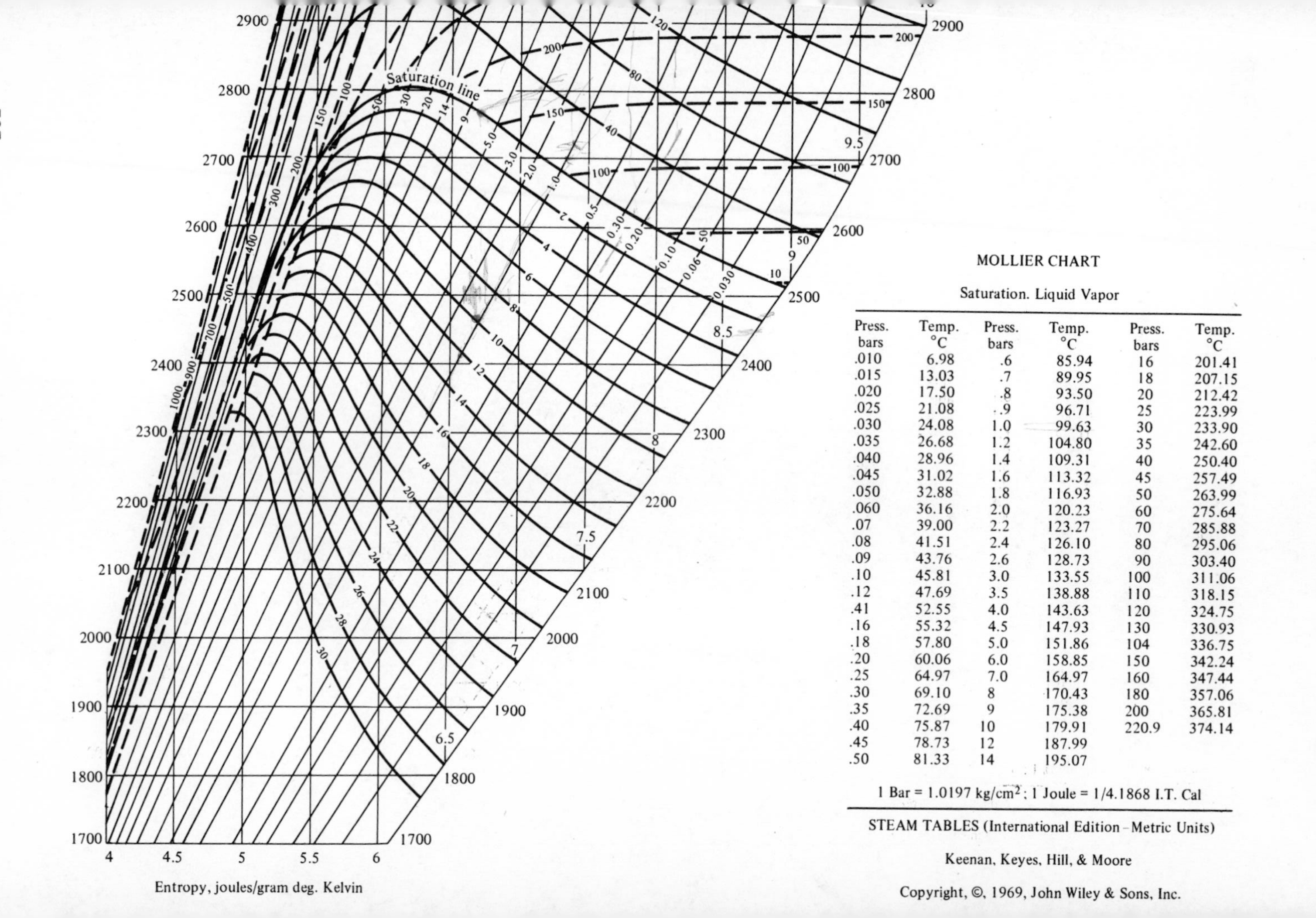

MOLLIER CHART

Saturation. Liquid Vapor

Press. bars	Temp. °C	Press. bars	Temp. °C	Press. bars	Temp. °C
.010	6.98	.6	85.94	16	201.41
.015	13.03	.7	89.95	18	207.15
.020	17.50	.8	93.50	20	212.42
.025	21.08	.9	96.71	25	223.99
.030	24.08	1.0	99.63	30	233.90
.035	26.68	1.2	104.80	35	242.60
.040	28.96	1.4	109.31	40	250.40
.045	31.02	1.6	113.32	45	257.49
.050	32.88	1.8	116.93	50	263.99
.060	36.16	2.0	120.23	60	275.64
.07	39.00	2.2	123.27	70	285.88
.08	41.51	2.4	126.10	80	295.06
.09	43.76	2.6	128.73	90	303.40
.10	45.81	3.0	133.55	100	311.06
.12	47.69	3.5	138.88	110	318.15
.41	52.55	4.0	143.63	120	324.75
.16	55.32	4.5	147.93	130	330.93
.18	57.80	5.0	151.86	104	336.75
.20	60.06	6.0	158.85	150	342.24
.25	64.97	7.0	164.97	160	347.44
.30	69.10	8	170.43	180	357.06
.35	72.69	9	175.38	200	365.81
.40	75.87	10	179.91	220.9	374.14
.45	78.73	12	187.99		
.50	81.33	14	195.07		

1 Bar = 1.0197 kg/cm^2; 1 Joule = 1/4.1868 I.T. Cal

STEAM TABLES (International Edition – Metric Units)

Keenan, Keyes, Hill, & Moore

B

steam tables

Table 1. Saturated Steam: Temperature Table

Temp Fahr t	Abs Press. Lb per Sq In. p	Specific Volume Sat. Liquid v_f	Specific Volume Evap v_{fg}	Specific Volume Sat. Vapor v_g	Enthalpy Sat. Liquid h_f	Enthalpy Evap h_{fg}	Enthalpy Sat. Vapor h_g	Entropy Sat. Liquid s_f	Entropy Evap s_{fg}	Entropy Sat. Vapor s_g	Temp Fahr t
32.0	0.08859	0.016022	3304.7	3304.7	0.0179	1075.5	1075.5	0.0000	2.1873	2.1873	**32.0**
34.0	0.09600	0.016021	3061.9	3061.9	1.996	1074.4	1076.4	0.0041	2.1762	2.1802	**34.0**
36.0	0.10395	0.016020	2839.0	2839.0	4.008	1073.2	1077.2	0.0081	2.1651	2.1732	**36.0**
38.0	0.11249	0.016019	2634.1	2634.2	6.018	1072.1	1078.1	0.0122	2.1541	2.1663	**38.0**
40.0	1.12163	0.016019	2445.8	2445.8	8.027	1071.0	1079.0	0.0162	2.1432	2.1594	**40.0**
42.0	0.13143	0.016019	2272.4	2272.4	10.035	1069.8	1079.9	0.0202	2.1325	2.1527	**42.0**
44.0	0.14192	0.016019	2112.8	2112.8	12.041	1068.7	1080.7	0.0242	2.1217	2.1459	**44.0**
46.0	0.15314	0.016020	1965.7	1965.7	14.047	1067.6	1081.6	0.0282	2.1111	2.1393	**46.0**
48.0	0.16514	0.016021	1830.0	1830.0	16.051	1066.4	1082.5	0.0321	2.1006	2.1327	**48.0**
50.0	0.17796	0.016023	1704.8	1704.8	18.054	1065.3	1083.4	0.0361	2.0901	2.1262	**50.0**
52.0	0.19165	0.016024	1589.2	1589.2	20.057	1064.2	1084.2	0.0400	2.0798	2.1197	**52.0**
54.0	0.20625	0.016026	1482.4	1482.4	22.058	1063.1	1085.1	0.0439	2.0695	2.1134	**54.0**
56.0	0.22183	0.016028	1383.6	1383.6	24.059	1061.9	1086.0	0.0478	2.0593	2.1070	**56.0**
58.0	0.23843	0.016031	1292.2	1292.2	26.060	1060.8	1086.9	0.0516	2.0491	2.1008	**58.0**
60.0	0.25611	0.016033	1207.6	1207.6	28.060	1059.7	1087.7	0.0555	2.0391	2.0946	**60.0**
62.0	0.27494	0.016036	1129.2	1129.2	30.059	1058.5	1088.6	0.0593	2.0291	2.0885	**62.0**
64.0	0.29497	0.016039	1056.5	1056.5	32.058	1057.4	1089.5	0.0632	2.0192	2.0824	**64.0**
66.0	0.31626	0.016043	989.0	989.1	34.056	1056.3	1090.4	0.0670	2.0094	2.0764	**66.0**
68.0	0.33889	0.016046	926.5	926.5	36.054	1055.2	1091.2	0.0708	1.9996	2.0704	**68.0**
70.0	0.36292	0.016050	868.3	868.4	38.052	1054.0	1092.1	0.0745	1.9900	2.0645	**70.0**
72.0	0.38844	0.016054	814.3	814.3	40.049	1052.9	1093.0	0.0783	1.9804	2.0587	**72.0**
74.0	0.41550	0.016058	764.1	764.1	42.046	1051.8	1093.8	0.0821	1.9708	2.0529	**74.0**
76.0	0.44420	0.016063	717.4	717.4	44.043	1050.7	1094.7	0.0858	1.9614	2.0472	**76.0**
78.0	0.47461	0.016067	673.8	673.9	46.040	1049.5	1095.6	0.0895	1.9520	2.0415	**78.0**
80.0	0.50683	0.016072	633.3	633.3	48.037	1048.4	1096.4	0.0932	1.9426	2.0359	**80.0**
82.0	0.54093	0.016077	595.5	595.5	50.033	1047.3	1097.3	0.0969	1.9334	2.0303	**82.0**
84.0	0.57702	0.016082	560.3	560.3	52.029	1046.1	1098.2	0.1006	1.9242	2.0248	**84.0**
86.0	0.61518	0.016087	227.5	527.5	54.026	1045.0	1099.0	0.1043	1.9151	2.0193	**86.0**
88.0	0.65551	0.016093	496.8	496.8	56.022	1043.9	1099.9	0.1079	1.9060	2.0139	**88.0**
90.0	0.69813	0.016099	468.1	468.1	58.018	1042.7	1100.8	0.1115	1.8970	2.0086	**90.0**
92.0	0.74313	0.016105	441.3	441.3	60.014	1041.6	1101.6	0.1152	1.8881	2.0033	**92.0**
94.0	0.79062	0.016111	416.3	416.3	62.010	1040.5	1102.5	0.1188	1.8792	1.9980	**94.0**
96.0	0.84072	0.016117	392.8	392.9	64.006	1039.3	1103.3	0.1224	1.8704	1.9928	**96.0**
98.0	0.89356	0.016123	370.9	370.9	66.003	1038.2	1104.2	0.1260	1.8617	1.9876	**98.0**
100.0	0.94924	0.016130	350.4	350.4	67.999	1037.1	1105.1	0.1295	1.8530	1.9825	**100.0**
102.0	1.00789	0.016137	331.1	331.1	69.995	1035.9	1105.9	0.1331	1.8444	1.9775	**102.0**
104.0	1.06965	0.016144	313.1	313.1	71.992	1034.8	1106.8	0.1366	1.8358	1.9725	**104.0**
106.0	1.1347	0.016151	296.16	296.18	73.99	1033.6	1107.6	0.1402	1.8273	1.9675	**106.0**
108.0	1.2030	0.016158	280.28	280.30	75.98	1032.5	1108.5	0.1437	1.8188	1.9626	**108.0**
110.0	1.2750	0.016165	265.37	265.39	77.98	1031.4	1109.3	0.1472	1.8105	1.9577	**110.0**
112.0	1.3505	0.016173	251.37	251.38	79.98	1030.2	1110.2	0.1507	1.8021	1.9528	**112.0**
114.0	1.4299	0.016180	238.21	238.22	81.97	1029.1	1111.0	0.1542	1.7938	1.9480	**114.0**
116.0	1.5133	0.016188	225.84	225.85	83.97	1027.9	1111.9	0.1577	1.7856	1.9433	**116.0**
118.0	1.6009	0.016196	214.20	214.21	85.97	1026.8	1112.7	0.1611	1.7774	1.9386	**118.0**
120.0	1.6927	0.016204	203.25	203.26	87.97	1025.6	1113.6	0.1646	1.7693	1.9339	**120.0**
122.0	1.7891	0.016213	192.94	192.95	89.96	1024.5	1114.4	0.1680	1.7613	1.9293	**122.0**
124.0	1.8901	0.016221	183.23	183.24	91.96	1023.3	1115.3	0.1715	1.7533	1.9247	**124.0**
126.0	1.9959	0.016229	174.08	174.09	93.96	1022.2	1116.1	0.1749	1.7453	1.9202	**126.0**
128.0	2.1068	0.016238	165.45	165.47	95.96	1021.0	1117.0	0.1783	1.7374	1.9157	**128.0**
130.0	2.2230	0.016247	157.32	157.33	97.96	1019.8	1117.8	0.1817	1.7295	1.9112	**130.0**
132.0	2.3445	0.016256	149.64	149.66	99.95	1018.7	1118.6	0.1851	1.7217	1.9068	**132.0**
134.0	2.4717	0.016265	142.40	142.41	101.95	1017.5	1119.5	0.1884	1.7140	1.9024	**134.0**
136.0	2.6047	0.016274	135.55	135.57	103.95	1016.4	1120.3	0.1918	1.7063	1.8980	**136.0**
138.0	2.7438	0.016284	129.09	129.11	105.95	1015.2	1121.1	0.1951	1.6986	1.8937	**138.0**
140.0	2.8892	0.016293	122.98	123.00	107.95	1014.0	1122.0	0.1985	1.6910	1.8895	**140.0**
142.0	3.0411	0.016303	117.21	117.22	109.95	1012.9	1122.8	0.2018	1.6534	1.8852	**142.0**
144.0	3.1997	0.016312	111.74	111.76	111.95	1011.7	1123.6	0.2051	1.6759	1.8810	**144.0**
146.0	3.3653	0.016322	106.58	106.59	113.95	1010.5	1124.5	0.2084	1.6684	1.8769	**146.0**
148.0	3.5381	0.016332	101.68	101.70	115.95	1009.3	1125.3	0.2117	1.6610	1.8727	**148.0**
150.0	3.7184	0.016343	97.05	97.07	117.95	1008.2	1126.1	0.2150	1.6536	1.8686	**150.0**
152.0	3.9065	0.016353	92.66	92.68	119.95	1007.0	1126.9	0.2183	1.6463	1.8646	**152.0**
154.0	4.1025	0.016363	88.50	88.52	121.95	1005.8	1127.7	0.2216	1.6390	1.8606	**154.0**
156.0	4.3068	0.016374	84.56	84.57	123.95	1004.6	1128.6	0.2248	1.6318	1.8566	**156.0**
158.0	4.5197	0.016384	80.82	80.83	125.96	1003.4	1129.4	0.2281	1.6245	1.8526	**158.0**
160.0	4.7414	0.016395	77.27	77.29	127.96	1002.2	1130.2	0.2313	1.6174	1.8487	**160.0**
162.0	4.9722	0.016406	73.90	73.92	129.96	1001.0	1131.0	0.2345	1.6103	1.8448	**162.0**
164.0	5.2124	0.016417	70.70	70.72	131.96	999.8	1131.8	0.2377	1.6032	1.8409	**164.0**
166.0	5.4623	0.016428	67.67	67.68	133.97	998.6	1132.6	0.2409	1.5961	1.8371	**166.0**
168.0	5.7223	0.016440	64.78	64.80	135.97	997.4	1133.4	0.2441	1.5892	1.8333	**168.0**
170.0	5.9926	0.016451	62.04	62.06	137.97	996.2	1134.2	0.2473	1.5822	1.8295	**170.0**
172.0	6.2736	0.016463	59.43	59.45	139.98	995.0	1135.0	0.2505	1.5753	1.8258	**172.0**
174.0	6.5656	0.016474	56.95	56.97	141.98	993.8	1135.8	0.2537	1.5684	1.8221	**174.0**
176.0	6.8690	0.016486	54.59	54.61	143.99	992.6	1136.6	0.2568	1.5616	1.8184	**176.0**
178.0	7.1840	0.016498	52.35	52.36	145.99	991.4	1137.4	0.2600	1.5548	1.8147	**178.0**

Table 1. Saturated Steam: Temperature Table—*Continued*

Temp Fahr t	Abs Press. Lb per Sq In. p	Specific Volume Sat. Liquid v_f	Specific Volume Evap v_{fg}	Specific Volume Sat. Vapor v_g	Enthalpy Sat. Liquid h_f	Enthalpy Evap h_{fg}	Enthalpy Sat. Vapor h_g	Entropy Sat. Liquid s_f	Entropy Evap s_{fg}	Entropy Sat. Vapor s_g	Temp Fahr t
180.0	7.5110	0.016510	50.21	50.22	148.00	990.2	1138.2	0.2631	1.5480	1.8111	**180.0**
182.0	7.850	0.016522	48.172	18.189	150.01	989.0	1139.0	0.2662	1.5413	1.8075	**182.0**
184.0	8.203	0.016534	46.232	46.249	152.01	987.8	1139.8	0.2694	1.5346	1.8040	**184.0**
186.0	8.568	0.016547	44.383	44.400	154.02	986.5	1140.5	0.2725	1.5279	1.8004	**186.0**
188.0	8.947	0.016559	42.621	42.638	156.03	985.3	1141.3	0.2756	1.5213	1.7969	**188.0**
190.0	9.340	0.016572	40.941	40.957	158.04	984.1	1142.1	0.2787	1.5148	1.7934	**190.0**
192.0	9.747	0.016585	39.337	39.354	160.05	982.8	1142.9	0.2818	1.5082	1.7900	**192.0**
194.0	10.168	0.016598	37.808	37.824	162.05	981.6	1143.7	0.2848	1.5017	1.7865	**194.0**
196.0	10.605	0.016611	36.348	36.364	164.06	980.4	1144.4	0.2879	1.4952	1.7831	**196.0**
198.0	11.058	0.016624	34.954	34.970	166.08	979.1	1145.2	0.2910	1.4888	1.7798	**198.0**
200.0	11.526	0.016637	33.622	33.639	168.09	977.9	1146.0	0.2940	1.4824	1.7764	**200.0**
204.0	12.512	0.016664	31.135	31.151	172.11	975.4	1147.5	0.3001	1.4697	1.7698	**204.0**
208.0	13.568	0.016691	28.862	28.878	176.14	972.8	1149.0	0.3061	1.4571	1.7632	**208.0**
212.0	14.696	0.016719	26.782	26.799	180.17	970.3	1150.5	0.3121	1.4447	1.7568	**212.0**
216.0	15.901	0.016747	24.878	24.894	184.20	967.8	1152.0	0.3181	1.4323	1.7505	**216.0**
220.0	17.186	0.016775	23.131	23.148	188.23	965.2	1153.4	0.3241	1.4201	1.7442	**220.0**
224.0	18.556	0.016805	21.529	21.545	192.27	962.6	1154.9	0.3300	1.4081	1.7380	**224.0**
228.0	20.015	0.016834	20.056	20.073	196.31	960.0	1156.3	0.3359	1.3961	1.7320	**228.0**
232.0	21.567	0.016864	18.701	18.718	200.35	957.4	1157.8	0.3417	1.3842	1.7260	**232.0**
236.0	23.216	0.016895	17.454	17.471	204.40	954.8	1159.2	0.3476	1.3725	1.7201	**236.0**
240.0	24.968	0.016926	16.304	16.321	208.45	952.1	1160.6	0.3533	1.3609	1.7142	**240.0**
244.0	26.826	0.016958	15.243	15.260	212.50	949.5	1162.0	0.3591	1.3494	1.7085	**244.0**
248.0	28.796	0.016990	14.264	14.281	216.56	946.8	1163.4	0.3649	1.3379	1.7028	**248.0**
252.0	30.883	0.017022	13.358	13.375	220.62	944.1	1164.7	0.3706	1.3266	1.6972	**252.0**
256.0	33.091	0.017055	12.520	12.538	224.69	941.4	1166.1	0.3763	1.3154	1.6917	**256.0**
260.0	35.427	0.017089	11.745	11.762	228.76	938.6	1167.4	0.3819	1.3043	1.6862	**260.0**
264.0	37.894	0.017123	11.025	11.042	232.83	935.9	1168.7	0.3876	1.2933	1.6808	**264.0**
268.0	40.500	0.017157	10.358	10.375	236.91	933.1	1170.0	0.3932	1.2823	1.6755	**268.0**
272.0	43.249	0.017193	9.738	9.755	240.99	930.3	1171.3	0.3987	1.2715	1.6702	**272.0**
276.0	46.147	0.017228	9.162	9.180	245.08	927.5	1172.5	0.4043	1.2607	1.6650	**276.0**
280.0	49.200	0.017264	8.627	8.644	249.17	924.6	1173.8	0.4098	1.2501	1.6599	**280.0**
284.0	52.414	0.01730	8.1280	8.1453	253.3	921.7	1175.0	0.4154	1.2395	1.6548	**284.0**
288.0	55.795	0.01734	7.6634	7.6807	257.4	918.8	1176.2	0.4208	1.2290	1.6498	**288.0**
292.0	59.350	0.01738	7.2301	7.2475	261.5	915.9	1177.4	0.4263	1.2186	1.6449	**292.0**
296.0	63.084	0.01741	6.8259	6.8433	265.6	913.0	1178.6	0.4317	1.2082	1.6400	**296.0**
300.0	67.005	0.01745	6.4483	6.4658	269.7	910.0	1179.7	0.4372	1.1979	1.6351	**300.0**
304.0	71.119	0.01749	6.0955	6.1130	273.8	907.0	1180.9	0.4426	1.1877	1.6303	**304.0**
308.0	75.433	0.01753	5.7655	5.7830	278.0	904.0	1182.0	0.4479	1.1776	1.6256	**308.0**
312.0	79.953	0.01757	5.4566	5.4742	282.1	901.0	1183.1	0.4533	1.1676	1.6209	**312.0**
316.0	84.688	0.01761	5.1673	5.1849	286.3	897.9	1184.1	0.4586	1.1576	1.6162	**316.0**
320.0	89.643	0.01766	4.8961	4.9138	290.4	894.8	1185.2	0.4640	1.1477	1.6116	**320.0**
324.0	94.826	0.01770	4.6418	4.6595	294.6	891.6	1186.2	0.4692	1.1378	1.6071	**324.0**
328.0	100.245	0.01774	4.4030	4.4208	298.7	888.5	1187.2	0.4745	1.1280	1.6025	**328.0**
332.0	105.907	0.01779	4.1788	4.1966	302.9	885.3	1188.2	0.4798	1.1183	1.5981	**332.0**
336.0	111.820	0.01783	3.9681	3.9859	307.1	882.1	1189.1	0.4850	1.1086	1.5936	**336.0**
340.0	117.992	0.01787	3.7699	3.7878	311.3	878.8	1190.1	0.4902	1.0990	1.5892	**340.0**
344.0	124.430	0.01792	3.5834	3.6013	315.5	875.5	1191.0	0.4954	1.0894	1.5849	**344.0**
348.0	131.142	0.01797	3.4078	3.4258	319.7	872.2	1191.1	0.5006	1.0799	1.5806	**348.0**
352.0	138.138	0.01801	3.2423	3.2603	323.9	868.9	1192.7	0.5058	1.0705	1.5763	**352.0**
356.0	145.424	0.01806	3.0863	3.1044	328.1	865.5	1193.6	0.5110	1.0611	1.5721	**356.0**
360.0	153.010	0.01811	2.9392	2.9573	332.3	862.1	1194.4	0.5161	1.0517	1.5678	**360.0**
364.0	160.903	0.01816	2.8002	2.8184	336.5	858.6	1195.2	0.5212	1.0424	1.5637	**364.0**
368.0	169.113	0.01821	2.6691	2.6873	340.8	855.1	1195.9	0.5263	1.0332	1.5595	**368.0**
372.0	177.648	0.01826	2.5451	2.5633	345.0	851.6	1196.7	0.5314	1.0240	1.5554	**372.0**
376.0	186.517	0.01831	2.4279	2.4462	349.3	848.1	1197.4	0.5365	1.0148	1.5513	**376.0**
380.0	195.729	0.01836	2.3170	2.3353	353.6	844.5	1198.0	0.5416	1.0057	1.5473	**380.0**
384.0	205.294	0.01842	2.2120	2.2304	357.9	840.8	1198.7	0.5466	0.9966	1.5432	**384.0**
388.0	215.220	0.01847	2.1126	2.1311	362.2	837.2	1199.3	0.5516	0.9876	1.5392	**388.0**
392.0	225.516	0.01853	2.0184	2.0369	366.5	833.4	1199.9	0.5567	0.9786	1.5352	**392.0**
396.0	236.193	0.01858	1.9291	1.9477	370.8	829.7	1200.4	0.5617	0.9696	1.5313	**396.0**
400.0	247.259	0.01864	1.8444	1.8630	375.1	825.9	1201.0	0.5667	0.9607	1.5274	**400.0**
404.0	258.725	0.01870	1.7640	1.7827	379.4	822.0	1201.5	0.5717	0.9518	1.5234	**404.0**
408.0	270.600	0.01875	1.6877	1.7064	383.8	818.2	1201.9	0.5766	0.9429	1.5195	**408.0**
412.0	282.894	0.01881	1.6152	1.6340	388.1	814.2	1202.4	0.5816	0.9341	1.5157	**412.0**
416.0	295.617	0.01887	1.5463	1.5651	392.5	810.2	1202.8	0.5866	0.9253	1.5118	**416.0**
420.0	308.780	0.01894	1.4808	1.4997	396.9	806.2	1203.1	0.5915	0.9165	1.5080	**420.0**
424.0	322.391	0.01900	1.4184	1.4374	401.3	802.2	1203.5	0.5964	0.9077	1.5042	**424.0**
428.0	336.463	0.01906	1.3591	1.3782	405.7	798.0	1203.7	0.6014	0.8990	1.5004	**428.0**
432.0	351.00	0.01913	1.30266	1.32179	410.1	793.9	1204.0	0.6063	0.8903	1.4966	**432.0**
436.0	366.03	0.01919	1.24887	1.26806	414.6	789.7	1204.2	0.6112	0.8816	1.4928	**436.0**
440.0	381.54	0.01926	1.19761	1.21687	419.0	785.4	1204.4	0.6161	0.8729	1.4890	**440.0**
444.0	397.56	0.01933	1.14874	1.16806	423.5	781.1	1204.6	0.6210	0.8643	1.4853	**444.0**
448.0	414.09	0.01940	1.10212	1.12152	428.0	776.7	1204.7	0.6259	0.8557	1.4815	**448.0**
452.0	431.14	0.01947	1.05764	1.07711	432.5	772.3	1204.8	0.6308	0.8471	1.4778	**452.0**
456.0	448.73	0.01954	1.01518	1.03472	437.0	767.8	1204.8	0.6356	0.8385	1.4741	**456.0**

Table 1. Saturated Steam: Temperature Table—*Continued*

Temp Fahr t	Abs Press. Lb per Sq In. p	Specific Volume Sat. Liquid v_f	Evap v_{fg}	Sat. Vapor v_g	Enthalpy Sat. Liquid h_f	Evap h_{fg}	Sat. Vapor h_g	Entropy Sat. Liquid s_f	Evap s_{fg}	Sat. Vapor s_g	Temp Fahr t
460.0	466.87	0.01961	0.97463	0.99424	441.5	763.2	1204.8	0.6405	0.8299	1.4704	**460.0**
464.0	485.56	0.01969	0.93588	0.95557	446.1	758.6	1204.7	0.6454	0.8213	1.4667	**464.0**
468.0	504.83	0.01976	0.89885	0.91862	450.7	754.0	1204.6	0.6502	0.8127	1.4629	**468.0**
472.0	524.67	0.01984	0.86345	0.88329	455.2	749.3	1204.5	0.6551	0.8042	1.4592	**472.0**
476.0	545.11	0.01992	0.82958	0.84950	459.9	744.5	1204.3	0.6599	0.7956	1.4555	**476.0**
480.0	566.15	0.02000	0.79716	0.81717	464.5	739.6	1204.1	0.6648	0.7871	1.4518	**480.0**
484.0	587.81	0.02009	0.76613	0.78622	469.1	734.7	1203.8	0.6696	0.7785	1.4481	**484.0**
488.0	610.10	0.02017	0.73641	0.75658	473.8	729.7	1203.5	0.6745	0.7700	1.4444	**488.0**
492.0	633.03	0.02026	0.70794	0.72820	478.5	724.6	1203.1	0.6793	0.7614	1.4407	**492.0**
496.0	656.61	0.02034	0.68065	0.70100	483.2	719.5	1202.7	0.6842	0.7528	1.4370	**496.0**
500.0	680.86	0.02043	0.65448	0.67492	487.9	714.3	1202.2	0.6890	0.7443	1.4333	**500.0**
504.0	705.78	0.02053	0.62938	0.64991	492.7	709.0	1201.7	0.6939	0.7357	1.4296	**504.0**
508.0	731.40	0.02062	0.60530	0.62592	497.5	703.7	1201.1	0.6987	0.7271	1.4258	**508.0**
512.0	757.72	0.02072	0.58218	0.60289	502.3	698.2	1200.5	0.7036	0.7185	1.4221	**512.0**
516.0	784.76	0.02081	0.55997	0.58079	507.1	692.7	1199.8	0.7085	0.7099	1.4183	**516.0**
520.0	812.53	0.02091	0.53864	0.55956	512.0	687.0	1199.0	0.7133	0.7013	1.4146	**520.0**
524.0	841.04	0.02102	0.51814	0.53916	516.9	681.3	1198.2	0.7182	0.6926	1.4108	**524.0**
528.0	870.31	0.02112	0.49843	0.51955	521.8	675.5	1197.3	0.7231	0.6839	1.4070	**528.0**
532.0	900.34	0.02123	0.47947	0.50070	526.8	669.6	1196.4	0.7280	0.6752	1.4032	**532.0**
536.0	931.17	0.02134	0.46123	0.48257	531.7	663.6	1195.4	0.7329	0.6665	1.3993	**536.0**
540.0	962.79	0.02146	0.44367	0.46513	536.8	657.5	1194.3	0.7378	0.6577	1.3954	**540.0**
544.0	995.22	0.02157	0.42677	0.44834	541.8	651.3	1193.1	0.7427	0.6489	1.3915	**544.0**
548.0	1028.49	0.02169	0.41048	0.43217	546.9	645.0	1191.9	0.7476	0.6400	1.3876	**548.0**
552.0	1062.59	0.02182	0.39479	0.41660	552.0	638.5	1190.6	0.7525	0.6311	1.3837	**552.0**
556.0	1097.55	0.02194	0.37966	0.40160	557.2	632.0	1189.2	0.7575	0.6222	1.3797	**556.0**
560.0	1133.38	0.02207	0.36507	0.38714	562.4	625.3	1187.7	0.7625	0.6132	1.3757	**560.0**
564.0	1170.10	0.02221	0.35099	0.37320	567.6	618.5	1186.1	0.7674	0.6041	1.3716	**564.0**
568.0	1207.72	0.02235	0.33741	0.35975	572.9	611.5	1184.5	0.7725	0.5950	1.3675	**568.0**
572.0	1246.26	0.02249	0.32429	0.34678	578.3	604.5	1182.7	0.7775	0.5859	1.3634	**572.0**
576.0	1285.74	0.02264	0.31162	0.33426	583.7	597.2	1180.9	0.7825	0.5766	1.3592	**576.0**
580.0	1326.17	0.02279	0.29937	0.32216	589.1	589.9	1179.0	0.7876	0.5673	1.3550	**580.0**
584.0	1367.7	0.02295	0.28753	0.31048	594.6	582.4	1176.9	0.7927	0.5580	1.3507	**584.0**
588.0	1410.0	0.02311	0.27608	0.29919	600.1	574.7	1174.8	0.7978	0.5485	1.3464	**588.0**
592.0	1453.3	0.02328	0.26499	0.28827	605.7	566.8	1172.6	0.8030	0.5390	1.3420	**592.0**
596.0	1497.8	0.02345	0.25425	0.27770	611.4	558.8	1170.2	0.8082	0.5293	1.3375	**596.0**
600.0	1543.2	0.02364	0.24384	0.26747	617.1	550.6	1167.7	0.8134	0.5196	1.3330	**600.0**
604.0	1589.7	0.02382	0.23374	0.25757	622.9	542.2	1165.1	0.8187	0.5097	1.3284	**604.0**
608.0	1637.3	0.02402	0.22394	0.24796	628.8	533.6	1162.4	0.8240	0.4997	1.3238	**608.0**
612.0	1686.1	0.02422	0.21442	0.23865	634.8	524.7	1159.5	0.8294	0.4896	1.3190	**612.0**
616.6	1735.9	0.02444	0.20516	0.22960	640.8	515.6	1156.4	0.8348	0.4794	1.3141	**616.0**
620.0	1786.9	0.02466	0.19615	0.22081	646.9	506.3	1153.2	0.8403	0.4689	1.3092	**620.0**
624.0	1839.0	0.02489	0.18737	0.21226	653.1	496.6	1149.8	0.8458	0.4583	1.3041	**624.0**
628.0	1892.4	0.02514	0.17880	0.20394	659.5	486.7	1146.1	0.8514	0.4474	1.2988	**628.0**
632.0	1947.0	0.02539	0.17044	0.19583	665.9	476.4	1142.2	0.8571	0.4364	1.2934	**632.0**
636.0	2002.8	0.02566	0.16226	0.18792	672.4	465.7	1138.1	0.8628	0.4251	1.2879	**636.0**
640.0	2059.9	0.02595	0.15427	0.18021	679.1	454.6	1133.7	0.8686	0.4134	1.2821	**640.0**
644.0	2118.3	0.02625	0.14644	0.17269	685.9	443.1	1129.0	0.8746	0.4015	1.2761	**644.0**
648.0	2178.1	0.02657	0.13876	0.16534	692.9	431.1	1124.0	0.8806	0.3893	1.2699	**648.0**
652.0	2239.2	0.02691	0.13124	0.15816	700.0	418.7	1118.7	0.8868	0.3767	1.2634	**652.0**
656.0	2301.7	0.02728	0.12387	0.15115	707.4	405.7	1113.1	0.8931	0.3637	1.2567	**656.0**
660.0	2365.7	0.02768	0.11663	0.14431	714.9	392.1	1107.0	0.8995	0.3502	1.2498	**660.0**
664.0	2431.1	0.02811	0.10947	0.13757	722.9	377.7	1100.6	0.9064	0.3361	1.2425	**664.0**
668.0	2498.1	0.02858	0.10229	0.13087	731.5	362.1	1093.5	0.9137	0.3210	1.2347	**668.0**
672.0	2566.6	0.02911	0.09514	0.12424	740.2	345.7	1085.9	0.9212	0.3054	1.2266	**672.0**
676.0	2636.8	0.02970	0.08799	0.11769	749.2	328.5	1077.6	0.9287	0.2892	1.2179	**676.0**
680.0	2708.6	0.03037	0.08080	0.11117	758.5	310.1	1068.5	0.9365	0.2720	1.2086	**680.0**
684.0	2782.1	0.03114	0.07349	0.10463	768.2	290.2	1058.4	0.9447	0.2537	1.1984	**684.0**
688.0	2857.4	0.03204	0.06595	0.09799	778.8	268.2	1047.0	0.9535	0.2337	1.1872	**688.0**
692.0	2934.5	0.03313	0.05797	0.09110	790.5	243.1	1033.6	0.9634	0.2110	1.1744	**692.0**
696.0	3013.4	0.03455	0.04916	0.08371	804.4	212.8	1017.2	0.9749	0.1841	1.1591	**696.0**
700.0	3094.3	0.03662	0.03857	0.07519	822.4	172.7	995.2	0.9901	0.1490	1.1390	**700.0**
702.0	3135.5	0.03824	0.03173	0.06997	835.0	144.7	979.7	1.0006	0.1246	1.1252	**702.0**
704.0	3177.2	0.04108	0.02192	0.06300	854.2	102.0	956.2	1.0169	0.0876	1.1046	**704.0**
705.0	3198.3	0.04427	0.01304	0.05730	873.0	61.4	934.4	1.0329	0.0527	1.0856	**705.0**
705.47*	3208.2	0.05078	0.00000	0.05078	906.0	0.0	906.0	1.0612	0.0000	1.0612	**705.47***

*Critical temperature

Table 2: Saturated Steam: Pressure Table *Critical pressure

Abs Press. Lb/Sq In. p	Temp Fahr t	Specific Volume Sat. Liquid v_f	Evap v_{fg}	Sat. Vapor v_g	Enthalpy Sat. Liquid h_f	Evap h_{fg}	Sat. Vapor h_g	Entropy Sat. Liquid s_f	Evap s_{fg}	Sat. Vapor s_g	Abs Press. Lb/Sq In. p
0.08865	32.018	0.016022	3302.4	3302.4	0.0003	1075.5	1075.5	0.0000	2.1872	2.1872	**0.08865**
0.25	59.323	0.016032	1235.5	1235.5	27.382	1060.1	1087.4	0.0542	2.0425	2.0967	**0.25**
0.50	79.586	0.016071	641.5	641.5	47.623	1048.6	1096.3	0.0925	1.9446	2.0370	**0.50**
1.0	101.74	0.016136	333.59	333.60	69.73	1036.1	1105.8	0.1326	1.8455	1.9781	**1.0**
5.0	162.24	0.016407	73.515	73.532	130.20	1000.9	1131.1	0.2349	1.6094	1.8443	**5.0**
10.0	193.21	0.016592	38.404	38.420	161.26	982.1	1143.3	0.2836	1.5043	1.7879	**10.0**
14.696	212.00	0.016719	26.782	26.799	180.17	970.3	1150.5	0.3121	1.4447	1.7568	**14.696**
15.0	213.03	0.016726	26.274	26.290	181.21	969.7	1150.9	0.3137	1.4415	1.7552	**15.0**
20.0	227.96	0.016834	20.070	20.087	196.27	960.1	1156.3	0.3358	1.3962	1.7320	**20.0**
30.0	250.34	0.017009	13.7266	13.7436	218.9	945.2	1164.1	0.3682	1.3313	1.6995	**30.0**
40.0	267.25	0.017151	10.4794	10.4965	236.1	933.6	1169.8	0.3921	1.2844	1.6765	**40.0**
50.0	281.02	0.017274	8.4967	8.5140	250.2	923.9	1174.1	0.4112	1.2474	1.6586	**50.0**
60.0	292.71	0.017383	7.1562	7.1736	262.2	915.4	1177.6	0.4273	1.2167	1.6440	**60.0**
70.0	302.93	0.017482	6.1875	6.2050	272.7	907.8	1180.6	0.4411	1.1905	1.6316	**70.0**
80.0	312.04	0.017573	5.4536	5.4711	282.1	900.9	1183.1	0.4534	1.1675	1.6208	**80.0**
90.0	320.28	0.017659	4.8779	4.8953	290.7	894.6	1185.3	0.4643	1.1470	1.6113	**90.0**
100.0	327.82	0.017740	4.4133	4.4310	298.5	888.6	1187.2	0.4743	1.1284	1.6027	**100.0**
110.0	334.79	0.01782	4.0306	4.0484	305.8	883.1	1188.9	0.4834	1.1115	1.5950	**110.0**
120.0	341.27	0.01789	3.7097	3.7275	312.6	877.8	1190.4	0.4919	1.0960	1.5879	**120.0**
130.0	347.33	0.01796	3.4364	3.4544	319.0	872.8	1191.7	0.4998	1.0815	1.5813	**130.0**
140.0	353.04	0.01803	3.2010	3.2190	325.0	868.0	1193.0	0.5071	1.0681	1.5752	**140.0**
150.0	358.43	0.01809	2.9958	3.0139	330.6	863.4	1194.1	0.5141	1.0554	1.5695	**150.0**
160.0	363.55	0.01815	2.8155	2.8336	336.1	859.0	1195.1	0.5206	1.0435	1.5641	**160.0**
170.0	368.42	0.01821	2.6556	2.6738	341.2	854.8	1196.0	0.5269	1.0322	1.5591	**170.0**
180.0	373.08	0.01827	2.5129	2.5312	346.2	850.7	1196.9	0.5328	1.0215	1.5543	**180.0**
190.0	377.53	0.01833	2.3847	2.4030	350.9	846.7	1197.6	0.5384	1.0113	1.5498	**190.0**
200.0	381.80	0.01839	2.2689	2.2873	355.5	842.8	1198.3	0.5438	1.0016	1.5454	**200.0**
210.0	385.91	0.01844	2.16373	2.18217	359.9	839.1	1199.0	0.5490	0.9923	1.5413	**210.0**
220.0	389.88	0.01850	2.06779	2.08629	364.2	835.4	1199.6	0.5540	0.9834	1.5374	**220.0**
230.0	393.70	0.01855	1.97991	1.99846	368.3	831.8	1200.1	0.5588	0.9748	1.5336	**230.0**
240.0	397.39	0.01860	1.89909	1.91769	372.3	828.4	1200.6	0.5634	0.9665	1.5299	**240.0**
250.0	400.97	0.01865	1.82452	1.84317	376.1	825.0	1201.1	0.5679	0.9585	1.5264	**250.0**
260.0	404.44	0.01870	1.75548	1.77418	379.9	821.6	1201.5	0.5722	0.9508	1.5230	**260.0**
270.0	407.80	0.01875	1.69137	1.71013	383.6	818.3	1201.9	0.5764	0.9433	1.5197	**270.0**
280.0	411.07	0.01880	1.63169	1.65049	387.1	815.1	1202.3	0.5805	0.9361	1.5166	**280.0**
290.0	414.25	0.01885	1.57597	1.59482	390.6	812.0	1202.6	0.5844	0.9291	1.5135	**290.0**
300.0	417.35	0.01889	1.52384	1.54274	394.0	808.9	1202.9	0.5882	0.9223	1.5105	**300.0**
350.0	431.73	0.01912	1.30642	1.32554	409.8	794.2	1204.0	0.6059	0.8909	1.4968	**350.0**
400.0	444.60	0.01934	1.14162	1.16095	424.2	780.4	1204.6	0.6217	0.8630	1.4847	**400.0**
450.0	456.28	0.01954	1.01224	1.03179	437.3	767.5	1204.8	0.6360	0.8378	1.4738	**450.0**
500.0	467.01	0.01975	0.90787	0.92762	449.5	755.1	1204.7	0.6490	0.8148	1.4639	**500.0**
550.0	476.94	0.01994	0.82183	0.84177	460.9	743.3	1204.3	0.6611	0.7936	1.4547	**550.0**
600.0	486.20	0.02013	0.74962	0.76975	471.7	732.0	1203.7	0.6723	0.7738	1.4461	**600.0**
650.0	494.89	0.02032	0.68811	0.70843	481.9	720.9	1202.8	0.6828	0.7552	1.4381	**650.0**
700.0	503.08	0.02050	0.63505	0.65556	491.6	710.2	1201.8	0.6928	0.7377	1.4304	**700.0**
750.0	510.84	0.02069	0.58880	0.60949	500.9	699.8	1200.7	0.7022	0.7210	1.4232	**750.0**
800.0	518.21	0.02087	0.54809	0.56896	509.8	689.6	1199.4	0.7111	0.7051	1.4163	**800.0**
850.0	525.24	0.02105	0.51197	0.53302	518.4	679.5	1198.0	0.7197	0.6899	1.4096	**850.0**
900.0	531.95	0.02123	0.47968	0.50091	526.7	669.7	1196.4	0.7279	0.6753	1.4032	**900.0**
950.0	538.39	0.02141	0.45064	0.47205	534.7	660.0	1194.7	0.7358	0.6612	1.3970	**950.0**
1000.0	544.58	0.02159	0.42436	0.44596	542.6	650.4	1192.9	0.7434	0.6476	1.3910	**1000.0**
1050.0	550.53	0.02177	0.40047	0.42224	550.1	640.9	1191.0	0.7507	0.6344	1.3851	**1050.0**
1100.0	556.28	0.02195	0.37863	0.40058	557.5	631.5	1189.1	0.7578	0.6216	1.3794	**1100.0**
1150.0	561.82	0.02214	0.35859	0.38073	564.8	622.2	1187.0	0.7647	0.6091	1.3738	**1150.0**
1200.0	567.19	0.02232	0.34013	0.36245	571.9	613.0	1184.8	0.7714	0.5969	1.3683	**1200.0**
1250.0	572.38	0.02250	0.32306	0.34556	578.8	603.8	1182.6	0.7780	0.5850	1.3630	**1250.0**
1300.0	577.42	0.02269	0.30722	0.32991	585.6	594.6	1180.2	0.7843	0.5733	1.3577	**1300.0**
1350.0	582.32	0.02288	0.29250	0.31537	592.3	585.4	1177.8	0.7906	0.5620	1.3525	**1350.0**
1400.0	587.07	0.02307	0.27871	0.30178	598.8	576.5	1175.3	0.7966	0.5507	1.3474	**1400.0**
1450.0	591.70	0.02327	0.26584	0.28911	605.3	567.4	1172.8	0.8026	0.5397	1.3423	**1450.0**
1500.0	596.20	0.02346	0.25372	0.27719	611.7	558.4	1170.1	0.8085	0.5288	1.3373	**1500.0**
1550.0	600.59	0.02366	0.24235	0.26601	618.0	549.4	1167.4	0.8142	0.5182	1.3324	**1550.0**
1600.0	604.87	0.02387	0.23159	0.25545	624.2	540.3	1164.5	0.8199	0.5076	1.3274	**1600.0**
1650.0	609.05	0.02407	0.22143	0.24551	630.4	531.3	1161.6	0.8254	0.4971	1.3225	**1650.0**
1700.0	613.13	0.02428	0.21178	0.23607	636.5	522.2	1158.6	0.8309	0.4867	1.3176	**1700.0**
1750.0	617.12	0.02450	0.20263	0.22713	642.5	513.1	1155.6	0.8363	0.4765	1.3128	**1750.0**
1800.0	621.02	0.02472	0.19390	0.21861	648.5	503.8	1152.3	0.8417	0.4662	1.3079	**1800.0**
1850.0	624.83	0.02495	0.18558	0.21052	654.5	494.6	1149.0	0.8470	0.4561	1.3030	**1850.0**
1900.0	628.56	0.02517	0.17761	0.20278	660.4	485.2	1145.6	0.8522	0.4459	1.2981	**1900.0**
1950.0	632.22	0.02541	0.16999	0.19540	666.3	475.8	1142.0	0.8574	0.4358	1.2931	**1950.0**
2000.0	635.80	0.02565	0.16266	0.18831	672.1	466.2	1138.3	0.8625	0.4256	1.2881	**2000.0**
2100.0	642.76	0.02615	0.14885	0.17501	683.8	446.7	1130.5	0.8727	0.4053	1.2780	**2100.0**
2200.0	649.45	0.02669	0.13603	0.16272	695.5	426.7	1122.2	0.8828	0.3848	1.2676	**2200.0**
2300.0	655.89	0.02727	0.12406	0.15133	707.2	406.0	1113.2	0.8929	0.3640	1.2569	**2300.0**
2400.0	662.11	0.02790	0.11287	0.14076	719.0	384.8	1103.7	0.9031	0.3430	1.2460	**2400.0**
2500.0	668.11	0.02859	0.10209	0.13068	731.7	361.6	1093.3	0.9139	0.3206	1.2345	**2500.0**
2600.0	673.91	0.02938	0.09172	0.12110	744.5	337.6	1082.0	0.9247	0.2977	1.2225	**2600.0**
2700.0	679.53	0.03029	0.08165	0.11194	757.3	312.3	1069.7	0.9356	0.2741	1.2097	**2700.0**
2800.0	684.96	0.03134	0.07171	0.10305	770.7	285.1	1055.8	0.9468	0.2491	1.1958	**2800.0**
2900.0	690.22	0.03262	0.06158	0.09420	785.1	254.7	1039.8	0.9588	0.2215	1.1803	**2900.0**
3000.0	695.33	0.03428	0.05073	0.08500	801.8	218.4	1020.3	0.9728	0.1891	1.1619	**3000.0**
3100.0	700.28	0.03681	0.03771	0.07452	824.0	169.3	993.3	0.9914	0.1460	1.1373	**3100.0**
3200.0	705.08	0.04472	0.01191	0.05663	875.5	56.1	931.6	1.0351	0.0482	1.0832	**3200.0**
3208.2*	705.47	0.05078	0.00000	0.05078	906.0	0.0	906.0	1.0612	0.0000	1.0612	**3208.2***

Table 3. Superheated Steam

Abs Press. Lb/Sq In. (Sat. Temp)		Sat. Water	Sat. Steam	Temperature – Degrees Fahrenheit 200	250	300	350	400	450	500	600	700	800	900	1000	1100	1200
1 (101.74)	Sh			98.26	148.26	198.26	248.26	298.26	348.26	398.26	498.26	598.26	698.26	798.26	898.26	998.26	1098.26
	v	0.01614	333.6	392.5	422.4	452.3	482.1	511.9	541.7	571.5	631.1	690.7	750.2	809.8	869.4	929.1	988.7
	h	69.73	1105.8	1150.2	1172.9	1195.7	1218.7	1241.8	1265.1	1288.6	1336.1	1384.5	1431.0	1480.8	1531.4	1583.0	1635.4
	s	0.1326	1.9781	2.0509	2.0841	2.1152	2.1445	2.1722	2.1985	2.2237	2.2708	2.3144	2.3512	2.3892	2.4251	2.4592	2.4918
5 (162.24)	Sh			37.76	87.76	137.76	187.76	237.76	287.76	337.76	437.76	537.76	637.76	737.76	837.76	937.76	1037.76
	v	0.01641	73.53	78.14	84.21	90.24	96.25	102.24	108.23	114.21	126.15	138.08	150.01	161.94	173.86	185.78	197.70
	h	130.20	1131.1	1148.6	1171.7	1194.8	1218.0	1241.3	1264.7	1288.2	1335.9	1384.3	1433.6	1483.7	1534.7	1586.7	1639.6
	s	0.2349	1.8443	1.8716	1.9054	1.9369	1.9664	1.9943	2.0208	2.0460	2.0932	2.1369	2.1776	2.2159	2.2521	2.2866	2.3194
10 (193.21)	Sh			6.79	56.79	106.79	156.79	206.79	256.79	306.79	406.79	506.79	606.79	706.79	806.79	906.79	1006.79
	v	0.01659	38.42	38.84	41.93	44.98	48.02	51.03	54.04	57.04	63.03	69.00	74.98	80.94	86.91	92.87	98.84
	h	161.26	1143.3	1146.6	1170.2	1193.7	1217.1	1240.6	1264.1	1287.8	1335.5	1384.0	1433.4	1483.5	1534.6	1586.6	1639.5
	s	0.2836	1.7879	1.7928	1.8273	1.8593	1.8892	1.9173	1.9439	1.9692	2.0166	2.0603	2.1011	2.1394	2.1757	2.2101	2.2430
14.696 * (212.00)	Sh				38.00	88.00	138.00	188.00	238.00	288.00	388.00	488.00	588.00	688.00	788.00	888.00	988.00
	v	0.0167	26.828		28.44	30.52	32.61	34.65	36.73	38.75	42.83	46.91	50.97	55.03	59.09	63.19	67.25
	h	180.07	1150.4		1169.2	1192.0	1215.4	1238.9	1262.1	1285.4	1333.0	1381.4	1430.5	1480.4	1531.1	1582.7	1635.1
	s	0.3120	1.7566		1.7838	1.8148	1.8446	1.8727	1.8989	1.9238	1.9709	2.0145	2.0551	2.0932	2.1292	2.1634	2.1960
15 (213.03)	Sh				36.97	86.97	136.97	186.97	236.97	286.97	386.97	486.97	586.97	686.97	786.97	886.97	986.97
	v	0.01673	26.290		27.837	29.899	31.939	33.963	35.977	37.985	41.986	45.978	49.964	53.946	57.926	61.905	65.882
	h	181.21	1150.9		1168.7	1192.5	1216.2	1239.9	1263.6	1287.3	1335.2	1383.8	1433.2	1483.4	1534.5	1586.5	1639.4
	s	0.3137	1.7552		1.7809	1.8134	1.8437	1.8720	1.8988	1.9242	1.9717	2.0155	2.0563	2.0946	2.1309	2.1653	2.1982
20 (227.96)	Sh				22.04	72.04	122.04	172.04	222.04	272.04	372.04	472.04	572.04	672.04	772.04	872.04	972.04
	v	0.01683	20.087		20.788	22.356	23.900	25.428	26.946	28.457	31.466	34.465	37.458	40.447	43.435	46.420	49.405
	h	196.27	1156.3		1167.1	1191.4	1215.4	1239.2	1263.0	1286.9	1334.9	1383.5	1432.9	1483.2	1534.3	1586.3	1639.3
	s	0.3358	1.7320		1.7475	1.7805	1.8111	1.8397	1.8666	1.8921	1.9397	1.9836	2.0244	2.0628	2.0991	2.1336	2.1665
25 (240.07)	Sh				9.93	59.93	109.93	159.93	209.93	259.93	359.93	459.93	559.93	659.93	759.93	859.93	959.93
	v	0.01693	16.301		16.558	17.829	19.076	20.307	21.527	22.740	25.153	27.557	29.954	32.348	34.740	37.130	39.518
	h	208.52	1160.6		1165.6	1190.2	1214.5	1238.5	1262.5	1286.4	1334.6	1383.3	1432.7	1483.0	1534.2	1586.2	1639.2
	s	0.3535	1.7141		1.7212	1.7547	1.7856	1.8145	1.8415	1.8672	1.9149	1.9588	1.9997	2.0381	2.0744	2.1089	2.1418
30 (250.34)	Sh					49.66	99.66	149.66	199.66	249.66	349.66	449.66	549.66	649.66	749.66	849.66	949.66
	v	0.01701	13.744			14.810	15.859	16.892	17.914	18.929	20.945	22.951	24.952	26.949	28.943	30.936	32.927
	h	218.93	1164.1			1189.0	1213.6	1237.8	1261.9	1286.0	1334.2	1383.0	1432.5	1482.8	1534.0	1586.1	1639.0
	s	0.3682	1.6995			1.7334	1.7647	1.7937	1.8210	1.8467	1.8946	1.9386	1.9795	2.0179	2.0543	2.0888	2.1217
35 (259.29)	Sh					40.71	90.71	140.71	190.71	240.71	340.71	440.71	540.71	640.71	740.71	840.71	940.71
	v	0.01708	11.896			12.654	13.562	14.453	15.334	16.207	17.939	19.662	21.379	23.092	24.803	26.512	28.220
	h	228.03	1167.1			1187.8	1212.7	1237.1	1261.3	1285.5	1333.9	1382.8	1432.3	1482.7	1533.9	1586.0	1638.9
	s	0.3809	1.6872			1.7152	1.7468	1.7761	1.8035	1.8294	1.8774	1.9214	1.9624	2.0009	2.0372	2.0717	2.1046
40 (267.25)	Sh					32.75	82.75	132.75	182.75	232.75	332.75	432.75	532.75	632.75	732.75	832.75	932.75
	v	0.01715	10.497			11.036	11.838	12.624	13.398	14.165	15.685	17.195	18.699	20.199	21.697	23.194	24.689
	h	236.14	1169.8			1186.6	1211.7	1236.4	1260.8	1285.0	1333.6	1382.5	1432.1	1482.5	1533.7	1585.8	1638.8
	s	0.3921	1.6765			1.6992	1.7312	1.7608	1.7883	1.8143	1.8624	1.9065	1.9476	1.9860	2.0224	2.0569	2.0899
45 ** (274.43)	Sh					25.57	75.57	125.57	175.57	225.57	325.57	425.57	525.57	625.57	725.57	825.57	925.57
	v	0.01722	9.403			9.782	10.503	11.206	11.897	12.584	13.939	15.284	16.623	17.959	19.292	20.623	21.954
	h	243.47	1172.1			1185.4	1210.8	1235.7	1260.2	1284.6	1333.3	1382.3	1432.0	1482.4	1533.6	1585.7	1638.8
	s	0.4021	1.6671			1.6949	1.7174	1.7472	1.7749	1.8010	1.8492	1.8934	1.9345	1.9730	2.0094	2.0439	2.0769
50 (281.02)	Sh					18.98	68.98	118.98	168.98	218.98	318.98	418.98	518.98	618.98	718.98	818.98	918.98
	v	0.1727	8.514			8.769	9.424	10.062	10.688	11.306	12.529	13.741	14.947	16.150	17.350	18.549	19.746
	h	250.21	1174.1			1184.1	1209.9	1234.9	1259.6	1284.1	1332.9	1382.0	1431.7	1482.2	1533.4	1585.6	1638.6
	s	0.4112	1.6586			1.6720	1.7048	1.7349	1.7628	1.7890	1.8374	1.8816	1.9227	1.9613	1.9977	2.0322	2.0652
55 ** (287.07)	Sh					12.93	62.93	112.93	162.93	212.93	312.93	412.93	512.93	612.93	712.93	812.93	912.93
	v	0.01733	7.787			7.947	8.550	9.134	9.706	10.270	11.385	12.489	13.587	14.682	15.775	16.865	17.954
	h	256.42	1176.0			1182.9	1208.9	1234.3	1259.1	1283.6	1132.6	1381.8	1431.6	1482.0	1533.3	1585.5	1638.5
	s	0.4196	1.6510			1.6602	1.6934	1.7238	1.7518	1.7781	1.8267	1.8710	1.9123	1.9507	1.9871	2.0217	2.0546
60 (292.71)	Sh					7.29	57.29	107.29	157.29	207.29	307.29	407.29	507.29	607.29	707.29	807.29	907.29
	v	0.1738	7.174			7.257	7.815	8.354	8.881	9.400	10.425	11.438	12.446	13.450	14.452	15.452	16.450
	h	262.21	1177.6			1181.6	1208.0	1233.5	1258.5	1283.2	1332.3	1381.5	1431.3	1481.8	1533.2	1585.3	1638.4
	s	0.4273	1.6440			1.6492	1.6934	1.7134	1.7417	1.7681	1.8168	1.8612	1.9024	1.9410	1.9774	2.0120	2.0450
65 (297.98)	Sh					2.02	52.02	102.02	152.02	202.02	302.02	402.02	502.02	602.02	702.02	802.02	902.02
	v	0.01743	6.653			6.675	7.195	7.697	8.186	8.667	9.615	10.552	11.484	12.412	13.337	14.261	15.183
	h	267.63	1179.1			1180.3	1207.0	1232.7	1257.9	1282.7	1331.9	1381.3	1431.1	1481.6	1533.0	1585.2	1638.3
	s	0.4344	1.6375			1.6390	1.6731	1.7040	1.7324	1.7590	1.8077	1.8522	1.8935	1.9321	1.9685	2.0031	2.0361
70 (302.93)	Sh						47.07	97.07	147.07	197.07	297.07	397.07	497.07	597.07	697.07	797.07	897.07
	v	0.01748	6.205				6.664	7.133	7.590	8.039	8.922	9.793	10.659	11.522	12.382	13.240	14.097
	h	272.74	1180.6				1206.0	1232.0	1257.3	1282.2	1331.6	1381.0	1430.9	1481.5	1532.9	1585.1	1638.2
	s	0.4411	1.6316				1.6640	1.6951	1.7237	1.7504	1.7993	1.8439	1.8852	1.9238	1.9603	1.9949	2.0279
75 (307.61)	Sh						42.39	92.39	142.39	192.39	292.39	392.39	492.39	592.39	692.39	792.39	892.39
	v	0.01753	5.814				6.204	6.645	7.074	7.494	8.320	9.135	9.945	10.750	11.553	12.355	13.155
	h	277.56	1181.9				1205.0	1231.2	1256.7	1281.7	1331.3	1380.7	1430.7	1481.3	1532.7	1585.0	1638.1
	s	0.4474	1.6260				1.6554	1.6868	1.7156	1.7424	1.7915	1.8361	1.8774	1.9161	1.9526	1.9872	2.0202

Sh = superheat, F
v = specific volume, cu ft per lb
h = enthalpy, Btu per lb
s = entropy, Btu per F per lb

*Values from STEAM TABLES, Properties of Saturated and Superheated Steam Published by COMBUSTION ENGINEERING, INC., Copyright 1940
**Values interpolated from ASME STEAM TABLES

Table 3. Superheated Steam—*Continued*

Abs Press. Lb/Sq In. (Sat. Temp)		Sat. Water	Sat. Steam	350	400	450	500	550	600	700	800	900	1000	1100	1200	1300	1400
				Temperature – Degrees Fahrenheit													
80	Sh			37.96	87.96	137.96	187.96	237.96	287.96	387.96	487.96	587.96	687.96	787.96	887.96	987.96	1087.96
(312.04)	v	0.01757	5.471	5.801	6.218	6.622	7.018	7.408	7.794	8.560	9.319	10.075	10.829	11.581	12.331	13.081	13.829
	h	282.15	1183.1	1204.0	1230.5	1256.1	1281.3	1306.2	1330.9	1380.5	1430.5	1481.1	1532.6	1584.9	1638.0	1692.0	1746.8
	s	0.4534	1.6208	1.6473	1.6790	1.7080	1.7349	1.7602	1.7842	1.8289	1.8702	1.9089	1.9454	1.9800	2.0131	2.0446	2.0750
85	Sh			33.74	83.74	133.74	183.74	233.74	283.74	383.74	483.74	583.74	683.74	783.74	883.74	983.74	1083.74
(316.26)	v	0.01762	5.167	5.445	5.840	6.223	6.597	6.966	7.330	8.052	8.768	9.480	10.190	10.898	11.604	12.310	13.014
	h	286.52	1184.2	1203.0	1229.7	1255.5	1280.8	1305.8	1330.6	1380.2	1430.3	1481.0	1532.4	1584.7	1637.9	1691.9	1746.8
	s	0.4590	1.6159	1.6396	1.6716	1.7008	1.7279	1.7532	1.7772	1.8220	1.8634	1.9021	1.9386	1.9733	2.0063	2.0379	2.0682
90	Sh			29.72	79.72	129.72	179.72	229.72	279.72	379.72	479.72	579.72	679.72	779.72	879.72	979.72	1079.72
(320.28)	v	0.01766	4.895	5.128	5.505	5.869	6.223	6.572	6.917	7.600	8.277	8.950	9.621	10.290	10.958	11.625	12.290
	h	290.69	1185.3	1202.0	1228.9	1254.9	1280.3	1305.4	1330.2	1380.0	1430.1	1480.8	1532.3	1584.6	1637.8	1691.8	1746.7
	s	0.4643	1.6113	1.6323	1.6646	1.6940	1.7212	1.7467	1.7707	1.8156	1.8570	1.8957	1.9323	1.9669	2.0000	2.0316	2.0619
95	Sh			25.87	75.87	125.87	175.87	225.87	275.87	375.87	475.87	575.87	675.87	775.87	875.87	975.87	1075.87
(324.13)	v	0.01770	4.651	4.845	5.205	5.551	5.889	6.221	6.548	7.196	7.838	8.477	9.113	9.747	10.380	11.012	11.643
	h	294.70	1186.2	1200.9	1228.1	1254.3	1279.8	1305.0	1329.9	1379.7	1429.9	1480.6	1532.1	1584.5	1637.7	1691.7	1746.6
	s	0.4694	1.6069	1.6253	1.6580	1.6876	1.7149	1.7404	1.7645	1.8094	1.8509	1.8897	1.9262	1.9609	1.9940	2.0256	2.0559
100	Sh			22.18	72.18	122.18	172.18	222.18	272.18	372.18	472.18	572.18	672.18	772.18	872.18	972.18	1072.18
(327.82)	v	0.01774	4.431	4.590	4.935	5.266	5.588	5.904	6.216	6.833	7.443	8.050	8.655	9.258	9.860	10.460	11.060
	h	298.54	1187.2	1199.9	1227.4	1253.7	1279.3	1304.6	1329.6	1379.5	1429.7	1480.4	1532.0	1584.4	1637.6	1691.6	1746.5
	s	0.4743	1.6027	1.6187	1.6516	1.6814	1.7088	1.7344	1.7586	1.8036	1.8451	1.8839	1.9205	1.9552	1.9883	2.0199	2.0502
105	Sh			18.63	68.63	118.63	168.63	218.63	268.63	368.63	468.63	568.63	668.63	768.63	868.63	968.63	1068.63
(331.37)	v	0.01778	4.231	4.359	4.690	5.007	5.315	5.617	5.915	6.504	7.086	7.665	8.241	8.816	9.389	9.961	10.532
	h	302.24	1188.0	1198.8	1226.6	1253.1	1278.8	1304.2	1329.2	1379.2	1429.4	1480.3	1531.8	1584.2	1637.5	1691.5	1746.4
	s	0.4790	1.5988	1.6122	1.6455	1.6755	1.7031	1.7288	1.7530	1.7981	1.8396	1.8785	1.9151	1.9498	1.9828	2.0145	2.0448
110	Sh			15.21	65.21	115.21	165.21	215.21	265.21	365.21	465.21	565.21	665.21	765.21	865.21	965.21	1065.21
(334.79)	v	0.01782	4.048	4.149	4.468	4.772	5.068	5.357	5.642	6.205	6.761	7.314	7.865	8.413	8.961	9.507	10.053
	h	305.80	1188.9	1197.7	1225.8	1252.5	1278.3	1303.8	1328.9	1379.0	1429.2	1480.1	1531.7	1584.1	1637.4	1691.4	1746.4
	s	0.4834	1.5950	1.6061	1.6396	1.6698	1.6975	1.7233	1.7476	1.7928	1.8344	1.8732	1.9099	1.9446	1.9777	2.0093	2.0397
115	Sh			11.92	61.92	111.92	161.92	211.92	261.92	361.92	461.92	561.92	661.92	761.92	861.92	961.92	1061.92
(338.08)	v	0.01785	3.881	3.957	4.265	4.558	4.841	5.119	5.392	5.932	6.465	6.994	7.521	8.046	8.570	9.093	9.615
	h	309.25	1189.6	1196.7	1225.0	1251.8	1277.9	1303.3	1328.6	1378.7	1429.0	1479.9	1531.6	1584.0	1637.2	1691.4	1746.3
	s	0.4877	1.5913	1.6001	1.6340	1.6644	1.6922	1.7181	1.7425	1.7877	1.8294	1.8682	1.9049	1.9396	1.9727	2.0044	2.0347
120	Sh			8.73	58.73	108.73	158.73	208.73	258.73	358.73	458.73	558.73	658.73	758.73	858.73	958.73	1058.73
(341.27)	v	0.01789	3.7275	3.7815	4.0786	4.3610	4.6341	4.9009	5.1637	5.6813	6.1928	6.7006	7.2060	7.7096	8.2119	8.7130	9.2134
	h	312.58	1190.4	1195.6	1224.1	1251.2	1277.4	1302.9	1328.2	1378.4	1428.8	1479.8	1531.4	1583.9	1637.1	1691.3	1746.2
	s	0.4919	1.5879	1.5943	1.6286	1.6592	1.6872	1.7132	1.7376	1.7829	1.8246	1.8635	1.9001	1.9349	1.9680	1.9996	2.0300
130	Sh			2.67	52.67	102.67	152.67	202.67	252.67	352.67	452.67	552.67	652.67	752.67	852.67	952.67	1052.67
(347.33)	v	0.01796	3.4544	3.4699	3.7489	4.0129	4.2672	4.5151	4.7589	5.2384	5.7118	6.1814	6.6486	7.1140	7.5781	8.0411	8.5033
	h	318.95	1191.7	1193.4	1222.5	1249.9	1276.4	1302.1	1327.5	1377.9	1428.4	1479.4	1531.1	1583.6	1636.9	1691.1	1746.1
	s	0.4998	1.5813	1.5833	1.6182	1.6493	1.6775	1.7037	1.7283	1.7737	1.8155	1.8545	1.8911	1.9259	1.9591	1.9907	2.0211
140	Sh				46.96	96.96	146.96	196.96	246.96	346.96	446.96	546.96	646.96	746.96	846.96	946.96	1046.96
(353.04)	v	0.01803	3.2190		3.4661	3.7143	3.9526	4.1844	4.4119	4.8588	5.2995	5.7364	6.1709	6.6036	7.0349	7.4652	7.8946
	h	324.96	1193.0		1220.8	1248.7	1275.3	1301.3	1326.8	1377.4	1428.0	1479.1	1530.8	1583.4	1636.7	1690.9	1745.9
	s	0.5071	1.5752		1.6085	1.6400	1.6686	1.6949	1.7196	1.7652	1.8071	1.8461	1.8828	1.9176	1.9508	1.9825	2.0129
150	Sh				41.57	91.57	141.57	191.57	241.57	341.57	441.57	541.57	641.57	741.57	841.57	941.57	1041.57
(358.43)	v	0.01809	3.0139		3.2208	3.4555	3.6799	3.8978	4.1112	4.5298	4.9421	5.3507	5.7568	6.1612	6.5642	6.9661	7.3671
	h	330.65	1194.1		1219.1	1247.4	1274.3	1300.5	1326.1	1376.9	1427.6	1478.7	1530.5	1583.1	1636.5	1690.7	1745.7
	s	0.5141	1.5695		1.5993	1.6313	1.6602	1.6867	1.7115	1.7573	1.7992	1.8383	1.8751	1.9099	1.9431	1.9748	2.0052
160	Sh				36.45	86.45	136.45	186.45	236.45	336.45	436.45	536.45	636.45	736.45	836.45	936.45	1036.45
(363.55)	v	0.01815	2.8336		3.0060	3.2288	3.4413	3.6469	3.8480	4.2420	4.6295	5.0132	5.3945	5.7741	6.1522	6.5293	6.9055
	h	336.07	1195.1		1217.4	1246.0	1273.3	1299.6	1325.4	1376.4	1427.2	1478.4	1530.3	1582.9	1636.3	1690.5	1745.6
	s	0.5206	1.5641		1.5906	1.6231	1.6522	1.6790	1.7039	1.7499	1.7919	1.8310	1.8678	1.9027	1.9359	1.9676	1.9980
170	Sh				31.58	81.58	131.58	181.58	231.58	331.58	431.58	531.58	631.58	731.58	831.58	931.58	1031.58
(368.42)	v	0.01821	2.6738		2.8162	3.0288	3.2306	3.4255	3.6158	3.9879	4.3536	4.7155	5.0749	5.4325	5.7888	6.1440	6.4983
	h	341.24	1196.0		1215.6	1244.7	1272.2	1298.8	1324.7	1375.8	1426.8	1478.0	1530.0	1582.6	1636.1	1690.4	1745.4
	s	0.5269	1.5591		1.5823	1.6152	1.6447	1.6717	1.6968	1.7428	1.7850	1.8241	1.8610	1.8959	1.9291	1.9608	1.9913
180	Sh				26.92	76.92	126.92	176.92	226.92	326.92	426.92	526.92	626.92	726.92	826.92	926.92	1026.92
(373.08)	v	0.01827	2.5312		2.6474	2.8508	3.0433	3.2286	3.4093	3.7621	4.1084	4.4508	4.7907	5.1289	5.4657	5.8014	6.1363
	h	346.19	1196.9		1213.8	1243.4	1271.2	1297.9	1324.0	1375.3	1426.3	1477.7	1529.7	1582.4	1635.9	1690.2	1745.3
	s	0.5328	1.5543		1.5743	1.6078	1.6376	1.6647	1.6900	1.7362	1.7784	1.8176	1.8545	1.8894	1.9227	1.9545	1.9849
190	Sh				22.47	72.47	122.47	172.47	222.47	322.47	422.47	522.47	622.47	722.47	822.47	922.47	1022.47
(377.53)	v	0.01833	2.4030		2.4961	2.6915	2.8756	3.0525	3.2246	3.5601	3.8889	4.2140	4.5365	4.8572	5.1766	5.4949	5.8124
	h	350.94	1197.6		1212.0	1242.0	1270.1	1297.1	1323.3	1374.8	1425.9	1477.4	1529.4	1582.1	1635.7	1690.0	1745.1
	s	0.5384	1.5498		1.5667	1.6006	1.6307	1.6581	1.6835	1.7299	1.7722	1.8115	1.8484	1.8834	1.9166	1.9484	1.9789
200	Sh				18.20	68.20	118.20	168.20	218.20	318.20	418.20	518.20	618.20	718.20	818.20	918.20	1018.20
(381.80)	v	0.01839	2.2873		2.3598	2.5480	2.7247	2.8939	3.0583	3.3783	3.6915	4.0008	4.3077	4.6128	4.9165	5.2191	5.5209
	h	355.51	1198.3		1210.1	1240.6	1269.0	1296.2	1322.6	1374.3	1425.5	1477.0	1529.1	1581.9	1635.4	1689.8	1745.0
	s	0.5438	1.5454		1.5593	1.5938	1.6242	1.6518	1.6773	1.7239	1.7663	1.8057	1.8426	1.8776	1.9109	1.9427	1.9732

Sh = superheat, F
v = specific volume, cu ft per lb
h = enthalpy, Btu per lb
s = entropy, Btu per F per lb

Table 3. Superheated Steam—*Continued*

Abs Press. Lb/Sq In. (Sat. Temp)		Sat. Water	Sat. Steam	Temperature—Degrees Fahrenheit													
				400	450	500	550	600	700	800	900	1000	1100	1200	1300	1400	1500
	Sh			14.09	64.09	114.09	164.09	214.09	314.09	414.09	514.09	614.09	714.09	814.09	914.09	1014.09	1114.09
210	v	0.01844	2.1822	2.2364	2.4181	2.5880	2.7504	2.9078	3.2137	3.5128	3.8080	4.1007	4.3915	4.6811	4.9695	5.2571	5.5440
(385.91)	h	359.91	1199.0	1208.02	1239.2	1268.0	1295.3	1321.9	1373.7	1425.1	1476.7	1528.8	1581.6	1635.2	1689.6	1744.8	1800.8
	s	0.5490	1.5413	1.5522	1.5872	1.6180	1.6458	1.6715	1.7182	1.7607	1.8001	1.8371	1.8721	1.9054	1.9372	1.9677	1.9970
	Sh			10.12	60.12	110.12	160.12	210.12	310.12	410.12	510.12	610.12	710.12	810.12	910.12	1010.12	1110.12
220	v	0.01850	2.0863	2.1240	2.2999	2.4638	2.6199	2.7710	3.0642	3.3504	3.6327	3.9125	4.1905	4.4671	4.7426	5.0173	5.2913
(389.88)	h	364.17	1199.6	1206.3	1237.8	1266.9	1294.5	1321.2	1373.2	1424.7	1476.3	1528.5	1581.4	1635.0	1689.4	1744.7	1800.6
	s	0.5540	1.5374	1.5453	1.5808	1.6120	1.6400	1.6658	1.7128	1.7553	1.7948	1.8318	1.8668	1.9002	1.9320	1.9625	1.9919
	Sh			6.30	56.30	106.30	156.30	206.30	306.30	406.30	506.30	606.30	706.30	806.30	906.30	1006.30	1106.30
230	v	0.01855	1.9985	2.0212	2.1919	2.3503	2.5008	2.6461	2.9276	3.2020	3.4726	3.7406	4.0068	4.2717	4.5355	4.7984	5.0606
(393.70)	h	368.28	1200.1	1204.4	1236.3	1265.7	1293.6	1320.4	1372.7	1424.2	1476.0	1528.2	1581.1	1634.8	1689.3	1744.5	1800.5
	s	0.5588	1.5336	1.5385	1.5747	1.6062	1.6344	1.6604	1.7075	1.7502	1.7897	1.8268	1.8618	1.8952	1.9270	1.9576	1.9869
	Sh			2.61	52.61	102.61	152.61	202.61	302.61	402.61	502.61	602.61	702.61	802.61	902.61	1002.61	1102.61
240	v	0.01860	1.9177	1.9268	2.0928	2.2462	2.3915	2.5316	2.8024	3.0661	3.3259	3.5831	3.8385	4.0926	4.3456	4.5977	4.8492
(397.39)	h	372.27	1200.6	1202.4	1234.9	1264.6	1292.7	1319.7	1372.1	1423.8	1475.6	1527.9	1580.9	1634.6	1689.1	1744.3	1800.4
	s	0.5634	1.5299	1.5320	1.5687	1.6006	1.6291	1.6552	1.7025	1.7452	1.7848	1.8219	1.8570	1.8904	1.9223	1.9528	1.9822
	Sh				49.03	99.03	149.03	199.03	299.03	399.03	499.03	599.03	699.03	799.03	899.03	999.03	1099.03
250	v	0.01865	1.8432		2.0016	2.1504	2.2909	2.4262	2.6872	2.9410	3.1909	3.4382	3.6837	3.9278	4.1709	4.4131	4.6546
(400.97)	h	376.14	1201.1		1233.4	1263.5	1291.8	1319.0	1371.6	1423.4	1475.3	1527.6	1580.6	1634.4	1688.9	1744.2	1800.2
	s	0.5679	1.5264		1.5629	1.5951	1.6239	1.6502	1.6976	1.7405	1.7801	1.8173	1.8524	1.8858	1.9177	1.9482	1.9776
	Sh				45.56	95.56	145.56	195.56	295.56	395.56	495.56	595.56	695.56	795.56	895.56	995.56	1095.56
260	v	0.01870	1.7742		1.9173	2.0619	2.1981	2.3289	2.5808	2.8256	3.0663	3.3044	3.5408	3.7758	4.0097	4.2427	4.4750
(404.44)	h	379.90	1201.5		1231.9	1262.4	1290.9	1318.2	1371.1	1423.0	1474.9	1527.3	1580.4	1634.2	1688.7	1744.0	1800.1
	s	0.5722	1.5230		1.5573	1.5899	1.6189	1.6453	1.6930	1.7359	1.7756	1.8128	1.8480	1.8814	1.9133	1.9439	1.9732
	Sh				42.20	92.20	142.20	192.20	292.20	392.20	492.20	592.20	692.20	792.20	892.20	992.20	1092.20
270	v	0.01875	1.7101		1.8391	1.9799	2.1121	2.2388	2.4824	2.7186	2.9509	3.1806	3.4084	3.6349	3.8603	4.0849	4.3087
(407.80)	h	383.56	1201.9		1230.4	1261.2	1290.0	1317.5	1370.5	1422.6	1474.6	1527.1	1580.1	1634.0	1688.5	1743.9	1800.0
	s	0.5764	1.5197		1.5518	1.5848	1.6140	1.6406	1.6885	1.7315	1.7713	1.8085	1.8437	1.8771	1.9090	1.9396	1.9690
	Sh				38.93	88.93	138.93	188.93	288.93	388.93	488.93	588.93	688.93	788.93	888.93	988.93	1088.93
280	v	0.01880	1.6505		1.7665	1.9037	2.0322	2.1551	2.3909	2.6194	2.8437	3.0655	3.2855	3.5042	3.7217	3.9384	4.1543
(411.07)	h	387.12	1202.3		1228.8	1260.0	1289.1	1316.8	1370.0	1422.1	1474.2	1526.8	1579.9	1633.8	1688.4	1743.7	1799.8
	s	0.5805	1.5166		1.5464	1.5798	1.6093	1.6361	1.6841	1.7273	1.7671	1.8043	1.8395	1.8730	1.9050	1.9356	1.9649
	Sh				35.75	85.75	135.75	185.75	285.75	385.75	485.75	585.75	685.75	785.75	885.75	985.75	1085.75
290	v	0.01885	1.5948		1.6988	1.8327	1.9578	2.0772	2.3058	2.5269	2.7440	2.9585	3.1711	3.3824	3.5926	3.8019	4.0106
(414.25)	h	390.60	1202.6		1227.3	1258.9	1288.1	1316.0	1369.5	1421.7	1473.9	1526.5	1579.6	1633.5	1688.2	1743.6	1799.7
	s	0.5844	1.5135		1.5412	1.5750	1.6048	1.6317	1.6799	1.7232	1.7630	1.8003	1.8356	1.8690	1.9010	1.9316	1.9610
	Sh				32.65	82.65	132.65	182.65	282.65	382.65	482.65	582.65	682.65	782.65	882.65	982.65	1082.65
300	v	0.01889	1.5427		1.6356	1.7665	1.8883	2.0044	2.2263	2.4407	2.6509	2.8585	3.0643	3.2688	3.4721	3.6746	3.8764
(417.35)	h	393.99	1202.9		1225.7	1257.7	1287.2	1315.2	1368.9	1421.3	1473.6	1526.2	1579.4	1633.3	1688.0	1743.4	1799.6
	s	0.5882	1.5105		1.5361	1.5703	1.6003	1.6274	1.6758	1.7192	1.7591	1.7964	1.8317	1.8652	1.8972	1.9278	1.9572
	Sh				29.64	79.64	129.64	179.64	279.64	379.64	479.64	579.64	679.64	779.64	879.64	979.64	1079.64
310	v	0.01894	1.4939		1.5763	1.7044	1.8233	1.9363	2.1520	2.3600	2.5638	2.7650	2.9644	3.1625	3.3594	3.5555	3.7509
(420.36)	h	397.30	1203.2		1224.1	1256.5	1286.3	1314.5	1368.4	1420.9	1473.2	1525.9	1579.2	1633.1	1687.8	1743.3	1799.4
	s	0.5920	1.5076		1.5311	1.5657	1.5960	1.6233	1.6719	1.7153	1.7553	1.7927	1.8280	1.8615	1.8935	1.9241	1.9536
	Sh				26.69	76.69	126.69	176.69	276.69	376.69	476.69	576.69	676.69	776.69	876.69	976.69	1076.69
320	v	0.01899	1.4480		1.5207	1.6462	1.7623	1.8725	2.0823	2.2843	2.4821	2.6774	2.8708	3.0628	3.2538	3.4438	3.6332
(423.31)	h	400.53	1203.4		1222.5	1255.2	1285.3	1313.7	1367.8	1420.5	1472.9	1525.6	1578.9	1632.9	1687.6	1743.1	1799.3
	s	0.5956	1.5048		1.5261	1.5612	1.5918	1.6192	1.6680	1.7116	1.7516	1.7890	1.8243	1.8579	1.8899	1.9206	1.9500
	Sh				23.82	73.82	123.82	173.82	273.82	373.82	473.82	573.82	673.82	773.82	873.82	973.82	1073.82
330	v	0.01903	1.4048		1.4684	1.5915	1.7050	1.8125	2.0168	2.2132	2.4054	2.5950	2.7828	2.9692	3.1545	3.3389	3.5227
(426.18)	h	403.70	1203.6		1220.9	1254.0	1284.4	1313.0	1367.3	1420.0	1472.5	1525.3	1578.7	1632.7	1687.5	1742.9	1799.2
	s	0.5991	1.5021		1.5213	1.5568	1.5876	1.6153	1.6643	1.7079	1.7480	1.7855	1.8208	1.8544	1.8864	1.9171	1.9466
	Sh				21.01	71.01	121.01	171.01	271.01	371.01	471.01	571.01	671.01	771.01	871.01	971.01	1071.01
340	v	0.01908	1.3640		1.4191	1.5399	1.6511	1.7561	1.9552	2.1463	2.3333	2.5175	2.7000	2.8811	3.0611	3.2402	3.4186
(428.99)	h	406.80	1203.8		1219.2	1252.8	1283.4	1312.2	1366.7	1419.6	1472.2	1525.0	1578.4	1632.5	1687.3	1742.8	1799.0
	s	0.6026	1.4994		1.5165	1.5525	1.5836	1.6114	1.6606	1.7044	1.7445	1.7820	1.8174	1.8510	1.8831	1.9138	1.9432
	Sh				18.27	68.27	118.27	168.27	268.27	368.27	468.27	568.27	668.27	768.27	868.27	968.27	1068.27
350	v	0.01912	1.3255		1.3725	1.4913	1.6002	1.7028	1.8970	2.0832	2.2652	2.4445	2.6219	2.7980	2.9730	3.1471	3.3205
(431.73)	h	409.83	1204.0		1217.5	1251.5	1282.4	1311.4	1366.2	1419.2	1471.8	1524.7	1578.2	1632.3	1687.1	1742.6	1798.9
	s	0.6059	1.4968		1.5119	1.5483	1.5797	1.6077	1.6571	1.7009	1.7411	1.7787	1.8141	1.8477	1.8798	1.9105	1.9400
	Sh				15.59	65.59	115.59	165.59	265.59	365.59	465.59	565.59	665.59	765.59	865.59	965.59	1065.59
360	v	0.01917	1.2891		1.3285	1.4454	1.5521	1.6525	1.8421	2.0237	2.2009	2.3755	2.5482	2.7196	2.8898	3.0592	3.2279
(434.41)	h	412.81	1204.1		1215.8	1250.3	1281.5	1310.6	1365.6	1418.7	1471.5	1524.4	1577.9	1632.1	1686.9	1742.5	1798.8
	s	0.6092	1.4943		1.5073	1.5441	1.5758	1.6040	1.6536	1.6976	1.7379	1.7754	1.8109	1.8445	1.8766	1.9073	1.9368
	Sh				10.39	60.39	110.39	160.39	260.39	360.39	460.39	560.39	660.39	760.39	860.39	960.39	1060.39
380	v	0.01925	1.2218		1.2472	1.3606	1.4635	1.5598	1.7410	1.9139	2.0825	2.2484	2.4124	2.5750	2.7366	2.8973	3.0572
(439.61)	h	418.59	1204.4		1212.4	1247.7	1279.5	1309.0	1364.5	1417.9	1470.8	1523.8	1577.4	1631.6	1686.5	1742.2	1798.5
	s	0.6156	1.4894		1.4982	1.5360	1.5683	1.5969	1.6470	1.6911	1.7315	1.7692	1.8047	1.8384	1.8705	1.9012	1.9307

Sh = superheat, F
v = specific volume, cu ft per lb
h = enthalpy, Btu per lb
s = entropy, Btu per F per lb

Table 3. Superheated Steam—*Continued*

Abs Press. Lb/Sq In. (Sat. Temp)		Sat. Water	Sat. Steam	Temperature—Degrees Fahrenheit 450	500	550	600	650	700	800	900	1000	1100	1200	1300	1400	1500
	Sh			5.40	55.40	105.40	155.40	205.40	255.40	355.40	455.40	555.40	655.40	755.40	855.40	955.40	1055.40
400	v	0.01934	116.10	1.1738	1.2841	1.3836	1.4763	1.5646	1.6499	1.8151	1.9759	2.1339	2.2901	2.4450	2.5987	2.7515	2.9037
(444.60)	h	424.17	1204.6	1208.8	1245.1	1277.5	1307.4	1335.9	1363.4	1417.0	1470.1	1523.3	1576.9	1631.2	1686.2	1741.9	1798.2
	s	0.6217	1.4847	1.4894	1.5282	1.5611	1.5901	1.6163	1.6406	1.6850	1.7255	1.7632	1.7988	1.8325	1.8647	1.8955	1.9250
	Sh			.60	50.60	100.60	150.60	200.60	250.60	350.60	450.60	550.60	650.60	750.60	850.60	950.60	1050.60
420	v	0.01942	1.1057	1.1071	1.2148	1.3113	1.4007	1.4856	1.5676	1.7258	1.8795	2.0304	2.1795	2.3273	2.4739	2.6196	2.7647
(449.40)	h	429.56	1204.7	1205.2	1242.4	1275.4	1305.8	1334.5	1362.3	1416.2	1469.4	1522.7	1576.4	1630.8	1685.8	1741.6	1798.0
	s	0.6276	1.4802	1.4808	1.5206	1.5542	1.5835	1.6100	1.6345	1.6791	1.7197	1.7575	1.7932	1.8269	1.8591	1.8899	1.9195
	Sh				45.97	95.97	145.97	195.97	245.97	345.97	445.97	545.97	645.97	745.97	845.97	945.97	1045.97
440	v	0.01950	1.0554		1.1517	1.2454	1.3319	1.4138	1.4926	1.6445	1.7918	1.9363	2.0790	2.2203	2.3605	2.4998	2.6384
(454.03)	h	434.77	1204.8		1239.7	1273.4	1304.2	1333.2	1361.1	1415.3	1468.7	1522.1	1575.9	1630.4	1685.5	1741.2	1797.7
	s	0.6332	1.4759		1.5132	1.5474	1.5772	1.6040	1.6286	1.6734	1.7142	1.7521	1.7878	1.8216	1.8538	1.8847	1.9143
	Sh				41.50	91.50	141.50	191.50	241.50	341.50	441.50	541.50	641.50	741.50	841.50	941.50	1041.50
460	v	0.01959	1.0092		1.0939	1.1852	1.2691	1.3482	1.4242	1.5703	1.7117	1.8504	1.9872	2.1226	2.2569	2.3903	2.5230
(458.50)	h	439.83	1204.8		1236.9	1271.3	1302.5	1331.8	1360.0	1414.4	1468.0	1521.5	1575.4	1629.9	1685.1	1740.9	1797.4
	s	0.6387	1.4718		1.5060	1.5409	1.5711	1.5982	1.6230	1.6680	1.7089	1.7469	1.7826	1.8165	1.8488	1.8797	1.9093
	Sh				37.18	87.18	137.18	187.18	237.18	337.18	437.18	537.18	637.18	737.18	837.18	937.18	1037.18
480	v	0.01967	0.9668		1.0409	1.1300	1.2115	1.2881	1.3615	1.5023	1.6384	1.7716	1.9030	2.0330	2.1619	2.2900	2.4173
(462.82)	h	444.75	1204.8		1234.1	1269.1	1300.8	1330.5	1358.8	1413.6	1467.3	1520.9	1574.9	1629.5	1684.7	1740.6	1797.2
	s	0.6439	1.4677		1.4990	1.5346	1.5652	1.5925	1.6176	1.6628	1.7038	1.7419	1.7777	1.8116	1.8439	1.8748	1.9045
	Sh				32.99	82.99	132.99	182.99	232.99	332.99	432.99	532.99	632.99	732.99	832.99	932.99	1032.99
500	v	0.01975	0.9276		0.9919	1.0791	1.1584	1.2327	1.3037	1.4397	1.5708	1.6992	1.8256	1.9507	2.0746	2.1977	2.3200
(467.01)	h	449.52	1204.7		1231.2	1267.0	1299.1	1329.1	1357.7	1412.7	1466.6	1520.3	1574.4	1629.1	1684.4	1740.3	1796.9
	s	0.6490	1.4639		1.4921	1.5284	1.5595	1.5871	1.6123	1.6578	1.6990	1.7371	1.7730	1.8069	1.8393	1.8702	1.8998
	Sh				28.93	78.93	128.93	178.93	228.93	328.93	428.93	528.93	628.93	728.93	828.93	928.93	1028.93
520	v	0.01982	0.8914		0.9466	1.0321	1.1094	1.1816	1.2504	1.3819	1.5085	1.6323	1.7542	1.8746	1.9940	2.1125	2.2302
(471.07)	h	454.18	1204.5		1228.3	1264.8	1297.4	1327.7	1356.5	1411.8	1465.9	1519.7	1573.9	1628.7	1684.0	1740.0	1796.7
	s	0.6540	1.4601		1.4853	1.5223	1.5539	1.5818	1.6072	1.6530	1.6943	1.7325	1.7684	1.8024	1.8348	1.8657	1.8954
	Sh				24.99	74.99	124.99	174.99	224.99	324.99	424.99	524.99	624.99	724.99	824.99	924.99	1024.99
540	v	0.01990	0.8577		0.9045	0.9884	1.0640	1.1342	1.2010	1.3284	1.4508	1.5704	1.6880	1.8042	1.9193	2.0336	2.1471
(475.01)	h	458.71	1204.4		1225.3	1262.5	1295.7	1326.3	1355.3	1410.9	1465.1	1519.1	1573.4	1628.2	1683.6	1739.7	1796.4
	s	0.6587	1.4565		1.4786	1.5164	1.5485	1.5767	1.6023	1.6483	1.6897	1.7280	1.7640	1.7981	1.8305	1.8615	1.8911
	Sh				21.16	71.16	121.16	171.16	221.16	321.16	421.16	521.16	621.16	721.16	821.16	921.16	1021.16
560	v	0.01998	0.8264		0.8653	0.9479	1.0217	1.0902	1.1552	1.2787	1.3972	1.5129	1.6266	1.7388	1.8500	1.9603	2.0699
(478.84)	h	463.14	1204.2		1222.2	1260.3	1293.9	1324.9	1354.2	1410.0	1464.4	1518.6	1572.9	1627.8	1683.3	1739.4	1796.1
	s	0.6634	1.4529		1.4720	1.5106	1.5431	1.5717	1.5975	1.6438	1.6853	1.7237	1.7598	1.7939	1.8263	1.8573	1.8870
	Sh				17.43	67.43	117.43	167.43	217.43	317.43	417.43	517.43	617.43	717.43	817.43	917.43	1017.43
580	v	0.02006	0.7971		0.8287	0.9100	0.9824	1.0492	1.1125	1.2324	1.3473	1.4593	1.5693	1.6780	1.7855	1.8921	1.9980
(482.57)	h	467.47	1203.9		1219.1	1258.0	1292.1	1323.4	1353.0	1409.2	1463.7	1518.0	1572.4	1627.4	1682.9	1739.1	1795.9
	s	0.6679	1.4495		1.4654	1.5049	1.5380	1.5668	1.5929	1.6394	1.6811	1.7196	1.7556	1.7898	1.8223	1.8533	1.8831
	Sh				13.80	63.80	113.80	163.80	213.80	313.80	413.80	513.80	613.80	713.80	813.80	913.80	1013.80
600	v	0.02013	0.7697		0.7944	0.8746	0.9456	1.0109	1.0726	1.1892	1.3008	1.4093	1.5160	1.6211	1.7252	1.8284	1.9309
(486.20)	h	471.70	1203.7		1215.9	1255.6	1290.3	1322.0	1351.8	1408.3	1463.0	1517.4	1571.9	1627.0	1682.6	1738.8	1795.6
	s	0.6723	1.4461		1.4590	1.4993	1.5329	1.5621	1.5884	1.6351	1.6769	1.7155	1.7517	1.7859	1.8184	1.8494	1.8792
	Sh				5.11	55.11	105.11	155.11	205.11	305.11	405.11	505.11	605.11	705.11	805.11	905.11	1005.11
650	v	0.02032	0.7084		0.7173	0.7954	0.8634	0.9254	0.9835	1.0929	1.1969	1.2979	1.3969	1.4944	1.5909	1.6864	1.7813
(494.89)	h	481.89	1202.8		1207.6	1249.6	1285.7	1318.3	1348.7	1406.0	1461.2	1515.9	1570.7	1625.9	1681.6	1738.0	1794.9
	s	1.6828	1.4381		1.4430	1.4858	1.5207	1.5507	1.5775	1.6249	1.6671	1.7059	1.7422	1.7765	1.8092	1.8403	1.8701
	Sh					46.92	96.92	146.92	196.92	296.92	396.92	496.92	596.92	696.92	796.92	896.92	996.92
700	v	0.02050	0.6556			0.7271	0.7928	0.8520	0.9072	1.0102	1.1078	1.2023	1.2948	1.3858	1.4757	1.5647	1.6530
(503.08)	h	491.60	1201.8			1243.4	1281.0	1314.6	1345.6	1403.7	1459.4	1514.4	1569.4	1624.8	1680.7	1737.2	1794.3
	s	0.6928	1.4304			1.4726	1.5090	1.5399	1.5673	1.6154	1.6580	1.6970	1.7335	1.7679	1.8006	1.8318	1.8617
	Sh					39.16	89.16	139.16	189.16	289.16	389.16	489.16	589.16	689.16	789.16	889.16	989.16
750	v	0.02069	0.6095			0.6676	0.7313	0.7882	0.8409	0.9386	1.0306	1.1195	1.2063	1.2916	1.3759	1.4592	1.5419
(510.84)	h	500.89	1200.7			1236.9	1276.1	1310.7	1342.5	1401.5	1457.6	1512.9	1568.2	1623.8	1679.8	1736.4	1793.6
	s	0.7022	1.4232			1.4598	1.4977	1.5296	1.5577	1.6065	1.6494	1.6886	1.7252	1.7598	1.7926	1.8239	1.8538
	Sh					31.79	81.79	131.79	181.79	281.79	381.79	481.79	581.79	681.79	781.79	881.79	981.79
800	v	0.02087	0.5690			0.6151	0.6774	0.7323	0.7828	0.8759	0.9631	1.0470	1.1289	1.2093	1.2885	1.3669	1.4446
(518.21)	h	509.81	1199.4			1230.1	1271.1	1306.8	1339.3	1399.1	1455.8	1511.4	1566.9	1622.7	1678.9	1735.7	1792.9
	s	0.7111	1.4163			1.4472	1.4869	1.5198	1.5484	1.5980	1.6413	1.6807	1.7175	1.7522	1.7851	1.8164	1.8464
	Sh					24.76	74.76	124.76	174.76	274.76	374.76	474.76	574.76	674.76	774.76	874.76	974.76
850	v	0.02105	0.5330			0.5683	0.6296	0.6829	0.7315	0.8205	0.9034	0.9830	1.0606	1.1366	1.2115	1.2855	1.3588
(525.24)	h	518.40	1198.0			1223.0	1265.9	1302.8	1336.0	1396.8	1454.0	1510.0	1565.7	1621.6	1678.0	1734.9	1792.3
	s	0.7197	1.4096			1.4347	1.4763	1.5102	1.5396	1.5899	1.6336	1.6733	1.7102	1.7450	1.7780	1.8094	1.8395
	Sh					18.05	68.05	118.05	168.05	268.05	368.05	468.05	568.05	668.05	768.05	868.05	968.05
900	v	0.02123	0.5009			0.5263	0.5869	0.6388	0.6858	0.7713	0.8504	0.9262	0.9998	1.0720	1.1430	1.2131	1.2825
(531.95)	h	526.70	1196.4			1215.5	1260.6	1298.6	1332.7	1394.4	1452.2	1508.5	1564.4	1620.6	1677.1	1734.1	1791.6
	s	0.7279	1.4032			1.4223	1.4659	1.5010	1.5311	1.5822	1.6263	1.6662	1.7033	1.7382	1.7713	1.8028	1.8329

Sh = superheat, F
v = specific volume, cu ft per lb
h = enthalpy, Btu per lb
s = entropy, Btu per F per lb

Table 3. Superheated Steam – *Continued*

Abs Press Lb/Sq In. (Sat. Temp)		Sat Water	Sat Steam	Temperature – Degrees Fahrenheit 550	600	650	700	750	800	850	900	1000	1100	1200	1300	1400	150
	Sh			11.61	61.61	111.61	161.61	211.61	261.61	311.61	361.61	461.61	561.61	661.61	761.61	861.61	961.
950	v	0.02141	0.4721	0.4883	0.5485	0.5993	0.6449	0.6871	0.7272	0.7656	0.8030	0.8753	0.9455	1.0142	1.0817	1.1484	1.21
(538.39)	h	534.74	1194.7	1207.6	1255.1	1294.4	1329.3	1361.5	1392.0	1421.5	1450.3	1507.0	1563.2	1619.5	1676.2	1733.3	179
	s	0.7358	1.3970	1.4098	1.4557	1.4921	1.5228	1.5500	1.5748	1.5977	1.6193	1.6595	1.6967	1.7317	1.7649	1.7965	1.82
	Sh			5.42	55.42	105.42	155.42	205.42	255.42	305.42	355.42	455.42	555.42	655.42	755.42	855.42	955.
1000	v	0.02159	0.4460	0.4535	0.5137	0.5636	0.6080	0.6489	0.6875	0.7245	0.7603	0.8295	0.8966	0.9622	1.0266	1.0901	1.15
(544.58)	h	542.55	1192.9	1199.3	1249.3	1290.1	1325.9	1358.7	1389.6	1419.4	1448.5	1505.4	1561.9	1618.4	1675.3	1732.5	179
	s	0.7434	1.3910	1.3973	1.4457	1.4833	1.5149	1.5426	1.5677	1.5908	1.6126	1.6530	1.6905	1.7256	1.7589	1.7905	1.82
	Sh				49.47	99.47	149.47	199.47	249.47	299.47	349.47	449.47	549.47	649.47	749.47	849.47	949
1050	v	0.02177	0.4222		0.4821	0.5312	0.5745	0.6142	0.6515	0.6872	0.7216	0.7881	0.8524	0.9151	0.9767	1.0373	1.09
(550.53)	h	550.15	1191.0		1243.4	1285.7	1322.4	1355.8	1387.2	1417.3	1446.6	1503.9	1560.7	1617.4	1674.4	1731.8	178
	s	0.7507	1.3851		1.4358	1.4748	1.5072	1.5354	1.5608	1.5842	1.6062	1.6469	1.6845	1.7197	1.7531	1.7848	1.81
	Sh				43.72	93.72	143.72	193.72	243.72	293.72	343.72	443.72	543.72	643.72	743.72	843.72	943.
1100	v	0.02195	0.4006		0.4531	0.5017	0.5440	0.5826	0.6188	0.6533	0.6865	0.7505	0.8121	0.8723	0.9313	0.9894	1.04
(556.28)	h	557.55	1189.1		1237.3	1281.2	1318.8	1352.9	1384.7	1415.2	1444.7	1502.4	1559.4	1616.3	1673.5	1731.0	178
	s	0.7578	1.3794		1.4259	1.4664	1.4996	1.5284	1.5542	1.5779	1.6000	1.6410	1.6787	1.7141	1.7475	1.7793	1.80
	Sh				39.18	89.18	139.18	189.18	239.18	289.18	339.18	439.18	539.18	639.18	739.18	839.18	939.
1150	v	0.02214	0.3807		0.4263	0.4746	0.5162	0.5538	0.5889	0.6223	0.6544	0.7161	0.7754	0.8332	0.8899	0.9456	1.00
(561.82)	h	564.78	1187.0		1230.9	1276.6	1315.2	1349.9	1382.2	1413.0	1442.8	1500.9	1558.1	1615.2	1672.6	1730.2	178
	s	0.7647	1.3738		1.4160	1.4582	1.4923	1.5216	1.5478	1.5717	1.5941	1.6353	1.6732	1.7087	1.7422	1.7741	1.80
	Sh				32.81	82.81	132.81	182.81	232.81	282.81	332.81	432.81	532.81	632.81	732.81	832.81	932.
1200	v	0.02232	0.3624		0.4016	0.4497	0.4905	0.5273	0.5615	0.5939	0.6250	0.6845	0.7418	0.7974	0.8519	0.9055	0.95
(567.19)	h	571.85	1184.8		1224.2	1271.8	1311.5	1346.9	1379.7	1410.8	1440.9	1499.4	1556.9	1614.2	1671.6	1729.4	178
	s	0.7714	1.3683		1.4061	1.4501	1.4851	1.5150	1.5415	1.5658	1.5883	1.6298	1.6679	1.7035	1.7371	1.7691	1.79
	Sh				22.58	72.58	122.58	172.58	222.58	272.58	322.58	422.58	522.58	622.58	722.58	822.58	922
1300	v	0.02269	0.3299		0.3570	0.4052	0.4451	0.4804	0.5129	0.5436	0.5729	0.6287	0.6822	0.7341	0.7847	0.8345	0.88
(577.42)	h	585.58	1180.2		1209.9	1261.9	1303.9	1340.8	1374.6	1406.4	1437.1	1496.3	1554.3	1612.0	1669.8	1727.9	178
	s	0.7843	1.3577		1.3860	1.4340	1.4711	1.5022	1.5296	1.5544	1.5773	1.6194	1.6578	1.6937	1.7275	1.7596	1.79
	Sh				12.93	62.93	112.93	162.93	212.93	262.93	312.93	412.93	512.93	612.93	712.93	812.93	912
1400	v	0.02307	0.3018		0.3176	0.3667	0.4059	0.4400	0.4712	0.5004	0.5282	0.5809	0.6311	0.6798	0.7272	0.7737	0.81
(587.07)	h	598.83	1175.3		1194.1	1251.4	1296.1	1334.5	1369.3	1402.0	1433.2	1493.2	1551.8	1609.9	1668.0	1726.3	178
	s	0.7966	1.3474		1.3652	1.4181	1.4575	1.4900	1.5182	1.5436	1.5670	1.6096	1.6484	1.6845	1.7185	1.7508	1.78
	Sh				3.80	53.80	103.80	153.80	203.80	253.80	303.80	403.80	503.80	603.80	703.80	803.80	903
1500	v	0.02346	0.2772		0.2820	0.3328	0.3717	0.4049	0.4350	0.4629	0.4894	0.5394	0.5869	0.6327	0.6773	0.7210	0.76
(596.20)	h	611.68	1170.1		1176.3	1240.2	1287.9	1328.0	1364.0	1397.4	1429.2	1490.1	1549.2	1607.7	1666.2	1724.8	178
	s	0.8085	1.3373		1.3431	1.4022	1.4443	1.4782	1.5073	1.5333	1.5572	1.6004	1.6395	1.6759	1.7101	1.7425	1.77
	Sh					45.13	95.13	145.13	195.13	245.13	295.13	395.13	495.13	595.13	695.13	795.13	895
1600	v	0.02387	0.2555			0.3026	0.3415	0.3741	0.4032	0.4301	0.4555	0.5031	0.5482	0.5915	0.6336	0.6748	0.71
(604.87)	h	624.20	1164.5			1228.3	1279.4	1321.4	1358.5	1392.8	1425.2	1486.9	1546.6	1605.6	1664.3	1723.2	178
	s	0.8199	1.3274			1.3861	1.4312	1.4667	1.4968	1.5235	1.5478	1.5916	1.6312	1.6678	1.7022	1.7347	1.76
	Sh					36.87	86.87	136.87	186.87	236.87	286.87	386.87	486.87	586.87	686.87	786.87	886
1700	v	0.02428	0.2361			0.2754	0.3147	0.3468	0.3751	0.4011	0.4255	0.4711	0.5140	0.5552	0.5951	0.6341	0.67
(613.13)	h	636.45	1158.6			1215.3	1270.5	1314.5	1352.9	1388.1	1421.2	1483.8	1544.0	1603.4	1662.5	1721.7	178
	s	0.8309	1.3176			1.3697	1.4183	1.4555	1.4867	1.5140	1.5388	1.5833	1.6232	1.6601	1.6947	1.7274	1.75
	Sh					28.98	78.98	128.98	178.98	228.98	278.98	378.98	478.98	578.98	678.98	778.98	878
1800	v	0.02472	0.2186			0.2505	0.2906	0.3223	0.3500	0.3752	0.3988	0.4426	0.4836	0.5229	0.5609	0.5980	0.63
(621.02)	h	648.49	1152.3			1201.2	1261.1	1307.4	1347.2	1383.3	1417.1	1480.6	1541.4	1601.2	1660.7	1720.1	177
	s	0.8417	1.3079			1.3526	1.4054	1.4446	1.4768	1.5049	1.5302	1.5753	1.6156	1.6528	1.6876	1.7204	1.75
	Sh					21.44	71.44	121.44	171.44	221.44	271.44	371.44	471.44	571.44	671.44	771.44	871
1900	v	0.02517	0.2028			0.2274	0.2687	0.3004	0.3275	0.3521	0.3749	0.4171	0.4565	0.4940	0.5303	0.5656	0.60
(628.56)	h	660.36	1145.6			1185.7	1251.3	1300.2	1341.4	1378.4	1412.9	1477.4	1538.8	1599.1	1658.8	1718.6	177
	s	0.8522	1.2981			1.3346	1.3925	1.4338	1.4672	1.4960	1.5219	1.5677	1.6084	1.6458	1.6808	1.7138	1.74
	Sh					14.20	64.20	114.20	164.20	214.20	264.20	364.20	464.20	564.20	664.20	764.20	864
2000	v	0.02565	0.1883			0.2056	0.2488	0.2805	0.3072	0.3312	0.3534	0.3942	0.4320	0.4680	0.5027	0.5365	0.56
(635.80)	h	672.11	1138.3			1168.3	1240.9	1292.6	1335.4	1373.5	1408.7	1474.1	1536.2	1596.9	1657.0	1717.0	177
	s	0.8625	1.2881			1.3154	1.3794	1.4231	1.4578	1.4874	1.5138	1.5603	1.6014	1.6391	1.6743	1.7075	1.73
	Sh					7.24	57.24	107.24	157.24	207.24	257.24	357.24	457.24	557.24	657.24	757.24	857
2100	v	0.02615	0.1750			0.1847	0.2304	0.2624	0.2888	0.3123	0.3339	0.3734	0.4099	0.4445	0.4778	0.5101	0.54
(642.76)	h	683.79	1130.5			1148.5	1229.8	1284.9	1329.3	1368.4	1404.4	1470.9	1533.6	1594.7	1655.2	1715.4	177
	s	0.8727	1.2780			1.2942	1.3661	1.4125	1.4486	1.4790	1.5060	1.5532	1.5948	1.6327	1.6681	1.7014	1.73
	Sh					55	50.55	100.55	150.55	200.55	250.55	350.55	450.55	550.55	650.55	750.55	850
2200	v	0.02669	0.1627			0.1636	0.2134	0.2458	0.2720	0.2950	0.3161	0.3545	0.3897	0.4231	0.4551	0.4862	0.51
(649.45)	h	695.46	1122.2			1123.9	1218.0	1276.8	1323.1	1363.3	1400.0	1467.6	1530.9	1592.5	1653.3	1713.9	177
	s	0.8828	1.2676			1.2691	1.3523	1.4020	1.4395	1.4708	1.4984	1.5463	1.5883	1.6266	1.6622	1.6956	1.72
	Sh						44.11	94.11	144.11	194.11	244.11	344.11	444.11	544.11	644.11	744.11	844
2300	v	0.02727	0.1513				0.1975	0.2305	0.2566	0.2793	0.2999	0.3372	0.3714	0.4035	0.4344	0.4643	0.49
(655.89)	h	707.18	1113.2				1205.3	1268.4	1316.7	1358.1	1395.7	1464.2	1528.3	1590.3	1651.5	1712.3	177
	s	0.8929	1.2569				1.3381	1.3914	1.4305	1.4628	1.4910	1.5397	1.5821	1.6207	1.6565	1.6901	1.72

Sh = superheat, F
v = specific volume, cu ft per lb
h = enthalpy, Btu per lb
s = entropy, Btu per F per lb

Table 3. Superheated Steam—*Continued*

ɔs Press. b/Sq In. (at. Temp)		Sat Water	Sat Steam	Temperature – Degrees Fahrenheit 700	750	800	850	900	950	1000	1050	1100	1150	1200	1300	1400	1500
	Sh			37.89	87.89	137.89	187.89	237.89	287.89	337.89	387.89	437.89	487.89	537.89	637.89	737.89	837.89
2400	v	0.02790	0.1408	0.1824	0.2164	0.2424	0.2648	0.2850	0.3037	0.3214	0.3382	0.3545	0.3703	0.3856	0.4155	0.4443	0.4724
ɔ62.11)	h	718.95	1103.7	1191.6	1259.7	1310.1	1352.8	1391.2	1426.9	1460.9	1493.7	1525.6	1557.0	1588.1	1649.6	1710.8	1771.8
	s	0.9031	1.2460	1.3232	1.3808	1.4217	1.4549	1.4837	1.5095	1.5332	1.5553	1.5761	1.5959	1.6149	1.6509	1.6847	1.7167
	Sh			31.89	81.89	131.89	181.89	231.89	281.89	331.89	381.89	431.89	481.89	531.89	631.89	731.89	831.89
2500	v	0.02859	0.1307	0.1681	0.2032	0.2293	0.2514	0.2712	0.2896	0.3068	0.3232	0.3390	0.3543	0.3692	0.3980	0.4259	0.4529
ɔ68.11)	h	731.71	1093.3	1176.7	1250.6	1303.4	1347.4	1386.7	1423.1	1457.5	1490.7	1522.9	1554.6	1585.9	1647.8	1709.2	1770.4
	s	0.9139	1.2345	1.3076	1.3701	1.4129	1.4472	1.4766	1.5029	1.5269	1.5492	1.5703	1.5903	1.6094	1.6456	1.6796	1.7116
	Sh			26.09	76.09	126.09	176.09	226.09	276.09	326.09	376.09	426.09	476.09	526.09	626.09	726.09	826.09
2600	v	0.02938	0.1211	0.1544	0.1909	0.2171	0.2390	0.2585	0.2765	0.2933	0.3093	0.3247	0.3395	0.3540	0.3819	0.4088	0.4350
ɔ73.91)	h	744.47	1082.0	1160.2	1241.1	1296.5	1341.9	1382.1	1419.2	1454.1	1487.7	1520.2	1552.2	1583.7	1646.0	1707.7	1769.1
	s	0.9247	1.2225	1.2908	1.3592	1.4042	1.4395	1.4696	1.4964	1.5208	1.5434	1.5646	1.5848	1.6040	1.6405	1.6746	1.7068
	Sh			20.47	70.47	120.47	170.47	220.47	270.47	320.47	370.47	420.47	470.47	520.47	620.47	720.47	820.47
2700	v	0.03029	0.1119	0.1411	0.1794	0.2058	0.2275	0.2468	0.2644	0.2809	0.2965	0.3114	0.3259	0.3399	0.3670	0.3931	0.4184
ɔ79.53)	h	757.34	1069.7	1142.0	1231.1	1289.5	1336.3	1377.5	1415.2	1450.7	1484.6	1517.5	1549.8	1581.5	1644.1	1706.1	1767.8
	s	0.9356	1.2097	1.2727	1.3481	1.3954	1.4319	1.4628	1.4900	1.5148	1.5376	1.5591	1.5794	1.5988	1.6355	1.6697	1.7021
	Sh			15.04	65.04	115.04	165.04	215.04	265.04	315.04	365.04	415.04	465.04	515.04	615.04	715.04	815.04
2800	v	0.03134	0.1030	0.1278	0.1685	0.1952	0.2168	0.2358	0.2531	0.2693	0.2845	0.2991	0.3132	0.3268	0.3532	0.3785	0.4030
ɔ84.96)	h	770.69	1055.8	1121.2	1220.6	1282.2	1330.7	1372.8	1411.2	1447.2	1481.6	1514.8	1547.3	1579.3	1642.2	1704.5	1766.5
	s	0.9468	1.1958	1.2527	1.3368	1.3867	1.4245	1.4561	1.4838	1.5089	1.5321	1.5537	1.5742	1.5938	1.6306	1.6651	1.6975
	Sh			9.78	59.78	109.78	159.78	209.78	259.78	309.78	359.78	409.78	459.78	509.78	609.78	709.78	809.78
2900	v	0.03262	0.0942	0.1138	0.1581	0.1853	0.2068	0.2256	0.2427	0.2585	0.2734	0.2877	0.3014	0.3147	0.3403	0.3649	0.3887
ɔ90.22)	h	785.13	1039.8	1095.3	1209.6	1274.7	1324.9	1368.0	1407.2	1443.7	1478.5	1512.1	1544.9	1577.0	1640.4	1703.0	1765.2
	s	0.9588	1.1803	1.2283	1.3251	1.3780	1.4171	1.4494	1.4777	1.5032	1.5266	1.5485	1.5692	1.5889	1.6259	1.6605	1.6931
	Sh			4.67	54.67	104.67	154.67	204.67	254.67	304.67	354.67	404.67	454.67	504.67	604.67	704.67	804.67
3000	v	0.03428	0.0850	0.0982	0.1483	0.1759	0.1975	0.2161	0.2329	0.2484	0.2630	0.2770	0.2904	0.3033	0.3282	0.3522	0.3753
ɔ95.33)	h	801.84	1020.3	1060.5	1197.9	1267.0	1319.0	1363.2	1403.1	1440.2	1475.4	1509.4	1542.4	1574.8	1638.5	1701.4	1763.8
	s	0.9728	1.1619	1.1966	1.3131	1.3692	1.4097	1.4429	1.4717	1.4976	1.5213	1.5434	1.5642	1.5841	1.6214	1.6561	1.6888
	Sh				49.72	99.72	149.72	199.72	249.72	299.72	349.72	399.72	449.72	499.72	599.72	699.72	799.72
3100	v	0.03681	0.0745		0.1389	0.1671	0.1887	0.2071	0.2237	0.2390	0.2533	0.2670	0.2800	0.2927	0.3170	0.3403	0.3628
700.28)	h	823.97	993.3		1185.4	1259.1	1313.0	1358.4	1399.0	1436.7	1472.3	1506.6	1539.9	1572.6	1636.7	1699.8	1762.5
	s	0.9914	1.1373		1.3007	1.3604	1.4024	1.4364	1.4658	1.4920	1.5161	1.5384	1.5594	1.5794	1.6169	1.6518	1.6847
	Sh				44.92	94.92	144.92	194.92	244.92	294.92	344.92	394.92	444.92	494.92	594.92	694.92	794.92
3200	v	0.04472	0.0566		0.1300	0.1588	0.1804	0.1987	0.2151	0.2301	0.2442	0.2576	0.2704	0.2827	0.3065	0.3291	0.3510
705.08)	h	875.54	931.6		1172.3	1250.9	1306.9	1353.4	1394.9	1433.1	1469.2	1503.8	1537.4	1570.3	1634.8	1698.3	1761.2
	s	1.0351	1.0832		1.2877	1.3515	1.3951	1.4300	1.4600	1.4866	1.5110	1.5335	1.5547	1.5749	1.6126	1.6477	1.6806
	Sh																
3300	v				0.1213	0.1510	0.1727	0.1908	0.2070	0.2218	0.2357	0.2488	0.2613	0.2734	0.2966	0.3187	0.3400
	h				1158.2	1242.5	1300.7	1348.4	1390.7	1429.5	1466.1	1501.0	1534.9	1568.1	1632.9	1696.7	1759.9
	s				1.2742	1.3425	1.3879	1.4237	1.4542	1.4813	1.5059	1.5287	1.5501	1.5704	1.6084	1.6436	1.6767
	Sh																
3400	v				0.1129	0.1435	0.1653	0.1834	0.1994	0.2140	0.2276	0.2405	0.2528	0.2646	0.2872	0.3088	0.3296
	h				1143.2	1233.7	1294.3	1343.4	1386.4	1425.9	1462.9	1498.3	1532.4	1565.8	1631.1	1695.1	1758.5
	s				1.2600	1.3334	1.3807	1.4174	1.4486	1.4761	1.5010	1.5240	1.5456	1.5660	1.6042	1.6396	1.6728
	Sh																
3500	v				0.1048	0.1364	0.1583	0.1764	0.1922	0.2066	0.2200	0.2326	0.2447	0.2563	0.2784	0.2995	0.3198
	h				1127.1	1224.6	1287.8	1338.2	1382.2	1422.2	1459.7	1495.5	1529.9	1563.6	1629.2	1693.6	1757.2
	s				1.2450	1.3242	1.3734	1.4112	1.4430	1.4709	1.4962	1.5194	1.5412	1.5618	1.6002	1.6358	1.6691
	Sh																
3600	v				0.0966	0.1296	0.1517	0.1697	0.1854	0.1996	0.2128	0.2252	0.2371	0.2485	0.2702	0.2908	0.3106
	h				1108.6	1215.3	1281.2	1333.0	1377.9	1418.6	1456.5	1492.6	1527.4	1561.3	1627.3	1692.0	1755.9
	s				1.2281	1.3148	1.3662	1.4050	1.4374	1.4658	1.4914	1.5149	1.5369	1.5576	1.5962	1.6320	1.6654
	Sh																
3800	v				0.0799	0.1169	0.1395	0.1574	0.1729	0.1868	0.1996	0.2116	0.2231	0.2340	0.2549	0.2746	0.2936
	h				1064.2	1195.5	1267.6	1322.4	1369.1	1411.2	1450.1	1487.0	1522.4	1556.8	1623.6	1688.9	1753.2
	s				1.1888	1.2955	1.3517	1.3928	1.4265	1.4558	1.4821	1.5061	1.5284	1.5495	1.5886	1.6247	1.6584
	Sh																
4000	v				0.0631	0.1052	0.1284	0.1463	0.1616	0.1752	0.1877	0.1994	0.2105	0.2210	0.2411	0.2601	0.2783
	h				1007.4	1174.3	1253.4	1311.6	1360.2	1403.6	1443.6	1481.3	1517.3	1552.2	1619.8	1685.7	1750.6
	s				1.1396	1.2754	1.3371	1.3807	1.4158	1.4461	1.4730	1.4976	1.5203	1.5417	1.5812	1.6177	1.6516
	Sh																
4200	v				0.0498	0.0945	0.1183	0.1362	0.1513	0.1647	0.1769	0.1883	0.1991	0.2093	0.2287	0.2470	0.2645
	h				950.1	1151.6	1238.6	1300.4	1351.2	1396.0	1437.1	1475.5	1512.2	1547.6	1616.1	1682.6	1748.0
	s				1.0905	1.2544	1.3223	1.3686	1.4053	1.4366	1.4642	1.4893	1.5124	1.5341	1.5742	1.6109	1.6452
	Sh																
4400	v				0.0421	0.0846	0.1090	0.1270	0.1420	0.1552	0.1671	0.1782	0.1887	0.1986	0.2174	0.2351	0.2519
	h				909.5	1127.3	1223.3	1289.0	1342.0	1388.3	1430.4	1469.7	1507.1	1543.0	1612.3	1679.4	1745.3
	s				1.0556	1.2325	1.3073	1.3566	1.3949	1.4272	1.4556	1.4812	1.5048	1.5268	1.5673	1.6044	1.6389

Sh = superheat, F
v = specific volume, cu ft per lb
h = enthalpy, Btu per lb
s = entropy, Btu per F per lb

Table 3. Superheated Steam – *Continued*

Abs Press. Lb/Sq In. (Sat. Temp)		Sat Water	Sat Steam	Temperature – Degrees Fahrenheit 750	800	850	900	950	1000	1050	1100	1150	1200	1250	1300	1400	1500
	Sh																
4600	v			0.0380	0.0751	0.1005	0.1186	0.1335	0.1465	0.1582	0.1691	0.1792	0.1889	0.1982	0.2071	0.2242	0.240
	h			883.8	1100.0	1207.3	1277.2	1332.6	1380.5	1423.7	1463.9	1501.9	1538.4	1573.8	1608.5	1676.3	1742
	s			1.0331	1.2084	1.2922	1.3446	1.3847	1.4181	1.4472	1.4734	1.4974	1.5197	1.5407	1.5607	1.5982	1.633
	Sh																
4800	v			0.0355	0.0665	0.0927	0.1109	0.1257	0.1385	0.1500	0.1606	0.1706	0.1800	0.1890	0.1977	0.2142	0.229
	h			866.9	1071.2	1190.7	1265.2	1323.1	1372.6	1417.0	1458.0	1496.7	1533.8	1569.7	1604.7	1673.1	1740
	s			1.0180	1.1835	1.2768	1.3327	1.3745	1.4090	1.4390	1.4657	1.4901	1.5128	1.5341	1.5543	1.5921	1.627
	Sh																
5000	v			0.0338	0.0591	0.0855	0.1038	0.1185	0.1312	0.1425	0.1529	0.1626	0.1718	0.1806	0.1890	0.2050	0.220
	h			854.9	1042.9	1173.6	1252.9	1313.5	1364.6	1410.2	1452.1	1491.5	1529.1	1565.5	1600.9	1670.0	1737
	s			1.0070	1.1593	1.2612	1.3207	1.3645	1.4001	1.4309	1.4582	1.4831	1.5061	1.5277	1.5481	1.5863	1.621
	Sh																
5200	v			0.0326	0.0531	0.0789	0.0973	0.1119	0.1244	0.1356	0.1458	0.1553	0.1642	0.1728	0.1810	0.1966	0.211
	h			845.8	1016.9	1156.0	1240.4	1303.7	1356.6	1403.4	1446.2	1486.3	1524.5	1561.3	1597.2	1666.8	1734
	s			0.9985	1.1370	1.2455	1.3088	1.3545	1.3914	1.4229	1.4509	1.4762	1.4995	1.5214	1.5420	1.5806	1.616
	Sh																
5400	v			0.0317	0.0483	0.0728	0.0912	0.1058	0.1182	0.1292	0.1392	0.1485	0.1572	0.1656	0.1736	0.1888	0.203
	h			838.5	994.3	1138.1	1227.7	1293.7	1348.4	1396.5	1440.3	1481.1	1519.8	1557.1	1593.4	1663.7	1732
	s			0.9915	1.1175	1.2296	1.2969	1.3446	1.3827	1.4151	1.4437	1.4694	1.4931	1.5153	1.5362	1.5750	1.610
	Sh																
5600	v			0.0309	0.0447	0.0672	0.0856	0.1001	0.1124	0.1232	0.1331	0.1422	0.1508	0.1589	0.1667	0.1815	0.195
	h			832.4	975.0	1119.9	1214.8	1283.7	1340.2	1389.6	1434.3	1475.9	1515.2	1552.9	1589.6	1660.5	1729
	s			0.9855	1.1008	1.2137	1.2850	1.3348	1.3742	1.4075	1.4366	1.4628	1.4869	1.5093	1.5304	1.5697	1.605
	Sh																
5800	v			0.0303	0.0419	0.0622	0.0805	0.0949	0.1070	0.1177	0.1274	0.1363	0.1447	0.1527	0.1603	0.1747	0.188
	h			827.3	958.8	1101.8	1201.8	1273.6	1332.0	1382.6	1428.3	1470.6	1510.5	1548.7	1585.8	1657.4	1726
	s			0.9803	1.0867	1.1981	1.2732	1.3250	1.3658	1.3999	1.4297	1.4564	1.4808	1.5035	1.5248	1.5644	1.600
	Sh																
6000	v			0.0298	0.0397	0.0579	0.0757	0.0900	0.1020	0.1126	0.1221	0.1309	0.1391	0.1469	0.1544	0.1684	0.181
	h			822.9	945.1	1084.6	1188.8	1263.4	1323.6	1375.7	1422.3	1465.4	1505.9	1544.6	1582.0	1654.2	1724
	s			0.9758	1.0746	1.1833	1.2615	1.3154	1.3574	1.3925	1.4229	1.4500	1.4748	1.4978	1.5194	1.5593	1.596
	Sh																
6500	v			0.0287	0.0358	0.0495	0.0655	0.0793	0.0909	0.1012	0.1104	0.1188	0.1266	0.1340	0.1411	0.1544	0.166
	h			813.9	919.5	1046.7	1156.3	1237.8	1302.7	1358.1	1407.3	1452.2	1494.2	1534.1	1572.5	1646.4	1717
	s			0.9661	1.0515	1.1506	1.2328	1.2917	1.3370	1.3743	1.4064	1.4347	1.4604	1.4841	1.5062	1.5471	1.584
	Sh																
7000	v			0.0279	0.0334	0.0438	0.0573	0.0704	0.0816	0.0915	0.1004	0.1085	0.1160	0.1231	0.1298	0.1424	0.154
	h			806.9	901.8	1016.5	1124.9	1212.6	1281.7	1340.5	1392.2	1439.1	1482.6	1523.7	1563.1	1638.6	1711
	s			0.9582	1.0350	1.1243	1.2055	1.2689	1.3171	1.3567	1.3904	1.4200	1.4466	1.4710	1.4938	1.5355	1.573
	Sh																
7500	v			0.0272	0.0318	0.0399	0.0512	0.0631	0.0737	0.0833	0.0918	0.0996	0.1068	0.1136	0.1200	0.1321	0.143
	h			801.3	889.0	992.9	1097.7	1188.3	1261.0	1322.9	1377.2	1426.0	1471.0	1513.3	1553.7	1630.8	1704
	s			0.9514	1.0224	1.1033	1.1818	1.2473	1.2980	1.3397	1.3751	1.4059	1.4335	1.4586	1.4819	1.5245	1.563
	Sh																
8000	v			0.0267	0.0306	0.0371	0.0465	0.0571	0.0671	0.0762	0.0845	0.0920	0.0989	0.1054	0.1115	0.1230	0.133
	h			796.6	879.1	974.4	1074.3	1165.4	1241.0	1305.5	1362.2	1413.0	1459.6	1503.1	1544.5	1623.1	1698
	s			0.9455	1.0122	1.0864	1.1613	1.2271	1.2798	1.3233	1.3603	1.3924	1.4208	1.4467	1.4705	1.5140	1.553
	Sh																
8500	v			0.0262	0.0296	0.0350	0.0429	0.0522	0.0615	0.0701	0.0780	0.0853	0.0919	0.0982	0.1041	0.1151	0.125
	h			792.7	871.2	959.8	1054.5	1144.0	1221.9	1288.5	1347.5	1400.2	1448.2	1492.9	1535.3	1615.4	1691
	s			0.9402	1.0037	1.0727	1.1437	1.2084	1.2627	1.3076	1.3460	1.3793	1.4087	1.4352	1.4597	1.5040	1.543
	Sh																
9000	v			0.0258	0.0288	0.0335	0.0402	0.0483	0.0568	0.0649	0.0724	0.0794	0.0858	0.0918	0.0975	0.1081	0.117
	h			789.3	864.7	948.0	1037.6	1125.4	1204.1	1272.1	1333.0	1387.5	1437.1	1482.9	1526.3	1607.9	1685
	s			0.9354	0.9964	1.0613	1.1285	1.1918	1.2468	1.2926	1.3323	1.3667	1.3970	1.4243	1.4492	1.4944	1.534
	Sh																
9500	v			0.0254	0.0282	0.0322	0.0380	0.0451	0.0528	0.0603	0.0675	0.0742	0.0804	0.0862	0.0917	0.1019	0.111
	h			786.4	859.2	938.3	1023.4	1108.9	1187.7	1256.6	1318.9	1375.1	1426.1	1473.1	1517.3	1600.4	1679
	s			0.9310	0.9900	1.0516	1.1153	1.1771	1.2320	1.2785	1.3191	1.3546	1.3858	1.4137	1.4392	1.4851	1.526
	Sh																
10000	v			0.0251	0.0276	0.0312	0.0362	0.0425	0.0495	0.0565	0.0633	0.0697	0.0757	0.0812	0.0865	0.0963	0.105
	h			783.8	854.5	930.2	1011.3	1094.2	1172.6	1242.0	1305.3	1362.9	1415.3	1463.4	1508.6	1593.1	1672
	s			0.9270	0.9842	1.0432	1.1039	1.1638	1.2185	1.2652	1.3065	1.3429	1.3749	1.4035	1.4295	1.4763	1.518
	Sh																
10500	v			0.0248	0.0271	0.0303	0.0347	0.0404	0.0467	0.0532	0.0595	0.0656	0.0714	0.0768	0.0818	0.0913	0.100
	h			781.5	850.5	923.4	1001.0	1081.3	1158.9	1228.4	1292.4	1351.1	1404.7	1453.9	1500.0	1585.8	1666
	s			0.9232	0.9790	1.0358	1.0939	1.1519	1.2060	1.2529	1.2946	1.3371	1.3644	1.3937	1.4202	1.4677	1.510

Sh = superheat, F
v = specific volume, cu ft per lb
h = enthalpy, Btu per lb
s = entropy, Btu per F per lb

Table 3. Superheated Steam—*Continued*

Abs Press. Lb/Sq In. (Sat. Temp)		Sat. Water	Sat. Steam	Temperature—Degrees Fahrenheit 750	800	850	900	950	1000	1050	1100	1150	1200	1250	1300	1400	1500
11000	v			0.0245	0.0267	0.0296	0.0335	0.0386	0.0443	0.0503	0.0562	0.0620	0.0676	0.0727	0.0776	0.0868	0.0952
	h			779.5	846.9	917.5	992.1	1069.9	1146.3	1215.9	1280.2	1339.7	1394.4	1444.6	1491.5	1578.7	1660.6
	s			0.9196	0.9742	1.0292	1.0851	1.1412	1.1945	1.2414	1.2833	1.3209	1.3544	1.3842	1.4112	1.4595	1.5023
11500	v			0.0243	0.0263	0.0290	0.0325	0.0370	0.0423	0.0478	0.0534	0.0588	0.0641	0.0691	0.0739	0.0827	0.0909
	h			777.7	843.8	912.4	984.5	1059.8	1134.9	1204.3	1268.7	1328.8	1384.4	1435.5	1483.2	1571.8	1654.7
	s			0.9163	0.9698	1.0232	1.0772	1.1316	1.1840	1.2308	1.2727	1.3107	1.3446	1.3750	1.4025	1.4515	1.4949
12000	v			0.0241	0.0260	0.0284	0.0317	0.0357	0.0405	0.0456	0.0508	0.0560	0.0610	0.0659	0.0704	0.0790	0.0869
	h			776.1	841.0	907.9	977.8	1050.9	1124.5	1193.7	1258.0	1318.5	1374.7	1426.6	1475.1	1564.9	1648.8
	s			0.9131	0.9657	1.0177	1.0701	1.1229	1.1742	1.2209	1.2627	1.3010	1.3353	1.3662	1.3941	1.4438	1.4877
12500	v			0.0238	0.0256	0.0279	0.0309	0.0346	0.0390	0.0437	0.0486	0.0535	0.0583	0.0629	0.0673	0.0756	0.0832
	h			774.7	838.6	903.9	971.9	1043.1	1115.2	1184.1	1247.9	1308.8	1365.4	1418.0	1467.2	1558.2	1643.1
	s			0.9101	0.9618	1.0127	1.0637	1.1151	1.1653	1.2117	1.2534	1.2918	1.3264	1.3576	1.3860	1.4363	1.4808
13000	v			0.0236	0.0253	0.0275	0.0302	0.0336	0.0376	0.0420	0.0466	0.0512	0.0558	0.0602	0.0645	0.0725	0.0799
	h			773.5	836.3	900.4	966.8	1036.2	1106.7	1174.8	1238.5	1299.6	1356.5	1409.6	1459.4	1551.6	1637.4
	s			0.9073	0.9582	1.0080	1.0578	1.1079	1.1571	1.2030	1.2445	1.2831	1.3179	1.3494	1.3781	1.4291	1.4741
13500	v			0.0235	0.0251	0.0271	0.0297	0.0328	0.0364	0.0405	0.0448	0.0492	0.0535	0.0577	0.0619	0.0696	0.0768
	h			772.3	834.4	897.2	962.2	1030.0	1099.1	1166.3	1229.7	1291.0	1348.1	1401.5	1451.8	1545.2	1631.9
	s			0.9045	0.9548	1.0037	1.0524	1.1014	1.1495	1.1948	1.2361	1.2749	1.3098	1.3415	1.3705	1.4221	1.4675
14000	v			0.0233	0.0248	0.0267	0.0291	0.0320	0.0354	0.0392	0.0432	0.0474	0.0515	0.0555	0.0595	0.0670	0.0740
	h			771.3	832.6	894.3	958.0	1024.5	1092.3	1158.5	1221.4	1283.0	1340.2	1393.8	1444.4	1538.8	1626.5
	s			0.9019	0.9515	0.9996	1.0473	1.0953	1.1426	1.1872	1.2282	1.2671	1.3021	1.3339	1.3631	1.4153	1.4612
14500	v			0.0231	0.0246	0.0264	0.0287	0.0314	0.0345	0.0380	0.0418	0.0458	0.0496	0.0534	0.0573	0.0646	0.0714
	h			770.4	831.0	891.7	954.3	1019.6	1086.2	1151.4	1213.8	1275.4	1332.9	1386.4	1437.3	1532.6	1621.1
	s			0.8994	0.9484	0.9957	1.0426	1.0897	1.1362	1.1801	1.2208	1.2597	1.2949	1.3266	1.3560	1.4087	1.4551
15000	v			0.0230	0.0244	0.0261	0.0282	0.0308	0.0337	0.0369	0.0405	0.0443	0.0479	0.0516	0.0552	0.0624	0.0690
	h			769.6	829.5	889.3	950.9	1015.1	1080.6	1144.9	1206.8	1268.1	1326.0	1379.4	1430.3	1526.4	1615.9
	s			0.8970	0.9455	0.9920	1.0382	1.0846	1.1302	1.1735	1.2139	1.2525	1.2880	1.3197	1.3491	1.4022	1.4491
15500	v			0.0228	0.0242	0.0258	0.0278	0.0302	0.0329	0.0360	0.0393	0.0429	0.0464	0.0499	0.0534	0.0603	0.0668
	h			768.9	828.2	887.2	947.8	1011.1	1075.7	1139.0	1200.3	1261.1	1319.6	1372.8	1423.6	1520.4	1610.8
	s			0.8946	0.9427	0.9886	1.0340	1.0797	1.1247	1.1674	1.2073	1.2457	1.2815	1.3131	1.3424	1.3959	1.4433

Sh = superheat, F
v = specific volume, cu ft per lb
h = enthalpy, Btu per lb
s = entropy, Btu per F per lb

Table 1. Saturation: Temperatures

		Specific Volume		Internal Energy			Enthalpy			Entropy		
Temp. °C	Press. Bars	Sat. Liquid	Sat. Vapor	Sat. Liquid	Evap.	Sat. Vapor	Sat. Liquid	Evap.	Sat. Vapor	Sat Liquid	Evap.	Sat. Vapor
T	P	v_l	v_g	u_l	u_{l-g}	u_g	h_l	h_{l-g}	h_g	s_l	s_{l-g}	s_g
.01	.006113	.0010002	206.136	.00	2375.3	2375.3	.01	2501.3	2501.4	.0000	9.1562	9.1562
1	.006567	.0010002	192.577	4.15	2372.6	2376.7	4.16	2499.0	2503.2	.0152	9.1147	9.1299
5	.008721	.0010001	147.120	20.97	2361.3	2382.3	20.98	2489.6	2510.6	.0761	8.9496	9.0257
10	.012276	.0010004	106.379	42.00	2347.2	2389.2	42.01	2477.7	2519.8	.1510	8.7498	8.9008
15	.017051	.0010009	77.926	62.99	2333.1	2396.1	62.99	2465.9	2528.9	.2245	8.5569	8.7814
20	.02339	.0010018	57.791	83.95	2319.0	2402.9	83.96	2454.1	2538.1	.2966	8.3706	8.6672
25	.03169	.0010029	43.360	104.88	2304.9	2409.8	104.89	2442.3	2547.2	.3674	8.1905	8.5580
30	.04246	.0010043	32.894	125.78	2290.8	2416.6	125.79	2430.5	2556.3	.4369	8.0164	8.4533
35	.05628	.0010060	25.216	146.67	2276.7	2423.4	146.68	2418.6	2565.3	.5053	7.8478	8.3531
40	.07384	.0010078	19.523	167.56	2262.6	2430.1	167.57	2406.7	2574.3	.5725	7.6845	8.2570
45	.09593	.0010099	15.258	188.44	2248.4	2436.8	188.45	2394.8	2583.2	.6387	7.5261	8.1648
50	.12349	.0010121	12.032	209.32	2234.2	2443.5	209.33	2382.7	2592.1	.7038	7.3725	8.0763
55	.15758	.0010146	9.568	230.21	2219.9	2450.1	230.23	2370.7	2600.9	.7679	7.2234	7.9913
60	.19940	.0010172	7.671	251.11	2205.5	2456.6	251.13	2358.5	2609.6	.8312	7.0784	7.9096
65	.2503	.0010199	6.197	272.02	2191.1	2463.1	272.06	2346.2	2618.3	.8935	6.9375	7.8310
70	.3119	.0010228	5.042	292.95	2176.6	2469.6	292.98	2333.8	2626.8	.9549	6.8004	7.7553
75	.3858	.0010259	4.131	313.90	2162.0	2475.9	313.93	2321.4	2635.3	1.0155	6.6669	7.6824
80	.4739	.0010291	3.407	334.86	2147.4	2482.2	334.91	2308.8	2643.7	1.0753	6.5369	7.6122
85	.5783	.0010325	2.828	355.84	2132.6	2488.4	355.90	2296.0	2651.9	1.1343	6.4102	7.5445
90	.7014	.0010360	2.361	376.85	2117.7	2494.5	376.92	2283.2	2660.1	1.1925	6.2866	7.4791
95	.8455	.0010397	1.9819	397.88	2102.7	2500.6	397.96	2270.2	2668.1	1.2500	6.1659	7.4159
100	1.0135	.0010435	1.6729	418.94	2087.6	2506.5	419.04	2257.0	2676.1	1.3069	6.0480	7.3549
105	1.2082	.0010475	1.4194	440.02	2072.3	2512.4	440.15	2243.7	2683.8	1.3630	5.9328	7.2958
110	1.4327	.0010516	1.2102	461.14	2057.0	2518.1	461.30	2230.2	2691.5	1.4185	5.8202	7.2387
115	1.6906	.0010559	1.0366	482.30	2041.4	2523.7	482.48	2216.5	2699.0	1.4734	5.7100	7.1833
120	1.9853	.0010603	.8919	503.50	2025.8	2529.3	503.71	2202.6	2706.3	1.5276	5.6020	7.1296
125	2.321	.0010649	.7706	524.74	2009.9	2534.6	524.99	2188.5	2713.5	1.5813	5.4962	7.0775
130	2.701	.0010697	.6685	546.02	1993.9	2539.9	546.31	2174.2	2720.5	1.6344	5.3925	7.0269
135	3.130	.0010746	.5822	567.35	1977.7	2545.0	567.69	2159.6	2727.3	1.6870	5.2907	6.9777
140	3.613	.0010797	.5089	588.74	1961.3	2550.0	589.13	2144.7	2733.9	1.7391	5.1908	6.9299
145	4.154	.0010850	.4463	610.18	1944.7	2554.9	610.63	2129.6	2740.3	1.7907	5.0926	6.8833
150	4.758	.0010905	.3928	631.68	1927.9	2559.5	632.20	2114.3	2746.5	1.8418	4.9960	6.8379
155	5.431	.0010961	.3468	653.24	1910.8	2564.1	653.84	2098.6	2752.4	1.8925	4.9010	6.7935
160	6.178	.0011020	.3071	674.87	1893.5	2568.4	675.55	2082.6	2758.1	1.9427	4.8075	6.7502
165	7.005	.0011080	.2727	696.56	1876.0	2572.5	697.34	2066.2	2763.5	1.9925	4.7153	6.7078
170	7.917	.0011143	.2428	718.33	1858.1	2576.5	719.21	2049.5	2768.7	2.0419	4.6244	6.6663
175	8.920	.0011207	.2168	740.17	1840.0	2580.2	741.17	2032.4	2773.6	2.0909	4.5347	6.6256
180	10.021	.0011274	.19405	762.09	1821.6	2583.7	763.22	2015.0	2778.2	2.1396	4.4461	6.5857
185	11.227	.0011343	.17409	784.10	1802.9	2587.0	785.37	1997.1	2782.4	2.1879	4.3586	6.5465
190	12.544	.0011414	.15654	806.19	1783.8	2590.0	807.62	1978.8	2786.4	2.2359	4.2720	6.5079

Table 1. Saturation: Temperatures—Continued

Temp. °C T	Press. Bars P	Specific Volume Sat. Liquid v_l	Sat. Vapor v_g	Internal Energy Sat. Liquid u_l	Evap. u_{l-g}	Sat. Vapor u_g	Enthalpy Sat. Liquid h_l	Evap. h_{l-g}	Sat. Vapor h_g	Entropy Sat. Liquid s_l	Evap. s_{l-g}	Sat. Vapor s_g
195	13.978	.0011488	.14105	828.37	1764.4	2592.8	829.98	1960.0	2790.0	2.2835	4.1863	6.4698
200	15.538	.0011565	.12736	850.65	1744.7	2595.3	852.45	1940.7	2793.2	2.3309	4.1014	6.4323
210	19.062	.0011726	.10441	895.53	1703.9	2599.5	897.76	1900.7	2798.5	2.4248	3.9337	6.3585
220	23.18	.0011900	.08619	940.87	1661.5	2602.4	943.62	1858.5	2802.1	2.5178	3.7683	6.2861
230	27.95	.0012088	.07158	986.74	1617.2	2603.9	990.12	1813.8	2804.0	2.6099	3.6047	6.2146
240	33.44	.0012291	.05976	1033.21	1570.8	2604.0	1037.32	1766.5	2803.8	2.7015	3.4422	6.1437
250	39.73	.0012512	.05013	1080.39	1522.0	2602.4	1085.36	1716.2	2801.5	2.7927	3.2802	6.0730
260	46.88	.0012755	.04221	1128.39	1470.6	2599.0	1134.37	1662.5	2796.9	2.8838	3.1181	6.0019
270	54.99	.0013023	.03564	1177.36	1416.3	2593.7	1184.51	1605.2	2789.7	2.9751	2.9551	5.9301
280	64.12	.0013321	.03017	1227.46	1358.7	2586.1	1235.99	1543.6	2779.6	3.0668	2.7903	5.8571
290	74.36	.0013656	.02557	1278.92	1297.1	2576.0	1289.07	1477.1	2766.2	3.1594	2.6227	5.7821
300	85.81	.0014036	.02167	1332.0	1231.0	2563.0	1344.0	1404.9	2749.0	3.2534	2.4511	5.7045
310	98.56	.0014474	.018350	1387.1	1159.4	2546.4	1401.3	1326.0	2727.3	3.3493	2.2737	5.6230
320	112.74	.0014988	.015488	1444.6	1080.9	2525.5	1461.5	1238.6	2700.1	3.4480	2.0882	5.5362
330	128.45	.0015607	.012996	1505.3	993.7	2498.9	1525.3	1140.6	2665.9	3.5507	1.8909	5.4417
340	145.86	.0016379	.010797	1570.3	894.3	2464.6	1594.2	1027.9	2622.0	3.6594	1.6763	5.3357
350	165.13	.0017403	.008813	1641.9	776.6	2418.4	1670.6	893.4	2563.9	3.7777	1.4335	5.2112
360	186.51	.0018925	.006945	1725.2	626.3	2351.5	1760.5	720.5	2481.0	3.9147	1.1379	5.0526
370	210.3	.002213	.004925	1844.0	384.5	2228.5	1890.5	441.6	2332.1	4.1106	.6865	4.7971
374.136	220.9	.003155	.003155	2029.6	0	2029.6	2099.3	0	2099.3	4.4298	0	4.4298

P = Bars = 10^5 N/m²
t = °C
u, h = kJ/kg
v = m³/kg³
s = kJ/kg °k

Condensed from Keenan, Keyes, Hill, and Moore *Steam Tables—International Edition,* J. Wiley & Sons, New York (1969) Used by Permission.

Table 2. Saturation: Pressures

Press. Bars	Temp. °C	Specific Volume Sat. Liquid	Sat. Vapor	Internal Energy Sat. Liquid	Evap.	Sat. Vapor	Enthalpy Sat. Liquid	Evap.	Sat. Vapor	Entropy Sat. Liquid	Evap.	Sat. Vapor
P	T	v_l	v_g	u_l	u_{l-g}	u_g	h_l	h_{l-g}	h_g	s_l	s_{l-g}	s_g
.006113	.01	.0010002	206.136	.00	2375.3	2375.3	.01	2501.3	2501.4	.0000	9.1562	9.1562
.010	6.98	.0010002	129.208	29.30	2355.7	2385.0	29.30	2484.9	2514.2	.1059	8.8697	8.9756
.015	13.03	.0010007	87.980	54.71	2338.6	2393.3	54.71	2470.6	2525.3	.1957	8.6322	8.8279
.020	17.50	.0010013	67.004	73.48	2326.0	2399.5	73.48	2460.0	2533.5	.2607	8.4629	8.7237
.040	28.96	.0010040	34.800	121.45	2293.7	2415.2	121.46	2432.9	2554.4	.4226	8.0520	8.4746
.060	36.16	.0010064	23.739	151.53	2273.4	2425.0	151.53	2415.9	2567.4	.5210	7.8094	8.3304
.080	41.51	.0010084	18.103	173.87	2258.3	2432.2	173.88	2403.1	2577.0	.5926	7.6361	8.2287
.10	45.81	.0010102	14.674	191.82	2246.1	2437.9	191.83	2392.8	2584.7	.6493	7.5009	8.1502
.15	53.97	.0010141	10.022	225.92	2222.8	2448.7	225.94	2373.1	2599.1	.7549	7.2536	8.0085
.20	60.06	.0010172	7.649	251.38	2205.4	2456.7	251.40	2358.3	2609.7	.8320	7.0766	7.9085
.40	75.87	.0010265	3.993	317.53	2159.5	2477.0	317.58	2319.2	2636.8	1.0259	6.6441	7.6700
.60	85.94	.0010331	2.732	359.79	2129.8	2489.6	359.86	2293.6	2653.5	1.1453	6.3867	7.5320
.80	93.50	.0010386	2.087	391.58	2107.2	2498.8	391.66	2274.1	2665.8	1.2329	6.2017	7.4346
1.00	99.63	.0010432	1.6940	417.36	2088.7	2506.1	417.46	2258.0	2675.5	1.3026	6.0568	7.3594
1.25	105.99	.0010483	1.3749	444.19	2069.3	2513.5	444.32	2241.0	2685.4	1.3740	5.9104	7.2844
1.50	111.37	.0010528	1.1593	466.94	2052.7	2519.7	467.11	2226.5	2693.6	1.4336	5.7897	7.2233
1.75	116.06	.0010568	1.0036	486.80	2038.1	2524.9	486.99	2213.6	2700.6	1.4849	5.6868	7.1717
2.00	120.23	.0010605	.8857	504.49	2025.0	2529.5	504.70	2201.9	2706.7	1.5301	5.5970	7.1271
2.25	124.00	.0010640	.7933	520.47	2013.1	2533.6	520.72	2191.3	2712.1	1.5706	5.5173	7.0878
2.50	127.44	.0010672	.7187	535.10	2002.1	2537.2	535.37	2181.5	2716.9	1.6072	5.4455	7.0527
2.75	130.60	.0010703	.6573	548.59	1991.9	2540.5	548.89	2172.4	2721.3	1.6408	5.3801	7.0209
3.00	133.55	.0010732	.6058	561.15	1982.4	2543.6	561.47	2163.8	2725.3	1.6718	5.3201	6.9919
3.25	136.30	.0010759	.5620	572.90	1973.5	2546.4	573.25	2155.8	2729.0	1.7006	5.2646	6.9652
3.50	138.88	.0010786	.5243	583.95	1965.0	2548.9	584.33	2148.1	2732.4	1.7275	5.2130	6.9405
3.75	141.32	.0010811	.4914	594.40	1956.9	2551.3	594.81	2140.8	2735.6	1.7528	5.1647	6.9175
4.0	143.63	.0010836	.4625	604.31	1949.3	2553.6	604.74	2133.8	2738.6	1.7766	5.1193	6.8959
5.0	151.86	.0010926	.3749	639.68	1921.6	2561.2	640.23	2108.5	2748.7	1.8607	4.9606	6.8213
6.0	158.85	.0011006	.3157	669.90	1897.5	2567.4	670.56	2086.3	2756.8	1.9312	4.8288	6.7600
7.0	164.97	.0011080	.2729	696.44	1876.1	2572.5	697.22	2066.3	2763.5	1.9922	4.7158	6.7080
8.0	170.43	.0011148	.2404	720.22	1856.6	2576.8	721.11	2048.0	2769.1	2.0462	4.6166	6.6628
9.0	175.38	.0011212	.2150	741.83	1838.6	2580.5	742.83	2031.1	2773.9	2.0946	4.5280	6.622
10.0	179.91	.0011273	.19444	761.68	1822.0	2583.6	762.81	2015.3	2778.1	2.1387	4.4478	6.5865
11.0	184.09	.0011330	.17753	780.09	1806.3	2586.4	781.34	2000.4	2781.7	2.1792	4.3744	6.553
12.0	187.99	.0011385	.16333	797.29	1791.5	2588.8	798.65	1986.2	2784.8	2.2166	4.3067	6.523
13.0	191.64	.0011438	.15125	813.44	1777.5	2591.0	814.93	1972.7	2787.6	2.2515	4.2438	6.495
14.0	195.07	.0011489	.14084	828.70	1764.1	2592.8	830.30	1959.7	2790.0	2.2842	4.1850	6.469
15.0	198.32	.0011539	.13177	843.16	1751.3	2594.5	844.89	1947.3	2792.2	2.3150	4.1298	6.4448
16.0	201.41	.0011587	.12380	856.94	1739.0	2596.0	858.79	1935.2	2794.0	2.3442	4.0776	6.421
17.0	204.34	.0011634	.11673	870.09	1727.2	2597.3	872.06	1923.6	2795.7	2.3718	4.0282	6.400
18.0	207.15	.0011679	.11042	882.69	1715.7	2598.4	884.79	1912.4	2797.1	2.3981	3.9812	6.379

Table 2. Saturation: Pressures—Continued

Press. Bars	Temp. °C	Specific Volume: Sat. Liquid	Specific Volume: Sat. Vapor	Internal Energy: Sat. Liquid	Internal Energy: Evap.	Internal Energy: Sat. Vapor	Enthalpy: Sat. Liquid	Enthalpy: Evap.	Enthalpy: Sat. Vapor	Entropy: Sat Liquid	Entropy: Evap.	Entropy: Sat. Vapor
P	T	v_l	v_g	u_l	u_{l-g}	u_g	h_l	h_{l-g}	h_g	s_l	s_{l-g}	s_g
19.0	209.84	.0011724	.10475	894.79	1704.6	2599.4	897.02	1901.4	2798.4	2.4233	3.9364	6.3597
20.0	212.42	.0011767	.09963	906.44	1693.8	2600.3	908.79	1890.7	2799.5	2.4474	3.8935	6.3409
25	223.99	.0011973	.07998	959.11	1644.0	2603.1	962.11	1841.0	2803.1	2.5547	3.7028	6.2575
30	233.90	.0012165	.06668	1004.78	1599.3	2604.1	1008.42	1795.7	2804.2	2.6457	3.5412	6.1869
35	242.60	.0012347	.05707	1045.43	1558.3	2603.7	1049.75	1753.7	2803.4	2.7253	3.4000	6.1253
40	250.40	.0012522	.04978	1082.31	1520.0	2602.3	1087.31	1714.1	2801.4	2.7964	3.2737	6.0701
50	263.99	.0012859	.03944	1147.81	1449.3	2597.1	1154.23	1640.1	2794.3	2.9202	3.0532	5.9734
60	275.64	.0013187	.03244	1205.44	1384.3	2589.7	1213.35	1571.0	2784.3	3.0267	2.8625	5.8892
70	285.88	.0013513	.02737	1257.55	1323.0	2580.5	1267.00	1505.1	2772.1	3.1211	2.6922	5.8133
80	295.06	.0013842	.02352	1305.57	1264.2	2569.8	1316.64	1441.3	2758.0	3.2068	2.5364	5.743[illegible]
90	303.40	.0014178	.02048	1350.51	1207.3	2557.8	1363.26	1378.9	2742.1	3.2858	2.3915	5.6772
100	311.06	.0014524	.018026	1393.04	1151.4	2544.4	1407.56	1317.1	2724.7	3.3596	2.2544	5.6141
110	318.15	.0014886	.015987	1433.7	1096.0	2529.8	1450.1	1255.5	2705.6	3.4295	2.1233	5.5527
120	324.75	.0015267	.014263	1473.0	1040.7	2513.7	1491.3	1193.6	2684.9	3.4962	1.9962	5.4924
140	336.75	.0016107	.011485	1548.6	928.2	2476.8	1571.1	1066.5	2637.6	3.6232	1.7485	5.3717
160	347.44	.0017107	.009306	1622.7	809.0	2431.7	1650.1	930.6	2580.6	3.7461	1.4994	5.2455
180	357.06	.0018397	.007489	1698.9	675.4	2374.3	1732.0	777.1	2509.1	3.8715	1.2329	5.1044
200	365.81	.002036	.005834	1785.6	507.5	2293.0	1826.3	583.4	2409.7	4.0139	.9130	4.9269
210	369.89	.002207	.004952	1842.1	388.5	2230.6	1888.4	446.2	2334.6	4.1075	.6938	4.8013
220.9	374.14	.003155	.003155	2029.6	0	2029.6	2099.3	0	2099.3	4.4298	0	4.4298

P = Bars = 10^5 N/m^2
t = °C
u, h = kJ/kg
v = m^3/kg
s = kJ/kg °K

Table 3. Super heated Vapor

Abs Press. Bar (Sat. Temp °C)		Temperature °C									
		50	**75**	**100**	**120**	**150**	**170**	**200**	**220**	**250**	**270**
	v	14.869	16.034	17.196	18.123	19.512	20.438	21.825	22.749	24.136	25.059
0.10	u	2.4439	2.4796	2.5155	2.5444	2.5879	2.6171	2.6613	2.6910	2.7360	2.7662
(45.81)	h	2592.6	2640.0	2687.5	2725.6	2783.0	2821.5	2879.5	2918.5	2977.3	3016.8
	s	8.1749	8.3162	8.4479	8.5474	8.6882	8.7770	8.9038	8.9844	9.1002	9.1743
	v		6.393	6.863	7.237	7.795	8.167	8.723	9.094	9.649	10.019
0.25	u		2.4777	2.5141	2.5432	2.5870	2.6164	2.6607	2.6905	2.7356	2.7659
(64.97)	h		2637.5	2685.6	2724.1	2781.9	2820.6	2878.8	2917.9	2976.8	3016.4
	s		7.8877	8.0212	8.1216	8.2633	8.3526	8.4798	8.5606	8.6766	8.7508
	v			3.418	3.608	3.889	4.076	4.356	4.542	4.820	5.006
0.50	u			2.5116	2.5413	2.5856	2.6152	2.6599	2.6898	2.7350	2.7654
(81.33)	h			2682.5	2721.6	2780.1	2819.1	2877.7	2916.9	2976.0	3015.7
	s			7.6947	7.7968	7.9401	8.0300	8.1580	8.2392	8.3556	8.4300

Abs Press. Bar (Sat. Temp °C)		**300**	**340**	**400**	**440**	**500**	**600**	**700**	**750**	**800**	**850**
	v	26.445	28.292	31.063	32.909	35.679	40.295	44.911	47.218	49.526	51.834
0.10	u	2.8121	2.8741	2.9689	3.0335	3.1323	3.3025	3.4796	3.5708	3.6638	3.7586
(45.81)	h	3076.5	3157.0	3279.6	3362.6	3489.1	3705.4	3928.7	4043.0	4159.0	4276.9
	s	9.2813	9.4170	9.6077	9.7275	9.8978	10.1608	10.4028	10.5174	10.6281	10.7354
	v	10.574	11.314	12.423	13.162	14.270	16.117	17.963	18.887	19.810	20.733
0.25	u	2.8118	2.8738	2.9688	3.0333	3.1322	3.3024	3.4795	3.5708	3.6638	3.7585
(64.97)	h	3076.2	3156.7	3279.3	3362.4	3488.9	3705.3	3928.6	4043.0	4159.0	4276.9
	s	8.8579	8.9937	9.1845	9.3044	9.4748	9.7378	9.9799	10.0945	10.2052	10.3125
	v	5.284	5.654	6.209	6.579	7.134	8.057	8.981	9.443	9.904	10.366
0.50	u	2.8113	2.8735	2.9685	3.0331	3.1320	3.3022	3.4794	3.5706	3.6636	3.7584
(81.33)	h	3075.5	3156.2	3278.9	3362.0	3488.7	3705.1	3928.5	4042.8	4158.9	4276.7
	s	8.5373	8.6732	8.8642	8.9841	9.1546	9.4178	9.6599	9.7744	9.8852	9.9926

Abs Press. Bar (Sat. Temp °C)		**900**	**950**	**1000**	**1100**	**1200**
	v	54.141	56.449	58.757	63.372	67.987
0.10	u	3.8550	3.9532	4.0530	4.2575	4.4679
(45.81)	h	4396.4	4517.7	4640.6	4891.2	5147.8
	s	10.8396	10.9408	11.0393	11.2287	11.4091
	v	21.656	22.579	23.502	25.349	27.195
0.25	u	3.8550	3.9532	4.0530	4.2574	4.4679
(64.97)	h	4396.4	4517.6	4640.5	4891.2	5147.8
	s	10.4167	10.5179	10.6163	10.8058	10.9862
	v	10.828	11.289	11.751	12.674	13.597
0.50	u	3.8549	3.9531	4.0529	4.2574	4.4678
(81.33)	h	4396.3	4517.6	4640.5	4891.1	5147.7
	s	10.0967	10.1979	10.2964	10.4859	10.6662

Table 3. Super heated Vapor—Continued

Abs Press. Bar (Sat. Temp °C)		Temperature °C									
		100	**120**	**150**	**170**	**200**	**220**	**250**	**270**	**300**	**340**
	v	2.270	2.398	2.587	2.713	2.900	3.025	3.211	3.335	3.520	3.768
0.75	u	2.5092	2.5393	2.5842	2.6141	2.6590	2.6890	2.7344	2.7648	2.8109	2.8731
(91.78)	h	2679.4	2719.1	2778.2	2817.5	2876.5	2915.9	2975.2	3014.9	3074.9	3155.7
	s	7.5009	7.6046	7.7496	7.8403	7.9690	8.0505	8.1673	8.2418	8.3493	8.4855
	v	1.6958	1.7929	1.9364	2.0311	2.172	2.266	2.406	2.499	2.639	2.824
1.0	u	2.5067	2.5373	2.5828	2.6129	2.6581	2.6882	2.7337	2.7643	2.8104	2.8727
(99.63)	h	2676.2	2716.6	2776.4	2816.0	2875.3	2914.8	2974.3	3014.2	3074.3	3155.2
	s	7.3614	7.4668	7.6134	7.7048	7.8343	7.9162	8.0333	8.1080	8.2158	8.3521
	v		1.4904	1.6108	1.6902	1.8083	1.8866	2.004	2.081	2.198	2.353
1.2	u		2.5357	2.5816	2.6119	2.6573	2.6876	2.7332	2.7638	2.8101	2.8724
(104.80)	h		2714.5	2774.9	2814.8	2874.3	2914.0	2973.7	3013.6	3073.8	3154.8
	s		7.3785	7.5265	7.6185	7.7486	7.8308	7.9482	8.0231	8.1310	8.2675
		400	**440**	**500**	**540**	**600**	**640**	**700**	**750**	**800**	**850**
	v	4.138	4.385	4.755	5.001	5.371	5.617	5.987	6.295	6.603	6.910
0.75	u	2.9682	3.0328	3.1318	3.1991	3.3021	3.3721	3.4793	3.5705	3.6636	3.7584
(91.78)	h	3278.5	3361.7	3488.4	3574.2	3704.9	3793.4	3928.3	4042.6	4158.7	4276.6
	s	8.6767	8.7967	8.9672	9.0754	9.2305	9.3296	9.4727	9.5872	9.6980	9.8054
	v	3.103	3.288	3.565	3.750	4.028	4.213	4.490	4.721	4.952	5.183
1.0	u	2.9679	3.0326	3.1316	3.1989	3.3019	3.3720	3.4792	3.5704	3.6635	3.7583
(99.63)	h	3278.2	3361.4	3488.1	3574.0	3704.7	3793.3	3928.2	4042.5	4158.6	4276.5
	s	8.5435	8.6636	8.8342	8.9424	9.0976	9.1967	9.3398	9.4544	9.5652	9.6725
	v	2.585	2.739	2.971	3.125	3.356	3.510	3.741	3.934	4.126	4.319
1.2	u	2.9677	3.0324	3.1314	3.1988	3.3018	3.3719	3.4791	3.5704	3.6634	3.7582
(104.80)	h	3277.8	3361.1	3487.9	3573.8	3704.6	3793.1	3928.1	4042.4	4158.6	4276.5
	s	8.4590	8.5791	8.7498	8.8581	9.0133	9.1125	9.2555	9.3701	9.4810	9.5883
		900	**950**	**1000**	**1100**	**1200**					
	v	7.218	7.526	7.834	8.449	9.065					
0.75	u	3.8549	3.9530	4.0529	4.2573	4.4678					
(91.78)	h	4396.2	4517.5	4640.4	4891.0	5147.6					
	s	9.9095	10.0107	10.1092	10.2987	10.4791					
	v	5.414	5.644	5.875	6.337	6.799					
1.0	u	3.8548	3.9530	4.0528	4.2573	4.4677					
(99.63)	h	4396.1	4517.4	4640.3	4891.0	5147.6					
	s	9.7767	9.8779	9.9764	10.1659	10.3463					
	v	4.511	4.704	4.896	5.281	5.666					
1.2	u	3.8547	3.9529	4.0527	4.2572	4.4677					
(104.80)	h	4396.1	4517.3	4640.3	4890.9	5147.5					
	s	9.6925	9.7937	9.8922	10.0817	10.2621					

u, h = kJ/kg; v = m³/kg; s = kJ/kg °K

Table 3. Super heated Vapor—Continued

Abs Press. Bar (Sat. Temp °C)		Temperature °C									
		120	**150**	**170**	**200**	**220**	**250**	**270**	**300**	**340**	**400**
	v	1.1878	1.2853	1.3493	1.4443	1.5073	1.6012	1.6636	1.7570	1.8812	2.067
1.5	u	2.5333	2.5798	2.6105	2.6562	2.6867	2.7325	2.7632	2.8095	2.8720	2.9673
(111.37)	h	2711.4	2772.6	2812.9	2872.9	2912.8	2972.7	3012.7	3073.1	3154.1	3277.4
	s	7.2693	7.4193	7.5123	7.6433	7.7259	7.8438	7.9189	8.0270	8.1638	8.3555
	v		.9596	1.0083	1.0803	1.1279	1.1988	1.2458	1.3162	1.4096	1.5493
2.0	u		2.5769	2.6081	2.6544	2.6851	2.7312	2.7621	2.8086	2.8712	2.9667
(120.23)	h		2768.8	2809.8	2870.5	2910.7	2971.0	3011.2	3071.8	3153.1	3276.6
	s		7.2795	7.3740	7.5066	7.5899	7.7086	7.7841	7.8926	8.0298	8.2218
	v		.6339	.6673	.7163	.7485	.7964	.8280	.8753	.9380	1.0315
3.0	u		2.5708	2.6031	2.6507	2.6820	2.7287	2.7598	2.8067	2.8697	2.9656
(133.55)	h		2761.0	2803.3	2865.6	2906.5	2967.6	3008.3	3069.3	3151.1	3275.0
	s		7.0778	7.1757	7.3115	7.3963	7.5166	7.5928	7.7022	7.8401	8.0330
	v		.4708	.4966	.5342	.5588	.5951	.6191	.6548	.7021	.7726
4.0	u		2.5645	2.5981	2.6468	2.6787	2.7261	2.7576	2.8048	2.8682	2.9644
(143.63)	h		2752.8	2796.7	2860.5	2902.3	2964.2	3005.2	3066.8	3149.0	3273.4
	s		6.9299	7.0312	7.1706	7.2570	7.3789	7.4559	7.5662	7.7049	7.8985
		440	**500**	**540**	**600**	**640**	**700**	**750**	**800**	**850**	**900**
	v	2.191	2.376	2.500	2.685	2.808	2.993	3.147	3.301	3.455	3.609
1.5	u	3.0321	3.1312	3.1986	3.3017	3.3717	3.4790	3.5702	3.6633	3.7581	3.8546
(111.37)	h	3360.7	3487.6	3573.5	3704.3	3792.9	3927.9	4042.3	4158.4	4276.3	4396.0
	s	8.4757	8.6466	8.7549	8.9101	9.0093	9.1524	9.2670	9.3779	9.4853	9.5894
	v	1.6422	1.7814	1.8741	2.013	2.106	2.244	2.360	2.475	2.591	2.706
2.0	u	3.0316	3.1308	3.1982	3.3014	3.3715	3.4788	3.5701	3.6631	3.7580	3.8545
(120.23)	h	3360.0	3487.1	3573.0	3704.0	3792.6	3927.6	4042.0	4158.2	4276.2	4395.8
	s	8.3423	8.5133	8.6217	8.7770	8.8763	9.0194	9.1341	9.2449	9.3524	9.4566
	v	1.0937	1.1867	1.2486	1.3414	1.4031	1.4957	1.5729	1.6499	1.7270	1.8041
3.0	u	3.0306	3.1300	3.1975	3.3008	3.3710	3.4784	3.5698	3.6629	3.7577	3.8542
(133.55)	h	3358.7	3486.0	3572.1	3703.2	3791.9	3927.1	4041.6	4157.8	4275.8	4395.4
	s	8.1538	8.3251	8.4337	8.5892	8.6886	8.8319	8.9467	9.0576	9.1650	9.2692
	v	.8194	.8893	.9359	1.0055	1.0519	1.1215	1.1794	1.2372	1.2951	1.3529
4.0	u	3.0296	3.1292	3.1968	3.3002	3.3705	3.4779	3.5693	3.6624	3.7574	3.8539
(143.63)	h	3357.4	3484.9	3571.2	3702.4	3791.3	3926.5	4041.0	4157.3	4275.4	4395.1
	s	8.0196	8.1913	8.3001	8.4558	8.5553	8.6987	8.8134	8.9244	9.0319	9.1362
		950	**1000**	**1100**	**1200**						
	v	3.763	3.917	4.225	4.532						
1.5	u	3.9528	4.0527	4.2571	4.4676						
(111.37)	h	4517.2	4640.2	4890.8	5147.5						
	s	9.6907	9.7892	9.9787	10.1591						

Table 3. Super heated Vapor—Continued

Abs Press. Bar (Sat. Temp) °C		Temperature °C									
		950	**1000**	**1100**	**1200**						
	v	2.822	2.937	3.168	3.399						
2.0	u	3.9527	4.0525	4.2570	4.4675						
(120.23)	h	4517.1	4640.0	4890.7	5147.3						
	s	9.5578	9.6563	9.8458	10.0262						
	v	1.8811	1.9581	2.1121	2.2661						
3.0	u	3.9524	4.0523	4.2568	4.4672						
(133.55)	h	4516.7	4639.7	4890.4	5147.1						
	s	9.3705	9.4690	9.6585	9.8389						
	v	1.4107	1.4685	1.5840	1.6996						
4.0	u	3.9522	4.0520	4.2565	4.4670						
(143.63)	h	4516.4	4639.4	4890.2	5146.8						
	s	9.2375	9.3360	9.5256	9.7060						
		170	**200**	**220**	**250**	**270**	**300**	**340**	**400**	**440**	**500**
	v	.3942	.4249	.4449	.4744	.4937	.5226	.5606	.6173	.6548	.7109
5.0	u	2.5928	2.6429	2.6755	2.7235	2.7553	2.8029	2.8666	2.9632	3.0286	3.1284
(151.86)	h	2789.9	2855.4	2897.9	2960.7	3002.2	3064.2	3146.9	3271.9	3356.0	3483.9
	s	6.9162	7.0592	7.1473	7.2709	7.3487	7.4599	7.5994	7.7938	7.9152	8.0873
		540	**600**	**640**	**700**	**750**	**800**	**850**	**900**	**950**	**1000**
	v	.7482	.8041	.8412	.8969	.9433	.9896	1.0359	1.0822	1.1284	1.1747
5.0	u	3.1961	3.2996	3.3700	3.4775	3.5689	3.6621	3.7570	3.8536	3.9519	4.0518
(151.86)	h	3570.2	3701.7	3790.6	3925.9	4040.5	4156.9	4275.0	4394.7	4516.1	4639.1
	s	8.1962	8.3522	8.4517	8.5952	8.7101	8.8211	8.9287	9.0329	9.1343	9.2328
		1100	**1200**								
	v	1.2672	1.3596								
5.0	u	4.2563	4.4668								
(151.86)	h	4889.9	5146.6								
	s	9.4224	9.6029								
		170	**200**	**220**	**250**	**270**	**300**	**340**	**400**	**440**	**500**
	v	.2610	.2830	.2971	.3176	.3311	.3509	.3771	.4158	.4413	.4795
7.4	u	2.5795	2.6332	2.6673	2.7171	2.7498	2.7983	2.8629	2.9604	3.0262	3.1265
(167.23)	h	2772.6	2842.6	2887.1	2952.2	2994.7	3058.0	3141.9	3268.1	3352.8	3481.3
	s	6.7045	6.8573	6.9496	7.0776	7.1575	7.2709	7.4124	7.6086	7.7309	7.9039
	v		.2060	.2169	.2327	.2429	.2579	.2776	.3066	.3257	.3541
10.0	u		2.6219	2.6580	2.7099	2.7436	2.7932	2.8588	2.9573	3.0236	3.1244
(179.91)	h		2827.9	2874.9	2942.6	2986.5	3051.2	3136.4	3263.9	3349.3	3478.5
	s		6.6940	6.7914	6.9247	7.0069	7.1229	7.2667	7.4651	7.5883	7.7622

h = kJ/kg; v = m³/kg; s = kJ/kg °K

Table 3. Super heated Vapor—Continued

Abs Press. Bar (Sat. Temp °C)		Temperature °C									
		540	**600**	**640**	**700**	**750**	**800**	**850**	**900**	**950**	**1000**
	v	.5048	.5427	.5679	.6056	.6370	.6683	.6997	.7310	.7623	.7936
7.4	u	3.1945	3.2983	3.3687	3.4764	3.5680	3.6613	3.7563	3.8530	3.9512	4.0512
(167.23)	h	3568.0	3699.8	3789.0	3924.6	4039.4	4155.8	4274.0	4393.9	4515.3	4638.4
	s	8.0132	8.1696	8.2694	8.4132	8.5283	8.6394	8.7471	8.8514	8.9528	9.0514
	v	.3729	.4011	.4198	.4478	.4711	.4943	.5175	.5407	.5639	.5871
10.0	u	3.1926	3.2968	3.3674	3.4753	3.5670	3.6604	3.7555	3.8522	3.9505	4.0505
(179.91)	h	3565.6	3697.9	3787.2	3923.1	4038.1	4154.7	4273.0	4392.9	4514.5	4637.6
	s	7.8720	8.0290	8.1290	8.2731	8.3883	8.4996	8.6074	8.7118	8.8133	8.9119

Abs Press. Bar (Sat. Temp °C)		**1100**	**1200**
	v	.8561	.9186
7.4	u	4.2557	4.4662
(167.23)	h	4889.2	5146.0
	s	9.2411	9.4215
	v	.6335	.6798
10.0	u	4.2551	4.4656
(179.91)	h	4888.6	5145.4
	s	9.1017	9.2822

Abs Press. Bar (Sat. Temp °C)		**200**	**220**	**250**	**270**	**300**	**340**	**400**	**440**	**500**	**540**
	v	.16195	.17117	.18427	.19269	.2050	.2210	.2444	.2598	.2827	.2979
12.5	u	2.6104	2.6487	2.7028	2.7374	2.7882	2.8549	2.9543	3.0211	3.1224	3.1909
(189.84)	h	2812.8	2862.6	2933.1	2978.3	3044.5	3131.1	3259.9	3345.9	3475.8	3563.2
	s	6.5658	6.6689	6.8077	6.8925	7.0111	7.1572	7.3576	7.4817	7.6566	7.7669
15.0	v	.13248	.14059	.15195	.15918	.16966	.18322	.2030	.2160	.2352	.2478
	u	2.5981	2.6388	2.6953	2.7311	2.7831	2.8508	2.9513	3.0185	3.1203	3.1891
(198.32)	h	2796.8	2849.7	2923.3	2969.9	3037.6	3125.7	3255.8	3342.5	3473.1	3560.9
	s	6.4546	6.5641	6.7090	6.7965	6.9179	7.0664	7.2690	7.3940	7.5698	7.6805

Abs Press. Bar (Sat. Temp °C)		**600**	**640**	**700**	**750**	**800**	**850**	**900**	**950**	**1000**	**1100**
	v	.3205	.3355	.3580	.3766	.3953	.4139	.4325	.4510	.4696	.5067
12.5	u	3.2953	3.3661	3.4742	3.5660	3.6595	3.7547	3.8515	3.9499	4.0499	4.2545
(189.84)	h	3695.9	3785.5	3921.7	4036.8	4153.6	4272.0	4392.1	4513.7	4636.9	4887.9
	s	7.9243	8.0246	8.1690	8.2844	8.3958	8.5037	8.6082	8.7097	8.8085	8.9982
	v	.2668	.2793	.2981	.3137	.3292	.3448	.3603	.3758	.3913	.4222
15.0	u	3.2939	3.3648	3.4731	3.5651	3.6587	3.7539	3.8508	3.9492	4.0492	4.2539
(198.32)	h	3694.0	3783.8	3920.2	4035.6	4152.5	4271.1	4391.2	4512.9	4636.1	4887.2
	s	7.8385	7.9391	8.0837	8.1993	8.3109	8.4189	8.5235	8.6251	8.7238	8.9137

Table 3. Super heated Vapor—Continued

Abs Press. Bar (Sat. Temp °C)		Temperature °C: 1200
	v	.5438
12.5	u	4.4650
(189.84)	h	5144.8
	s	9.1788
	v	.4531
15.0	u	4.4644
(198.32)	h	5144.1
	s	9.0942

Abs Press. Bar (Sat. Temp °C)		220	250	270	300	340	400	440	500	540	600
	v	.11867	.12883	.13521	.14442	.15624	.17340	.18460	.2012	.2121	.2284
17.5	u	2.6284	2.6876	2.7246	2.7779	2.8468	2.9483	3.0160	3.1183	3.1873	3.2924
(205.76)	h	2836.1	2913.1	2961.2	3030.6	3120.2	3251.8	3339.0	3470.3	3558.5	3692.1
	s	6.4711	6.6227	6.7131	6.8375	6.9886	7.1933	7.3192	7.4960	7.6072	7.7657

Abs Press. Bar (Sat. Temp °C)		640	700	800	850	900	950	1000	1100	1200
	v	.2392	.2553	.2821	.2954	.3087	.3220	.3353	.3619	.3884
17.5	u	3.3635	3.4720	3.6578	3.7532	3.8501	3.9485	4.0486	4.2533	4.4638
(205.76)	h	3.7821	3918.8	4151.4	4270.1	4390.3	4512.1	4635.4	4886.6	5143.5
	s	7.8665	8.0115	8.2390	8.3470	8.4517	8.5534	8.6522	8.8421	9.0227

Abs Press. Bar (Sat. Temp °C)		220	250	270	300	340	400	440	500	540	600
	v	.10215	.11144	.11721	.12547	.13600	.15120	.16108	.17568	.18529	.19960
20.0	u	2.6174	2.6796	2.7179	2.7726	2.8426	2.9452	3.0134	3.1162	3.1856	3.2909
(212.42)	h	2821.7	2902.5	2952.3	3023.5	3114.6	3247.6	3335.5	3467.6	3556.1	3690.1
	s	6.3861	6.5453	6.6388	6.7664	6.9201	7.1271	7.2540	7.4317	7.5434	7.7024
	v	.08923	.09788	.10319	.11071	.12025	.13392	.14279	.15585	.16444	.17721
22.5	u	2.6057	2.6713	2.7110	2.7671	2.8384	2.9422	3.0108	3.1142	3.1838	3.2895
(218.45)	h	2806.4	2891.5	2943.2	3016.3	3109.0	3243.5	3332.1	3464.8	3553.8	3688.2
	s	6.3068	6.4744	6.5713	6.7023	6.8588	7.0681	7.1959	7.3747	7.4868	7.6463
	v		.08700	.09195	.09890	.10764	.12010	.12815	.13998	.14776	.15930
25.0	u		2.6626	2.7039	2.7616	2.8342	2.9391	3.0082	3.1121	3.1820	3.2880
(223.99)	h		2880.1	2933.7	3008.8	3103.3	3239.3	3328.6	3462.1	3551.4	3686.3
	s		6.4085	6.5091	6.6438	6.8031	7.0148	7.1436	7.3234	7.4360	7.5960

Abs Press. Bar (Sat. Temp °C)		640	700	800	850	900	950	1000	1100	1200
	v	.20908	.2232	.2467	.2584	.2700	.2817	.2933	.3166	.3398
20.0	u	3.3622	3.4709	3.6570	3.7524	3.8493	3.9479	4.0480	4.2527	4.4633
(212.42)	h	3780.4	3917.4	4150.3	4269.1	4389.4	4511.2	4634.6	4885.9	5142.9
	s	7.8035	7.9487	8.1765	8.2847	8.3895	8.4912	8.5901	8.7800	8.9607

u, h = kJ/kg; v = m^3/kg; s = kJ/kg °K

Table 3. Super heated Vapor—Continued

Abs Press. Bar (Sat. Temp °C)		Temperature °C									
		640	**700**	**800**	**850**	**900**	**950**	**1000**	**1100**	**1200**	
	v	.18567	.19828	.2192	.2296	.2400	.2503	.2607	.2814	.3021	
22.5	u	3.3610	3.4698	3.6561	3.7516	3.8486	3.9472	4.0473	4252.1	4.4627	
(218.45)	h	3778.7	3916.0	4149.2	4268.1	4388.5	4510.4	4633.9	4885.2	5142.3	
	s	7.7477	7.8933	8.1214	8.2297	8.3345	8.4363	8.5352	8.7252	8.9059	
	v	.16694	.17832	.19716	.20654	.21590	.22524	.2346	.2532	.2718	
25.0	u	3.3596	3.4687	3.6553	3.7508	3.8479	3.9465	4.0467	4.2515	4.4621	
(223.99)	h	3777.0	3914.5	4148.2	4267.2	4387.6	4509.6	4633.1	4884.6	5141.7	
	s	7.6976	7.8435	8.0720	8.1804	8.2853	8.3871	8.4861	8.6762	8.8569	
		250	**270**	**300**	**340**	**400**	**440**	**500**	**540**	**600**	**640**
	v	.07807	.08273	.08922	.09732	.10879	.11617	.12700	.13411	.14465	.15161
27.5	u	2.6535	2.6965	2.7559	2.8299	2.9359	3.0056	3.1100	3.1802	3.2865	3.3583
(229.12)	h	2868.2	2924.0	3001.2	3097.5	3235.1	3325.0	3459.3	3549.0	3684.3	3775.3
	s	6.3464	6.4511	6.5896	6.7519	6.9661	7.0959	7.2767	7.3898	7.5503	7.6522
		700	**800**	**850**	**900**	**950**	**1000**	**1100**	**1200**		
	v	.16199	.17915	.18769	.19621	.20472	.2132	.2302	.2471		
27.5	u	3.4676	3.6544	3.7500	3.8472	3.9459	4.0460	4.2509	4.4615		
(229.12)	h	3913.1	4147.0	4266.2	4386.8	4508.8	4632.4	4883.9	5141.1		
	s	7.7984	8.0272	8.1357	8.2407	8.3426	8.4416	8.6318	8.8125		
		250	**270**	**300**	**340**	**400**	**440**	**500**	**540**	**600**	**640**
	v	.07058	.07502	.08114	.08871	.09936	.10619	.11619	.12273	.13243	.13884
30.0	u	2.6440	2.6889	2.7501	2.8255	2.9328	3.0029	3.1080	3.1784	3.2850	3.3570
(233.90)	h	2855.8	2913.9	2993.5	3091.6	3230.9	3321.5	3456.5	3546.6	3682.3	3773.5
	s	6.2872	6.3963	6.5390	6.7045	6.9212	7.0520	7.2338	7.3474	7.5085	7.6106
		700	**800**	**850**	**900**	**950**	**1000**	**1100**	**1200**		
	v	.14838	.16414	.17198	.17980	.18761	.19541	.21098	.22652		
30.0	u	3.4665	3.6535	3.7493	3.8465	3.9452	4.0454	4.2503	4.4609		
(233.90)	h	3911.7	4145.9	4265.2	4385.9	4508.0	4631.6	4883.3	5140.5		
	s	7.7571	7.9862	8.0948	8.1999	8.3019	8.4009	8.5912	8.7720		
		250	**270**	**300**	**340**	**400**	**440**	**500**	**540**	**600**	**640**
	v	.05872	.06286	.06842	.07517	.08453	.09050	.09918	.10486	.11324	.11877
35.0	u	2.6237	2.6728	2.7380	2.8165	2.9264	2.9976	3.1038	3.1747	3.2821	3.3544
(242.60)	h	2829.2	2892.8	2977.5	3079.6	3222.3	3314.4	3450.9	3541.7	3678.4	3770.1
	s	6.1749	6.2942	6.4461	6.6184	6.8405	6.9734	7.1572	7.2717	7.4339	7.5366

Table 3. Super heated Vapor—Continued

Abs Press. Bar (Sat. Temp °C)		Temperature °C									
		700	**800**	**850**	**900**	**950**	**1000**	**1100**	**1200**		
	v	.12699	.14056	.14730	.15402	.16073	.16743	.18080	.19415		
35.0	u	3.4643	3.6518	3.7477	3.8450	3.9438	4.0441	4.2492	4.4598		
(242.60)	h	3908.8	4143.7	4263.2	4384.1	4506.4	4630.1	4881.9	5139.3		
	s	7.6837	7.9134	8.0223	8.1276	8.2297	8.3288	8.5192	8.7000		
		270	**300**	**340**	**400**	**440**	**500**	**540**	**600**	**640**	**700**
	v	.05365	.05884	.06499	.07341	.07872	.08643	.09145	.09885	.10372	.11095
40.0	u	2.6555	2.7253	2.8073	2.9199	2.9922	3.0995	3.1711	3.2791	3.3518	3.4621
(250.40)	h	2870.1	2960.7	3067.3	3213.6	3307.1	3445.3	3536.9	3674.4	3766.6	3905.9
	s	6.1990	6.3615	6.5413	6.7690	6.9041	7.0901	7.2056	7.3688	7.4720	7.6198
		800	**850**	**900**	**950**	**1000**	**1100**	**1200**			
	v	.12287	.12879	.13469	.14057	.14645	.15817	.16987			
40.0	u	3.6500	3.7461	3.8436	3.9425	4.0429	4.2480	4.4586			
(250.40)	h	4141.5	4261.3	4382.3	4504.8	4628.7	4880.6	5138.1			
	s	7.8502	7.9593	8.0647	8.1669	8.2662	8.4567	8.6376			
		270	**300**	**340**	**400**	**440**	**500**	**540**	**600**	**640**	**700**
	v	.04641	.05135	.05706	.06475	.06956	.07651	.08102	.08765	.09201	.09847
45.0	u	2.6367	2.7120	2.7978	2.9133	2.9868	3.0953	3.1674	3.2760	3.3491	3.4599
(257.49)	h	2845.6	2943.1	3054.6	3204.7	3299.8	3439.6	3531.9	3670.5	3763.1	3903.0
	s	6.1079	6.2828	6.4710	6.7047	6.8420	7.0301	7.1467	7.3110	7.4148	7.5631
		800	**850**	**900**	**950**	**1000**	**1100**	**1200**			
	v	.10911	.11439	.11965	.12489	.13013	.14056	.15098			
45.0	u	3.6483	3.7446	3.8422	3.9412	4.0416	4.2468	4.4575			
(257.49)	h	4139.3	4259.3	4380.6	4503.2	4627.2	4879.3	5136.9			
	s	7.7942	7.9035	8.0091	8.1115	8.2108	8.4015	8.5825			
		270	**300**	**340**	**400**	**440**	**500**	**540**	**600**	**640**	**700**
	v	.04054	.04532	.05070	.05781	.06223	.06857	.07267	.07869	.08264	.08849
50.0	u	2.6162	2.6980	2.7880	2.9066	2.9813	3.0910	3.1637	3.2730	3.3464	3.4576
(263.99)	h	2818.9	2924.5	3041.5	3195.7	3292.4	3433.8	3527.0	3666.5	3759.6	3900.1
	s	6.0189	6.2084	6.4058	6.6459	6.7855	6.9759	7.0934	7.2589	7.3632	7.5122
		800	**850**	**900**	**950**	**1000**	**1100**	**1200**			
	v	.09811	.10287	.10762	.11235	.11707	.12648	.13587			
50.0	u	3.6466	3.7430	3.8407	3.9398	4.0404	4.2456	4.4563			
(263.99)	h	4137.1	4257.4	4378.8	4501.6	4625.7	4878.0	5135.7			
	s	7.7440	7.8535	7.9593	8.0618	8.1612	8.3520	8.5331			

h = kJ/kg; v = m^3/kg; s = kJ/kg °K

Table 3. Super heated Vapor—Continued

Abs Press. Bar (Sat. Temp °C)		Temperature °C									
		300	**340**	**400**	**440**	**500**	**540**	**600**	**640**	**700**	**800**
	v	.04034	.04548	.05213	.05622	.06207	.06584	.07136	.07498	.08033	.08911
55.0	u	2.6831	2.7779	2.8998	2.9757	3.0866	3.1599	3.2700	3.3437	3.4554	3.6448
(270.02)	h	2905.0	3028.0	3186.5	3284.9	3428.0	3522.0	3662.4	3756.1	3897.2	4134.9
	s	6.1369	6.3446	6.5915	6.7335	6.9262	7.0448	7.2114	7.3163	7.4659	7.6984
		850	**900**	**950**	**1000**	**1100**	**1200**				
	v	.09345	.09778	.10209	.10639	.11496	.12351				
55.0	u	3.7414	3.8393	3.9385	4.0391	4.2444	4.4552				
(270.02)	h	4255.4	4377.1	4500.0	4624.2	4876.7	5134.5				
	s	7.8081	7.9141	8.0167	8.1163	8.3072	8.4883				
		300	**340**	**400**	**440**	**500**	**540**	**600**	**640**	**700**	**800**
	v	.03616	.04111	.04739	.05122	.05665	.06015	.06525	.06859	.07352	.08160
60.0	u	2.6672	2.7674	2.8929	2.9700	3.0822	3.1561	3.2669	3.3410	3.4531	3.6431
(275.64)	h	2884.2	3014.0	3177.2	3277.3	3422.2	3517.0	3658.4	3752.6	3894.2	4132.7
	s	6.0674	6.2866	6.5408	6.6853	6.8803	6.9999	7.1677	7.2731	7.4234	7.6566
	v	.02947	.03421	.03993	.04334	.04814	.05120	.05565	.05855	.06283	.06981
70.0	u	2.6322	2.7452	2.8786	2.9585	3.0734	3.1485	3.2607	3.3356	3.4485	3.6395
(285.88)	h	2838.4	2984.7	3158.1	3261.9	3410.3	3506.9	3650.3	3745.4	3888.3	4128.2
	s	5.9305	6.1776	6.4478	6.5976	6.7975	6.9193	7.0894	7.1960	7.3476	7.5822
		850	**900**	**950**	**1000**	**1100**	**1200**				
	v	.08560	.08958	.09354	.09749	.10536	.11321				
60.0	u	3.7398	3.8378	3.9372	4.0378	4.2433	4.4540				
(275.64)	h	4253.4	4375.3	4498.4	4622.7	4875.4	5133.3				
	s	7.7666	7.8727	7.9755	8.0751	8.2661	8.4474				
	v	.07326	.07669	.08010	.08350	.09027	.09703				
70.0	u	3.7367	3.8350	3.9345	4.0353	4.2409	4.4517				
(285.88)	h	4249.5	4371.8	4495.2	4619.8	4872.8	5130.9				
	s	7.6926	7.7991	7.9021	8.0020	8.1933	8.3747				
		340	**400**	**440**	**500**	**540**	**600**	**640**	**700**	**800**	**850**
	v	.02897	.03432	.03742	.04175	.04448	.04845	.05102	.05481	.06097	.06401
80.0	u	2.7213	2.8638	2.9467	3.0643	3.1408	3.2544	3.3301	3.4439	3.6360	3.7335
(295.06)	h	2953.1	3138.3	3246.1	3398.3	3496.7	3642.0	3738.3	3882.4	4123.8	4245.6
	s	6.0747	6.3634	6.5190	6.7240	6.8481	7.0206	7.1282	7.2812	7.5173	7.6282
		900	**950**	**1000**	**1100**	**1200**					
	v	.06702	.07002	.07301	.07896	.08489					
80.0	u	3.8321	3.9318	4.0328	4.2386	4.4495					
(295.06)	h	4368.3	4492.0	4616.9	4870.3	5128.5					
	s	7.7351	7.8384	7.9384	8.1300	8.3115					

Table 3. Super heated Vapor—Continued

Abs Press. Bar (Sat. Temp °C)		Temperature °C									
		340	**400**	**440**	**500**	**540**	**600**	**640**	**700**	**800**	**850**
	v	.02484	.02993	.03281	.03677	.03926	.04285	.04517	.04857	.05409	.05681
90.0	u	2.6953	2.8484	2.9346	3.0552	3.1329	3.2481	3.3245	3.4393	3.6325	3.7303
(303.40)	h	2918.8	3117.8	3229.9	3386.1	3486.3	3633.7	3731.0	3876.5	4119.3	4241.6
	s	5.9751	6.2854	6.4471	6.6576	6.7839	6.9589	7.0679	7.2221	7.4596	7.5710
		900	**950**	**1000**	**1100**	**1200**					
	v	.05950	.06218	.06485	.07016	.07544					
90.0	u	3.8292	3.9292	4.0303	4.2363	4.4472					
(303.40)	h	4364.8	4488.8	4614.0	4867.7	5126.2					
	s	7.6783	7.7818	7.8821	8.0740	8.2556					
		340	**400**	**440**	**500**	**540**	**600**	**640**	**700**	**800**	**850**
	v	.021472	.02641	.02911	.03279	.03508	.03837	.04048	.04358	.04859	.05105
100.0	u	2.6666	2.8324	2.9221	3.0458	3.1250	3.2417	3.3189	3.4347	3.6289	3.7272
(311.06)	h	2881.4	3096.5	3213.2	3373.7	3475.8	3625.3	3723.7	3870.5	4114.8	4237.7
	s	5.8763	6.2120	6.3805	6.5966	6.7254	6.9029	7.0131	7.1687	7.4077	7.5196
		900	**950**	**1000**	**1100**	**1200**					
	v	.05349	.05591	.05832	.06312	.06789					
100.0	u	3.8263	3.9265	4.0278	4.2340	4.4449					
(311.06)	h	4361.2	4485.6	4611.0	4865.1	5123.8					
	s	7.6272	7.7311	7.8315	8.0237	8.2055					
		400	**440**	**500**	**540**	**600**	**640**	**700**	**800**	**850**	**900**
	v	.015649	.017933	.02080	.02251	.02491	.02643	.02861	.03210	.03379	.03546
150.0	u	2.7407	2.8542	2.9966	3.0835	3.2086	3.2901	3.4109	3.6109	3.7112	3.8119
(342.24)	h	2975.5	3123.2	3308.6	3421.3	3582.3	3686.5	3840.1	4092.4	4218.0	4343.8
	s	5.8811	6.0945	6.3443	6.4864	6.6776	6.7943	6.9572	7.2040	7.3184	7.4279
		950	**1000**	**1100**	**1200**						
	v	.03711	.03875	.04200	.04523						
150.0	u	3.9133	4.0154	4.2226	4.4338						
(342.24)	h	4469.9	4596.6	4852.6	5112.3						
	s	7.5332	7.6348	7.8283	8.0108						
		400	**440**	**500**	**540**	**600**	**640**	**700**	**800**	**850**	**900**
	v	.009942	.012223	.014768	.016213	.018178	.019397	.02113	.02385	.02516	.02645
200.0	u	2.6193	2.7749	2.9429	3.0393	3.1740	3.2602	3.3864	3.5927	3.6951	3.7975
(365.81)	h	2818.1	3019.4	3238.2	3363.5	3537.6	3648.1	3809.0	4069.7	4198.3	4326.4
	s	5.5540	5.8450	6.1401	6.2982	6.5048	6.6286	6.7993	7.0544	7.1715	7.2830

u, h = kJ/kg; v = m^3/kg; s = kJ/kg °K

Table 3. Super heated Vapor—Continued

Abs Press. Bar (Sat. Temp °C)		Temperature °C			
		950	**1000**	**1100**	**1200**
	v	.02771	.02897	.03145	.03391
200.0	u	3.9001	4.0031	4.2113	4.4228
(365.81)	h	4454.3	4582.5	4840.2	5101.0
	s	7.3898	7.4925	7.6874	7.8707

Abs Press. Bar (Sat. Temp °C)		**400**	**440**	**500**	**540**	**600**	**640**	**700**	**800**	**850**	**900**
	v	.002790	.006228	.008678	.009891	.011446	.012374	.013661	.015623	.016548	.017448
300.0	u	2.0674	2.5634	2.8207	2.9425	3.1005	3.1975	3.3358	3.5555	3.6626	3.7685
	h	2151.1	2750.3	3081.1	3239.2	3443.9	3568.7	3745.6	4024.2	4159.0	4291.9
	s	4.4728	5.3434	5.7905	5.9900	6.2331	6.3729	6.5606	6.8332	6.9560	7.0718

Abs Press. Bar (Sat. Temp °C)		**950**	**1000**	**1100**	**1200**
	v	.018329	.019196	.020903	.022589
300.0	u	3.8738	3.9788	4.1892	4.4013
	h	4423.6	4554.7	4816.3	5079 0
	s	7.1817	7.2867	7.4845	7.6692

Abs Press. Bar (Sat. Temp °C)		**400**	**440**	**500**	**540**	**600**	**640**	**700**	**800**	**850**	**900**
	v	.0019077	.003206	.005622	.006737	.008094	.008878	.009941	.011523	.012255	.012962
400.0	u	1.8546	2.2661	2.6784	2.8357	3.0226	3.1319	3.2836	3.5178	3.6297	3.7394
	h	1930.9	2394.3	2903.3	3105.2	3346.4	3487.0	3.6812	3978.7	4120.0	4257.9
	s	4.1135	4.7809	5.4700	5.7248	6.0114	6.1689	6.3750	6.6662	6.7948	6.9150

Abs Press. Bar (Sat. Temp °C)		**950**	**1000**	**1100**	**1200**
	v	.013650	.014324	.015642	.016940
400.0	u	3.8475	3.9546	4.1674	4.3801
	h	4393.6	4527.6	4793.1	5057.7
	s	7.0282	7.1356	7.3364	7.5224

P = Bars = 10^5 N/m^2
T = °C
u, h = kJ/kg
v = m^3/kg
s = kJ/kg °K

reduced property correlation charts

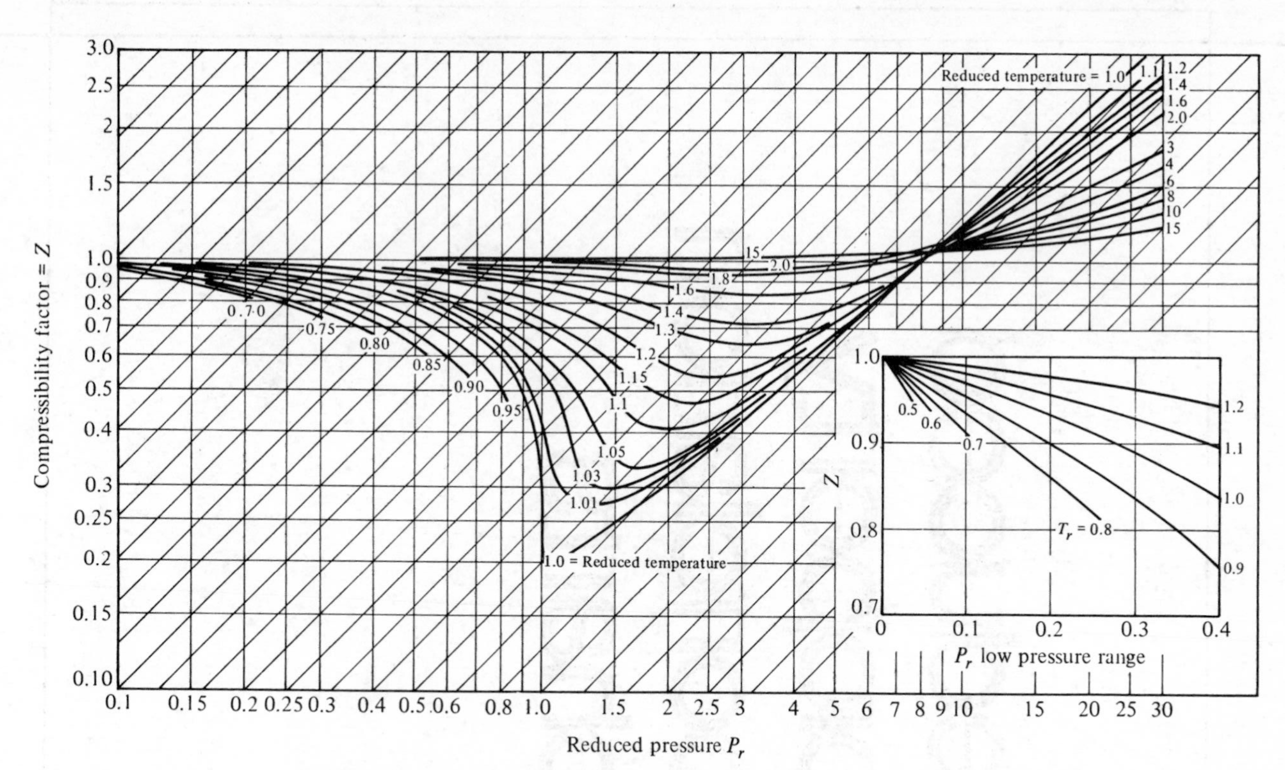

Figure C-1. Compressibility factors of gases and vapors. (From O. A. Hougen and K. M. Watson, *Chemical Process Principles Charts*, 1st ed., John Wiley & Sons, New York, 1946.)

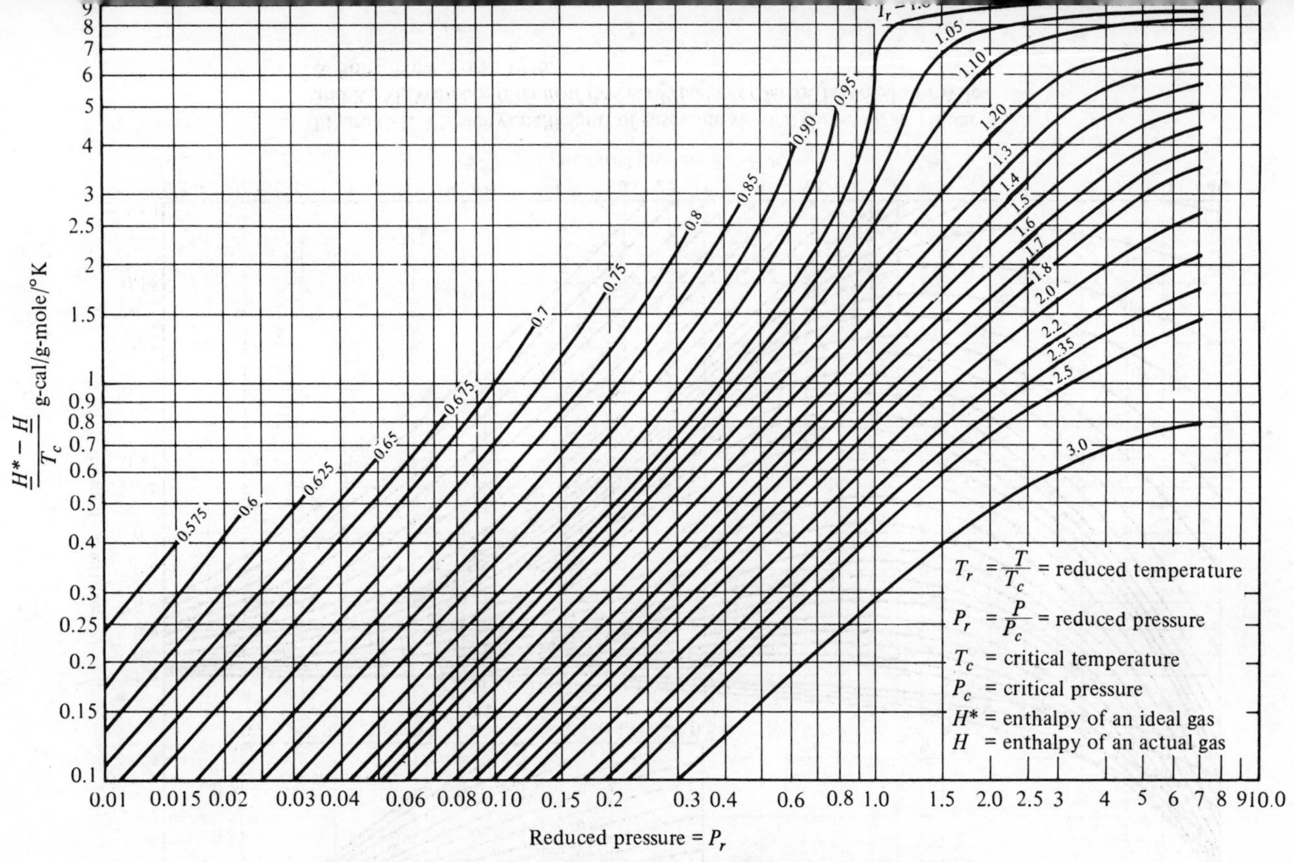

Figure C-2. Enthalpy correction of gases and vapors due to lack of ideal behavior. (From O. A. Hougen and K. M. Watson, *Chemical Process Principles Charts*, 1st ed., John Wiley & Sons, New York, 1946.)

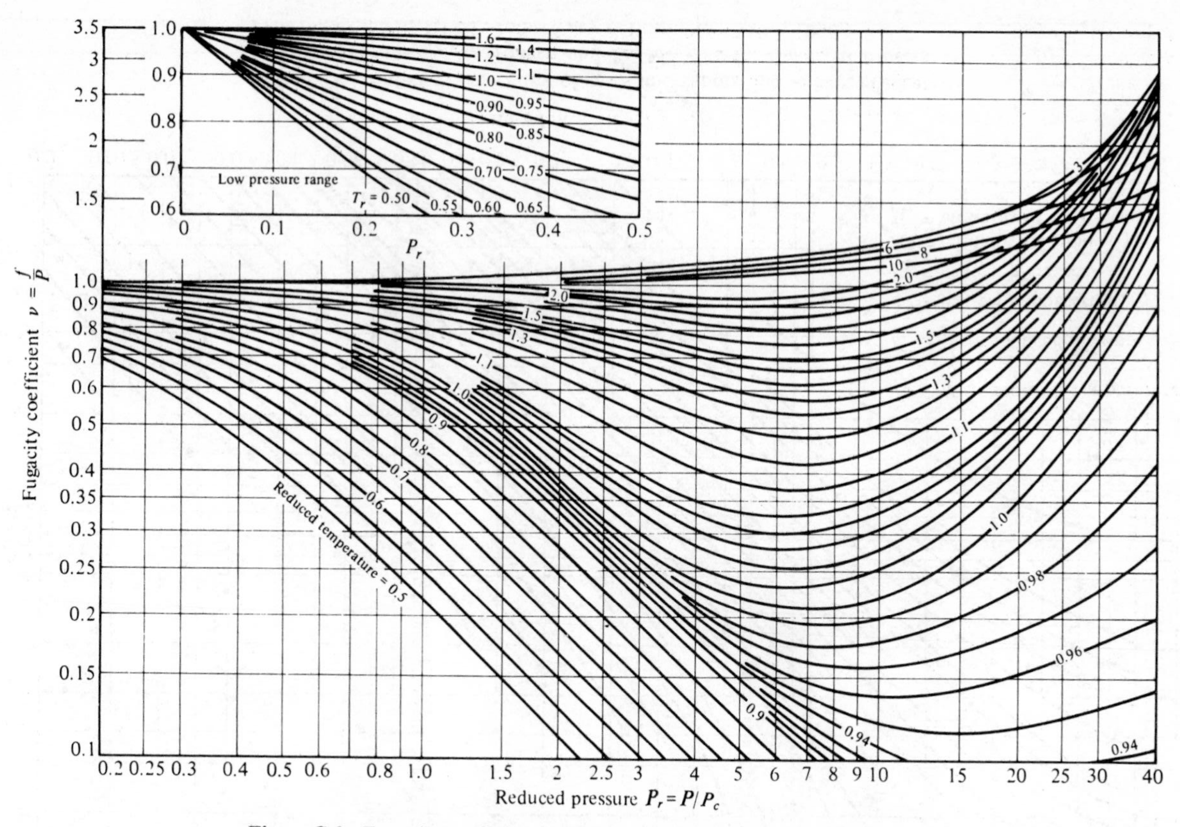

Figure C-3. Fugacity coefficients of gases and vapors. (From O. A. Hougen and K. M. Watson, *Chemical Process Principles Charts*, 1st ed., John Wiley & Sons, New York, 1946.)

D

property charts for selected compounds

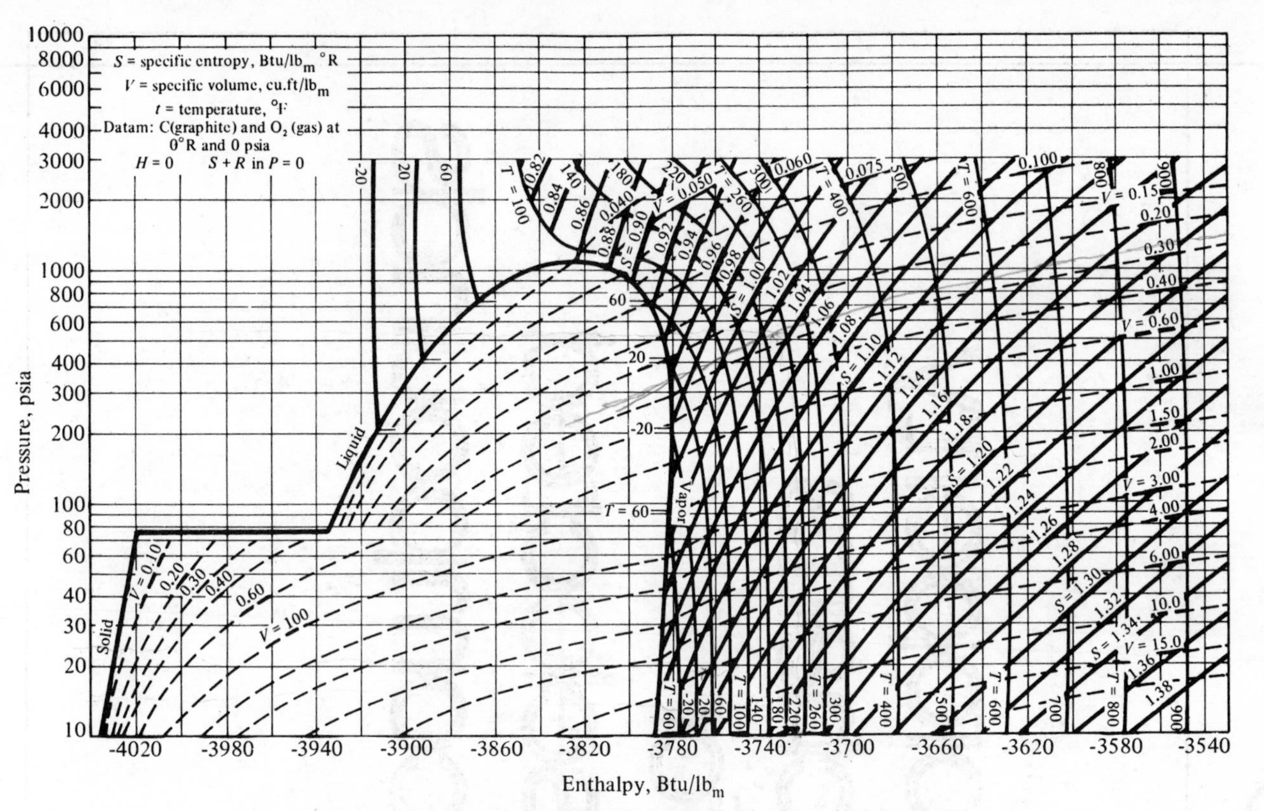

Figure D-1. Carbon dioxide pressure-enthalpy diagram. (Reproduced by permission from L. N. Canjar and F. S. Manning, *Thermodynamic Properties and Reduced Correlations for Gases*, copyright Gulf Publishing Co., Houston, 1967.)

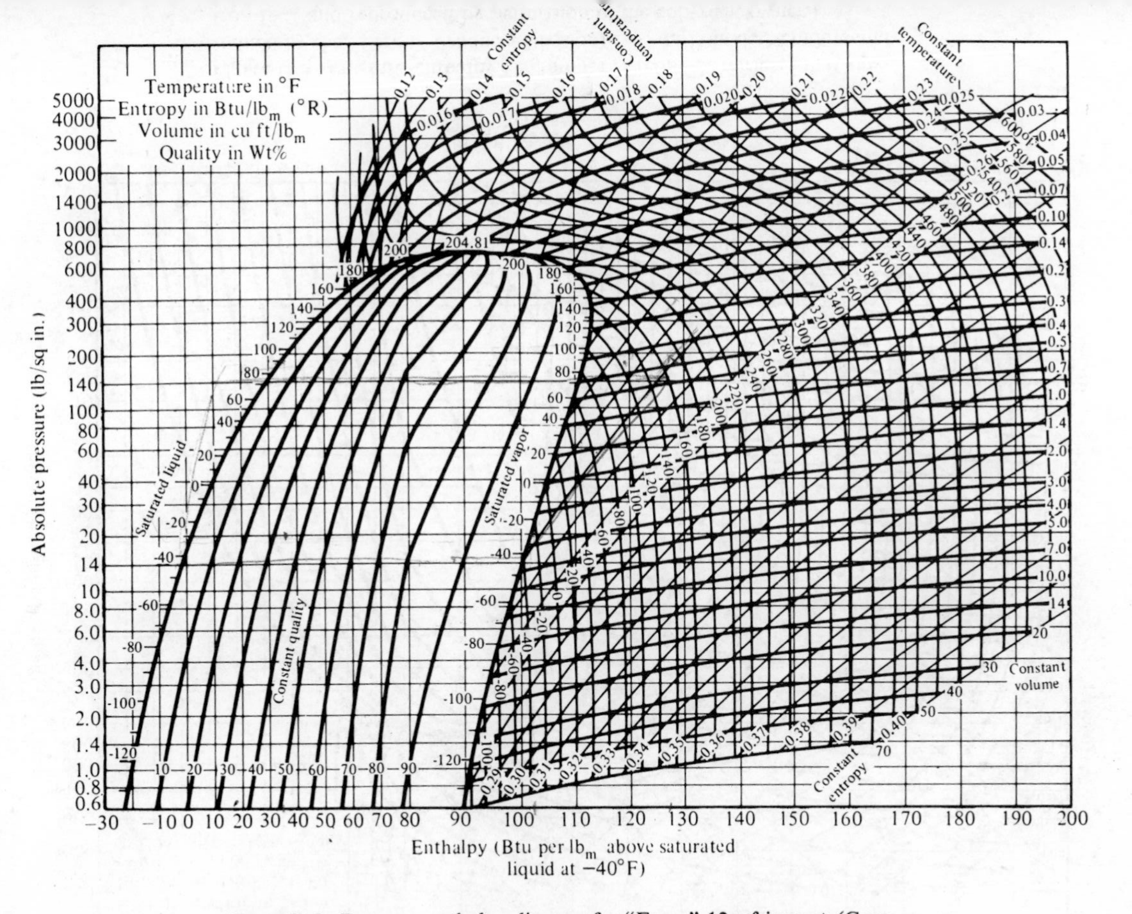

Figure D-2. Pressure–enthalpy diagram for "Freon"-12 refrigerant. (Copyrighted by the "Freon" Products Division, E. I. du Pont de Nemours and Co., Inc., and reproduced by permission of the copyright owner.)

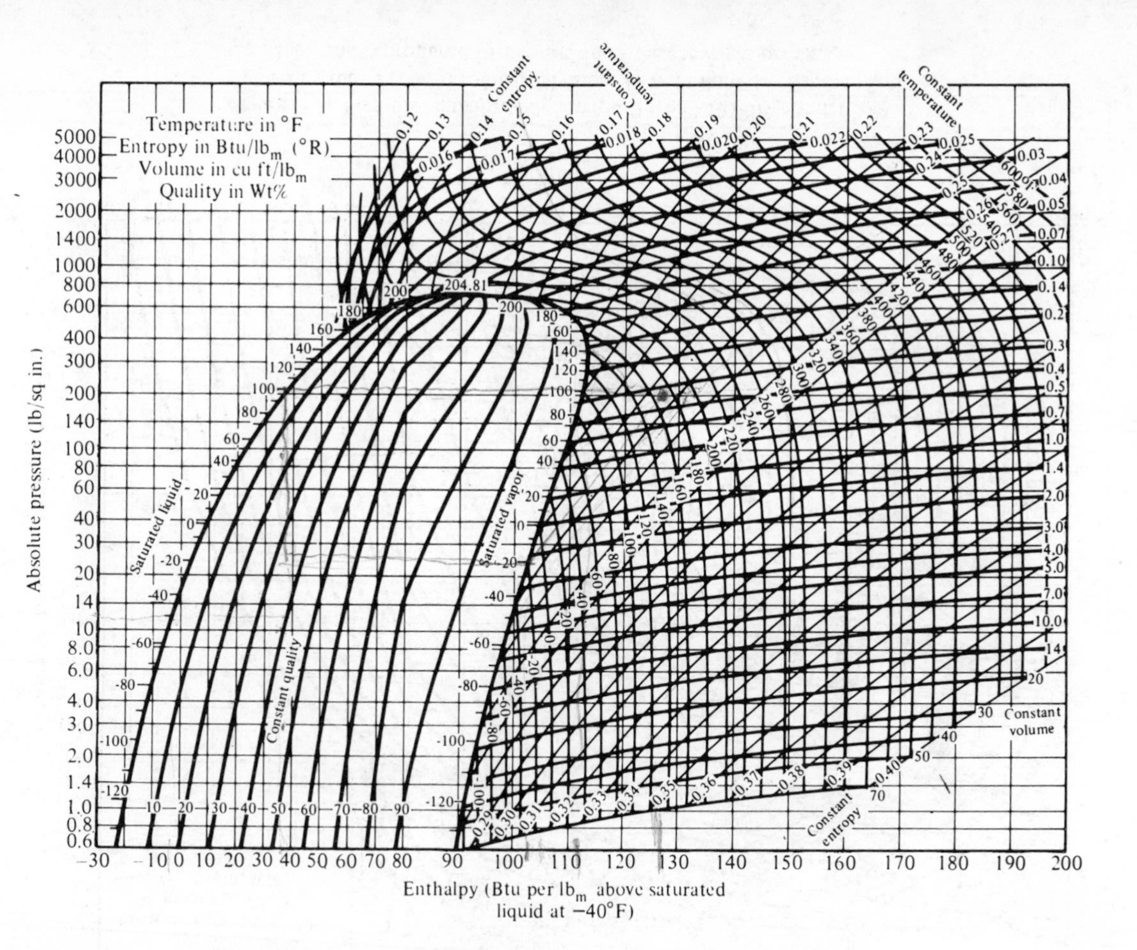

Figure D-3. Pressure–enthalpy diagram for "Freon"-22 refrigerant. (Copyrighted by the "Freon" Products Division, E. I. du Pont de Nemours and Co., Inc., and reproduced by permission of the copyright owner.)

E

properties of air at low pressures

Table E. Properties of Air at Low Pressures, per Pound

T, °R	h, Btu/lb	p_r	u, Btu/lb	v_r	ϕ Btu/(lb)(°R)
360	85.97	0.3363	61.29	396.6	0.50369
380	90.75	0.4061	64.70	346.6	0.51663
400	95.53	0.4858	68.11	305.0	0.52890
420	100.32	0.5760	71.52	270.1	0.54058
440	105.11	0.6776	74.93	240.6	0.55172
460	109.90	0.7913	78.36	215.33	0.56235
480	114.69	0.9182	81.77	193.65	0.57255
500	119.48	1.0590	85.20	174.90	0.58233
520	124.27	1.2147	88.62	158.58	0.59173
537	128.10	1.3593	91.53	146.34	0.59945
540	129.06	1.3860	92.04	144.32	0.60078
560	133.86	1.5742	95.47	131.78	0.60950
580	138.66	1.7800	98.90	120.70	0.61793
600	143.47	2.005	102.34	110.88	0.62607
620	148.28	2.249	105.78	102.12	0.63395
640	153.09	2.514	109.21	94.30	0.64159
660	157.92	2.801	112.67	87.27	0.64902
680	162.73	3.111	116.12	80.96	0.65621
700	167.56	3.446	119.58	75.25	0.66321
720	172.39	3.806	123.04	70.07	0.67002
740	177.23	4.193	126.51	65.38	0.67665
760	182.08	4.607	129.99	61.10	0.68312
780	186.94	5.051	133.47	57.20	0.68942
800	191.81	5.526	136.97	53.63	0.69558
820	196.69	6.033	140.47	50.35	0.70160
840	201.56	6.573	143.98	47.34	0.70747
860	206.46	7.149	147.50	44.57	0.71323
880	211.35	7.761	151.02	42.01	0.71886
900	216.26	8.411	154.57	39.64	0.72438
920	221.18	9.102	158.12	37.44	0.72979
940	226.11	9.834	161.68	35.41	0.73509

T, °R	h, Btu/lb	p_r	u, Btu/lb	v_r	ϕ Btu/(lb)(°R)
1460	358.63	50.34	258.54	10.743	0.84704
1480	363.89	53.04	262.44	10.336	0.85062
1500	369.17	55.86	266.34	9.948	0.85416
1520	374.47	58.78	270.26	9.578	0.85767
1540	379.77	61.83	274.20	9.226	0.86113
1560	385.08	65.00	278.13	8.890	0.86456
1580	390.40	68.30	282.09	8.569	0.86794
1600	395.74	71.73	286.06	8.263	0.87130
1620	401.09	75.29	290.04	7.971	0.87462
1640	406.45	78.99	294.03	7.691	0.87791
1660	411.82	82.83	298.02	7.424	0.88116
1680	417.20	86.82	302.04	7.168	0.88439
1700	422.59	90.95	306.06	6.924	0.88758
1720	428.00	95.24	310.09	6.690	0.89074
1740	433.41	99.69	314.13	6.465	0.89387
1760	438.83	104.30	318.18	6.251	0.89697
1780	444.26	109.08	322.24	6.045	0.90003
1800	449.71	114.03	326.32	5.847	0.90308
1820	455.17	119.16	330.40	5.658	0.90609
1840	460.63	124.47	334.50	5.476	0.90908
1860	466.12	129.95	338.61	5.302	0.91203
1880	471.60	135.64	342.73	5.134	0.91497
1900	477.09	141.51	346.85	4.974	0.91788
1920	482.60	147.59	350.98	4.819	0.92076
1940	488.12	153.87	355.12	4.670	0.92362
1960	493.64	160.37	359.28	4.527	0.92645
1980	499.17	167.07	363.43	4.390	0.92926
2000	504.71	174.00	367.61	4.258	0.93205
2020	510.26	181.16	371.79	4.130	0.93481
2040	515.82	188.54	375.98	4.008	0.93756

Table E. Properties of Air at Low Pressures, per Pound (Continued)

T, °R	h, Btu/lb	p_r	u, Btu/lb	v_r	ϕ Btu/(lb)(°R)
960	231.06	10.610	165.26	33.52	0.74030
980	236.02	11.430	168.83	31.76	0.74540
1000	240.98	12.298	172.43	30.12	0.75042
1020	245.97	13.215	176.04	28.59	0.75536
1040	250.95	14.182	179.66	27.17	0.76019
1060	255.96	15.203	183.29	25.82	0.76496
1080	260.97	16.278	186.93	24.58	0.76964
1100	265.99	17.413	190.58	23.40	0.77426
1120	271.03	18.604	194.25	22.30	0.77880
1140	276.08	19.858	197.94	21.27	0.78326
1160	281.14	21.18	201.63	20.293	0.78767
1180	286.21	22.56	205.33	19.377	0.79201
1200	291.30	24.01	209.05	18.514	0.79628
1220	296.41	25.53	212.78	17.700	0.80050
1240	301.52	27.13	216.53	16.932	0.80466
1260	306.65	28.80	220.28	16.205	0.80876
1280	311.79	30.55	224.05	15.518	0.81280
1300	316.94	32.39	227.83	14.868	0.81680
1320	322.11	34.31	231.63	14.253	0.82075
1340	327.29	36.31	235.43	13.670	0.82464
1360	332.48	38.41	239.25	13.118	0.82848
1380	337.68	40.59	243.08	12.593	0.83229
1400	342.90	42.88	246.93	12.095	0.83604
1420	348.14	45.26	250.79	11.622	0.83975
1440	353.37	47.75	254.66	11.172	0.84341

T, °R	h, Btu/lb	p_r	u, Btu/lb	v_r	ϕ Btu/(lb)(°R)
2060	521.39	196.16	380.18	3.890	0.94026
2080	526.97	204.02	384.39	3.777	0.94296
2100	532.55	212.1	388.60	3.667	0.94564
2150	546.54	233.5	399.17	3.410	0.95222
2200	560.59	256.6	409.78	3.176	0.95868
2250	574.69	281.4	420.46	2.961	0.96501
2300	588.82	308.1	431.16	2.765	0.97123
2350	603.00	336.8	441.91	2.585	0.97732
2400	617.22	367.6	452.70	2.419	0.98331
2450	631.48	400.5	463.54	2.266	0.98919
2500	645.78	435.7	474.40	2.125	0.99497
2550	660.12	473.3	485.31	1.9956	1.00064
2600	674.49	513.5	496.26	1.8756	1.00623
2650	688.90	556.3	507.25	1.7646	1.01172
2700	703.35	601.9	518.26	1.6617	1.01712
2750	717.83	650.4	529.31	1.5662	1.02244
2800	732.33	702.0	540.40	1.4775	1.02767
2850	746.88	756.7	551.52	1.3951	1.03282
2900	761.45	814.8	562.66	1.3184	1.03788
2950	776.05	876.4	573.84	1.2469	1.04288
3000	790.68	941.4	585.04	1.1803	1.04779
3500	938.40	1829.3	698.48	0.7087	1.09332
4000	1088.26	3280	814.06	0.4518	1.13334
4500	1239.86	5521	931.39	0.3019	1.16905
5000	1392.87	8837	1050.12	0.20959	1.20129
6000	1702.29	20120	1291.00	0.11047	1.25769
6500	1858.44	28974	1412.87	0.08310	1.28268

Source: V. M. Faires, "Thermodynamics," 4th ed., The Macmillan Company, New York, 1962. Data from J. H. Keenan and J. Kaye, "Gas Tables," John Wiley & Sons, Inc., New York, 1945.

f

vapor pressures of selected compounds

For vapor pressures between 10 and 1500 mm Hg Rossini[1] suggests the following form

$$\log_{10} P' = A - \frac{B}{T + C}$$

log P' = vapor pressure in mm Hg
T = temperature in °C

Values of the constants are given below:

Compound	A	B	C	Temp. range of applicability
Methane	6.61184	389.93	266.0	−183 to −152.5
Ethane	6.80266	656.40	256.0	−142 to − 75
Ethylene	6.74756	585.00	255.0	−153 to − 91
Propylene	6.81960	785.0	247.0	−112 to − 32
Propane	6.82973	813.20	248.0	−108 to − 25
1-Butene	6.84290	926.1	240.0	− 81 to + 12.5
n-Butane	6.83029	945.9	240.0	− 77 to + 19
n-Pentane	6.85221	1064.64	232.0	− 50 to 57
iso-Pentane	6.80380	1027.25	234.0	− 57 to 49
Benzene	6.89745	1206.35	220.237	− 5.5 to 104
n-Hexane	6.87773	1171.53	224.366	− 25 to 92
Cyclohexane	6.84498	1203.526	222.863	6.6 to 105
Tolume	6.95334	1343.943	219.377	6 to 136
n-Heptane	6.90319	1268.586	216.954	− 2 to 123
n-Octane	6.92374	1355.126	209.517	19 to 152

Although these values may be used to predict vapor pressures greater than 1500 mm Hg, such usage is likely to lead to errors—especially for vapor

[1]F. D. Rossini et al., A.P.I. Research Project 44, Nat. Bur. Stand. Circular C461 (1947).

pressures approaching the critical pressure. For such higher pressures where accurate predictions are needed, the following values of Thodos[2] should be used

$$\log_{10} P' = A + \frac{B}{T} + \frac{C}{T^2} + D\left(\frac{T}{T_d} - 1\right)^n$$

where P' is in mm Hg, T is in °K, and the last term is neglected for $T < T_d$, the various constants are given below:

Compound	T_d(°K)	A	B	C	D	n
Methane	118.83	6.18025	− 296.1	− 8,000	0.257	1.32
Ethane	204.74	6.73244	− 624.24	− 15,912	0.1842	1.963
Propane	261.20	6.80064	− 785.6	− 27,800	0.2102	2.236
n-Butane	312.30	6.78880	− 902.4	− 44,493	0.4008	2.40
n-Pentane	357.79	6.77767	− 988.6	− 66,936	0.6550	2.46
n-Hexane	398.79	6.75933	−1054.9	− 92,720	0.9692	2.49
n-Heptane	436.34	6.74242	−1108.0	−121,489	1.3414	2.50
n-Octane	471.00	6.72908	−1151.6	−152,835	1.7706	2.50
n-Nonane	503.14	6.72015	−1188.2	−186,342	2.2438	2.50
n-Decane	533.13	6.71506	−1219.3	−221,726	2.7656	2.50
n-Dodecane	587.61	6.71471	−1269.7	−296,980	3.9302	2.50

[2]G. Thodos, *Ind. Eng. Chem.*, **42**, 8, p. 1514 (1950).

G

microscopic introduction to entropy

The concept of entropy was first suggested by the German physicist Clausius. Entropy was conceived as a measure of the change in the ability of the universe to produce work in the future, as a result of past or presently occurring transformations or processes. Initially it was considered in a purely macroscopic sense and was related quantitatively to heat, work, and temperature. Subsequent advances in our understanding of the microscopic nature of matter and the application of statistical and quantum mechanical principles to the structure of matter have led to a microscopic interpretation of entropy which is considerably more revealing than that given by macroscopic considerations alone. The introduction to entropy that follows utilizes a microscopic treatment to establish entropy as a state property and attempts to give the student a physical appreciation of its significance. Equally important in this introduction to entropy is the interrelationship it possesses to the fundamental state properties: mass, volume, and energy.

We will begin our discussion by considering the microscopic characterization of matter. We quickly observe that a true microscopic characterization is impossible—the amount of information involved is simply too vast. On the other hand we will observe that although microscopic characterization of individual particles is not possible, the large numbers of particles involved makes the use of statistical, or average characterizations extremely valuable. After that, the relationships between statistical characterizations and the entropy function will be examined. Finally, we will close the discussion by considering the relationships between entropy and the macroscopic concepts of heat and work.

Microscopic Characterization of Matter

The complete description of a collection of monatomic molecules in the gaseous state would require the specification of a tremendously large number of coordinate positions and particle velocities. In normal systems where the number of molecules (particles) exceeds billions of billions, a procedure for completely describing such a system is difficult to imagine, even in the age of computers. Furthermore, since the position and velocity of each particle are continuously changing, the problem becomes even more complex. The problem in completely describing a polyatomic molecular system is orders of magnitude more difficult because of the need to account for rotational and vibrational behavior as well. In addition, the Heisenberg uncertainty principle tells us that it is impossible to know exactly what is happening to any single particle at a given instant of time. However, the application of statistical procedures to an entire collection can provide much insight into the behavior of a "typical" particle. Just as insurance companies employ statistical analyses to estimate the number of deaths among a certain element of our population without knowing just who will die, the scientist may use such procedures to predict the position and velocity distribution possessed by a system without relating it to specific particles. The insurance statistician needs certain pieces of information that characterize the population before making such an analysis. Similarly, the statistical characterization of a system of many particles requires the specification of certain parameters: the number of particles, the volume in which they are free to move, and the total energy they possess.

The specification of the number of particles and the volume of the container in which they are free to move is relatively straightforward. If a simple monatomic species is studied under conditions where rotational, vibrational, nuclear, and electronic energy changes are either absent or ignored, the problem of describing the energy of the collection of particles reduces to one in which only the translational contributions are of importance. Under these conditions internal energy changes can be directly related to changes in particle velocities and a much simplified problem results. For this initial exposure to entropy such a simplification is highly desirable.

The specification of internal energy as a system parameter then serves to limit the total translational energy. Without such a limit a given number of particles in a given volume could possess any distribution of velocities (or momenta) and a unique, or most likely, distribution would be impossible to specify. Thus by limiting the internal energy of the entire collection, we have said in effect that the sum of the individual particle energies must total a given amount. Each particle could have the same energy (equal to the total internal energy, U, divided by the number of particles, N), or some presumably could have none and others correspondingly greater amounts.

Ultimately entropy will be shown to serve as an indicator of the equilibrium, or most probable, distribution among a system of particles given values of U, V, and N for the system as a whole. However, its relationship to microscopic behavior can be illustrated equally well, and with greater clarity, for the undergraduate student by considering initially just the spatial distributions of many-particle systems in space. The development that follows considers the placement of a small number of particles (molecules) in a small number of boxes (volume cells in space). The numbers are intentionally small so that the student can follow clearly each step of the development which relates entropy to the most probable configuration.

Consider first the many possible positional arrangements that exist for six particles, numbered 1 through 6, in two volume elements, A and B. Table G-1 identifies each of the 64 distinguishable distributions that result for such a system. The number of distinguishable distributions may be predicted directly by means of equation (G-1), which is obtained from the mathematics of permutations and combinations:

$$\Omega = M^N \tag{G-1}$$

where Ω = number of distinguishable arrangements
M = number of distinguishable cells
N = number of distinguishable particles

Equation (G-1) illustrates the effect that increasing the number of particles or volume elements has on the number of distinguishable distributions realizable. For example, if we take 1 lb-mole of a gas at standard conditions of temperature and pressure, it will occupy approximately 359 ft^3. If we divide the volume into 1-$in.^3$ elements, there would be 620,352 1-$in.^3$ volume elements into which approximately 2.5×10^{26} molecules could distribute themselves. Application of equation (G-1) yields 620,352 raised to the 2.5×10^{26} power different ways in which such a system could distribute itself. Needless to say, that number is virtually beyond our comprehension, but it is characteristic of the enormous numbers that we encounter when we attempt to describe systems on a microscopic basis.

In statistical mechanics the term *microstate* is used to characterize an individual, distinguishable distribution. The six particle–two box example possessed a total of 64 such microstates. The term *macrostate* is used to designate a group of microstates with common characteristics. In this case we may choose to consider a macrostate as consisting of all microstates with the same number of particles in compartment A and the same number in compartment B. Thus all microstates that have any 3 particles in compartment A and the other 3 in B would constitute one macrostate. All those microstates with 4 particles in A and 2 in B would constitute a second macrostate; 4 particles in B and 2 in A would represent still a third macrostate.

Table G-1 shows that the 3-3 macrostate consists of 20 microstates, while the 4-2 and 2-4 macrostates each include 15 microstates. On the other hand, the 6-0 and 0-6 macrostates contain only a single microstate each. It should be noted that different arrangements (permutations) of the same particles within a given box do not constitute additional microstates. Thus for the $0A$-$6B$ microstate, the 123456 arrangement is considered identical to the 234561 arrangement.

Table G-1 Distribution of Six Particles in Two Boxes

Distributions	Compartment A	Compartment B
0–6		1–2–3–4–5–6
1–5	1	2–3–4–5–6
1–5	2	1–3–4–5–6
1–5	3	1–2–4–5–6
1–5	4	1–2–3–5–6
1–5	5	1–2–3–4–6
1–5	6	1–2–3–4–5
2–4	1–2	3–4–5–6
2–4	1–3	2–4–5–6
2–4	1–4	2–3–5–6
2–4	1–5	2–3–4–6
2–4	1–6	2–3–4–5
2–4	2–3	1–4–5–6
2–4	2–4	1–3–5–6
2–4	2–5	1–3–4–6
2–4	2–6	1–3–4–5
2–4	3–4	1–2–5–6
2–4	3–5	1–2–4–6
2–4	3–6	1–2–4–5
2–4	4–5	1–2–3–6
2–4	4–6	1–2–3–5
2–4	5–6	1–2–3–4
3–3	1–2–3	4–5–6
3–3	1–2–4	3–5–6
3–3	1–2–5	3–4–6
3–3	1–2–6	3–4–5
3–3	1–3–4	2–5–6
3–3	1–3–5	2–4–6
3–3	1–3–6	2–4–5
3–3	1–4–5	2–3–6
3–3	1–4–6	2–3–5
3–3	1–5–6	2–3–4
3–3	2–3–4	1–5–6
3–3	2–3–5	1–4–6
3–3	2–3–6	1–4–5
3–3	2–4–5	1–2–6
3–3	2–4–6	1–3–5
3–3	2–5–6	1–3–4
3–3	3–4–5	1–2–6
3–3	3–4–6	1–2–5
3–3	3–5–6	1–2–4
3–3	4–5–6	1–2–3
4–2 5–1 0–6	These are mirror images of the first 3 distributions.	

The number of combinations for placing N particles in two cells with M in one and $N - M$ in a second is given by the expression

$$C_N^{M,N-M} = \frac{N!}{(N-M)!\,M!} \tag{G-2}$$

Thus for $N = 6$ we see that if $M = 5$,

$$C_6^{5,1} = \frac{6!}{1!\,5!} = 6$$

We observe indeed that there are 6 microstates in the macrostate $5A$-$1B$ and 6 in the macrostate $1A$-$5B$.

SAMPLE PROBLEM G-1. Deduce a general expression for the number of combinations for placing N particles in three cells with M_1 in one cell, M_2 in a second cell (where $M_1 + M_2 \leq N$). Extend this result to k cells with M_1, $M_2, \ldots, M_k$ particles in each of the cells.

Solution: Let us consider, first, the ways by which the $1A$-$5B$ macrostate in the preceding discussion might be formed. Assume that the first ball picked will go in A and the remaining 5 in B. We may pick any of the 6 for the first, any of the remaining 5 next—and so forth—for a total of $6 \cdot 5 \cdot 4 \cdot 3 \cdot 2 \cdot 1 = 6!$ ways of placing the 6 balls in the two boxes. (Had we chosen to put the second ball drawn in A, then we would have had 6 choices for the first ball in B, 5 for the one in A, 4, 3, 2, and 1 for the remaining choices in B. Again a total of 6!. If we choose any other draw to go in A, clearly we still end up with 6! total permutations possible. Similarly, the result would have been the same for any other macrostate chosen.) However, of these 6! possible orderings, many just involve perturbations of the order in which particles are placed in the boxes, not the particles that finally end up in the two boxes. Thus if we examine the microstate where ball 1 is in box A, and balls 2, 3, 4, 5, and 6 are in B, the orderings

$$1 \quad A \quad 2\ 3\ 4\ 5\ 6 \quad B$$

$$1 \quad A \quad 3\ 4\ 5\ 6\ 2 \quad B$$

are clearly the same microstate. The number of such permutations with particle 1 in A and 2, 3, 4, 5 and 6 in B is

$$1 \cdot (5 \cdot 4 \cdot 3 \cdot 2 \cdot 1) = 1!\,5!$$

In a like manner, every other microstate making up the $1A$-$5B$ macrostate can be formed in 1!5! permutations. The number of microstates in turn is given by the total number of permutations of the six balls divided by the number of permutations in each microstate that has 1 ball in A and 5 in B. Thus

$$\text{no. microstates} = \frac{6!}{5!\,1!} = C_6^{5,1}$$

This result may easily be generalized to show that the number of microstates in the M_1A-M_2B macrostate ($M_1 + M_2 = 6$) is

$$\text{no. microstates} = C_N^{M_1, M_2} = \frac{N!}{M_1!\,M_2!}$$

If we have cells with M_1, M_2, and M_3 balls in each ($M_1 + M_2 + M_3 = N$), the total number of microstates in the $M_1 - M_2 - M_3$ macrostate is simply

$$\text{no. microstates} = \frac{N!}{M_1!\,M_2!\,M_3!}$$

For k cells we simply add additional factorial terms to the denominator for each additional cell as follows:

$$C_N^{M_1 \cdots M_k} = \frac{N!}{M_1!\,M_2!\cdots M_{k-1}!\,M_k!}$$

or, in terms of the continued product, Π,

$$C_N^{M_1 \cdots M_k} = \frac{N!}{\prod_{i=1}^{k} M_i!}$$

If each of the six particles has an equal probability of going into either compartment A or B, then any one of the 64 distributions (or microstates) would have an equal probability of occurring each time the six particles are distributed in the two cells. (It is assumed that the order in which particles are arranged in the cells is irrelevant.) However, it can be seen that there is considerably less likelihood of finding macrostate $6A$-$0B$ than $3A$-$3B$, because it occurs in just 1 of the 64 possibilities. Its probability is thus $\frac{1}{64}$, as compared with a probability of $\frac{20}{64}$ for $3A$-$3B$ or $\frac{15}{64}$ for either $2A$-$4B$ or $4A$-$2B$. The probability of finding any given macrostate under these conditions is directly related to the number of microstates it contains: The larger the number of microstates the greater the likelihood of finding that particular macrostate. If the particles are free to fluctuate between the two compartments, we would expect the system to spend a greater fraction of its time in the $3A$-$3B$ macrostate than in any of the other six, as shown in Table G-2.

An examination of Table G-2 indicates that of the seven possible macrostates, the 3-3 or uniform distribution contains the largest number of microstates and would exist $\frac{5}{16}$ of the time. Inclusion of the 2-4 and 4-2 distributions with the 3-3 arrangement raises to $\frac{25}{32}$ the probability of finding the particles in a "more or less" uniform distribution.

Table G-3 represents the behavior of a twenty particle–two cell problem which contains 1,048,576 microstates. One begins to get an appreciation for what happens as the number of particles increases. Note that although the most probable macrostate occurs with a probability of only 0.176, the com-

Table G-2 Macrostates for a Six Particle–Two Box System

Macrostate A	Macrostate B	No. of Microstates	Fraction of Time in Macrostate
6	0	1	1/64
5	1	6	3/32
4	2	15	15/64
3	3	20	5/16
2	4	15	15/64
1	5	6	3/32
0	6	1	1/64

Table G-3 Macrostates for a Twenty Particle–Two Box System

Particles in Cell A	Particles in Cell B	No. of Microstates	Fraction of Total Micro-states	Fraction of Microstates At Least This Evenly Distributed	Fraction of Microstates Less Evenly Distributed
20	0	1	0.000001		
19	1	20	0.00002		
18	2	190	0.00018	0.99998	0.00002
17	3	1,140	0.00109	0.99962	0.00038
16	4	4,845	0.00462	0.99744	0.00256
15	5	15,504	0.01479	0.98820	0.01180
14	6	38,760	0.03696	0.95862	0.04138
13	7	77,520	0.07393	0.88470	0.11530
12	8	125,970	0.12014	0.73684	0.26316
11	9	167,960	0.16018	0.49656	0.50344
10	10	184,756	0.17620	0.17620	0.82380
9	11	167,960	0.16018		
8	12	125,970	0.12014		
7	13	77,520	0.07393		
6	14	38,760	0.03696		
5	15	15,504	0.01479		
4	16	4,845	0.00462		
3	17	1,140	0.00109		
2	18	190	0.00018		
1	19	20	0.00002		
0	20	1	0.000001		
		1,048,576 = 2^{20}	1.00002		

bined probability of the three most likely equals 0.496. Thus such a system would spend approximately half its time in these three macrostates.

If this type of analysis is extended to systems containing many particles—say the order of Avogadro's number (10^{23})—we find that the probability of obtaining this "more or less" evenly distributed arrangement becomes overwhelming.

Although slight variations from the most probable distribution are continually occurring, these variations are so small that we have no instruments sensitive enough to measure them. The likelihood of any measurable deviation from an essentially uniform distribution is so small that it may essentially be discounted! For example, Denbigh[1] reports that the probability of observing a 0.001 per cent variation in the density of 1 cm^3 of air is less than 10^{-10^8} and is not likely to be observed in trillions of years. Thus we find that the spontaneous appearance of all the gaseous molecules in one portion of the room is *not totally impossible*; it is just so unlikely that the chances of it occurring are negligibly small.

Sample Problem G-2. What is the probability of finding all the molecules in a particular cubic inch of space if we have a 1 lb-mole of gas occupying 359 ft^3 of space at standard temperature and pressure? If the molecules are uniformly distributed, how many would each 1-$in.^3$ cell contain? How many microstates would make up the latter macrostate? (*Hint:* Stirling's approximation for evaluation of $N!$ when N is large is helpful: $\ln N! \cong N \ln N - N$.)

Solution:

$$359 \text{ ft}^3 = 620{,}352 \text{ in.}^3 = \text{number of cells}$$

$$1 \text{ lb-mole} \cong 2.7 \times 10^{26} \text{ molecules}$$

From equation (G-1): $\Omega = 620{,}352^{2.7 \times 10^{26}}$ = total number of microstates. If all of the particles are to be in *one particular* cell, there is only one microstate that can exist. The probability, P, of finding the desired macrostate is given by the number of microstates in that macrostate divided by the total number of microstates. Thus

$$P = \frac{1}{(620{,}352)^{2.7 \times 10^{26}}}$$

If we allow the molecules to collect in any one of the 620,352 cubic inches, then there are 620,352 possible microstates and the probability is given by

$$P = \frac{620{,}352}{(620{,}352)^{2.7 \times 10^{26}}}$$

A uniform distribution of N particles among M cells would find (N/M) particles in each cell. The number of microstates in such a uniform distribution would be given by

$$C_N^{(N/M)_1 \cdots (N/M)_M} = \frac{N!}{\prod_{i=1}^{M} \left(\frac{N}{M}\right)!} = C$$

[1]K. Denbigh, *The Principles of Chemical Equilibrium*, Cambridge University Press, New York, 1966, p. 59.

or upon taking natural logarithms

$$\ln C = \ln (N!) - M \ln \left(\frac{N}{M}\right)!$$

Assuming N is large so Stirling's approximation is applicable, we find

$$\ln C = N \ln N - N - M\left(\frac{N}{M} \ln \frac{N}{M} - \frac{N}{M}\right)$$

$$\ln C = N \ln M$$

which gives upon exponentiation

$$C = M^N$$

Thus if we consider 2.7×10^{26} particles in 620,352 cells, we find from Stirling's approximation that the number of ways we can form a uniform particle distribution is

$$C = (620{,}352)^{2.7 \times 10^{26}}$$

The probability of finding this macrostate then is given by

$$P = \frac{C}{\Omega} = \frac{(620{,}352)^{2.7 \times 10^{26}}}{(620{,}352)^{2.7 \times 10^{26}}} = 1.0$$

However, we know this result is seriously in error: The actual probability of finding an *exactly even* distribution is negligible, even though the probability of a *more or less even* distribution is, as we have seen, overwhelming. The Stirling's approximation we have used is not accurate enough to distinguish between the many "almost evenly" distributed arrangements. Rather, the approximation lumps all of these together, and hence we obtain the result that the exactly even distribution is *apparently* always obtained. (Had we examined any of the other nearly even distributions with Stirling's approximation we would have obtained the same result.)

In addition to a spatial distribution, it is necessary to consider the energy distribution of individual molecules in the various energy levels to which the molecules have access. If the energy, U, of the collection is specified along with the number of particles, N, many possibilities exist regarding the energy that any one molecule might possess. For example, the energy possessed by an individual polyatomic molecule may be distributed among its various energy modes, such as translation, vibration, and rotation. A spectrum of energies will be found for each of the modes among the molecules making up the system.

Quantum principles state that in any of these modes only certain discrete energy levels are permissible. Stated differently, this means that a translating molecule cannot posses any velocity but is limited to particular values that depend on system parameters, specifically its volume and energy. Transitions from one energy state to another involve the liberation or absorption of a finite amount of energy that is exactly the amount of energy needed to shift from one quantum state to the next. The process is similar to an electron which jumps from one discrete energy level to an adjacent one by emitting or absorbing a fixed amount of energy.

For any given system of molecules with N, V, and U specified, there is a finite number of permissible energy levels available to each of the molecules. In our earlier examples involving spatial distributions, we have considered a given number of particles and a finite number of volume elements, M, within a given volume, V. The number M was rather arbitrarily selected, and it could be made to approach infinity if one chose to select infinitesimally small elements. Similarly, our consideration of energy distributions will involve the placement of a given number of particles, N, into a large, but finite, number of "energy cells." Each cell (or level, as it is commonly termed) has a specific energy associated with it. It, too, can accommodate varying numbers of particles, as could a volume element. However, if we place a total energy, U, on our system of particles, only distributions in which the sum of particle energies totals U are permitted. Thus, if we define arbitrary energy units (e.u.), such that the permitted energy levels have energies of 0, 1, 2, . . . , 100 e.u., and if we consider the distribution of 10 particles such that the total system energy equals 20 e.u., we immediately recognize that certain distributions are unacceptable. For example, no particle could occupy a level above that with a value of 20 e.u., or the system energy exceeds the prescribed value regardless of which levels the other particles occupy. Such levels are said to be inaccessible for the total energy specified, even though at higher energies some of them might be populated.

In like manner, all 10 particles could not reside in the energy level of 1 e.u., for a system energy of 10 e.u. does not meet the total energy condition of 20 e.u. The reader can immediately identify numerous other distributions that are inconsistent with the stipulated value for U.

The most obvious of the many distributions that would satisfy the energy requirement is the placement of each of the 10 particles in energy level 2 such that the system energy totals 20 e.u. Equally sufficient from a total energy point of view, but far less probable (as we might expect from our earlier discussion), is the distribution of 1 particle in the 20-e.u. level and the remaining 9 in the 0-e.u. level.

As was the case with spatial considerations, the macrostate that contains the greatest number of microstates will have the largest probability of existing. As the number of particles and hence microstates becomes very large, it will have properties that will approach the average properties of all the microstates. Thus the macrostate representing the most probable energy distribution can be used to characterize the equilibrium macroscopic behavior of a system, even though the system may seldom exist in precisely that state. (It should be pointed out again that a many-particle system at equilibrium is continuously fluctuating between many microstates, most of which differ only insignificantly from *the* most probable one.)

SAMPLE PROBLEM G-3. Six molecules, A through F, are to be distributed among 21 energy levels ranging from 0 to 20 e.u. with no limit on the number of particles per level. Calculate the number of microstates in each possible macrostate if the total energies, U, of the system are 5 and 10 e.u.

Solution: Let us begin by identifying the macrostates that satisfy the conditions for each of the two total energies specific. If we systematically consider all possibilities that exist by first assigning 6 molecules to the 0 level, then 5, 4, 3, 2, 1, and 0 in succession, we minimize the likelihood of omitting any possibilities. Clearly all molecules cannot be in the 0 level or the total energy requirement is not met. However, if we place 5 in the 0 level and 1 in either the 5 or 10 level, we can satisfy the conditions. Since we have six different molecules that could be chosen to place in the 5 or 10 level, there are six microstates comprising each of these macrostates. Recalling Sample Problem G-2,

$$C_N^{M_1 \cdots M_k} = \frac{N!}{\prod_{i=0}^{k} M_i!}$$

For $N = 6$ and $M_0 = 5$, M_5 or $M_{10} = 1$,

$$C_6^{5,1} = \frac{6!}{5!\,1!} = 6$$

If we then consider all possibilities in which four molecules remain the 0 level for each total energy requirement, we shall find two configurations for $U =$ 5 e.u. and five for $U = 10$ e.u. Applying the same formula to each distribution permits us to calculate the number of ways each particular configuration (macrostate) can be achieved. Tables SG-3a and SG-3b give all possible arrangements for $U = 5$ and 10 e.u.

Table SG-3a Distribution of Six Particles in Various Energy Levels (0, 1, 3, 4, . . . Energy Units)–System Energy = 5 e.u.

	Energy Level													No. of	
Macrostate	0	1	2	3	4	5	6	7	8	9	10	11	12	Microstates	Probablity, %
1	5	0	0	0	0	1								6	2.4
2	4	0	1	1	0	0								30	11.9
3	4	1	0	0	1	0								30	11.9
4	3	1	2	0	0	0								60	23.8
5	3	2	0	1	0	0								60	23.8
6	2	3	1	0	0	0								60	23.8
7	1	5	0	0	0	0								6	2.4
														252	

Table SG-3b Distribution of Six Particles in Various Energy Levels (0, 1, 3, 4, . . . Energy Units)–System Energy = 10 e.u.

Macrostate	Energy Level													No. of	Probability, %
	0	1	2	3	4	5	6	7	8	9	10	11	12	Microstates	
1	5	0	0	0	0	0	0	0	0	0	1			6	0.2
2	4	1	0	0	0	0	0	0	0	1	0			30	1.0
3	4	0	1	0	0	0	0	0	1	0	0			30	1.0
4	4	0	0	1	0	0	0	1	0	0	0			30	1.0
5	4	0	0	0	1	0	1	0	0	0	0			30	1.0
6	4	0	0	0	0	2	0	0	0	0	0			15	0.5
7	3	2	0	0	0	0	0	0	1	0	0			60	2.0
8	3	1	1	0	0	0	0	1	0	0	0			120	4.1
9	3	1	0	1	0	0	1	0	0	0	0			120	4.1
10	3	1	0	0	1	1	0	0	0	0	0			120	4.1
11	3	0	2	0	0	0	1	0	0	0	0			60	2.0
12	3	0	1	1	0	1	0	0	0	0	0			120	4.1
13	3	0	1	0	2	0	0	0	0	0	0			60	2.0
14	3	0	0	2	1	0	0	0	0	0	0			60	2.0
15	2	3	0	0	0	0	0	1	0	0	0			60	2.0
16	2	2	1	0	0	0	1	0	0	0	0			180	6.1
17	2	2	0	1	0	1	0	0	0	0	0			180	6.1
18	2	2	0	0	2	0	0	0	0	0	0			90	3.1
19	2	1	2	0	0	1	0	0	0	0	0			180	6.1
20	2	1	1	1	1	0	0	0	0	0	0			360	12.2
21	2	1	0	3	0	0	0	0	0	0	0			60	2.0
22	2	0	2	2	0	0	0	0	0	0	0			90	3.1
23	1	4	0	0	0	0	1	0	0	0	0			30	1.0
24	1	3	1	0	0	1	0	0	0	0	0			120	4.1
25	1	3	0	1	1	0	0	0	0	0	0			120	4.1
26	1	2	2	0	1	0	0	0	0	0	0			180	6.1
27	1	1	3	1	0	0	0	0	0	0	0			120	4.1
28	1	0	5	0	0	0	0	0	0	0	0			6	0.2
29	1	2	1	2	0	0	0	0	0	0	0			180	6.1
30	0	5	0	0	0	1	0	0	0	0	0			6	0.2
31	0	4	1	0	1	0	0	0	0	0	0			30	1.0
32	0	4	0	2	0	0	0	0	0	0	0			15	0.5
33	0	3	2	1	0	0	0	0	0	0	0			60	2.0
34	0	2	4	0	0	0	0	0	0	0	0			15	0.5
														2943	

As Tables SG-3a and SG-3b show, the given internal energy of the collection can be achieved by many different distributions within the permissible energy levels. Just as was the case with spatial distributions, certain energy distributions occur with overwhelming probability (for systems with large numbers of particles and energy levels), because they can be achieved in a great many ways. Under conditions where the system is permitted to remain undisturbed for a sufficient period of time, those distributions that are most probable are the only ones that will be observed macroscopically. As was the case with spatial distributions, the most probable distributions tend to be those with the most uniform distribution of particles among the energy levels.

Although an in-depth treatment of statistical thermodynamics would show entropy to be related to the most probable energy distribution, as well as the spatial distributions discussed earlier, this development intentionally avoids such a detailed treatment because of complicating factors which arise, both conceptually and mathematically, without a commensurate contribution to our understanding of entropy. Since our goal in this treatment is the latter and not one of teaching statistical thermodynamics, we refer the reader to any of several fine textbooks which cover this subject in greater detail.[2]

Relationship of Entropy to Microscopic Character of Matter

At this point you may logically ask: What is the connection between these arguments about the probability of a given macrostate and the concept of entropy? In answer to such a question we might make the following observations: As an isolated system moves toward equilibrium, it moves successively from less probable (more ordered) states, where macroscopic potential differences exist, to more probable ones; eventually the system will reach its most probable or equilibrium condition at which point all macroscopic potential differences within the system have vanished. At this point work may no longer be obtained from the (isolated) system. Thus it seems logical to attempt to express entropy in terms of the thermodynamic probability of a system's macrostate, because both approach a maximum as the system approaches equilibrium. Before attempting to relate the two, it is helpful to define the *thermodynamic probability*, p_i, of any macrostate, i, as the number of microstates it possesses:

$$
\begin{aligned}
p_i &= \text{number of microstates in the } i \text{ macrostate} \\
p_{\max} &= \text{number of microstates in the most probable or} \\
&= \text{equilibrium macrostate}
\end{aligned}
$$

The functional relationship between entropy and thermodynamic probability was first proposed by Boltzmann and may be formulated by recognizing the following conditions, which the functional form must satisfy. We know that entropy, as originally formulated by Clausius and developed by others, is an extensive property. If we combine two systems that are in the same state, the total entropy must be the sum of the entropies of the two sys-

[2]Knuth, Eldon, *Introduction to Statistical Thermodynamics*, McGraw-Hill Inc., New York (1966). Desloge, Edward, *Statistical Physics*, Holt, Rinehart, & Winston (1966). M. T. Howerton, *Engineering Thermodynamics*, D. van Nostrand Co., Inc. New York (1962).

tems. However, the thermodynamic probability, (p_{1+2}), for the combined system is the product of the thermodynamic probabilities of the two original systems, (p_1) and (p_2). Thus, although the entropies must be additive, the thermodynamic probabilities of the combined system are multiplicative. The only functional relationship that satisfies these conditions is the logarithmic function. With these considerations in mind it is postulated that the entropy of a given configuration, or macrostate, is related to its thermodynamic probability, p_i, by the simple expression

$$S_i = k \ln p_i \tag{G-3}$$

The proportionality constant, k, is the *Boltzmann constant*, which is equal to the ideal gas constant, R, divided by Avogadro's number. This relationship between thermodynamic probability and entropy is now generally accepted and serves as the basis for the development of statistical thermodynamics.

Although it is not our goal in this book to develop this subject, it should be observed that as our understanding of atomic and molecular behavior is increased, statistical thermodynamics becomes an increasingly valuable tool for the engineer. Its utilization requires an understanding of all the energy modes and the values of energy levels to which particles have access in each of the energy macrostates. Today such information is available for only the simplest of molecules in the gaseous state. If in the future the behavior of more complex molecules can also be described by scientists, then in theory it would be possible to calculate thermodynamic properties directly from a knowledge of the fundamental parameters. Once established, such procedures could eliminate the need for scientists and engineers to devote sizable efforts to obtain experimentally the property data needed to make engineering calculations. However, that day has not yet arrived, and engineers will continue in the foreseeable future to determine much of the property data required for their analyses experimentally.

The value of having introduced equation (G-3) in this development relates to the increased insight one gains with regard to entropy. For each macrostate we can compute an entropy from equation (G-3) provided only that we can determine the number of microstates that comprise the macrostate. Although the number of macrostates for given values of N, V, and U in any real system is enormous, it is, nevertheless, a finite number. Of these, we know that certain macrostates will occur with far greater frequency because of the overwhelming number of microstates which they contain. Figure G-3 shows the data of Table G-3 with p_i for each macrostate. The ordinate can be interpreted as either the thermodynamic probability, p_i or entropy, S_i, for the ith macrostate with an appropriate modification of the scale utilized. Although these data originate from just a twenty particle–two cell distribution, it clearly shows the tendency for the system to cluster about the most probable configurations. As implied earlier, when the number of cells and particles is increased to a level that compares with the number of molecules in a system and the number of energy levels they can occupy, the distribution peaks considerably more sharply about the most probable value.

The macrostates with the largest thermodynamic probability, or largest values of entropy, are those which exist once the system reaches equilibrium. As previously indicated, fluctuations about the equilibrium state occur, but these fluctuations are so small that for practical purposes the system may be considered to be in its equilibrium state. Thus we might associate the value of entropy, labeled S_{max}, with the equilibrium state corresponding to the values of N, V, and U for which the plot applies. A change in any of these parameters would produce a corresponding change in the curve in Fig. G-1 as well as the system's equilibrium entropy.

One very important point should be learned from the foregoing discussion: that system with a given N, V, and U possesses a unique value of S only if considered to be in its equilibrium state. If we treat thermodynamics as a study of equilibrium states, entropy retains its state character. However, a system in a perturbed or nonequilibrium condition can be thought of as possessing some value of entropy less than its equilibrium value, as seen in Fig. G-1. The nonequilibrium system would correspond to a less probable macrostate than that at the peak of the curve, and the entropy of any such state is clearly less than S_{max}. Energy exchange between various energy macrostates or molecules in a nonequilibrium isolated system would move it from less probable to more probable distributions until the system finally reaches the most probable, or equilibrium, state at the peak of the curve.

A three-dimensional plot of entropy, S, plotted against the internal energy, U, and volume, V, of the system is shown in Fig. G-2. The surface that results represents all the equilibrium states for the possible combinations of U and V for a given value of N. All nonequilibrium states possible would fall beneath the surface, inasmuch as such states would possess entropies less than those corresponding to the equilibrium values.

Two equilibrium states are designated on the surface. Any path connecting the two points and lying wholly on the surface would represent a reversible path, because each intermediate state is itself an equilibrium state. There are an infinite number of such paths possible for any system in going from one state to another. (*Note:* Processes that occur along the equilibrium U–V–S surface are also quasistatic processes!)

Similarly, there are an infinite number of paths that depart from the surface which could also be used in taking a system from state 1 to state 2. Such processes represent irreversible and nonquasistatic processes. The difficulties in analyzing such a process should take on added significance in view of the earlier discussion on entropy. Since these processes pass through nonequilibrium states, any given value of U and V can have many different values of S.

In Fig. G-2 it should be noted that a decrease in U at constant V or a

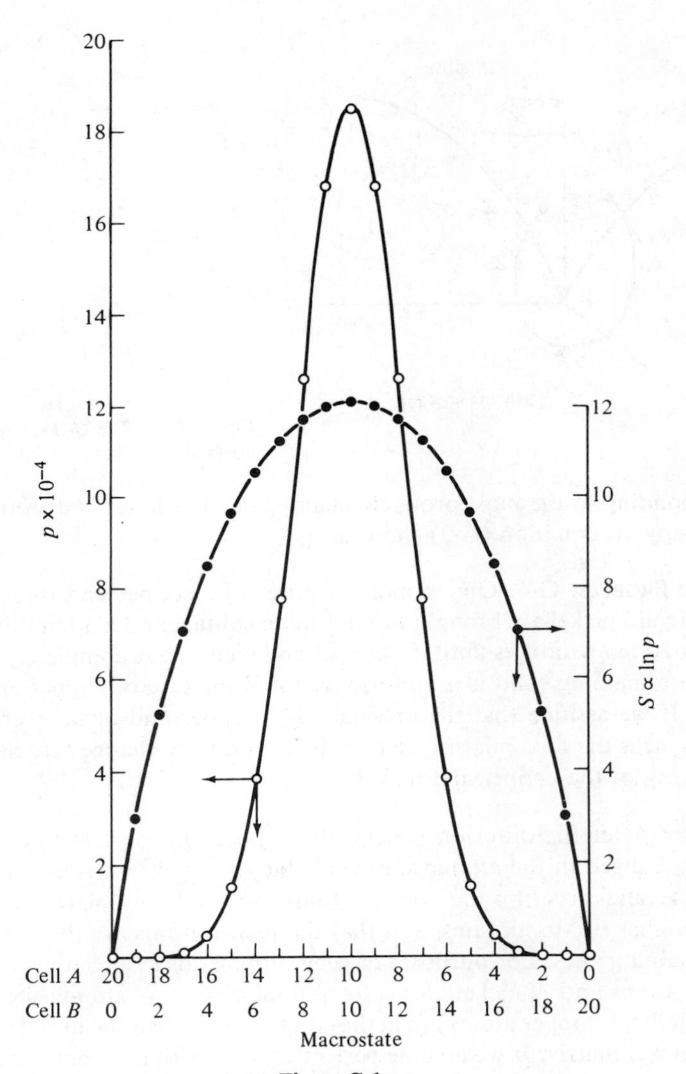

Figure G-1.

decrease in V at constant U both produce a decrease in S. In view of our previous discussion, this would be expected, because a reduction of either U or V reduces the number of microstates accessible to the system. The greater the energy, the greater the number of ways in which it may be distributed among the various particles (see Sample Problem G-3). Similarly, as V is increased (equivalent to increasing M, the number of cells, in our earlier discussion), the total number of cells increases, as does the thermodynamic probability

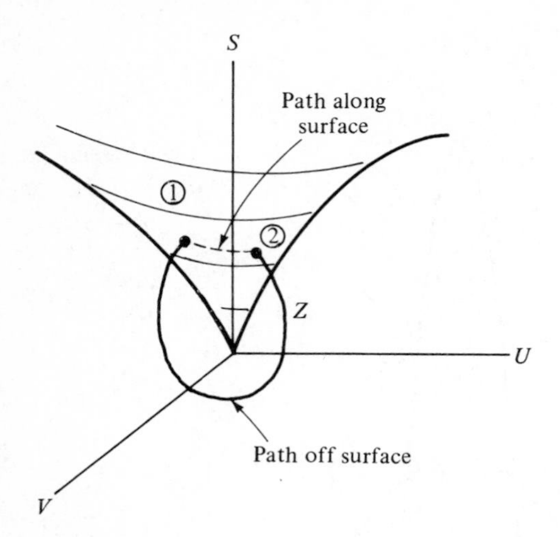

Figure G-2. The U–V–S equilibrium surface.

corresponding to the most probable macrostate. The latter is directly related to entropy, as equation (G-3) showed.

SAMPLE PROBLEM G-4. One lb-mole of pure solid copper and three lb-mole of pure solid nickel are brought into intimate contact and held in this fashion at elevated temperatures until the copper and nickel have completely diffused, and the remaining solid is a uniform, random mixture of copper and nickel atoms. If we assume that the original solid copper and nickel were perfect crystals, as is the final mixture, determine the entropy change associated with the mixing of the copper and nickel.

Solution: After the diffusion occurs, the copper and nickel atoms are randomly arranged in the atomic sites available to them. Let us assume that in our molecular ordering the copper atoms are randomly placed among the total number of atomic sites, and that the nickel atoms are then used to fill the remaining sites. For purposes of generality, let us assume that we have C copper atoms and N nickel atoms for a total of $C + N$ atomic sites.

Initially all copper atoms are in the copper matrix and the nickel atoms are in the nickel matrix. If we assume perfect crystals with no atomic movement, then there is only one possible microstate which satifies the initial conditions and the initial entropy is

$$S = k \ln p_i = 0$$

Now after the diffusion, the C copper atoms are randomly dispersed between $C + N$ sites. The total number of ways in which these C atoms can be so arranged is found by observing that the first copper atom can go in any of the $(N + C)$ sites, the second in the $(N + C - 1)$ remaining, and so on until we have

$$(N + C)(N + C - 1)\ldots(N + 1) = \frac{(N + C)!}{N!}$$

arrangements. But of these, $C!$ are simply rearrangements of the C atoms in the same atomic sites, and hence do not represent new microstates. Thus the total number of distinguishable microstates is given by

$$p_i = \frac{(N + C)!}{N!\,C!}$$

and the entropy of the mixture is given by

$$S = k \ln p_i = k \ln\left[\frac{(N + C)!}{N!\,C!}\right]$$

Applying Stirling's approximation and simplifying gives

$$S = k[N + C]\left[\ln (N + C) - \frac{N}{N + C}\ln N - \frac{C}{N + C}\ln C\right]$$

but

$$\frac{N}{(N + C)} = x_{\text{Ni}} \quad \text{and} \quad \frac{C}{(N + C)} = x_{\text{Cu}}$$

where x represents mole fractions. Thus

$$S = k[N + C]\,[\ln (N + C) - x_{\text{Ni}} \ln N - x_{\text{Cu}} \ln C]$$

but

$$x_{\text{Ni}} + x_{\text{Cu}} = 1$$

so that

$$S = -k[N + C]\left[x_{\text{Ni}} \ln\left(\frac{N}{N + C}\right) + x_{\text{Cu}} \ln\left(\frac{C}{N + C}\right)\right]$$

$$= -k[N + C][x_{\text{Ni}} \ln x_{\text{Ni}} + x_{\text{Cu}} \ln x_{\text{Cu}}]$$

but

$$N + C = n_T A,$$

where n_T = Total number of moles
A = Avogadro's number
R = Ideal gas constant

and

$$R = Ak$$

so that

$$S = -Rn_T[x_{\text{Ni}} \ln x_{\text{Ni}} + x_{\text{Cu}} \ln x_{\text{Cu}}]$$

$$= (-1.987)(4.0)[0.75\,(\ln 0.75) + 0.25\,(\ln 0.25)]\text{Btu/°R}$$

$$= +4.5\ \text{Btu/°R}$$

and we observe, as expected, that the entropy has increased.

notation index

Listed here are most of the symbols used in this text together with a brief descriptive name. Where applicable the dimensions are also given in terms of mass (M), length (L), time (t), temperature (T), and force (F). Because of the common use of thermal energy units such as Btu and cal, the symbolism TEU is substituted for the dimension $F \cdot L$ in many places. In general the dimension M can be construed as either mass or moles except where one or the other is specifically stated. Those symbols of particular importance or having unique use in this text are further identified with the number of the equation in which they are first introduced.

Latin Symbols

Symbol	Description
a	Acceleration (L/T^2)
a_i	Activity of species i in a mixture (dimensionless)
a_{ji}	Number of atoms of element j in chemical species i
A	Area (L^2)
A	Chemical affinity (TEU/M)
A	Helmholtz free energy (TEU)
a	Specific Helmholtz free energy (TEU/M)
$\bar{A}_i$	Partial molar Helmholtz free energy (TEU/M)
A_i	ith chemical species in a chemical reaction
b_j	Total number of gram atoms of element j
C	Number of chemical species (components) in a mixture
C	Molar concentration (Moles/L^3)
C	Sonic Velocity (L/T)
C_P	Heat capacity at constant pressure (TEU/$M \cdot T$)
C_V	Heat capacity at constant volume (TEU/$M \cdot T$)
$\Delta\mathcal{C}_P^\circ$	Net heat capacity of stoichiometric reaction mixture (TEU/$M \cdot T$)

$(\mathcal{C}_P^\circ)_{exit}$	Total heat capacity of product stream (TEU/$M \cdot T$)
COP	Coefficient of performance for refrigeration
D	Diameter (L)
D	Diffusivity (L^2/t)
E	Total energy (TEU)
e	Specific total energy (TEU/M)
E	Electrical potential difference (volts)
f	Degrees of freedom
f	Fanning friction factor (dimensionless)
f	Fugacity (F/L^2)
$\bar{f}_i$	Partial fugacity (F/L^2)
F	Force
F	Faraday constant (coulombs/equivalent)
g	Acceleration of gravity (L/t^2)
g_c	Universal conversion factor ($M \cdot L/F \cdot t^2$)
G	Mass flow rate per unit area ($M/L^2 \cdot t$)
G	Gibbs free energy (TEU)
g	Specific Gibbs free energy (TEU/M)
$\bar{G}_i$	Partial molar Gibbs free energy (TEU/M)
Δg_f°	Standard state Gibbs free energy of formation (TEU/mole)
$\underline{\mathcal{G}}$	Gibbs free energy per mole of species 1 entering a reactor (TEU/mole)
$\Delta\underline{\mathcal{G}}^\circ$	Standard state Gibbs free energy change of reaction(TEU/mole)
h	Height (L)
H	Humidity
H	Enthalpy (TEU)
h	Specific enthalpy (TEU/M)
$\bar{H}_i$	Partial molar enthalpy (TEU/M)
Δh_f°	Standard-state enthalpy of formation (TEU/M)
$\mathcal{H}^\circ$	Enthalpy per mole of species 1 entering a reactor (TEU/M)
$\Delta\underline{\mathcal{H}}^\circ$	Standard state enthalpy change (heat) of reaction (TEU/M)
J_a	Activity ratio (dimensionless)
k	Boltzmann constant
k	Ratio of specific heats
K_i	Equilibrium vaporization ratio
K_a	Equilibrium constant
KE	Kinetic energy ($F \cdot L$ or TEU)
L	Length
LW	Total lost work ($F \cdot L$ or TEU)
lw	Lost work per unit mass flowing (TEU/M)
lw	Local lost work per unit volume (TEU/L^3)
m	Molality

M	Molecular weight
M	Mass
$\dot{M}$	Mass flow rate (M/t)
M	Mach Number
n	Number of moles
N	Number of competing chemical reactions
$\underline{\mathfrak{N}}$	Number of moles of electrons liberated per mole of species 1 reacted
p or P	Number of coexisting phases
P	Pressure (F/L^2)
P'	Vapor pressure (F/L^2)
P_i	Partial pressure (F/L^2)
PE	Potential Energy ($F \cdot L$ or TEU)
q	Electrical charge transferred (coulombs)
$\boldsymbol{q}$	Local heat transferred per unit area of boundary (TEU/L^2)
Q	Thermal energy flux or heat (TEU)
q	Heat flow per unit of material flowing (TEU/M)
$\dot{Q}$	Rate of thermal energy flux (TEU/t)
$\underline{\mathfrak{Q}}$	Heat per mole of species 1 entering a reactor (TEU/mole)
$\bar{R}$	Ideal gas constant
RH	Relative Humidity
S	Entropy (TEU/T)
s	Specific entropy (TEU/$T \cdot M$)
$\bar{S}_i$	Partial molar entropy (TEU/$T \cdot M$)
T	Temperature
t	Time
V	Velocity (L/t)
U	Internal energy ($F \cdot L$ or TEU)
u	Specific internal energy (TEU/M)
$\bar{U}_i$	Partial molar internal energy (TEU/M)
V	Volume (L^3)
v	Specific volume (L^3/M)
$\bar{V}_i$	Partial molar volume (L^3/M)
W	Weight (F)
W	Mechanical energy flux or work transfer ($F \cdot L$ or TEU)
w	Work transfer per unit of material flowing (TEU/M)
$\dot{W}$	Rate of mechanical energy flux (TEU/t)
$\underline{\mathfrak{W}}$	Work per mole of species 1 entering a reactor (TEU/mole)
x_i	Mole fraction of chemical specie i
X	Measure of length or position (L)
X	Extent of reaction (moles)
$\underline{X}$	Fractional extent of reaction (dimensionless)
y_i	Mole fraction, particularly in gaseous phase

Z	Length or distance (L)
Z	Compressibility facor (dimensionless)

Greek Symbols

α_i	Stoichiometric coefficient (dimensionless)
α_{ji}	Stoichiometric coefficient for ith species in jth competing chemical reaction (dimensionless)
γ_i	Activity coefficient (dimensionless)
ζ	Euken coefficient ($L^3 \cdot T/M$)
η	Efficiency factor, actual quantity/ideal quantity
η	Thermal efficiency, fraction of thermal energy supplied to a heat engine that is converted to work
η_i	Portion of species i passing through a reactor unreacted
θ	Angle (radians)
μ	Joule–Thomson coefficient ($L^2 \cdot T/F$)
μ	Viscosity ($M/L \cdot t$ or $F \cdot t/L^2$)
μ_i	Chemical potential (TEU/M)
ν	Fugacity coefficient (dimensionless)
ν	Kinematic viscosity (L^2/t)
ρ	Density (M/L^3)

Operators

δ	Denotes infinitesimal change in path function
d	Denotes infinitesimal change in a state property
Δ	Finite change in state property, final minus initial value
d	Total derivative operator
∂	Partial derivative operator
$\int$	Integral operator
$\oint$	Integral operator for closed path
ln	Natural logarithm operator (base e)
log	Common logarithm operator (base 10)
$\prod$	Cumulative product operator
$\sum$	Cumulative summation operator

Subscripts

acc	Accumulation
c	Denotes value at critical state
gen	Generation

i, j, k, etc. Component identification
in Addition to system
H High temperature reservior
L Low temperature reservoir
m With operator Δ denotes change upon mixing
m or mix Mixture
out Removal from system
prod Product
r Reduced property
react Reactant
res Resistance of surroundings to change in system
rev Reversible
soln Solution or mixture
surr Surroundings
sys System

Superscripts

E With operator Δ denotes excess change in property upon mixing
L, V, G,
I, II etc. Phase identification
$\bar{}$ (overbar) Denotes partial molar quantity
$\dot{}$ (overdot) Time rate of change
* Denotes value of property in limit as $P \longrightarrow 0$ ($V \longrightarrow \infty$)
$\circ$ Reference state

index